Dynamics of Curved Fronts

PERSPECTIVES IN PHYSICS

Huzihiro Araki, J.A. Libchaber, and Giorgio Parisi, editors

Dynamics of Curved Fronts

Edited by

Pierre Pelcé
Laboratoire de Recherche en Combustion
Université de Provence—St. Jerome
Marseille, France

ACADEMIC PRESS, INC.
Harcourt Brace Jovanovich, Publishers

Boston San Diego New York
Berkeley London Sydney
Tokyo Toronto

ACADEMIC PRESS, INC.
1250 Sixth Avenue, San Diego, CA 92101

United Kingdom Edition published by
ACADEMIC PRESS INC. (LONDON) LTD.
24-28 Oval Road, London NW1 7DX

Library of Congress Cataloging-in-Publication Data

Pelcé, Pierre. 1959–
Dynamics of curved fronts / Pierre Pelcé.
p. cm.—(Perspectives in physics)
Bibliography: p.
Includes index.
ISBN 0-12-550355-5
1. Surfaces (Physics) 2. Surfaces (Technology) 3. Hydrodynamics.
4. Metallurgy. 5. Combustion. I. Title. II. Series.
QC173.4.S94P45 1988
530.4′1--dc19 87-28916
CIP

Printed in the United States of America
88 89 90 91 9 8 7 6 5 4 3 2 1

A mes parents
A ma femme Chantal

Contents

Part II. Collected Papers

Planar Front Propagation

Dragging of a Liquid by a Moving Plate

Saffman-Taylor Finger

Dendrites

Premixed Flames

Foreword

In 1917 D'Arcy Thompson wrote a stimulating and classical book entitled "On Growth and Form," where he defined a program and a hope:

> The waves of the sea, the little ripples on the shore, the sweeping curve of the sandy bay between the headlands, the outline of the hills, the shape of the clouds, all these are so many riddles of form, so many problems of morphology, and all of them the physicist can more or less easily read and adequately solve....
>
> The terms Growth and Form, which make up the title of this book, are to be understood, as I need hardly say, in their relation to the study of organisms. We want to see how, in some cases at least, the forms of living things, and of the parts of living things, can be explained by physical considerations, and to realise that in general no organic forms exist save such as are in conformity with physical and mathematical laws.

Pierre Pelcé's book fits naturally in this ambitious program. It is divided into two parts. The first one is a clear and lucid introduction to the problem of selection and stability of moving fronts, from the flat interface to the complex dendrite. The second part is a compendium of the relevant articles on the subject, very useful, given the dispersion in time, language and journals for this subject which covers so many disciplines. This book is an important contribution to the field of nonlinear dynamics.

A. LIBCHABER
Professor of Physics
University of Chicago

Preface

In the last ten years, much progress has been made in the understanding of interface dynamics in various fields (hydrodynamics, metallurgy, combustion). The field has grown rapidly, and there is no need today to stress its importance. This book consists of an introduction aimed at highlighting the unifying ideas of the field, and a collection of papers representing some of the most important contributions in the field.

This book will be useful for researchers and graduate students in chemical engineering, mechanical engineering, physics and mathematics. Its aim is not to be complete but to show that problems apparently unique to one discipline are in fact similar to those found in other disciplines and, moreover, that insights from all the areas are essential to resolve these questions of interface dynamics.

I am particularly indebted to A. Libchaber who suggested to me to gather this collection of articles and introduce it by what was in part my Thèse d'Etat "Dynamique des fronts courbes" (in French). I thank him too for the valuable advice he gave me during the preparation of this work.

I thank also the many people with whom the point of view was developed in the monograph. Among them are G. Albinet, J. Bechhoefer, M. Benamar, D. Bensimon, B. Billia, L. Boyer, P. Clavin, T. Dombre, V. Hakim, G. Joulin, A. Karma, A. Linan, C. Nicoli, Y. Pomeau, A. Pumir, J. Quinard, G. Searby, B. Shaw, B. Shraiman, P. Tabeling and G. Zocchi.

I also thank J. Bechhoefer who reread the manuscript and corrected the numerous errors of English.

The manuscript was written with the help of the McCormick Fellowship given by the University of Chicago and the grant NSF DMR MRL 85-19460.

Part I. Introduction

1. Introduction

1.1. Examples of Interface Propagation

When a viscous fluid (oil for instance) contained in a Hele-Shaw cell is pushed by an air overpressure applied at one of its extremities, the planar air-oil interface moving with constant velocity is unstable. Many fingers of air are first formed. Then, one of them develops more than the others and, after some time, moves with constant velocity and stationary shape (Fig. 1.1). For larger air pressure, localised perturbations of the shape appear which move from the tip to the rear of the finger. When air pressure is further increased secondary fingers are formed either on the sides of the main finger by amplification of these last perturbations or by splitting of the tip (see Tabeling *et al.* [1]).

When a solid nucleus is introduced in a supercooled liquid (succinonitrile for instance), this nucleus grows and forms a crystal of dendritic shape whose dendrites grow in each crystallographic direction (six in the case of succinonitrile, whose crystallographic structure is cubic). At the tip of a primary dendrite, secondary dendrites are generated which grow on the sides, at rest in the laboratory frame (Fig. 1.2). Experiments have shown (Glicksman *et al.* [2]) that primary dendrites grow with a well defined velocity, function of the succinonitrile supercooling. Velocity measurements are in good agreement with a similarity law such as

$$U = \frac{D}{d_0} f(\Delta), \tag{1.1}$$

where U is the growth velocity of the crystal, D the thermal diffusion coefficient, and d_0 a capillary length. Δ is the dimensionless supercooling defined as $\Delta = c_p(T_s - T_\infty)/Q$, where T_s is the crystallisation temperature of the planar interface, T_∞ the temperature of the melt far ahead of the crystal, Q the latent heat and c_p the specific heat per unit volume of solid.

If one considers a vertical cylindrical closed tube filled with water whose bottom is suddenly opened, the freely falling water will be gradually replaced by air raising in the tube. An interface between air and water is formed, whose shape is stationary and rises with constant velocity (Fig. 1.3). Experiments have shown (Davies and Taylor [3]) that interface velocity measurements are

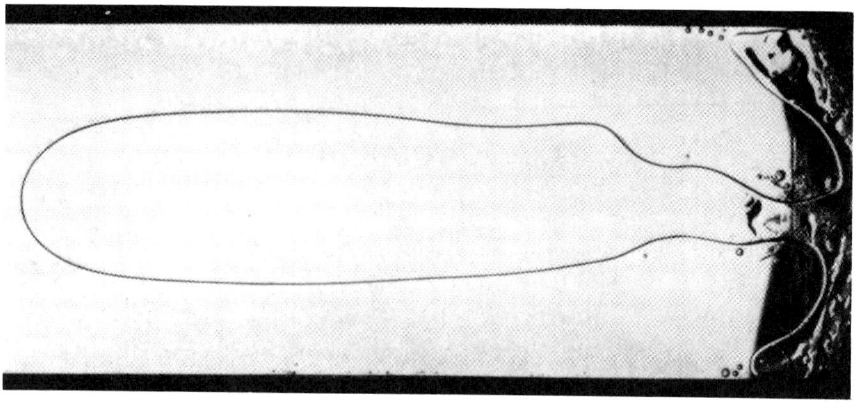

Fig. 1.1. An air finger advancing into glycerine. (Reprinted from plate 2 of Saffman and Taylor [9] with the permission of the Royal Society.)

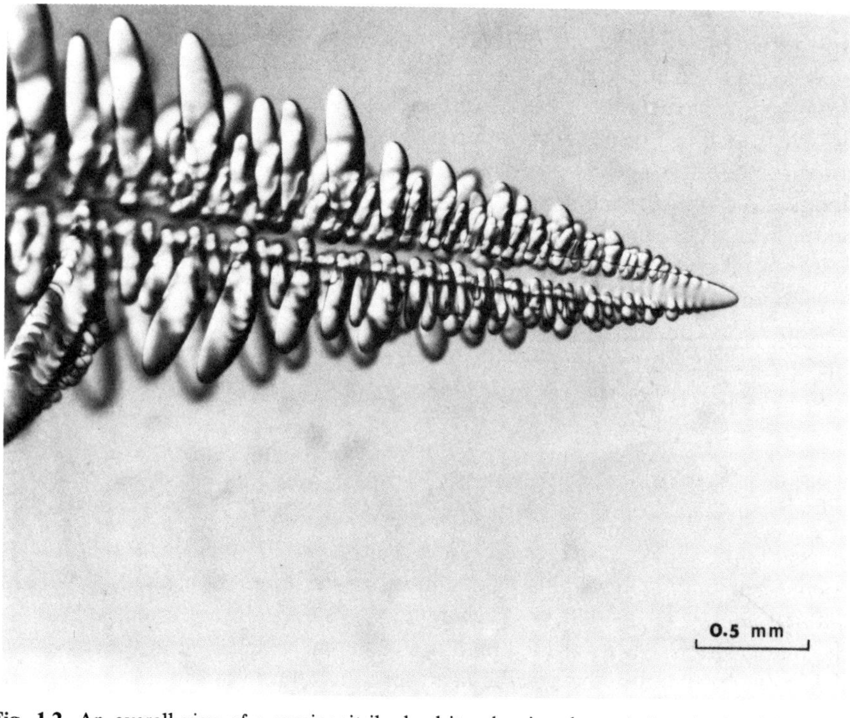

Fig. 1.2. An overall view of a succinonitrile dendrite, showing the variations in the sidebranch morphology with distance from the dendrite tip. (Reprinted with permission from Fig. 2 of Huang and Glicksman [41]. © 1981 Pergamon Journals, Ltd.)

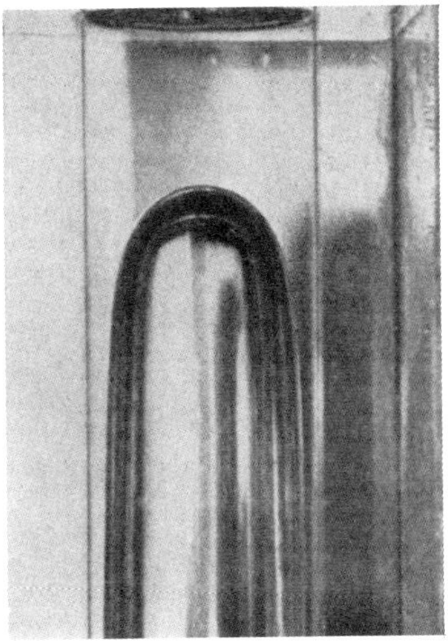

Fig. 1.3. Emptying a glass tube 7.9 cm diameter. (Reprinted from Davies and Taylor [3] with the permission of the Royal Society.)

in good agreement with the law

$$U = .46\sqrt{gR}\,, \tag{1.2}$$

where g is the acceleration of gravity and R the tube radius.

The last example we will consider is the case of premixed flames propagating in tubes. Here there are many rich configurations where flame shape and propagation velocity depend upon various control parameters such as the dilution and equivalence ratio of the reactive mixture, and upon orientation of the tube with the vertical. For instance, a downwards propagating flame is flat at low velocity (Quinard *et al.* [4], Quinard [5]). Flames that propagate faster are cellular and can have an autoturbulent characteristic: wrinkles spontaneously appear on the burning interface and move chaotically even if the flow in which the flame propagates is laminar. Another configuration uses a flame that propagates up the tube (Von Lavante and Strehlow [6], Pelcé-Savornin *et al.* [7]). The flame shape is in this case stationary, curved towards the fresh mixture and seems very stable (Fig. 1.4). For larger propagation velocity a time-dependent process develops in which many individual curved flames compete.

These experimental observations, apparently disparate, since they represent distinct fields of research, (hydrodynamics, crystal growth, combustion) can be grouped together under one class of phenomena: The propagation of interfaces. We will see in the following that a real unity ties together these physical

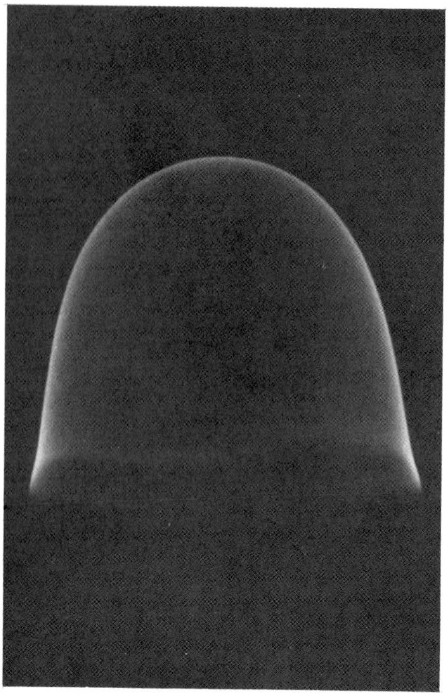

Fig. 1.4. Curved flame dynamically stabilized in a tube. The reactive gas flows downwards, fresh and burned mixture are above and below respectively. (Courtesy of Pelce-Savornin, Quinard, Searby.)

phenomena and that it is useful to study them together. In this domain, there are two basic issues:

- The problem of interface morphology (planar or curved interface, cellular structure; nonstationary shape, chaotic, turbulent) as a function of the control parameters.
- The problem of propagation velocity, or growth velocity, as a function of the same control parameters.

These questions belong to the program of D'Arcy Thompson [8] to study many problems of growth and forms in different fields of physics and biology.

1.2. Interface Propagation Considered as a Dynamical System

Interface motion is equivalent to the solution of a free boundary problem. The question is to determine a solution for a scalar field (pressure, temperature, concentration of an impurity) or a vector field (such as the fluid velocity field) satisfying a partial differential equation (diffusion equation, Euler or Navier-Stokes equation) with boundary conditions applied on the interface. These last are determined by a study of the interface structure and by conservation relations.

An example of a free boundary problem is the one posed by Saffman and Taylor [9]. The question is to determine the motion of an interface between two fluids of different viscosities in a Hele-Shaw cell. The field considered is the pressure p of the fluid satisfying the Laplace equation[1]

$$\Delta p = 0. \tag{1.3a}$$

The interface must satisfy two boundary conditions. One of them, the kinematic condition, requires the non-penetrability of the two fluids in contact, that is to say equality between normal velocity of the interface $\mathbf{v}_i \cdot \mathbf{n}$ and that of the fluid. As, in this case, the fluid velocity field is proportional to the pressure gradient (Darcy's law) this condition can be written as

$$\mathbf{v}_i \cdot \mathbf{n} = -\frac{b^2}{12\mu} \nabla p \cdot \mathbf{n}. \tag{1.3b}$$

Here, μ is the viscosity of the driven fluid and b the thickness between the plates of the cell. The other, the dynamical condition, gives the pressure of the fluid at the interface. The relation usually applied is the Laplace law[2]

$$p_{\text{air}} - p = \frac{\sigma}{R}, \tag{1.3c}$$

where σ is the surface tension between the two fluids and R the radius of curvature at a given point of the interface.

It is interesting to compare this free boundary problem with the classical problem of Neumann where one must solve $\Delta f = 0$ inside a volume limited by a fixed surface on which the normal gradient of f is given. In the Saffman-Taylor problem, the interface shape is not given in advance but is the solution of the free boundary problem. So it is necessary to specify another boundary condition (the dynamical one) in contrast to the Neumann problem.

[1] One considers the simplified case where the less viscous fluid is air. As its viscosity is considerably smaller than that of oil, air pressure can be considered as constant. The pressure p is then that of oil.

[2] In the case where the fluid completely wets the horizontal walls of the Hele-Shaw cell, films of fluids are left behind the finger along the walls. They modify the pressure jump at the interface (see Section 2, relation (2.13)).

1.2.1. Stability of the Planar Interface

The simplest fixed point (stationary solution) of such a dynamical system is the planar interface moving with constant velocity V. The determination of these stationary solutions does not present any difficulty, since the field considered depends only on the coordinate normal to the interface and satisfies in this case an ordinary differential equation.

The linear stability of these planar interfaces is now well understood. The method consists in studying the time evolution of a small perturbation of the planar interface. Because of the translational invariance in the direction parallel to the interface, all the quantities (field considered, interface shape) are proportional to $\exp(\Omega t + iky)$, y being the coordinate parallel to the front, Ω the growth rate of the perturbation, and k its wave number. This kind of perturbation is a solution of the interface dynamics only if Ω and k are related by a dispersion relation $\Omega(k)$. If there exists a value of k for which the real part of Ω is positive, the perturbation is amplified and the planar interface is unstable. The linear stability analysis of the Saffman-Taylor interface was done by Chuoke *et al.* [10]. We consider the case where the air of negligible viscosity pushes oil of viscosity μ with a constant velocity V. If film effects due to the wetting of the oil along the walls are neglected, the dispersion relation is

$$\Omega = V|k|\left(1 - \frac{b^2}{12Ca}k^2\right), \tag{1.4a}$$

where $Ca = \mu V/\sigma$ is the capillary number and b the thickness between the plates. This relation shows the destabilising effect of the interface motion ($\Omega = V|k|$), which leads to positive Ω and the stabilising effect of surface tension at large wave number. The planar interface is thus unstable, with a most unstable wavelength

$$\lambda_c = \frac{\pi b}{\sqrt{Ca}}, \tag{1.4b}$$

which would be the most represented in the interface structure a short time after the appearance of the perturbation. When the Hele-Shaw cell is vertical, air being above oil, the acceleration of gravity stabilises large wavelengths. Then the moving planar interface is stable if its velocity is sufficiently small.

In the case of crystallisation, the planar liquid-solid interface moving with constant velocity V is a solution of the free boundary problem for a determined value of the supercooling Δ. The latent heat that is released during crystallisation heats the liquid from T_∞ to T_s, the crystallisation temperature, which leads to the relation $T_s - T_\infty = Q/c_p$ or $\Delta = 1$. So, the stability analysis of this particular solution would seem to be academic. However it shows up the mechanisms of an important instability, called the Mullins-Sekerka instability [11], which in many cases is the origin of pattern formations on crystal shapes. In a simplified model of crystallisation, which assumes that the temperature field is quasistationary (see for instance Langer [12]), the corresponding dispersion relation is

$$\Omega = V|k|\left(1 - \frac{D}{V} d_0 k^2\right), \tag{1.5a}$$

where D is the thermal diffusive coefficient of the liquid and d_0 a length built with liquid-solid surface tension and other dimensional quantities taking part in the interface dynamics. So, the interface is unstable with the most unstable wavelength

$$\lambda_c = 2\pi\sqrt{3d_0 \frac{D}{V}}, \tag{1.5b}$$

whose typical size is about 10 microns. The dispersion relation (1.5a) is very similar to the relation (1.4a) obtained in the Saffman-Taylor case. This is because the free boundary problems corresponding to these two situations, linearised around the planar interface, are similar, the pressure field in the Hele-Shaw cell being associated with the quasistationary temperature field in the case of crystallisation. A similar phenomenon occurs when crystal growth is limited by diffusion of an impurity. The dispersion relation for perturbations of the planar interface is then similar to (1.5a) where D is now the coefficient of diffusion of the solute. This instability is of great importance in directional solidification of alloys or binary mixtures (see for instance Jackson and Hunt [13], Trivedi [14], Kurz and Fisher [15]). The interface is pulled by a temperature gradient imposed along the sample, at constant velocity. At small velocity, the interface is stable, stabilised both by the temperature gradient and capillary effects. At larger velocity, the instability develops, giving birth to cellular and dendritic structures.

The linear stability of a planar horizontal interface between two fluids of different densities, at rest in the acceleration of a gravity field, was first analysed by Stokes (see Lamb [16]). If one considers an interface between air and a fluid of much larger density ρ lying above it, the corresponding dispersion relation is

$$\Omega^2 = g|k|\left(1 - \frac{\sigma}{\rho g} k^2\right). \tag{1.6a}$$

The interface is unstable, the most unstable wavelength being

$$\lambda_c = 2\pi\sqrt{3\frac{\sigma}{\rho g}}, \tag{1.6b}$$

whose typical size is about one centimeter.

In the case of premixed flames, the front instability is due to the gas expansion across the flame. It was first studied by Darrieus [17] and Landau [18]. When the flame is assumed infinitely thin, the dispersion relation associated to this instability is

$$\Omega = f(\alpha) V |k|, \tag{1.7a}$$

where V is the propagation velocity of the planar flame, $\alpha = \rho_b / \rho_0$ the ratio between burned and fresh gas density (expansion ratio) and f a function which increases from 0 when $\alpha = 1$, to infinity when $\alpha = 0$.

When effects due to flame thickness d are taken into account, this dispersion relation is noticeably modified for wavelength λ of the order of d. One can compute analytically this modification by expanding in large Peclet number λ / d, so that the flame structure can be considered as quasi-planar. This calculation involves a singular perturbation method worked out in successive studies starting from that of Markstein [19]: Einbinder [20], Eckhaus [21], Frankel and Sivashinsky [22], and Pelcé and Clavin [23]. (For a review of these works see Clavin [24].) It shows that in general, diffusion of reactant and heat inside the flame thickness is a globally stabilising effect[3]. The dispersion relation obtained at large Peclet number is

$$\Omega = V|k|(f(\alpha) - h(\alpha, \beta, Le) d |k|), \tag{1.7b}$$

where h is a function of the gas expansion parameter α, of the reduced activation energy of the chemical reaction β and of the Lewis number Le, the ratio between thermal and limiting reactant diffusive coefficient. The most amplified wavelength is

$$\lambda_c = 4\pi \frac{h}{f} d, \tag{1.7c}$$

which is usually of the order of a centimeter. When the effect of the acceleration of gravity is taken into account it stabilises the large wavelengths in the case of downwards propagation (burned gas, lighter, lies above the fresh mixture). Then the flat flame can be stable if its propagation velocity is sufficiently small ($\approx$ 10 cm/s) [4], [23].

1.2.2. Curved Interfaces of Stationary Shape

When the planar interface between two viscous fluids moving in the Hele-Shaw cell becomes unstable, it soon wrinkles with a typical wavelength λ_c (see relation (1.4b)). Then a transient dynamics develops where some of the

[3] In the case where the reactant limiting the reaction diffuses much faster than heat (lean hydrogen-oxygen flame for instance) the flame could be submitted to a thermodiffusive instability (Zeldovich [25]).

wrinkles grow faster than others, screening them, until one single finger finally remains with stationary shape and moving with constant velocity U. This finger is stable over a large range of the control parameter, the capillary number $Ca = \mu U/\sigma$, [9], [1].

An important theoretical point is that a continuum of exact solutions of fingers can be found to the related free boundary problem in which surface tension effects have been neglected (fluid pressure is constant on the interface) [9]. This continuum is parametrised either by the relative width λ or by the velocity U which product is constant for a given velocity of the fluid far ahead of the finger. This situation contradicts experiment, where a well-defined finger width is observed as a function of the capillary number. Saffman and Taylor [9] have proposed to take into account surface tension effects between the two fluids in order to resolve this contradiction.

The determination of stationary symmetric shapes of a finger moving with constant velocity, surface tension effects included, was initiated by McLean and Saffman [26]. They obtain an integral equation for the interface which depends only upon a single dimensionless parameter, which characterizes the effects of surface tension:

$$\kappa = \left(\frac{b}{w}\right)^2 \frac{1}{Ca} \frac{1}{(1-\lambda)^2}, \tag{1.8}$$

where w is the cell width. When $\kappa = 0$, one finds again the solutions without surface tension. When $\kappa \neq 0$ their equation must be solved numerically. The method used by McLean and Saffman gives one single branch of solution $\lambda(\kappa)$ which starts at $\frac{1}{2}$ when κ goes to 0 and tends to 1 for large κ (large surface tension). This result is very attractive since it helps one to understand why in experiments the finger seems to have a relative width close to $\frac{1}{2}$ at large velocity. Then, Romero [27] found two other branches of solutions. Soon after, Vanden-Broeck [28], using another numerical method, found a countable set of solutions $\lambda_n(\kappa)$. For a given value of κ, a shape is computed which in general has a nonzero slope at the tip. The vanishing of this slope, which is required for a symmetric finger, appears as a solvability condition which leads to a quantification of the allowed widths of the finger. All these branches of solutions meet at $\lambda = \frac{1}{2}$ when κ goes to 0.

The analytical study of these solutions in the neighbourhood of the point ($\kappa = 0$, $\lambda = \frac{1}{2}$) was done by Dombre *et al.* [29], Shraiman [30], Hong and Langer [31], Combescot *et al.* [32] and Tanveer [33] who used singular perturbation methods. They show that in general the slope of the interface at the tip is exponentially small and not 0 as required for a symmetric finger. The vanishing of this slope leads to a quantification of the allowed widths of the finger, which confirms the Vanden-Broeck findings.

The example described above belongs to a class of problems where one must determine the propagation velocity of nonlinear waves. This class belongs to the more general problem of finding self-similar solutions of partial differential equations, the propagation velocity of the nonlinear waves being a particular case of self-similarity exponent (Barenblatt [34]). These exponents are eigenvalues of nonlinear equations. Their spectrum can be either continuous as in the case of reaction diffusion wave (Kolmogorov *et al.* [35]) or discrete as mentioned above.

Experiments [1] don't agree with the saturation of the relative size of the finger at $\lambda = \frac{1}{2}$. It appears that, in addition to surface tension effects, the effects of films of oil left by the finger along the plates are important. This was early noticed by Saffman and Taylor [9] and experimentally verified by Tabeling and Libchaber [36]. These effects were introduced by changing the boundary conditions for the jump of pressure at the interface (Park and Homsy [37], Reinelt and Saffman [38], Reinelt [39].

The case of free dendrites growing at constant velocity is somewhat different since their shape is usually time dependent. Only the tip has a stationary shape, roughly close to a paraboloid of revolution. At some distance from the tip secondary dendrites (sidebranching) grow on the sides, at rest in the laboratory frame. However the study of stationary shapes of needle crystals has been developed, as a first step to an understanding of time-dependent patterns.

As in the case of the Saffman-Taylor finger, a continuum of exact solutions of needle crystals growing with constant velocity U has been found by Ivantsov [40] who assumed that the interface is at the constant temperature T_s. The solutions form a family of paraboloids of revolution whose radius of curvature ρ is related to the velocity U by the Ivantsov law:

$$\Delta = P \exp P \int_P^{+\infty} \frac{\exp(-u)}{u}\, du, \tag{1.9}$$

where P is the Peclet number $\rho U/2D$. This formula does not determine the velocity of the crystal since the supercooling Δ fixes only the Peclet number P. It says only that, for a given supercooling, thin needle crystals grow faster than thick ones.

Experiments on growth of free dendrites of succinonitrile (Glicksman *et al.* [2], Huang and Glicksman [41]) and ice crystals (Fujioka [42]) have shown that dendrites grow with a well-defined velocity as a function of the supercooling. As in the Saffman-Taylor finger case, a velocity selection problem is posed. Effects due to liquid-solid surface tension have been introduced in order to remove this degeneracy. In this case the interface temperature is no longer constant but depends on the curvature according to the thermodynamic Gibbs-Thomson law (see Chapter 3). Many approximate models have been built where the paraboloid shape is conserved but where one of the boundary conditions has been applied at the tip only (Sekerka *et al.* [43], Glicksman and Schaefer [44], Trivedi [45]). These models determine a curve $U = f(\rho)$ which has a maximum. It has been conjectured for a long time that the dendritic growth may occur at maximum velocity. A more precise study was proposed by Nash and Glicksman [46] where they determine an integral equation for the interface shape and determine numerically a maximum velocity. At small supercooling, all these theories lead to a velocity law such as

$$U \approx \frac{D}{d_0}\Delta^{\alpha}, \tag{1.10}$$

where $\alpha \approx 2.5$.

Another point of view was adopted by Langer and Muller-Krumbhaar [47] where they deal with the assumption that the growth velocity is dynamically selected. They introduce a marginal stability criterion which says that the selected radius of curvature is proportional to the most unstable wavelength of the planar front (relation (1.5a)), or that the ratio

$$\sigma = 2\frac{Dd_0}{\rho^2 U} \tag{1.11}$$

is a constant σ_c. The proportionality constant is determined numerically and corresponds to a state where the tip is marginally stable.

An intensive study of simplified models for interface dynamics was undertaken (geometrical model: Brower *et al.* [48], Kessler *et al.* [49]; boundary layer model: Ben Jacob *et al.* [50]) in order to understand more generally the problem of velocity selection of a needle crystal. In the first model, for instance, the interface is a closed curve whose normal velocity at a given point is a function of the local curvature and its second derivative with respect to the arc length (the derivative term mimics the surface tension effect). These models share common features with the realistic models, such as the existence of a continuum of needle crystals moving with arbitrary velocity when the higher derivative terms are neglected (corresponding to zero surface tension). They enable us to understand analytically how the continuum of solutions without surface tension can be broken when a small surface tension effect is added (Kruskal and Segur [51], Langer [52]).

Then the study of fully non-local problem was undertaken. An analytical study of the Nash-Glicksman equation was carried out by Pelcé and Pomeau [53] for the limit of small supercooling, which is usually the case in experiments. The other limit ($\Delta \approx 1$) was studied too (Caroli *et al.* [54]). In the limit of small supercooling, the shape of the crystal can be divided into two regions.

Far from the tip, the surface tension effects are negligible. The interface shape is an Ivantsov paraboloid of tip radius ρ which satisfies relation (1.9) at small supercooling, i.e.,

$$\frac{\rho U}{D} \approx -\frac{\Delta}{\log \Delta}. \tag{1.12}$$

Close to the tip, the initial integral equation for the shape can be reduced to a nonlinear eigenvalue problem. The eigenvalue is the number $C = 8/\sigma$, where σ is defined by relation (1.11), with the difference that ρ is not the tip radius of the crystal, but the tip radius of the Ivantsov paraboloid towards which the shape is asymptotic at large distance from the tip. Eliminating ρ in relations (1.11) and (1.12) leads to

$$U \approx \frac{1}{C}\frac{D}{d_0}\left(\frac{\Delta}{\log \Delta}\right)^2. \tag{1.13}$$

Thus, the possible velocities of the crystal are determined up to a constant which is an eigenvalue of a nonlinear integral equation.

Numerical work on the two-dimensional version of the Nash-Glicksman equation was taken over by Meiron [55] and Kessler *et al.* [56] and it was found that, as in the case of geometrical models and contrary to the results of the Nash-Glicksman study, no solution exists when surface tension is assumed isotropic. But when surface tension anisotropy is introduced some solution can be found.

This strange property was confirmed by both numerical integration of the Pelcé-Pomeau eigenvalue equation (Benamar and Moussallam [57]) and by singular perturbation analysis similar to the one developed in the framework of the geometrical and boundary layer models (Benamar and Pomeau [58], Barbieri *et al.* [59]). For isotropic surface tension, no eigenvalue is found. When surface tension anisotropy is added, one finds a discrete set of eigenvalues.

One can also consider dendritic growth in a capillary tube (Honjo and Sawada [60]). From a theoretical point of view, Pelcé and Pumir [61], Pelcé [62] and Kessler *et al.* [63] have shown that a two-dimensional needle crystal, growing sufficiently slowly inside the capillary tube, has a shape of a Saffman-Taylor finger. This shows more precisely the connection that exists between the motion of a bubble and the growth of a crystal.

The study of stationary shapes of the interface between air and water that is formed when the bottom of a tube initially filled with water is opened was undertaken by Davies and Taylor [3]. Contrary to the previous cases, no exact solution of interface moving with constant velocity was found when surface tension effects are neglected. However the authors determined an approximate solution, the interface velocity being determined by relation (1.2), in good agreement with experimental results. The method consists of choosing an upstream potential flow that depends upon free parameters. The parameters are fixed by the boundary conditions for the velocity field applied at particular points along the interface. Then a more precise study was done by Birkhoff and Carter [64] in which they determined an integral equation for the interface, surface tension effects being neglected. Garabedian [65] showed that the flow is not uniquely determined by the tube radius R and the acceleration of gravity g; the velocity U of the bubble must be given too. Furthermore, he suggested that solutions exist if the Froude number $F = U/\sqrt{gR}$ is smaller than a critical value F_c. By using the same numerical method as the one developed in the Saffman-Taylor case, Vanden-Broeck [66] determined a unique solution of the integral equation for each Froude number less than .36, this confirming Garabedian's findings. When surface tension is taken into account there exists a countably infinite number of solutions, each of these corresponding to a different value of the Froude number (Vanden-Broeck [67]).

The propagation of curved flames in tubes has been the subject of numerous studies, experimental (Uberoi [68], Maxworthy [69]) as well as theoretical (Ball [70], Zeldovich *et al.* [71]). As in the previous problem, no exact stationary solution for a curved flame is known even if curvature effects are neglected; this means that the normal combustion velocity u_L is assumed constant along the flame front. However, important qualitative considerations have been developed by Zeldovich *et al.* [71] leading to the determination of approximate solutions. The method is similar to the one adopted by Davies and Taylor [3]. The flow upstream from the flame is potential. The potential is chosen to depend on free parameters that are determined using the global conservation of mass and momentum of the flow. These approximate solutions do not satisfy all the boundary conditions for the pressure and the velocity field at all points of the interface. The propagation velocity of a curved flame given by this model is

$$U = u_L F(\alpha), \tag{1.14}$$

where F is a function of the gas expansion α that must be determined numerically, whose value is 1 when $\alpha = 1$ (planar flame), and which increases when α decreases.

Gravity effects were introduced in this model by Pelcé [72]. In this case, the velocity of a flame propagating in a vertical tube becomes

$$U = u_L G(\alpha, Fr), \tag{1.15}$$

where $Fr = \pm u_L^2/gR$ is the Froude number positive (resp. negative) for downwards (resp. upwards) propagation. The flame can behave either as a flame in zero gravity when the Froude number is large, and its velocity is given by relation (1.14) or as a Davies-Taylor bubble of hot gas rising in a less dense gaseous medium, for small negative Froude number [6]; the relation (1.15) describes intermediate cases.

1.2.3. Stability of Stationary Curved Interfaces

The Stability of the curved, stationary interfaces determined in the previous paragraph presents some common characteristics.

Secondary fingering. This phenomenon is mainly observed in dendritic growth [2] and in the case of the Saffman-Taylor finger [1]. Perturbations of the stationary shape appear close to the tip of the curved interface, go away from it and grow altogether. Perturbations are observed on the Saffman-Taylor shape when the control paramcter $1/\kappa$ is sufficiently large. Numerical simulations (DeGregoria and Schwartz [73] and Bensimon [74]) and experiments [1] show that this large critical value depends upon the amplitude of the noise. For instance Couder *et al.* [75], [76] have shown that when a smooth Saffman-Taylor interface comes up against a small air bubble, one observes secondary fingering. The same phenomenon occurs when the fingers move between two plates on which a regular array of streaks have been drawn (Ben-Jacob *et al.* [77]). In directional solidification of pivalic acid, Bechhoefer and Libchaber [78] have shown that when the wavelength of the cells cannot readjust (because of inhibition of tip-splitting) secondary sidebranching is periodically emitted from the tip. In the case of free dendrites, these perturbations are in general always observed for any value of the supercooling [2].

Tip splitting. When the control parameter reaches some large value which still seems to depend upon the amplitude of the noise or of the turbulence in the system, the tip of the curved interface splits into two parts, leading in general to an instationary process where many curved shapes seem to compete. This process is often observed in the case of Saffman-Taylor fingers [1] and flames rising in a vertical tube [6].

A qualitative stability analysis of a curved interface was proposed by Zeldovich *et al.* [71] for the case of curved flames propagating in tubes. When the front is curved, a tangential non-uniform velocity field is set up in the frame where the flame is at rest. This has two kinds of consequences. First, the overall effect of the tangential field is to advect any perturbations of the front from the tip of the flame to the tube wall while they grow because of the Darrieus-Landau instability.

Second, the wavelength of the perturbation stretches because of the increase of the tangential velocity field from the tip where it vanishes, to the wall. If a localised perturbation of wavelength λ_c, the most unstable wavelength of the planar front, appears close to the tip, it is advected towards the walls of the tube by the tangential velocity field. At the same time, the wavelength of the perturbation increases and so its growth rate decreases (see relation (1.7b)). If the amplitude of the perturbation is sufficiently small and is advected sufficiently rapidly towards the wall of the tube, it will not have enough time to grow in order to destabilise the stationary curved shape.

These arguments can be made more precise when the tip radius of the flame is large compared to λ_c. Assume that a perturbation whose wavelength is λ_c appears near the tip. The total growth of the amplitude of this perturbation can be computed via a W.K.B. approximation that takes advantage of the fact that the wavelength of the perturbation is very small compared to the typical radius of curvature of the front. The growth factor Γ is found to be proportional to the Reynolds number. By using the formula

$$A_{\mathrm{f}} \approx A_{\mathrm{i}} \exp \Gamma, \tag{1.16}$$

where A_{i} (resp. A_{f}) is the initial (resp. final) amplitude of the perturbation and assuming that the amplitude of the perturbation which is necessary to destabilise the flame is of order R, the radius of the tube, Zeldovich *et al.* [71] find the stability criterion

$$Re_{\mathrm{c}} \approx -\log\left(\frac{A_{\mathrm{i}}}{R}\right). \tag{1.17}$$

These ideas can be applied to other interfaces. In the case of the Saffman-Taylor finger, Bensimon *et al.* [79] and Pelcé [62] have determined a stability threshold of the form

$$\frac{w}{b\sqrt{Ca}} \approx -\log\left(\frac{A_{\mathrm{i}}}{w}\right). \tag{1.18}$$

This relation is in good agreement with numerical simulation [74]. This picture given by Zeldovich *et al.* [71] is especially well observed in the experiments of Tabeling *et al.* [1] and in the numerical simulations of DeGregoria and Schwartz [73]. There they observe that a localised perturbation starts from the tip, grows, and finally decays upon arrival at the rear of the finger where the front is stable to perturbations. In the case of the needle crystal, Pelcé and Clavin [80] showed that, at small supercooling, a needle crystal associated with a large eigenvalue C (see relation (1.11)) is stable if the criterion

$$C \approx -\log\left(\frac{A_{\mathrm{i}}}{\rho}\right) \tag{1.19}$$

is satisfied, where ρ is the radius of curvature of the tip. In this approximate theory, only disturbances with initial wavelength close to λ_c are considered since they seem the more dangerous for the stability of the curved interface.

A more rigorous and complete way to solve the stability problem of a curved front is to determine the growth rates of the eigenmodes of the free boundary problem linearised around the stationary solution. This can be done analytically for the Saffman-Taylor solutions without surface tension (Taylor and Saffman [81]) and for the isothermal Ivantsov paraboloid (Langer and Muller-Krumbhaar [47]). Both of these stationary solutions are unstable.

Surface tension effects are introduced by the parameter $\epsilon = \lambda_c/\rho$ ($\epsilon = 0$ is the case without surface tension).[4] The problem is to know for which value of ϵ the curved interface becomes stable. First, a numerical approach was taken by Langer and Muller-Krumbhaar [47] for the case of dendrites. They wrote an evolution equation for a perturbation of the Ivantsov paraboloid and introduced surface tension effects only in the linearised equations. Two kinds of eigenmodes are found according to the value of ϵ. If ϵ is larger than a critical value ϵ_c, a continuum of complex growth rates corresponding to sidebranching is found. If ϵ is smaller than ϵ_c the same continuum is found but in addition there appears a real, positive growth rate corresponding to the splitting of the tip.

In the Saffman-Taylor case, Kessler and Levine [82] and Bensimon [74] have studied the linearised problem around the Saffman-Taylor solution ($\lambda = \frac{1}{2}$) without surface tension. They found numerically growth rates only with negative real part, for values of ϵ larger than .3, leading to linear stability.

Then Kessler and Levine [83] studied the linearised free boundary problem around the exact stationary solutions, both for the Saffman-Taylor finger and the needle crystal. The growth rate spectrum is similar in both cases: a complex continuum of extended modes with negative real part. In addition, there are real discrete localised modes. The nth branch of solutions has $n - 1$ such modes with positive growth rate. Thus, at least for values of ϵ larger than the one studied in this numerical study ($\approx .5$), only the first branch of solutions is stable. The study of these localised modes was done analytically in the W.K.B. limit on an approximate model by Bensimon *et al.* [84] which confirms qualitatively the Kessler and Levine findings. Then, Tanveer [85] did a complete stability analysis of the Saffman-Taylor finger and found linear stability for the first branch of solutions whatever the value of ϵ. This is in agreement with the qualitative theory proposed by Zeldovich *et al.* [71]. The curved interface is linearly stable with respect to disturbances of infinitesimal amplitude.

[4] In the various works on this subject, authors use different parameters, but all are related to ϵ.

1.3. Time-dependent Interface Shapes

A few works were devoted to the determination of nontrivial time-dependent solutions for interface motion. In the case of the Saffman-Taylor interface, Shraiman and Bensimon [86] and Howinson [87] determined an equation for the motion of the interface by using conformal mapping techniques. The exterior of the interface is mapped into the interior of the unit disc of the complex plane. The interface becomes a fixed boundary (the unit disc) and its evolution is determined by an equation for the conformal mapping. In the absence of surface tension, this equation has solutions that have polar decomposition. The only time dependence of the solutions appears in poles which move in the complex plane according to a dynamical system. Some solutions lead to cusps after a finite time (Shraiman and Bensimon [86], Sarkar [88]) while others do not (Howinson [89]). More complicated solutions can be obtained by considering hierarchical structures of poles leading to an infinite cascade of tip splitting (Bensimon and Pelcé [90], Howinson [89]).

Another approach was developed by Sivashinsky [91] in the context of flame propagation. He derives a nonlinear time-dependent equation for the flame front position $\xi(x, t)$ which, in its more general form, can be written

$$\frac{\partial \xi}{\partial t} = \mathrm{P}\int \frac{\partial_{x'}\xi}{x - x'}\, dx' - \epsilon \frac{\partial^2 \xi}{\partial x^2} - \nu \frac{\partial^4 \xi}{\partial x^4} - \frac{1}{2}\left(\frac{\partial \xi}{\partial x}\right)^2, \qquad (1.20)$$

where the first term of the right-hand side represents the Darrieus-Landau hydrodynamic instability, the second term represents a thermodiffusive instability of the flame thickness (if ϵ is positive), the third term represents diffusive effects stabilising the large wave numbers, and the nonlinear term comes from the cosine of the angle between the normal to the interface and the direction of propagation. When the first term of the right-hand side is not present and when ϵ is positive (diffusive instability only) one obtains the so-called Kuramoto-Sivashinsky equation, whose solution can be autoturbulent [92]. When $\nu = 0$ and ϵ is negative, one obtains the so-called Michelson-Sivashinsky equation. Numerical simulation shows solutions with large stationary cells matched by cusps (Michelson and Sivashinsky [93], Pumir [94]). Lee and Chen [95], who studied a similar equation in another context, and Thual *et al.* [96] have shown that these solutions can be obtained as a polar decomposition of the equation.

In the following, we will present in detail several aspects of the selection mechanism for stationary propagation of curved fronts. This will be done in the next three sections of this monograph. In the first we discuss the Saffman-Taylor finger, in the second the growing needle crystal, and in the third the curved flame propagating in a tube. The last chapter is devoted to the stability of these curved fronts. In each section we emphasize the physical aspects of the problem and try to bring out the unifying ideas. Thus some specifics particular to each subject will inevitably be left out, and for more complete information the reader is invited to consult, for instance, the following review articles: On the Saffman-Taylor finger, Bensimon *et al.* [79], Saffman [97]; on the dendritic growth, Langer [12], [98]; and on flame propagation, Clavin [24], Sivashinsky [99].

The second part of this book is a collection of articles which contributed to the present understanding of the dynamics of curved fronts. First are two important articles on propagation of flat fronts by Kolmogorov *et al.* [35] and Zeldovich and Frank-Kamenetskii [100], which are at the origin of the field. These are followed by articles on the Saffman-Taylor finger and dendritic growth, interfaces which can be considered as prototypes for the understanding of the dynamics of curved fronts. Finally, there are articles on directional solidification, bubbles rising in a vertical tube, and flame propagation, interfaces whose study is important for applications.

2. The Saffman-Taylor Finger

2.1. Model for the Interface Propagation

Let an interface between two fluids of different viscosities move at small Reynolds number in a horizontal Hele-Shaw cell. A Hele-Shaw cell is a long thin channel that is formed by two rectangular plates of glass whose dimensions (length ≈ 1 m, width $w \approx 10$ cm) are considerably larger than the distance $b \approx 1$ mm between them (Fig. 2.1). We consider the case where the driven fluid is an oil of viscosity μ that wets the plates. The driving fluid is air whose viscosity can be neglected. The air is thus at constant pressure p_{air}. The dynamics of the driven fluid follows the Stokes equation

$$\nabla p = \mu \, \Delta \mathbf{v} \tag{2.1a}$$

and the mass conservation equation

$$\nabla \cdot \mathbf{v} = 0 \tag{2.1b}$$

with the boundary conditions

$$\mathbf{v} = 0 \tag{2.2a}$$

on the plates ($z = \pm b/2$), and

$$\begin{aligned} \mathbf{v}_\text{i} \cdot \mathbf{n} &= \mathbf{v} \cdot \mathbf{n} \\ \mathbf{t} \cdot \mathbf{T} \cdot \mathbf{n} &= 0 \\ p_{\text{air}} - \mathbf{n} \cdot \mathbf{T} \cdot \mathbf{n} &= \frac{\sigma}{R} \end{aligned} \tag{2.2b}$$

on the interface whose equation is $z = h(x, y)$. Here $\mathbf{v}_\text{i}$ is the velocity of the interface, $\mathbf{T}$ the stress tensor, $\mathbf{t}$ and $\mathbf{n}$ the unit tangent and normal vectors to the interface, and σ the surface tension. The longitudinal direction is x, and z and y are the transverse directions, respectively perpendicular and parallel to the plates. We will see now that this three-dimensional free boundary problem can be reduced to a two-dimensional one, by taking advantage of the fact that the vertical dimension is very small compared to the horizontal one. Furthermore, when the capillary number $Ca = \mu U/\sigma$ is small, the boundary conditions at the interface of this new free boundary problem can be determined analytically.

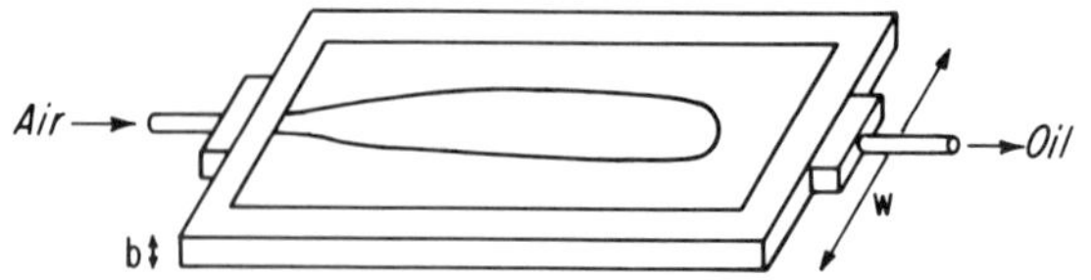

Fig. 2.1. Sketch of the Hele-Shaw cell apparatus.

2.1.1. Flow and Shape of the Interface in the Direction Perpendicular to the Plates at a Small Capillary Number

We assume that the interface is moving at constant velocity U in the x direction and that its shape is flat in the y direction (Fig. 2.2).

Far ahead of the finger, the flow is parallel to the x direction and the velocity profile parabolic (Poiseuille flow). The mean velocity of the flow is $\bar{v}$ and is related to the uniform pressure gradient by a Darcy law:

$$\bar{v} = -\frac{b^2}{12\mu}\frac{dp}{dx}. \tag{2.3}$$

Close to the interface the streamlines are curved and films of fluid are left behind the finger, along the plates. Three regions can be distinguished on the interface. The first region, located around the tip of the finger, of size b, is hydrostatic, in the sense that the flow does not influence the shape of the interface. In this region, the typical variation of pressure on the interface due to the flow is

$$\delta p_{\text{flow}} \approx \frac{\mu U}{b}. \tag{2.4a}$$

The typical variation of pressure due to surface tension is

$$\delta p_{\text{s.t.}} \approx \frac{\sigma}{b}. \tag{2.4b}$$

So at small capillary number, $\delta p_{\text{s.t.}}$ is very large compared to δp_{flow} and the shape is nearly hydrostatic, i.e., a half-circle of radius $b/2$. Far in the tail of the finger, a film of fluid at rest in the laboratory frame remains, of constant

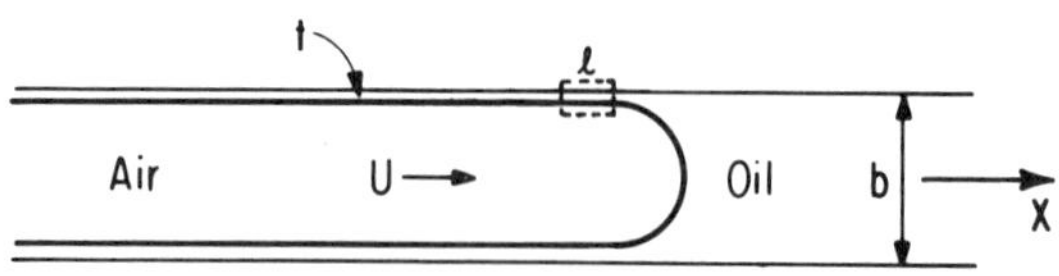

Fig. 2.2. Shape of the interface in the direction perpendicular to the plates at small capillary number. U is the velocity of the finger, and t is the thickness of the film left behind it.

thickness t. Between these two regions exists a transition region, where the variation of pressure due to the flow is of the same order as the variation of pressure due to surface tension. This region was first described by Landau and Levich [101] for the problem of the dragging of a fluid by a moving plate. It is essentially a rectangle close to the plate of length l and thickness t (Fig. 2.2). One can evaluate l and t by the following argument. The variation δp_{flow} is now

$$\delta p_{\text{flow}} \approx \frac{\mu U l}{t^2}, \tag{2.5a}$$

and the variation of capillary pressure is

$$\delta p_{\text{s.t.}} \approx \frac{\sigma t}{l^2}, \tag{2.5b}$$

since in this region the slope of the shape is very small. Furthermore, the jump of pressure between the ends of this region is σ/b: In the film region the shape is flat, so the pressure is p_{air}, and in the hydrostatic region the pressure of the fluid is $p \approx p_{\text{air}} + \sigma/b$. From the relations

$$\frac{\mu U l}{t^2} \approx \frac{\sigma t}{l^2} \approx \frac{\sigma}{b} \tag{2.6}$$

one gets the size of the transition region

$$\begin{aligned} l &\approx b(Ca)^{1/3} \\ t &\approx b(Ca)^{2/3}. \end{aligned} \tag{2.7}$$

The main effect of the film is to reduce the tip radius of the meniscus and thus to increase the pressure drop at the interface by capillary effects. The corresponding correction of the pressure drop (Bretherton [102], Park and Homsy [37]) is

$$\delta p \approx \frac{\sigma t}{b^2} \approx \frac{\sigma}{b}(Ca)^{2/3}. \tag{2.8}$$

When the Hele-Shaw cell is horizontal and effects due to the acceleration of gravity are taken into account, the buoyancy force tends to push the meniscus towards the upper plate leading to an upper film thinner than the bottom film. When the Bond number $B = \rho g b^2/\sigma$ is larger than one, the meniscus leaves the bottom plate and the bottom film disappears. The thickness of the meniscus becomes of the order of the capillary length. The new correction to the static pressure drop (Jensen *et al.* [103]) is

$$\delta p \approx \sqrt{\rho g \sigma}\,(Ca)^{2/3}. \tag{2.9}$$

For larger capillary number the film thickness saturates (Reinelt and Saffman [38], Reinelt [39]).

2.1.2. Reduction of the Interface Dynamics to a Two-Dimensional Free-Boundary Problem

It appears that, in a direction parallel to the plates, the interface is curved on the scale w, the width of the plates, which is considerably larger than the thickness between the plates b, the dimension of the meniscus. So, the local transverse structure of the flow in a direction normal to the interface is the same as the one described previously, U being changed by $\mathbf{v}_{\mathrm{i}} \cdot \mathbf{n}$, the normal velocity of the interface. For instance in the case of a finger with stationary shape moving with constant velocity U, the local thickness of the film left behind the finger is given by relation (2.7) where Ca is changed by $Ca \cos \theta$; here, θ is the angle between the normal to the interface and the direction of propagation. Such a relation has been verified experimentally by interferometric methods (Tabeling and Libchaber [36]). As the local transverse structure of the flow is known, the interface dynamics can be reduced to a bidimensional free boundary problem. The new interface is the curve formed by the tips of the local meniscus and the flow $\mathbf{v}$ is the average in the transverse small dimension of the local Poiseuille flow. It satisfies the relation (2.3), i.e.,

$$\mathbf{v} = -\frac{b^2}{12\mu}\nabla p, \tag{2.10}$$

with the mass conservation

$$\nabla \cdot \mathbf{v} = 0. \tag{2.11}$$

Two boundary conditions must be satisfied on the interface: the kinematic condition,

$$\mathbf{v}_{\mathrm{i}} \cdot \mathbf{n} = \mathbf{v} \cdot \mathbf{n}, \tag{2.12}$$

which requires the non-penetrability of the two fluids in contact, and the dynamical condition,

$$p_{\mathrm{air}} - p = \frac{\sigma}{b}\left(1 + 3.8\left(\frac{\mu \mathbf{v}_{\mathrm{i}} \cdot \mathbf{n}}{\sigma}\right)^{2/3}\right) + \frac{\pi}{4}\frac{\sigma}{R}, \tag{2.13}$$

which gives the pressure of the fluid at the interface. Here, R is the radius of curvature of the interface in a plane parallel to the plates[5]. On the walls ($y = \pm a$), where $a = w/2$, the normal velocity of the fluid vanishes, and at infinity, far ahead of the interface, the flow is uniform, parallel to the x axis, and of magnitude V.

[5] The coefficient of σ/R, $\pi/4$, was computed by Park and Homsy [37].

2.2. The Problem Without Surface Tension

Saffman and Taylor found a continuum of stationary solutions to a simplified version of the free boundary problem mentioned above, in which surface tension is neglected and the pressure along the interface is taken as constant. These solutions are fingers with stationary shape that move with constant velocity U. As shown by relation (2.10) the velocity flow derives from the potential $\Phi = -b^2/12\mu(p - p_{\text{air}})$. As the flow is two-dimensional and obeys Laplaces' equation, the complex potential $W = \Phi + i\Psi$ is an analytic function of the complex number $z = x + iy$. (Here x and y are scaled against the cell half-width a). Ψ is the stream function defined as

$$\frac{\partial \Phi}{\partial x} = \frac{\partial \Psi}{\partial y}, \qquad \frac{\partial \Phi}{\partial y} = -\frac{\partial \Psi}{\partial x}. \tag{2.14}$$

On the interface, $\Phi = 0$, and $\Psi = Uy$, as can be deduced from (2.12). On the walls ($y = \pm 1$), $\Psi = \pm V$, and at infinity in front of the interface, $\Phi \approx Vx$. The trick is to take Φ and Ψ as new variables and find z as a function of W. As a matter of fact, in these variables, the problem is no longer a free boundary problem since the interface shape is known (the segment $\Phi = 0$). It remains to determine the analytical function $z(W)$ with prescribed simple values on fixed boundaries. This function is found (Saffman and

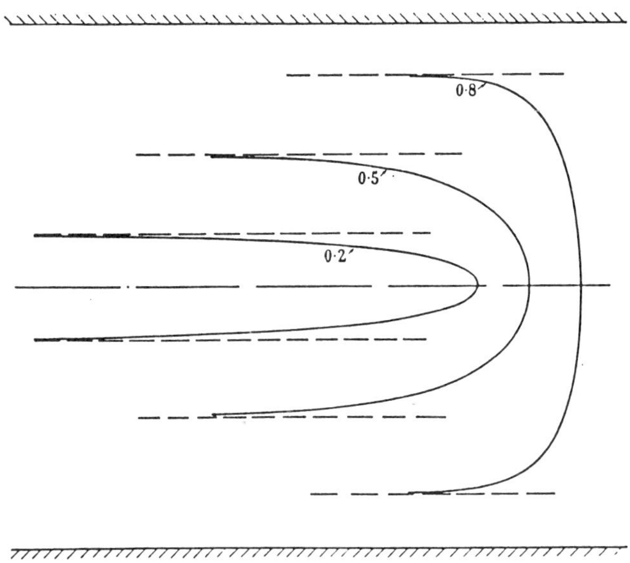

Fig. 2.3. Calculated profiles for $\lambda = 0.2$, 0.5, and 0.8. (Reprinted from Fig. 7 of Saffman and Taylor [9] with the permission of the Royal Society.)

Taylor [9]) to be

$$z = \frac{W}{V} + \frac{2}{\pi}(1 - \lambda)\log\left(\frac{1 + \exp(-\pi W/V)}{2}\right), \tag{2.15}$$

where $\lambda = V/U$. The shape of the interface is obtained by putting $\Phi = 0$ and $\Psi = Uy$ in relation (15):

$$x = \frac{(1 - \lambda)}{\pi}\log\left(\frac{1 + \cos(\pi y/\lambda)}{2}\right), \tag{2.16}$$

i.e., the shape of a finger of thickness 2λ (Fig. 2.3).

These solutions form a continuous family parametrised by λ which can vary between 0 and 1. In experiments [9], [1] only one λ is observed for a given capillary number. At large velocity, the observed shape is in very good agreement with the calculated shape given by (2.16) with λ close to $\frac{1}{2}$. This poses a serious selection problem. More than this, Saffman and Taylor [104] have found asymmetric solutions of fingers moving with constant velocity. In experiments, only the axisymmetric ones are observed. An important step towards an understanding of this was taken by McLean and Saffman [26], who took into account surface tension effects.

2.3. Effects of Surface Tension

2.3.1. Integral Equations

The starting point of the McLean and Saffman study [26] is the determination of an integral equation for the shape of the interface. They take into account the surface tension effects only for the large dimension of the cell. The effects due to the film's wetting are omitted. We give here the main steps in the derivation of this equation. The frame of reference is chosen to be at rest with respect to the finger. The velocity potential and the stream function are now defined as

$$\hat{\Phi} = \frac{\Phi - Ux}{(1 - \lambda)Ua}, \qquad \hat{\Psi} = \frac{\Psi - Uy}{(1 - \lambda)Ua}. \tag{2.17}$$

The conformal map

$$\sigma = s + it = \exp\left(-\left(\hat{W} - \hat{\Phi}_0\right)\right) \tag{2.18}$$

maps the potential plane into the upper half plane (Fig. 2.4). Here Φ_0 is the velocity potential at the tip. By applying Cauchy's integral theorem to the logarithm of the complex velocity $u - iv = q\exp(-i\theta)$, one obtains the integral relation

$$\log(q(s)) = -\frac{1}{\pi}\mathrm{P}\int_0^1 \frac{\theta(s') - \pi}{s - s'}\, ds', \tag{2.19}$$

which comes solely from kinematic considerations. Here q is the modulus of the velocity and θ the angle between the velocity vector and the x axis. The dynamics is introduced by relation (2.13) where film effects are omitted. If one differentiates this relation with respect to s along the interface one obtains a second relation between q and θ:

$$\frac{\sigma b^2 \pi^2}{12\mu U a^2} qs \frac{d}{ds}\left(qs \frac{d\theta}{ds} \right) - (1-\lambda) q = \cos\theta, \qquad 0 < s < 1, \tag{2.20}$$

with the appropriate boundary conditions:

$$\begin{aligned} &\theta = \pi/2, q = 0 && \text{at the tip } (s = 1), \\ &\theta = \pi, q = \frac{1}{1-\lambda} && \text{in the tail } (s = 0). \end{aligned} \tag{2.21}$$

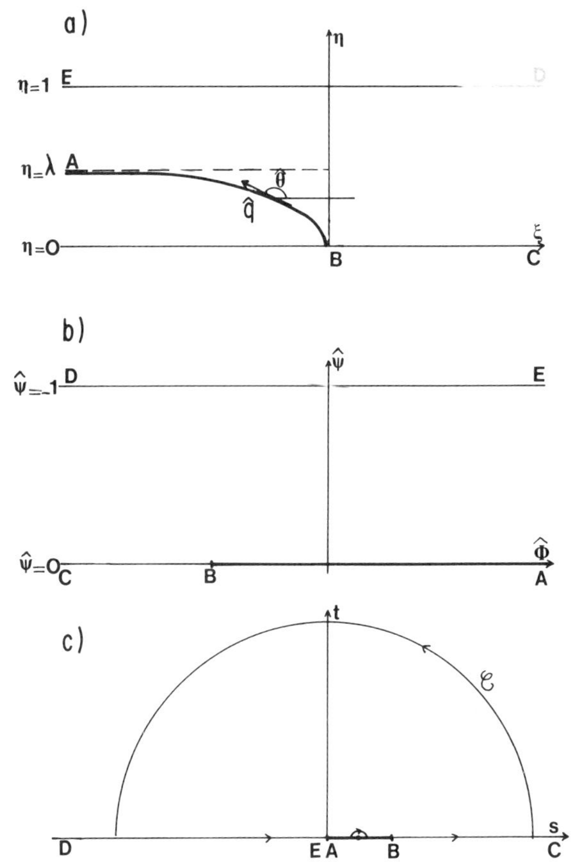

Fig. 2.4. Conformal mapping of the half cell region: a) real space, b) flow variables, c) s, t variables. The finger is represented by the thick line AB.

One can rewrite these equations in a form which does not explicitly involve λ. One notices from relations (2.19) and (2.21) that

$$\log(1-\lambda)=\frac{1}{\pi}\mathrm{P}\int_0^1\frac{\theta-\pi}{s'}\,ds'. \tag{2.22}$$

So, by redefining $\theta \equiv \theta - \pi$ and $q \equiv (1-\lambda)q$, one obtains from (2.19)

$$\log(q(s))=-\frac{s}{\pi}\mathrm{P}\int_0^1\frac{\theta(s')}{s'(s'-s)}\,ds', \tag{2.23}$$

and from (2.20)

$$\kappa qs\frac{d}{ds}\left(qs\frac{d\theta}{ds}\right)-q=-\cos\theta, \tag{2.24}$$

with

$$\kappa=\frac{\sigma b^2\pi^2}{12\mu Ua^2(1-\lambda)^2} \tag{2.25}$$

and the appropriate boundary conditions

$$\theta(0)=0,\quad q(0)=1,\quad \theta(1)=-\pi/2,\quad q(1)=0 \tag{2.26}$$

The problem is to find the number of solutions of the system of equations (2.23), (2.24), with boundary conditions (2.26) for a given κ. It appears as a nonlinear eigenvalue problem in the sense of Barenblatt [34]. When a solution is found, the corresponding λ is computed from relation (2.22). Then one can find the relation between λ and the capillary number by using relation (2.25).

When $\kappa = 0$ one regains the Saffman-Taylor continuum of solutions which in these variables are

$$q=\sqrt{\frac{1-s}{1+s\alpha}},\qquad \theta=\cos^{-1}q \tag{2.27}$$

where $\alpha=(2\lambda-1)/(1-\lambda)^2$ is arbitrary.

Solutions can be determined numerically when $\kappa \neq 0$ and analytically when κ is close to zero or to infinity.

2.3.2. Numerical Solutions

Many attempts have been made to determine numerically the solutions of the McLean and Saffman equations. First, McLean and Saffman [26] determined one branch of solutions $\lambda(\kappa)$ starting from $\frac{1}{2}$ for small κ and going to 1 for κ large. Romero [27] found two other branches of solutions. Then Vanden-Broeck [28] found a countably infinite set of solutions $\lambda_n(\kappa)$ (Fig. 2.5). After relaxing the condition $\theta(1)=0$, he found a unique solution for all values of λ and κ. For a fixed κ, $\theta(1)$ was found to be zero for a countably infinite set of values $\lambda_n(\kappa)$, this condition giving the acceptable solutions. All these branches of

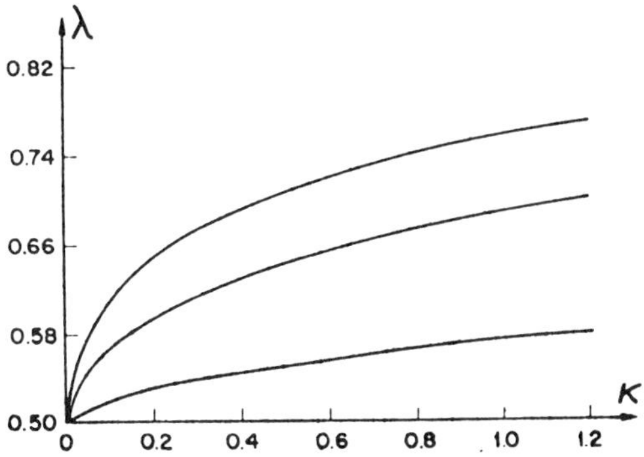

Fig. 2.5. Values of λ versus κ. (Reprinted from Fig. 4 of Vanden-Broeck [28] with the permission of the American Institute of Physics.)

solutions are increasing functions of κ, starting from $\kappa = 0$, $\lambda = \frac{1}{2}$ (small surface tension) and tending to $\lambda = 1$ when κ goes to infinity (large surface tension).

The large κ limit (slow fingers) was studied analytically by Pomeau [105] and Dombre and Hakim [106]. The results obtained in the limit of small κ (fast fingers) are rather surprising. The flat front is unstable with a typical wavelength λ_c which behaves as $U^{-1/2}$, so which decreases when the velocity increases. In general, the nonlinear structures which develop at large time after the beginning of the instability are roughly scaled with the most unstable wavelength of the instability. This is not the case here since, at large velocity, the thickness of the finger saturates at the value $w/2$ while the most unstable wavelength tends to zero. This is why we will now present in detail the analytical study of the McLean-Saffman-Romero-Vanden-Broeck solutions in the limit of small κ.

2.3.3. Singular Perturbation Analysis in the Limit $\kappa \ll 1$ ($\lambda \approx \frac{1}{2}$)

A formal inner-outer perturbation expansion was performed by McLean and Saffman [26]. In each region (the inner region is where s is close to zero) the variables q and θ are expanded in powers of κ and matched at each order of the expansion, whatever the value of λ. This result is in complete disagreement with the numerical study of Vanden-Broeck where values of λ close to $\frac{1}{2}$ are selected.

They suggested that this selection is due to transcendentally small terms of the expansion which they were not able to compute at that time. This argument was confirmed by a numerical calculation by Kessler and Levine [82] who showed that, when λ does not belong to the set of allowed values, the slope of the tip, which in this case is not zero, is exponentially small. The vanishing of the slope appears as a solvability condition that they called the microsolvability mechanism.

The analytical evaluation of these transcendentally small terms was done almost simultaneously by many groups (Dombre *et al.* [29], Shraiman [30], Hong and Langer [31], Combescot *et al.* [32], and Tanveer [33]) by using a Wentzel-Kramers-Brillouin (W.K.B.) method developed first in the case of the geometrical models by Kruskal and Segur [51] and Langer [52]. We present here the main steps which lead to the analytical determination of the solvability condition.

The first step is rather usual in perturbation expansions (see Pelcé and Pomeau [53] and Shraiman [30]). Let $X_0(l)$ (for instance a Saffman-Taylor shape parametrized by the curvilinear coordinate l) be a solution of the equation

$$F(X_0(l)) = 0. \tag{2.28}$$

One wants to find a solution $X(l)$ of the equation

$$F(X(l)) + \kappa G(X(l)) = 0 \tag{2.29}$$

for κ small (F and G are in general nonlinear integral operators). Such a form of the equations can be obtained by writing q as a function of θ by using relation (2.23) and then replacing it in equation (2.24). If one writes the solution as $X = X_0 + X_1$, where X_1 is a small perturbation, X_1 satisfies the equation

$$LX_1 = -\kappa G(X_0), \tag{2.30}$$

where L is the linear operator $(dF/dX + \kappa\, dG/dX)|_{X_0}$. Equation (2.30) has a nontrivial solution if the solvability condition

$$\int_{-\infty}^{+\infty} Y_1(l) G(X_0(l))\, dl = 0 \tag{2.31}$$

is satisfied. Here Y_1 is any element of the kernel of L^+, the adjoint of L. It is clear that L has nontrivial elements in its kernel since Equation (2.28) has a continuum of solutions. If X_0 is one of these solutions a nearby solution is $X_0 + \epsilon X_1$ where ϵ is small and $LX_1 = 0$. But since L is not self-adjoint in general, this does not help to find the kernel of L^+.

To get nontrivial solvability conditions, one must consider elements Y_1 of the kernel of L^+ which vary on the fast scale (λ_c the most unstable wavelength of the planar front) compared to the spatial scale (w the cell width) of the finger. We present here a way to compute these elements which, despite its lack of mathematical rigor, gives more physical insight into the problem and can be applied to other kinds of curved fronts (Pelcé and Bensimon [107]). The main idea is that as far as fast modes are concerned, there is a link between the operator L and the linear stability analysis of the planar interface.

Let us begin with a simple example, a planar front moving with constant velocity V. An element of the kernel of L is a perturbation of the interface which is marginal (it does not evolve with time). In this case, the marginal modes are of the form $\exp(ikl)$, where k is a zero of the dispersion relation

$$\omega = |k|\left(V - \frac{b^2\sigma}{12\mu}k^2\right), \tag{2.32a}$$

with $\omega = 0$. There are three solutions. One is a constant ($k = 0$) which corresponds to a simple translation of the front in the direction of propagation. The other modes are of the form $\xi = \cos(k_{\mathrm{m}}l)$, $\xi = \sin(k_{\mathrm{m}}l)$ where

$$k_{\mathrm{m}} = \frac{1}{b}\sqrt{12Ca}\,. \tag{2.32b}$$

The operator L is so naturally deduced with the right-hand side of relation (2.32a). To the Fourier component $\xi(k)$ of the perturbation $\xi(l)$ it associates the number $|k|(V - b^2\sigma/12\mu k^2)\xi(k)$. In the real space, L is defined as

$$L: \xi(l) \to \mathrm{P}\int_{-\infty}^{+\infty} \frac{dl'}{l - l'}\left(V\frac{\partial\xi}{\partial l'} + \frac{b^2\sigma}{12\mu}\frac{\partial^3\xi}{\partial l'^3}\right). \tag{2.33}$$

This operator is self-adjoint. The elements of the kernel of L^+ are thus the same as those mentioned above.

Consider now the case of the planar front moving with constant velocity U in the x direction but tilted at an angle θ. In a frame fixed with the interface there is now a uniform tangential flow of magnitude $-U\sin\theta$. In this frame, the dispersion relation is simply deduced from relation (2.32a) by a change of galilean frame, i.e.,

$$\omega = iU\sin\theta k + |k|\left(U\cos\theta - \frac{b^2\sigma}{12\mu}k^2\right). \tag{2.34}$$

The nontrivial marginal perturbations ($\omega = 0$) cannot be found with k real. If one allows complex k one must adapt the previous formula

$$iU\sin\theta k + \{k\}\left(U\cos\theta - \frac{b^2\sigma}{12\mu}k^2\right) = 0, \tag{2.35}$$

where $\{k\} = k$ if $\mathrm{Re}(k) > 0$ and $\{k\} = -k$ if $\mathrm{Re}(k) < 0$[6]. The corresponding linear operator L associates to the Fourier component $\xi(k)$ the left-hand

[6] $\{k\}$ appears as a generalisation of $|k|$ for k complex. It comes in relation (2.35) because the perturbation of the potential flow is proportional to $\exp(iks - qn)$. Here n is the normal coordinate with increasing value towards the fluid, and $q^2 = k^2$. Since the flow cannot diverge when $n \to +\infty$, one must choose $q = \{k\}$.

side of the relation (2.35) multiplied by $\xi(k)$ itself. It is no longer self-adjoint because of the presence of the term ik in relation (2.35) which corresponds to a first derivative in real space. Thus, in this simple formulation, it appears that the term which breaks the self-adjoint property of the linear operator L is the tangential velocity of the flow, in the frame of the finger. One can build the adjoint operator simply by changing ik to $-ik$. The marginal modes of this new operator have a wave number k which now satisfies

$$-iU\sin\theta k + \{k\}\left(U\cos\theta - \frac{b^2\sigma}{12\mu}k^2\right) = 0, \tag{2.36}$$

i.e., they are

$$\begin{aligned} &\cos(k_{\mathrm{m}}\cos(\theta/2)l)\exp(k_{\mathrm{m}}\sin(\theta/2)l),\\ &\sin(k_{\mathrm{m}}\cos(\theta/2)l)\exp(k_{\mathrm{m}}\sin(\theta/2)l). \end{aligned} \tag{2.37}$$

When θ is between 0 and $\pi/2$ (resp. between $-\pi/2$ and 0) these functions increase exponentially as l goes to $+\infty$ (resp. to $-\infty$) and decrease exponentially as l goes to $-\infty$ (resp. to $+\infty$).

We consider now the case where the interface is a finger. When κ is small, the spatial scale of variation of some element of the kernel of L^+, which is of order λ_c is very small compared to w, the spatial scale of the finger. At a scale of order λ_c, the front can be considered as planar, the local wave number being fixed by θ. As is usual in the W.K.B. approximation, when a solution Ψ of a wave equation for k constant is $\exp(ikl)$, the corresponding solution for k slowly varying is, to lowest order,

$$\Psi \approx \exp\left(\int_0^l ik(l')\,dl'\right). \tag{2.38}$$

Thus, the elements of the kernel of L^+ which vary rapidly in space can be built from (2.37) as

$$\begin{aligned} &\cos\left(k_{\mathrm{m}}\int_0^l \cos\left(\frac{\theta}{2}\right)dl'\right)\exp\left(k_{\mathrm{m}}\int_0^l \sin\left(\frac{\theta}{2}\right)dl'\right),\\ &\sin\left(k_{\mathrm{m}}\int_0^l \cos\left(\frac{\theta}{2}\right)dl'\right)\exp\left(k_{\mathrm{m}}\int_0^l \sin\left(\frac{\theta}{2}\right)dl'\right). \end{aligned} \tag{2.39}$$

As opposed to the previous case, these functions decrease exponentially as l goes to $\pm\infty$ and thus can lead to convergent integrals when they are used in the solvability condition (2.31).

One can now write the solvability condition associated with these two elements as

$$\int_{-\infty}^{+\infty} dl\, G(X_0(l))\exp\left(-ik_{\mathrm{m}}\int_0^l \exp\left(\frac{i\theta}{2}\right)dl'\right) = 0. \tag{2.40}$$

The main interest of this formulation is that it is adaptable to other kinds of fronts. One needs the continuum of solutions $X_0(l)$ (the Saffman-Taylor shapes) from which the function $\theta(l)$ is deduced, the inhomogeneous term G (the curvature term) and the dispersion relation for the perturbations on the planar front.

We now compute explicitly the solvability condition (2.40). One needs first the relation between θ and l for a Saffman-Taylor shape. From the relation

$$\frac{dl}{ds} = -\frac{a(1-\lambda)}{\pi s q}, \tag{2.41}$$

where q is given by relation (2.27), one obtains

$$l = \frac{a(1-\lambda)}{\pi}\left(\log\left(\frac{1-\sin\theta}{1+\sin\theta}\right) - 2\sqrt{\alpha}\,\mathrm{Arctg}\left(\sqrt{\alpha}\sin\theta\right)\right). \tag{2.42}$$

By changing variable to $\eta = \tan\theta$, one can rewrite the solvability condition (2.40) as

$$\int_{-\infty}^{+\infty} G(\eta)\exp\left(\frac{\Psi(\eta)}{\sqrt{\kappa}}\right) d\eta = 0. \tag{2.43}$$

Here, G is a slowly varying function of η, and $\Psi(\eta)$ is defined as

$$\Psi(\eta) = 2i(1+\alpha)\int_0^{\eta} \frac{(1+i\eta)^{3/4}(1-i\eta)^{1/4}}{1+(1+\alpha)\eta^2}\, d\eta. \tag{2.44}$$

Relations (2.43) and (2.44) are the same as the ones determined by Hong and Langer [31]. The integral (2.43) is evaluated by steepest descent methods (see the chapter "Asymptotic Expansion of Integrals" in the book by Bender and Orzsag [108]). The function Ψ has two stationary phase points at $\eta = i$ and $\eta = -i$. One can show that the contribution of the point $\eta = i$ to the integral (2.43) dominates. To evaluate this contribution, one deforms the original contour (the real axis) to a contour of steepest descent which goes through the point $\eta = i$. Now, two cases must be considered, depending upon the sign of $\lambda - \frac{1}{2}$.

When λ is less than $\frac{1}{2}$, there is no singularity below the steepest descent path and the resulting value of the integral (2.43) is of the form

$$A\kappa^m \exp\left(\frac{\Psi(i)}{\sqrt{\kappa}}\right). \tag{2.45}$$

Here A is a constant, m a power which depends upon the expansion of Ψ around $\eta = i$ and $\Psi(i)$ can be written as

$$\Psi(i) = -2(1+\alpha)\int_0^1 \frac{(1-x)^{3/4}(1+x)^{1/4}}{1-(1+\alpha)x^2}\, dx, \tag{2.46}$$

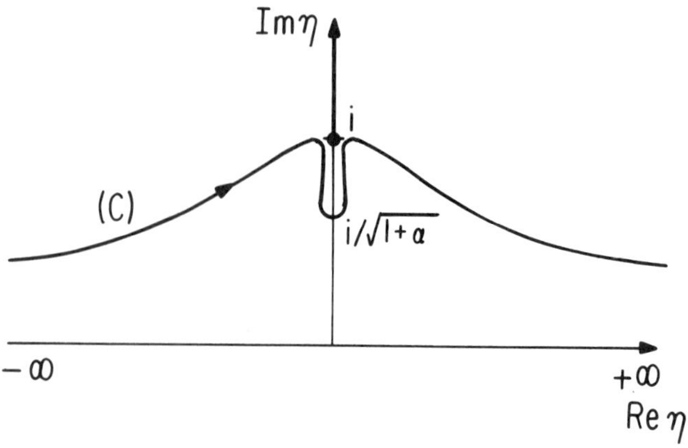

Fig. 2.6. Contour (C) of integration.

which is a negative real number. So the quantity (2.45) is exponentially small and cannot vanish: The solvability condition (2.43) cannot be satisfied up to exponentially small terms.

When λ is larger than $\frac{1}{2}$, a logarithmic singularity now lies on the imaginary axis between 0 and i, at $\eta = i/\sqrt{(1+\alpha)}$, and the steepest descent path must turn around it (Fig. 2.6). This gives an oscillatory factor to the exponentially small resulting value of the integral of the form $\cos(A\alpha^{3/4}/\sqrt{\kappa})$. The solvability condition (2.43) can now vanish for some discrete values of λ. These selected values satisfy the following scaling law:

$$\lambda - \tfrac{1}{2} \approx n^{3/4}\kappa^{2/3}, \tag{2.47}$$

which is in good agreement with the numerical results of Vanden-Broeck [28]. A more rigorous method, developed by Kruskal and Segur [51] in the context of the geometrical model [48], consists of solving an analytical continuation of the equations in the complex plane. It has been applied by Combescot *et al.* [32] and Tanveer [33] to the Saffman-Taylor problem. This method leads to the same scaling as in relation (2.47).

2.4. Comparison with Experiments and Numerical Simulations

2.4.1. Comparison with Experiments

The most recent results are those of Tabeling *et al.* [1] who carefully redid the original experiment of Saffman and Taylor [9]. In this experiment air is

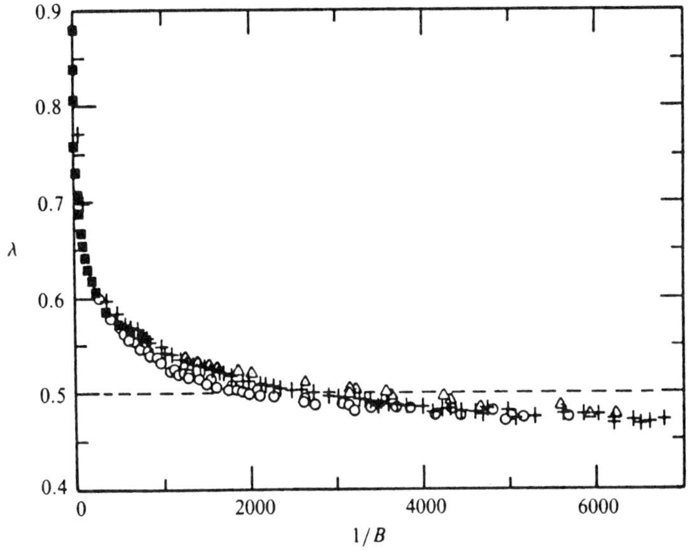

Fig. 2.7. Plot of λ versus $1/B$ for three different aspect ratios: $w/b = 112.5$ (Δ), $w/b = 65(+)$, and $w/b = 34(0)$. Here $1/B = 4\pi^2/(1-\lambda)\kappa$. (Reprinted with permission from Fig. 8 of Tabeling *et al.* [1]. © 1987 Cambridge University Press.)

pushing silicone oil in Hele-Shaw cells of various aspect ratios ($w/b = 34$ to 112.5). The graph of λ versus the control parameter $1/B \approx 1/(1-\lambda)^2\kappa$ lies almost on the same curve (Fig. 2.7) which is not far from the theoretical curve of McLean and Saffman (the first of the branches found by Vanden-Broeck). However there is one important discrepancy. For large $1/B$, the experimental curve goes slightly below $\frac{1}{2}$, which is in disagreement with the theory of stationary states of McLean, Saffman and Vanden-Broeck. This discrepancy may be due to the film effect which is not taken into account in the theory. The only remaining question, then, is why only the first branch of solutions is observed in experiments. The answer lies in the stability of the solutions, which will be taken up in the last chapter.

2.4.2. Comparison with Numerical Simulations

There have been several time-dependent numerical calculations simulating the motion of a Saffman-Taylor interface. They use the same two-dimensional free boundary problem given by relations (10), (11), (12), (13), where film effects have been neglected. Different numerical methods are used. Simulations by Meng and Thomson [109], and Tryggvason and Aref [110], [111] use a "vortex in cell" method and can reproduce the McLean and Saffman steady-state results for relatively large surface tension. DeGregoria and Schwartz [73] use a boundary-integral approach and reproduce steady-state results for lower surface tension. Similar results were obtained by Bensimon [74], by a pole dynamics method, and by Liang [112], by a method using random walkers. In a more recent work, DeGregoria and Schwartz [113] obtain steady fingers with relative width slightly less than $\frac{1}{2}$ at very low surface tension. This is apparently in contradiction with the steady-state results except if the fingers produced in [113] are quasi-steady. This point is not very clear for the moment.

3. Stationary Shapes of a Needle Crystal Growing From a Supercooled Liquid

3.1. The Model of Growth

The liquid-solid interface is a region that is in general very small (some angstroms) compared to the macroscopic scale. During crystal growth, it is here that the latent heat is released, that the density of the material and the concentration of an impurity vary rapidly. In order to study the dynamic of growth, one can consider the interface as a discontinuity surface for the fields of temperature, flow velocity of the liquid, and concentration of an impurity. At the interface are satisfied boundary conditions determined by an analysis of the interface structure and by conservation laws.

3.1.1. The Interface

The interface is assumed to be rough, that is to say formed of many crystalline steps, whose thickness is on the order of molecular dimensions (Burton and Cabrera [114]). In this case the kinetic time of transfer of molecules between solid and liquid is very fast compared to the characteristic diffusion time of heat or of an impurity. Thus, one can consider the limiting case where the conditions of growth are out of equilibrium because of the presence of temperature or concentration gradient in the medium, the interface being itself at equilibrium. In these conditions, a thermodynamic relation, the Gibbs-Thomson relation, gives the temperature at the interface as a function of the local curvature of the front.

3.1.2. The Gibbs-Thomson Relation

Consider the case of the pure liquid. Let the solid be in equilibrium with the melt in a vessel in which the liquid is in contact with a reservoir at constant pressure P_1. The temperature T of the system is uniform. When the surface of the crystal whose equation is $z = \alpha(x, y)$ is slightly deformed, the variation of the volume of the solid (resp. the liquid) is δV_s (resp δV_1). The variation of the number of molecules of the solid is δN_s and the one of the liquid $-\delta N_s$ since the total number of molecules is conserved. During this small deformation, the exchange of work due to volume variations is

$$\delta W = -P_l(\delta V_l + \delta V_s). \tag{3.1}$$

Thus, the equilibrium shape will be determined by the minimum of the thermodynamic potential $G = F + P_l V$, F being the Helmholtz free energy of the system and V the total volume occupied by the two phases. The variation of this potential is

$$dG = (P_l - P_s)\delta V_s + (\mu_s - \mu_l)\delta N_s + \delta \int_{(S)} \sigma(\mathbf{n})\, dS, \tag{3.2}$$

where $\sigma(\mathbf{n})$ is the anisotropic liquid-solid surface tension, $\mathbf{n}$ the unit vector normal to the interface and S the surface of the crystal[7]. Let R_1 and R_2, the principal radii of curvature of the interface, be positive when they are directed towards the solid side. Let ϕ_1 and ϕ_2 be the angles of the normal to the interface with some fixed direction inside each main plan. One can show (see, for instance, Chernov [115]) that

$$\delta \int_{(S)} \sigma(\mathbf{n})\, dS = \int_{(S)} \left\{ \left(\sigma + \frac{\partial^2 \sigma}{\partial \phi_1^2} \right) \frac{1}{R_1} + \left(\sigma + \frac{\partial^2 \sigma}{\partial \phi_2^2} \right) \frac{1}{R_2} \right\} \delta\alpha\, dS, \tag{3.3}$$

a relation obtained for the first time by Herring [116]. As

$$\delta V_s = \int_{(S)} \delta\alpha\, dS, \tag{3.4}$$

the minimization of G with respect to the independent variables δV_s and δN_s leads to the relations

$$P_s - P_l = \left(\sigma + \frac{\partial^2 \sigma}{\partial \phi_1^2} \right) \frac{1}{R_1} + \left(\sigma + \frac{\partial^2 \sigma}{\partial \phi_2^2} \right) \frac{1}{R_2}, \tag{3.5}$$

which appears as a generalization of the Laplace law to the case of anisotropic surface tension, and

$$\mu_l(P_l, T) = \mu_s(P_s, T). \tag{3.6}$$

One expands relation (3.6) around a point (P^*, T^*) of the curve of coexistence of the two phases of the planar interface, and obtain

$$(P_l - P^*)v_l - (T - T^*)s_l = (P_s - P^*)v_s - (T - T^*)s_s, \tag{3.7}$$

where v_l (resp. v_s) is the specific volume of the liquid (resp. the solid) and s_l (resp. s_s) is the specific entropy of the liquid (resp. the solid). Using relation

[7] One assumes that the stress tensor $\sigma_{\mu\nu}$ of the solid is simply hydrostatic, that is to say, $\sigma_{\mu\nu} = -P_s\delta_{\mu\nu}$. For instance, effects due to elastic deformations are neglected.

(3.5), one obtains

$$\delta p\left(\frac{v_1}{v_s} - 1\right) - \frac{\delta T}{T^*}Q = \left(\sigma + \frac{\partial^2\sigma}{\partial\phi_1^2}\right)\frac{1}{R_1} + \left(\sigma + \frac{\partial^2\sigma}{\partial\phi_2^2}\right)\frac{1}{R_2}, \tag{3.8}$$

where $Q = T^*(s_1 - s_s)/v_s$ is the latent heat per unit volume of solid and δp (resp. δT) is the variation of the liquid pressure (resp. the temperature) with respect to P^* (resp. T^*). One considers only the case where the pressure of the liquid is constant and uniform. One can thus take it as P^* (which corresponds to $\delta p = 0$) and pick out the temperature variation δT to the corresponding T^*. One then obtains the Gibbs-Thomson law, which relates the temperature at the interface function of the local radius of curvature

$$\frac{\delta T}{T^*}Q = -\left(\sigma + \frac{\partial^2\sigma}{\partial\phi_1^2}\right)\frac{1}{R_1} - \left(\sigma + \frac{\partial^2\sigma}{\partial\phi_2^2}\right)\frac{1}{R_2}. \tag{3.9}$$

This law says essentially that when a bulge of solid penetrates inside the melt, the temperature at the tip of the bulge is smaller than the fusion temperature of the planar interface.

3.1.3. The Diffusion Field

The metastable liquid is assumed to be free of all impurities, whose very slow diffusion would limit the crystal growth. Instead, the growth is limited by diffusion of the latent heat in the medium, which is released at the interface. The relevant field is the one of temperature, which satisfies the diffusion equation

$$\frac{\partial T}{\partial t} = D_1 \Delta T \tag{3.10}$$

in the liquid, and

$$\frac{\partial T}{\partial t} = D_s \Delta T \tag{3.11}$$

in the solid. D_1 and D_s are respectively the diffusion coefficients for heat in the liquid and the solid.

3.1.4. Conservation Relation

Conservation of energy at the interface is written by matching the normal components to the interface of the energy density fluxes, in a frame where the interface is at rest:

$$\mathbf{j}_1 \cdot \mathbf{n} = \rho_1 h_1 \mathbf{v}_1 \cdot \mathbf{n} - \lambda_1 \nabla T \cdot \mathbf{n} = \mathbf{j}_s \cdot \mathbf{n} = \rho_s h_s \mathbf{v}_s \cdot \mathbf{n} - \lambda_s \nabla T \cdot \mathbf{n}. \tag{3.12}$$

By using the conservation of mass,

$$\rho_s \mathbf{v}_s \cdot \mathbf{n} = \rho_1 \mathbf{v}_1 \cdot \mathbf{n} = -\rho_s \mathbf{v} \cdot \mathbf{n}, \tag{3.13}$$

one obtains

$$(h_1 - h_s)\rho_s \mathbf{v} \cdot \mathbf{n} = \left(\lambda_s(\nabla T)_s - \lambda_1(\nabla T)_1\right) \cdot \mathbf{n}, \tag{3.14}$$

where h_1 (resp. h_s) is the enthalpy per unit mass of liquid (resp. of solid), λ_1 (resp. λ_s) is the thermal conductivity of the liquid (resp. the solid) and $\mathbf{v} \cdot \mathbf{n}$ is the normal velocity of the interface. Since the latent heat per unit volume of solid is $Q = (h_1 - h_s)\rho_s$ and the diffusion constant is $D = \lambda/c_p$, where c_p is the specific heat per unit volume, one can express the conservation of energy as

$$Q\mathbf{v} \cdot \mathbf{n} = \left(D_s c_{ps}(\nabla T)_s - D_1 c_{p1}(\nabla T)_1\right) \cdot \mathbf{n}. \tag{3.15}$$

3.1.5. The Free Boundary Problem

In the model explored in the following, we will assume that diffusion coefficients and specific heat per unit volume are equal in both phases (the symmetric model). The dynamics of the liquid-solid interface is determined by the following free boundary problem.

In both liquid and solid, the temperature satisfies the diffusion equation

$$\frac{\partial T}{\partial t} = D\Delta T \tag{3.16}$$

with, at infinity in the liquid, $T = T_\infty$. At the solid-liquid interface $z = \alpha(x, y)$, the following two boundary conditions hold: the relation of conservation of energy,

$$Q\mathbf{v} \cdot \mathbf{n} = Dc_p((\nabla T)_s - (\nabla T)_1) \cdot \mathbf{n}, \tag{3.17}$$

and the Gibbs-Thomson relation,

$$T - T_s = -\frac{T_s}{Q}\left\{\left(\sigma + \frac{\partial^2\sigma}{\partial\phi_1^2}\right)\frac{1}{R_1} + \left(\sigma + \frac{\partial^2\sigma}{\partial\phi_2^2}\right)\frac{1}{R_2}\right\}. \tag{3.18}$$

3.2. The Problem Without Surface Tension

3.2.1. The Ivantsov Paraboloid

Ivantsov [40] has shown that the free boundary problem has a family of exact solutions when surface tension is neglected, i.e., when the interface is assumed isothermal. The corresponding crystal shapes are paraboloids of revolution with tip radius ρ and moving with constant velocity U in the direction Oz, the axis of revolution (Fig. 3.1). Let us introduce the parabolic coordinates $\xi = (r - z)/\rho$, $\eta = (r + z)/\rho$ and θ the angle around the Oz axis; r is the distance $(x^2 + y^2 + z^2)^{1/2}$.

In these variables, the equation of the paraboloid with tip radius ρ is simply $\eta = 1$. The stationary diffusion equation is, in the frame moving with constant

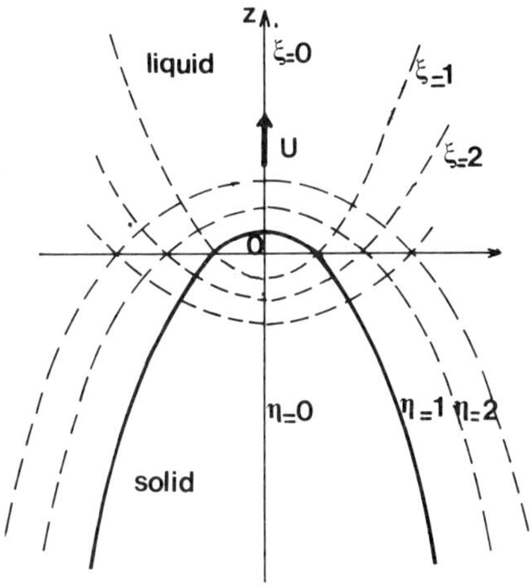

Fig. 3.1. Parabolic coordinates.

velocity U,

$$-P\frac{1}{\xi+\eta}\left(\eta\frac{\partial T}{\partial\eta}-\xi\frac{\partial T}{\partial\xi}\right)=\frac{1}{\xi+\eta}\left(\frac{\partial}{\partial\eta}\eta\frac{\partial T}{\partial\eta}+\frac{\partial}{\partial\xi}\xi\frac{\partial T}{\partial\xi}\right)+\frac{1}{4\eta\xi}\frac{\partial^2 T}{\partial\theta^2}, \tag{3.19}$$

where P is the Peclet number $P = \rho U/2D$.

On the interface $\eta = \eta(\xi, \theta)$ the boundary conditions are $T = T_s$ and

$$\frac{Q}{c_p}\left(\eta+\xi\frac{\partial\eta}{\partial\xi}\right)=\frac{1}{P}(G_s-G_1) \tag{3.20}$$

where

$$G=\eta\frac{\partial T}{\partial\eta}-\xi\frac{\partial\eta}{\partial\xi}\frac{\partial T}{\partial\xi}-\frac{\eta+\xi}{4\eta\xi}\frac{\partial\eta}{\partial\theta}\frac{\partial T}{\partial\theta}. \tag{3.21}$$

One sees that the paraboloid $\eta = 1$ is a solution of the free boundary problem. The corresponding temperature field depends only upon η ($T = T(\eta)$) outside the paraboloid ($\eta > 1$) and is constant and equal to T_s inside it ($\eta < 1$). By applying relation (3.19) and the boundary conditions, one obtains

$$T(\eta)=T_\infty+(T_s-T_\infty)\frac{\displaystyle\int_{p\eta}^{+\infty}\frac{\exp(-u)}{u}\,du}{\displaystyle\int_{P}^{+\infty}\frac{\exp(-u)}{u}\,du} \tag{3.22}$$

and

$$\Delta = P\exp(P)\int_{p}^{+\infty}\frac{\exp(-u)}{u}\,du, \tag{3.23}$$

where $\Delta = c_p(T_s - T_\infty)/Q$ is the dimensionless supercooling. Thus, at this level, the velocity of the crystal is not determined since at given Δ, only the product ρU is determined.

3.2.2. Dimensional Analysis

One could anticipate the relation (3.23) by a dimensional analysis. In the class of systems M (mass), L (length), T (time), θ (temperature) (see for instance Barenblatt [34]), the dimension of the parameters which are involved are

$(U) = L/T$, for the velocity of the crystal
$(D) = L^2/T$, for the diffusive coefficient of heat
$(Q) = M/LT^2$, for the latent heat per unit volume of solid
$(T_s) = \theta$, for the fusion temperature of the planar interface
$(T_\infty) = \theta$, for the temperature at infinity in the liquid
$(c_p) = M/LT^2\theta$, for the specific heat per unit volume.

With these parameters, one can build only two dimensionless expressions, which are Δ and c_pT_s/Q. Thus, there is no way to relate the velocity of the crystal U to the dimensionless supercooling Δ. If solutions exist, they must belong to a family of surfaces parametrized by a characteristic length ρ. With this new length, one can build the Peclet number $P = \rho U/2D$ in which the velocity is involved. As T_s does not play any particular role with respect to T_∞, one must relate Δ and P, as in relation (3.23).

The parabolic shape of the crystal could be anticipated by reasoning typical of boundary layer problems. When the crystal grows with constant velocity, the heat released at the interface is advected along the axis Oz according to the law $z = Ut$. Along a transverse direction, it diffuses, according to the law $y = (Dt)^{1/2}$. An isotherm (the interface for instance) is obtained by eliminating the time between these two relations, that is to say $z = y^2U/D$. A simple characteristic length is the radius of curvature ρ of the paraboloid.

More generally, Horway and Cahn [117] have shown that elliptic paraboloids are solutions too.

3.3. Effects of Surface Tension

The surface tension introduces a new length in the problem: σ/Q. We will assume for the moment that the surface tension is isotropic. With the physical quantities mentioned above, one can build three dimensionless quantities, which are

$$DQ/U\sigma, \qquad c_pT_s/Q, \qquad \text{and } \Delta. \tag{3.24}$$

Thus, it is no longer necessary to introduce an additional length ρ in order to determine the crystal velocity. One must write the velocity as

$$U = DQ/\sigma f\left(c_p T_s/Q, \Delta\right). \tag{3.25}$$

3.3.1. Phenomenological Models

Numerous attempts have been made to determine the velocity of the crystal when surface tension is taken into account. Many of them assume that the shape of the crystal is still a paraboloid with a radius of curvature ρ. The boundary conditions cannot be satisfied at all points of the interface, since now the temperature is no longer constant but depends upon the curvature of the interface. For example the "modified Ivantsov" method introduced by Sekerka *et al.* [43] and Glicksman and Schaefer [44] assumes that the temperature field satisfies relation (3.17) at all points but satisfies the Gibbs-Thomson relation at the tip only. Other methods have been used, as summarized in Table I. Each of these results in a relation between the Peclet number $P = \rho U/2D$ and the dimensionless numbers given by (3.24). The elimination of the curvature radius at the tip ρ is realized by a selection principle. One of them used for a long time is the following: The curve $V(\rho)$, other quantities being fixed, has a maximum for some critical value ρ^*. This is not true when surface tension is not taken into account. The selection principle says that the observed crystal shape is the one of maximum velocity.

3.3.2. The Nash-Glicksman Approach

A more rigorous approach was developed by Nash and Glicksman [46]. They determined a nonlinear integral equation for the shape of the interface which can be written as

$$\Delta - \frac{1}{R} = u\int_{-\infty}^{+\infty} \frac{d\tau}{(4\pi\tau)^{3/2}} \int_{-\infty}^{+\infty} dx_1 \int_{-\infty}^{+\infty} dy_1$$
$$\times \exp\left[-\frac{1}{4\tau}\left((x-x_1)^2 + (y-y_1)^2 + (\alpha - \alpha_1 + u\tau)^2\right)\right]. \tag{3.26}$$

(A simple derivation is given by Langer [118].) Here, lengths are scaled by the capillary length $d_0 = \sigma c_p T_s/Q^2$, $1/R$ is the dimensionless curvature and $u = Ud_0/D$ is the dimensionless velocity. The left-hand side of the equation, which is the temperature at the point $(x, y, \alpha(x, y))$ of the interface, is determined by the sum of the contributions of the heat sources distributed at the points of coordinates $(x_1, y_1, \alpha_1(x_1, y_1))$ (the right-hand side of the equation). They determine numerically a continuum of solutions of this equation with one of maximum velocity. By applying the principle of maximum velocity, they determine the velocity of the crystal as

$$U = .128(D/d_0)\Delta^{2.65} \tag{3.27}$$

when the supercooling Δ is small, which is usually the case in experiments.

Table I. Steady–State Theories of Dendritic Growth
(Reprinted with permission from Table 1 of Glicksman *et al.* [2] (a column giving references has been deleted). Glicksman, M.E., Schaefer, R.J., Ayers, J.D., *Metallurgical Transactions A*, vol. 7A, pgs. 1747–1759, American Society for Metals, Metals Park, Ohio, 1976.)

Theory	Dendrite Shape	Gibbs-Thomson Boundary Condition	Thermal Flux, Balance	Molecular Attachment, Boundary Condition	Comment
Bolling-Tiller	Paraboloid of revolution	Tip only	Tip only	Tip only	
Fisher	Cylinder with hemispherical cap	Cap only	Approximated by macroscopic theory	No	$V \alpha \Delta T^2$
Glicksman-Schaefer	Paraboloid of revolution	Tip only	All points	Tip only	"Modified Ivantsov"
Holtzmann	Paraboloid of revolution	All points	Tip only	All points	
Horvay-Cahn	Elliptical paraboloids	No	All points	No	Isothermal theory
Ivantsov	Paraboloid of revolution	No	All points	No	Isothermal theory
Nash-Glicksman	Unconstrained body of revolution	All points	All points	No	Non-linear theory
Sekerka, *et al.*	Paraboloid of revolution	Tip only	All points	No	"Modified Ivantsov"
Tarshis-Kotler	Paraboloid of revolution	All points	Tip only	All points	
Temkin	Paraboloid of revolution	All points	Tip only	All points	
Trivedi	Paraboloid of revolution	All points	Tip only	All points	"Modified Temkin"

3.3.3. Asymptotic Analysis at Small Supercooling

Pelcé and Pomeau [53] have undertaken the Nash and Glicksman approach from an analytical point of view. In the small supercooling limit, they show that the velocity of an axisymmetric needle crystal takes the form

$$U = \frac{4}{C}\frac{D}{d_0}\left(\frac{\Delta}{\log \Delta}\right)^2. \tag{3.28}$$

Here C is an eigenvalue of a nonlinear integral equation that is independent of the dimensionless parameters. Before dealing with the axisymmetric case, let us begin with the less realistic but simpler case of a two-dimensional needle crystal.

Two-Dimensional Case. As the shape depends only upon the coordinate x, one can integrate the right-hand side of equation (3.26) with respect to the coordinates y_1 and τ to obtain

$$\Delta + \frac{\alpha_{xx}}{\left(1+\alpha_x^2\right)^{3/2}} = \frac{u}{2\pi}\int_{-\infty}^{+\infty} dx_1 \exp\left(-\frac{u}{2}(\alpha-\alpha_1)\right) \times K_0\left(\frac{u}{2}\sqrt{(x-x_1)^2+(\alpha-\alpha_1)^2}\right) \tag{3.29}$$

Here K_0 is the modified Bessel function of zero order.

It can be verified [53] that the parabola $\alpha = -x^2/2\rho$ is solution of equation (29), in which the term of curvature of the left-hand side has been omitted, if the condition

$$\Delta = \frac{2\rho u}{\sqrt{\pi}}\int_0^{+\infty} dx\, \frac{\exp(-x^2/4)}{x^2+2\rho u} \tag{3.30}$$

is satisfied. This is another way to find the solution of Ivantsov, Horway, and Cahn. This relation defines the tip radius ρ of a parabola moving with the same velocity u of a stationary solution of equation (3.29) and with the same supercooling Δ. When Δ is small, the relation (3.30) can be written as

$$\rho = \frac{2}{\pi}\Delta^2\frac{1}{u}. \tag{3.31}$$

Thus one can define two regions of very different size. The outer region, of size $1/u$, much larger than the size ρ of the inner region localised around the tip. The ratio between these two lengths is the small parameter of the asymptotic expansion: $\epsilon = 2\rho u = 4\Delta^2/\pi$.

The outer region. Let us define a new spatial variable of order unity in this region, $\xi = ux$, and rescale α as $\alpha_e = u\alpha$. The solution can be expanded in this region as

$$\alpha_{\mathrm{e}} = \frac{1}{\epsilon}\alpha_{\mathrm{e}}^{(0)}(\xi) + o(1/\epsilon), \tag{3.32}$$

where $\alpha_{\mathrm{e}}^{(0)}$ satisfies the equation

$$\left(\frac{\pi}{4}\epsilon\right)^{1/2} + u\epsilon^2\frac{\alpha_{\mathrm{e}\xi\xi}^{(0)}}{\left(\alpha_{\mathrm{e}\xi}^{(0)}\right)^3} = \frac{1}{2\pi}\int_{-\infty}^{+\infty} d\xi_1 \exp\left[-\frac{1}{2\epsilon}\left(\alpha_{\mathrm{e}}^{(0)} - \alpha_{\mathrm{e1}}^{(0)}\right)\right]$$
$$\times K_0\left(\frac{1}{2\epsilon}|\alpha_{\mathrm{e}}^{(0)} - \alpha_{\mathrm{e1}}^{(0)}|\right). \tag{3.33}$$

One anticipates that $u\epsilon^2$ is very small compared to $\epsilon^{1/2}$, which will be verified at the end of the calculation. Thus, the curvature term in (3.33) is negligible and the solution is the Ivantsov parabola

$$\alpha_{\mathrm{e}}^{(0)}(\xi) = -\xi^2. \tag{3.34}$$

The inner region. The spatial variable is defined here as $\eta = \xi/\epsilon$ and is of order unity in the region of size ρ. The inner solution is expanded as

$$\alpha_{\mathrm{i}}(\eta) = \epsilon\alpha_{\mathrm{i}}^{(0)}(\eta) + o(\epsilon), \tag{3.35}$$

where $\alpha_{\mathrm{i}}^{(0)}$ satisfies the equation

$$\left(\frac{\pi}{4}\epsilon\right)^{1/2} + \frac{u}{\epsilon}\frac{\alpha_{\mathrm{i}\eta\eta}^{(0)}}{\left(1 + \left(\alpha_{\mathrm{i}\eta}^{(0)}\right)^2\right)^{3/2}} = \frac{\epsilon}{2\pi}\int_{-\infty}^{+\infty} d\eta_1$$
$$\times K_0\left(\frac{\epsilon}{2}\sqrt{\left(\alpha_{\mathrm{i}}^{(0)} - \alpha_{\mathrm{i1}}^{(0)}\right)^2 + (\eta - \eta_1)^2}\right). \tag{3.36}$$

Since $\alpha_{\mathrm{i}}^{(0)}(\eta) = -\eta^2$ is a solution of equation (3.36) in which the curvature term has been omitted, and since $K_0(\epsilon) \approx -\log\epsilon$ when ϵ is small, one obtains by subtraction

$$\frac{\alpha_{\mathrm{i}\eta\eta}^{(0)}}{\left(1 + \left(\alpha_{\mathrm{i}\eta}^{(0)}\right)^2\right)^{3/2}} = -\frac{C}{4\pi}\int_{-\infty}^{+\infty}\log\left(\frac{\left(\alpha_{\mathrm{i}}^{(0)} - \alpha_{\mathrm{i1}}^{(0)}\right)^2 + (\eta - \eta_1)^2}{\left(-\eta^2 + \eta_1^2\right)^2 + (\eta - \eta_1)^2}\right) d\eta_1. \tag{3.37}$$

Here $C = \epsilon^2/u$ appears as a nonlinear eigenvalue of equation (3.37).

Matching of the solutions. One can show [53] that the inner and outer solutions can be matched if the condition

$$\int_{-\infty}^{+\infty} d\eta_1\left(\eta_1^2 + \alpha_{\mathrm{i}}^{(0)}(\eta_1)\right) = 0 \tag{3.38}$$

is satisfied. At the beginning of the analysis, we have assumed that $u\epsilon^{3/2}$ is very small compared to unity. Since $u = \epsilon^2/C$ (deduced from the analysis of the inner region), this condition is certainly satisfied.

Thus, when the dimensionless supercooling Δ is small (which is the case in most experimental situations), the problem of determining the stationary shapes of two-dimensional needle crystals is reduced to solving a nonlinear eigenvalue problem. If C denotes one of these eigenvalues, the velocity of the crystal is given by

$$U = \frac{16}{\pi^2 C}\frac{D}{d_0}\Delta^4. \tag{3.39}$$

The shape of the crystal is a parabola far from the tip and is a solution of (3.37) in the vicinity of it. We will return later to the problem of the determination of the eigenvalues C.

The Axisymetric Case. The asymptotic analysis is similar to the one developed previously. Let (r, θ, z) be cylindrical coordinates. For an axisymmetric solution $\alpha(r)$, the equation (3.26) can be written as

$$\begin{aligned}\Delta + \frac{\alpha_{rr} + \frac{1}{r}\left(\alpha_r^3 + \alpha_r\right)}{\left(1 + \alpha_r^2\right)^{3/2}} &= \frac{u}{4\pi}\int_0^{\pi} d\Phi \int_0^{+\infty} dr_1 \\ &\times \frac{r_1 \exp\left[-\frac{u}{2}(\alpha - \alpha_1)\right]\exp\left[-\frac{u}{2}\sqrt{(\alpha-\alpha_1)^2 + (r-r_1)^2 + 4rr_1\sin^2\Phi}\right]}{\sqrt{(\alpha-\alpha_1)^2 + (r-r_1)^2 + 4rr_1\sin^2\Phi}}.\end{aligned} \tag{3.40}$$

The Ivantsov paraboloid $\alpha = \quad r^2/2\rho$ is solution of this equation, in which the curvature term of the left-hand side has been omitted, if the condition

$$\Delta = \rho u \int_0^{+\infty} dr\, \frac{r\exp(-r^2/4)}{r^2 + 2\rho u} \tag{3.41}$$

is satisfied. When Δ is small, one obtains

$$\rho = -\frac{\Delta}{\log \Delta}\frac{1}{u}. \tag{3.42}$$

As in the two-dimensional case, one can distinguish two regions, the outer region of size $1/u$ being much larger than ρ, the size of the inner region located around the tip. The small parameter of the asymptotic expansion is the ratio between these two lengths: $\epsilon = 2\rho u = -2\Delta/\log \Delta$. One can show [53] that in the outer region, the shape of the crystal is a paraboloid with a radius of curvature at the tip equal to ρ. In the inner region, the shape is a solution of a nonlinear equation with eigenvalues $C = \epsilon^2/u$, as in the two-dimensional case. Thus, if C is one of these eigenvalues, the corresponding velocity of the crystal is

$$U = \frac{4}{C}\frac{D}{d_0}\left(\frac{\Delta}{\log \Delta}\right)^2. \tag{3.43}$$

It is interesting to compare these predictions with the theory of marginal stability of Langer and Muller-Krumbhaar [47]. This theory says that the dendritic growth is such that the tip of the crystal is marginally stable. This operating point corresponds to a constant value of the dimensionless number $\sigma = 2Dd_0/\rho^2 U$ where ρ is the radius of curvature of the tip. In the theory presented previously, a needle crystal is associated to an eigenvalue $C = 4\rho^2 U/Dd_0$ where ρ is the radius of curvature of the Ivantsov paraboloid towards which the shape of the crystal is asymptotic far from the tip. Thus the two theories are very similar as far as their predictions are concerned. For large eigenvalues C they are qualitatively identical since the two radii of curvature are the same. As one can see from relation (3.37), when C is large the shape of the crystal is very close to the Ivantsov parabola.

The Eigenvalue Problem. The determination of the eigenvalues C is a problem of mathematics. A numerical approach (Benamar and Moussallam [57]) leads to the conclusion that there are no solutions to the eigenvalue equation either in the bidimensional or the axisymmetric case (see also Benamar [119]). This is a confirmation of previous numerical results obtained by Meiron [55] in the two-dimensional and Kessler *et al.* [56] in the axisymetric case, on the full Nash-Glicksman equation (3.26) and where no stationary solutions of needle crystals were found when surface tension is isotropic. They found that the introduction of anisotropy in surface tension leads to stationary solutions. This idea came from a similar situation that occurs in the geometrical model of Brower *et al.* [48], Kessler *et al.* [49] and the boundary layer model of Ben-Jacob *et al.* [50]. Stationary solutions of needle crystals can be found only if anisotropy in the surface tension is introduced.

If we come back now to the eigenvalue problem, the situation is the same. Introduction of crystalline anisotropy leads to a discrete set of eigenvalues. Some analytical work can be done when the W.K.B. approximation applies (Benamar, Pomeau [58], Barbieri *et al.* [59]). From relation (1.5b), the most unstable wavelength of the planar front is $\lambda_c \approx (d_0 D/U)^{1/2}$, so that the eigenvalue C defined previously can be written as $C \approx (\rho/\lambda_c)^2$. Thus, the W.K.B. approximation which applies when ρ is very large compared to λ_c is valid in the limit of large C. In this configuration, as previously mentioned, the shape of the crystal is very close to a paraboloid (or to a parabola in the two-dimensional case). We will present here the determination of the eigenvalues C for the two-dimensional problem by the method presented previously in the Saffman-Taylor case (Pelcé and Bensimon [107]).

When anisotropy in surface tension is introduced, Equation (3.37) becomes

$$(1-\beta\cos 4\theta)\frac{\alpha_{\mathrm{i}\eta\eta}^{(0)}}{\left(1+\left(\alpha_{\mathrm{i}\eta}^{(0)}\right)^2\right)^{3/2}}$$

$$=-\frac{C}{4\pi}\int_{-\infty}^{+\infty}\log\left(\frac{\left(\alpha_{\mathrm{i}}^{(0)}-\alpha_{\mathrm{i}1}^{(0)}\right)^2+\left(\eta-\eta_1\right)^2}{\left(-\eta^2+\eta_1^2\right)^2+\left(\eta-\eta_1\right)^2}\right)d\eta_1. \qquad (3.44)$$

Here, a four-fold symmetry for the crystal has been assumed, so that the capillary length takes the form $d_0(1-\beta\cos 4\theta)$, where θ is the angle between the normal at the interface and the direction of propagation. Equation (3.44) can be written with the same form as equation (2.29):

$$F(X(l))+\frac{1}{C}G(X(l))=0. \qquad (3.45)$$

Here F is the integral operator of the right-hand side of equation (3.37) and G the curvature term. The Ivantsov parabola $X_0(l)$ is solution of equation (3.45) when C is infinite. As already shown in Section 2, one can get a non-trivial solvability condition, with elements of the adjoint of the linearised operator which vary on a scale λ_c, which is very fast compared to the radius of curvature of the needle crystal. For this, one needs the dispersion relation for perturbations of a tilted planar front moving with a constant velocity U, in a frame where the interface is at rest. From relation (1.5a), one obtains

$$\omega=iUk\sin\theta+|k|\left(U\cos\theta-Dd_0(1-\beta\cos 4\theta)k^2\right). \qquad (3.46)$$

As already mentioned this dispersion relation is very close to the one obtained for the Saffman-Taylor interface (see relation (2.34)), so that the following will be similar to what has been done in Section 2. As shown previously, the nontrivial element of the kernel of the adjoint of the linearized operator takes the form

$$\Psi\approx\exp\left(\int_0^l ik(l')\,dl'\right). \qquad (3.47)$$

Here k is the marginal mode of the relation (3.46), where ik has replaced $-ik$, and $|k|$ has replaced $\{k\}$ (see relation (2.35)). Thus, one can write the nontrivial solvability condition as

$$\int_{-\infty}^{+\infty}dl\,G(X_0(l))\exp\left(-i\sqrt{\frac{U}{Dd_0}}\int_0^l dl'\frac{\exp(i\theta/2)}{\sqrt{1-\beta\cos 4\theta}}\right)=0. \qquad (3.48)$$

Here $\theta(l)$ is computed for the Ivantsov parabola.

We now compute explicitly the solvability condition (3.48). The relation $\theta(l)$ for the Ivantsov parabola is easily obtained from

$$l=-\frac{\rho}{2}\left(\frac{\tan\theta}{\cos\theta}+\log\left(\frac{1}{\cos\theta}+\tan\theta\right)\right), \qquad (3.49)$$

where ρ is the radius of curvature at the tip. By changing variable to $\eta = \tan\theta$, one can rewrite the solvability condition (3.48) as

$$\int_{-\infty}^{+\infty} G(\eta)\exp\left(\sqrt{C}\,\Psi(\eta)\right) d\eta = 0. \tag{3.50}$$

Here, G is a slowly varying function of η, and $\Psi(\eta)$ is defined as

$$\Psi(\eta) = \frac{i}{2}\int_0^{\eta} \frac{(1+i\eta)^{7/4}(1-i\eta)^{5/4}}{\sqrt{(1+\eta^2)^2(1-\beta)+8\beta\eta^2}}\, d\eta. \tag{3.51}$$

Relations (3.50) and (3.51) are the same as those obtained by Barbieri *et al.* [59]. The integral (3.50) is evaluated by steepest descent methods (see Bender and Orzsag [108]). The function $\Psi(\eta)$ has two stationary phase points at $\eta = \pm i$. One can show that the contribution of the point $\eta = i$ to the integral (3.50) dominates. To evaluate this contribution one deforms the original contour (the real axis) to a contour of steepest descent which goes close to the point $\eta = i$ and turns around the branch cut which begins at $\eta = i(1-\sqrt{2\beta})$ (see Fig. 3.2). On the loop which turns around the cut, Ψ is imaginary, so that in front of the exponentially small integral will be an oscillating factor of the form

$$\cos\left(A\sqrt{C}\,\beta^{7/8}\right), \tag{3.52}$$

where A is a numerical constant. This factor vanishes for a discrete set of selected values of C,

$$C \approx \frac{n^2}{\beta^{7/4}}, \tag{3.53}$$

where n is an arbitrary integer.

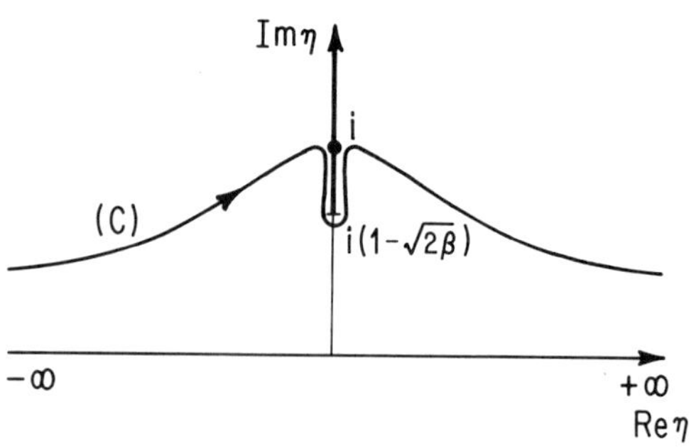

Fig. 3.2. Contour (C) of integration.

3.4. Comparison With Experiments

Experiments on growth of dendrites have been performed on succinonitrile by Huang and Glicksman [41] and on ice by Fujioka [42]. It appears that, in this case, needle crystals of stationary shapes are never observed. At some distance from the tip there appears a persistent sidebranching at rest in the laboratory frame, so that the whole shape is not stationary. Only the tip has a stationary shape.

Results on growth velocity are summarized in Fig. 3.3. As mentioned by Langer [12], the data agree well with the "marginal stability" theory. It says

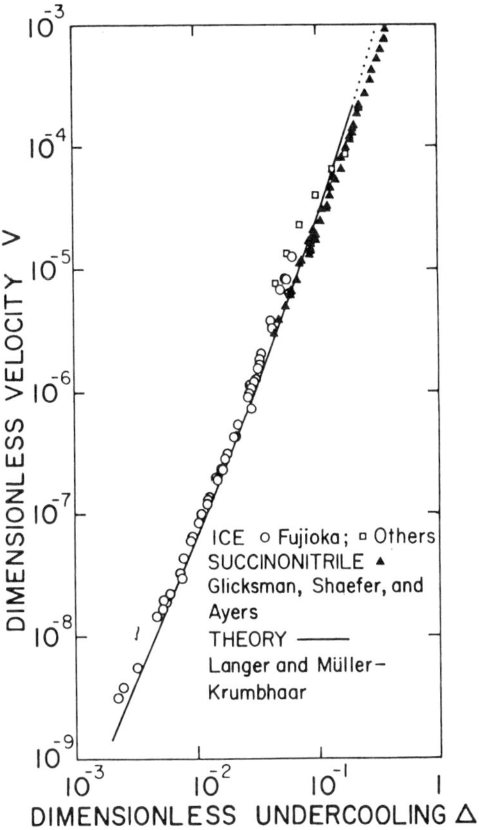

Fig. 3.3. Dimensionless growth velocity, $V = Ud_0/2D$, as a function of dimensionless undercooling Δ. Experimental points are for ice and succinonitrile as shown. The square points summarize a large body of older data (see [12] for references). The theoretical curve is drawn for $\sigma = 0.025$. The dotted extension to this curve identifies the region where the Peclet number p is greater than 0.1, so that the assumptions of the calculations may be inaccurate. (Reprinted from Fig. 15 of Langer [12] with the permission of the American Physical Society.)

that the number $\sigma = 2Dd_0/\rho^2 U$ is a constant approximately equal to .02 (ρ is the tip radius). When one eliminates ρ by using the Ivantsov relation (3.42) one obtains

$$U = .012\frac{D}{d_0}\left(\frac{\Delta}{\log \Delta}\right)^2, \tag{3.54}$$

which is the same formula as the one given by the "eigenvalue" theory (see relation (3.43)) with $C = 330$, which is large compared to unity. We have seen that the shapes of crystals corresponding to large eigenvalues are close to paraboloids. This fact is well observed in experiments. Thus, both scaling of the crystal velocity versus supercooling and the crystal shapes predicted by the theory are in good agreement with experiments.

On the other hand, the prediction that eigenvalues are selected by anisotropy of liquid-solid surface tension does not seem to be confirmed yet by experiments. Experiments performed by Glicksman and Singh [120] on succinonitrile (weak anisotropy) and pivalic acid (strong anisotropy) lead to the same values of σ and so of C. It is possible that the discrepancy between theory and experiments comes from the presence of the persistent sidebranching. It is also possible that other effects could lead to the selection of the eigenvalues C.

3.5. Analogy Between Saffman-Taylor Finger and Dendrite

There seem to be strong similarities between a Saffman-Taylor finger and a dendrite. The similarities become more striking when one considers the growth of a dendrite in a capillary of radius a. Some experiments relative to this configuration were realised by Honjo and Sawada [60], in which a solid grows in an isothermal supersaturated solution of ammonium bromine, and by Bechhoefer *et al.* [121], in which a solid grows in a binary succinonitrile-acetone mixture. Here, the limiting process of growth is the diffusion of solute, much slower than the diffusion of heat released at the interface. As in the pure thermal case, one will assume the interface at thermal equilibrium.

3.5.1. The Free-Boundary Problem

The Interface. The variation of the thermodynamical potential $G = F + P_1 V$ is

$$dG = (P_1 - P_s)\delta V_s + (\mu_s - \mu_1)\delta N_s + (\mu'_s - \mu'_1)\delta n_s + \delta\int_{(S)} \sigma(\mathbf{n})\, dS. \tag{3.55}$$

Here, N_s is the number of molecules of solvent in solid phase, n_s the number of molecules of solute in solid phase, and μ (resp. μ') the chemical potential of the solvent (resp. of the solute). As with the case at the beginning of the section, the minimisation of G with respect to δV_s, δN_s and δn_s leads to the relations

$$P_s - P_1 = \left(\sigma + \frac{\partial^2\sigma}{\partial\phi_1^2}\right)\frac{1}{R_1} + \left(\sigma + \frac{\partial^2\sigma}{\partial\phi_2^2}\right)\frac{1}{R_2}, \tag{3.56}$$

$$\mu_1(P_1, T, c_1) = \mu_s(P_s, T, c_s), \tag{3.57}$$

$$\mu_1'(P_1, T, c_1) = \mu_s'(P_s, T, c_s), \tag{3.58}$$

where c_1 (resp. c_s) is the concentration of solute in the liquid (resp. in the solid). We will assume in the following that the solutions are dilute. In this case (see Landau and Lifchitz [122]) one has

$$\begin{aligned} \mu &= \mu(P, T) - kTc, \\ \mu' &= \Psi(P, T) + kT \log c. \end{aligned} \tag{3.59}$$

The linearisation of relation (3.57) around the equilibrium situation where the planar interface separates the two phases of the pure solvent leads to

$$\delta p\left(\frac{v_1}{v_s} - 1\right) - \frac{\delta T}{T^*}Q = \left(\sigma + \frac{\partial^2\sigma}{\partial\phi_1^2}\right)\frac{1}{R_1} + \left(\sigma + \frac{\partial^2\sigma}{\partial\phi_2^2}\right)\frac{1}{R_2} + \frac{kT^*}{v_S}(c_1 - c_s). \tag{3.60}$$

Furthermore, the equality of chemical potentials of the solute leads to

$$\begin{aligned} kT \log\left(\frac{c_s}{c_1}\right) &= \Psi_1(P_1, T) - \Psi_s(P_s, T) \\ &\approx \Delta\Psi(P^*, T^*), \end{aligned} \tag{3.61}$$

from which one deduces

$$c_s = Kc_1, \tag{3.62}$$

where K, the partition coefficient, is

$$K = \exp\left(\frac{\Delta\Psi(P^*, T^*)}{kT^*}\right). \tag{3.63}$$

If one takes as P^* the constant pressure of the liquid, one obtains a relation between temperature and solute concentration in the liquid at the interface

$$\frac{\delta T}{T^*}Q = \frac{mc_1}{T^*}Q - \left(\sigma + \frac{\partial^2\sigma}{\partial\phi_1^2}\right)\frac{1}{R_1} - \left(\sigma + \frac{\partial^2\sigma}{\partial\phi_2^2}\right)\frac{1}{R_2}. \tag{3.64}$$

Here, m is the slope of the liquidus line in the binary mixture phase diagram (Fig. 3.4).

The Diffusion Field. As the solute diffuses much slower than the latent heat, the field limiting the growth is the concentration of solute in the liquid. It satisfies the diffusion equation

$$\frac{\partial c_1}{\partial t} = D\,\Delta c_1, \tag{3.65}$$

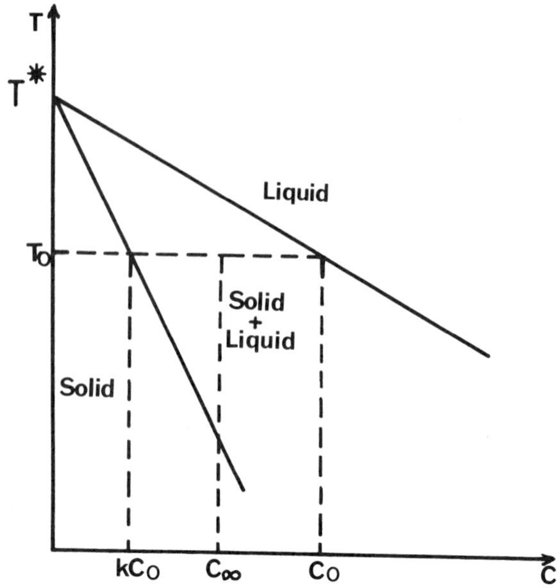

Fig. 3.4. Part of the phase diagram of a binary mixture.

where D is the coefficient of diffusion in the liquid. The diffusion of solute in the solid is assumed here negligible.

Conservation relation. Conservation of number of molecules of solute at the interface is written by matching the normal components to the interface of the concentration fluxes, in a frame where the interface is at rest:

$$\mathbf{j}_l \cdot \mathbf{n} = \rho_l c_l \mathbf{v}_l \cdot \mathbf{n} - \rho_l D \nabla c_l \cdot \mathbf{n} = \mathbf{j}_s \cdot \mathbf{n} = \rho_s c_s \mathbf{v}_s \cdot \mathbf{n}. \qquad (3.66)$$

Then, by using conservation of mass (see relation (3.13)), one obtains

$$c_l(1 - K)\mathbf{v} \cdot \mathbf{n} = -D\frac{\rho_l}{\rho_s} \nabla c_l \cdot \mathbf{n}. \qquad (3.67)$$

In the following, one assumes crystal and melt densities equal.

The Stationary Free Boundary Problem. Let us look at the two-dimensional problem of the growth at constant velocity U along the axis of the channel of a needle crystal with stationary shape.

In a frame fixed with the crystal, the stationary diffusion equation for the concentration in the liquid is

$$-U\frac{\partial c_l}{\partial z} = D\Delta c_l \qquad (3.68)$$

in the strip of width $2a$ with the boundary conditions

$$\begin{aligned} c &= c_\infty && \text{at infinity,} \\ \partial c/\partial y &= 0 && \text{on the walls } (y = \pm a), \end{aligned} \qquad (3.69)$$

and, on the interface,

$$D \nabla c_1 \cdot \mathbf{n} = -c_1(1-K)U\cos\theta, \tag{3.70}$$

$$c_1 = c_0 + \frac{T^*}{mQ}\left(\sigma + \frac{d^2\sigma}{d\theta^2}\right)\frac{1}{R}. \tag{3.71}$$

Here $c_0 = \delta T/m$ is the equilibrium concentration at the planar interface corresponding to the supercooling δT.

3.5.2. Asymptotic Analysis at Small Peclet Number

Pelcé and Pumir [61] and Pelcé [62] have shown that in the small Peclet number limit $P = aU/D$, the shape of the crystal is the same as a Saffman-Taylor finger. In this limit, the concentration field varies on two different scales. In the outer region, of size D/U, in front of the crystal, the concentration field is longitudinal. One must solve the one-dimensional diffusion equation in a semi-infinite medium with boundary conditions applied at the crystal tip position, determined by the study of the inner region, of size a, located around the tip of the crystal. In this region, the field is deformed in the transverse direction, and one must solve a two-dimensional free-boundary problem. Far from the tip, the concentration becomes uniform and equal to c_0 the temperature of crystallisation at temperature T_0 (Fig. 3.5). In the following, we scale all the lengths by a, the radius of the capillary, and omit the subscript l for the concentration.

The Inner Region. In this region, the rescaled concentration $c_i = (c - c_0)/P$ varies on a scale a, so that the Laplacien term in equation (3.68) will dominate the advective term $-P\partial c/\partial z$. Thus, one must solve the Laplace equation $\Delta c_i = 0$. The boundary conditions at the interface are simply given by:

$$\nabla c_i \cdot \mathbf{n} = -c_0(1-K)\cos\theta \tag{3.72}$$

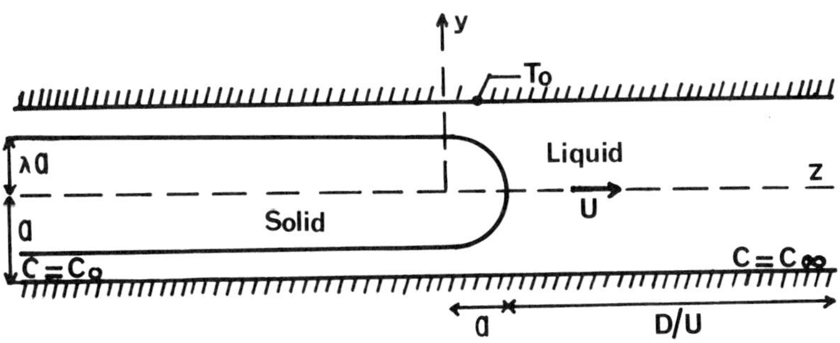

Fig. 3.5. Shape of the needle crystal growing in a capillary tube at small Peclet number.

and

$$c_i = \frac{T^*}{PmQa}\left(\sigma + \frac{d^2\sigma}{d\theta^2}\right)\frac{d\theta}{dl}. \tag{3.73}$$

Here, l is the dimensionless curvilinear coordinate on the interface. Far away from the crystal (but still in the inner region) one anticipates that the concentration gradient is uniform ($c_i \approx -Sz$) as z goes to infinity. In the case of isotropic surface tension, the free-boundary problem so posed is identical to the Saffman-Taylor problem (see relations (2.10), (2.11), (2.12) and (2.13) in which film effects have been neglected). One must identify the concentration of solute with the pressure, the concentration gradient $-S$ with the velocity V of the fluid far ahead of the finger and the quantity $-c_0(1-K)$ with the velocity of the finger. Thus the shape of the crystal is a Saffman-Taylor finger of relative half-width $\lambda(\kappa)$, where κ is determined by analogy with relation (2.25), i.e.,

$$\kappa = \frac{Dd_0}{a^2U}\frac{\pi^2}{(1-\lambda)^2}, \tag{3.74}$$

where d_0 is the capillary length which can be written as

$$d_0 = \frac{\sigma T^*}{Qmc_0(K-1)}. \tag{3.75}$$

As this inner problem is formally the same as the one studied by McLean and Saffman [26], one gets a countably infinite set of branches of solutions $\lambda_n(\kappa)$. The concentration gradient $-S$ is then deduced to be $S = \lambda c_0(1-K)$.

The Outer Region. In this region, the concentration field c_e depends only upon the z coordinate. If one assumes a growth at a constant velocity, one obtains simply from (3.68)

$$c_e = c_\infty + A\exp - Pz \tag{3.76}$$

where A is for the moment an undetermined constant.

Matching of the Solutions. The matching of the solutions requires that at $z = 0$ (around the tip), $c_e = c_0$ and $dc_e/dz = -PS$. (the concentration flux which leaves the inner region is equal to the concentration flux which enters the outer region [61]). This leads to the relation $A = (c_0 - c_\infty) = S$, from which one deduces $\lambda = \Delta = (c_0 - c_\infty)/c_0(1-K)$ the dimensionless supersaturation. In contrast to the Saffman-Taylor case, where the width of the finger was determined by a solvability condition, the width is here fixed, by conservation of impurities, to be the control parameter Δ. Far ahead of the crystal, the concentration is uniform and equal to c_∞. Far behind it, the solid is in equilibrium with the liquid at the temperature T_0. As the fraction of solid

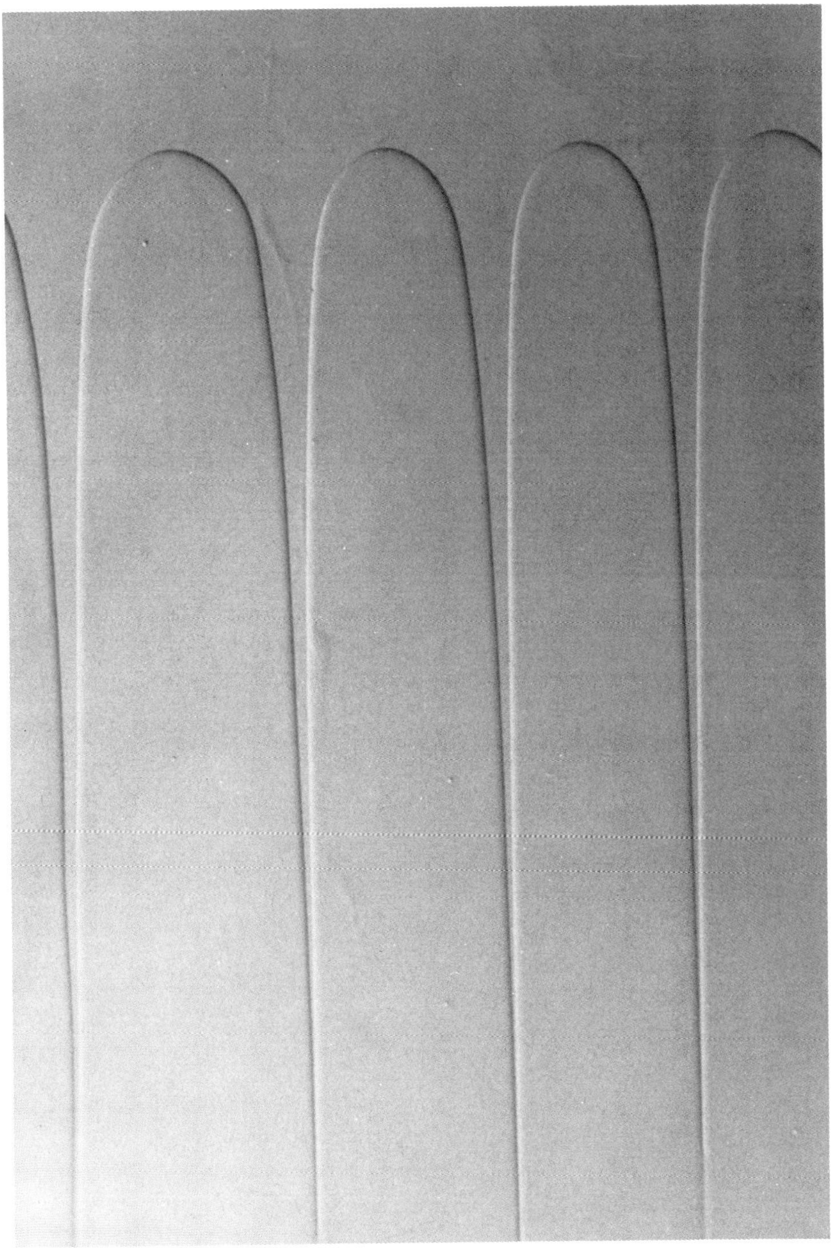

Fig. 3.6. Cell shape in directional solidification of succinonitrile. The Peclet number $P = 2aU/D$ is equal to 0.3. (Courtesy of J. Bechhoefer.)

is λ and the one of liquid, $1 - \lambda$, one obtains

$$c_\infty = \lambda K c_0 + (1 - \lambda) c_0, \tag{3.77}$$

from which, one gets $\lambda = \Delta$. Thus, no solution of this kind exists for Δ less than $\frac{1}{2}$. Such solutions can be obtained if one introduces anisotropy in surface tension (Kessler *et al.* [63], Dorsey and Martin [123]).

Similar qualitative results were obtained by Karma [124] in the case of the symmetric model (coefficients of diffusion equal in liquid and solid). Deviations from the small Peclet number limit were studied by Kessler *et al.* [63]. When a temperature gradient is applied to the system one returns to the problem of directional solidification at small Peclet number (in the regime where elongated cells are observed) (Fig. 3.6). U becomes a control parameter; $2a$, the periodicity of the cells, and Δ, the local supersaturation at the tip, become free parameters. The previous study determines a relation between these quantities, this leading to a one-parameter continuum of solutions. The effect of the temperature gradient introduces a new length in the problem. Contrary to the conclusion of the work of Karma [125], this effect does not break the continuum. (Dombre and Hakim [106], Benamar and Moussallam [146]).

4. Stationary Shapes of a Curved Flame Propagating in a Channel

4.1. Model for Flame Propagation

4.1.1. The Flat Flame

A premixed flame propagating in a reactive mixture can be considered as an interface separating fresh and burned gas. Its thickness d is of order .1 mm, which is very small compared to other relevant dimensions such as the radius of the tube in which it propagates. The structure of this interface is complex. A large number of chemical reactions take place (about 300 in common hydrocarbon-oxygen combustion) releasing heat that both diffuses and sustains the propagation. Nevertheless, measurements of flame velocity seem to indicate that the macroscopic dynamics of a flame do not reflect this huge underlying complexity. Thus, insofar as the large-scale motion of a flame is concerned (determination of its shape or the flow around it), the interface structure can be considerably simplified. The simplest model consists of a single exothermic chemical reaction:

$$A \to B + Q, \tag{4.1}$$

where A is the limiting reactant, B the product and Q the heat released by the reaction. The reaction rate follows an Arrhenius law:

$$\frac{1}{\tau} \approx \frac{1}{\tau_0} Y \exp\left(-\frac{E}{kT}\right), \tag{4.2}$$

where T is the temperature, Y the mass fraction of A, E the activation energy and τ_0 the mean collision time between molecules. The structure of the flat premixed flame was first described by Zeldovich and Frank-Kamenetskii [100]. Because of the high activation energy of the chemical reaction, a flame has two adjacent layers where different balances occur. One of these, close to the burned side (Fig. 4.1), is the reaction zone. In this layer, the combustion effectively occurs at a temperature close to the final temperature of the burned gas $T_b = T_0 + Y_0 Q/c_p$. Here T_0 is the temperature of the fresh mixture and Y_0 the initial mass fraction of the limiting reactant. The second layer, between the

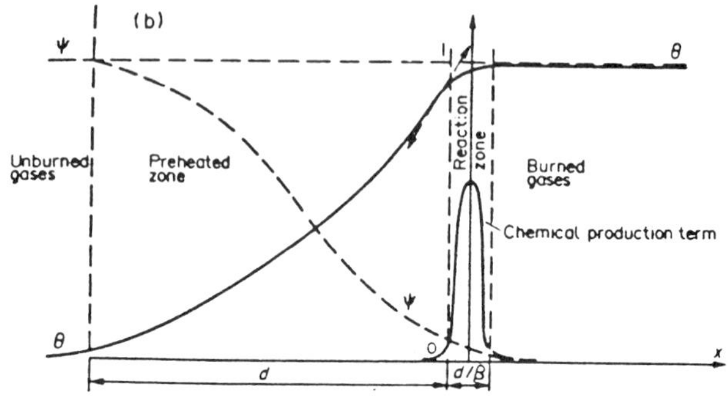

Fig. 4.1. Flame structure associated with the one-step overall reaction model. The variable θ is the dimensionless temperature defined as $\theta = (T - T_0)/(T_b - T_0)$, and $\psi = Y/Y_0$. (Reprinted with permission from Fig. 2b of Clavin [24]. © 1980 Pergamon Journals, Ltd.)

fresh mixture and the reaction zone, is the diffusive zone. In this layer, the heat released by the reaction zone heats the fresh mixture by diffusion up to T_b but no reaction occurs since the temperature is too low to sustain it. The respective thicknesses of these layers are d/β and $d = D/u_L$. Here, u_L is the flame propagation velocity, D the coefficient of diffusion of heat and $\beta = E(T_b - T_0)/RT_b^2$, the reduced activation energy (or Zeldovich number), which is in general a large number. The velocity u_L is an eigenvalue of the problem which is found to be

$$u_L \approx \frac{1}{\beta}\sqrt{\frac{D}{\tau_0}\exp\left(-\frac{E}{RT_b}\right)}\,. \tag{4.3}$$

It is of order $(D/\tau)^{1/2}$, where τ is a mean reaction time inside the combustion zone. Because of the exponentially small Arrhenius factor, τ can be much larger than τ_0, the mean collision time. Then u_L can be very small compared to $(D/\tau_0)^{1/2} \approx c$, the sound velocity in the gas. This explains that this slow combustion regime (deflagration) involves flow at small Mach number. Flame velocities of 10 cm/s are usual in very dilute mixtures. As $D \approx .1\ \mathrm{cm}^2/\mathrm{s}$, one has $d \approx .1$ mm, the thickness of the reaction zone being ten times smaller ($\beta \approx 10$). Thus, to a good approximation, the flame can be considered as an interface of thickness d between two incompressible gases, which moves relatively to the fresh mixture at constant normal velocity u_L.

4.1.2. The Curved Flame

Flat flames that propagate freely in a tube are rather rare. They can be observed when they propagate downwards at sufficiently low velocities (between 5 and 15 cm/s) [4], [5]. In general, flames are either curved or wrinkled. Observed patterns usually have wavelengths of wrinkling that are much larger than the flame thickness $d(\lambda \approx 1\,\text{cm})$. This means that, to a first approximation, a flame can be considered as a surface of discontinuity between two incompressible fluids (the fresh mixture of density ρ_0 and the burned mixture of density ρ_b), which moves relative to the fresh mixture with a constant normal velocity u_L. The effects of deformation of the flame structure come at first order in the large Peclet number limit $P = \lambda/d$. Because of transverse transport of heat and mass that appear inside the flame thickness when the flame is wrinkled, the normal burning velocity of the flame is no longer constant but depends on the front curvature and the deformation of the flow just ahead of the flame. Markstein [19] developed a phenomenological model in which the normal burning velocity was a function of the front curvature. Then, Eckhaus [21], using perturbative techniques at large Peclet number on the original equations, derived an expression for the flame velocity valid for slightly curved fronts. His result is that the burning velocity depends upon its time derivative, the front curvature and the velocity gradient at the flame boundary. Clavin and Williams [126] clarified this result by using correct boundary conditions at the reaction zone (see for instance Joulin and Clavin [127]). Later, Clavin and Joulin [128] extended the method to the case of fronts with large amplitude. The formula obtained at first order in $1/P$ for the normal burning velocity can be written as

$$\mathbf{v}_i \cdot \mathbf{n} = u_L\left(1 + L\left(\frac{1}{R} + \frac{1}{u_L}\mathbf{n} \cdot \nabla \mathbf{w} \cdot \mathbf{n}\right)\right). \tag{4.4}$$

Here, $1/R$ is the curvature of the interface, $\mathbf{n}$ the unit vector normal to it, $\nabla \mathbf{w}$ the strain tensor of the flow, and L the Markstein length. L is proportional to d, the flame thickness and is a function of the expansion ratio $\alpha = \rho_b/\rho_0$, of the reduced activation energy β and the Lewis number of the limiting component of the reaction $Le = D_T/D_M$ where D_T (resp. D_M) is the coefficient of diffusion of heat (resp. of the reactant).

4.1.3. The Free Boundary Problem

Since many experiments on curved flame propagation are conducted at large Peclet number, we will neglect its effect. This leads to two main approximations: first, the normal burning velocity is constant, and second, that the gas is inviscid. In fact, for gases, a large Peclet number implies a large Reynolds number, since the coefficients of diffusion of heat and viscosity have nearly the same values. This leads to the following boundary problem:

The velocity field $\mathbf{w}$ of the flow satisfies the incompressible Euler equation

$$\nabla \cdot \mathbf{w} = 0 \tag{4.5}$$

for conservation of mass, and

$$\rho\left(\frac{\partial \mathbf{w}}{\partial t} + (\mathbf{w}\cdot\nabla)\mathbf{w}\right) = -\nabla p \tag{4.6}$$

for the dynamics, where p is the pressure, and ρ is equal to ρ_0 in the fresh gas and to ρ_b in burned gas. At the interface, the following four boundary conditions must be satisfied:

$$\rho_0(\mathbf{w}_{-f} - \mathbf{v}_i)\cdot\mathbf{n} = \rho_b(\mathbf{w}_{+f} - \mathbf{v}_i)\cdot\mathbf{n} \tag{4.7}$$

for mass conservation,

$$\mathbf{w}_{-f}\cdot\boldsymbol{\tau} = \mathbf{w}_{+f}\cdot\boldsymbol{\tau} \tag{4.8}$$

showing the equality of tangential velocities,

$$p_{-f} + \rho_0\left((\mathbf{w}_{-f} - \mathbf{v}_i)\cdot\mathbf{n}\right)^2 = p_{+f} + \rho_b\left((\mathbf{w}_{+f} - \mathbf{v}_i)\cdot\mathbf{n}\right)^2 \tag{4.9}$$

for conservation of the momentum component normal to the interface, and

$$(\mathbf{w}_{-f} - \mathbf{v}_i)\cdot\mathbf{n} = u_L \tag{4.10}$$

showing that the normal burning velocity is constant and equal to u_L. Here, $\mathbf{n}$ is the unit vector normal to the interface, directed towards the burned mixture, $\boldsymbol{\tau}$ is the unit vector tangent to the interface, and indices $-$f and $+$f mean that the values of the velocity field are taken on the fresh side (resp. the burned side). We look at situations where the flame propagates in a tube of radius R. Thus, on the walls, the normal component of the velocity field must vanish.

Many authors (Uberoi [68], Maxworthy [69], von Lavante and Strehlow [6], Pelcé-Savornin *et al.* [7]) have observed in a tube the propagation of a curved flame, with stationary shape and constant velocity. This configuration is important because it is intermediate between the planar flame, in general unstable because of the Darrieus-Landau instability and turbulent forms of combustion. Furthermore, it has a high degree of stability since it is observed in a large range of Reynolds numbers. Below, we discuss the determination of stationary shapes of curved flames. We first present the case of the two-dimensional inviscid problem and then discuss the effects of geometry, the acceleration of gravity and the flame thickness on the shape and the velocity of the curved flame.

4.2. The Inviscid Problem

4.2.1. The Two-dimensional Case

In order to determine the flame shape, Ball [70] looked at the hydrodynamics around the flame in a frame where it is at rest. He made the following hypotheses: The flow is inviscid, incompressible, and two-dimensional. The normal burning velocity is constant and equal to u_L. Thus the effects of the curvature of the flame on the burning velocity are neglected. The procedure is similar in the problems of the Saffman-Taylor finger and of the needle crystal, where one begins by neglecting surface tension effects. Nevertheless, even in this simplified problem, no exact stationary solution can be found, in contrast to the previous cases. Ball assumes that the flame velocity is a known multiple of u_L. He solves the Euler equations by a relaxation method and obtains flame shape and streamlines in good agreement with observations. A more recent approach of Zeldovich *et al.* [71] consists in writing an integral relation for the flow.

The Structure of the Flow. We describe the flow in a frame where the flame is at rest. Since there is no vorticity far ahead in the fresh mixture (The fresh gas is at rest in the laboratory frame), and since vorticity is conserved along the streamlines, the upstream flow is potential. The flow at infinity is parallel and uniform with velocity U, the propagation velocity of the flame. At the interface, the density jump due to the heating of the gas by the exothermic chemical reaction generates vorticity which is carried along the streamlines so that the downstream flow is rotational. At infinity in the burned mixture, the flow is parallel but non-uniform. Furthermore, as is observed in experiments, the flame front is convex and meets the wall at a fixed angle. As the streamlines are deflected towards the center of the tube because of the density jump, a stagnation zone is created from the intersection of the flame with the wall and lies in the burned mixture (Fig. 4.2).

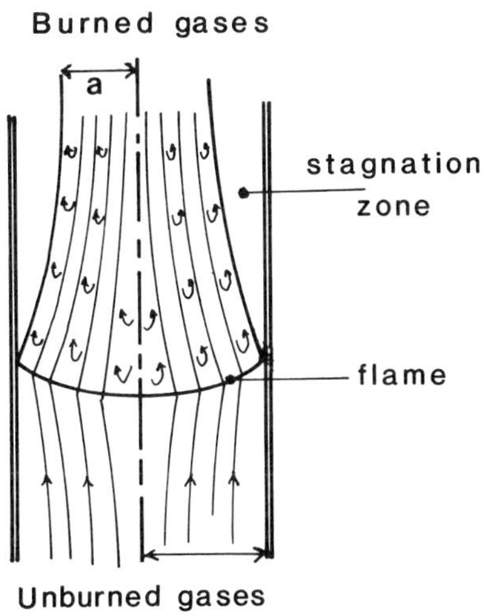

Fig. 4.2. The structure of the flow.

Integral Relation. The conservation of mass between the sections of the channel located at $x = -\infty$ and $x = +\infty$ can be written as

$$\rho_0 UR = \rho_b \int_0^a u_{+\infty}(y)\,dy, \tag{4.11}$$

where a ($a < R$) is the radius of the part of the channel filled with the fluid in motion far at infinity in the burned mixture. If one introduces the stream function defined as

$$\frac{\partial \Psi}{\partial y} = \rho u, \qquad \frac{\partial \Psi}{\partial x} = -\rho v, \tag{4.12}$$

one can rewrite (4.11) as $\rho_0 UR = \Psi_{+\infty}(a)$.
The conservation of momentum can be written as

$$(p_{-\infty} + \rho_0 U^2)R = p_{+\infty}R + \rho_b \int_0^a u_{+\infty}^2(y)\,dy, \tag{4.13}$$

or, using the stream function, as

$$(p_{-\infty} - p_{+\infty} + \rho_0 U^2)R = \int_0^{\Psi_{+\infty}(a)} u_{+\infty}(\Psi)\,d\Psi. \tag{4.14}$$

Because of the presence of the stagnation zones in the burned mixture, where the fluid is at rest with respect to the flame, one can express $p_{-\infty} - p_{+\infty}$ and $u_{+\infty}$ in terms of the upstream flow quantities. Thus, one can obtain a closed integral relation for the upstream potential flow. In order to do this, we follow a loop in the flow field along which different relations are used. For instance, in computing $p_{-\infty} - p_{+\infty}$, one uses three steps (Fig. 4.3):

- The Bernoulli law in the upstream potential flow between $x = -\infty$ and the point where the flame meets the tube at the wall.
- The jump relations across the flame front.
- The constancy of pressure inside the stagnation zone.

Then, one can deduce

$$p_{-\infty} - p_{+\infty} = \rho_0 u_L^2\left(\frac{1-\alpha}{\alpha}\right) - \frac{1}{2}\rho_0 U^2 + \frac{1}{2}\rho_0 u_{-f}^2(R). \tag{4.15}$$

In a similar way, the calculation of $u_{+\infty}(\Psi)$ involves six steps (Fig. 4.3):

- The Bernoulli law along the streamline where the stream function takes the value Ψ, between $x = +\infty$ and just behind the front.
- The jump relations across the flame front.
- The Bernoulli law in the upstream flow between just before the front and the point where the flame meets the tube at the wall.
- Again, the jump relations across the flame front.
- The constancy of pressure inside the stagnation zone.
- The uniformity of pressure at infinity.

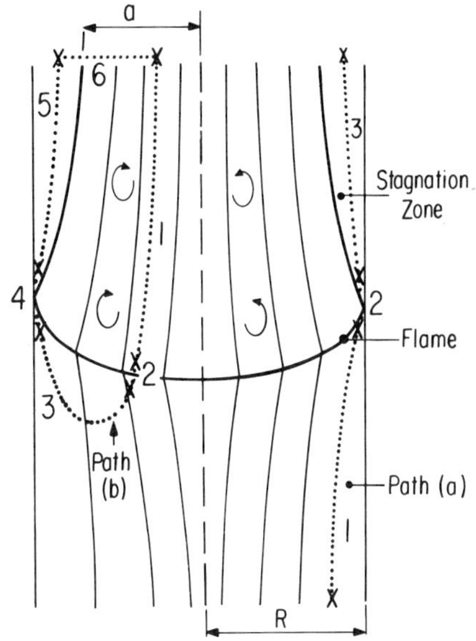

Fig. 4.3. Paths used (dotted lines) for the determination of the integral relations. Path (a) is used to determine $p_{-\infty} - p_{+\infty}$; path (b) is used to determine $u_{+\infty}(\Psi)$.

From these steps, one gets the relation

$$u_{+\infty}(\Psi) = \frac{1}{\alpha}\left((1-\alpha^2)u_L^2 + \alpha u_{-f}^2(R) - \alpha(1-\alpha)w_{-f}^2\right)^{1/2}. \qquad (4.16)$$

Thus relation (4.14) can be written as

$$\begin{aligned} &U^2 + u_{-f}^2(R) + 2\left(\frac{1-\alpha}{\alpha}\right)u_L^2 \\ &= \frac{2}{\alpha\rho_0 R}\int_0^{\rho_0 UR} d\Psi_{-f}\left((1-\alpha^2)u_L^2 + \alpha u_{-f}^2(R) - \alpha(1-\alpha)w_{-f}^2\right)^{1/2} \end{aligned} \qquad (4.17)$$

Furthermore, from relation (4.12), one can deduce the half thickness a as

$$a = \int_0^{\Psi_{+\infty}(a)} \frac{d\Psi}{\rho_b u_{+\infty}(\Psi)} \qquad (4.18)$$

where $u_{+\infty}(\Psi)$ is given by relation (4.16).

Approximate Shapes and Velocities. One can use relation (4.17) to determine approximate shapes and velocities of the curved flame. By using the mass conservation law, the potential Φ that determines the upstream flow satisfies $\Delta\Phi = 0$. As the normal component of the velocity vanishes at the wall, $\partial\Phi/\partial y = 0$ for $y = R$. The simplest potential for a non-uniform flow which

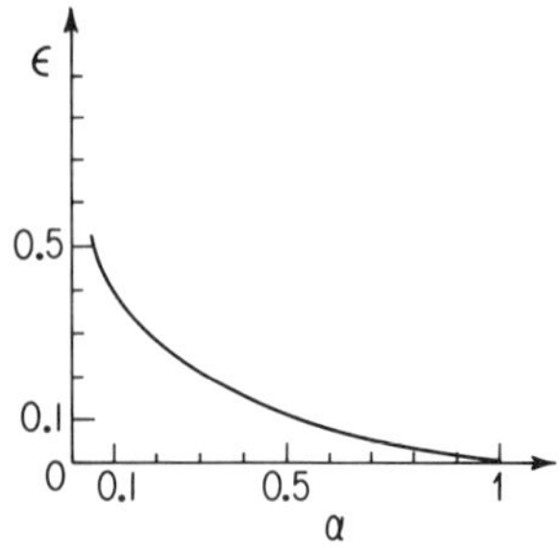

Fig. 4.4. Relative increment of the flame velocity $\epsilon = (U - u_L)/u_L$ as a function of the gas expansion in the flame-front. (Reprinted with permission from Zeldovich *et al.* [71]. © 1980 Gordon and Breach Science Publishers Inc.)

satisfies these conditions can be written as

$$\Phi = Ux + A \cos ky \exp kx, \tag{4.19}$$

where $k = \pi/R$. As at the tip of the flame ($x = 0$), $u = u_L$, one has $A = (u_L - U)/k$. Thus, one obtains a one-parameter upstream flow defined as

$$\begin{aligned} \frac{u_-}{u_L} &= 1 + \epsilon(1 - \cos ky \exp kx), \\ \frac{v_-}{u_L} &= \epsilon \sin ky \exp kx, \end{aligned} \tag{4.20}$$

where $\epsilon = (U - u_L)/u_L$. The upstream flow being given by relation (4.20), one finds the flame shape $\xi(y)$ by integrating equation (4.10), i.e.,

$$u_- - v_- \frac{\partial \xi}{\partial y} = u_L \sqrt{1 + \left(\frac{\partial \xi}{\partial y}\right)^2} \tag{4.21}$$

with the condition that $\xi(0) = 0$. By inserting the solution of (4.21) and relation (4.20) in relation (4.17), one obtains a relation $\epsilon(\alpha)$. Of course $\epsilon = 0$, which corresponds to the flat flame, is a solution for any value of the expansion ratio α. But another nontrivial branch of solutions for curved flames exists starting from the point $\epsilon = 0$ where $\alpha = 1$ and increasing when α decreases (Fig. 4.4). Thus, the smaller α is (the jump of densities at the interface is large) the larger ϵ becomes, i.e., the more the flame is curved, since the velocity of the flame is proportional to its area.

4.2.2. The Axisymmetric Curved Flame

In the more realistic case of an axisymetric curved flame the method is the same; only the formulation of the relations changes (Zeldovich *et al.* [71]). The mass conservation is now

$$\Psi_{+\infty}(a) = \frac{1}{2}\rho_0 U R^2 = \frac{1}{2\pi}\rho_b \int_0^a 2\pi r u_{+\infty}(r)\, dr, \tag{4.22}$$

where Ψ is the stream function defined as

$$\frac{\partial \Psi}{\partial r} = \rho u r, \qquad \frac{\partial \Psi}{\partial x} = -\rho v r. \tag{4.23}$$

The global conservation of momentum is now

$$\left(p_{-\infty} + \rho_0 U^2\right)\pi R^2 = p_{+\infty}\pi R^2 + \rho_b \int_0^a 2\pi r u_{+\infty}^2(r)\, dr. \tag{4.24}$$

As the expressions of $p_{-\infty} - p_{+\infty}$ and of $u_{+\infty}(\Psi)$ do not depend on the geometry, one can use relations (4.15), (4.16), and (4.24) to obtain

$$\begin{aligned} U^2 + u_{-f}^2(R) + 2\left(\frac{1-\alpha}{\alpha}\right)u_L^2 \\ = \frac{4}{\alpha\rho_0 R^2}\int_0^{\rho_0 U R^2/2}\Big((1-\alpha^2)u_L^2 + \alpha u_{-f}^2(R) \\ -\alpha(1-\alpha)w_{-f}^2\Big)^{1/2} d\Psi. \end{aligned} \tag{4.25}$$

The simplest potential one can choose for the upstream flow is now

$$\Phi = Ux - \frac{u_L}{k}\epsilon J_0(kr)\exp kx, \tag{4.26}$$

with $k = \lambda_1/R$, where λ_1 is the first zero of the Bessel function J_1. As in the two-dimensional case, one first finds the flame shape which corresponds to the flow (4.26) (the curve which satisfies in all points normal velocity equal to the constant u_L) and then computes relation (4.25). The resulting curve $\epsilon(\alpha)$ is qualitatively similar to the one obtained in the two-dimensional case. Quantitatively, the flame velocities are twice as large.

4.3. Effect of the Acceleration of Gravity

In some experiments on flame propagation, the effect of gravity is important. The reason is that the fresh and burned mixtures have different densities ($\alpha = .2$ in general) and that the Froude number $Fr = u_L^2/gR$ is of order unity. For instance, in the case of a methane-air flame (7% methane) propagating in a vertical tube of radius 2.5 cm, $u_L \approx 10$ cm/s and thus $Fr \approx .1$ (gravity effects actually dominate) [6]. Thus, in order to compare experimental results with theory, one must include the effects of the acceleration of gravity in theory (Pelcé [72]).

For this, one must consider the free boundary problem defined by relations (4.5) to (4.10), but where the pressure is changed by the effective pressure $P = p + \rho g x$. We assume here that the channel is vertical and that positive g correspond to downwards propagation and negative g to upward propagation (the x-axis is always oriented in the direction of propagation). We look here only at the two-dimensional case. As in section 4.2.1 one writes an integral relation on the upstream flow as

$$U^2 + u_{-\mathrm{f}}^2(R) + 2\left(\frac{1-\alpha}{\alpha}\right)u_L^2 + 2(\alpha-1)\frac{g}{R}\int_0^R (\xi(y) - \xi(R))\,dy$$
$$= \frac{2}{\alpha\rho_0 R}\int_0^{\rho_0 UR} d\Psi\big((1-\alpha^2)u_L^2 + \alpha u_{-\mathrm{f}}^2(R) - \alpha(1-\alpha)w_{-\mathrm{f}}^2 \qquad (4.27)$$
$$-2\alpha(1-\alpha)g(\xi(y) - \xi(R))\big)^{1/2}$$

(compare with relation (4.17)). The potential upstream flow again takes always the form given by relations (4.20). By using the same method as in the previous case, one obtains from relation (4.27) the value of ϵ, now a function of two parameters: the expansion ratio α and the Froude number Fr. The curve $\epsilon(E)$ ($E = 1/Fr$) is drawn for a value of α equal to .2, a value typically found in experiments (Fig. 4.5). Negative (resp. positive) Froude numbers correspond to upwards (resp. downwards) propagation. It appears that ϵ is a decreasing function of the inverse of the Froude number. In the case of negative Froude numbers (upwards propagation), effects of Darrieus-Landau and Rayleigh-Taylor instabilities add together so that the amplitude and thus the area of the curved flame is larger. Since the flame velocity is proportional to the flame area, ϵ is correspondingly large. On the other hand, for positive Froude numbers, effects of the acceleration of gravity become stabilizing and oppose the Darrieus-Landau instability. Consequently, the flame amplitude is smaller and thus so is its velocity. There exists a critical Froude number F_c under which only the flat flame is a solution. In other words, for given u_L and g, there exists a maximal size for a curved flame propagating downwards in a tube: $R_c = \pi u_L^2/gF_c$.

In the experiment of Pelcé-Savornin *et al.* [7], the flame is maintained at rest in the vertical tube by a feedback system using laser tomography (Boyer [129]). Fresh mixture is injected by the top extremity of the tube and burned gases are drained off at the bottom. Thus, the burned gases are below the fresh ones and

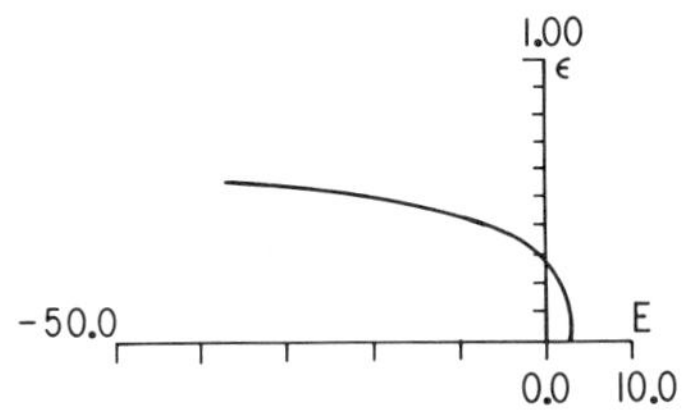

Fig. 4.5. Relative increment of the flame velocity $\epsilon = (U - u_L)/u_L$ as a function of the Froude number, for a gas expansion equal to 0.2.

therefore, the Froude number is negative, i.e., Rayleigh-Taylor instability adds to Darrieus-Landau instability. Stationary shapes of curved flames have been observed such as in Fig. 1.4.

4.4. Effects of Flame Thickness

Experiments show that flames curved on a scale much larger than the flame thickness can propagate with stationary shape and constant velocity. Thus, the effects of the variation of combustion velocity with the front curvature are weak and can at first be neglected when determining solutions of curved flames. For instance, von Lavante and Strehlow [6] note the existence of curved stationary flames propagating in a tube with square cross section of size 10 cm, in a methane-air mixture (6% methane). In this case, the normal combustion velocity is $u_L \approx 10\,\mathrm{cm/s}$ and the propagation velocity $U \approx 30\,\mathrm{cm/s}$. In order to evaluate the relative importance of the flame thickness effects one uses formula (4.4), i.e.,

$$\frac{\delta u_L}{u_L} \approx \frac{d}{R}\left(1 + \frac{U}{u_L}\right) \approx 5 \cdot 10^{-2}. \tag{4.28}$$

Thus, one can think that the model with constant burning velocity gives a good evaluation of the shape and the propagation velocity of the curved flame. This is not evident if one considers the results obtained in the Saffman-Taylor problem. One knows that in this case, the problem without surface tension admits a continuum of solutions and that this continuum of solutions is broken when small surface tension is added (singular perturbation). Here the question of multiplicity of solutions for the model with u_L constant is not yet solved. If there is a continuum, it is possible that it will be broken by the effects of flame thickness, even if they are small. The answers to these questions seem for the moment out of reach.

Occasionally, effects of flame thickness are important. For instance, when the normal velocity of combustion is sufficiently slow, which is the case when the reactive mixture is very dilute, the flame thickness becomes larger and its stabilizing effects more important. This is the reason that a flat flame propagating downwards can be stable [23]. The condition of small burning velocity is not sufficient to ensure important thermodiffusive effects. One sees from (4.28) that if U/u_L is large, the relative variation of normal burning velocity can be important even at large Peclet number. This is the case for instance when a flame propagates upwards at small Froude number. As was shown by von Lavante and Strehlow [6], the flame then behaves as a bubble of hot gas rising in a more dense fluid. Thus its propagation velocity is $U \approx \sqrt{gR}$. The ratio U/u_L is about $\sqrt{gR}/u_L^2 \approx \sqrt{1/F}$, which is very large compared to unity. Close to flame extinction, thermodiffusive effects can even dominate. As was suggested by Peters and Smooke [130] the competition between chain branching reactions and radical recombination which appears close to the extinction limit leads to a vanishing burning velocity. This effect considerably thickens the flame thickness and thus increases the thermodiffusive effects.

In all these cases, the theory developed in this chapter is not valid, and thermodiffusive effects must be taken into account.

4.5. Analogy Between Saffman-Taylor Finger and Retrocombustion Front

Consider the combustion of a piece of balsa wood (Britten *et al.* [131]) which, as is the case for a Hele-Shaw cell, is thin compared to its lateral dimensions. Air flow is injected at a constant rate at one extremity and combustion is initiated at the other (Fig. 4.6). Then a retrocombustion front propagates along the sample and reaches after some time a state of constant velocity. This experiment is an idealisation of a method called reverse combustion, widely used in past underground coal gasification field tests, to enhance the permeability of the coal seam prior to forward gasification. The balsa wood can be considered as a porous medium of permeability κ for the flow of air of viscosity μ. When the wood burns, its structure changes so that the permeability of the burned material is much smaller than for the initial material. Thus, as the combustion zone is thin compared to the lateral dimensions of the sample, the retrocombustion front can be considered as a discontinuity surface for the air flow between two media of different permeabilities. The flow of air $\mathbf{v}$ obeys the Darcy law

$$\mathbf{v} = -\frac{\kappa}{\mu}\nabla p \qquad (4.29)$$

on both sides of the interface, with two boundary conditions applied on it. As the permeability of the burned material is much smaller than the initial one, the pressure of the flow in the burned side can be considered constant. Thus if the interface is assumed infinitely thin, the pressure is constant at the interface

$$p = p_0. \qquad (4.30)$$

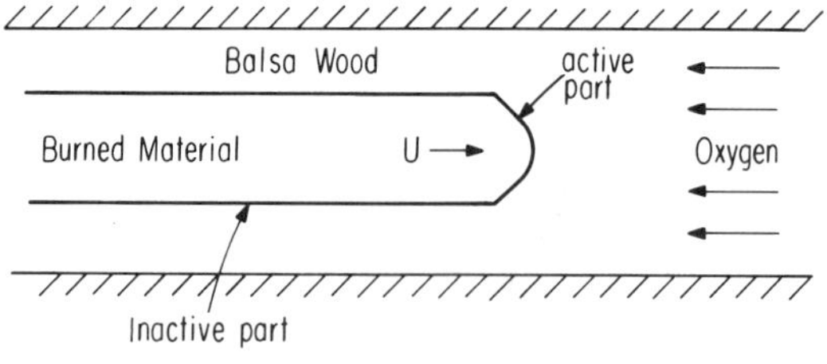

Fig. 4.6. Retrocombustion front in a piece of balsa wood.

The second boundary condition determines the normal velocity of the interface function of the normal component of the flux of air. It takes the form

$$\rho_s \mathbf{v}_i \cdot \mathbf{n} = f(\rho_g \mathbf{v} \cdot \mathbf{n}), \tag{4.31}$$

where ρ_s and ρ_g are respectively the densities of the wood and the gas and f is a function which must be determined from an analysis of the interface structure. This function has been obtained numerically from a model of the interface by Kotowski and Gunn [132] and by asymptotic methods by Britten [133] and Britten and Krantz [134]. Qualitatively, the interface velocity is an increasing function of the gas flux. In order to simplify the problem, one can approximate this curve by a linear relation, i.e.,

$$\rho_s \mathbf{v}_i \cdot \mathbf{n} = -\alpha \rho_g \mathbf{v} \cdot \mathbf{n}, \tag{4.32}$$

where α is some positive constant (the minus sign comes because the motion of the interface and the flow of air are in opposite directions). Far ahead of the finger, the velocity of the flow is uniform: $v = -Vx$. The sides of the sample are isolated so that no air flow can go out from the piece of balsa wood. This leads to the boundary condition $\mathbf{v} \cdot \mathbf{n} = 0$ for $y = \pm R$. We look now for stationary shapes of interfaces moving with constant velocity U, solutions of this free boundary problem. As in the Saffman-Taylor problem (see Section 2) one can define a velocity potential $\Phi = -\kappa/\mu(p - p_0)$ and a stream function Ψ. Far ahead of the finger, $\Phi = -Vx$, $\Psi = -Vy$ so that on the sides of the sample, which are streamlines, $\Psi = \pm V$. On the interface, $\Phi = 0$ and $\Psi = -\rho_s Uy/\alpha\rho_g$. If one comes back now to the Saffman-Taylor problem, one sees that the equations and the boundary conditions are the same if one applies the correspondence $V \to -V$ and $U \to -\rho_s U/\alpha\rho_g$ (the left-hand side of the arrow corresponds to the Saffman-Taylor problem and the right-hand side to the reverse combustion problem. Thus one can conclude that the shape of the retrocombustion interface is a Saffman-Taylor finger of relative thickness

$$\lambda = \alpha \frac{\rho_g V}{\rho_s U}. \tag{4.33}$$

Thus there is a strong analogy between a Saffman-Taylor finger and a retrocombustion front (Britten [133]). These shapes are observed in experiments. The real situation, however, seems more complicated. When the normal velocity of the flow becomes too small (at some distance from the tip of the finger) the reaction cannot be sustained and the wood does not burn. The front is then composed of two parts: an active part (where reaction effectively occurs) around the tip of the finger and two straight lines (inactive) which mark the trace left by the tip. Solutions corresponding to such fronts can be found (Clavin *et al.* [135]).

5. Stability of Curved Fronts

In contrast to the problem of determining the stationary states of curved interfaces, the problem of their stability is not yet completely understood. We present here the current status of the question. The problem consists of determining how an initial perturbation of the stationary shape evolves with time. If all perturbations decay with time, the curved interface is stable. From a mathematical point of view, the problem is well posed. One linearises the equations of motion for the interface around the stationary shape moving with constant velocity and determines the normal modes; i.e., one looks for solutions of the form $f(s)\exp \Omega t$ where f is a function of the curvilinear coordinate of the unperturbed shape, which vanishes at infinity far from the tip. Ω is in general a complex number. If all the growth rates Ω have a negative real part, the curved interface is stable. Even if the mathematical problem does not reflect completely the experimental situation (there is often a problem in meeting the exact boundary conditions, for instance), one takes into account the experimental observations before tackling the mathematical problem.

5.1. Experimental Observations

Either in experiments on propagation of Saffman-Taylor finger, dendrites or flames, a few modes of instability of the curved shape are observed.

5.1.1. Modes of Instability

Secondary Fingering. One of these modes can be called secondary fingering. It appears close to the tip and develops on the side of the curved front. This mode is observed on the Saffman-Taylor finger (Fig. 5.1) as well as the needle crystal (Fig. 5.2). In both cases the perturbation stays at rest in the laboratory frame so that in the frame of the interface it propagates from the tip to the tail

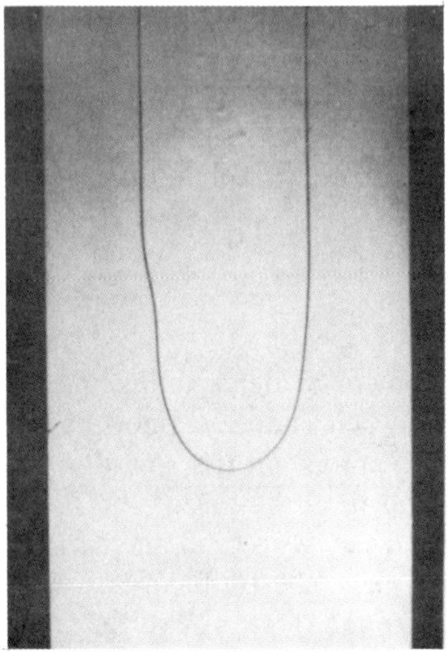

Fig. 5.1. Picture of the first instability $1/B \approx 7000$. Here, $1/B = 4\pi^2/(1 - \lambda)\kappa$. (Reprinted with permission from Fig. 10 of Tabeling *et al.* [1]. © 1987 Cambridge University Press.)

at a tangential velocity $U \sin\theta$[8]. Here U is the velocity of the curved front and θ is the angle between the axis of propagation and the normal to the interface. Huang and Glicksman [41] illustrated the development of this mode nicely by taking a multiexposure photograph of the dendrite (Fig. 5.2) where it is clear that it grows at rest in the laboratory frame. Thus, this mode is spatially extended and associated with a complex growth rate Ω, whose imaginary part corresponds to propagation from the tip to the tail. A difference between the Saffman-Taylor finger and the dendrite is that in the first case the deformations are generated by single bumps (whose origin is not yet clear), while in the case of the dendrite, the perturbations appear nearly periodically. However, when one streaks evenly spaced grooves on the plates of the Hele-Shaw cell [77] or puts a small bubble at the tip of the Saffman-Taylor finger [76], periodic sidebranching is generated. The typical wavelength of the perturbation (or its size) is of the order of the radius of curvature of the tip of the curved interface.

[8]In the case of the Saffman-Taylor finger, the tangential velocity field is proportional to the tangential gradient of pressure on the interface. In experiments, disturbances of the shape appear when the finger is sufficiently fast, i.e., when the shape is not affected by surface tension effects. Thus, the pressure is constant on the interface, and consequently, the tangential velocity field vanishes in the frame of the cell.

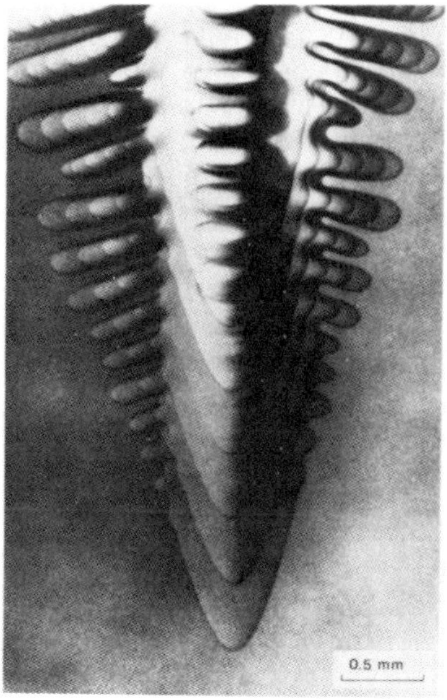

Fig. 5.2. Superimposition of time-lapsed photographs of a growing dendritic crystal, showing the sidebranch evolution process on the {100} branching sheets. (Reprinted with permission from Fig. 5 of Huang and Glicksman [41]. © 1981 Pergamon Journals, Ltd.)

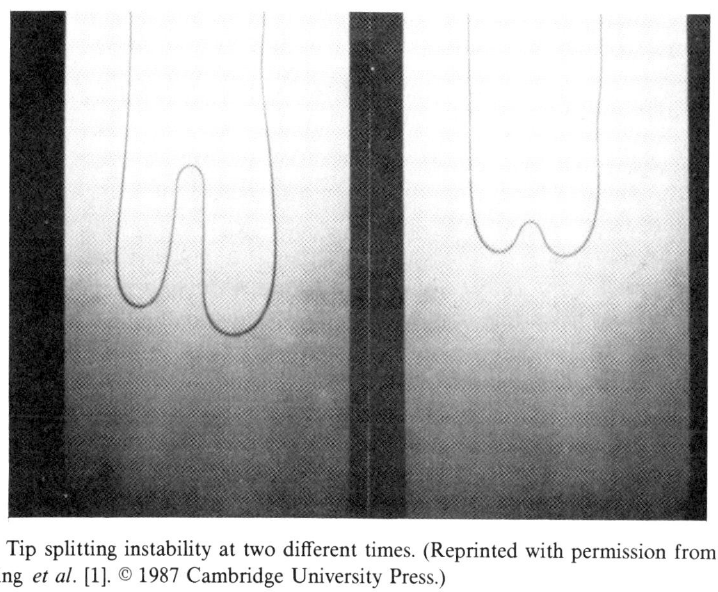

Fig. 5.3. Tip splitting instability at two different times. (Reprinted with permission from Fig. 11 of Tabeling *et al.* [1]. © 1987 Cambridge University Press.)

Tip Splitting. Another mode which is observed is tip splitting. The tip splits in two, leading in general to a non-stationary process where many curved shapes seem to compete. This process is observed for the Saffman-Taylor finger [1] (Fig. 5.3) as well as for curved flames propagating in tubes. In contrast to the previous mode, this one seems localised around the tip. In terms of the growth rate, it must be associated with a real and positive Ω.

5.1.2. Sensitivity to Noise

At least in the case of the Saffman-Taylor finger [1], the stability threshold seems to depend on noise amplitude. There is a clear variation of the stability threshold when the finger propagates in a dirty or a clean cell, the latter one improving stability.

5.2. Qualitative Theory for the Problem of Stability

Zeldovich *et al.* [71] took up the problem of the stability of a curved flame by evaluating the growth rate of a perturbation initially localised at the tip, of typical wavelength λ_c, the most unstable wavelength of the planar front. These perturbations seem to be the most dangerous for the stability of the curved stationary shape. Similar methods can be applied to other curved fronts, such as the Saffman-Taylor finger and the needle crystal [79], [62], [80]. From the evaluation of this growth rate a stability criterion can be deduced relating a dimensionless number characterizing the stationary curved front and the amplitude of the fluctuations in the medium.

5.2.1. Case of the Curved Flame

Theory of Zeldovich et al. [71]. When the front is curved, a tangential non-uniform velocity field is set up in the frame where the flame is at rest. This effect is stabilizing for two reasons. First, as a perturbation grows because of the Darrieus-Landau instability, it is carried along the interface by the tangential flow, from the tip, where it appears, to the extremities of the flame (where the flame meets the walls of the tube). Thus, if the initial perturbation (thermal noise, turbulence) is not too large, it will not have time to grow sufficiently before disappearing and the flame will be considered to be stable. The second reason is that the wavelength of the perturbation is stretched by the gradient of the tangential field. Thus, since the growth rate of the Darrieus-Landau instability is of the form $\Omega = u_L|k|$, where u_L is the normal burning velocity and k the wave number of the perturbation, a stretching of the wavelength corresponds to a decrease in the growth rate of the perturbation. These two effects would explain why one can observe flames curved on a

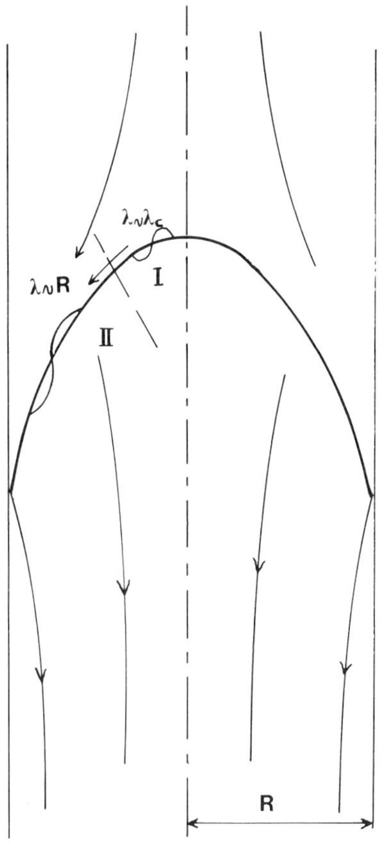

Fig. 5.4. Evolution of a perturbation of a curved flame-front.

scale (the tube radius R) larger than the most unstable wavelength of the planar front λ_c. For this, one evaluates the growth rate Γ of an initial perturbation, localised, on a scale λ_c, which appears close to the tip at a distance of the same order.

At the beginning of its evolution (Phase I of Fig. 5.4), the wavelength of the perturbation is much smaller than the radius of curvature of the flame. Its growth rate Γ_I can be evaluated by a W.K.B. method, i.e.,

$$\Gamma_I(l) \approx \int_{\lambda_c}^{l} \frac{\Omega}{u_T}\, dl', \tag{5.1}$$

where $\Omega(k)$ is the local growth rate of the perturbation determined by the linear stability analysis of the planar front (see relation (1.7b)), and l is the curvilinear coordinate along the interface.

The wavelength is stretched according to the law

$$\frac{\delta\lambda}{\lambda} \approx \frac{\delta u_\tau}{u_\tau}, \tag{5.2}$$

from which one deduces

$$\lambda = \lambda_c \frac{u_\tau(l)}{u_\tau(\lambda_c)}, \tag{5.3}$$

since the initial condition is that for $l = \lambda_c$, $\lambda = \lambda_c$. Thus, around the tip, where the tangential velocity field can be approximated by $u_\tau'(0) \cdot l$, the wavelength grows linearly with the distance l. As Zeldovich [136] remarks, this reminds one of the stretching of the wavelength of radiation in an expanding Friedmann universe, where the expansion velocity follows Hubble's law, i.e., it grow linearly with distance.

If one writes the stationary flow field in dimensionless form, in the frame where the flame is at rest, as

$$u_\tau(l) = u_L \bar{u}(s), \tag{5.4}$$

where $s = l/R$ is the dimensionless curvilinear coordinate, then relation (5.3) can be written as

$$\lambda = R \frac{\bar{u}_\tau(s)}{\bar{u}_\tau'(0)}. \tag{5.5}$$

When s is of order unity (l of order R), λ becomes of the order of the radius of curvature of the flame front, and the W.K.B. approximation is no longer valid (Phase II in Fig. 5.1). But since the growth of the perturbation is largest at the tip (it stays for a long time since the tangential velocity field is weak), the main contribution to the growth rate is given by relation (5.1) where l is chosen arbitrarily to be smaller than R but much larger than λ_c.

Using relation (1.7b), one can write relation (5.1) as

$$\Gamma_{\mathrm{I}}(l) = \int_{\lambda_c}^{l} \frac{u_L f |k| \left(1 - \dfrac{|k|\lambda_c}{4\pi}\right)}{u_\tau}\, dl. \tag{5.6}$$

Defining the dimensionless variable

$$x = \frac{\bar{u}_\tau(s)}{\epsilon}, \tag{5.7}$$

where

$$\epsilon = \bar{u}_\tau\left(\frac{\lambda_c}{R}\right) \approx \bar{u}_\tau'(0)\frac{\lambda_c}{R}, \tag{5.8}$$

one obtains

$$\Gamma_{\mathrm{I}}(l) = \frac{\bar{u}_\tau'(0)}{\epsilon} \int_l^{\bar{u}_\tau(s)/\epsilon} dx \frac{2\pi f\left(1 - \dfrac{1}{2x}\right)}{x^2 \left.\dfrac{d\bar{u}_\tau}{ds}\right|_{s=\bar{u}_\tau^{-1}(\epsilon x)}}, \tag{5.9}$$

which, to leading order in $1/\epsilon$, is

$$\Gamma_{\mathrm{I}} \approx \frac{3\pi f}{2\epsilon}. \tag{5.10}$$

From the approximate velocity field defined by relation (4.20), one obtains

$$\bar{u}_\tau'(0) = \pi \frac{(U - u_L)}{u_L}. \tag{5.11}$$

Thus by using relation (1.7c), one deduces

$$\epsilon = \frac{4\pi^2 h}{f}\left(\frac{U - u_L}{u_L}\right)\frac{1}{Re}, \tag{5.12}$$

since d the flame thickness is $d = D/u_L \approx v/u_L$. Then, one can write the growth rate of the perturbation in the first region as

$$\Gamma_{\mathrm{I}} \approx \frac{3}{8\pi}\frac{f^2}{h}\frac{u_L}{U - u_L} Re. \tag{5.13}$$

The contribution Γ_{II} to the growth rate coming from the part of the flame exterior to the tip region cannot be calculated by the W.K.B. method since the wavelength of the perturbation is of the order of the radius of curvature of the flame. Nevertheless, one can evaluate it from dimensional analysis by writing $\Gamma_{\mathrm{II}} \approx \omega t$, where $t \approx R/U$ is the transit time of the perturbation across this region and $\omega \approx u_L/\lambda \approx U/R$ from the Darrieus-Landau formula. This gives $\Gamma_{\mathrm{II}} \approx 1$ which is negligible compared to Γ_{I} since the Reynolds number is assumed large compared to unity. Thus, the total growth rate is well evaluated by formula (5.13).

Relation (5.13) allows one to calculate the final amplitude of the perturbation as a function of the initial one by the relation

$$A_{\mathrm{f}} \approx A_{\mathrm{i}} \exp \Gamma. \tag{5.14}$$

One can determine a stability criterion for the curved flame by fixing a final amplitude likely to destabilize the flame. This amplitude must be determined from a nonlinear theory for the dynamics of the perturbation, a calculation that has not yet been done[9]. For the moment, one can choose $A_f \approx R$ [10]. By using relation (5.14), one deduces

$$\Gamma \approx -\log\left(\frac{A_i}{R}\right), \tag{5.15}$$

which can be written (by using relation (5.13)) as

$$Re \approx -\frac{8\pi}{3}\frac{h}{f^2}\frac{(U-u_L)}{u_L}\log\left(\frac{A_i}{R}\right). \tag{5.16}$$

Typical values for a flame are

the expansion ratio $\alpha = .2$,
the reduced activation energy $E(T_b - T_0)/RT_b^2 = 15$, and
the Lewis number $Le = 1.2$.

By using these data, one finds from Fig. 4.3, $(U - u_L)/u_L = .27$, and from [23], $f \approx .25$ and $h \approx 7.25$. Thus, relation (5.16) becomes

$$Re \approx -200\log\left(\frac{A_i}{R}\right). \tag{5.17a}$$

From relation (1.7c) one deduces $\lambda_c = 350d$. Thus, (5.17a) can also be written as

$$\frac{R}{\lambda_c} \approx -.5\log\left(\frac{A_i}{R}\right). \tag{5.17b}$$

This relation indicates that curved flames can be theoretically stable in tubes of radius much larger than the most unstable wavelength of the planar front, if the amplitude of the noise is sufficiently small. In experiments, the ratio R/λ_c is about some units.

Model Equation. This behaviour can be described exactly by a model equation for the perturbation. If one includes the effect of uniform stretch in the linear part of the Michelson-Sivashinsky equation (see equation (1.20)), one obtains

$$\frac{\partial \xi}{\partial t} + x\frac{\partial \xi}{\partial x} = \mathrm{P}\int^x \frac{\partial_{x'}\xi}{x - x'}\,dx' + \frac{1}{Re}\frac{\partial^2 \xi}{\partial x^2}. \tag{5.18}$$

Here, lengths are scaled against R the tip radius of the curved flame and time by R/U where U is the flame propagation velocity. If the stretch effect

[9] A particular class of localised solutions of a nonlinear model equation for stretched disturbances was determined by Joulin [137].

[10] Because of the presence of the logarithm in the right-hand side of relation (5.15), the determination of the exact value of A_f has little importance, even from a quantitative point of view.

$x\,\partial\xi/\partial x$ is absent one obtains the linear evolution of a perturbation on a planar front whose dispersion relation is

$$\Omega = |k|\left(1 - \frac{1}{Re}|k|\right), \tag{5.19}$$

a relation similar to (1.7b).

Introducing a new spatial coordinate $y = x\exp(-t)$, equation (5.18) becomes

$$\frac{\partial\xi}{\partial t} = \exp(-t)\mathrm{P}\int^{y}\frac{\partial_{y'}\xi}{y-y'}\,dy' + \frac{1}{Re}\exp(-2t)\frac{\partial^2\xi}{\partial y^2}, \tag{5.20}$$

whose solutions take the form

$$\xi_q(y,t) = \frac{A_i}{R}\exp\left\{(1-\exp(-t))|q| - \frac{1}{2Re}(1-\exp(-2t))q^2\right\}\exp(iqy). \tag{5.21}$$

Thus, an initial modulation of wave number q and amplitude A_i/R has a final amplitude $A_f/R = A_i/R\exp(|q| - (1/2Re)q^2)$. The largest one is obtained for $q = q_m = Re$ the marginal mode of the dispersion relation (5.19). The corresponding final amplitude is $(A_i/R)\exp(Re/2)$. Thus the stability criterion for the curved flame is now $A_f/R \approx 1$, i.e., $Re \approx -2\log(A_i/R)$, which is of the same form as (5.17).

5.2.2. Case of the Saffman-Taylor Finger

A similar argument can be developed here [79], [62]. As in the previous case, one evaluates the growth rate of a localised perturbation, of size λ_c, which appears close to the tip at a distance of the same order. The wavelength is always stretched according to relations (5.2) and (5.3) where $\bar{u}_\tau(s)$ is the dimensionless velocity field introduced by the relation

$$u_\tau(l) = U\bar{u}_\tau(s). \tag{5.22}$$

Here, $s = l/a$ is the dimensionless curvilinear coordinate (not the McLean-Saffman coordinate used in Section 2), a the half-width of the Hele-Shaw cell and U the velocity of the finger. One can distinguish three phases in the evolution of the amplitude of the perturbation.

First, there is an initial phase (phase I of Fig. 5.5) where the perturbation is close to the tip and where the wavelength is small compared to the radius of curvature of the finger ($s \ll 1$, $\lambda \ll a$). One can use the W.K.B. method in order to calculate the growth rate of the perturbation in this region.

Second, there is an intermediate phase (phase II in Fig. 5.5) where the wavelength is of the order of the radius of curvature of the finger ($s \approx 1$, $\lambda \approx a$) and where the W.K.B. method cannot be applied.

Third, there is a final phase (phase III in Fig. 5.5) where the wavelength of the perturbation saturates at a value of order a, since the tangential field

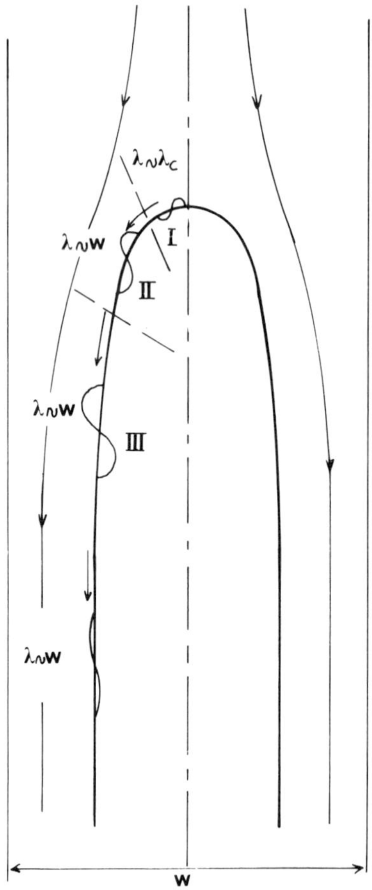

Fig. 5.5. Evolution of a perturbation of a Saffman-Taylor finger.

becomes nearly uniform ($s \gg 1$, $\lambda \approx a$). Even if the interface is nearly straight in this region, one cannot use the growth rate given by relation (1.4), since the velocity field is perturbed on a distance of the order of the wavelength, i.e., a, and thus interacts with the wall of the cell. Nevertheless, to have an idea of what occurs in this region, one will evaluate the growth rate of the perturbation by neglecting the interaction of the perturbed velocity field with the wall. The amplitude of the perturbation grows only if the wavelength is larger than the marginal wavelength of the planar front. Using relation (1.4), and writing $V = U\cos\theta$, the marginal wavelength is

$$\lambda_m = \frac{1}{\sqrt{3}} \frac{\lambda_c}{\sqrt{\cos\theta}}. \tag{5.23}$$

Here, θ is the angle between the normal to the interface and the direction of propagation and λ_c the most unstable wavelength of the planar front at the tip ($\theta = 0$). There exists an angle θ_f for which $\lambda_m = \lambda$ and thus from which the amplitude of the perturbation begins to decay. This angle is determined by the relation

$$\lambda_c \frac{\bar{u}_\tau(s_f)}{\epsilon} = \frac{1}{\sqrt{3}} \frac{\lambda_c}{\sqrt{\cos\theta}}. \tag{5.24}$$

Thus, one has the following picture where the amplitude of the perturbation begins to grow close to the tip, reaches a maximum and then decays in the tail since the instability is weaker there. This behaviour is in good agreement with the numerical simulations of DeGregoria and Schwartz [73] and the observations of Tabeling *et al.* [1].

In the first phase of its evolution, the growth rate of the perturbation is given by (5.1), i.e.,

$$\Gamma_I(l) = \int_{\lambda_c}^{l} \frac{Uk\left(\cos\theta - \left(k\lambda_c/2\pi\sqrt{3}\right)^2\right)}{u_\tau}\, dl, \tag{5.25}$$

where l is large compared to λ_c but small compared to a. By using the change of variable given by (5.7), it can be written as

$$\Gamma_I(l) = \frac{\bar{u}_\tau'(0)}{\epsilon} \int_1^{\bar{u}_\tau(s)/\epsilon} \frac{2\pi\left(\cos\theta - \dfrac{1}{3x^2}\right)}{x^2 \left.\dfrac{d\bar{u}_\tau}{ds}\right|_{s=\bar{u}_\tau^{-1}(\epsilon x)}}\, dx, \tag{5.26}$$

which, to leading order in $1/\epsilon$ is

$$\Gamma_I \approx \frac{2\pi}{\epsilon} \int_1^{\infty} \frac{3x^2 - 1}{3x^4}\, dx. \tag{5.27}$$

In experiment [1], the finger becomes unstable when its velocity is sufficiently large. The width of the finger is then about $\frac{1}{2}$ of the cell width and surface tension has very weak influence on the shape of the finger. Thus, one can compute the velocity field $u_\tau(s)$ by using the ($\lambda = \frac{1}{2}$) solution [9]. Then, one has

$$\begin{aligned} \bar{u}_\tau(s) &= \sin\theta = \operatorname{th}\left(\frac{\pi}{2}s\right), \\ \bar{u}_\tau'(0) &= \frac{\pi}{2}. \end{aligned} \tag{5.28}$$

Thus from relation (5.27) and relation (1.4a), one obtains

$$\Gamma_I \approx \frac{16}{9\pi} \frac{w}{b} \sqrt{Ca}\,. \tag{5.29}$$

In the intermediate phase (phase II), one evaluates the growth rate by dimensional analysis, i.e., $\Gamma_{\mathrm{II}} \approx \Omega t$ with $t \approx R/U$ and $\Omega \approx U/\lambda \approx U/R$ which leads to $\Gamma_{\mathrm{II}} \approx 1$.

In the final phase (Phase III), if one neglects the effect of the wall on the perturbed velocity field, one obtains

$$\Gamma_{\mathrm{III}} \approx \int_{l'}^{l_{\mathrm{f}}} \frac{Uk\left(\cos\theta - \left(k\lambda_c/2\pi\sqrt{3}\right)^2\right)}{u_\tau} \, dl, \tag{5.30}$$

or, by using the variable $y = u_\tau(s)$,

$$\Gamma_{\mathrm{III}} \approx \bar{u}_\tau'(0) \int_{\bar{u}_\tau(s')}^{\bar{u}_\tau(s_{\mathrm{f}})} \frac{2\pi\left(\cos\theta - \dfrac{\epsilon^2}{3y^2}\right)}{y^2 \left.\dfrac{d\bar{u}_\tau}{ds}\right|_{s=\bar{u}_\tau^{-1}(y)}} \, dy. \tag{5.31}$$

Here, l' (resp. s') is small compared to l_{f} (resp. s_{f}) but large compared to a (resp. 1). From relation (5.28), one deduces

$$\begin{aligned} \bar{u}_\tau'(s) &= \frac{\pi}{2}\left(1 - \mathrm{th}\left(\frac{\pi}{2}s\right)\right) \\ &= \frac{\pi}{2}(1 - y^2). \end{aligned} \tag{5.32}$$

The upper bound of the integral (5.31) is determined by relations (5.24) and (5.28), i.e.,

$$\bar{u}_\tau(s_{\mathrm{f}}) = \sin\theta_{\mathrm{f}} = 1 - \frac{1}{18}\epsilon^4, \tag{5.33}$$

which leads to

$$\Gamma_{\mathrm{III}} \approx 4\int_{\bar{u}_\tau(s')}^{1-1/18\epsilon^4} \frac{\left(\sqrt{1-y^2} - \dfrac{\epsilon^2}{3y^2}\right)}{y^2(1-y^2)} \, dy, \tag{5.34}$$

whose dominant order in ϵ is a number of order unity, i.e., $\Gamma_{\mathrm{III}} \approx 1$.

Thus, the total growth rate of the perturbation is mainly determined by the contribution of the tip (relation (5.29)). A stability criterion for the Saffman-Taylor finger is then

$$\frac{w}{b}\sqrt{Ca} \approx -\frac{9\pi}{16}\log\left(\frac{A_{\mathrm{i}}}{R}\right), \tag{5.35a}$$

where A_{i} is the initial amplitude of the perturbation. For a given amplitude of noise, a finger is unstable either if its velocity is large or if the Hele-Shaw cell is of large width and thin thickness. A relation of this kind was anticipated by the numerical simulation of Bensimon [74]. Thus, the heuristic argument of Zeldovich *et al.* [71] gives good predictions on the example of a curved interface whose analytical, numerical, and experimental results are the most advanced. From the analytical expression of the $\lambda = \frac{1}{2}$ solution (see relation (2.16)), one gets the tip radius of the finger, $\rho = w/2\pi$. Thus, by using the analytical expression of λ_c given by relation (1.4a) one can write the criterion (5.35a) as

$$\frac{\rho}{\lambda_c} \approx -\frac{9}{32\pi}\log\left(\frac{A_i}{R}\right). \tag{5.35b}$$

In the W.K.B. method used above, it is assumed that ρ/λ_c is large, i.e., relation (5.35b) is valid for weak amplitude of noise. In this sense noise amplitude is not so weak in numerical simulations [73], [74] or experiments [1] mentioned above, since the largest values of ρ/λ_c observed do not exceed a factor 2. As a result, there is not yet clear experimental evidence of an anomalous stability of the Saffman-Taylor finger.

5.2.3. Case of the Free Needle Crystal

The case of free needle crystals is different from the case of curved fronts propagating in channels (Saffman-Taylor finger, flame) in that the stationary shape is never observed, except perhaps in helium (Rolley *et al.* [138]). What is observed is a dendrite, i.e., a crystal whose shape is stationary only in the vicinity of the tip (on a length of a few tip radii). Near the tip, the interface is approximately a paraboloid. At some distance from the tip sidebranches develop, almost at rest in the laboratory frame. Thus, in the frame of the crystal they propagate away from the tip towards the tail with a tangential velocity $u_\tau = U \sin\theta$, implying that the shape of the dendrite is not stationary. In order to understand the origin of this persistent sidebranching, it is useful to study first the stability of the needle crystal. In the same way as in the previous cases, one evaluates the growth rate Γ of a perturbation, starting from the tip and with initial wavelength λ_c[11]. We look at the case where the W.K. B. method applies, i.e., when λ_c is very small compared to the tip radius ρ, or as mentioned in Section 3, when the eigenvalue C characterizing the stationary state is large (weak crystalline anisotropy for instance). As in the case of the Saffman-Taylor finger, one can notice three phases in the evolution of the perturbation.

In the first phase (Phase I in Fig. 5.6) the wavelength of the perturbation is small compared to the tip radius. Thus, the W.K.B. method applies in this

[11] More complete analyses study the evolution of a wave packet in the W.K.B. method by using the original equations for the shape of the crystal (Pieters and Langer [139] in the frame of the boundary layer model [50], Caroli *et al.* [140] and Barbieri *et al.* [141] by using the time-dependent Nash-Glicksman equation).

region and the growth rate takes the form

$$\Gamma_{\mathrm{I}}(l) = \int_{\lambda_c}^{l} \frac{Uk\left(\cos\theta - \left(k\lambda_c/2\pi\sqrt{3}\right)^2\right)}{u_\tau}\, dl, \tag{5.36}$$

where l is large compared to λ_c but small compared to the tip radius ρ. For this, one used the linear growth rate of a perturbation on the planar interface given by relation (1.5). This relation is similar to relation (5.25) obtained in the case of the Saffman-Taylor finger. Thus, the development of the calculations is the same and one obtains

$$\Gamma_{\mathrm{I}} \approx \frac{16\pi}{9\epsilon}. \tag{5.37}$$

Here,

$$\epsilon = \bar{u}_\tau\left(\frac{\lambda_c}{\rho}\right) \approx \frac{4\pi\sqrt{3}}{\sqrt{C}}\bar{u}'_\tau(0), \tag{5.38}$$

and $u_\tau(s) = \sin\theta$, the dimensionless tangential drift of the perturbation, is a function of the dimensionless curvilinear coordinate $s = l/\rho$. As already mentioned in Section 3, large values of C correspond to stationary states close to the paraboloid of tip radius ρ which can be associated with it. The equation of this paraboloid is $z = -r^2/\rho$, from which one deduces

$$\begin{aligned} \bar{u}_\tau(s) &= \sin\theta = \frac{r}{\sqrt{r^2+\rho^2}}, \\ \bar{u}'_\tau(0) &= 1. \end{aligned} \tag{5.39}$$

Thus, in the first phase, the growth rate is given by

$$\Gamma_{\mathrm{I}} \approx \frac{4}{9\sqrt{3}}\sqrt{C}. \tag{5.40}$$

In this initial phase, the wavelength is stretched to a value of order ρ. At this step, the growth rate is determined dimensionally (Phase II in Fig. 5.6), i.e., $\Gamma_{\mathrm{II}} \approx \Omega t \approx (U/\rho)\cdot(\rho/U) \approx 1$.

In the final phase, (Phase III in Fig. 5.6) the curvature radii of the paraboloid become very large compared to the wavelength of the perturbation (of order ρ). Thus the growth rate can be computed by the W.K.B. method, i.e.,

$$\Gamma_{\mathrm{III}} \approx \int_{l'}^{l_f} \frac{Uk\left(\cos\theta - \left(k\lambda_c/2\pi\sqrt{3}\right)^2\right)}{u_\tau}\, dl, \tag{5.41}$$

or, by using the variable $y = u_\tau(s)$,

$$\Gamma_{\mathrm{III}} \approx \bar{u}'_\tau(0)\int_{\bar{u}_\tau(s')}^{\bar{u}_\tau(s_f)} \frac{2\pi\left(\cos\theta - \dfrac{\epsilon^2}{3y^2}\right)}{y^2 \left.\dfrac{d\bar{u}_\tau}{ds}\right|_{s=\bar{u}_\tau^{-1}(y)}}\, dy. \tag{5.42}$$

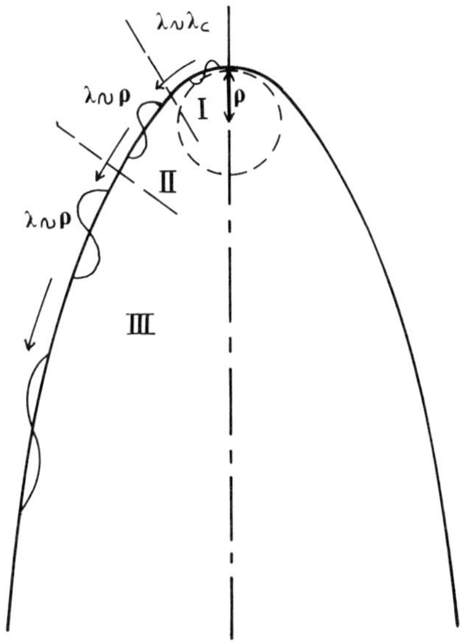

Fig. 5.6. Evolution of a perturbation of a needle crystal.

The upper bound of the integral (5.42) is determined by the place where the marginal wavelength of the planar front becomes larger than the wavelength of the perturbation, i.e., where the amplitude of the perturbation begins to decay. As these wavelengths have the same dependence in θ as in the case of the Saffman-Taylor finger, this upper bound is determined by relation (5.33). The length l' (resp. s') is small compared to l_{f} (resp. s_{f}) but large compared to ρ (resp. 1). From relation (5.39), one deduces

$$\cos\theta = \sqrt{1 - y^2} \tag{5.43}$$

and

$$\bar{u}'_\tau(s) = \frac{\rho^4}{r^2 + \rho^2} = \left(1 - y^2\right)^2. \tag{5.44}$$

Thus, one obtains

$$\Gamma_{\text{III}} \approx \int_{\bar{u}_\tau(s')}^{1-\epsilon^4/18} \frac{\left(\sqrt{1 - y^2} - \dfrac{\epsilon^2}{3y^2}\right)}{y^2\left(1 - y^2\right)^2}\, dy \tag{5.45}$$

(compare with relation (5.34)), whose dominant order in $1/\epsilon$ is

$$\Gamma_{\text{III}} \approx \frac{3}{2\epsilon^2} = \frac{1}{32\pi^2}C. \tag{5.46}$$

Thus, the situations differs from the Saffman-Taylor case since the major part of the growth is now in the tail and not at the tip. This can be understood in that the wall of the Hele-Shaw cell constrains the interface to be parallel to the direction of propagation at a shorter distance from the tip than for the freely growing needle crystal. As the growth rate of the instability depends directly on cos θ, the perturbation has less time to grow in the tail in the confined case than in the free case.

Thus, the stability criterion for the needle crystal can be written as

$$C = -32\pi^2 \log\left(\frac{A_i}{\rho}\right). \tag{5.47}$$

Here, A_i is the amplitude of the initial perturbation. This is a link between the eigenvalue characterizing the needle crystal and the amplitude of the fluctuations in the medium. There is an important difference with the Saffman-Taylor case. Here, the stability limits are very weakly dependent on the control parameter of the growth, the supercooling Δ. It appears only in the logarithm of $\rho(\rho \approx d_0/\Delta$ from Section 3). This control parameter does not enter in the left-hand side of the stability criterion as in the case of the Saffman-Taylor finger. Thus, at least in this configuration, there is no control parameter to change in order to observe a stability threshold. This is in agreement with experiment in the sense that no such threshold has yet been observed. Needle crystals are in general unstable, the dominant mode of instability being the sidebranching. The origin of this sidebranching, i.e., the nature of the initial perturbations, remains unclear. Either it comes from fluctuations of the medium (thermal noise, fluctuation of concentration of some impurity), as seems to be observed by Dougherty *et al.* [142], or it is of intrinsic nature (dynamical coupling between the tip and the sidebranches which could lead to a limit cycle).

5.3. Linear Stability Analysis

Although the linear stability problem is well posed, it is nontheless difficult to solve. We will discuss here only the cases of the Saffman-Taylor finger and of the needle crystal for which linear stability analyses were proposed. An important first step is the stability analysis of the exact analytical solutions without surface tension, i.e., the Saffman-Taylor solution and the Ivantsov paraboloid. It is remarkable that in these cases the analysis can be done analytically.

5.3.1. The Case Without Surface Tension

The Saffman-Taylor Finger. The first stability analysis was done by Taylor and Saffman [81]. It can be found in more recent works by McLean and

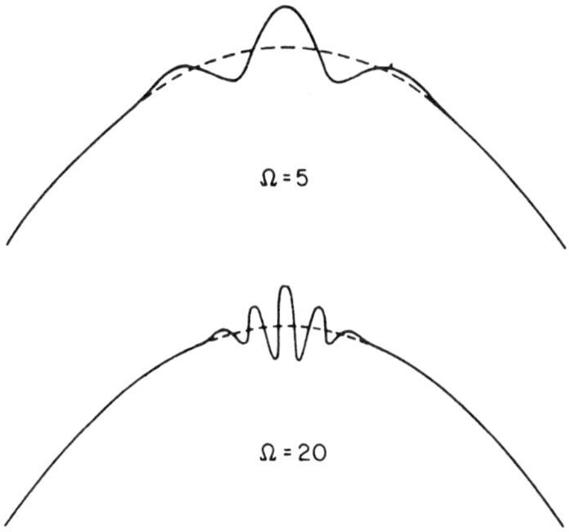

Fig. 5.7. Unstable modes in the two-dimensional model. (Reprinted with permission from Fig. 2 of Langer and Muller-Krumbhaar [47]. © 1978 Pergamon Journals, Ltd.)

Saffman [26] and McLean [143]. The growth rates Ω which are found are real, positive and correspond to localised modes (tip splitting). Some of these correspond to symmetrical disturbances, $\Omega = 2n\pi U/w$, where n is a positive integer and U the velocity of the finger, and others correspond to asymmetrical disturbances, $\Omega = (2n + 1)\pi U/w$. This result is identical to the case of the planar front with boundary conditions at the wall. The only difference is that the analytical expression of the shape of the disturbance is more complicated in the case of the finger because of the curved geometry. Thus the curved interface is still completely unstable when surface tension effects are neglected. A recent work by Tanveer [85] shows that this spectrum is not complete but is a subset of a continuum of modes with $\text{Re}(\Omega) > 0$ and arbitrary imaginary part. The physical interpretation of these modes is still lacking.

The Ivantsov Paraboloid. The analysis was done by Langer and Muller-Krumbhaar [47]. Either in the two-dimensional or in the three-dimensional case, one finds analytically, in the small Peclet number limit (small supercooling), a continuum of growth rates real and positive corresponding to localised modes (tip splitting). Some of these modes are shown in Fig. 5.7 in the two-dimensional case. Thus, here again the situation is analogous to the planar front. The curved interface is still fully unstable. As in the Saffman-Taylor case, it is possible that these modes will form a part of the whole spectrum. This is not known for the moment.

5.3.2. Effect of Surface Tension

For clarity, we will introduce one dimensionless number characterizing the effect of surface tension, which will be $\epsilon = \lambda_c/\rho$, where ρ is the tip radius of the curved interface. All the parameters used in the various existing stability analyses are directly related to this number.

The first attempt to determine the spectrum of growth rate of perturbations of a curved interface, when surface tension effects are included, was done by Langer and Muller-Krumbhaar [47] for the case of the two-dimensional needle crystal. They use the fact that when ϵ is small, the stationary shape of the needle crystal is not very different from the Ivantsov parabola, and write an integral linear equation for small disturbances of the Ivantsov stationary field. Surface tension effects are introduced only in the linearised boundary conditions at the interface. The spectrum of this linear equation is determined numerically. First the marginal mode ($\Omega = 0$) corresponding to the whole translation of the needle crystal is found. Second, in general complex growth rates are found which correspond to propagative modes (sidebranching modes). They are stable ($\mathrm{Re}(\Omega) < 0$) above some critical value of ϵ. It seems that these modes are the ones studied in the qualitative theory developed in Section 5.2.3 and which were found linearly stable by the heuristic argument adapted from Zeldovich *et al.* whatever the value of ϵ. Thus, a first discrepancy occurs there, which may be due to the fact that these modes are extended and thus, in the spectral analysis, may be very sensitive to the boundary conditions applied at some finite distance from the tip. Third, an isolated tip splitting mode (Ω real and positive) seems to have been resolved for very small value of ϵ.

One can already notice two important consequences of inclusion of surface tension effects. First, complex growth rates appear which correspond to propagative modes (sidebranching). Second, the tip splitting modes become discrete. Without surface tension, this part of the spectrum is continuous; with surface tension, it seems discrete since one isolated mode is found.

Many attempts at stability analysis of the Saffman-Taylor finger were made by McLean and Saffman [26], Kessler and Levine [82] and Bensimon [74]. In the two last studies (which use different methods), the stationary shape is assumed very close to the $\lambda = \frac{1}{2}$ solution, and surface tension is introduced only in the linearised equation (in this sense, the approach is analogous to the one used by Langer and Muller-Krumbhaar for the needle crystal). They conclude that the interface is linearly stable (all modes with $\mathrm{Re}(\Omega) < 0$). As before, complex growth rates were found (corresponding to propagative modes) but tip splitting modes were not found. It is not certain that these modes cannot be present. As a matter of fact one can expect tip splitting only if the tip radius contains at least a few most unstable wavelengths λ_c, i.e., if ϵ is much smaller than unity. But, because of some difficulty encountered in the numerical study (singular behaviour at small surface tension), previous analyses were performed for ϵ not smaller than .3.

A unified picture has been proposed by Kessler and Levine [83]. In both cases (Saffman-Taylor finger and needle crystal) the free boundary problem admits a countably infinite set of branches of stationary solutions indexed by a positive integer n. For each branch, they find a continuum of complex growth rates with negative imaginary part, and a discrete set of localised tip splitting modes (real and positive growth rate). These last modes are ordered as the following: The stationary solutions which belong to the nth branch have $n - 1$ tip splitting modes. Thus they are all unstable except the first one. This property is remarkable in that the selection problem which was posed by the multiplicity of stationary solutions is now solved.

The same property was shown analytically by Bensimon *et al.* [84], by using a W.K.B. method, similar to the one developed by Pokrovskii and Khalatnikov [144] for a problem of quantum mechanics, on a model equation for disturbances of a curved interface.

As before, the numerical study of Kessler and Levine [83] was done for not so small values of ϵ and it is not clear that the solutions of the first branch of solutions could not become unstable for smaller values of ϵ. A complete stability analysis of the Saffman-Taylor finger was done analytically by Tanveer [85] in the limit of small surface tension, which confirms the results obtained by Kessler and Levine [83] even at infinitesimally small ϵ. This result is rather paradoxical: When ϵ is very small, the interface can be locally considered as planar and thus ought to break in many pieces. This point gives some weight to the qualitative theory of Zeldovich *et al.* [71]. The curved interface is linearly stable, but can be destabilised by a finite amplitude of noise.

Thus there is an anomalous stability of a curved interface. This fact was noticed early for the case of flame propagation illustrated by this passage from the book on hydrodynamics by Landau and Lifshitz [145]: "The experimental data, on the other hand, show that the 'self-turbulence' of the flame does not occur up to very large Reynolds numbers are reached. This means that there must be other factors which stabilize the flame front and postpone its becoming turbulent until very large Reynolds numbers. It is possible that the change in the normal velocity when the front is deformed is of importance: where the front is concave, u_L increases (since the heat transfer into the unburnt mixture in the concavity is improved), while where it is convex u_L is reduced (Ya.B. Zeldovich). This important problem has not yet been resolved."

Conclusion

Interface dynamics plays an important role in spontaneous pattern formation in nature. It determines the shapes of bodies, and thus its application extends to many fields: hydrodynamics (shapes of boundaries between fluids), metallurgy (shapes of crystals), geology (shapes of rocks), biology (shapes of cells, bacteria, leaves, among others). Out of this huge domain, we have isolated a small group of interfaces which seem to behave similarly.

These interfaces are characterized by a planar front that is linearly unstable, with a most highly amplified wavelength λ_c. The wavelength λ_c also characterizes the order of magnitude of the length below which perturbations of the interface shape are damped.

The instability develops and, depending on the value of the control parameter driving the instability, the interface can restabilise in a convex curved shape. It has typical dimension ρ (the tip radius for instance) and moves with constant velocity. For a given value of the control parameter, a discrete set of stationary shapes are allowed, each of them being associated with a value of the parameter $\epsilon = \lambda_c/\rho$ and a corresponding propagation velocity.

Only one of the solutions is linearly stable, the one of largest ϵ. But it can be destabilised by a finite amplitude A of noise, a typical relation characterizing this effect being $\epsilon \approx -\log(A/\rho)$.

The results obtained here seem very weak in comparison with what is observed in nature, but they are an important step in the comprehension of the effects of nonlinearities in instabilities in general. From a theoretical point of view, this field is privileged. In two cases (Saffman-Taylor finger and needle crystal) we have at our disposal an exact solution of curved interface of large amplitude. This is very important since usually nonlinearities are introduced by an expansion at small amplitude around the unstable state with no control on the radius of convergence of the series.

These exact stationary solutions correspond to the particular value $\epsilon = 0$ (no surface tension) and are degenerate since the steady-state problem admits a continuum of solutions. This continuum breaks down for nonzero value of ϵ. The expansion at small ϵ is rather subtle since the discretisation of the steady state is due to exponentially small terms of the expansion. The spectrum of values of ϵ can be very "model dependent." For instance, in the case of the needle crystal when surface tension anisotropy factor is zero, there is no solution, but when it is nonzero, a discrete set of ϵ is found. One needs to understand more completely the link between the ingredients that one puts in a model and the corresponding spectrum of values of ϵ.

These two ideal cases can be useful to understand more difficult problems such as bubbles or flames, where analytical approaches are unsuccessful.

The problem of the determination of the values of ϵ is similar to the determination of similarity exponents of self-similar solutions. In fact, fronts propagating with constant velocity can be considered as a special case of self similar solutions, their velocity being a similarity exponent.

As far as stability is concerned, the results mentioned above are demonstrated completely from a theoretical point of view in the case of the Saffman-Taylor finger and the needle crystal. But the relation between ϵ and the amplitude of noise can be derived in all cases. This relation is derived rigorously on condition that ϵ be small; nevertheless, it is in good agreement with numerical simulations of the Saffman-Taylor finger dynamics, even when ϵ is of order unity. Up to now, neither numerical simulations nor experiments have exhibited situations where ϵ is reasonably small. Thus the anomalous stability of a curved interface is not yet experimentally confirmed.

In this kind of matter, complete agreement between theory and experiments is difficult to obtain. The main reason is the following: In experiments, many phenomena that are difficult to avoid can perturb the idealised case that one wants to observe. The model, in general, can take into account only a few of these phenomena. The question which then is posed is, are the phenomena which are neglected relevant or not? Are they singular perturbations? That is, even if the dimensionless parameter which characterizes them is small, are they essential? In very simple systems, like the Saffman-Taylor finger, the agreement between theory and experiment is complete. But this seems an exceptional situation. It is not the same, for instance, in crystallisation or combustion processes. One needs time to solve these questions.

References

[1] Tabeling, P., Zocchi, G., and Libchaber, A. (1987). *J. Fluid. Mech.* **177**, 67–82. An experimental study of the Saffman-Taylor instability.

[2] Glicksman, M.E., Shaefer, R.J., and Ayers, J.D. (1976). *Metall. Trans.* **A7**, 1747–1757. Dendritic growth—A test of theory.

[3] Davies, R.M. and Taylor, G.I. (1950). *Proc. Roy. Soc. A* **200**, 375–388. The mechanics of large bubbles rising through extended liquids and through liquids in tubes.

[4] Quinard, J., Searby, G., and Boyer, L. (1983). Ninth International Colloquium on Gasdynamics of Explosions and Reactive Systems, Poitiers, France. *Progress in Astronautics and Aeronautics* **95**, 129–141. Stability limits and critical size of structures in premixed flames.

[5] Quinard, J. (1984). Thèse d'Etat Presentée à l'Université de Provence, Marseille, France, (in French). Limites de stabilité et structures cellulaires dans les flammes prémélangées—Etude experimentale.

[6] von Lavante, E. and Strehlow, R.A. (1983). *Comb. and Flame* **49**, 123–140. The mechanism of lean limit flame extinction.

[7] Pelcé-Savornin, C., Quinard, J., and Searby, G. (1988). To appear in *Comb. Sci. Tech.* The flow field of a curved flame propagating freely upwards.

[8] Thompson, D'Arcy W., "On Growth and Form." Cambridge University Press, vol. 1 and 2, edition of 1952.

[9] Saffman, P.G. and Taylor, G.I. (1958). *Proc. Roy Soc. A* **245**, 312–329. The penetration of a fluid into a porous medium or Hele-Shaw cell containing a more viscous liquid.

[10] Chuoke, R.L., Van Meurs, P., and Van der Pol, C. (1959). *Trans. AIME* **216**, 188–194. The instability of slow immiscible viscous liquid-liquid displacements in permeable media.

[11] Mullins, W.W. and Sekerka, R.F. (1964). *J. Appl. Phys.* **3**, 444–451. Stability of a planar interface during solidification of a dilute binary alloy.

[12] Langer, J.S. (1980). *Reviews of Modern Physics* **52**, 1–28. Instabilities and pattern formation in crystal growth.

[13] Jackson, K.A. and Hunt, J.D. (1965). *Acta Metal.* **13**, 1212–1215. Transparent compounds that freeze like metals.

[14] Trivedi, R. (1984). *Metal. Trans.* **15A**, 977–982. A comparison of theory and experiment.

[15] Kurz, W. and Fisher, D.J. (1984). Fundamentals of dendritic solidification. Trans Tech Publications, Aedermannsdorf, Switzerland.

[16] Lamb, H. (1932). "Hydrodynamics," sixth edition, Cambridge University Press.

[17] Darrieus, G. (1938). Propagation d'un front de flamme. Essai de théorie des vitesses anormales de déflagration par développement spontané de la turbulence. Unpublished works presented at "La Technique Moderne" 1938 and at "Congrès de Mécanique Appliquée" 1945.

[18] Landau, L.D. (1944). *Acta Physicochimica URSS* **19**, 77–85. On the theory of slow combustion.

[19] Markstein, G.H. (1951). *J. Aero. Sci.* **18**, 199–209. Experimental and theoretical studies of flame-front stability.

[20] Einbinder, H. (1961). *J. Chem. Phys.* **21**, 480–489. The hydrodynamic stability of flame-fronts.

[21] Eckhaus, W. (1961). *J. Fluid. Mech.* **10**, 80–100. Theory of flame-front stability.

[22] Frankel, M.L. and Sivashinsky, G.J. (1982). *Comb. Sci. Tech.* **20**, 207–229. The effect of viscosity on hydrodynamic stability of a plane flame front.

[23] Pelcé, P. and Clavin, P. (1982). *J. Fluid. Mech.* **124**, 219–237. Influence of hydrodynamics and diffusion upon the stability limits of laminary premixed flames.

[24] Clavin, P. (1985). *Prog. Energy Combust. Sci.* **11**, 1–59. Dynamic behavior of premixed flame-fronts in laminar and turbulent flows.

[25] Zeldovich, Ya.B. (1944). Theory of Combustion and Detonation of Gases. Acad. Sci. USSR, Moscou.

[26] McLean, J.W. and Saffman, P.G. (1981). *J. Fluid. Mech.* **102**, 455–469. The effect of surface tension on the shape of fingers in a Hele-Shaw cell.

[27] Romero, L. (1982). Ph.D. Thesis, California Institute of Technology.

[28] Vanden-Broeck, J.M. (1983). *Phys. Fluids* **26** (8), 2033–2034. Fingers in a Hele-Shaw cell with surface tension.

[29] Dombre, T., Hakim, V., and Pomeau, Y. (1986). *C.R. Acad. Sci.* **302**, 803–808. Séléction de la largeur des doigts dans l'instabilité de Saffman-Taylor (in French).

[30] Shraiman, B.I. (1986). *Phys. Rev. Lett.* **56**, 2028–2031. Velocity selection in the Saffman-Taylor problem.

[31] Hong, D.C. and Langers, J.S. (1986). *Phys. Rev. Lett.* **56**, 2032–2035. Analytical theory of the selection mechanism in the Saffman-Taylor problem.

[32] Combescot, R., Dombre, T., Hakim, V., Pomeau, Y., and Pumir, A. (1986). *Phys. Rev. Lett.* **56**, 2036–2039. Shape selection of Saffman-Taylor fingers.

[33] Tanveer, S. (1987). To appear in *Phys. Fluids*. Analytical theory for the selection of symmetric Saffman-Taylor fingers in a Hele-Shaw cell.

[34] Barenblatt, G.I. (1979). Similarity, Self Similarity and Intermediate Asymptotics. Consultants Bureau, New York.

[35] Kolmogorov, A.N., Petrovskii, I.G., and Piskunov, N.S. (1937). Study of the diffusion equation with growth of the quantity of matter and its application to a

biological problem. Translated from *Bulletin de l'Université d'Etat a Moscou, serie internationale, Section A*, **1**, Mathématique et Mécanique, 1–25.

[36] Tabeling, P. and Libchaber, A. (1986). *Phys. Rev. A* **33**, 794–796. Film draining and the Saffman-Taylor problem.

[37] Park, C.W. and Homsy, G.M. (1984). *J. Fluid. Mech.* **139**, 291–308. Two-phase displacement in Hele-Shaw cells: theory.

[38] Reinelt, D.A. and Saffman, P.G. (1985). *SIAM J. Sci. Stat. Comput.* **6**, 542–561. The penetration of a finger into a viscous fluid in a channel and tube.

[39] Reinelt, D.A. (1987). *J. Fluid. Mech.* **183**, 219–234. Interface condition for two-phase displacement in Hele-Shaw cells.

[40] Ivantsov, G.P. (1947). Translated From *Doklady Akad. Nauk* **58** (4), 567–569. Temperature field around spheroidal, cylindrical and acicular crystal growing in a supercooled melt.

[41] Huang, S.C. and Glicksman, M.E. (1980). *Acta Metall.* **29**, 701–734. Fundamentals of dendritic solidification—I. Steady state tip growth; II. Development of sidebranch structure.

[42] Fujioka, T. (1978). Ph.D. Thesis, Carnegie Mellon University.

[43] Sekerka, R.F., Seidensticker, R.G., Hamilton, D.R., and Harrison, J.D. (1967). "Investigation of desalination by freezing," Westinghouse Research Lab Report, chapter 3.

[44] Glicksman, M.E. and Schaefer, R.J. (1968). *J. Crystal Growth* **2**, 239–242. Comments on theoretical analysis of isenthalpic solidification.

[45] Trivedi, R. (1970). *Acta Metall.* **18**, 287. Growth of dendritic needles from a supercooled melt.

[46] Nash, G.E. and Glicksman, M.E. (1974). *Acta Metall.* **22**, 1283–1299. Capillarity-limited steady-state dendritic growth—I. Theoretical development; II. Numerical results.

[47] Langer, J.S. and Muller-Krumbhaar, H. (1978). *Acta Metall.* **26**, 1681–1708. Theory of dendritic growth—I. Elements of a stability analysis; II. Instabilities in the limit of vanishing surface tension; III. Effect of surface tension.

[48] Brower, R., Kessler, D., Koplik, J., and Levine, H. (1983). *Phys. Rev. Lett.* **51**, 1111–1114. Geometrical approach to moving-interface dynamics; (1984) *Phys. Rev. A* **29**, 1335–1342. Geometrical models of interface evolution.

[49] Kessler, D., Koplik, J., and Levine, H. (1984). *Phys. Rev. A* **30**, 3161–3174. Geometrical models of interface evolution—II. Numerical simulations; (1985) *Phys. Rev. A* **31**, 1712–1717. Geometrical models of interface evolution—III. Theory of dendritic growth.

[50] Ben-Jacob, E., Goldenfeld, N.D., Langer, J.S., and Schon, G. (1983). *Phys. Rev. Lett.* **51**, 1930–1932. Dynamics of interfacial pattern formation; (1984) *Phys. Rev. A* **29**, 330–340. Boundary-layer model of pattern formation in solidification.

[51] Kruskal, M. and Segur, H. (1985). Aeronautical Res. Associates of Princeton Technical Memo, 85–25.

[52] Langer, J.S. (1986). *Phys. Rev. A* **33**, 435–441. Existence of needle crystals in local models of solidification.

[53] Pelce, P. and Pomeau, Y. (1986). *Studies in Appl. Math.* **74**, 245–258. Dendrites in the small undercooling limit.

[54] Caroli, B., Caroli, C., Roulet, B., and Langer, J.S. (1986). *Phys. Rev. A* **33**, 442–452. Solvability condition for needle crystals at large undercooling in a nonlinear model of solidification.

[55] Meiron, D. (1986). *Phys. Rev. A* **33**, 2704–2715. Selection of steady states in the two-dimensional symmetric model of dendritic growth.

[56] Kessler, D., Koplik, J., and Levine, H. (1986). *Phys. Rev. A* **33**, 3352–3357. Steady-state dendritic crystal growth.

[57] Benamar, M. and Moussalam, B. (1986). *Physica D* **25** (1–3), 155–164. Numerical results on two-dimensional dendritic solid.

[58] Benamar, M. and Pomeau, Y. (1986). *Euro. Phys. Lett.* **2** (4), 307–314. Theory of dendritic growth in a weakly undercooled melt.

[59] Barbieri, A., Hong, D.C., and Langer, J.S. (1986). *Phys. Rev. A* **35**, 1802–1808. Velocity selection in the symmetric model of dendritic crystal growth.

[60] Honjo, H. and Sawada, Y. (1982). *J. Crystal. Growth* **58**, 297–303. Quantitative measurements on the morphology of a NH_4Br dendritic crystal growth in a capillary.

[61] Pelcé, P. and Pumir, A. (1985). *J. Crystal. Growth* **73**, 337–342. Cell shape in directional solidification in the small Peclet number limit.

[62] Pelcé, P. (1986). Thèse d'Etat: "Dynamique des fronts courbes" présentée à l'Université de provence, Marseille, France (in French).

[63] Kessler, D., Koplik, J., and Levine, H. (1987). *Phys. Rev. A* **34**, 4980–4987. Dendritic growth in a channel.

[64] Birkhoff, G. and Carter, D. (1957). *J. of Math. and Mech.* **6**, 769–779. Rising plane bubbles.

[65] Garabedian, P.R. (1957). *Proc. Roy. Soc. A* **241**, 423–431. On steady-states bubbles generated by Taylor instability.

[66] Vanden-Broeck, J.M. (1984). *Phys. Fluids* **27** (5), 1090. Bubbles rising in a tube and jets falling from a nozzle.

[67] Vanden-Broeck, J.M. (1984). Technical Summary Report #2705, University of Wisconsin-Madison. Rising bubbles in a two-dimensional tube with surface tension.

[68] Uberoi, M.S. (1959). *Phys. Fluids* **2**, 72–78. Flow field of a flame in a channel.

[69] Maxworthy, T. (1962). *Phys. Fluids* **5**, 407–417. Flame propagation in tubes.

[70] Ball, G.A. (1951). Ph.D. Thesis, Harvard University.

[71] Zeldovich, Ya.B., Istratov, A.G., Kidin, N.I., and Librovich, V.B. (1980), *Comb. Science and Tech.* **24**, 1–13. Flame propagation in tubes: Hydrodynamics and stability.

[72] Pelcé, P. (1985). *J. Physique* **46**, 503–510. Effect of gravity on the propagation of flames in tubes.

[73] DeGregoria, A.J. and Schwartz, L.W. (1986). *J. Fluid Mech.* **164**, 383–400. A boundary-integral method for two-phase displacement in Hele-Shaw cells.

[74] Bensimon, D. (1986). *Phys. Rev. A* **33** 1302–1308. Stability of viscous fingering.

[75] Couder, Y., Cardoso, O., Dupuy, D., and Tavernier, P. (1986). *Europhys. Lett.* **2**, 437–443. Dendritic growth in the Saffman-Taylor experiment.

[76] Couder, Y., Gerard, N., and Rabaud, M. (1986). *Phys. Rev. A* **34**, 5175–5178. Narrow fingers in the Saffman-Taylor instability.

[77] Ben-Jacob, E., Godbey, R., Goldenfeld, N.D., Koplik, J., Levine, H., Mueller, T., and Sander, L.M. (1985). *Phys. Rev. Lett.* **55**, 1315–1318. Experimental demonstration of the role of anisotropy in interfacial pattern formation.

[78] Bechhoefer, J. and Libchaber, A. (1987). *Phys. Rev. B* **35**, 1393–1396. Testing shape selection in directional solidification.

[79] Bensimon, D., Kadanoff, L.P., Liang, S., Shraiman, B.I., and Tang, C. (1986). *Rev. Mod. Phys.* **58**, 977–999. Viscous flow in two dimensions.

[80] Pelcé, P. and Clavin, P. (1987). *Europhys. Lett.* **3** (8) 907–913. The stability of curved fronts.

[81] Taylor, G.I. and Saffman, P.G. (1958). Proc. 2nd Symp. on Naval Hydrodynamics, pp. 277–291.

[82] Kessler, D. and Levine, H. (1986). *Phys. Rev. A* **33**, 2621–2639. Theory of the Saffman-Taylor "finger" pattern.

[83] Kessler, D. and Levine, H. (1986). *Phys. Rev. Lett.* **57**, 3069–3072. Stability of dendritic crystals.

[84] Bensimon, D., Pelcé, P., and Shraiman, B.I. (1987). *J. Physique* **48**, 2081–2087. Dynamics of curved fronts and pattern selection.

[85] Tanveer, S. (1987). To appear in *Phys. Fluids*. Analytical theory for the linear stability of the Saffman-Taylor finger.

[86] Shraiman, B.I. and Bensimon, D. (1984). *Phys. Rev. A* **30**, 2840–2842. Singularities in non local interface dynamics.

[87] Howinson, S.D. (1986). *J. Fluid. Mech.* **167**, 439–453. Fingering in Hele-Shaw cells.

[88] Sarkar, S. (1984). *Phys. Rev. A* **31**, 3468–3469. Generalization of singularities in non local interface dynamics.

[89] Howinson, S.D. (1986). *SIAM J. Appl. Math.* **46**, 20–. Cusp development in Hele-Shaw flow with free surface.

[90] Bensimon, D. and Pelcé, P. (1986). *Phys. Rev. A* **33**, 4477–4478. Tip-splitting solutions to a Stefan problem.

[91] Sivashinsky, G.I. (1977). *Acta Astronautica* **4**, 1177–1206. Nonlinear analysis of hydrodynamic instability in laminar flames—I. Derivation of basic equations.

[92] Kuramoto, Y. and Yamada, (1976). *Progress of theoretical physics* **56**, 679–.

[93] Michelson, D.M. and Sivashinsky, G.I. (1977). *Acta Astronautica* **4**, 1207–1221. Nonlinear analysis of hydrodynamic instability in laminar flames—II. Numerical experiments.

[94] Pumir, A. (1985). *Phys. Rev. A* **31**, 543–546. Equation describing wrinkled flame fronts.

[95] Lee, Y.C. and Chen, H.H. (1982). *Phys. Scr.* **T2**, 41–47. Nonlinear dynamical models for plasma turbulence.

[96] Thual, O., Frisch, U., and Henon, M. (1985). *J. Physique* **46**, 1485–1494. Application of pole decomposition to an equation governing wrinkled flame fronts.

[97] Saffman, P.G. (1986). *J. Fluid. Mech.* **173**, 73–94. Viscous fingering in Hele-Shaw cells.

[98] Langer, J.S. (1986). Lectures in the theory of Pattern formation. Les Houches Summer School.

[99] Sivashinsky, G.I. (1983). *A. Rev. Fluid. Mech.* **15**, 179–199. Instabilities, pattern formation and turbulence in flames.

[100] Zeldovich, Ya.B and Frank-Kamenetskii, D.A. (1938). *Acta Physicochimica URSS* **2**, 341–350. A theory of thermal propagation of flame.

[101] Landau, L.D. and Levich, V.G. (1942). *Acta Physicochimica URSS* **17**, 42–54. Dragging of a liquid by a moving plate.

[102] Bretherton, F.P. (1961). *J. Fluid Mech.* **10**, 166–188. The motion of long bubbles in tubes.

[103] Jensen, H., Libchaber, A., Pelcé, P., and Zocchi, G. (1987). *Phys. Rev. A* **35**, 2221–2227. Effect of gravity on the Saffman-Taylor meniscus: Theory and experiment.

[104] Saffman, P.G. and Taylor, G.I. (1959). *Quarterly J. Mech. and Appl. Math.* **104**, 330–343. A note on the motion of bubbles in a Hele-Shaw cell and porous medium.

[105] Pomeau, Y. (1985). *Studies in Appl. Math.* **73**, 75–89. The Saffman-Taylor fingers in the limit of very large surface tension.

[106] Dombre, T. and Hakim, V. (1987). *Phys. Rev. A* **36**, 2811–2817. Saffman-Taylor fingers and directional solidification at low velocity.

[107] Pelcé, P. and Bensimon, D. (1987). *Nuclear Phys. B* (Proc. Suppl) **2**, 259–270.

[108] Bender, C.M. and Orszag, S.A. (1979). Advanced Mathematical Methods for Scientists and Engineers. McGraw-Hill, New York.

[109] Meng, J.C.S. and Thomson, J.A.L. (1978). *J. Fluid. Mech.* **84**, 433–453. Numerical studies of some nonlinear hydrodynamic problems by discrete vortex element methods.

[110] Tryggvason, G. and Aref, H. (1983). *J. Fluid. Mech.* **136**, 1–30. Numerical experiments on Hele-Shaw flow with a sharp interface.

[111] Tryggvason, G. and Aref, H. (1985). *J. Fluid. Mech.* **154**, 287–301. Finger-interaction mechanisms in stratified Hele-Shaw flow.

[112] Liang, S. (1986). *Phys. Rev. A* **33**, 2663–2674. Random walk simulations of flow in Hele-Shaw cells.

[113] DeGregoria, A.J. and Schwartz, L.W. (1987). *Phys. Rev. Lett.* **58**, 1742–1744. Saffman-Taylor finger width at low interfacial tension.

[114] Burton, W.K. and Cabrera, N. (1949). *Discussions Faraday Society* **5**, 33–48. Crystal growth and surface structure.

[115] Chernov, A.A. (1984). Modern crystallography III. Springer, Berlin.

[116] Herring, C. (1953). "Structures and properties of Solid Surfaces," edited by R. Gomer and C.S. Smith. University of Chicago Press.

[117] Horway, G. and Cahn, J.W. (1961). *Acta Metall.* **9**, 695–705. Dendritic and spheroidal growth.

[118] Langer, J.S. (1977). *Acta Metall.* **25**, 1121–1137. Studies in the theory of interfacial stability—II. Moving symmetric model.

[119] Benamar, M. (1988). To be published in *Physica D*. Theory of the needle crystal.

[120] Glicksman, M.E. and Singh, B. (1984). Workshop "Solidification and Fluid Dynamics in the Earth's and the Space Laboratory," p. 182.

[121] Bechhoefer, J., Guido, H., and Libchaber, A. (1988).*C. R. Acad. Sci.* Paris, t. **306**, Série II, 619–625. Solidification dans un capillaire rectangulaire.

[122] Landau, L.D. and Lifshitz, E. Statistical Physics. Translated from Russian by E. Peierls and R.F. Peierls. Pergamon Press, London.

[123] Dorsey, A.T. and Martin, O. (1987). *Phys. Rev. A* **35**, 3989–3992. Saffman-Taylor fingers with anisotropic surface tension.

[124] Karma, A. (1986). *Phys. Rev. A* **34**, 4353–4362. Solidification cells at low velocity: The moving symmetric model.

[125] Karma, A. (1986). *Phys. Rev. Lett.* **57**, 858–861. Wavelength selection in directional solidification.

[126] Clavin, P. and Williams, F.A. (1981). *J. Fluid. Mech.* **116**, 251–282. Effect of molecular diffusion and of thermal expansion on the structure and dynamics of premixed flames in turbulent flows of large scale and low intensity.

[127] Joulin, G. and Clavin, P. (1979). *Comb. and Flame* **35**, 139–153. Linear stability analysis of nonadiabatic flames; diffusional-thermal model.

[128] Clavin, P. and Joulin, G. (1983). *J. Phys. Lett.* **44**, L–1, L–12. Premixed flames in large scale and high intensity turbulent flows.

[129] Boyer, L. (1980). *Comb. and Flame* **39**, 321–323. Laser tomographic method for flame front movement studies.

[130] Peters, N. and Smooke, M.D. (1985). *Comb. and Flame* **60**, 171–182. Fluid dynamic-chemical interaction at the lean flammability limit.

[131] Britten, J.A., Krantz, W.B., Visser, W., and Gunn, R.D. (1985). Internal Report University of Colorado, Boulder. Theoretical and experimental studies of reverse combustion.

[132] Kotowski, M.D. and Gunn, R.D. (1976). Laramie Energy Research Report LERC/RI-76/4 Laramie, WY.

[133] Britten, J.A. (1984). Ph.D. Thesis, University of Colorado. Activation energy-asymptotic modelling of reverse combustion instabilities in porous media.

[134] Britten, J.A. and Krantz, W.B. (1985). *Comb. and Flame* **60**, 125–140. Linear stability of planar reverse combustion in porous media.

[135] Clavin, P., Denet, B., Monteiller, J., and Pelcé, P. (1986). *J. Meca. Theo. et Appl.*, Numero special, 173–192. Study of reaction-diffusion waves representative of reverse combustion in porous combustible media.

[136] Zeldovich, Ya.B. (1980). *Comb. and Flame* **40**, 225–234. Structure and stability of steady laminar flame at moderately large Reynolds number.

[137] Joulin, G. (1988) Submitted to *J. Physique*. On the Hydrodynamic stability of curved premixed flames.

[138] Rolley, E., Balibar, S. and Gallet, F. (1986). *Europhys. Lett.* **2** (3) 247–255. The first roughening transition of ^{3}He crystals.

[139] Pieters, R. and Langer, J.S. (1986). *Phys. Rev. Lett.* **56**, 1948–

[140] Caroli, B., Caroli, C., and Roulet, B. (1987). *J. Physique*, **48**, 1423–1437. On the linear stability of needle crystals: evolution of a Zel'dovich localised front deformation.

[141] Barbieri, A., Barder, and Langer, J.S. (1987). Submitted.

[142] Dougherty, A., Kaplan, P.D., and Gollub, J.P. (1987). *Phys. Rev. Lett.* **58**, 1652–1655. Development of sidebranching in dendritic crystal growth.

[143] McLean, J.W. (1981). Ph.D. Thesis, California Institute of Technology.

[144] Pokrovskii, V.L. and Khalatnikov, I.M. (1961). *Soviet Physics JETP* **13** (6), 1207–1210. On the problem of above-barrier reflection of high-energy particles.

[145] Landau, L.D. and Lifshitz, E.M. Fluid Mechanics. Translated from Russian by J.B. Sykes and W.H. Reid. Pergamon Press, London.

[146] Benamar, M. and Moussallam, B. (1988). *Phys. Rev. Lett.* **60**, 317–320. Absence of selection in directional solidification.

Part II. Collected Papers

Planar Front Propagation

Study of the Diffusion Equation with Growth of the Quantity of Matter and its Application to a Biology Problem

A. Kolmogoroff, I. Petrovsky, and N. Piscounoff

Introduction

We start off from the diffusion equation, and for greater simplicity in just two dimensions:

$$\frac{\partial v}{\partial t} = k\left(\frac{\partial^2 v}{\partial x^2} + \frac{\partial^2 v}{\partial y^2}\right), \qquad k > 0, \tag{1}$$

x and y denoting the plane coordinates of a point, t time, and v the density of the matter at (x, y) at time t. Now we suppose that besides diffusion there is a growth of the quantity of matter whose rate around a given point at the given time depends on the density already present. We then obtain the equation

$$\frac{\partial v}{\partial t} = k\left(\frac{\partial^2 v}{\partial x^2} + \frac{\partial^2 v}{\partial y^2}\right) + F(v). \tag{2}$$

Of course we are only interested in the values of $F(v)$ for $v \geq 0$. We suppose $F(v)$ is a continuous function, having all the necessary derivatives below with respect to v and also satisfying the conditions

$$F(0) = F(1) = 0; \tag{3}$$
$$F(v) > 0, \qquad 0 < v < 1; \tag{4}$$
$$F'(0) = a > 0; \qquad F'(v) < a \text{ for } 0 < v \leq 1. \tag{5}$$

Thus we are supposing that, v being infinitely small, the rate $F(v)$ of growth of v is proportional to v (with proportionality factor α), and that a state of saturation occurs when v approaches 1, at which time the growth of v ceases.

Translated from the French by Protrans. Originally published in *Bulletin de l'université d'état à Moscou, Série internationale*, section A, Vol. 1, 1937

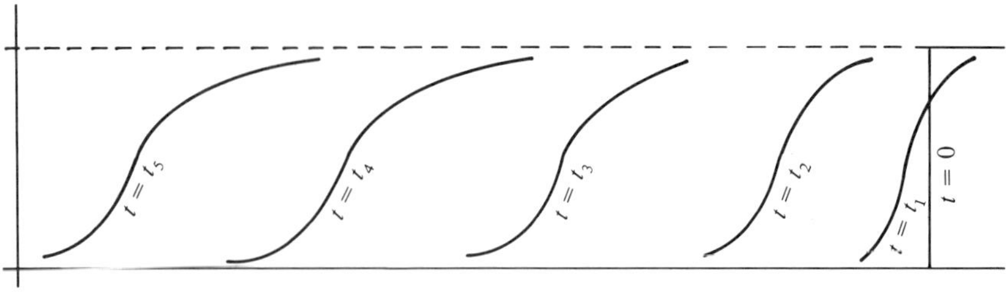

Fig. 1.

Accordingly, we only consider the solutions of (2) satisfying

$$0 \le v \le 1. \tag{6}$$

The arbitrary initial values of v for $t = 0$ that satisfy (6) define a unique solution of (2) for $t > 0$ subject to the same condition (6).[1]

We suppose after that, that the density v does not depend on the y coordinate. In this case the fundamental equation (2) will have the form

$$\frac{\partial v}{\partial t} = k\frac{\partial^2 v}{\partial x^2} + F(v). \tag{7}$$

Now suppose that at the initial time $t = 0$ the density vanishes for $x < a$ and attains its greatest possible value $v = 1$ for $x > b \ge a$.

Of course the domain of densities close to 1 spreads out as t increases from right to left, pushing back the domain of small densities to the left. In the particular case $a = b$, the picture is roughly that shown on Fig. 1. The portion of the density curve (as a function of x) which accounts for the principal part of the decrease of the density from 1 to 0 shifts with time from right to left. One sees that the shape of the density curve, as $t \to \infty$, approaches a limiting shape. The problem is to find this limit shape of the density curve and the limiting rate of its displacement from right to left. One can show that the limit rate to be found is equal to

$$\lambda_0 = 2\sqrt{k\alpha}\,, \tag{8}$$

and that the limit shape of the density curve is given by the solution of the equation

$$\lambda_0 \frac{dv}{dx} = k\frac{d^2 v}{dx^2} + F(v), \tag{9}$$

which vanishes for $x = -\infty$ and approaches 1 for $x = +\infty$. Such a solution

[1] See the proof in Section 3.

always exists and is unique to within a transformation $x' = x + c$, not changing the shape of the curve. Note that (9) can be obtained as follows. Let us look for a solution of (7) with the property that as t varies the shape of the curve representing the dependence between v and x does not change while the curve itself moves from right to left at the rate λ. Such a solution will have the form

$$v(x, t) = v(x + \lambda t). \tag{10}$$

By now considering v as a function of a single variable $z = x + \lambda t$, we obtain the equation

$$\lambda \frac{dv}{dz} = k \frac{d^2 v}{dz^2} + F(v).$$

This equation has solutions for $\lambda \geq \lambda_0$ satisfying the stated conditions relative to (9). But only for $\lambda = \lambda_0$ do we obtain the limit shape that we are interested in for the density curve under the initial conditions indicated above. To understand this fact, strange at first glance, of the existence of the solutions for (7) of the form (10) for $\lambda > \lambda_0$, solutions for which the shift of the domain of large (near to 1) densities takes place at a rate greater than λ_0, we examine the limiting case $k = 0$. In this case there is no diffusion and (7) is easily integrated. Under our initial conditions, at the points $x < a$ where the initial density was zero, it will remain zero for all $t > 0$. However, it is easy to calculate that one can find for any positive λ solutions of (7) of the form (10) satisfying all the conditions named above. The apparent shift of the matter from right to left will actually be caused there by the growth of the density at each point, occurring completely independently of what goes on at other points.

The facts presented in the introduction are applied in Section 1 to the study of some biology problems, and proved in Sections 2 and 3.

Section 1

Now let us consider a certain territory populated by an arbitrary species. First we suppose that a certain dominant gene A is spread out over this territory with a constant concentration p $(0 \leq p \leq 1)$. Suppose further that individuals having character A (i.e., belonging to the genotypes AA and Aa) are privileged in the struggle for survival over individuals not having this character (belonging to the genotype aa): We shall suppose precisely that the survivability probability ratio for an individual with A to one without A is

$$1 + \alpha,$$

α being a small positive number. Then, to within terms of order α^2, we obtain for the increase of the concentration due to a single generation the formula[2]

$$\Delta p = \alpha p(1 - p)^2. \tag{11}$$

Now suppose that the concentration p is different at different points of the territory occupied by our species, i.e. p depends on the x and y coordinates of the plane point. If also the individuals of our species remained immobile, linked to fixed points, (11) would hold as before. But suppose that in the interval between birth and reproduction, each individual moves in a random direction (all directions being equiprobable) a certain distance. Let $f(r)\,dr$ be the probability of a move whose distance lies between r and $r + dr$ and

$$\rho = \sqrt{\int_0^\infty r^2 f(r)\,dr}$$

be the mean square displacement. Instead of (11) we then get

$$\begin{aligned}\Delta p(x, y) &= \int_{-\infty}^{+\infty}\int_{-\infty}^{+\infty} p(\xi, \eta)\frac{f(r)}{2\pi r}\,d\xi\,d\eta - p(x, y)\\ &\quad + ap(x, y)\{1 - p(x, y)\}^2,\\ r &= \sqrt{(x - \xi)^2 + (y - \eta)^2}.\end{aligned} \tag{12}$$

We now suppose that we can differentiate p with respect to x, y and t (counting t in generations), that α and ρ are very small and that the third moment

$$d^3 = \int_0^\infty |r|^3 f(r)\,dr$$

is small with respect to ρ^2. In this case, by expanding $p(\xi, \eta)$ in (12) in a Taylor series in $\xi - x$ and $\eta - y$ and confining ourselves to the second-order terms (the first-order terms are disappearing), we approximately obtain for p the differential equation[3]

$$\frac{\partial p}{\partial t} = \frac{\rho^2}{4}\left(\frac{\partial^2 p}{\partial x^2} + \frac{\partial^2 p}{\partial y^2}\right) + ap(1 - p)^2. \tag{13}$$

All the considerations relating to (2) apply to this equation.

Let us again recall the hypotheses that were made. We have supposed that the concentration p varies continuously as a function of place and time (p being differentiable with respect to x, y and t), that these variations are caused by selection in the ratio $(1 + \alpha) : 1$ in favor of the dominant criterion A

[2] See R.S. Fisher, The Genetical Theory of Natural Selection, Oxford University Press, 1930.

[3] As to getting from (12) to (13), see the similar considerations, e.g., in A. Khintchine, "Assymptotische Gesetze der Wahrscheinlichkeitsrechnung," Springer, 1933.

and by accidental displacements of particular individuals, such as the mean square displacement of a single individual counted from birth to reproduction, or ρ; and, finally, that α and ρ are small (ρ with respect to the distances over which considerable changes in the concentration p occur). In this case, one obtains (13) by measuring t in generations.

Let us examine the case where a vast territory is already occupied by gene A with a concentration p close to 1. There must naturally be along the boundary of this domain a strip of intermediate concentrations. We suppose that beyond this strip p is close to zero. In view of positive selection, the domain occupied by gene A is going to spread, or restated, its boundary is going to move to territories not yet occupied by A, but along this boundary there will always remain a strip of intermediate concentrations. Our first problem is to find *the rate of propagation of the gene* A, i.e., the rate of displacement of the boundary. Formula (8) gives an immediate answer to this question: since in the case at hand $k = \rho^2/4$, the desired rate is

$$\lambda = \rho\sqrt{\alpha}\,. \tag{14}$$

In the second place, one must naturally find the width of the intermediate strip. By (9), the concentration p satisfies, in the direction of the normal to the boundary, the equation

$$\lambda\frac{dp}{dn} = \frac{\rho^2}{4}\frac{d^2p}{dn^2} + ap(1-p)^2$$

or, dividing by α and applying (14) one has

$$\frac{\rho}{\sqrt{a}}\frac{dp}{dn} = \frac{1}{4}\frac{\rho^2}{a}\frac{d^2p}{dn^2} + p(1-p)^2.$$

By introducing a new variable

$$\nu = \frac{\sqrt{a}}{\rho}n \tag{15}$$

we obtain the equation

$$\frac{dp}{d\nu} = \frac{1}{4}\frac{d^2p}{d\nu^2} + p(1-p)^2 \tag{16}$$

not containing α or ρ. The boundary conditions for (16) are the same as for (9):

$$p(+\infty) = 1, \qquad p(-\infty) = 0.$$

We deduce from (15) that the width of the intermediate strip is proportional to

$$L = \frac{\rho}{\sqrt{a}}. \tag{17}$$

e.g., for $\rho = 1$ and $\alpha = 0.0001$, one has $\lambda = 0.01$ and $L = 100$.

Section 2

In this section we consider the equation

$$\lambda \frac{dv}{dx} = k \frac{d^2v}{dx^2} + F(v), \tag{18}$$

λ and k being positive and $F(v)$ satisfying the conditions stated in the introduction. Our goal is to find relations between λ, k and $\alpha = F'(0)$ for which this equation would have a solution satisfying

$$0 \leq v(x) \leq 1, \qquad v(x) \to 1 \text{ as } x \to +\infty, \qquad \text{and } v(x) \to 0 \text{ as } x \to -\infty.$$

Putting $dv/dx = p$ we have

$$\frac{d^2v}{dx^2} = \frac{dp}{dv} \cdot \frac{dv}{dx} = \frac{dp}{dv} \cdot p.$$

Substituting this into (18), we find that

$$\frac{dp}{dv} = \frac{\lambda p - F(v)}{kp}. \tag{19}$$

We are interested in the integral curves of this equation that in the (p, v) plane pass between the lines $v = 0$ and $v = 1$. Of these there are generally the following types.

1. Integral curves that do not approach one of the lines $v = 0, v = 1$ at a distance less than a certain $\epsilon > 0$.
2. Integral curves that draw infinitely away from the v axis, asymptotically approaching one of the lines $v = 0, v = 1$.
3. Interval curves that intersect one of these lines at a point not on the v axis.
4. Integral curves not of any of the preceding types, approaching the points $v = 0, p = 0$ and $v = 1, p = 0$.

It is easy to see that the integral curves of the first type cannot correspond to solutions of (18) satisfying the stated conditions, since for them v cannot be made as close to 0 and 1 as one desires.

The integral curves of the second type do not generally exist for they would have to have points for which, $|p|$ being very large, the quantity $|dp/dv|$ would be very large. But the quotient $(\lambda p - F(v))/kp$, for very large values of $|p|$, is roughly equal to λ/k (since $F(v)$ is bounded on the interval $(0, 1)$.

The integral curves of the third type correspond to solutions of (18) that do not always remain between 0 and 1. In effect, suppose for example that some curve of this type approaches the point $v = 1$, $p = p_1 \neq 0$. Near the line $v = 1$,

$$\frac{dp}{dv} \approx \frac{\lambda}{k \neq 0}.$$

Consequently, here one may consider p as a function of v, say $p = \phi(v)$. Since $\phi(1) = p_1 \neq 0$, $|\phi(v)|$ would remain, on a certain small interval $(1 - \epsilon, 1 + \epsilon)$, greater than a certain positive constant C. Let x_0 denote the value of x for which $v = 1 - \epsilon$. Then, integrating the equation $dv/dx = \phi(v)$, we obtain

$$\int_{x_0}^{x} dx = x - x_0 = \int_{1-\epsilon}^{v} \frac{dv}{\phi(v)}.$$

It results that when v varies from $1 - \epsilon$ to $1 + \epsilon$, the variations of x do not exceed in modulus the number $2\epsilon/C$. This is why when x varies from x_0 to $x_0 + 2\epsilon/C$, v must pass through 1.

Now for the fourth type. Each of the points $v = 0$, $p = 0$ and $v = 1$, $p = 0$ is a singular point of (19). The integral curve of the fourth type must approach each of these points without intersecting the lines $v = 0$ and $v = 1$ elsewhere. Thus, in order for such curves to exist, it would be necessary that the characteristic equation for each of these points have real roots. Let us put $F(v)$ in the form

$$F(v) = \alpha v + \phi_1(v).$$

Then obviously $\phi_1(v) = o(v)$. Thus the characteristic equation at $v = 0$, $p = 0$ has the form

$$\begin{vmatrix} \lambda - \rho & -a \\ k & -\rho \end{vmatrix} = 0,$$

i.e.,

$$\rho^2 - \lambda\rho - ak = 0. \tag{20}$$

For this equation to have real roots, it would be necessary that

$$\lambda^2 \geq \alpha k.$$

To obtain the characteristic equation at $v = 1$, $p = 0$, we change the variables, $v = 1 - u$. We obtain

$$\frac{dp}{du} = \frac{-\lambda p + \Phi(u)}{kp}$$
$$\Phi(u) = F(1 - u).$$

We obviously have $F'(1) \leq 0$ and $\Phi'(0) = -F'(1) = A \geq 0$. Consequently,

$$\Phi(u) = Au + o(u),$$

and the characteristic equation at $v = 1$, $p = 0$ takes the form

$$\begin{vmatrix} -\lambda - \rho & A \\ k & -\rho \end{vmatrix} = 0,$$

i.e.,

$$\rho^2 + \lambda\rho - Ak = 0. \tag{21}$$

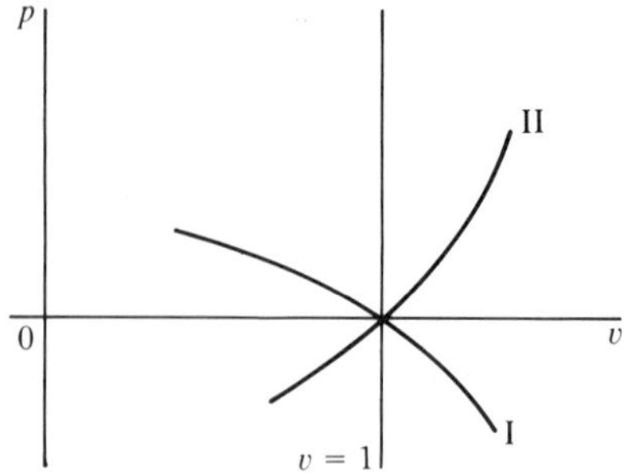

Fig. 2.

For this equation to have real roots, it would be necessary that

$$\lambda^2 \geq -4Ak.$$

Since $\alpha > 0$, (20) has real roots of the same sign. Consequently $(0,0)$ is a node. All the curves in a tight enough neighborhood of the point pass through it. This is why only two curves pass through this point following perfectly determined directions; namely, the directions given by the equations

$$\begin{aligned} m_1 u + n_1 p &= 0 \\ m_2 u + n_2 p &= 0. \end{aligned} \tag{22}$$

It is known[4] that the coefficients m_1, n_1, m_2, n_2 are determined by the equations

$$\begin{aligned} km_1 - \rho_1 n_1 &= 0 \\ km_2 - \rho_2 n_2 &= 0, \end{aligned} \tag{23}$$

where ρ_1 and ρ_2 are the roots of the characteristic equation (21). As these roots have different signs, the slopes of the lines (22) also have different signs.[5] This is why in each of the angles formed by the intersection of the lines $v = 1$ and $p = 0$ there passes only one integral curve of (19) going through $v = 1$, $p = 0$. On Fig. 2 below is a sketch of these curves.

The curve II cuts the p axis below the origin, for (19) shows that $dp/dv > 0$ in the part of the band between the lines $v = 0$, $v = 1$ below the v axis. For this reason, we eliminate curve II from consideration. Now for curve I.[6]

---

[4] E.g., see Bendixon, *Acta Mathematica*, **24**, 1900.

[5] One can also show that the slopes of the tangents to the integral curves of equation (19) are positive at the origin.

[6] If $A = 0$, one can only say that there is at least one integral curve of type I approaching $(1,0)$ in the direction with negative slope. (See I. Petrowsky, Über das Verhalten der Integralkurven eines Systems gewöhnlicher Differentialgleichungen, *Recueil Mathématique* **41**, pp. 107–156, 1934).

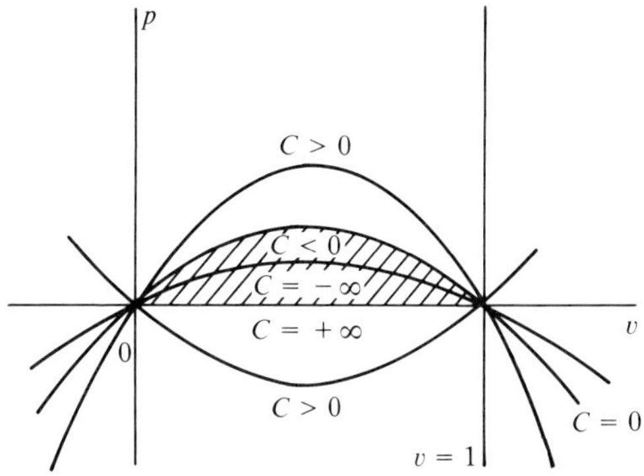

Fig. 3.

We wish to show that the curves I cut the p axis at the origin. Let us first show that these curves cannot cut the p axis below the origin. For this we consider the isoclines of (19). The equation for this family has the form

$$\frac{\lambda p - F(v)}{kp} = C, \tag{24}$$

where C is the value of dp/dv at (v, p). Hence

$$p = \frac{F(v)}{\lambda - Ck}. \tag{24'}$$

Equation (24) is a family of curves passing through $(0,0)$ and $(1,0)$. See Fig. 3. We see beside each curve the value of C corresponding to it. The thick line is the curve for $C = 0$. As the summits of the curves rise, C increases, approaching λ/k, which corresponds to the lines $v = 0$, $v = 1$; in the domain between $C = 0$ and the v axis (hatched on Fig. 3) we have $C < 0$, where C is very large in absolute value at points close to the v axis. Below the v axis, $C > 0$ and the summit of the curve drops from the level of the v axis to $-\infty$ while C decreases from $+\infty$ to λ/k.

Now we easily see that the integral curve of type 1 (Fig. 2) cannot intersect the p axis below the origin. In effect, in this case the curve I would have to intersect the v axis. Since dp/dv is equal to $-\infty$ above this axis and to $+\infty$ below, the convexity of the integral curve I at the point of intersection with the v axis is turned toward the line $v = 1$. Thus, for this curve to reach the point $(1,0)$ it would be necessary that dp/dv become ∞ above the v axis, which is impossible. Therefore, the integral curves I cannot intersect the line $v = 1$ above the v axis.

Now let us prove that the integral curve I cannot cut the p axis above the origin. For this it suffices to show that there is a halfline passing through the origin and crossing the first quadrant, that is not cut by any of the integral curves meeting the p axis in its positive part. Let us go now to the proof of the existence of such a halfline.

From (24) we obtain[7]

$$\left(\frac{d\bar{p}}{dv}\right)_{v=0} = \frac{a}{\lambda - Ck}.$$

Let us find C so that $(d\bar{p}/dv)_{v=0} = C$. For it we have

$$\frac{a}{\lambda - Ck} = C,$$

i.e.,

$$kC^2 - C\lambda + a = 0.$$

Hence

$$C = \frac{\lambda \pm \sqrt{\lambda^2 - 4ak}}{2k}. \tag{25}$$

Since we supposed that

$$\lambda^2 \geq 4\alpha k,$$

the two values of C from (25) are real and positive. Let C_0 be one of them and draw the line

$$p = C_0 v. \tag{26}$$

We easily see that for all points of the strip between the lines $v = 0$ and $v = 1$ above the line (26) or even on it (only the origin being excluded) we have[8]

$$\frac{dp}{dv} > C_0.$$

Thus no integral curve passing through any point of the p axis above the origin will ever intersect the part of the line (26) above the v axis. We have thus shown that each integral curve of type I (Fig. 2) passes through the origin.

We are going to show that there exists only one integral curve of type I. (This demonstration is certainly necessary only in the case $A = 0$.) In effect, we have shown that all the curves of type I pass through the origin. On the other hand, it follows from (19) that for $p > 0$, dp/dv increases with p, while v remains invariable. Thus, two integral curves emerging from the origin cannot pass through $(1, 0)$.

Now let us show that the curve I corresponds to a solution of (18) satisfying the conditions stated at the outset. Note first that no perpendicular to the v

[7] Here $d\bar{p}/dv$ denotes the derivative of $p = p(v)$ defined by (24′).
[8] p being linked to v by (18).

axis intersects the integral curve I of equation (19) except in a single point, for otherwise dp/dv would become ∞ above the v axis. This is why p is a function of v, $p = \phi(v)$, along this curve. Recall again that the curve I cuts the v axis at the point $(1, 0)$ at an angle whose tangent is negative, and at the origin at an angle whose tangent is positive. Thus for small values of v,

$$p = k_1 v + o(v) \tag{27}$$

and for small values of $1 - v$,

$$p = k_2(1 - v) + o(1 - v), \tag{28}$$

where k_1 and k_2 are positive.

Now recall that $p = dv/dx$. Thus $dv/dx = \phi(v)$ or $dx = dv/\phi(v)$. Integrating the latter, we have

$$x - x_0 = \int_{v_0}^{v} \frac{dv}{\phi(v)}, \qquad 0 < v_0 < 1.$$

From (27) and (28), as $v \to 0$, $x \to -\infty$, and as $v \to 1$, $x \to +\infty$, QED.

Section 3

Instead of equation (7), of which we spoke in the introduction, we shall consider in this section the equation

$$\frac{\partial v}{\partial t} - \frac{\partial^2 v}{\partial x^2} = F(v) \tag{29}$$

where the function $F(v)$ satisfies the following conditions:

$$F(0) = F(1) = 0; \tag{30}$$

$$F(v) > 0, \qquad 0 < v < 1; \tag{31}$$

$$F'(0) = 1; \tag{32}$$

$$F'(v) < 1, \qquad 0 < v \leq 1. \tag{33}$$

$F'(v)$ is bounded and continuous on the interval $(0, 1)$. Besides that, we suppose that we can differentiate $F(v)$ a number of times. This is precisely the form to which we can always reduce the general equation (7) given in the introduction with the aid of the change of variables

$$x = \sqrt{\frac{k}{a}}\,\bar{x}, \qquad t = \frac{\bar{t}}{u}.$$

The main goal of this section is to show that as $t \to +\infty$, the portion of the density curve $v(x, t)$ (as a function of x) on which occurs the fundamental part of the drop from 1 to 0 moves with time from right to left at a rate that approaches 2 from below; and the shape of the density curve approaches that of the curve that is the solution $u(x)$ of the equation

$$\frac{d^2u}{dx^2} - 2\frac{du}{dx} + F(u) = 0 \tag{34}$$

which becomes 0 as $x \to -\infty$ and 1 as $x \to +\infty$; the existence of such a solution was proved in Section 2.

Before undertaking the proof of the fundamental propositions of this section, we are going to prove the existence of a solution of the equation

$$\frac{\partial v}{\partial t} - \frac{\partial^2 v}{\partial x^2} = F(x, t, v)$$

of which a particular case is equation (29) taking on given values for $t = 0$, and we shall study the properties of it.

Theorem 1. *Let*

$$\frac{\partial v}{\partial t} - \frac{\partial^2 v}{\partial x^2} = F(x, t, v) \tag{35}$$

where the bounded and continuous $F(x, t, v)$ satisfies the Lipschitz condition with respect to v and x, i.e.,

$$|F(x_2, t, v_2) - F(x_1, t, v_1)| < k|x_2 - x_1| + k|v_2 - v_1|, \tag{36}$$

where k is a constant independent of x, t and v. Let $f(x)$ be a bounded function defined for all values of x. For greater simplicity, we shall suppose that $f(x)$ has only a finite number of points of discontinuity. Then there exists a unique bounded function $v(x, t)$ for bounded t that for $t > 0$ satisfies (35) *and that for $t = 0$ is equal to $f(x)$ at all points of continuity of this function.*

Below for greater brevity, in saying that $v(x, t)$ becomes equal to $f(x)$ for $t = 0$, we shall mean that this refers only to the points of continuity of this function.

Proof. Let $v_0(x, t)$ be a bounded function satisfying for $t > 0$ the equation

$$\frac{\partial v}{\partial t} - \frac{\partial^2 v}{\partial x^2} = 0 \tag{37}$$

and equal to $f(x)$ for $t = 0$. By substituting this function instead of v in the second member of (35), we find according to the formula

$$\tilde{v}_1(x, t) = \frac{1}{2\sqrt{\pi}} \int_0^t d\eta \int_{-\infty}^{+\infty} \frac{e^{-\frac{(x-\xi)^2}{4(t-\eta)}}}{\sqrt{t-\eta}} F(\xi, \eta, v_0(\xi, \eta))\, d\xi \tag{38}$$

the solution of this equation taking on the value 0 at $t = 0$ (see Gevrey,

Journal de math. pures et appliqueés, 1914, p. 308). The function

$$v_1(x, t) = v_0(x, t) + \tilde{v}_1(x, t)$$

for $t = 0$ is equal to $f(x)$ and for $t > 0$ satisfies the equation

$$\frac{\partial v_1}{\partial t} - \frac{\partial^2 v_1}{\partial x^2} = F(x, t, v_0(x, t)).$$

In general, according to the formula

$$v_{i+1}(x, t) = v_0(x, t) + \frac{1}{2\sqrt{\pi}} \int_0^t d\eta \int_{-\infty}^{+\infty} \frac{e^{-\frac{(x-t)^2}{4(t-\eta)}}}{\sqrt{t-\eta}} F(\xi, \eta, v_1(\xi, \eta))\, d\xi \quad (39)$$

we obtain a function $v_{i+1}(x, t)$ satisfying for $t > 0$ the equation

$$\frac{\partial v_{i+1}}{\partial t} - \frac{\partial^2 v_{i+1}}{\partial x^2} = F(x, t, v_i) \quad (40)$$

and reducing to $f(x)$ for $t = 0$.

Let us show that the sequence of functions $v_i(x, t)$ converges uniformly. Indeed, noting (36), we find from (39) that

$$\begin{aligned} M_{i+1}(t) &= \limsup_{\eta \le t} |v_{i+1}(x, \eta) - v_i(x, \eta)| \\ &\le \frac{1}{2\sqrt{\pi}} \int_0^t d\eta \int_{-\infty}^{+\infty} \frac{e^{-\frac{(x-\xi)^2}{4(t-\eta)}}}{\sqrt{t-\eta}} \\ &\quad \times |F(\xi, \eta, v_i(\xi, \eta)) - F(\xi, \eta, v_{i-1}(\xi, \eta))|\, d\xi \\ &\le \int_0^t k M_i(\eta)\, d\eta \end{aligned} \quad (41)$$

Because

$$\int_{-\infty}^{+\infty} \frac{e^{-\frac{(x-\xi)^2}{4(t-\eta)}}}{\sqrt{t-\eta}} d\xi = 2\sqrt{\pi}.$$

Now, denoting by M_0 the upper limit of the values of $|f(x)|$ and $|F(x, t, 0)|$, we obtain

$$|v_0(x, t)| \le M_0,$$

hence, with the aid of (38),

$$M_1 \le \int_0^t (k + 1) M_0\, dt = (k + 1) M_0 t = Mt.$$

Using (41), we easily deduce from this that

$$M_i \le M \frac{k^{i-1} t^i}{i!},$$

and hence we obtain very simply the uniform convergence of v_i.

Put

$$v(x, t) = \lim_{i \to \infty} v_i(x, t).$$

The function $v(x, t)$ becomes $f(x)$ for $t = 0$. On the other hand, as one can easily see, it satisfies the equation

$$v(x, t) = v_0(x, t) + \frac{1}{2\sqrt{\pi}} \int_0^t d\eta \int_{-\infty}^{+\infty} \frac{e^{-\frac{(x-\xi)^2}{4(t-\eta)}}}{\sqrt{t - \eta}} \times F(\xi, \eta, v(\xi, \eta))\, d\xi. \tag{42}$$

It follows that $v(x, t)$ is a continuous function of x and t for $t > 0$. In the already cited memoire of Gevrey one finds (pp. 343–344) a proof that for any bounded function F the second term of the right-hand part of the last inequality has a bounded derivative with respect to x. In view of condition (36), it turns out that the function $F(x, t, v(x, t))$ has finite derivative numbers with respect to x for $t > 0$; this is why $v(x, t)$ (see Gevrey, loc. cit., p. 351) satisfies (35).

The uniqueness of the bounded solution is proved as follows. Let us suppose there exist two bounded functions $v_1(x, t)$ and $v_2(x, t)$ having the same values for $t = 0$. Then they must satisfy the equation

$$v_2(x, t) - v_1(x, t) = \frac{1}{2\sqrt{\pi}} \int_0^t d\eta \int_{-\infty}^{+\infty} \frac{e^{-\frac{(x-\xi)^2}{4(t-\eta)}}}{\sqrt{t - \eta}} \times \left[F(\xi, \eta, v_2) - F(\xi, \eta, v_1)\right] d\xi. \tag{43}$$

Put

$$M(t) = \limsup_{t \geq \eta} |v_2(x, \eta) - v_1(x, \eta)|;$$

then with the aid of (36) we obtain from (43) that

$$M(t) \leq k \int_0^t M(\eta)\, d\eta,$$

which is impossible. ■

Remark. It is proved in the cited memoire of Gevrey that in a domain bounded on the left and right by arbitrary lines $x = \phi_1(t)$, $x = \phi_2(t)$, and from above and below by straight lines $t = t_0$ and $t = t_1 > t_0$, there exists a unique bounded function satisfying, in the interior of this domain, equation (35) and having bounded and continuous prescribed values on $x = \phi_1(t)$,

$x = \phi_2(t)$ and $t = t_0$. One could prove similarly the existence and uniqueness of the bounded function satisfying (35) in the interior of the domain G bounded only on one side by a curve $x = \phi(t)$ and from above and below by straight lines $t = t_0$ and $t = t_1 > t_0$ and having prescribed bounded and continuous values on $x = \phi(t)$ and on $t = t_0$.

Theorem 2. *By replacing the function $F(x, t, v)$ by another $F_1(x, t, v)$ such that everywhere*

$$F_1(x, t, v) \geq F(x, t, v)$$

one does not diminish the function $v(x, t)$, provided the initial conditions do not change.

Remark. If we interpret (35) as a heat-propagation equation, then $F(x, t, v)$ characterizes the intensity of heat creation, and the theorem becomes physically obvious.

Proof. Let the functions $v(x, t)$ and $v_1(x, t)$ satisfy respectively equation (35) and equation

$$\frac{\partial v_1}{\partial t} - \frac{\partial^2 v_1}{\partial x^2} = F_1(x, t, v_1).$$

Subtracting term by term, we see that the function

$$w(x, t) = v_1(x, t) - v(x, t)$$

satisfies the equation

$$\frac{\partial w}{\partial t} - \frac{\partial^2 w}{\partial x^2} = F_1(x, t, v_1) - F(x, t, v).$$

Put

$$w(x, t) = \bar{w}(x, t) e^{-kt}.$$

where k is the same as in (36). Then

$$\frac{\partial \bar{w}}{\partial t} - \frac{\partial^2 \bar{w}}{\partial x^2} = k\bar{w} + e^{kt}[F_1(x, t, v_1) - F(x, t, v)],$$

whence

$$\begin{aligned}
\overline{w}(x,t) &= \frac{1}{2\sqrt{\pi}}\int_0^t d\eta \int_{-\infty}^{+\infty} \frac{e^{-\frac{(x-\xi)^2}{4(t-\eta)}}}{\sqrt{t-\eta}} \\
&\quad \times \left\{ k\overline{w} + e^{k\eta}\left[F_1(\xi,\eta,v_1) - F(\xi,\eta,v)\right]\right\} d\xi \\
&= \frac{1}{2\sqrt{\pi}}\int_0^t d\eta \int_{-\infty}^{+\infty} \frac{e^{-\frac{(x-\xi)^2}{4(t-\eta)}}}{\sqrt{t-\eta}} \\
&\quad \times \left\{ k\overline{w} + e^{k\eta}\left[F_1(\xi,\eta,v_1) - F(\xi,\eta,v_1)\right.\right. \\
&\qquad\qquad \left.\left. + F(\xi,\eta,v_1) - F(\xi,\eta,v)\right]\right\} d\xi \qquad (44) \\
&\geq \frac{1}{2\sqrt{\pi}}\int_0^t d\eta \int_{-\infty}^{+\infty} \frac{e^{-\frac{(x-\xi)^2}{4(t-\eta)}}}{\sqrt{t-\eta}} \\
&\quad - \left\{ k\overline{w} + e^{k\eta}\left[F(\xi,\eta,v_1) - F(\xi,\eta,v)\right]\right\} d\xi \\
&\geq \frac{1}{2\sqrt{\pi}}\int_0^t d\eta \int_{-\infty}^{+\infty} \frac{e^{-\frac{(x-\xi)^2}{4(t-\eta)}}}{\sqrt{t-\eta}} (k\overline{w} - k|\overline{w}|)\, d\xi.
\end{aligned}$$

The expression in square brackets is equal to zero if $\overline{w} \geq 0$ and to $-2k\overline{w}$ if $w \leq 0$. Let $-m(t)$ denote the lower limit of the values of $\overline{w}(\xi,\eta) - |\overline{w}(\xi,\eta)|$ for $\eta \leq t$. To prove our theorem it obviously suffices to show that $m(t) \equiv 0$. For this, we note that the inequality (44) yields

$$w(x,t) \geq -k\int_0^t m(\eta)\, d\eta,$$

and so

$$m(t) \leq k\int_0^t m(\eta)\, d\eta,$$

which is possible only for $m(t) \equiv 0$, QED. ■

Theorem 3. *$v(x,t)$ cannot decrease when $f(x)$ increases.*

Physically, the sense of this theorem is as clear as that of the preceding theorem if we interpret (35) as the heat-propagation equation for a rod. The function $f(x)$ represents the initial temperature of the rod. When this temperature increases, $v(x,t)$ increases also.

Proof of Theorem 3. Let $v_1(x,t)$ and $v_2(x,t)$ be two functions satisfying equation (35) and for $t = 0$ becoming equal to $f_1(x)$ and $f_2(x)$ respectively, $f_2(x) \geq f_1(x)$. Let us show that $v_2 \geq v_1$.

The function $w = v_2 - v_1$ satisfies

$$\frac{\partial w}{\partial t} - \frac{\partial^2 w}{\partial x^2} = F(x, t, v_2) - F(x, t, v_1).$$

According to (36), we have

$$F(x, t, v_2) - F(x, t, v_1) \geq -k|w|.$$

Thus, according to Theorem 2, the function $w(x, t)$ is not less than the function $v^*(x, t)$ that for $t = 0$ becomes $f_2(x) - f_1(x) \geq 0$ and for $t > 0$ satisfies

$$\frac{\partial v^*}{\partial t} - \frac{\partial^2 v^*}{\partial x^2} = -k|v^*|.$$

The bounded solution for bounded t (unique according to Theorem 1) of this equation is given by the function $e^{-kt}v^{**}$, where $v^{**}(x, t)$ denotes the solution of (37) reducing to $f_2(x) - f_1(x)$ for $t = 0$ (obviously a non-negative solution). Consequently,

$$w = v_2 - v_1 \geq 0,$$

QED. ■

Theorem 4. *If $f(x) \geq 0$ everywhere and $F(x, t, 0) = 0$, one also has $v(x, t) \geq 0$.*

Proof. By Theorem 3 the function $v(x, t)$ cannot increase when $f(x)$ decreases. For $f(x) \equiv 0$ we have $v(x, t) \equiv 0$. Consequently, for $f(x) \geq 0$ one has $v(x, t) \geq 0$, QED. ■

Theorem 5. *If, in addition to the conditions of Theorem* 4, *one has $f(x) > 0$ on a certain interval of positive length, then for $t > 0$ one has $v(x, t) > 0$.*

The proof follows from that of Theorem 3 if we put $v_2 = v$ and $v_1 = 0$ and note that $v^{**}(x, t)$, being represented by a Poisson integral, has to be positive for $t > 0$.

Theorem 6. *If $F(x, t, 1) \equiv 0$ and $f(x) \leq 1$, then $v(x, t) \leq 1$.*

Proof. By Theorem 3, the function $v(x, t)$ cannot decrease when $f(x)$ increases. If $f(x) \equiv 1$, the function $v(x, t) \equiv 1$, QED. ■

Theorem 7. *If the function $v(x, t)$ coincides for $t = 0$ with the increasing and differentiable function $f(x)$ and if for $t > 0$ it satisfies the equation*

$$\frac{\partial v}{\partial t} - \frac{\partial^2 v}{\partial x^2} = F(t, v) \tag{45}$$

then for $t > 0$, $v(x, t)$ is a nondecreasing function of x.

Proof. By Theorem 1,

$$v(x, t) = v_0(x, t) + \frac{1}{2\sqrt{\pi}} \int_0^t d\eta \int_{-\infty}^{+\infty} \frac{F(\eta, v(\xi, \eta)) e^{-\frac{(x-\xi)^2}{4(t-\eta)}}}{\sqrt{t-\eta}} \, d\eta \quad (46)$$

where $v_0(x, t)$ satisfies

$$\frac{\partial v}{\partial t} - \frac{\partial^2 v}{\partial x^2} = 0 \quad (47)$$

for $t > 0$ and reduces to $f(x)$ for $t = 0$. If $f(x)$ is differentiable, then $v'_{0x}(x, t)$ approaches $f'(x)$ as (x, t) approaches $(x, 0)$.[9] On the other hand, the partial derivative with respect to x of the second component of the right-hand part of (46) does not exceed in absolute value the quantity $(4/\sqrt{\pi})Mt^{1/2}$ if $|F| \leq M$.[10] Consequently, $v'_x(x, t)$ approaches $f'(x)$ as $t \to 0$. If one supposes in addition that $v(x, t)$ has the derivatives $\partial^2 v/\partial t\,\partial x$ and $\partial^3 v/\partial x^3$, as is the case when $F(t, v)$ has three successive derivatives with respect to v, then the function $w(x, t) = v'_x(x, t)$ satisfies

$$\frac{\partial w}{\partial t} - \frac{\partial^2 w}{\partial x^2} = \frac{\partial F}{\partial v} w. \quad (48)$$

Then, applying Theorem 4, we find that $w(x, t) \geq 0$, QED. ■

Theorem 8. *If* $f^{(\epsilon)}(x) \underset{\epsilon \to 0}{\longrightarrow} f^{(0)}(x)$ *in such a way that*

$$\int_{-\infty}^{+\infty} |f^{(\epsilon)} - f^{(0)}| \, dx \to 0,$$

then for each $t > 0$ *the function* $v^{(\epsilon)}(x, t)$ *satisfying* (35) *for* $t > 0$ *and reducing to* $f^{(\epsilon)}(x)$ *for* $t = 0$ *approaches the function* $v^{(0)}(x, t)$ *that also satisfies* (35) *for* $t > 0$ *but reduces to* $f^{(0)}(x)$ *for* $t = 0$.

Proof. Let us determine $v^{(\epsilon)}(x, t)$ and $v^{(0)}(x, t)$ by the method of successive approximations as we did in the proof of Theorem 1. The functions $v_0^{(\epsilon)}$ and $v_0^{(0)}$ are determined respectively by the formulas

$$v_0^{(\epsilon)}(x, t) = \frac{1}{2\sqrt{\pi}} \int_{-\infty}^{+\infty} f^{(\epsilon)}(\xi) \frac{e^{-\frac{(x-\xi)^2}{4t}}}{\sqrt{t}} \, d\xi,$$

$$v_0^{(0)}(x, t) = \frac{1}{2\sqrt{\pi}} \int_{-\infty}^{+\infty} f^{(0)}(\xi) \frac{e^{-\frac{(x-\xi)^2}{4t}}}{\sqrt{t}} \, d\xi.$$

[9]Gevrey, *loc. cit.*, pp. 330–381.
[10]Gevrey, *loc. cit.*, p. 334.

We see at once that for $t > 0$,

$$v_0^{(\epsilon)}(x, t) \to v_0^{(0)}(x, t) \qquad \text{as } \epsilon \to 0.$$

The quantity $\tilde{v}_1^{(\epsilon)}(x, t) - \tilde{v}_1^{(0)}(x, t)$ (the notation is the same as in Theorem 1) is given by the formula

$$\frac{1}{2\sqrt{\pi}} \int_0^t d\eta \int_{-\infty}^{+\infty} \frac{e^{-\frac{(x-\xi)^2}{4(t-\eta)}}}{\sqrt{t-\eta}} \left[F\left(\xi, \eta, v_0^{(\epsilon)}\right) - F\left(\xi, \eta, v_0^{(0)}\right)\right] d\eta,$$

so

$$\begin{aligned} v_1^*(x, t) &= |\tilde{v}_1^{(\epsilon)}(x, t) - \tilde{v}_1^{(0)}(x, t)| \\ &\leq \frac{k}{2\sqrt{\pi}} \int_0^t d\eta \int_{-\infty}^{+\infty} \frac{e^{-\frac{(x-\xi)^2}{4(t-\eta)}}}{\sqrt{t-\eta}} - |v_0^{(\epsilon)}(\xi, \eta) - v_0^{(0)}(\xi, \eta)| \, d\xi. \end{aligned}$$

Put

$$v_0^*(x, t) = \frac{1}{2\sqrt{\pi}} \int_{-\infty}^{+\infty} \left| f^{(\epsilon)}(\xi) - f^{(0)}(\xi) \right| \frac{e^{-\frac{(x-\xi)^2}{4(t-\eta)}}}{\sqrt{t}} \, d\xi.$$

It is obvious that

$$v_0^*(x, t) \geq \left| v_0^{(\epsilon)}(x, t) - v_0^{(0)}(x, t) \right|,$$

so

$$\begin{aligned} v_1^*(x, t) &\leq \frac{k}{2\sqrt{\pi}} \int_0^t d\eta \int_{-\infty}^{+\infty} \frac{e^{-\frac{(x-\xi)^2}{4(t-\eta)}}}{\sqrt{t-\eta}} v_0^*(\xi, \eta) \, d\xi \\ &= k \int_0^t v_0^*(x, t) \, d\eta = ktv_0^*(x, t). \end{aligned}$$

The next-to-last equality holds because $v_0^*(x, t)$ satisfies (47). Thus,

$$v_1^*(x, t) \leq ktv_0^*(x, t).$$

In exactly the same way we find that

$$v_i^*(x, t) = \left| v_i^{(\epsilon)}(x, t) - v_i^{(0)}(x, t) \right| \leq \frac{(kt)^i}{i!} v_0^*(x, t).$$

It turns out that by choosing ϵ sufficiently small, we can make $\Sigma_{i=0}^{\infty} v_i(x, t)$ and thus also $|v^{(\epsilon)}(x, t) - v^{(0)}(x, t)|$ arbitrarily small, QED. ■

Theorem 9. *The function $v(x, t)$ that satisfies* (45) *for $t > 0$, vanishes for $t = 0$, $x < 0$, and becomes equal to 1 for $t = 0$, $x > 0$, is a nondecreasing function of x, and one has $v_x'(x, t) > 0$ for $t > 0$.*

***Proof*.** By the last theorem, we can consider $v(x, t)$ to be the limit as $\epsilon \to 0$ of the function $v^{(\epsilon)}(x, t)$ that takes on the same values as $v(x, t)$ on the x axis for $|x| \geq \epsilon$ and that is continuous on the whole x axis along with its derivative with respect to x, and is monotonic. Now we have just shown in Theorem 7 that the function $v^{(\epsilon)}(x, t)$ increases with x for $t > 0$. So the same holds true for $v(x, t)$.

Let us further show that for $t > 0$ one has $v_x'(x, t) > 0$. For this it suffices to prove that for $t > 0$ the equality $v_x'(x, t) = 0$ cannot hold. Here is why: the quantity $v_x'(x, t)$ for $t > 0$ satisfies (48); consequently, the quantity $\overline{w}(x, t) = e^{Mt}v_x'(x, t)$, where M is the upper bound of the values of $|\partial F/\partial v|$, satisfies the equation

$$\frac{\partial \overline{w}}{\partial t} - \frac{\partial^2 \overline{w}}{\partial x^2} = \left(\frac{\partial F}{\partial v} + M\right)\overline{w}.$$

Since $\partial F/\partial v + M \geq 0$ and by Theorem 2, the function $w(x, t)$, for $t > t_0 > 0$, is not less than the function $\overline{\overline{w}}(x, t)$ that, for $t = t_0$, coincides with $\overline{w}(x, t)$ and for $t > t_0$ satisfies the equation

$$\frac{\partial \overline{\overline{w}}}{\partial t} - \frac{\partial^2 \overline{\overline{w}}}{\partial x^2} = 0.$$

This latter function, for sufficiently small t_0, is positive for $t > t_0$ since for $t = t_0$ the function $\overline{w}(x, t)$ is not equal to zero identically. ■

In all that follows, we shall denote by $v(x, t)$ a function satisfying for $t > 0$ the equation (29), vanishing for $t = 0$, $x < 0$ and becoming equal to 1 for $t = 0$ and $x > 0$.

Theorem 10. *For every fixed $x < 0$, one has*

$$v(x - 2t, t) \to 0 \text{ as } t \to +\infty.$$

***Proof*.** The function $\bar{v}(x, t) \equiv v(x - 2t, t)$ satisfies the equation

$$\frac{\partial \bar{v}}{\partial t} - \frac{\partial^2 \bar{v}}{\partial x^2} = -2\frac{\partial \bar{v}}{\partial x} + F(\bar{v})$$

and the function[11] $v^*(x, t) = v(x - 2t, t)e^{-x}$ the equation

$$\frac{\partial v^*}{\partial t} - \frac{\partial^2 v^*}{\partial x^2} = (F(\bar{v}) - \bar{v})e^{-x}.$$

[11]It is easy to see that $v^*(x, t)$ remains bounded for bounded $t > 0$.

In view of conditions (32) and (33) imposed on $F(v)$, we have $F(v) - v \leq 0$. Consequently, $v^*(x, t)$ is less than the function that satisfies (37) for $t > 0$ and takes the value 0 for $t = 0$ when $x < 0$ and the value e^{-x} when $x > 0$. This last function now tends uniformly to zero with respect to x as $t \to +\infty$. ∎

Theorem 11. *For fixed t, let us consider the quantity $v_x'(x, t)$ as a function of v, which is possible by Theorem 9. Let*

$$v_x'(x, t) = \psi(v, t). \tag{49}$$

Then the function ψ cannot increase when t increases, v remaining invariable.

Proof. Consider the functions $v(x, t)$ and $v(x + c, t + t_0) = v_{t_0}(x, t)$, where c is some constant and $t_0 > 0$. Put

$$w(x, t) = v(x, t) - v_{t_0}(x, t).$$

Let M denote the set of points (x, t) where $w(x, t) \geq 0$. Let us show first that this set is only bounded on the left and moreover by a curve emerging from the origin along which the t coordinate never decreases. To show this, note that $w(x, t)$ satisfies

$$\frac{\partial w}{\partial t} - \frac{\partial^2 w}{\partial x^2} = k(x, t)w, \tag{50}$$

where $k(x, t)$ is a certain bounded function, i.e.,

$$k(x, t) = F'(\bar{u}(x, t)),$$

where $\bar{u}(x, t)$ is a certain number between $v(x, t)$ and $v_{t_0}(x, t)$. Thus the set M cannot include isolated pieces[12] separated from the x axis. Consequently, it only includes a single piece that obviously touches the half axis $x > 0$. To prove that this set is bounded on the left by a curve along which t never decreases, let us suppose the contrary, i.e., suppose that this curve includes a piece of the type shown on Fig. 4. E.g., suppose that this curve drops starting from the point A. Then the function $w(x, t)$ would take on negative values on the right of the segment OA, while on the segment OA itself it is equal to 0, and on the x axis, for $x > 0$, it takes on positive values. But by methods similar to those we have already used in proving Theorem 4, we can show that this is impossible.

We can similarly show that the set M is not bounded on the right.

[12] For a proof of a similar proposition in the case of finite pieces see I. Petrovsky, *Compositie Mathematica* **I** pp. 386–387, 1935. One can show that the same proposition is valid for infinite pieces. Compare the remark on Theorem 1.

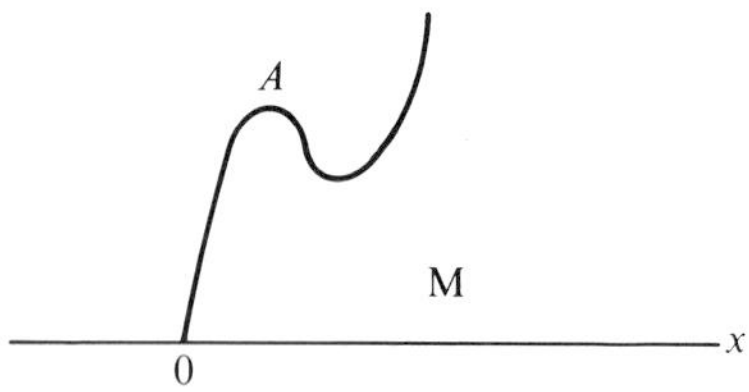

Fig. 4.

After these remarks, the theorem can be shown very simply. Indeed, by a suitable choice of c, we can make the values of $v(x_0, t)$ and $v_{t_0}(x_0, t)$ coincide for a certain x_0, t being given in advance. And then, in view of the arguments just cited, we shall have for all $x > x_0$ that $v(x, t) \geq v_{t_0}(x, t)$, and consequently $v_x'(x_0, t) \geq v_{t_0x}'(x_0, t)$, QED. ■

Theorem 12. *For any t we have*

$$v_x'(x, t) \geq u'(x)$$

if $v(x, t) = u(x)$. Here $u(x)$ denotes the solution of equation (34) *mentioned at the beginning of this section.*

The proof is absolutely the same as that of the preceding theorem, except we must take the function $u(x + c)$ instead of $v_{t_0}(x, t)$ and instead of the function $w(x, t)$, consider first the difference $v(x, t) - u(x + c)$.

Theorem 13. *Let*

$$v^*(x, t) = v(x + \phi(t), t),$$

where $\phi(t)$ is chosen so that everywhere

$$v^*(0, t) = c = \text{const.}$$

Then

$$v^*(x, t) \to v^*(x) \text{ as } t \to \infty$$

uniformly with respect to x.

Proof. From (49) we obtain

$$x = \int_c^{v^*} \frac{dv}{\phi(v, t)}. \tag{51}$$

The function under the integral sign increases monotonically as $t \to \infty$ by Theorem 11. In addition, by Theorem 12, the integral $\int_c^{v^*} dv/\psi(v, t)$ does not increase without bound. Thus in (51) we can pass to the limit under the integral sign. Let $\psi(v, t) \to \psi(v)$ as $t \to +\infty$. Then (51) becomes in the limit $x = \int_c^{v^*} dv/\psi(v)$. Since by Theorem 12 we have $\psi(v) > 0$, this equality defines a certain function v^* of x. It remains to prove the uniform convergence of $v^*(x, t)$ to $v^*(x)$. For this we note that (51) implies the uniform convergence of $x(v^*, t)$ to $x(v^*)$ on every interval $\epsilon < v^* < 1 - \epsilon$. If we now take into account that the function $\psi(v^*, t)$, by Theorem 11, remains bounded on such an interval, we deduce the uniform convergence of $v^*(x, t)$ to $v^*(x)$ for the values of x where $v^*(x)$ lies between ϵ and $1 - \epsilon$, where ϵ is arbitrary. Outside this interval ofvalues of x, $v^*(x, t) \to v^*(x)$ uniformly, because for sufficiently large t, the function $v^*(x, t)$ can only take on values not very different from 0 and 1. ■

Theorem 14. *As $t_0 \to +\infty$ the function*

$$v_{t_0}(x, t) = v(x + \phi(t_0), t + t_0)$$

converges uniformly in the domain $t \leq T = const$ to a certain solution $\bar{v}(x, t)$ of (29). *The function $\phi(t_0)$ is defined so that for every t_0,*

$$v_{t_0}(0, t_0) = c = \text{const}.$$

Proof. The function

$$w(x, t) = v_{t_0}(x, t) - v_{t_0+T}(x, t)$$

satisfies the equation

$$\frac{\partial w}{\partial t} - \frac{\partial^2 w}{\partial x^2} = F'(\bar{v})w, \tag{52}$$

where $\bar{v}(x, t)$ is between $v_{t_0}(x, t)$ and $v_{t_0+T}(x, t)$. By Theorem 13, for sufficiently large t we have

$$|w(x,0)| < \epsilon,$$

where $\epsilon > 0$ is arbitrarily small. By Theorems 2 and 3, $w(x, t)$ is less than the function $\overline{w}(x, t) = \epsilon e^{kt}$, where k is the upper limit of the values of $|F'(v)|$; for $t = 0$, $\overline{w}(x, t)$ takes on values no less than $w(x,0)$ and satisfies for $t > 0$ the equation

$$\frac{\partial \overline{w}}{\partial t} - \frac{\partial^2 \overline{w}}{\partial x^2} = k|\overline{w}|,$$

where the second member, for $w = \overline{w}$, is not less than the second member of (52). We can likewise show that

$$w(x, t) > -\epsilon e^{kt}.$$

Thus we have proved that the sequence of functions $v_{t_0}(x, t)$ converges as

$t_0 \to +\infty$ uniformly in a certain domain $t < T$ to a function that we shall denote by $\bar{v}(x, t)$. Let us show that $\bar{v}(x, t)$ satisfies (29).

Using (42), we write

$$v_{t_0}(x, t) = v_{t_0,0}(x, t) + \frac{1}{\sqrt{2\pi}} \int_0^t d\eta \int_{-\infty}^{+\infty} \frac{e^{-\frac{(x-\xi)^2}{4(t-\eta)}}}{\sqrt{t-\eta}} \times F\big(v_{t_0}(\xi, \eta)\big)\, d\xi. \tag{53}$$

In (53) we can pass to the limit by replacing v_{t_0} by $\bar{v}$. The function that satisfies (53) also satisfies (29), as we already brought out in the proof of Theorem 1. ■

Theorem 15. *As $t_0 \to +\infty$ the partial derivatives of first order of the function $v_{t_0}(x, t)$ with respect to x and t approach the corresponding partial derivatives of $\bar{v}(x, t)$, uniformly in every domain $\epsilon < t < T$, where ϵ and T are arbitrary positive constants.*

Proof. The uniform convergence of $\partial v_{t_0}/\partial x$ can be proved with (53). Indeed, the uniform convergence for $t > \epsilon$ of the partial with respect to x of the first component of the second member results from the fact that this component is represented by a Poisson integral. To prove the same with respect to the second component, we form the difference of its values for two values of t_0, say $\bar{t}_0$ and $\bar{\bar{t}}_0$; it is equal to

$$\frac{1}{2\sqrt{\pi}} \int_0^t d\eta \int_{-\infty}^{+\infty} \frac{e^{-\frac{(x-\xi)^2}{4(t-\eta)}}}{\sqrt{t-\eta}} \big[F(v_{\bar{t}_0}) - F(v_{\bar{\bar{t}}_0})\big]\, d\xi. \tag{54}$$

By Theorem 14, the quantity

$$F\big(v_{\bar{t}_0}(\xi, \eta)\big) - F\big(v_{\bar{\bar{t}}_0}(\xi, \eta)\big)$$

is arbitrarily small for sufficiently large $\bar{t}_0$ and $\bar{\bar{t}}_0$. Applying here Gevrey's result cited above, we find that the partial of (54) with respect to x, for sufficiently large $\bar{t}_0$ and $\bar{\bar{t}}_0$, becomes as small as desired uniformly with respect to x if $t < T$.

The function $w_{t_0}(x, t) = \partial v_{t_0}(x, t)/\partial x$ satisfies the equation

$$\frac{\partial w_{t_0}}{\partial t} - \frac{\partial^2 w_{t_0}}{\partial x^2} = F'(v_{t_0}) w_{t_0}.$$

We have already proved the uniform convergence for $\epsilon < t < T$ of the second member of this equation as $t_0 \to \infty$. The same procedures that we used to prove the uniform convergence of $\partial v_{t_0}/\partial x$ also apply to prove the uniform convergence of $\partial^2 v_{t_0}/\partial x^2$. And since the function v_{t_0} satisfies (29), we see that the convergence of $\partial v_{t_0}/\partial t$ is uniform. ■

Theorem 16. *Suppose that the function $v_{t_0}(x, t)$ (resp. $\bar{v}(x, t)$) remains constantly equal to c along the curve $x = \phi_{t_0}(t)$ (resp. $x = \phi(t)$). Then, uniformly with respect to t for $\epsilon < t < T$, we have*

$$\phi'_{t_0}(t) \to \phi'(t) \text{ as } t \to \infty.$$

Proof. The quantity $\phi'_{t_0}(t)$ (resp. $\phi'(t)$) is equal to

$$-\frac{\dfrac{\partial v_{t_0}}{\partial t}}{\dfrac{\partial v_{t_0}}{\partial x}} \left(\text{resp. } -\frac{\dfrac{\partial \bar{v}}{\partial t}}{\dfrac{\partial \bar{v}}{\partial x}} \right)$$

at the point $(\phi_{t_0}(t), t)$ (resp. $(\phi(t), t)$). By Theorems 12 and 14, we have for sufficiently large t_0 that $|\phi_{t_0}(t) - \phi(t)| < \epsilon_1$ in the domain $\bar{G}$ defined by the inequalities $\epsilon < t < T$, $\epsilon_1 > 0$ being arbitrarily small. By Theorem 15, the numerators and denominators of the quotients

$$\frac{\dfrac{\partial v_{t_0}}{\partial t}}{\dfrac{\partial v_{t_0}}{\partial x}} \quad \frac{\dfrac{\partial \bar{v}}{\partial t}}{\dfrac{\partial \bar{v}}{\partial x}} \tag{55}$$

differ as little as desired uniformly in $\bar{G}$. In addition, in the strip $\phi(t) - \epsilon_2 < x < \phi(t) + \epsilon_2$ the quantity $\partial \bar{v}/\partial x$ is greater than some positive constant. This is why the quotients (55) for sufficiently large t_0 differ by less than ϵ_3 in the strip $\epsilon < t < T$, $\phi(t) - \epsilon_2 < x < \phi(t) + \epsilon_2$. Since the quotient $(\partial \bar{v}/\partial t)/(\partial \bar{v}/\partial x)$ is uniformly continuous in this strip, its values at the points of this strip with the same t differ for sufficiently small ϵ_3 by as little as one desires, whence the theorem. QED. ■

Theorem 17. *For any t we have*

$$\bar{v}(x, t) = u(x + 2), \qquad d\phi/dt \to -2 \text{ as } t \to \infty.$$

(*The notation is the same as in Theorem* 14).

Proof. Consider the function

$$v^*(x, t) = \bar{v}(x + c_1(t), t),$$

where $c_1(t)$ is chosen so that $v^*(0, t) \equiv c \equiv$ const. Then

$$\frac{\partial v^*}{\partial t} = \frac{\partial^2 v^*}{\partial x^2} + c_1'(t)\frac{\partial v^*}{\partial x} + F(v^*).$$

But on the other hand, by the definition of $\bar{v}(x, t)$, the quantity $v^*(x, t)$ is independent of t for every x. Thus, $\partial v^*/\partial t = 0$, $c_1'(t) =$ const. According to

Section 2, this constant cannot be greater than -2. But by Theorem 10, this constant cannot be less than -2. Consequently it is equal to -2 and, by Theorem 16, $d\phi/dt \to -2$ as $t \to \infty$, QED. ■

Remark. Suppose that the initial values of $v(x, t)$ were not as we have thus far supposed but were

1) $v(x,0) = 1 \qquad (x \geq c_1)$,
2) $v(x,0) = 0 \qquad (x \leq c_2 < c_1)$,
3) $v(x,0)$ takes on arbitrary values between 0 and 1 for $c_2 < x < c_1$.

Then we can easily see that the rate of displacement to the left of the portion accounting for the principal part of the drop of v from 1 to 0 would still tend to 2 as $t \to \infty$, because in this case one would have

$$v(x - c_1, t) \leq \bar{v}(x, t) \leq v(x - c_2, t),$$

where $\bar{v}(x, t)$ denotes the solution of (29) satisfying the new initial conditions.

1938 ACTA PHYSICOCHIMICA U.R.S.S. VOL. IX. No. 2

A Theory of Thermal Propagation of Flame

By *J. B. Zeldowitsch* and *D. A. Frank-Kamenetzki*

It was already very long ago[1] that a conception was formulated of the thermal propagation of flame as the most usual mechanism of combustion connected with the successive ignition of the explosive mixture by the heat liberated during the reaction. The existing theories[2,3,4], however, are unsatisfactory[5] since they use the conception of the „ignition" temperature of the mixture.

At the same time it is well known[6,7] that the phenomenon of self-ignition is quite naturally explained with the assumption of a smooth increase in the rate of exothermic reaction with the increase of temperature if the heat balance of the system is considered; the ignition temperature itself proves to be dependent not only upon the properties of the explosive mixture, but also upon the size and form of the vessel in which the ignition temperature is measured and upon other factors. A rational theory must find an expression for the velocity of flame propagation in a mixture for which the kinetics of chemical reaction is given as a function of temperature and concentration of the reacting substances.

A simultaneous analysis of the equation of heat conduction allowing for the heat liberated owing to chemical reaction, toge-

1 Mallard et Le Chatelier, Ann. de Mines, **4**, 296 (1883).
2 Nusselt, Z. V. D. I., **59**, 872 (1915).
3 Jaugnet, C. R., **168**, 820 (1919).
4 Daniell, Proc. Roy. Soc., **A 126**, 393 (1930).
5 Jost u. Müffling, Z. physik. Chem., **A 181**, 208 (1937).
6 Van't Hoff, Études de dynamique chimique.
7 N. N. Semenoff, „Chemical Kinetics and Chain Reactions", Leningrad, 1935.

 Reprinted from *Acta Physicochimica U.R.S.S.*, Vol. IX, No. 2, 1938.

ther with the diffusion equation governing the concentrations of the reacting substances and reaction products — also allowing for the change of these concentrations due to chemical reaction — leads to conclusions which are very essential for the subsequent study[8].

Namely, the relative increase of temperature and the relative changes in the concentrations of all the reacting substances are identically equal:

$$\frac{T-T_0}{T_1-T_0}=\frac{a-a_0}{a_1-a_0}=\frac{b-b_0}{b_1-b_0}\,. \tag{1}$$

Here T_0, a_0 and b_0 etc. are the temperature, concentration of substance A, concentration of substance B, respectively, etc. In the initial mixture, T_1, a_1, b_1 etc. are the corresponding values for the final reaction products (T_1 is the theoretical temperature of combustion), and T, a, b, etc. are the temperature and the concentrations, respectively, at any given spot in the flame zone. T, a, b, etc. are functions of space coordinates and of time, while T_0, a_0, b_0 and T_1, a_1, b_1, etc. are constant values for any given mixture.

The meaning of formula (1) can be stated as the similarity of the fields of concentrations to the field of temperature in the flame.

This similarity of the fields of concentration and of that of the temperature means that the temperature and the concentrations are related to each other in the same manner as for an adiabatic reaction though the course of reaction is quite different.

In the derivation of formula (1) the following assumption must be made:

(1) The equality of the temperature conductivity (ratio of thermal conductivity to heat capacity per unit volume) and of the diffusion coefficients of the substances dealt with.

(2) The absence of heat transfer by radiation.

(3) The absence of heat transfer to the walls of the vessel in which the flame is propagated.

(4) The formula relates only to the initial and final substanses, only the gross reaction being present. This formula does not apply to the concentrations of the intermediate products, which must be

[8] The rôle of diffusion was pointed out in a qualitative manner in a recent investigation of Jost[5].

small as compared with those of the reacting substances. But the active rôle of the intermediate products or any centers in the course of a chemical reaction does not affect the correctness of the formula as applied to the initial and final products. Assumptions (2) and (3) have the result that formula (1) actually holds true only in the vicinity of the flame itself, and only here does it correctly express the relation between the rise in the temperature and the change in the mixture composition due to the chemical reaction and to the heat and diffusion currents in the flame.

Further, where the reaction products at distances exceeding the width of the flame, get cool without changing the composition, formula (1) holds true no longer.

For the derivation of formula (1), let us compare the equation of heat conduction

$$\frac{\partial T}{\partial t} = \varkappa \Delta T - (\mathfrak{v}\nabla) T + \frac{Q}{C},$$

where $\varkappa$ is the temperature conductivity (see above), $\mathfrak{v}$ is the gas velocity, C is the heat capacity per unit volume, the limiting conditions being $T = T_0$ in the initial substance, $T = T_1$ in the combustion products, with the equation of diffusion

$$\frac{\partial a}{\partial t} = D\Delta a - (\mathfrak{v}\nabla) a + M,$$

where D is the diffusion coefficient and M is the rate of formation of the substance A owing to reaction; the limiting conditions are:

$a = a_0$ in the initial substance,
$a = a_1$ in the combustion products.

Introducing

$$\theta = \frac{T - T_0}{T_1 - T_0} \quad \text{and} \quad \alpha = \frac{a - a_0}{a_1 - a_0}$$

we get

$$\frac{\partial \theta}{\partial t} = \varkappa \Delta \theta - (\mathfrak{v}\nabla) \theta + \frac{Q}{C(T_1 - T_0)},$$

where $\theta = 0$ in the initial substance, $\theta = 1$ in the combustion products, and

$$\frac{\partial \alpha}{\partial t} = D\Delta \alpha - (\mathfrak{v}\nabla) \alpha + \frac{M}{a_1 - a_0},$$

where $\alpha = 0$ in the initial substance, $\alpha = 1$ in the combustion products.

We evidently have

$$\frac{M}{a_1 - a_0} = \frac{Q}{C(T_1 - T_0)}; \quad \frac{Q}{M} = \frac{C(T_1 - T_0)}{a_1 - a_0}.$$

When only the gross reaction is present, the rate of heat liberation and the rate of change of concentration have the same ratio as the total amount of heat liberated $C(T_1 - T)$ and the total change in concentration $(a_1 - a_0)$ when the reaction proceeds to the end. But then, on the assumption that $D = \varkappa$, the differential equations as well as the limiting conditions for θ and α coincide:

$$\theta = \alpha,$$

$$\frac{T - T_0}{T_1 - T_0} = \frac{a - a_0}{a_1 - a_0}.$$

Using this assumption we can express the concentrations involved in the formula of the rate of chemical reaction in terms of temperature, and this will permit us to confine ourselves to the consideration of but one equation of heat conduction.

In the zone of flame all the temperatures lying in the interval between the initial temperature and the theoretical temperature of combustion, $T_0 \leqslant T \leqslant T_1$, are realized simultaneously. The rapid increase in the rate of reaction with the rise of temperature, obeying the exponential law of Arrhenius, leads to the result that the reaction practically proceeds just at high temperatures in the vicinity of T_1.

Comparing this conclusion with formula (1) we can see that the reaction proceeds in a mixture which contains a large amount of the final products of combustion and comparatively small amounts of the initial substances. Hence some qualitative conclusions may be drawn as to the character of the dependence between the velocity of flame propagation and various parameters. Thus, as compared with the classical reaction with the initial velocity equaling the maximum velocity of autocatalytic reaction, the auto-catalysis of the combustion reaction by the final products, which strongly retards the self- ignition of the mixture, must not affect the velocity and the character of the propagation of flame, since the reaction in the flame proceeds in the zone of high concentrations of the final products of reaction.

Various admixtures may also exert an appreciable influence on the velocity of flame propagation only in so far as they affect the rate of reaction at very high temparatures which approach T_1.

Finally, the initial temperature and the mixture composition must influence the rate of propagation principally through the change in the theoretical temperature of combustion T_1 [9].

The conception that the reaction proceeds principally at a temperature which is near to T_1 also permits us to find closed analytical expressions for the velocity of flame propagation. The heat liberated during the reaction is partly consumed by the heating of the reacting gas itself and partly carried away by heat conduction to the neighbouring elements of the gas volume. If the temperature of the zone in which the reaction actually proceeds is already near to T_1, the amount of the heat consumed by the heating of the reacting gas up to its final temperature beyond the front of the flame, T_1, is small as compared with the total heat of reaction liberated. Approximately we can assume that all the heat from the zone of reaction is carried away by conduction. The Fourier equation of heat conduction is [10]:

$$C\frac{dT}{dt} = C\frac{\partial T}{\partial t} + C(\mathfrak{v}\nabla T) = dw\,(k\nabla T) + Q. \tag{2}$$

Here C is the heat capacity, K is the thermal conductivity, $Q = Q(t)$ is the volume rate of heat liberation due to chemical reaction (in cals./cm.^{3} per sec.), while the dependence of the velocity of reaction upon the concentrations is reduced, with the aid of formula (1), to the dependence upon temperature; the thermal expansion is here also allowed for.

---

[9] Throughout the present investigation we consistently neglect the heat losses by radiation, by heat transmission through the cold parts of the apparatus etc. (cf. above).

[10] In the foregoing derivation of similarity we assumed that the heat capacity and conductivity were constant, which is no longer necessary here.

We neglect the heat consumed for the heating of the mixture in the reaction zone itself $\left(C\frac{dT}{dt}\right)$ and further consider the thermal conductivity in this zone to be constant and equal, for the sake of definitiveness, to the thermal conductivity, k_1, of the reaction products at T_1. Passing to the one-dimensional problem we obtain the equation

$$k_1\frac{d^2T}{dx^2} = -Q\,[T]. \tag{3}$$

This equation can be easily integrated by the method of quadratures.

Taking into account the limiting condition

$$\frac{dT}{dt} = 0 \text{ at } T = T_1 \tag{4}$$

we shall find

$$\frac{dT}{dx} = \sqrt{\frac{2}{k_1}\int_T^{T_1} Q(T)\,dT.} \tag{5}$$

To find $\frac{dT}{dt}$ at the somewhat conventional boundary of the reaction zone, we should rather substitute the temperature existing at this boundary for the lower limit of the integral in formula (5).

Owing to the rapid fall of $Q\,[T]$ with a decrease in temperature we shall find the limiting value of $\left(\frac{dT}{dx}\right)$ at the boundary of the zone of reaction according to the formula

$$\left(\frac{dT}{dx}\right)_z = \sqrt{\frac{2}{k_1}\int_0^{T_1} Q[T]dT,} \tag{6}$$

this value differing but little from the true value.

Finally, we equate the heat flow from the zone of reaction to the total amount of heat liberated in the flame per unit time

$$k_1\left(\frac{dT}{dx}\right)_z = u\lambda, \tag{7}$$

where u is the mass rate of combustion (in grams of mixture per unit·time per cm.² of the flame surface), and λ is the calorific value of the mixture in cals./gr.).

The specific reaction rate obeying the Arrhenius law decreases e times in the temperature interval $\frac{RT^2}{E}$, where E is the activation energy.

The statement that practically the entire reaction proceeds at temperatures close to T_1 is justified if

$$\frac{RT_1^2}{E} \ll T_1 - T_0.$$

By comparing (6) and (7) we shall get the final expression for the mass rate of combustion

$$u = \frac{1}{\lambda}\sqrt{2k_1\int_0^{T_1} Q[T]dT}. \tag{8}$$

In considering the latter expression it should be remembered that according to formula (1)

$$Q(T) = Q\,(T,\ T_0,\ T_1,\ a_0,\ a_1\ .\ .\ .\ .\ ), \tag{9}$$

and that the final temperature itself

$$T_1 = T_0 + \frac{\lambda}{C} \tag{10}$$

depends upon the initial temperature and calorific value of the mixture.

The mass rate of combustion is related to the linear velocity of flame propagation by a simple formula $u = v\rho$, where v is the linear velocity and ρ is the density.

Substituting the density of the initial mixture, we shall find the linear velocity of flame propagation relative to the initial mixture v_0, and substituting the density ρ_1 of the reaction products at T_1, we shall get the velocity v_1 as referred to the substance burnt.

For some simplest cases of mono- and bi-molecular reactions as well as reactions the rates of which do not depend upon concentration (so-called reactions of zero order), we can easily establish simple formulae of the velocity of flame propagation.

Neglecting the volume change due to reaction, we obtain the linear velocities of flame propagation as follows.

For the reaction of zero order, with the reaction rate equaling

$$\frac{dn}{dt} = Ce^{-\frac{E}{RT}},$$

where n is the concentration of the reacting substance (in grams per cm.³), we have the propagation velocity:

$$v_0 = \sqrt{\frac{2k_1 Ce^{-E/RT_1} RT_1^2}{LEn_0\rho_0}}.$$

The fact that the calorific value of the mixture L turns out to be in the denominator is not in disagreement with the usual conceptions; in fact, at a constant initial temperature T_0, the theoretical temperature of combustion T_1 involved in the formula increases with L, so that velocity v_0 increases as a whole. Conversely, an increase in L at constant T_1 involves a fall in T_0, so that the decrease in the velocity is quite natural.

For the first order, with the reaction rate equaling

$$\frac{dn}{dt} = C'e^{-\frac{E}{RT}} \cdot n,$$

we have the propagation velocity

$$v_0 = \sqrt{\frac{2k_1 C' e^{-\frac{E}{RT_1}}}{\rho_0 L (T_1 - T_0)} \left(\frac{RT_1^2}{E}\right)^2}.$$

For the second order reaction, with the reaction rate equaling

$$\frac{dn}{dt} = C'' e^{-\frac{E}{RT}} \cdot n^2,$$

the propagation velocity is

$$v_0 = \sqrt{\frac{4k_1 C'' e^{-\frac{E}{RT_1}} n_0}{\rho_0 L (T_1 - T_0)^2} \left(\frac{RT_1}{E}\right)^3}.$$

Let us define once more the meaning and dimensions of all the symbols used:

v_0 — linear velocity of flame propagation relative to the unburnt mixture, cm./sec.;

k_1 — thermal conductivity at the theoretical temperature of combustion, cals./cm. sec.°C;

ρ — initial density of the mixture, gr./cm.³;

n — concentration of the reacting substance, gr./cm.³;

n_0 — initial concentration of the reacting substance, gr./cm.³;

L — calorific value of the mixture, cals./gr.;

T_0 — initial temperature, °C;

T_1 — theoretical temperature of combustion, $T = T_0 + \frac{L}{C}$, °C;

E — heat of activation, cals./mol.;

R — gas constant, cals./mol., °C;

C — specific rate of zero order reaction gr./cm.³ sec.;

C' — specific rate of first order reaction, 1/sec.;

C'' — specific rate of second order reaction, cm.³/gr. sec.

In conclusion we must note that the evaluation of the velocity of flame propagation places extremely strict requirements on the study of the kinetics of the reaction of combustion. As has been shown above, the rate of reaction at extremely high temperatures of the order of 1000—2000°C at which the reaction proceeds in the course of 10^{-2}—10^{-6} sec., proves to be very essential.

The presence of parallel reactions, the formation of considerable amounts of intermediate products, the active participation (autocatalysis, chains) of intermediate products in the course of reaction, — all these deviations from the simplest kinetic schemes greatly restrict the range of applicability of the quantitative results of the theory proposed.

The detailed derivations and the problem of cold flame propagation here omitted will be published elsewhere.

The Physico-Chemical Laboratory of the
Academy of Sciences of USSR,
Leningrad.

Received
April 7, 1938.

Dragging of a Liquid by a Moving Plate

Dragging of a Liquid by a Moving Plate

By *L. Landau* and *B. Levich*

One of the methods of depositing a thin layer of liquid upon a surface of a solid, which is wetted by it, consists of pulling out an infinite film with constant speed v_0.

This method is applied, in particular, in the cinefilm manufacture with the purpose of depositing a uniform layer of photosensitive emulsion upon the cinefilm base.

From the trough containing the dissolved photosensitive emulsion an infinite film, wetted by the solution, is pulled out.

After the solvent has evaporated, a uniform layer of emulsion is left deposited upon the surface of the film base.

The problem of determining the thickness of the dragged layer as a function of the speed of the motion of the film and of parameters characteristic of the properties of the fluid (its viscosity η, its surface tension σ and its density ρ) is of essential interest for practice.

Numerous attempts at evaluating the thickness of dragged layer of fluid found in literature contain some incorrect assumptions in the very basis of the method of computation, thus leading to erroneous formulae for the value of this thickness.

In the present paper the thickness of the layer and the quantity of fluid carried along when pulling an infinite plate out of a vessel, which is sufficiently large to permit the neglecting of the effect of its walls and of the edges of the plate, is evaluated.

Let us choose a set of coordinates in such a way as to make the plate — a plane $x=0$ and the surface of the liquid undis-

Reprinted from *Acta Physicochimica U.R.S.S.*, Vol. XVII, No. 1–2, 1942.

turbed by the capillary meniscus (*i. e.* sufficiently far from the plate) — the plane $z=0$, with the z axis directed upward along the plate.

First of all, let us consider the case of low velocity of motion of the plate (it will be stated below which velocities can still be considered low).

In this case all the surface of the liquid may be separated into two independent regions: the region of the surface situated high above the meniscus and directly dragged by the plate, where the surface of liquid may be taken to be nearly parallel to the plate surface, and the region of the meniscus of liquid, which will be slightly deformed by the motion of the plate, hence the shape of the surface will nearly coincide with the shape of static meniscus.

Below we shall write down the solutions of hydrodynamical equations in both independent regions and then connect both of the solutions found.

Let us denote by h the thickness of a layer of liquid, when measured from the plate. We shall look for $h=h(z)$ in both of the independent regions. First of all, let us write down the equations for the thickness of the liquid film carried along by the moving plate, *i. e.* the equations for h in the first region.

Since the surface of the liquid in the firsf region is nearly flat, if is clear that the motion of the moving liquid in this region will also be nearly flat. In other words, the main component of the motion of the liquid in the first region will be the flowing down of it nearly parallel to the plate surface.

This peculiarity may be used in order to simplify the equations of hydrodynamics suitably for this case.

In fact, it is evident that the only component of the velocity of fluid which plays an essential part is the vertical (along the z axis) component u_z.

It is evident as well that the gradients of the velocity along the direction normal to the plate $\frac{\partial u_z}{\partial x}$ are large as compared with the gradients of velocity along tbe plate $\frac{\partial u_z}{\partial z}$.

Therefore the motion of a liquid carried along with the plate is described by the equations of the Prandtl boundary layer, which in the stationary case have the form:

$$\nu \frac{d^2u}{dz^2} = \frac{1}{\rho} \frac{\partial p}{\partial z} + g, \tag{1}$$

$$\frac{\partial p}{\partial x} = 0. \tag{2}$$

Here, for the sake of brevity, the z component of the velocity u_z is designated simply by u. Other symbols have their usual meaning.

As the boundary conditions for equations (1) and (2) the following conditions will serve: on the surface of the plate, no slip between the liquid and the plate occurs, hence

$$u = v_0 \quad \text{at} \quad x = 0. \tag{3}$$

Here v_0 is the velocity of motion of the plate.

On the free surface of the liquid, at $x = h(z)$, where h is the thickness of the liquid layer, the pressure inside the liquid must be equal to the capillary pressure p_σ and, moreover, the tangent stress must be absent, so that:

$$\left.\begin{aligned} p &= p_\sigma, \\ \eta \frac{\partial u}{\partial x} &= 0, \end{aligned}\right\} \quad x = h. \tag{4}$$

The capillary pressure is known to equal

$$p_\sigma = \frac{\sigma}{R},$$

σ being the surface tension, R — the radius of curvature of the surface.

Substituting the well-known expression for the radius of curvature R, we get:

$$p_\sigma = -\sigma \frac{\frac{d^2h}{dz^2}}{\left[1 + \left(\frac{dh}{dz}\right)^2\right]^{3/2}}. \tag{5}$$

As long as the thickness of the layer of the liquid carried along is very small, it is evident that the curvature of the surface of the liquid in the vicinity of the plate also will be very small. Therefore, the square of dh/dz in the denominator of equation (5) may be neglected, and the following equation for the capillary pressure in the region of the liquid carried along may be finally written down:

$$p_c \approx -\sigma \frac{d^2h}{dz^2}. \tag{5'}$$

Therefore, the first of the boundary conditions (4) may be rewritten as follows:

$$p = -\sigma \frac{d^2h}{dz^2} \quad \text{at} \quad z = h.$$

However, it follows from equation (2) that the pressure is constant along the thickness of the liquid layer carried along. Thus, not only on the free surface, but throughout inside the liquid pressure is also given by

$$p = -\sigma \frac{d^2h}{dz^2}. \tag{6}$$

The solution of equation (1) satisfying the boundary conditions (3) and (4) is:

$$u = v_0 + \left(\frac{1}{\eta}\frac{dp}{dz} + \frac{\rho g}{\eta}\right)\left(\frac{x^2}{2} - hx\right) =$$

$$= v_0 + \left(\frac{\rho g}{\eta} - \frac{\sigma}{\eta}\frac{d^3h}{dz^3}\right)\left(\frac{x^2}{2} - hx\right) \tag{7}$$

where the value of p is substituted from equation (6).

Let us finally make use of the continuity equation, in order to connect the thickness of the liquid layer with the flow of the fluid carried along by the plate.

For steady motion of the fluid, keeping in mind the incompressibility of fluid, we may, evidently, write down the continuity equation in the form:

$$j = \int_0^h u\, dh = \text{const},$$

where j is the flux of the fluid per unit of width of the plate. Substituting the value of u from equation (7), we obtain:

$$v_0 h - \left(\rho g - \sigma \frac{d^3h}{dz^3}\right)\frac{h^3}{3\eta} = j. \tag{8}$$

Equation (8) defines the thickness of liquid layer $h(z)$ far from the surface of the fluid (in the first region[1]).

In the second region, near the fluid surface, the thickness $h(z)$, as has been said above, is determined by the equation for the static meniscus.

Let us rewrite equation (8) in a more suitable form:

$$\frac{d^3h}{dz^3} = \frac{3\eta}{\sigma} \frac{(j - v_0 h)}{h^3} - \frac{\rho g}{\sigma}. \tag{8'}$$

Let us introduce a new dimensionless coordinate λ determined by the equation

$$\lambda = \left(\frac{\sigma}{3\eta}\right)^{1/3} \frac{j}{v_0^{4/3}} z, \tag{9}$$

and a dimensionless expression for the thickness of the liquid layer

$$\mu(\lambda) = \frac{v_0 h(z)}{j}. \tag{10}$$

Then, introducing λ and $\mu(\lambda)$ into equation (8'), we find:

$$\frac{d^3\mu}{d\lambda^3} - \frac{1-\mu}{\mu^3} - \frac{\rho g j^2}{3\eta v_0^3} = 0. \tag{11}$$

The order of magnitude of the last term in equation (11) which contains all the dimensional quantities involved in the problem, is determined by the kind of dependence of the flux j on the velocity of the plate v_0. If the flux j is proportional to v_0 in the power higher than $^2/_3$, then the last term in equation (11) will be simply proportional to v_0, and for sufficiently small values of velocity of elevation v_0 will be small as compared with unity.

We assume the flux j to depend on v_0 in the way mentioned above. Then the last term in equation (11) is in fact small as compared with both the first ones and may therefore be neglected. It will be shown later that this assumption turns out to hold within some regions of velocities of elevation of the plate. Thus, the region of validity of solutions of equation (11) obtained on the basis of this assumption will be defined.

[1] One of us (B. Levich) has been kindly informed by B. V. Derjaguin that he was first to obtain this equation. However, B. V. Derjaguin, has failed to derive from it any particular conclusions concerning the thickness of the film carried along.

Neglecting the last term in equation (11), we get finally:

$$\frac{d^3\mu}{d\lambda^3}=\frac{1-\mu}{\mu^3}. \tag{12}$$

The following conditions serve as the boundary conditions for equation (12): on the upper boundary of the region, for very large values of z (far from the surface of the liquid) the thickness of the liquid layer h must tend to a constant limit h_0, which evidently equals:

$$h_0 \to \frac{f}{v_0} \qquad \text{for } z \to \infty.$$

The derivatives dh/dz and d^2h/dz^2 must in this case tend to zero.

Therefore, in terms of dimensionless quantities λ and μ, the boundary conditions may on the upper boundary of the region be written down in the form:

$$\left.\begin{aligned} &\mu \to 1 \\ &\frac{d\mu}{d\lambda} \to 0 \\ &\frac{d^2\mu}{d\lambda^2} \to 0 \end{aligned}\right\} \quad \text{for } h \to \infty. \tag{13}$$

In order to determine the boundary conditions on the lower boundary of the region, at small values of z (near the meniscus), let us turn to the equations for the static meniscus, which determines the shape of the surface of the layer in the second region.

The equations for the static meniscus have, as is known, the following form:

$$\frac{\dfrac{d^2h}{dz^2}}{\left[1+\left(\dfrac{dh}{dz}\right)^2\right]^{3/2}}=\frac{\rho g z}{\sigma}. \tag{14}$$

Integrating equation (14), we find

$$\frac{\dfrac{dh}{dz}}{\left[1+\left(\dfrac{dh}{dz}\right)^2\right]^{1/2}}=\frac{\rho g z^2}{2\sigma}+c.$$

The constant of integration may be determined from the boundary conditions far away from the plate. In this case precisely

$z \to 0$ and $dh/dz \to 0$, *i. e.* the surface of the fluid is horizontal. Therefore $c = -1$ and

$$\frac{\frac{dh}{dz}}{\left[1+\left(\frac{dh}{dz}\right)^2\right]^{1/2}} = \frac{\rho g z^2}{2\varsigma} - 1. \tag{15}$$

The quantity $\left(\frac{\rho g}{\varsigma}\right)^{1/2}$ entering the equations (14) and (15) is the Laplace capillary constant, having dimensionality of the length. The capillary constant is the characteristic length of the problem of the static meniscus. If we shall designate it by a, then equation (14) and (15) may be written down in the form:

$$\frac{\frac{d^2h}{dz^2}}{\left[1+\left(\frac{dh}{dz}\right)^2\right]^{3/2}} = \frac{z}{a^2}, \tag{14'}$$

$$\frac{\frac{dh}{dz}}{\left[1+\left(\frac{dh}{dz}\right)^2\right]^{1/2}} = \frac{z^2}{2a^2} - 1. \tag{15'}$$

Equations (14) and (15) determine the thickness of the layer of liquid $h(z)$ in the second region close to the meniscus.

At small values of $h(z)$, small as compared with the capillary constant, and large values of z, the solution of the equality (14) must evidently go over into the solution of the equation (12) for the thickness of the film carried along. We must, therefore, chain together the solution of equation (12) with that of equation (11). The conditions of chaining together of both the equations will at the same time serve at the sought for boundary conditions of equation (11) at the lower boundary of the first region.

The conditions of chaining together of both solutions shall be obtained from equation (15′) with the aid of the transition to the limit of the small thicknesses $h \to 0$.

It is clear that with h tending to zero the quantity dh/dz in the formula (15′) will tend to zero as well. Since the surface of the liquid wetting the film in the vicinity of the film itself would be almost vertical, we find at the same time from equation (15) that to the thickness, tending to zero and to the almost vertical surface of the liquid, the finite altitude, tending to the limit value

$$z \to 2a \tag{16}$$

corresponds.

Next, with the aid of equation (14) we find at the same time that the second derivative of the thickness d^2h/dz^2 in the static solution tends to the limit

$$\frac{d^2h}{dz^2} \to \frac{2}{a}. \tag{17}$$

Going over to the dimensionless coordinate λ and to the thickness μ, we find, with the aid of equations (19), (10), (16), (17):

$$\mu \to 0 \quad \begin{cases} \lambda \to \left(\frac{\sigma}{3\eta}\right)^{1/3} \frac{2ja}{v_0^{4/3}}, \\ \frac{d^2\mu}{d\lambda^2} \to \frac{\sqrt{2}\, a\, \sigma^{2/3} j}{v_0^{5/3} (3\eta)^{2/3}}. \end{cases} \tag{18}$$

Thus, we see that the boundary with the first region, $\frac{d^2\mu}{d\lambda^2}$, tends to a constant limit, determined by equation (18).

Let us turn now to the boundary conditions of equation (11) at the lower boundary of the region. Here the thickness of the film of the liquid carried along will be large as compared with the limit thickness $\mu = 1$ at the upper boundary of the region. In other words, the valuea of μ tending to infinity correspond to the lower boundary of the region. Therefore, we must find the boundary conditions of equation (11) with $\mu \to \infty$. Thus, the boundaries of both independent regions overlap. $\mu \to \infty$ corresponds to the lower boundary of the first region, $\mu \to 0$ — to the upper boundary of the second region.

We shall require, as a condition for chaining together the solutions of both regions, the continuity of the second derivative $\frac{d^2\mu}{d\lambda^2}$.

From the point of view of geometry, the continuity of $\frac{d^2\mu}{d\lambda^2}$ expresses the continuity of the curvature of the surface in the region of small curvatures. If we shall designate by α the limit $\lim \left(\frac{d^2\mu}{d\lambda^2}\right)_{\mu\to\infty}$, where μ is the solution of equation (11), then with the aid of equation (18) the condition of chaining together the solutions for both regions may be written in the form:

$$\left(\frac{d^2\mu}{d\lambda^2}\right)_{\mu\to 0}=\frac{\sqrt{2}\,a\,\sigma^{2/3}j}{v_0^{5/3}(3\eta)^{2/3}}=\alpha. \tag{19}$$

Inasmuch as no dimensional quantities enter equation (11), α is a pure number. The quantity α may be found by means of the numerical integration of equation (11), which will be performed below.

Equation (19) gives us

$$j=\frac{\alpha}{\sqrt{2}}\,\frac{v_0^{5/3}(3\eta)^{2/3}}{\sigma^{2/3}a}. \tag{20}$$

The condition (20) will be discussed later on. In order to perform the numerical integration (11), it is necessary to investigate in more detail the character of the tending of the derivatives $d\mu/d\lambda$ and $d^2\mu/d\lambda^2$ to zero with λ increasing infinitely, *i. e.* at the upper boundary of the region [see equation (13)]. It may be established from the behaviour of the asymptotic solutions of equation (11).

Namely, for sufficiently large values of λ, the thickness μ may, evidently, be represented in the form:

$$\mu(\lambda)=1+\mu_1(\lambda) \tag{21}$$

with $\mu_1(\lambda)\ll 1$.

Then substituting the value $\mu(\lambda)$ from equation (21) into equation (11), we find, after neglecting the squares of small quantities, the linear equation for μ_1:

$$\frac{d^3\mu_1}{d\lambda^3}=-\mu_1. \tag{22}$$

As the boundary conditions for equation (22) serves:

$$\mu_1\to 0 \qquad \text{for} \quad \lambda\to\infty.$$

As a particular solution of equation (22), satisfying this boundary condition, may serve:

$$\mu_1=\text{const}\cdot e^{-\lambda}.$$

It is evident at the same time that

$$\frac{d\mu_1}{d\lambda} = -\text{const}\cdot e^{-\lambda} = -\mu_1,$$

$$\frac{d^2\mu_1}{d\lambda^2} = \text{const}\cdot e^{-\lambda} = \mu_1.$$

Thus, at large values of λ we have the following asymptotic equations for μ and for its derivatives:

$$\mu = 1 + \text{const}\cdot e^{-\lambda}, \tag{23}$$

$$\frac{d\mu}{d\lambda} = 1 - \mu, \tag{23'}$$

$$\frac{d^2\mu}{d\lambda^2} = \mu - 1. \tag{23''}$$

The values of the constant figuring in equation (23) turn out to be unessential for our purpose. For the direct performance of the numerical integration of equation (11), with boundary conditions(23), (23′) and (23″) kept in mind, it would be convenient to lower the order of the equation, choosing μ as the new variable quantity, and $(d\mu/d\lambda)^2$ as the new unknown function. If we shall designate $(d\mu/d\lambda)^2$ by ξ, then, after simple transformations, we find the following equation for ξ:

$$\frac{d^2\xi}{d\mu^2} = \frac{2(1-\mu)}{\mu^3\sqrt{\xi}}. \tag{24}$$

The boundary conditions (23)—(23″) may be now written down as

$$\left.\begin{aligned} &\xi \to (1-\mu)^2, \\ &\frac{d\xi}{d\mu} \to 2(\mu - 1) \end{aligned}\right\} \tag{25}$$

for $\mu \to 1$.

We are directly interested in the quantity α equal in the new designations to

$$\alpha = \lim\left(\frac{d^2\mu}{d\lambda^2}\right)_{\mu\to\infty} = \frac{1}{2}\lim\left(\frac{d\xi}{d\mu}\right)_{\mu\to\infty}.$$

The numerical analysis of equation (18) with boundary condition (27) gives for α the value

$$\alpha = 0.63\ldots$$

4*

Substituting this value of α into equation (20) for the flux of the fluid, we get finally

$$j = 2.29 \cdot \frac{\eta^{2/3} v_0^{5/3}}{c^{1/6} (\rho g)^{1/2}} . \tag{20'}$$

We see thus that the conditions of chaining together of the solutions in both regions allow to express the flux j by the characteristic quantities of our problem. It follows from equation (20) that the quantity of the fluid carried along by a slowly moving plate is directly proportional to the velocity of the movement v_0 to the power $^5/_3$, and to the viscosity of the fluid to the power $^2/_3$, and is inversely proportional to the tension to the power $^1/_6$ (*i. e.* it shows a very slight dependence on the surface tension).

The limit thickness of the layer of the liquid carried along by the plate far away from the meniscus of the fluid, *i. e.* at $\mu = 1$, would be [*cf.* (10)]:

$$h_0 = 2.29 \frac{(v_0 \eta)^{2/3}}{c^{1/2} \sqrt{\rho g}} . \tag{26}$$

We see, therefore, that the limit thickness of the layer of liquid carried along is proportional to the velocity of the elevation of the plate and to its viscosity to the power $^2/_3$, and shows a rather slight dependence on the surface tension, being inversely proportional to it to the power $^1/_6$.

Let us ascertain now the conditions of the applicability of the formulae received [equations (20) and (21)] for the consumption and for the thickness of the layer carried along.

In going over from (15) to (16), we omit the last term of equation (15), on the assumption that the quantity $\rho g j^2/3\eta v_0^3$ is small as compared with unity.

Substituting for j its value obtained from formula (22), we find that this quantity is really small as compared with unity, and our calculation is legitimate if the following inequality is fulfilled:

$$\left(\frac{\eta v_0}{c}\right)^{1/3} \ll 1, \quad i.\ e. \quad v_0 \ll \frac{c}{\eta} , \tag{27}$$

i. e. at sufficiently small values of the velocity of the plate.

The expressions obtained for j and h seem to be in good accordance with the experiment. Indeed, the experiment shows that the thickness of the layer carried along is proportional at small velocities to v_0 to the power 0.6, which agrees with the power index obtained by us.

In the opposite limiting case, when the velocity v_0 turns out to be greater than σ/η, the given calculation becomes inapplicable. Namely, the supposition that all the surface of the fluid may be split into two independent regions, which has led to all the expressions written above, does not hold here any longer. It is impossible to obtain in this case the exact expressions for j and h. However, on the basis of dimensionality considerations, the general character of the dependence of these quantities on the fundamental parameters v_0, η, ρ and σ at sufficiently large velocity may be pointed out.

Namely, it is clear that at high velocities the consumption j and the thickness h of the layer carried along should not depend on the surface tension. From the viewpoint of physics, this may be seen from the fact that at sufficiently high velocities of the plate, the shape of the entire surface of the fluid would be determined by the character of the process oi carrying along, and not by the static properties of the surface of the fluid. Therefore, at high vefocities of the plate, the thickness of the layer carried along must depend only on the quantities v_0, η, ρ and g. The only quantity of the dimensionality of length, which may be obtained from these quantities, is the quantity $(\eta v/\rho g)^{1/2}$.

Therefore, at sufficiently high velocities of the elevation of the plate, the thickness of the layer of liquid carried along must have the form:

$$h \approx A\left(\frac{\eta v_0}{\rho g}\right) \tag{28}$$

and the consumption

$$j \approx A\left(\frac{\eta}{\rho g}\right)^{1/2} v_0^{2/3}.$$

The numerical value of the constant A may be found only by means of experiment.

Finally, in the intermediate region of velocities, it may be seen out of dimensionality considerations that the thickness of the layer carried along must have the following form:

$$h = \left(\frac{\eta v_0}{\rho g}\right)^{1/2} f\left(\frac{v_0 \eta}{\sigma}\right) \tag{29}$$

where $f(v_0\eta/\sigma)$ is some function of the dimensionless parameter $v_0\eta/\sigma$, the form of which must be found by experiment.

In both the limit cases of large and small values of $v_0\eta/\sigma$ the function $f(v_0\eta/\sigma)$ tends accordingly to unity and to $(v_0\eta/\sigma)^{1/6}$, respectively, so that equation (29) goes over into equations (28) or (26).

In conclusion we should like to express our gratitude to M. M. Kussakov and B. V. Derjaguin, who have called our attention to the experimental and the technical interest of the problem treated here.

Academy of Sciences of the USSR,
Institute for Physical Problems.
Institute of Colloid Chemistry and Electrochemistry.

Received
June 15, 1942.

Saffman-Taylor Finger

The penetration of a fluid into a porous medium or Hele-Shaw cell containing a more viscous liquid

By P. G. Saffman and Sir Geoffrey Taylor, F.R.S.
Cavendish Laboratory, University of Cambridge

(*Received* 17 *January* 1958—*Read* 17 *April* 1958)

[Plates 2 and 3]

When a viscous fluid filling the voids in a porous medium is driven forwards by the pressure of another driving fluid, the interface between them is liable to be unstable if the driving fluid is the less viscous of the two. This condition occurs in oil fields. To describe the normal modes of small disturbances from a plane interface and their rate of growth, it is necessary to know, or to assume one knows, the conditions which must be satisfied at the interface. The simplest assumption, that the fluids remain completely separated along a definite interface, leads to formulae which are analogous to known expressions developed by scientists working in the oil industry, and also analogous to expressions representing the instability of accelerated interfaces between fluids of different densities. In the latter case the instability develops into round-ended fingers of less dense fluid penetrating into the more dense one. Experiments in which a viscous fluid confined between closely spaced parallel sheets of glass, a Hele-Shaw cell, is driven out by a less viscous one reveal a similar state. The motion in a Hele-Shaw cell is mathematically analogous to two-dimensional flow in a porous medium.

Analysis which assumes continuity of pressure through the interface shows that a flow is possible in which equally spaced fingers advance steadily. The ratio λ = (width of finger)/(spacing of fingers) appears as the parameter in a singly infinite set of such motions, all of which appear equally possible. Experiments in which various fluids were forced into a narrow Hele-Shaw cell showed that single fingers can be produced, and that unless the flow is very slow λ = (width of finger)/(width of channel) is close to $\frac{1}{2}$, so that behind the tips of the advancing fingers the widths of the two columns of fluid are equal. When $\lambda = \frac{1}{2}$ the calculated form of the fingers is very close to that which is registered photographically in the Hele-Shaw cell, but at very slow speeds where the measured value of λ increased from $\frac{1}{2}$ to the limit 1·0 as the speed decreased to zero, there were considerable differences. Assuming that these might be due to surface tension, experiments were made in which a fluid of small viscosity, air or water, displaced a much more viscous oil. It is to be expected in that case that λ would be a function of $\mu U/T$ only, where μ is the viscosity, U the speed of advance and T the interfacial tension. This was verified using air as the less viscous fluid penetrating two oils of viscosities 0·30 and 4·5 poises.

1. The stability of the interface between two fluids in a porous medium

It has been pointed out (Taylor 1950) and verified experimentally (Lewis 1950) that when two superposed fluids of different densities and negligible viscosities are accelerated in a direction perpendicular to their interface, this surface is stable or unstable for small deviations according as the acceleration is directed from the more dense to the less dense fluid or vice versa.

An analogous instability can occur when two superposed viscous fluids are forced by gravity and an imposed pressure gradient through a porous medium. If the steady state is one of uniform motion with velocity V vertically upwards and the

Reprinted with permission from *Proceedings of the Royal Society of London, Series A, Mathematical and Physical Sciences*, Vol. 245, 1958.

interface between the two fluids is horizontal, then it can be shown that the interface is stable for small deviations from the steady state if

$$\left(\frac{\mu_2}{k_2} - \frac{\mu_1}{k_1}\right) V + (\rho_2 - \rho_1) g > 0, \tag{1}$$

and unstable if

$$\left(\frac{\mu_2}{k_2} - \frac{\mu_1}{k_1}\right) V + (\rho_2 - \rho_1) g < 0, \tag{2}$$

where the suffix 1 refers to the upper fluid and the suffix 2 to the lower. The motion of the fluids through the medium is here supposed to be governed by Darcy's law which asserts that the velocity of the fluid is given by

$$\mathbf{u} = -\frac{k}{\mu} \operatorname{grad}(p + \rho g x) = \operatorname{grad} \phi, \quad \text{say}, \tag{3}$$

where $\mathbf{u}$ denotes the velocity, μ the viscosity and ρ the density, k is the permeability of the medium to the fluid, g the acceleration due to gravity, x the vertical height above some horizontal plane, and ϕ is called the velocity potential.

To describe a disturbance of the surface of separation, take rectangular coordinates (x, y, z), the instantaneous position of the undisturbed interface coinciding with the plane $x = 0$. Suppose the interface is deformed slightly into a wave-like corrugation of wavelength $2\pi/n$ described by

$$x = a\, e^{iny + \sigma t}. \tag{4}$$

Assuming that the fluids are incompressible and that the medium is of uniform porosity, the equation of continuity satisfied by the velocity field is $\operatorname{div} \mathbf{u} = 0$ and the velocity potential therefore satisfies Laplace's equation $\nabla^2 \phi = 0$.

It is now necessary to make some assumption about the nature of the motion in the vicinity of the interface because, in fact, a sharp interface between the two fluids does not exist but there is, rather, an ill-defined transition region in which the two fluids intermingle. This region is often not very thick and we shall assume that the fluids do not interpenetrate to any marked extent and that the width of the transition zone is small compared with the length scale of the motion. It is then reasonable to assume for the purpose of mathematical analysis that the two fluids are separated by a sharp interface, across which the normal component of velocity and the pressure are continuous (surface tension or any other similar effect is neglected).†

It follows from the continuity of normal velocity that the velocity potentials in the upper and lower fluid satisfy on $x = 0$, to the first order in the deviation

$$\frac{\partial \phi_1}{\partial x} = \frac{\partial \phi_2}{\partial x} = V + a\sigma\, e^{iny + \sigma t}. \tag{5}$$

† The assumption that one fluid completely expels the other may sometimes be relaxed and the analysis can be modified to treat cases in which a proportion of one fluid is left behind to be surrounded by the oncoming fluid. The continuity of normal velocity across the interface no longer holds, but provided that the mixture of fluids can be regarded as homogeneous and the proportion left behind the interface is constant, then it is found that the interface moves as if it separated two fluids whose viscosities and densities differ from those of the original two fluids but which completely expel one another. The values of these viscosities and densities depend upon the physical properties of the mixture and the proportion left behind. In this connexion, see also the Appendix.

Hence, $$\phi_1 = Vx - (a\sigma/n)\,e^{iny-nx+\sigma t}$$

and $$\phi_2 = Vx + (a\sigma/n)\,e^{iny+nx+\sigma t},$$

these being the appropriate solutions of $\nabla^2\phi = 0$ which satisfy (5) and for which the disturbance vanishes at infinity.

The pressure, p_1, in the upper fluid is $-(\mu_1/k_1)\,\phi_1 - \rho_1 gx$ and that, p_2, in the lower fluid is $-(\mu_2/k_2)\,\phi_2 - \rho_2 gx$. Equating the values of p_1 and p_2 on the interface (4), we find that, to the first order in the deviation, σ must satisfy

$$\frac{\sigma}{n}\left(\frac{\mu_1}{k_1}+\frac{\mu_2}{k_2}\right) = (\rho_1-\rho_2)\,g + \left(\frac{\mu_1}{k_1}-\frac{\mu_2}{k_2}\right)V. \tag{6}$$

If the right-hand side of (6) is positive, then σ is positive and the amplitude of the deviation increases at an exponential rate and the interface is then unstable to small disturbances. If the right-hand side is negative, the deviation is damped at an exponential rate and the motion is stable to small disturbances. Thus, the results (1) and (2) are true for all wavelengths and consequently for all types of small disturbance. They can also be put as follows. When two superposed fluids of different viscosities are forced through a porous medium in a direction perpendicular to their interface, this surface is stable or unstable to small deviations according as the direction of motion is directed from the more viscous to the less viscous fluid or vice versa, whatever the relative densities of the fluids, provided that the velocity is sufficiently large.

It appears that this result is not essentially new and that mining engineers and geologists have long been aware of it. In certain types of oil wells the oil taken out of the ground is replaced by encroaching water which comes from the expansion of a large water accumulation or seepage from the surface. Since oil is lighter than water but somewhat more viscous, it follows from (2) that when the velocity of extraction is too large the interface will become unstable. It is indeed observed in practice that when the velocity of extraction is too large, long tongues or cones of water penetrate the oil and it comes out of the well mixed with water. This phenomenon is known in the literature as 'water tonguing or coning'. However, earlier writers (e.g. Dietz 1953; Kidder 1956) do not appear to have considered explicitly the stability of the interface, but rather the conditions necessary in certain cases for a steady interface to exist.

2. An analogue for two-dimensional flow in a porous medium

The motion of fluid in a porous medium according to Darcy's law can be derived from a potential $\phi = -(k/\mu)\,(p+\rho gx)$. Motion in two dimensions can therefore be studied experimentally by means of an analogue devised by Hele-Shaw (1898). This makes use of the result that the motion of a viscous fluid, between two fixed parallel plates which are sufficiently close together, is such that the components of the mean velocity across the stratum are

$$u = -\frac{b^2}{12\mu}\left(\frac{\partial p}{\partial x}+\rho g\right), \quad v = -\frac{b^2}{12\mu}\frac{\partial p}{\partial y}, \tag{7}$$

where b denotes the distance between the plates (see, for example, Lamb 1932, § 330). The plates are here taken as vertical, the x-axis is vertically upwards, the y-axis parallel to the plates, and the z-axis perpendicular to the plates; u and v are the components of mean velocity in the x- and y-directions, respectively.

These are the equations satisfied by the velocity in a porous media of permeability $b^2/12$ and there is thus a direct analogy between two-dimensional flow in a porous medium and the flow between parallel plates, the velocity in the former case corresponding to the mean velocity in the latter. In this way, for example, the streamlines for the flow around bodies of arbitrary shape can be determined experimentally.

The analogue can also be used to reproduce experimentally the (two-dimensional) motion of the interface between two fluids in a porous medium. Consider the motion between parallel plates of two immiscible fluids of viscosities μ_1, μ_2 and densities ρ_1, ρ_2, respectively, and suppose the direction of motion is away from fluid 2 towards fluid 1. Now the fluid 1 is not necessarily completely expelled or replaced by fluid 2, a film of fluid 1 may wet the plates and adhere to them, while a tongue of fluid 2 advances along the middle of the gap between the plates. The thickness of the tongue may be taken as a fraction t, say, of the gap between the plates and the analysis which follows is applicable and the analogue valid provided that t is constant.

In experiments to be described later, one of the fluids (fluid 2) was air and in order to find out what proportion of fluid 1 was left behind after the interface (or the tip of the meniscus) has passed, subsidiary experiments were made in which a measured volume of air was blown centrally into the narrow space (0·09 cm) between the flat base of a metal vessel containing oil or glycerine and a circular flat glass disk. The rather irregular outline of the bubble so formed was photographed and the area contained within it determined. In several such trials it was found that the volume of the bubble divided by the area gave a thickness which was always less than 0·09 cm but was never more than $1\frac{1}{2}\%$ less. The rate at which the air was blown into the apparatus had to be kept low for otherwise the instability which gives rise to the irregular outlines of the interface made it difficult to measure the area. The velocity of the interface was, however, of the same order as those occurring in the experiments to be described later. In discussing those experiments, it is therefore legitimate to assume that $t = 1$ and the outlines which were observed and photographed represent interfaces completely separating the two fluids.

The thickness of the film of liquid left behind when a bubble moves in a capillary tube has been investigated (Fairbrother & Stubbs 1935) and shown to depend on the non-dimensional parameter $\mu U/T$, where μ is the viscosity, U is the velocity of the bubble and T the surface tension. The value of the parameter in our experiments was such that only a small fraction of the fluid would be expected to remain behind.

The mean velocity across the stratum of fluid 1 ahead of the interface is given by (assuming b to be sufficiently small)

$$\mathbf{u}_1 = -(b^2/12\mu_1)\,\mathrm{grad}\,(p+\rho_1 g x) = \mathrm{grad}\,\phi_1, \quad \text{say}, \tag{8}$$

taking the plates vertical and the x-axis vertically upwards. Taking $t = 1$ and fluid 1 to be completely expelled, the mean velocity across the stratum of fluid 2 is given by

$$\mathbf{u}_2 = -(b^2/12\mu_2)\,\mathrm{grad}\,(p+\rho_2 g x) = \mathrm{grad}\,\phi_2, \quad \text{say}. \tag{9}$$

Now when b is small, the width of the projection of the meniscus on to the plates is small, and expressions (8) and (9) can be supposed to hold (for the purposes of the analysis) right up to the interface which may be regarded as a sharp line. It follows from continuity that the components of $\mathbf{u}_1$ and $\mathbf{u}_2$ normal to the interface are continuous across the interface; if surface tension effects are negligible the pressure is constant across the interface (see, further, § 5 below), and the motion of the two fluids will then reproduce the two-dimensional motion of the interface between two fluids of viscosities μ_1 and μ_2 in a porous medium of permeability $b^2/12$.

The analogue is still valid when $t \neq 1$, provided that t is constant. For this case, the interface can be identified with the tip of the meniscus between the two fluids. The modifications that are required are given in the Appendix.

The considerations of § 1 apply to the motion in the Hele-Shaw cell and it will be noticed that when the cell is vertical there is a critical velocity for the interface, which separates unstable from stable conditions. This is given by (1) and (2) with k_1 and k_2 replaced by $b^2/12$. When the Hele-Shaw apparatus is set horizontal the analysis applies with g put equal to zero, and it follows that the interface is always unstable when the less viscous fluid is driving the more viscous.

3. Effect of surface tension on stability in the Hele-Shaw cell

The effect of surface tension on the stability of the interface may depend on a variety of physical conditions. The simplest assumption is to take the pressure drop through the interface as $T(2/b+1/R)$, where R is the radius of curvature of the projection on the planes bounding the cell of the tip of the meniscus. In discussing the stability of a plane interface in the Hele-Shaw apparatus, this may be taken as $T(2/b+\mathrm{d}^2x/\mathrm{d}y^2)$, where x is given by (4), and it is easily seen that (6) is thereby altered to

$$\frac{12}{b^2}\sigma(\mu_1+\mu_2) = \frac{2\pi}{l}\left\{\frac{12V}{b^2}(\mu_1-\mu_2)+g(\rho_1-\rho_2)\right\} - \frac{8\pi^3 T}{l^3}, \qquad (10)$$

where σ is the amplification factor of disturbances of wavelength $l = 2\pi/n$.

It will be seen that this effect of surface tension is to limit the range of disturbances which are unstable to those of wavelength greater than

$$l_{\text{crit.}} = 2\pi T^{\frac{1}{2}} b\{12V(\mu_1-\mu_2)+b^2 g(\rho_1-\rho_2)\}^{-\frac{1}{2}}. \qquad (11)$$

That surface tension should have this effect was pointed out to us by Dr Chuoke in a similar connexion, the amplification factor is a maximum for disturbances of wavelength $\sqrt{3}\, l_{\text{crit.}}$.

4. Experiments using the Hele-Shaw cell

It was shown experimentally (Lewis 1950) that the unstable accelerating interface between two fluids of different density develops in such a way that long fingers of the less dense fluid penetrate into the more dense one, and beyond the level to which these fingers have penetrated into the more dense fluid the acceleration has the same value it would have if the interface had remained plane. Analogous results are obtained when the Hele-Shaw apparatus contains two immiscible fluids.

In the first apparatus, two pieces of commercially flat plate glass were separated by strips of rubber 0·09 cm thick laid long their long edges. The channel 0·09 × 12 × 38 cm thus formed was connected at its two ends with vessels containing the two fluids. The pressure gradient along the channel was produced by applying air pressure or suction to the airspace above the fluid in one of the end vessels and the pressure at the other was maintained at that of the atmosphere. The apparatus which is shown in the sketch (figure 1) could be used either vertically or horizontally and the meniscus of the interface could be photographed as a sharp line. The three-cock shown in figure 1 on the left side of the top of the front view made it possible

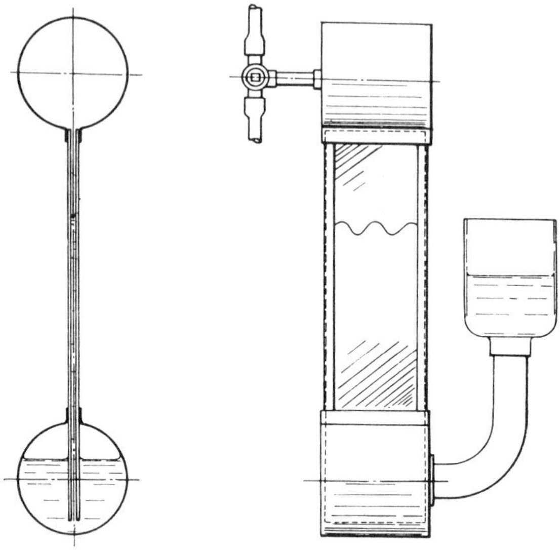

FIGURE 1. Sketch of Hele-Shaw cell.

to change the pressure in the air chamber rapidly from pressure to suction. The first experiments were made with the apparatus vertical and with glycerine as the more viscous and air as the less viscous fluid. In this case the critical velocity is downwards and the unstable disturbances are to be expected when the air is above if the velocity is downwards and greater than this. If the glycerine lies above the air the flow becomes stable when the downwards velocity is greater than the critical.

In the experiment shown in figure 2, plate 2, the fluid was sucked up to near the top of the Hele-Shaw channel and allowed to fall with a velocity less than the critical and then maintained at rest for a short time. The fluid left behind on the glass then gathered itself together to form vertical streaks which produced on the interface a small variation in level. The air pressure was then turned on; this is indicated by the mercury manometer seen to the left of the channel. The photograph was taken by a flash bulb after the glycerine had been forced downwards through a few centimetres. It will be seen that the instability has already manifested itself. In

20-2

this experiment V was 0·1 cm/s and the critical wavelength given by (11) was 1·2 cm; the average wavelength of the disturbance shown in figure 2 was 2·2 cm, so that instability was to be expected.

Figure 3, plate 2, shows a later stage of the instability (not in the same experiment as that of figure 2). In this case the pressure was turned on when the nearly straight interface was at the level of the top of the mark seen projecting from the left-hand side of the channel. A characteristic feature of the later stages of the growth of 'instability' into 'fingers', is shown in figure 3, namely the tendency of the fingers to space themselves so that the width of the air fingers and the columns of fluid between them are of approximately the same breadth. The development of these fingers is very similar to those recorded by Lewis for the later states of instability of an accelerated interface, but it differs from them in that the air fingers in Lewis's experiment were separated by very narrow columns of fluid.

The reason for the narrow columns in Lewis's experiment was that the water left behind after the passage of the front of 'fingers' is in a field of uniform pressure, and therefore moves uniformly at the speed at which it was passed by the front, while the fluid ahead of this front is accelerating away from it. There is thus a longitudinal rate of strain in the columns so that they must continually decrease in thickness as the front leaves them.

Figure 4, plate 2, shows another characteristic feature, the inhibiting effect on the growth of its neighbours that happens when the end of one of the fingers gets ahead of them. On the right-hand side of this photograph can be seen three fingers which started to grow at the same time. The middle finger, however, was slightly larger than its neighbours and at the stage shown in figure 4 has almost completely inhibited their growth and as it passed them it spread laterally. The inhibiting effect on the growth of neighbours by any finger which gets ahead of them also occurred in Lewis's experiments and was due to the same cause.

In attempting to form a mathematical description of the mechanics of the formation of 'fingers' we were naturally led to consider an infinite set of equal and equally spaced fingers all advancing at the same speed. Since each finger is then identical mathematically with all the others and the fluid on the straight lines halfway between neighbours has no transverse component of velocity, we considered only a single finger propagating itself in a channel of fixed width. The details of the analysis will be given in the next section.

Description of plate 2

Figure 2. Interface between air and glycerine at an early stage of the instability.

Figure 3. Development of instability.

Figure 4. Inhibiting effect of a finger which gets ahead of its neighbour.

Figure 8. An air finger advancing into glycerine.

Figure 9. Enlarged forward portion of the finger shown in figure 8. ○, points calculated from (17) using $\lambda = \frac{1}{2}$.

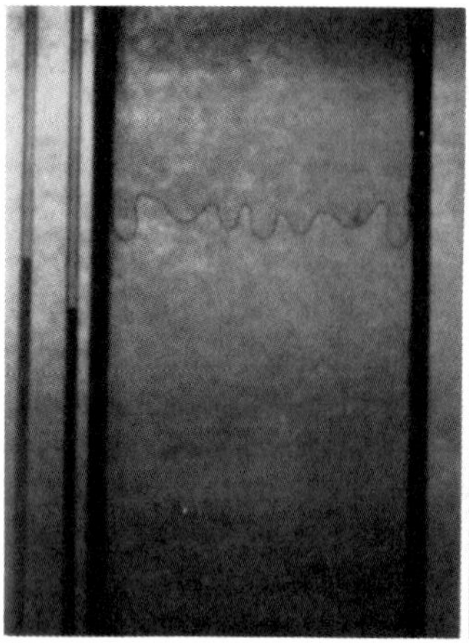

FIGURE 2

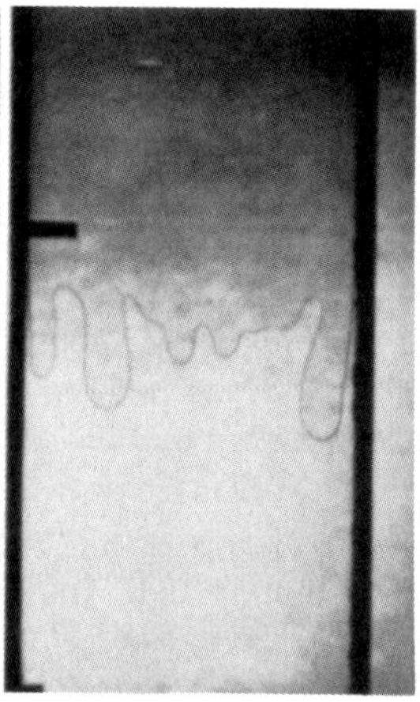

FIGURE 3

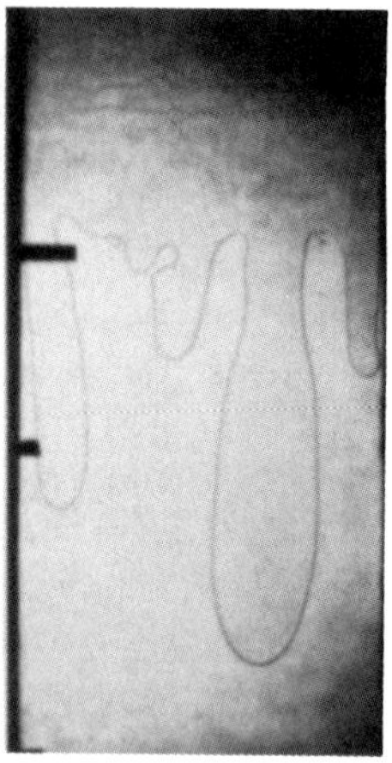

FIGURE 4

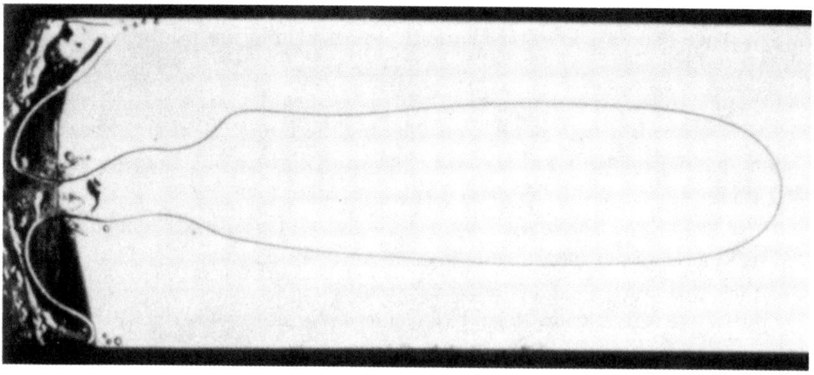

FIGURE 8

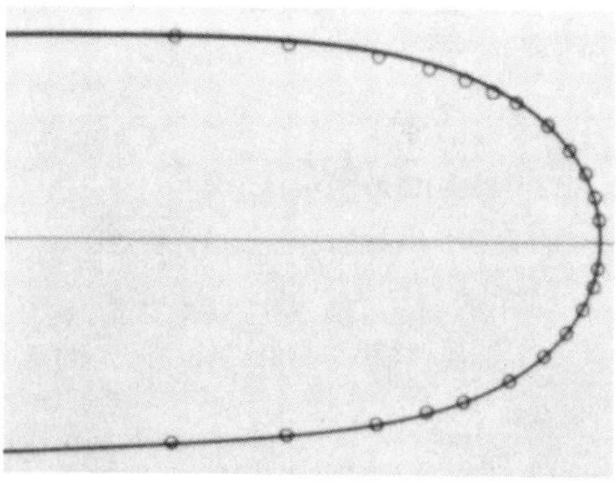

FIGURE 9

(*Facing p.* 318)

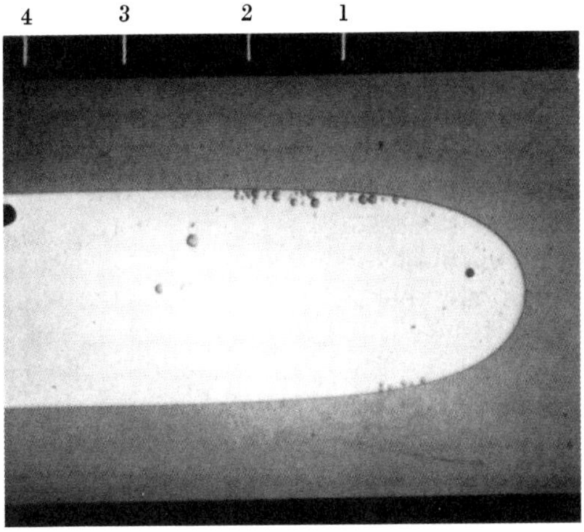

FIGURE 11. Finger of oil ($\mu = 4\cdot5$ P) penetrating glycerine ($\mu = 9$ P).

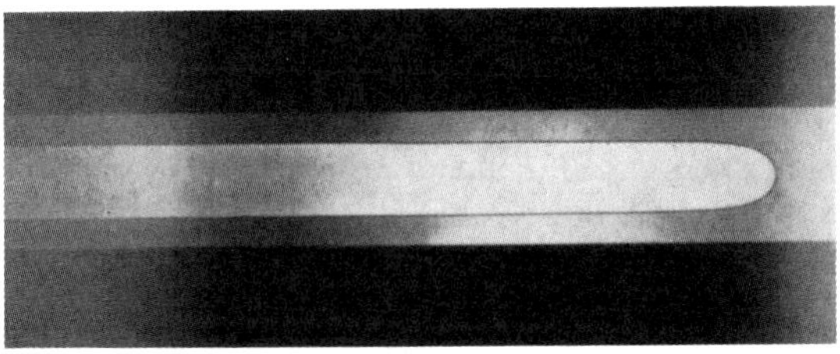

FIGURE 12. Finger of water penetrating oil (Shell Diala).

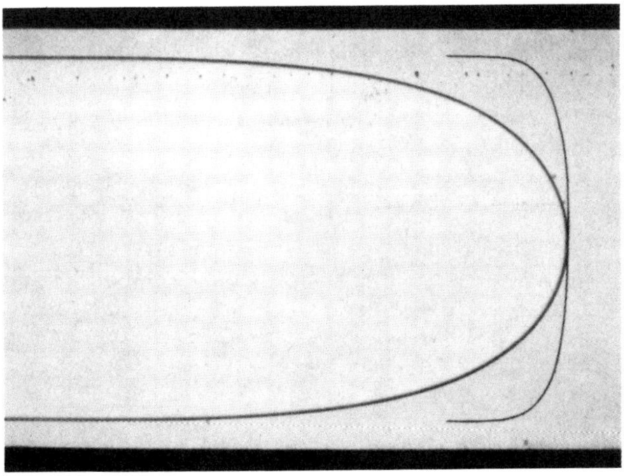

FIGURE 13. Water penetrating into oil at slow speed. Profile calculated from (17) for $\lambda = 0\cdot87$ superposed.

5. Penetration of a single 'finger' into a channel

In this section we consider the analysis of the motion of a long bubble or 'finger' of fluid moving through an infinite channel in the Hele-Shaw cell filled with a viscous fluid and bounded by straight parallel walls. As described in the previous section, the motion of these bubbles is connected with the mechanics of the formation and propagation of 'fingers' and, by virtue of the analogue, also bears on the question of how long it would take for a vertical tube or channel of saturated porous material to drain when closed at the top and open at the bottom. The analogous problem for inviscid liquids of the motion of bubbles through liquids in tubes and channels has been studied experimentally and theoretically in some detail (see, for example, Davies & Taylor 1950; Garabedian 1957). The present problem is of particular interest, since it is possible to obtain exact solutions of the equations of motion in closed form and compare them with experiment.

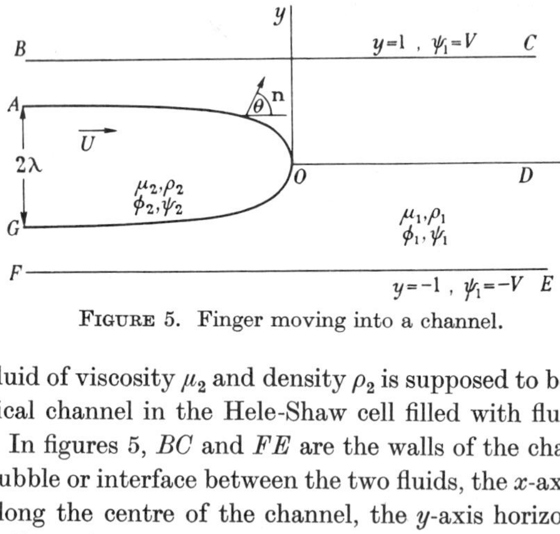

Figure 5. Finger moving into a channel.

A bubble of fluid of viscosity μ_2 and density ρ_2 is supposed to be moving steadily through a vertical channel in the Hele-Shaw cell filled with fluid of viscosity μ_1 and density ρ_1. In figures 5, BC and FE are the walls of the channel, AOG is the surface of the bubble or interface between the two fluids, the x-axis is taken vertically upwards along the centre of the channel, the y-axis horizontal and perpendicular to the walls, and the origin is at the nose of the bubble which is supposed to be of infinite extent and symmetrical about the centre of the channel. The velocity of the bubble is denoted by U and the velocity of the fluid at infinity in front of the bubble by V. The walls of the channel are taken as $y = \pm 1$, and the width of the bubble at infinity as 2λ, where λ is a parameter which is for the present unspecified except that it lies between 0 and 1. Suffixes 1 and 2 refer to quantities outside and inside the bubble, respectively.

The mean velocity across the stratum is given by equations (8) and (9) and the equation of continuity takes the form

$$\frac{\partial u}{\partial x} + \frac{\partial v}{\partial y} = 0,$$

where u and v are the components of mean velocity parallel to the axes. A stream function ψ can be defined by

$$u = \frac{\partial \phi}{\partial x} = \frac{\partial \psi}{\partial y}, \quad v = \frac{\partial \phi}{\partial y} = -\frac{\partial \psi}{\partial x},$$

and it follows from the Cauchy–Riemann equations that $w = \phi + i\psi$ is an analytic function of $z = x + iy$.

It will be assumed that the experiment described in § 2 is valid, i.e. one fluid completely expels the other, and that the meniscus separating the two fluids is a sharp line, so that the components of mean velocity normal to the interface are continuous, i.e.

$$\frac{\partial \phi_1}{\partial n} = \frac{\partial \phi_2}{\partial n} = U \cos\theta, \tag{12}$$

where θ is the angle between the x-axis and the outward normal $\mathbf{n}$ to the interface. Now $\partial\phi/\partial n = \partial\psi/\partial s$ where $\partial/\partial s$ denotes differentiation along the interface, and $\cos\theta = \partial y/\partial s$, from which it follows that

$$\psi_1 = \psi_2 = Uy \tag{13}$$

on the surface of the bubble.

If the pressure change due to surface tension at the interface is ignored, the equation for continuity of pressure is

$$(12/b^2)(\mu_1\phi_1 - \mu_2\phi_2) = g(\rho_2 - \rho_1)\,x. \tag{14}$$

The same equation is valid if the change in pressure on passing through the interface is constant, since an arbitrary constant may be added to the velocity potential. If the fluid wets the plane surfaces it might be expected that under static conditions the pressure drop would be $T(2/b + 1/R)$, where R is the radius of curvature of the meniscus on the plane of the two parallel sheets which bound the cell. The assumption that this is constant amounts to neglecting $1/R$ in comparison with $2/b$ and to assuming that the surface tension and curvature of the interface are identical with their static values. The range of validity of equation (14) will be discussed later.

The remaining conditions on the velocity potentials and stream functions are $\phi_1 \sim Vx$ as $x \to +\infty$, $\phi_2 \sim Ux$ as $x \to -\infty$, and $\psi_1 = \pm V$ on $y = \pm 1$, since the walls of the channel must be streamlines. (We neglect the edge effects which occur at solid boundaries in the Hele-Shaw cell and invalidate the equations of motion (8) and (9) within a distance of order b from these boundaries.)

We examine first the case in which the fluid inside the bubble is of negligible viscosity and density, and suppose also that gravity forces are negligible, i.e. that the imposed pressure gradient which moves the fluid is large compared with that due to gravity. This case corresponds to the experimental arrangement described in § 7 below when a bubble of air is blown through glycerine.

Equation (14) now reduces to $\phi_1 = 0$ on the bubble surface. Further, the flow becomes uniform a long way behind the nose of the bubble, i.e. as $x \to -\infty$, and since ϕ_1 is zero on the interface, $\phi_1 \to 0$ as $x \to -\infty$ and the fluid is at rest at $x = -\infty$. Hence, the stream function has the same value at A as at B (and also at G as at F) and it follows from (13) that

$$V = \lambda U. \tag{15}$$

Equation (15) gives the velocity of the bubble in terms of its width and the velocity at infinity; the latter velocity will be determined by the external means by which the motion is generated.

The shape of the interface is as yet unknown and to solve this free boundary problem, we transform into the ϕ, ψ plane and consider $x+\mathrm{i}y$ as an analytic function of $\phi+\mathrm{i}\psi$ (for brevity we drop the suffix 1).† In figure 6, corresponding points in the physical and potential planes are marked with the same letter; the exterior of the bubble transforms into the semi-infinite strip $\phi>0$, $-V<\psi<V$, the surface of the bubble to the ϕ axis between $\psi=\pm V$, and the walls of the channel to $\psi=\pm V$.

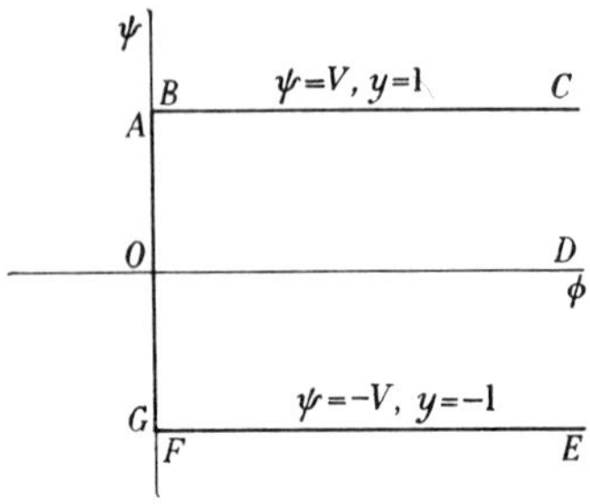

FIGURE 6. The potential plane for motion in a channel.

Now y is a harmonic function of ϕ and ψ which has the values -1 on FE, 0 on DO, $+1$ on BC, $y=\psi/U$ on GA (the surface of the bubble), and $y\to\psi/V$ as $\phi\to+\infty$.

Take, therefore, $$y=\frac{\psi}{V}+\sum_{1}^{\infty}A_n\sin\frac{n\pi\psi}{V}\exp\left(-\frac{n\pi\phi}{V}\right).$$

This is a harmonic function satisfying all the boundary conditions, provided

$$\frac{\psi}{U}=\frac{\psi}{V}+\sum_{1}^{\infty}A_n\sin\frac{n\pi\psi}{V},\quad\text{for}\quad -V<\psi<V.$$

Calculating the coefficients by the usual method of Fourier series, we find that

$$A_n=-\frac{2}{\pi}\left(1-\frac{V}{U}\right)=-\frac{2}{\pi}(1-\lambda),$$

and therefore $$z=\frac{w}{V}+\frac{2}{\pi}(1-\lambda)\ln\frac{1}{2}\left(1+\exp\left[-\frac{\pi w}{V}\right]\right). \tag{16}$$

This equation determines the complex potential implicitly as a function of z. The pressure then follows from the relation

$$p=-\frac{b^2}{12\mu}\phi+\text{const.}$$

The parametric equation of the interface is obtained by putting $\phi=0$ in equation (16) and it follows that the bubble surface is

$$x=\frac{1-\lambda}{\pi}\ln\frac{1}{2}\left(1+\cos\frac{\pi y}{\lambda}\right). \tag{17}$$

This completes the solution for the case of a bubble of fluid of negligible viscosity when effects due to gravity forces and departures from equation (14) are unimportant.

† We are grateful to Mr F. Ursell for first suggesting this transformation to us.

Before discussing this further, we give briefly the solution for the more general case in which the viscosity of the fluid in the bubble is not neglected and pressure gradients due to gravity are taken into account. The equations of motion for the fluid inside the bubble can be satisfied by taking

$$\phi_2 + i\psi_2 = U(x+iy). \tag{18}$$

The boundary condition for ϕ_1 then becomes $\phi_1 = -U^*x$ on the bubble, where

$$\frac{12\mu_1 U^*}{b^2} = g(\rho_1-\rho_2) - \frac{12\mu_2 U}{b^2}. \tag{19}$$

Define now $$W = \Phi + i\Psi = w_1 + U^*z. \tag{20}$$

On the interface, $\Phi = 0$, $\Psi = (U+U^*)y$; further $\Psi = \pm(V+U^*)$ on $y = \pm 1$ and $W \to (V+U^*)z$ as $x \to +\infty$. The problem is thus reduced to the simpler case considered above and the complex potential for the motion outside the bubble is given by (15) and (16) with V replaced by $V+U^*$, U by $U+U^*$, and w by W. The bubble surface corresponds to $\Phi = 0$ and is therefore given by (17) also. It is worth noticing that if the asymptotic width at infinity is fixed, then the shape of the bubble is independent of the physical properties of the fluids. The width 2λ of the bubble is given in terms of U (the velocity of the bubble), V (the velocity at infinity), and the physical properties of the fluids by

$$\lambda = \frac{V+U^*}{U+U^*}.$$

The maximum velocity of propagation again corresponds to $\lambda = 0$ and is

$$U_{\text{max.}} = \frac{\mu_1}{\mu_2} V + \frac{b^2 g}{12\mu_2}(\rho_1-\rho_2).$$

6. Non-uniqueness of the solution

There is nothing in the preceding mathematical analysis to determine the width of the bubble and the value of λ, i.e. the fraction of the channel occupied by the bubble after the nose has passed. In other words, if only the velocity at infinity ahead of the bubble is specified (this is equivalent to specifying the pressures driving the less viscous finger into the more viscous fluid), then the free-boundary problem does not have a unique solution and there are an infinite number of possible steady shapes, each with a different velocity of propagation. These shapes are the members of the family of curves given by (17) for values of λ between 0 and 1. The velocity of propagation of the bubble corresponding to each of these shapes is related to the velocity at infinity by (15). The two extreme members are given by $\lambda = 1$ when the interface extends in a straight line across the whole width of the channel, and $\lambda = 0$ when the bubble has zero width and propagates with infinite velocity. The calculated shapes for $\lambda = 0{\cdot}2$, $0{\cdot}5$ and $0{\cdot}8$ are shown in figure 7.

It was pointed out recently by Garabedian (1957) that the analogous free-boundary problem for the propagation of an air bubble through a vertical tube or channel containing an inviscid liquid also does not possess a unique solution, and that there are many possible 'equilibrium' shapes which a bubble can have, each again corresponding to a different velocity of rise. (Garabedian did not demonstrate explicitly the multiplicity of solutions and it is of some interest that this can be done for the problem we are considering.) He gave arguments based on the hypothesis of a maximum rate of loss of potential energy to show that in practice only one of these possible shapes would occur, the one occurring being that with the maximum velocity of propagation.

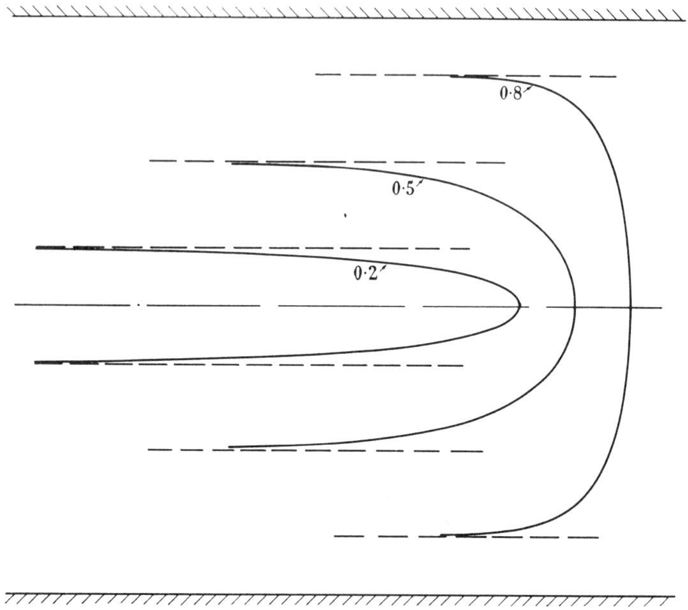

FIGURE 7. Calculated profiles for $\lambda = 0{\cdot}2$, $0{\cdot}5$ and $0{\cdot}8$.

These arguments do not apply to motion in a porous medium or a Hele-Shaw cell, since the motion is then dissipative and in any case the maximum velocity of propagation according to the analysis is infinite, but we should still expect on the grounds of physical experience that only one of all the possible shapes would occur in practice. It should be noted that the hypothesis of a maximum or minimum rate of dissipation of energy by viscosity does not determine a unique value for λ, since this rate for a channel of finite, but very large length is proportional to the product of the velocity at infinity and the pressure difference between the ends of the channel, and it follows from (16) that the value of this product becomes independent of λ as the length of the channel tends to infinity. (The terms involving λ, which tend to zero as the length increases, are monotonic in λ and a hypothesis of maximum or minimum dissipation would in any case give $\lambda = 1$ or $\lambda = 0$.)

Our experiments with the Hele-Shaw cell indicate that, as the speed of flow for any given fluid increases, λ rapidly decreases to $\lambda = \frac{1}{2}$ and remains close to this value over a large range of speeds, till at high speeds of flow the tongue or finger of the advancing fluid itself breaks down and divides into smaller fingers.

7. Experiments in channels

The apparatus with which the photographs of figures 2 to 4 were taken was not suitable for producing single steadily moving bubbles of the type considered in the analysis. It was not long enough to permit one of the fingers to grow at the expense of the others and thus be propagated as a single steadily moving column. The apparatus was therefore modified by inserting liners between the plates which limited the width of the viscous fluid channel to 5·6 cm, leaving the length (38 cm) and thickness (0·09 cm) unchanged, and it was used horizontally. With this width it was possible to obtain bubbles which for several inches behind the advancing meniscus were parallel sided. Figure 8, plate, 2 is a photograph of one. It will be seen that the 'finger' starting from an artificial initial disturbance, produced by blowing a small central air bubble close to the initially straight meniscus, swells out to a definite breadth and is then propagated without further change down the channel. The outer edge of the outline of the bubble represents the outer edge of the meniscus and, measuring the breadth of this at its widest section on the enlarged photograph of figure 8, it was found to be 5·10 cm. The two liners which limit the breadth of the channel appear black in figure 8 and the width between them measured on this photograph with a reading microscope was 10·29 cm, so that

$$\lambda = \frac{5\cdot10}{10\cdot29} = 0\cdot496.$$

In several experiments, λ was found to be within 2 % of 0·50.

To compare the shape of the bubble with the results of calculation, the photograph was enlarged till the distance between the walls was 20·0 cm and the points calculated from expression (17) for the case when $\lambda = 0\cdot50$ were pricked through transparent squared paper laid on the photograph. Small ink circles were then centred on these pinholes and the result reproduced in figure 9. It will be seen that the bubble profile agrees very well with the calculation when λ is taken as 0·50.

8. Experiments with pairs of viscous fluids

In the experiments so far described, air was used to drive out glycerine. The viscosity of air is only 1/50 000th of that of glycerine, but according to our calculations the same kind of instability would occur whatever the viscosity of the driving fluid, provided it is less than that of the fluid it is displacing. On the other hand, since we have no clue on theoretical grounds as to what determines the value of λ which will occur, it seemed possible that it might depend on the ratio of the viscosities, μ_2/μ_1. For this reason experiments were made in which a shell oil mixture of viscosity 2·75 P was used to drive coloured glycerine of viscosity 8 P through our channel. It was found that when the two viscosities are comparable the length of run required before the driving fluid penetrates into the more viscous fluid in the form of a steadily moving 'finger' was much greater than when one is air. For this reason, a new channel was built of Perspex $2\cdot54 \times 91 \times 0\cdot08$ cm. This is shown in figure 10. The camera A points vertically downwards at the channel B which is illuminated by a flash-bulb D placed under a tracing-paper screen C. The driving pressure was produced by raising a water vessel K so as to increase the air pressure in the bottle J. This raised the pressure above the fluid contained in the upstream reservoir L. The more viscous fluid was supplied from a vessel F, so as to fill the upstream reservoir E. If the pressure in E were maintained constant, as it would be if the pipe connecting E and F were left open, the velocity of the finger would increase as the length of the column of more viscous fluid decreased. To maintain a more constant velocity the cock G between E and F was closed during an experiment and the fluid in E escaped to the atmosphere through a needle valve H, the resistance of which was greater than that of the channel itself. By measuring the velocity of the head of the advancing finger before and after passing under the camera A, it was possible to obtain a good estimate of the velocity at the moment the photograph was taken.

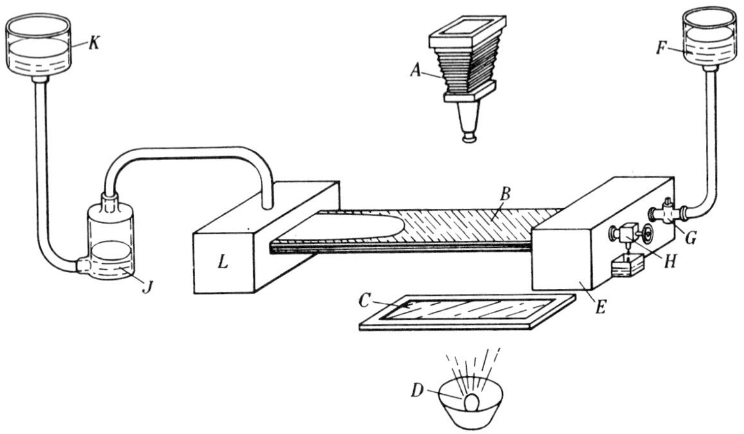

FIGURE 10. Arrangement of long Hele-Shaw cell for photographing fingers.

Figure 11, plate 3, shows a 'finger' of the oil penetrating at 1 mm/s into the tinted glycerine. It will be seen that the outline is very similar to that obtained with air and glycerine. Measurements of λ = (width of finger)/(width of channel) were made at four sections which are marked on the photograph. These gave the following results:

position	1	2	3	4
λ	0·485	0·502	0·508	0·514

It will be seen that after running a distance of 70 cm, i.e. 28 times the width of the channel, the finger has nearly, but not quite, settled down to a constant width of a little more than half the width of the channel.

Figure 12, plate 3, covers a larger part of the channel and shows water penetrating into oil in the form of a very uniform finger.

9. Effect on the shape of the bubble of surface stress at the interface

The good agreement between the photograph of the bubble shape (figure 9) and that calculated for the case when $\lambda = \frac{1}{2}$ is an indication that the boundary condition (14), which expresses the assumption that the pressure difference between the two sides of the interface is constant over its length, is justifiable at the flow speed used. When experiments were made at much lower speeds, however, it was found that the bubble was wider but that the wider forms do not conform to the corresponding profiles calculated for the same value of λ from equation (17). Figure 13, plate 3, for instance, shows water penetrating into an oil (Shell Diala) at a speed 0·226 cm/s in the channel of dimensions $0{\cdot}08 \times 2{\cdot}54 \times 91$ cm. The value of λ obtained by measuring this photograph is 0·87. The contour for $\lambda = 0{\cdot}87$ calculated using (17) is shown superposed on the photograph. The calculated contour is one along which the pressure in the fluid is constant. The fact that the observed contour is less flat than the calculated curve indicates intuitively that the pressure in the fluid increases on passing along it from the vertex. (It is hoped later to investigate quantitatively the actual pressure distribution and verify this directly.) Since the pressure inside the finger must be nearly constant along the contour, this means that the pressure difference which is supported by the interface must decrease on passing from the vertex towards the parallel portion. This pressure difference can only be attributed to surface tension or some equivalent surface stress. Under that general heading, however, various physical effects might be distinguished.

If, for instance, they are due to a constant surface tension of a fluid which wets the walls of the channel and leaves only a negligible amount of the penetrated fluid behind after the interface has passed, one would naturally look to the curvature of the meniscus in the plane midway between the parallel sheets. If, on the other hand, the interface has a finite angle of contact it might be necessary to study how that angle of contact varies when the interface moves over the plane surface. Possible differences between surface tension measured statically and that which acts over newly formed surfaces should also be studied. These things involve further experimental work, some of which is now being carried out at the Cavendish Laboratory, and it may be some time before the results are available. In the present note we propose to discuss only those which afford justification for the use of the boundary condition (14) in the theoretical discussion.

Confining attention to the case when a viscous fluid of interfacial tension T is driven slowly by a fluid of much smaller viscosity such as water or air, it is to be expected on dimensional grounds that the shape of the meniscus in a channel of given shape should be a function of $\mu U/T$ only, when the shape of the channel is fixed and the fluid wets the flat slides of the channel. It was found that neither water nor glycerine wet the Perspex which was used to construct the channel; accordingly two oils of very different viscosities, which have in static experiments nearly the same interfacial tension, were used to test whether λ is a function of $\mu U/T$ only.

The two oils used were Shell Diala ($\mu = 0{\cdot}30$ P, $\rho = 0{\cdot}875\,\mathrm{g/cm^3}$) and Shell Talpa ($\mu = 4{\cdot}5$ P, $\rho = 0{\cdot}90\,\mathrm{g/cm^3}$), and measurements were made using both water and air as the less viscous fluid. The viscosities were determined by measuring the flow through a Veridia accurate bore tube of 1·0 mm bore at 20 °C. To measure the interfacial tension between these oils and air a tube of 0·5 mm bore was dipped into the oil and the height to which the meniscus rose was observed. The surface tension of Diala was found to be between 27 and 30 dyn/cm at 20 °C, and that of Talpa slightly higher. The interfacial tension between the oils and water was measured by two methods. In the first a vertical short length of 0·5 mm tube was lowered through a thick layer of oil lying above water, till the upper end was submerged in the oil and the lower end penetrated into the water. The oil wetted the tube and a meniscus convex upwards was formed in the tube. The depth of this below the level of the oil/water interface was measured. The high viscosities of the oils and the comparatively small difference in density of the oil and water made the meniscus move slowly, but consistent results were obtained even with the more viscous oil when an hour or so was allowed to elapse between inserting the tube and measuring the height of the interface.

In this way interfacial tensions between 13·5 and 15 dyn/cm were measured for the Diala/water interface. The other method was to insert a longer piece of $\frac{1}{2}$ mm tube through the oil into the water. There were then two interfaces in the tube: an oil/air interface concave upwards and an oil/water interface downwards. The positions of these in relation to the levels of the flat oil and water surfaces outside the tube were measured. Knowing the densities of oil and water, and assuming that the oil wets the glass as it appeared to do, we obtain with this method the difference between the surface tensions at the two interfaces. Independent measurements gave the surface tension at the oil/air interface. This second method gave values for the interfacial tensions of Diala/water and Talpa/water of between 14 and 16·5.

The results of the experiments with water are shown in figure 14 in which the measured values of λ = (width of finger)/(width of channel) are plotted as ordinates and $\mu U/T$ as abscissae. It will be seen that in spite of a 15 : 1 ratio in the viscosities the points obtained with the two oils appear to fall nearly on the same curve. Some of the points were obtained by direct measurement after the finger had been formed for some time while others were obtained by measuring the photographs. It will be noticed that the direct measurement points, particularly for the larger values of $\mu U/T$, correspond with rather smaller values of λ than those derived photographically.This seems to be because at the higher values of $\mu U/T$ a small amount of the fluid is left behind after the passage of the air finger, and this fluid then flows slowly outwards towards the sides of the channel thus slightly reducing the width of the air finger.

The most interesting feature of the results exhibited in figure 14, and of similar sets of observations using air instead of water as the finger, or using oil penetrating into glycerine, is that in all cases the value of λ rapidly decreases as $\mu U/T$ increases till it reaches a value which is very close to $\frac{1}{2}$. We have never measured values of λ appreciably less than $\frac{1}{2}$.

As $\mu U/T$ increases, the effect of surface tension in determining the shape of the interface decreases relatively to that of viscous stress. The fact that λ tends rapidly to $\frac{1}{2}$ as $\mu U/T$ increases, is an indication that when the physical conditions are such that the boundary condition (14) can legitimately be used, the only one of the set of shapes described by (17) which can actually occur is that for which $\lambda = \frac{1}{2}$. The probability that this conclusion is correct is strengthened by the very close agreement between the shape of the contour shown in the photograph of figure 9 with that calculated using $\lambda = \frac{1}{2}$ in (17).

We have found no theoretical reason for this deduction from observation, but we have noticed a few purely analytical features which distinguish the particular shape corresponding with $\lambda = \frac{1}{2}$ from those corresponding with other values of λ. These have been omitted from the present paper owing to lack of any obvious physical meaning.

It is perhaps worth mentioning, in conclusion, that we have investigated theoretically the stability for infinitesimal disturbance of the shapes given by (17) for the case in which the fluid inside the bubble is of negligible viscosity. The not entirely unexpected result was found that if surface tension effects are neglected, i.e. if the velocity potential is taken as constant along the perturbed surface, then all the shapes are unstable, whatever the value of λ. The analysis gives no indication about why the shape with $\lambda = \frac{1}{2}$ is observed in practice.

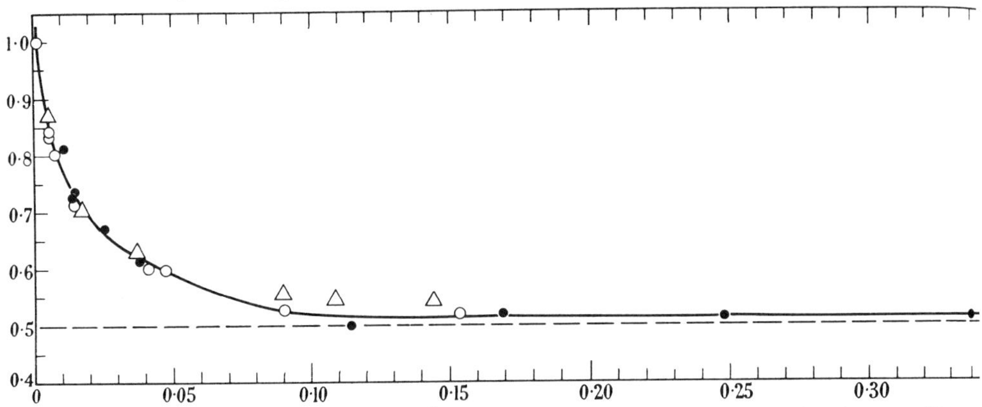

FIGURE 14. Measured values of λ for water penetrating into two oils: △, Diala (photographic measurements); ○, Diala (direct measurement); ●, Talpa (direct measurements).

APPENDIX

In this appendix we give for completeness the modifications of the analysis of § 2 which are required when the penetrating fluid (fluid 2) does not completely expel the other (fluid 1), but a tongue of fluid 2, occupying a constant fraction t of the stratum between the plates bounding the Hele-Shaw cell, advances along the gap. It is assumed that the gap is small so that the mean velocity may be calculated by assuming that the motion is everywhere parallel to the plates, and that derivatives of the velocity in directions other than normal to the plates can be neglected in comparison with those along the normal. Using the co-ordinate system introduced in § 2, the mean velocity in fluid 1 ahead of the tongue is

$$\mathbf{u}_1 = -\frac{b^2}{12\mu_1}\,\mathrm{grad}\,(p+\rho_1 gx). \tag{A 1}$$

After the passage of the tongue, the mean velocity of fluid 2 is

$$\mathbf{u}_2 = -\frac{b^2}{12\mu_2}\left(t^2+\frac{3\mu_2}{2\mu_1}(1-t^2)\right)\mathrm{grad}\,(p+\rho_2' gx), \tag{A 2}$$

where
$$\rho_2' = \rho_2\left[1+\frac{3\mu_2(1-t)^2\,(\rho_1-\rho_2)}{\rho_1\{2\mu_1 t^2+3\mu_2(1-t^2)\}}\right]. \tag{A 3}$$

The mean velocity of the film of fluid 1 left adhering to the plates is

$$\mathbf{u}_{12} = -\frac{b^2}{12\mu_1}(1-t)(1+\tfrac{1}{2}t)\,\mathrm{grad}\,(p+\rho_1 gx) - \frac{b^2 t(1-t)}{8\mu_1}\,\mathrm{grad}\,(\rho_2-\rho_1)\,gx. \quad \text{(A 4)}$$

The interface is taken as the tip of the meniscus, which advances with velocity $U = \mathbf{n}.\mathbf{u}_2$ normal to itself, where $\mathbf{n}$ is the outwards normal to the projection of the tip of the meniscus on the plates. The equation of continuity across the interface is

$$\mathbf{n}.\mathbf{u}_1 = t\mathbf{n}.\mathbf{u}_2 + (1-t)\,\mathbf{n}.\mathbf{u}_{12}. \quad \text{(A 5)}$$

If we define $\mathbf{u}_1'$ by

$$\mathbf{u}_1' = -\frac{b^2}{12\mu_1 . A}\,\mathrm{grad}\,(p+\rho_1(1+B)\,gx), \quad \text{(A 6)}$$

where

$$A = t + \frac{\mu_2(1-t)^2(1+\frac{1}{2}t)}{\mu_1 t + \frac{3}{2}\mu_2(1-t^2)}, \quad B = (1-t)^2\left\{(1+\tfrac{1}{2}t)\left(\frac{\rho_2'}{\rho_1}-1\right) - \tfrac{3}{2}t\left(\frac{\rho_2}{\rho_1}-1\right)\right\},$$

then it can be shown that

$$U = \mathbf{n}.\mathbf{u}_1',$$

at the interface.

Hence, the motion of the interface in the case in which a constant fraction t of fluid 2 penetrates fluid 1 corresponds, according to the analysis, with the motion of the interface between a fluid of viscosity μ_2 and density ρ_2', which completely expels a fluid of viscosity $A\mu_1$ and density $\rho_1(1+B)$. Thus, the kinematics of the motion are unaltered by a constant fraction of the penetrated fluid being left adhering to the plates bounding the cell.

References

Davies, R. M. & Taylor, Sir Geoffrey 1950 *Proc. Roy. Soc.* A, **200**, 375.
Dietz, D. N. 1953 *Proc. Acad. Sci. Amst.* B, **56**, 83.
Fairbrother, F. & Stubbs, A. E. 1935 *J. Chem. Soc.* p. 527.
Garabedian, P. R. 1957 *Proc. Roy. Soc.* A, **241**, 423.
Hele-Shaw, H. S. S. 1898 *Trans. Instn Nav. Archit., Lond.*, **40**, 21.
Kidder, R. E. 1956 *J. Appl. Phys.* **27**, 867.
Lamb, H. 1932 *Hydrodynamics*, 6th ed. Cambridge University Press.
Lewis, D. J. 1950 *Proc. Roy. Soc.* A, **202**, 81.
Taylor, Sir Geoffrey 1950 *Proc. Roy. Soc.* A, **201**, 192.

J. Fluid Mech. (1981), *vol.* 102, *pp.* 455–469
Printed in Great Britain

The effect of surface tension on the shape of fingers in a Hele Shaw cell

By J. W. McLEAN AND P. G. SAFFMAN

Applied Mathematics, California Institute of Technology, Pasadena, California 91125

(Received 24 June 1980)

The experimental results of Saffman & Taylor (1958) and Pitts (1980) on fingering in a Hele Shaw cell are modelled by two-dimensional potential flow with surface-tension effects included at the interface. Using free streamline techniques, the shape of the free surface is expressed as the solution of a nonlinear integro-differential equation. The equation is solved numerically and the solutions are compared with experimental results. The shapes of the profiles are very well predicted, but the dependence of finger width on surface tension is not quantitatively accurate, although the qualitative behaviour is correct. A conflict between the numerics and a formal singular perturbation analysis is noted but not resolved. The stability of the steady finger to small disturbances is also examined. Linearized stability analysis indicates that the two-dimensional fingers are not stabilized by the surface-tension effect, which disagrees with the experimental observations. A possible reason for the discrepancy between theory and experiment is suggested.

1. Introduction

It is well known that when two superposed fluids of different densities are accelerated in a direction perpendicular to the surface, the interface will be unstable if the acceleration is from the less dense to the more dense fluid. A similar instability may occur in a porous medium when a less viscous fluid drives a more viscous fluid (Saffman & Taylor 1958). This instability is responsible for water flooding of oil wells, and is of considerable importance to oil reservoir engineering (Wooding & Morel-Seytoux 1976).

In order to study this instability, experiments have been performed in a Hele Shaw cell: a channel formed by two closely spaced parallel glass plates which provides a model of two-dimensional flow through a porous medium (Saffman & Taylor 1958, Pitts 1980). It was observed that the unstable interface forms into a number of 'fingers'. As the flow develops, one of the fingers dominates and suppresses the growth of the others, and the flow reaches a steady state with a single finger propagating without change of shape.† Our interest will be in the steady-state shape of the finger.

We adopt the notation that U is the velocity of the finger, $2a$ the lateral width of the channel, b the transverse thickness or spacing ($b/a \ll 1$), T the interfacial tension between the driven (outside finger) and driving fluids, and μ is the viscosity of the driven fluid. The parameter λ will denote the ratio of the width of the finger to the

† Meng & Thomson (1978) present results of numerical simulations.

width of the channel. The adjective lateral will apply to variations in the x, y plane of the plates, whereas transverse will apply to changes across the gap in the z direction.

By neglecting surface tension effects at the free surface, Saffman & Taylor obtained a closed form solution for the shape of the finger; however, λ is undetermined by their analysis. Experiments where the surface tension is negligible ($T/\mu U \ll 1$) yield $\lambda = \frac{1}{2}$. Comparison of the observed shapes with the Saffman–Taylor solution in these experiments shows excellent agreement. As surface tension effects become important, the finger widens to fill the channel, and the agreement becomes poor.

Several attempts have been made to remove the non-uniqueness of the Saffman–Taylor result. For example, Taylor & Saffman (1958) gave a number of purely analytical features which singled out the $\lambda = \frac{1}{2}$ solution, but were unable to provide any physical basis for the arguments. Jaquard & Seguier (1962) presented an exact solution describing the growth of a small circular bubble into fingers with $\lambda = \frac{1}{2}$, but this depends on the initial shape being circular; an initial elliptical shape would give $\lambda \neq \frac{1}{2}$. To our knowledge, there is still no adequate mathematical theory for the width of fingers in a Hele Shaw cell.

A valuable contribution was recently made by Pitts (1980). He put forward the hypothesis that independent of λ the finger shapes satisfy $R \sin \hat{\theta} =$ constant, where R is the lateral radius of curvature and $\hat{\theta}$ is the angle between the tangent and the direction of advance of the finger. (It is easily verified that the Saffman–Taylor solution for $\lambda = \frac{1}{2}$ satisfies this relation.) The Saffman–Taylor experiment was repeated and the observed shapes were found to satisfy this relation closely for values of λ up to 0·77. Deviations for $\lambda = 0{\cdot}88$ were attributed to drainage effects and inadequate validity of the Hele Shaw approximations. Pitts proposed further that equating the change in pressure across the meniscus at the nose and sides due to variations of the transverse curvature with the change in pressure due to the flow leads to a unique relation between λ and $\mu Ua/Tb$.

In the experiments carried out by Pitts and most of those by Saffman & Taylor, the viscosity of the driving fluid was negligible and Pitts' relation is consistent with the observed dependence of λ on $\mu U/T$. In particular, $\lambda \to \frac{1}{2}$ as $\mu U/T$ increases and $\lambda \to 1$ as $\mu U/T$ decreases. Further experiments are, however, needed to check the a/b dependence. It should also be noted that Pitts' relation involves an unknown constant m describing the difference in transverse curvature at the nose and sides.

The object of this paper is to describe the results of an extension of the Saffman–Taylor analysis to include surface tension effects due to lateral curvature of the interface. Following the assumptions of Saffman & Taylor, variations in transverse curvature are neglected. The problem is formulated in a potential plane, and the equations for the interface are reduced to a pair of nonlinear integro-differential equations. For finite surface tension, these equations are replaced by a finite difference form and solved numerically using Newton's method. The numerical scheme converged quadratically in all cases to a unique solution for a given value of $\mu U/T$. It appears therefore that the degeneracy of the Saffman–Taylor solutions is removed and the equations have a unique solution for non-zero surface tension. The numerical profiles are compared with experimental photographs, and in all cases show excellent agreement. The qualitative dependence of λ on $\mu U/T$ is in agreement with observation; complete quantitative comparison cannot be made because the dependence of the pressure drop across the interface due to lateral curvature depends on the detailed structure of the transverse flow and is not known, but the indications are that the agreement is less than satisfactory which suggests that neglect of variations of transverse curvature is questionable. We shall return to this point in § 6.

The case of small surface tension has also been examined by conventional singular perturbation methods. The strange result is found that it is apparently possible to construct a formally consistent perturbation expansion for arbitrary finger widths, suggesting that the equations for the interface do not have a unique solution. Stringent checks of the numerical procedure were carried out and the internal consistency leads us to believe that the numerical results are correct and the perturbation expansion contains an undetected inconsistency. This conflict with the numerics has not been resolved and raises a problem of considerable mathematical interest.

The effect of the lateral curvature and surface tension on the stability of the steady motion to infinitesimal disturbances has also been examined. The analysis based on the Hele Shaw equations with surface tension neglected predicts that the fingers are unstable, a result which disagrees completely with observation (Taylor & Saffman 1958). The inclusion of surface tension does not resolve this discrepancy, which suggests that the Hele Shaw equations with the present surface-tension modification are not appropriate for unsteady motion.

2. Derivation of the equations

The slow flow of an incompressible fluid in a homogeneous porous medium is governed by Darcy's law and the equation of continuity:

$$\left.\begin{aligned} \mathbf{u} &= -\frac{k}{\mu}\nabla p = \nabla\phi, \\ \nabla.\mathbf{u} &= \nabla^2\phi = 0, \end{aligned}\right\} \tag{1}$$

where $\mathbf{u}$ is the average velocity vector, p the pressure, k is the permeability of the medium, and ϕ is the velocity potential.

Two-dimensional flow in a porous medium may be studied experimentally by means of an analogue developed by Hele Shaw (1898). The motion of an incompressible viscous fluid between two closely spaced parallel plates is such that the components of the mean velocity (averaged across the gap) satisfy:

$$\mathbf{u} = -\frac{b^2}{12\mu}\nabla p = \nabla\phi \tag{2}$$

(see Lamb 1932, § 330). Comparison with (1) shows that the motion in a Hele-Shaw cell is equivalent to two-dimensional flow in a porous medium of permeability $b^2/12$.

We now derive the equations for steady two-dimensional flow produced by a finger along a channel with surface tension effects due to lateral curvature included. It is assumed that the viscosity of the driving fluid is negligible, buoyancy forces are unimportant, and that the driven fluid is completely expelled from the gap by the advancing finger. Then the boundary conditions on the (harmonic) velocity potential ϕ and pressure p in the driven fluid are

$$\frac{\partial\phi}{\partial y} = 0 \quad \text{on} \quad y = \pm a, \tag{3}$$

$$\frac{\partial \phi}{\partial n} = U \sin \hat{\theta}, \quad p_0 - p = \frac{T}{R} + \frac{2T}{b} \cos \chi \quad \text{on the finger,} \tag{4}$$

$$\phi \sim 0 \quad \text{as} \quad x \to -\infty, \quad \lambda a < |y| < a, \tag{5}$$

$$\phi \sim U\lambda x \quad \text{as} \quad x \to \infty, \quad -a < y < a, \tag{6}$$

where χ is the contact angle of the meniscus, p_0 the constant pressure inside the finger, and $p = -12\mu\phi/b^2$. The harmonic conjugate of ϕ defines a stream function ψ. Following Saffman & Taylor, we assume further that $\chi = 0$, or at least is constant so that the variations of ϕ or p along the free boundary are proportional to the lateral curvature. Some of these assumptions can be questioned and some are in fact not valid for the actual experiments. These matters will be discussed later and we shall point out (§ 6) that the problem with more general assumptions about the flow in the gap can be formulated in this way, with appropriate changes in the values of μ and T, provided the finger occupies a constant fraction t of the stratum between the plates.

It is now convenient to work in a frame of reference moving with the finger with origin at the nose and introduce dimensionless variables (with a and $(1-\lambda)U$ as the units of length and velocity)

$$\left.\begin{aligned} \hat{x} = (x - Ut)/a, \quad \hat{y} = y/a, \quad \hat{R} = R/a, \\ \hat{\phi} = \frac{\phi - Ux}{(1-\lambda)Ua}, \quad \hat{\psi} = \frac{\psi - Uy}{(1-\lambda)Ua}. \end{aligned}\right\} \tag{7}$$

The free-boundary potential problem is illustrated in figure 1. Symmetry about the centre-line of the channel has been assumed. The centre-line and the interface is a streamline ($\hat{\psi} = 0$), as is also the wall $\hat{y} = 1 (\hat{\psi} = -1)$.

The flow region of figure 1 maps into the infinite strip in the potential plane $-\infty < \hat{\phi} < \infty$, $-1 < \hat{\psi} < 0$, and a natural formulation is to use Kirchhoff–Helmholtz free streamline theory to study the analytic function $-\log \hat{q} + i\hat{\theta}$, where $\hat{u} - i\hat{v} = \hat{q}e^{-i\hat{\theta}}$ is the (dimensionless) complex velocity relative to the finger, in terms of the complex potential $\hat{w} = \hat{\phi} + i\hat{\psi}$. Note that if $\hat{S}$ measures (dimensionless) arclength along the interface from the nose,

$$\frac{1}{\hat{R}} = \frac{d\hat{\theta}}{d\hat{S}} = \frac{\partial \hat{\theta}}{\partial \hat{\phi}} \frac{d\hat{\phi}}{d\hat{S}} = \hat{q} \frac{\partial \hat{\theta}}{\partial \hat{\phi}} \tag{8}$$

so the radius of curvature takes a simple form.

The conformal map

$$\sigma = s + it = e^{-(\hat{w} - \hat{\phi}_0)\pi} \tag{9}$$

maps the potential plane into the upper half σ-plane, with the upper wall DE going into $-\infty < s < 0$, the interface AB into $0 < s < 1$, and the centreline BC into $1 < s < \infty$. On BC and DE, $\hat{\theta} = \pi$. From a standard application of Cauchy's integral theorem, the speed $\hat{q}$ on the interface AB is related to the direction $\hat{\theta}$ by the Cauchy principal-value integral

$$\log(\hat{q}(s)) = -\frac{1}{\pi} P \int_0^1 \frac{\hat{\theta}(s') - \pi}{s' - s} ds'. \tag{10}$$

The constants are fixed by the requirement that $\hat{q} \sim 1$ as $\hat{\phi} \to -\infty$.

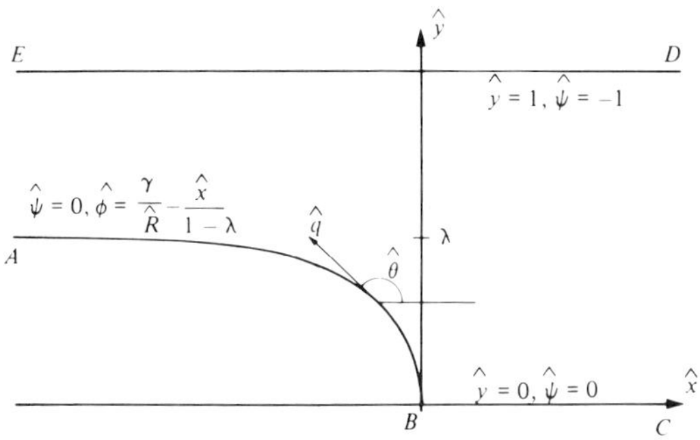

FIGURE 1. The free boundary potential problem in dimensionless variables. $\gamma = Tb^2/12\mu Ua^2(1-\lambda)$.

This relation for the unknowns $\hat{q}(s)$ and $\hat{\theta}(s)$ comes solely from kinematic considerations. The dynamics are introduced through (4) and (8), which gives

$$\frac{Tb^2}{12\mu Ua^2}\frac{\hat{q}}{1-\lambda}\frac{\partial\hat{\theta}}{\partial\hat{\phi}}-\frac{\hat{x}}{1-\lambda}=\hat{\phi}, \quad \hat{\phi}_0<\hat{\phi}<\infty. \tag{11}$$

Using s as the independent variable and the relations

$$\cos\hat{\theta}=\frac{d\hat{x}}{d\hat{S}}, \quad \frac{d\hat{\phi}}{d\hat{S}}=\hat{q}, \quad \frac{d\hat{\phi}}{ds}=-\frac{1}{\pi s}, \tag{12}$$

we differentiate (11) along the interface to obtain the second relation between $\hat{q}$ and $\hat{\theta}$,

$$\frac{Tb^2\pi^2}{12\mu Ua^2}\hat{q}s\frac{d}{ds}\left(\hat{q}s\frac{d\hat{\theta}}{ds}\right)-(1-\lambda)\hat{q}=\cos\hat{\theta}, \quad 0<s<1. \tag{13}$$

The appropriate boundary conditions for (10) and (13) are

$$\left.\begin{aligned}\hat{\theta}(0)&=\pi, & \hat{q}(0)&=1/(1-\lambda),\\ \hat{\theta}(1)&=\pi/2, & \hat{q}(1)&=0,\end{aligned}\right\} \tag{14}$$

corresponding to uniform flow at the tail A ($s=0$) and a stagnation point at the nose B ($s=1$).

Our task is to solve equations (10) and (13) subject to the boundary conditions (14) for given values of U, μ, T, a, b, and determine if λ is then uniquely determined. Since from (10) and (14)

$$\log(1-\lambda)=\frac{1}{\pi}\int_0^1\frac{\hat{\theta}-\pi}{s'}ds', \tag{15}$$

we can rewrite the equations in a form which does not explicitly involve λ. Introducing

$$\theta=\hat{\theta}-\pi, \quad q=(1-\lambda)\hat{q} \tag{16}$$

$$\kappa=\frac{Tb^2\pi^2}{12\mu Ua^2(1-\lambda)^2} \tag{17}$$

we have

$$\log q(s)=-\frac{s}{\pi}P\int_0^1\frac{\theta(s')}{s'(s'-s)}ds' \tag{18}$$

$$\kappa qs \frac{d}{ds}\left(qs\frac{d\theta}{ds}\right) - q = -\cos\theta, \tag{19}$$

with

$$\theta(0) = 0, \quad q(0) = 1, \quad \theta(1) = -\pi/2, \quad q(1) = 0. \tag{20}$$

If these equations have a unique solution for given κ, then the width of the finger is uniquely determined by (15) to give λ as a function of $Tb^2/\mu Ua^2$. Note that this result is qualitatively different from that of Pitts, whose argument gives λ as a function of $Tb/\mu Ua$. Further experiments with channels of varying gap should distinguish between the theories.

The shape of the interface follows from

$$\frac{d\hat{z}}{d\hat{w}} = \frac{e^{i\hat{\theta}}}{\hat{q}} \tag{21}$$

and is given by

$$\hat{x}(s) + i\hat{y}(s) = -\frac{(1-\lambda)}{\pi}\int_s^1 \frac{e^{i\theta}}{sq}\,ds. \tag{22}$$

The value of λ can also be obtained by evaluating $\hat{y}(0)$, and provides a check on the accuracy of the computations.

In these variables, the Saffman–Taylor solution is

$$q_0 = \left(\frac{1-s}{1+s\alpha}\right)^{\frac{1}{2}}, \quad \theta_0 = \cos^{-1} q_0, \tag{23}$$

where $\alpha = (2\lambda-1)/(1-\lambda)^2$ is arbitrary. This solution can be shown to satisfy (18) for $\kappa = 0$ (McLean 1980).

3. Endpoint singularity

Before proceeding with the numerical computations, it is necessary to analyse the possible singularities of the solution. For $\kappa = 0$, the solution has square-root singularities at both endpoints. For non-zero surface tension we seek a solution about $s = 0$ of the form

$$\left.\begin{aligned} \theta(s) &= a_1 s^\tau + a_2 s^{2\tau} + \dots, \\ q(s) &= 1 + b_1 s^\tau + b_2 s^{2\tau} + \dots \end{aligned}\right\} \tag{24}$$

This expansion is substituted into (18). Using the result

$$P\int_0^1 \frac{x^\tau}{x-y}\,dx = \frac{1}{\tau} + \frac{y}{\tau-1} - \pi y^\tau \cot\pi\tau + O(y^2)$$

for $0 < \tau < 1$, we equate the coefficients of the s^τ term, giving

$$b_1 = a_1 \cot\pi\tau.$$

From (19) we obtain

$$b_1 = \kappa\tau^2 a_1.$$

Eliminating b_1 and requiring that $a_1 \neq 0$, we find that the exponent τ is given by

$$\frac{1}{\tau^2}\cot\pi\tau = \kappa, \tag{25}$$

which has a root with $0 < \tau < 1$. The coefficients b_1, b_2, and a_2 can be found recursively as functions of a_1. As $\kappa \to 0$, $\tau \to \frac{1}{2}$ which agrees with the Saffman–Taylor solution (23).

It can be shown (McLean 1980) that terms linear in s in the expansion generate singularities from (18) of the form $s \log^n s$ which cannot satisfy the equation and so are not present. Since the cotangent is periodic, (25) has multiple solutions $\{\tau_n\}$. A proper expansion must account for these extra terms, and the full local expansion is of the form

$$\theta(s) = {}_1a_1 s^{\tau_1} + {}_1a_2 s^{2\tau_1} + \ldots + {}_2a_1 s^{\tau_2} + {}_2a_2 s^{2\tau_2} + \ldots . \tag{26}$$

At the nose ($s = 1$) an expansion in powers of $(1-s)^{\frac{1}{2}}$ can be shown to be consistent with the equations as required by symmetry. But in this case, the coefficients cannot be found recursively, but are given globally as integrals of θ over the entire interval. For details, see McLean (1980).

4. Numerical treatment

We change the independent variable so that q and θ will be differentiable at the endpoints. If

$$\zeta = (1-s^{\tau})^{\frac{1}{2}}, \quad s^{\tau} = 1-\zeta^2 \tag{27}$$

the variables have at least two derivatives with respect to ζ at the end points. To evaluate the principal value integral, the singular part of the integrand is subtracted off and (18) is written

$$\log q = \frac{1}{\pi}\left\{\int_0^1 \frac{\theta(s')}{s'}\,ds' - \int_0^1 \frac{\theta(s')-\theta(s)}{s'-s}\,ds' - \theta(s)\log\left(\frac{1-s}{s}\right)\right\}.$$

The system is written in terms of ζ with derivatives replaced by a three point centred difference quotient and the integral evaluated using the trapezoidal rule. The interval $[0, 1]$ is broken into N equal segments of length $h = 1/N$. Newton's method is used to solve the finite difference equations for θ at the $N-1$ interior mesh points. The numerical scheme was coded in vectorized FORTRAN and computed on the CDC CYBER 203 in Minneapolis. The Newton method is said to have converged when the maximum residual is less than 10^{-10} (generally 4–5 iterations, taking about 0·2 seconds CPU time for $N = 50$). After a converged solution is obtained, the Cartesian coordinates are given by evaluation of equation (22).

Newton's method is particularly valuable for this problem because of the singularity of the system when $\kappa = 0$. If the solutions are not unique for $\kappa > 0$, then the Jacobian of the Newton iteration should show singular behaviour. For the first attempt to find a solution for $\kappa \neq 0$, we used the Saffman–Taylor solution (23) for $\lambda = \frac{1}{2}$ as the first guess and took $\kappa = 0{\cdot}237$ ($\tau = 0{\cdot}48$). The Newton iteration converged quadratically with no sign of singular behaviour. Continuation was employed to construct a unique family of solutions for values of κ from 0·064 to 4×10^7 and the shapes were computed.

The numerical profiles are compared with the experimental results of Pitts (1980) in figures 2(*a*), 2(*b*), and 2(*c*), for values of κ chosen so that the calculated λ is equal to the observed λ. The agreement in profile shape is remarkably good.

Numerous tests were performed to check the numerical procedure. The number of mesh points was varied between 25 and 200. Examination of the results indicated that the error was $O(h^2)$ as expected from the trapezoidal rule. The results were independent

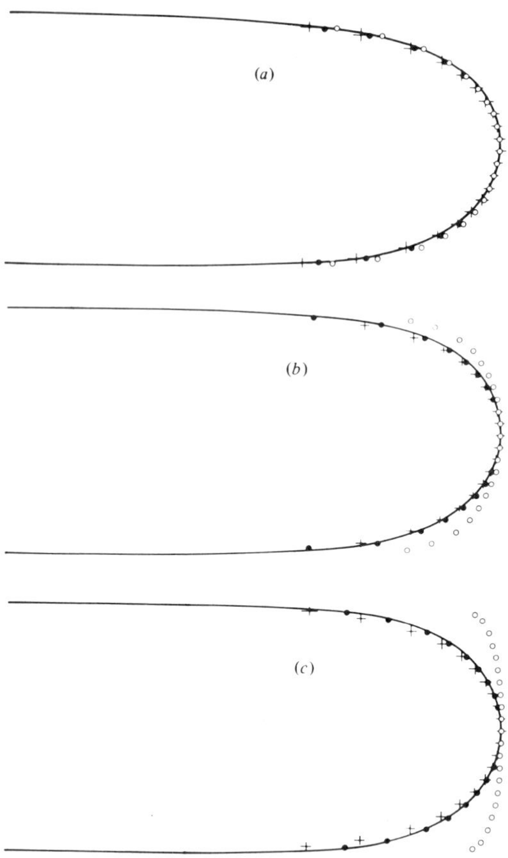

FIGURE 2. Comparison of theory and experiment. (*a*) $\lambda = 0{\cdot}54$, (*b*) $\lambda = 0{\cdot}68$, (*c*) $\lambda = 0{\cdot}88$. Solid line, experiment (Pitts 1980); ○, Saffman & Taylor solution; +, Pitts theory; ●, numerical results.

of the initial guess, provided this was not so bad that the iteration diverged. The convergence was always quadratic, provided κ was neither too small nor too large. The Jacobian determinant is shown in figure 3 for $N = 50$. The behaviour is consistent with isolated solutions for $\kappa > 0$ and singular behaviour when $\kappa = 0$.

Doubts raised by a conflict of these results with a perturbation expansion (§ 5) suggested some additional tests; e.g. fixing the slope at $s = 0$ (i.e. giving a_1 in (24)) and solving (18) and (19) on a reduced interval, or fixing λ and solving an inflated system based on (15), (18), and (19). Both these tests confirm the previous calculations.

In short, all indications of the numerical investigation are that the system defined by (18), (19), and (20) has a unique solution for $\kappa > 0$. Values of λ for various κ are listed in table 1 and plotted in figure 4. Also in table 1 are the corresponding values of $\mu U/T$ for the Saffman–Taylor and Pitts value of $a/b = 15{\cdot}875$. The comparison of calculated with observed values will be discussed later.

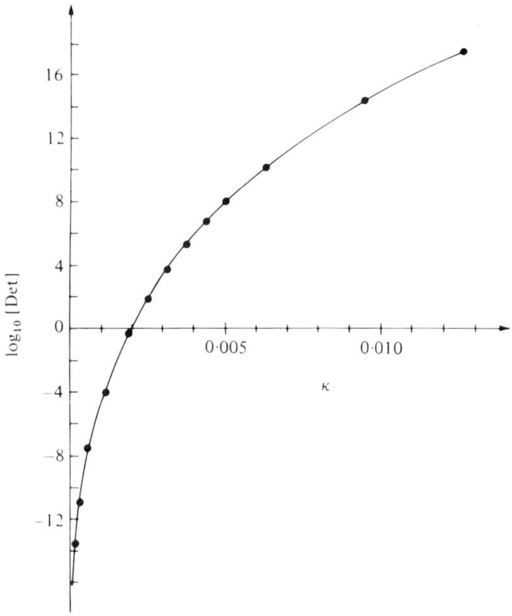

FIGURE 3. Determinant of the Jacobian versus the surface tension parameter, indicating singular behaviour as $\kappa \to 0$.

5. Perturbation expansions

For $\kappa \ll 1$, the system would appear suitable for investigation by standard singular perturbation methods. A formal perturbation expansion was constructed by the method outlined here. Since the behaviour of θ near $s = 0$ is non-uniform in κ, we construct an 'inner–outer' matched asymptotic expansion. The outer expansion is taken to be

$$\left.\begin{aligned} \theta(s;\kappa) &= \theta_0(s) + \kappa\theta_1(s) + \kappa^2\theta_2(s) + \dots, \\ q(s;\kappa) &= q_0(s) + \kappa q_1(s) + \kappa^2 q_2(s) + \dots, \end{aligned}\right\} \tag{28}$$

where the leading term (q_0, θ_0) is the Saffman–Taylor solution (23). Substituting the expansion (28) in the equations (18) and (19) and matching terms linear in κ, $\theta_1(s)$ is found to satisfy a singular linear integral equation

$$\frac{s}{\pi} P\int_0^1 \frac{\theta_1(s')}{s'(s'-s)}\,ds' = \theta_1 \tan\theta_0 - s\frac{d}{ds}\left(q_0 s\frac{d\theta_0}{ds}\right). \tag{29}$$

The solution of (29) can be calculated explicitly (Carrier, Krook & Pearson 1966) and is found to be

$$\theta_1(s) = \frac{(1+\alpha)^{\frac{1}{2}}}{\pi}\frac{(s(1-s))^{\frac{1}{2}}}{1+s\alpha}\left\{A + \frac{\pi(1+\alpha)^{\frac{1}{2}}}{4}\frac{s^{\frac{1}{2}}(1-2s\alpha)}{(1+s\alpha)^{\frac{5}{2}}} + \frac{1}{2(1+s\alpha)^2}\right.$$
$$\left.\left[\frac{1+2\alpha-s\alpha(\alpha+2)}{\alpha+1} - \frac{(1-s)^{\frac{1}{2}}(2\alpha s-1)}{2(1+\alpha s)^{\frac{1}{2}}}\log\frac{s(1+\alpha)}{2\{(1-s)(1+\alpha s)\}^{\frac{1}{2}} + (\alpha-1)s + 2}\right]\right\}, \tag{30}$$

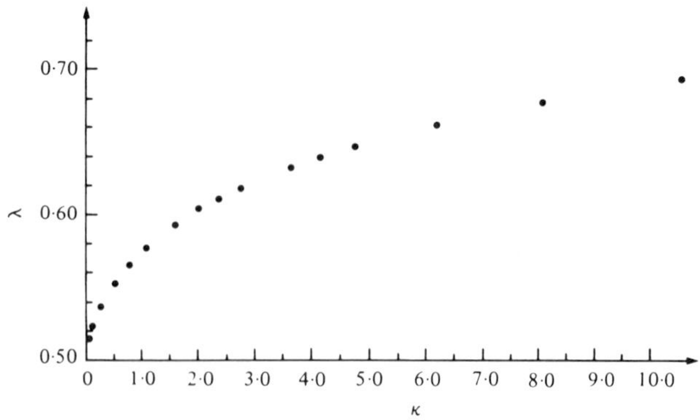

FIGURE 4. The finger width λ as a function of the surface-tension parameter κ.

τ	κ	$\mu U/T$	λ
0·495	0·069	0·2168	0·515
0·49	0·131	0·1098	0·524
0·48	0·273	0·0557	0·537
0·46	0·597	0·0279	0·557
0·43	1·21	0·0154	0·582
0·40	2·03	0·0102	0·604
0·35	3·09	0·0082	0·640
0·30	8·07	0·0039	0·678
0·25	16·0	0·0026	0·720
0·20	34·4	0·0017	0·767
0·16	71·1	0·0013	0·809
0·10	308	0·0008	0·881
0·05	2 526	0·0005	0·947
0·02	39 740	0·0003	0·984

TABLE 1

where the constant A is an arbitrary parameter. Each succeeding term in the expansion (28) satisfies on equation of the form (29) with a different inhomogeneous term. This equation can be solved provided the inhomogeneous term vanishes at $s = 0$, which was always found to be the case.

The outer expansion (28) is expanded about $s = 0$, and the inner expansion (24) is expanded in powers of κ. The coefficient a_1 is chosen to match the $s^{\frac{1}{2}}$ terms. The remaining terms are then compared to check consistency. We have verified that the $s^{\frac{1}{2}} \log s$ terms match at $O(\kappa)$, the $s^{\frac{1}{2}} \log^2 s$ terms match at $O(\kappa^2)$, the s terms match at $O(\kappa)$, and the $s^{n+\frac{1}{2}}$ terms match to all orders. To the order considered, α and A, and hence λ remain undetermined. For full details, see McLean (1980).

The perturbation analysis suggests that the system defined by equations (18), (19), and (20) does not have a unique solution, a conclusion in conflict with the numerical results. It is possible that the perturbation expansion is inconsistent at higher order or that transcendentally small terms are present, but we detected no reason why this should happen. The fact that the difficulty is subtle is suggested by the finding that the choice of A and α, to give a width in agreement with the numerics, leads to values of $\theta_0 + \kappa\theta_1$ which agree with the computed values of θ for small κ to within the accuracy of the computations. Despite considerable effort, the apparent failure of the perturbation expansion to fix a unique λ remains unexplained.

An attempt was also made to construct a perturbation expansion for $\kappa \gg 1$, in order to find the limiting solution as $\kappa \to \infty$ or determine if there is an upper bound on κ for solutions of the assumed type to exist. We were unable to construct a consistent expansion or answer this question. Numerical solutions were found for $\kappa = O(10^7)$; the values of λ were about 0·99 and the profiles were smooth to the eye. For larger values the numerics became unreliable due to the appearance of large gradients of q and θ at the endpoints.

6. The dependence of finger widths on surface tension

The equations that we have studied are based on the boundary conditions (4), which in turn are based on the hypothesis that the driving fluid completely expels the driven fluid. In reality, it is expected that a thin film of the driven fluid will be left adhering to the plates and that the driving fluid penetrates the gap in a tongue of thickness bt, where t can be expected to be a function of $\mu U/T$. For the case of a viscous fluid expelled from a tube by a finger of less viscous fluid advancing with speed U, it was found experimentally that (Fairbrother & Stubbs 1935; Taylor 1961)

$$t = 1-(\mu U/T)^{\frac{1}{2}}, \quad 0 < \mu U/T < 0{\cdot}09. \tag{31}$$

A similar dependence can be expected in a Hele-Shaw experiment. Now it was shown by Saffman & Taylor that the motion when the advancing finger fills a constant fraction t of the stratum is similar to that when the driven fluid is assumed completely expelled and has viscosity $A\mu$, where

$$A = t + \frac{\mu_2(1-t)^2(1+\frac{1}{2}t)}{\mu t + \frac{3}{2}\mu_2(1-t^2)} \tag{32}$$

and μ_2 is the viscosity of the driving fluid.

The pressure drop across the interface depends on the details of the flow near the nose of the tongue, and cannot apparently be found without a full analysis of the three-dimensional flow about the advancing tongue. All we can safely conclude, essentially from dimensional analysis, is that there should be a relation of the form

$$\Delta p = \frac{T}{b}f(t) + \frac{T}{R}g(t), \tag{33}$$

where f and g are unknown functions of the tongue width t which in turn is some function of $\mu U/T$. However, if we assume that t is constant, then the problem that we have solved is equivalent to the real one with

$$\kappa = \frac{g}{A}\frac{Tb^2\pi^2}{12\mu U a^2(1-\lambda)^2}. \tag{34}$$

In other words, finite tongue thickness is just equivalent to increasing the surface tension by a factor g/A, provided it is constant.

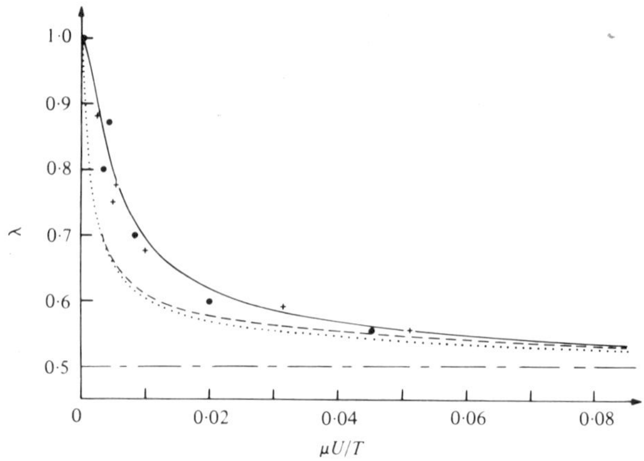

FIGURE 5. Finger width λ versus $\mu U/T$. +, Pitts' experimental results; ●, Saffman & Taylor experimental results; ——, Pitts' relation; ····, numerical results; ---, numerics adjusted for finite tongue thickness.

In figure 5, we plot the relation between λ and $\mu U/T$. The upper solid line is the empirical relation of Pitts (1980), which agrees well with both Pitts' and Saffman & Taylor's experimental observations.† The current results are indicated by the lower dotted line. The uncorrected theoretical values ($g/A = 1$) differ significantly from the experimental values. Corrected values, shown by the dashed line, were found as follows. From figure 4 or table 1, we have λ *vs.* κ. These were substituted into (34) with $g = 1$ and $A = t$ (the appropriate value for $\mu_2/\mu \ll 1$) and combined with (31) to give $\mu U/T$ as a function of λ. The improvement is slight. Thus, the quantitative dependence of λ on $\mu U/T$ is unsatisfactory, unless g is greater than 1, or t is less than the value suggested by (31) which is for experiments in tubes.

An alternative possibility is that $t \neq$ constant. Indeed, the U in (31) should strictly have been the normal velocity of the interface, which is not constant. Then the variations in $f(t)$ would control the shape. This, of course, is essentially Pitt's suggestion. The theoretical problem now becomes far more difficult, but it is hoped to study this possibility in further work. As mentioned earlier, experiments on the dependence of λ on $\mu U/T$ for various a/b would help establish the relative importance of the two terms in (33).

7. Stability analysis

Taylor & Saffman (1958) examined the stability of the fingers neglecting surface tension. Here, the lateral surface tension effects are included. Also, we suppose that the tongue completely fills the gap. In this section, t denotes time, and we use dimensional variables with ϕ and ψ denoting the velocity potential and streamfunction in coordinates moving with the finger.

† We are grateful to Dr Pitts for pointing out that the data of Saffman & Taylor were incorrectly plotted in figure 14 of the original paper and that the correct figure is to be found in vol. IV of Taylor's collected papers, p. 328.

In the physical plane, the steady finger is bounded by the streamline $\psi = 0$. The motion in oscillations about the steady shape is described in the potential plane by

$$x = x_0(\phi, \psi) + \epsilon\xi(\phi, \psi, t),$$
$$y = y_0(\phi, \psi) + \epsilon\eta(\phi, \psi, t).$$

The interface is given by an equation

$$G(\phi, \psi, t) \equiv \psi - \epsilon F(\phi, t) = 0,$$

where ξ, η, and F are the perturbations about the steady motion. The free surface $G = 0$ is a material surface moving with the fluid, so the kinematic condition is:

$$\frac{D}{Dt}(\psi - \epsilon F) = 0 \quad \text{on} \quad G = 0,$$

where D/Dt is the convective derivative. The stability analysis will be carried out in the potential plane, so the kinematic condition must be written with (ϕ, ψ) as independent variables (see Longuet-Higgins 1978). Retaining terms linear in the perturbations, the boundary condition becomes

$$F_t + q^2 F_\phi + q\cos\hat{\theta}\eta_t - q\sin\hat{\theta}\xi_t = 0 \tag{35}$$

on the undisturbed finger $\psi = 0$, $\phi_0 < \phi < \infty$.

The second condition at the interface is the pressure boundary condition. The lateral curvature has the parametric representation

$$\frac{1}{R} = \left|\frac{G_{xx}G_y^2 - 2G_x G_y G_{xy} + G_{yy}G_x^2}{(G_x^2 + G_y^2)^{\frac{3}{2}}}\right|.$$

In the potential plane, this becomes, to $O(\epsilon)$,

$$\frac{1}{R} = q\hat{\theta}_\phi + \epsilon(qF_{\phi\phi} - q_\phi F_\phi).$$

Retaining terms linear in the perturbation, the dynamic boundary condition to be satisfied on $\psi = 0$ becomes

$$\frac{Tb^2}{12\mu U}\{F(q\hat{\theta}_\phi^2 - q_{\phi\phi} + q_\phi^2/q) + qF_{\phi\phi} - q_\phi F_\phi\} - \xi + F\frac{\sin\hat{\theta}}{q} + g(t) = 0, \tag{36}$$

where $g(t)$ is an arbitrary function of time.

The entire potential plane $-Ua(1-\lambda) < \psi < Ua(1-\lambda)$ is mapped to a semi-infinite strip $-\pi/2 < u < \pi/2$, $v > 0$ through the transformations

$$w = \phi + i\psi = -\frac{Ua(1-\lambda)}{\pi}\log(1 - Z^2) + \phi_0,$$

$$Z = \sin\Omega = \sin(u + iv),$$

where the u axis corresponds to all the interface, both above and below the centre-line.

The perturbation $\xi + i\eta$ must be an analytic function of Ω. For a symmetric perturbation, we choose the representation

$$\xi + i\eta = e^{\sigma t}\sum_0^\infty a_n e^{2ni\Omega}, \quad W = F/q = e^{\sigma t}\sum_0^\infty b_n \cos(2n-1)u. \tag{37}$$

The real coefficients $\{a_n, b_n\}$ are chosen to satisfy (35) and (36) on $v = 0$, and the condition that this can be done determines the eigenvalue σ.

For no surface tension ($T = 0$), the equations can be solved in closed form. Using (37) in (35) and (36), and equating Fourier coefficients, one can show that the series is finite if σ is one of the values

$$\sigma_n = 2\pi n U/a, \quad n = 0, 1, 2, \dots . \tag{38}$$

The corresponding eigenfunction includes the first $n+1$ terms of (37). In particular, $n = 0$ corresponds to a translation of the finger, while $n = 1$ is the perturbation

$$\xi + i\eta = e^{2\pi t U/a}(1 + \tfrac{1}{3} e^{2i\Omega}).$$

This result was first obtained by Taylor & Saffman (1958). Since the eigenvalues σ_n are positive, the analysis suggests that the fingers are unstable, a conclusion in complete disagreement with the experimental observations, which show not the slightest trace of instability.

To follow these eigenvalues to finite surface tension, equations (35) and (36) are written using ζ as an independent variable. The series (37) is truncated at a finite number of terms N. The coefficients $\{a_n, b_n\}$ are chosen to satisfy (35) and (36) at N interior mesh points, yielding the linear system

$$(A_N - \sigma B_N)\begin{pmatrix} a_n \\ b_n \end{pmatrix} = 0.$$

A_N and B_N are square matrices of order $2N \times 2N$. Newton's method is used to compute the eigenvalues for $T > 0$. The truncation N is varied to gain confidence in the solution. Using $N = \{5, 6, 8, 10, 12, 15\}$, the lowest eigenvalue did not vary more than 1 %. The lowest eigenvalue was computed for $0 < \kappa < 1$. It remains real and positive, decreasing by about 10 % from its value of $2\pi U/a$ for $\kappa = 0$, indicating the finger is unstable and that surface tension associated with lateral curvature does not stabilize the finger. This is a further indication that variations in the gap may be crucial. For full details of the stability analysis, again see McLean (1980).

8. Other steady solutions

Although the inclusion of surface tension in the manner considered here has apparently removed the non-uniqueness of the solution (apart from the unresolved conflict between the numerics and the perturbation analysis), it is worth noting that there still exist a continuum of steady solutions for given U and T if the interface stretches entirely across the channel. For suppose in channel fixed co-ordinates the interface has the equation

$$f(x, y, t) \equiv Ut + \eta(y) - x = 0, \quad -a \leqslant y \leqslant a, \tag{39}$$

and let us again suppose the tongue thickness is constant so that we may assume the driving fluid completely expels the driven. Then Darcy's law and the kinematic and dynamic boundary conditions are satisfied by $\phi_x = U$ in each fluid if

$$(\mu - \mu_2)\, U\eta + \frac{Tb^2}{12}\frac{\eta_{yy}}{(1+\eta_y^2)^{\frac{3}{2}}} = fn\,(t).$$

Thus any $\eta(y)$ which satisfies the equation

$$\eta_{yy} + \frac{12U}{Tb^2}(\mu - \mu_2)(1+\eta_y^2)^{\frac{3}{2}}\eta = 0, \quad -a \leqslant y \leqslant a \tag{40}$$

is a possible interface, provided only that $|\eta| < \infty$. The curious feature here is that the degeneracy vanishes when $T = 0$, as then $\eta = 0$ is the only solution.

This work was supported by the Department of Energy (Office of Basic Energy Sciences). We also wish to thank Control Data Corporation for the granting of time on their CYBER 203 computer at the C.D.C. Service Center, Arden Hills, Minnesota, which enabled us to implement Newton's method for large matrices.

REFERENCES

CARRIER, G. F., KROOK, M. & PEARSON, C. E. 1966 *Functions of a Complex Variable*. McGraw-Hill.

FAIRBROTHER, F. & STUBBS, A. E. 1935 The 'bubble tube' method of measurement. *J. Chem. Soc.* **1**, 527–530.

HELE-SHAW, H. J. S. 1898 On the motion of a viscous fluid between two parallel plates. *Nature* **58**, 34–36.

JACQUARD, P. & SÉGUIER, P. 1962 Mouvement de deux fluides en contact dans un milieu poreux. *J. Méc.* **1**, 367–394.

LAMB, H. 1932 *Hydrodynamics*, 6th edn. Cambridge University Press.

LONGUET-HIGGINS, M. S. 1978 The instabilities of gravity waves of finite amplitude in deep water. *Proc. Roy. Soc.* A **360**, 471–488.

MCLEAN, J. W. 1980 The fingering problem in flow through porous media. Ph.D. dissertation. California Institute of Technology.

MENG, J. C. S. & THOMSON, J. A. L. 1978 Numerical studies of some nonlinear hydrodynamic problems by discrete vortex element methods. *J. Fluid Mech.* **84**, 433–453.

PITTS, E. 1980 Penetration of fluid into a Hele Shaw cell. *J. Fluid Mech.* **97**, 53–64.

SAFFMAN, P. G. & TAYLOR, G. I. 1958 The penetration of fluid into a porous medium or Hele-Shaw cell. *Proc. Roy. Soc.* A **245**, 312–329.

TAYLOR, G. I. 1961 Deposition of a viscous fluid on the wall of a tube. *J. Fluid Mech.* **10**, 161–165.

TAYLOR, G. I. & SAFFMAN, P. G. 1958 Cavity flows of viscous fluids in narrow spaces. *2nd Symp. on Naval Hydrodynamics*.

WOODING, R. A. & MOREL-SEYTOUX, H. J. 1976 Multiphase fluid flow through porous media. *Ann. Rev. Fluid Mech.* **8**, 233–274.

Fingers in a Hele–Shaw Cell with surface tension

Jean-Marc Vanden-Broeck
Department of Mathematics and Mathematics Research Center, University of Wisconsin-Madison, Madison, Wisconsin 53705

(Received 4 April 1983; accepted 17 May 1983)

McLean and Saffman's model for the fingering in a Hele–Shaw cell is solved numerically. The results suggest that a countably infinite number of solutions exist for nonzero surface tension. This set of solutions contains the solution previously obtained by McLean and Saffman.

In a recent paper McLean and Saffman[1] modeled the fingering in a Hele–Shaw Cell by a two-dimensional potential flow with surface tension included at the interface (see Fig. 1). They introduced the notation that U is the velocity of the finger, $2a$ the lateral width of the channel, b the transverse thickness, T the surface tension, and μ the velocity. In addition they denoted the ratio of the width of the finger to the width of the channel by a parameter λ.

Taking a as the unit length and $(1-\lambda)U$ as the unit velocity, they derived the following nonlinear integrodifferential equation for the unknown shape of the free surface:

$$\log q(S) = -\frac{S}{\pi}\int_0^1 \frac{\theta(S')}{S'(S'-S)}\,dS', \tag{1}$$

$$\kappa q S \frac{d}{dS}\left(q S \frac{d\theta}{dS}\right) - q = -\cos\theta, \tag{2}$$

$$\theta(0) = 0, \quad q(0) = 1, \tag{3}$$

$$\theta(1) = -\pi/2, \quad q(1) = 0. \tag{4}$$

Here

$$\theta = \hat{\theta} - \pi, \tag{5}$$

$$q = (1-\lambda)\hat{q}, \tag{6}$$

$$\kappa = Tb^2\pi^2/12\mu\, Ua^2(1-\lambda)^2. \tag{7}$$

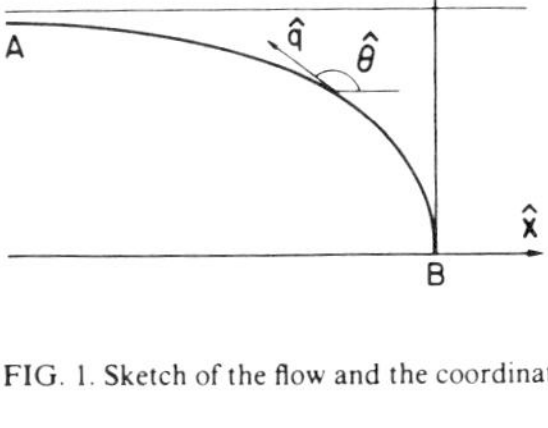

FIG. 1. Sketch of the flow and the coordinates.

The integral in (1) is of Cauchy principal value form. The variables $\hat{q}$ and $\hat{\theta}$ in (5) and (6) are defined in terms of the dimensionless complex velocity $\hat{u} - i\hat{v}$ by the relation

$$\hat{u} - i\hat{v} = \hat{q}e^{-i\hat{\theta}}. \tag{8}$$

The flow configuration is illustrated in Fig. 1. The $\hat{x}$ axis is parallel to the walls of the channel and is the axis of symmetry of the finger. The points A and B in Fig. 1 correspond to $s = 0$ and $s = 1$, respectively.

After a solution for θ and q is obtained, the shape of the finger is given by

$$\hat{x}(S) + i\hat{y}(S) = -\frac{1-\lambda}{\pi}\int_S^1 \frac{e^{i\theta}}{Sq}\,dS. \tag{9}$$

For $\kappa = 0$, Saffman and Taylor[2] obtained the following closed form solution:

$$q = \left(\frac{(1-S)(1-\lambda)^2}{(1-\lambda)^2 + S(2\lambda-1)}\right)^{1/2}, \tag{10}$$

$$\theta = \cos^{-1} q.$$

The solution (10) leaves the parameter λ undetermined.

McLean and Saffman[1] discretized the system (1)–(4) with $\kappa \neq 0$ and solved the resultant algebraic equations by Newton's method. They obtained one family of solutions. Romero[3] repeated the calculations with a different numerical scheme and different initial guesses. He obtained two other families of solutions.

In this paper we present a systematic numerical procedure to compute all the possible solutions of the system (1)–(4). We show that for each $\kappa \neq 0$ there exists a countably infinite number of solutions. This set of solutions contains the families of solutions previously obtained by McLean and Saffman[1] and Romero.[3] Our results show that the degeneracy of the Saffman–Taylor[2] solutions is not completely removed for $\kappa \neq 0$.

In order to find all the solutions of (1)–(4), we define a modified problem which has solutions for all values of λ and κ. This modified problem is obtained by replacing (4) by

 Reprinted with permission from *The Physics of Fluids*, Vol. 26, pp. 2033–2034 (1983).

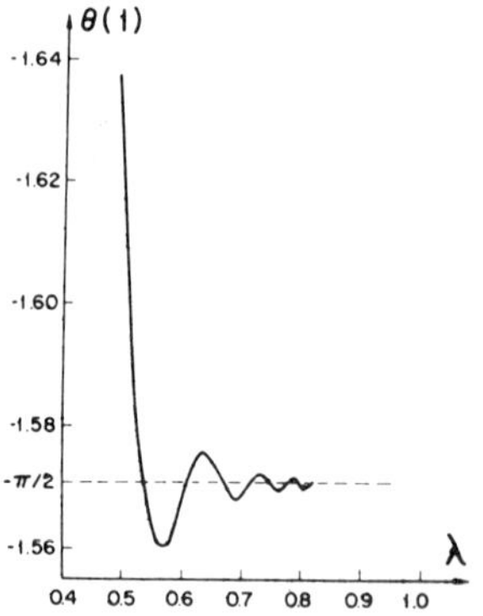

FIG. 2. Values of $\theta(1)$ vs λ for $\kappa = 0.273$.

$$q(1) = 0. \tag{11}$$

Therefore $\theta(1)$ becomes a free parameter to be found as part of the solution.

We will solve the modified problem defined by (1)–(3) and (11) and obtain solutions for all values of λ and κ. Then we will obtain the solutions of the original problem by selecting among the solutions of the modified problem those for which $\theta(1) = -\pi/2$.

Following McLean and Saffman[1] we introduce the change of variable

$$S^{\tau} = 1 - \zeta^{\gamma}. \tag{12}$$

Here τ is the smallest root of

$$(1/\tau^2)\cot \pi\tau = \kappa. \tag{13}$$

With (12), θ is twice differentiable with respect to ζ at both end points. McLean and Saffman[1] chose $\gamma = 2$ in (12). In order to solve the modified problem we choose $\gamma = 4$.

To solve the system (1)–(3) and (11) we introduce the N mesh points

$$\zeta_I = (I-1)/N, \quad I = 1,...,N. \tag{14}$$

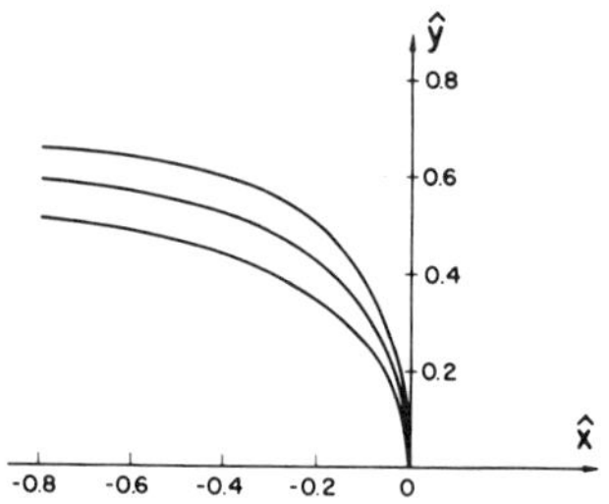

FIG. 3. Profiles of the finger for $\kappa = 0.273$. The values of λ are 0.54, 0.61, and 0.67.

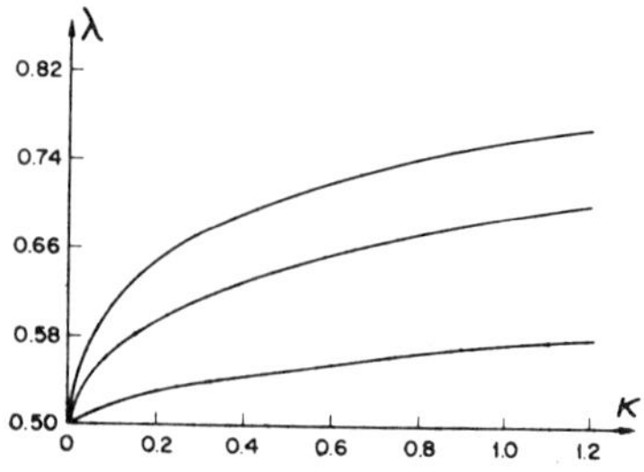

FIG. 4. Values of λ vs κ.

We also define the quantities

$$\theta_I = \theta\,[(1-\zeta_I^{\gamma})^{1/\tau}], \quad I = 1,...,N. \tag{15}$$

We discretize the system (1)–(3) and (11) by following the procedure outlined by McLean and Saffman.[1] Thus we obtain $N-1$ nonlinear algebriac equations for the $N-1$ unknowns θ_I, $I = 2,...,N$. For given values of λ and κ this system is solved by Newton's method.

In Fig. 2 we present numerical values of $\theta(1)$ vs λ for $\kappa = 0.273$. As λ approaches zero, $\theta(1) \to -\pi$ and the finger collapses on the negative $\hat{x}$ axis. As λ approaches one, $\theta(1)$ oscillates infinitely often around $-\pi/2$.

Figure 2 shows that there exists a countably infinite number of values of λ for which $\theta(1) = -\pi/2$. The solutions corresponding to these values of λ are the solutions of the original problem.

The three smallest values of λ for which $\theta(1) = -\pi/2$ are 0.54, 0.61, and 0.67. The corresponding profiles are shown in Fig. 3. The value $\lambda = 0.54$ corresponds to the solution obtained by McLean and Saffman.[1]

Similar results were found for other values of κ. As κ decreases, the amplitudes and the wavelengths of the oscillations in Fig. 2 decrease. In Fig. 4 we present the values of three smallest values of λ as functions of κ. The three curves approach $\lambda = \frac{1}{2}$ as κ tends to zero. The lowest of the three curves is the family of solutions previously obtained by McLean and Saffman.[1] The two other curves correspond to the families of solutions obtained by Romero.[3] The values of λ presented by Romero agree with our numerical results within 4% for $\kappa > 0.2$. However, Romero did not obtain reliable results for small values of κ.

I wish to thank Professor P. G. Saffman for bringing to my attention the work of Romero.

This work was sponsored by the United States Army under Contract No. DAAG29-80-C-0041 and the National Science Foundation under Grant No. MCS800-1960.

[1] J. W. McLean and P. G. Saffman, J. Fluid Mech. **102**, 455 (1981).
[2] P. G. Saffman and G. I. Taylor, Proc. R. Soc. London Ser. A **245**, 312 (1958).
[3] L. Romero, Ph.D. thesis, California Institute of Technology, 1982.

Shape Selection of Saffman-Taylor Fingers

Roland Combescot, Thierry Dombre, Vincent Hakim,[a] and Yves Pomeau[b]
Groupe de Physique des Solides, Ecole Normale Supérieure, F-75231 Paris Cédex 05, France

and

Alain Pumir[b]
Laboratory of Atomic and Solid State Physics, Cornell University, Ithaca, New York 14853
(Received 10 March 1986)

Among all problems of pattern selection the one posed by the Saffman-Taylor finger is yet unsolved, although numerics and experiments indicate the very simple property that the relative width of the finger tends to $\frac{1}{2}$ in the low–surface-tension limit. We explain this by performing an expansion beyond all orders leading to the formulation of a nonlinear eigenvalue problem with a discrete set of solutions. In particular we predict that the relative width of the finger tends to $\frac{1}{2}$ as the surface tension to the power $\frac{2}{3}$.

PACS numbers: 47.10.+g, 03.40.Gc, 68.10.−m

Pattern selection in nonequilibrium phenomena, such as dendritic solidification or multiphase fluid flow, is a subject of much recent interest. The determination of the width of Saffman-Taylor fingers is a primary example in that field. Saffman and Taylor[1] have studied the displacement of a viscous fluid by a less viscous one in a Hele-Shaw cell (a rectangular cavity between narrowly spaced glass plates) in an attempt to model displacement of oil by water in porous media. A single finger of the less-viscous fluid is eventually formed and propagates at constant velocity keeping a steady shape. In the absence of interfacial tension between the two fluids, Saffman and Taylor have computed explicitly the interface shape. They have obtained a continuous family of fingers parametrized by the relative width λ between the finger and the channel. What determined the actual shape of the finger was not understood. Then, McLean and Saffman[2] derived the equations for the finger profile in the presence of interfacial tension. Their numerical solutions[3] give a discrete set of possible relative finger widths which surprisingly all decrease to $\frac{1}{2}$ when the dimensionless surface tension goes to zero. This was quite mysterious because attempts treating interfacial tension as a conventional perturbation were unable to detect any sign of selection. Very recently, in the problem of velocity selection in dendritic growth, some evidence has been produced numerically,[4] and analytically in simplifed models,[5,6] that the selection is due to transcendentally small terms in the small parameter. It was proposed in a previous publication[7] that analogous effects are present in the Saffman-Taylor problem and are responsible for the selection. This was done by a partly heuristic method similar to the one of Ref. 5, but important nonlinear effects that cannot be neglected are missed by this linear approach, as explained below. In this Letter, we give a fully nonlinear treatment of the problem, free of ary arbitrary assumption, which is inspired by the work of Kruskal and Segur[6] on the existence of needle crystals in geometric models of crystallization. Our approach to the nonlocal Saffman-Taylor model resolves the aforementioned difficulties of the linear treatment. It opens the way to a clear understanding of realistic selection phenomena in the broad area of diffusion-controlled interface motion.[8]

We explain briefly our approach before proceeding to give more details on our analysis. We are interested in understanding how an arbitrarily small interfacial tension ($k << 1$) selects a discrete set of finger widths out of a continuum family. McLean and Saffman's equations [see Eqs. (1a) and (1b) below] are equations for a half-finger profile. We find it convenient to introduce a new coordinate along the interface which makes apparent their symmetry with respect to reflection along the center axis of the cell. Perturbation in powers of k gives an asymptotic expansion of the finger profile which respects this natural symmetry for any finger width. The crucial point is surprisingly that this does not give enough information to decide whether a given finger exists or not. Transcendentally small terms in k (i.e., lying beyond all orders of the asymptotic expansion), whose existence is qualitatively explainable by the fact that the small parameter is in front of the highest derivative, may make the symmetric continuation of the finger incompatible with a smooth behavior at its tip. We must therefore find a method to compute such terms. Generalizing the approach of Ref. 6 to our problem, we define it in the complex plane. In the neighborhood of one of its singularities, the asymptotic expansion breaks down and small terms become large enough to be noticed and therefore are under control. Here we find a couple of singularities and in their neighborhood we ob-

Reprinted with permission from *Physical Review Letters*, Vol. 56, No. 19, 1986.

tain a rescaled problem which is different from the original one and independent of the small parameter. It gives to leading order in its asymptotic expansion the effects that were previously beyond all orders. This internal region contains therefore the whole selection mechanism. It gives, for example, the new prediction that the width of selected fingers goes to $\frac{1}{2}$ as $k^{2/3}$ when $k \to 0$, because $a = \alpha^{3/2}/k$ is the only parameter in the internal problem. We should emphasize that, in contrast to Ref. 6, our internal problem still depends on the dimensionless quantity a, which appears as a nonlinear eigenvalue of an otherwise parameterless problem. The magnitude of the terms that prevent the half-profile to be continued by parity into a full smooth finger is a function of a. The nullification of this function gives the searched for solvability condition and its zeros are the selected finger widths. We emphasize that this solvability condition is exact (within the limit of k going to zero, a finite). For large values of a, the inner problem itself has a WKB-type structure and its asymptotic solution is obtained. The actual magnitude of transcendental corrections at the finger tip could be computed via a WKB transmission factor from the internal region to the real axis. This is the main difference from the previously proposed linear analysis.[7] The internal region was not defined and the magnitude of transcendental terms was approximately estimated via an inhomogeneous second member in the WKB equation.

By using conformal mapping techniques, McLean and Saffman[2] have reduced the determination of the finger shape to the solution of two coupled nonlinear integrodifferential equations:

$$\ln(q) = \frac{-s}{\pi} \mathscr{P}\int_0^1 \frac{\theta(s')}{s'(s'-s)} ds', \tag{1a}$$

$$kqs\frac{d}{ds}\left[qs\frac{d\theta}{ds}\right] - q = -\cos\theta, \tag{1b}$$

with the boundary conditions $\theta(0)=0$, $q(0)=1$, $\theta(1)=-\frac{1}{2}\pi$, and $q(1)=0$; $\mathscr{P}$ denotes the Cauchy principal value. The parameter k is proportional to the surface tension and inversely proportional to the velocity of the finger. The variable s parametrizes a half finger only; $s=1$ corresponds to the tip and $s=0$ to the trailing part of the finger.

In the $k=0$ case, the continuum of solutions discovered by Saffman and Taylor[1] is recovered: $q_0 = [(1-s)/(1+s\alpha)]^{1/2}$, $\theta_0 = \cos^{-1} q_0$, with $\alpha = (2\lambda - 1)/(1-\lambda)^2$, λ being the relative width of the finger.

Numerical analysis[2,3] suggests that the relative width of the finger decreases toward $\lambda = \frac{1}{2}$ when $k \to 0$. We thus focus on small values of α and k. We introduce the change of variables $s = \cosh^{-2}\frac{1}{2}v$, $v \in (-\infty, 0]$, as in Ref. 7. Although v is defined on $(-\infty, 0]$ only, the solutions can be extended to the whole v axis: $v<0$ describes one half of the finger and $v>0$ the other half. It is physically clear that the solutions $\theta(v) + \frac{1}{2}\pi$ and $q(v)$ must be odd functions of v. One can indeed extend the McLean-Saffman equations to the whole v axis and check that the determination of the shape of steady fingers is equivalent to the existence of odd solutions of these extended equations.

Let us start off with a solution of the Saffman-Taylor problem ($k=0$), corresponding to a relative width λ, which defines a value of α, and look for a solution of the complete problem nearby. The perturbation analysis has been carried out by McLean and Saffman but does not give any indication of a selection phenomenon. There is a difficulty at the point $s_0 = -1/\alpha$, where the expansion is increasingly divergent. However, this point does not lay in the physical region, $0<s<1$. The existence of singularities in the complex domain generates exponentially small terms that can be computed[6] only by investigation of the neighborhood of these singularities. In the present problem, we need first to continue analytically Eqs. (1), and then to use a standard inner-outer expansion[9] in the vicinity of the singularity.

Equation (1a) is readily extended in the upper-half complex v plane:

$$\ln(q) = \frac{-s}{\pi}\int_0^1 \frac{\theta(s')}{s'(s'-s)} ds' + i\theta(s)$$
$$= I(v) + i\theta(v), \tag{2}$$

where s is understood as a function of v. The closest singularities to the real axis are $v = i\pi + w_0$, where w_0 is a solution of $\sinh\frac{1}{2}w_0 = \sqrt{\alpha}$. The roots close to 0 of this equation are real for $\alpha>0$ and imaginary for $\alpha<0$. In the vicinity of the singularities ("inner problem"), $I(v)$ is a regular analytical function and can be represented by its Taylor series. In the limit $k \to 0$, it is enough to retain the lowest-order terms. Thus, Eq. (2) relates in a single fashion the varying parts of $q(v)$ and $\theta(v)$ near their singularities, allowing us to transform Eqs. (1a) and (1b) into a single differential equation for the inner problem. Moreover, one can check that $I(v)$ can be safely evaluated to zeroth order in k (i.e., by use of the original Saffman-Taylor solution). This leads to

$$ikw\frac{d}{dw}\left[w\frac{dq}{dw}\right] = -q_0^{-2} + q^{-2}$$
$$= \alpha - \frac{w^2}{4} + q^{-2}, \tag{3}$$

where $w = v - i\pi$. Rescaling w and q as $w = 2\sqrt{|\alpha|}x$ and $q = Q/\sqrt{|\alpha|}$, one obtains

$$ix\frac{d}{dx}\left[x\frac{dQ}{dx}\right] = \mathrm{a}(\epsilon - x^2 + Q^{-2}), \tag{4}$$

with $\epsilon = \text{sgn}(\alpha)$ and $a = |\alpha|^{3/2}/k$. As Eq. (4) is valid in the inner region only, we need to match its solution with the limit of the outer problem when w is small (although large compared to the size of the inner region). In this limit the asymptotic expansion is still valid and provides the boundary condition on Q:

$$Q \approx -1/x \text{ for } \text{Re}(x) \to -\infty. \tag{5}$$

The McLean-Saffman equations have an odd solution on the real axis only if the solution of Eqs. (4) and (5) is purely imaginary on the imaginary axis. We are thus lead to a nonlinear eigenvalue problem for the parameter a whose solution gives ultimately the selection mechanism. This is the searched for exact solvability condition. In order to get more insight, we investigate the large-x behavior of the solutions of Eqs. (4) and (5). When x is large, an asymptotic expansion in powers of $1/x$ can be generated. It is expected to be correct up to small corrections. The possible form of these corrections is computed by linearization of (4) around the asymptotic value of Q, and satisfy

$$ix\frac{d}{dx}\left(x\frac{dQ_1}{dx}\right) = 2ax^3Q_1, \tag{6}$$

where $Q = -1/x + \ldots + Q_1$. (By the ellipses we mean an asymptotic series when $|x| \to \infty$.) The only possible asymptotic behavior of Q_1 on the negative imaginary axis is

$$Q_1 = [C(a)/(ix)^{3/4}]\exp\int(-2iax')^{1/2}dx'. \tag{7}$$

It would usually be meaningless to add exponentially small terms to a truncated asymptotic series. Here, the asymptotic series is purely imaginary on the imaginary axis. Thus, its real part trivially converges to zero and can be compared to the real part of Q [the solution of Eqs. (4) and (5)] giving sense to the real part of Q_1. The integrand in Eq. (7) is real on the half imaginary axis $\text{Im}(x) < 0$, so that the phase of Q_1 is given by the phase of $C(a)$. Whenever $C(a)$ is purely imaginary, we obtain an odd solution of the McLean-Saffman equations. The solvability condition is translated in the nullification of the real part of $C(a)$ whose zeros give the selected widths.

This function must in general be computed numerically by solution of the nonlinear problem [Eq. (4)] with the given boundary condition [Eq. (5)]. In the large-a limit we evaluate it as follows. We first consider the case $\alpha > 0$ (relative finger width larger than $\frac{1}{2}$). In the neighborhood of the left singularity ($x = -1 + y$, $y \ll 1$), the internal problem, Eq. (4), simplifies to

$$i\,d^2Q/dy^2 = a(2y + Q^{-2}). \tag{8}$$

As before, the small parameter $1/a$ can be eliminated by a rescaling of the function Q and variable y: $Q = a^{1/7}F$, $y = a^{-2/7}r$. In terms of the new variable and function we get

$$i\,d^2F/dr^2 = 2r + F^{-2}. \tag{9}$$

This defines a new inner region of extension $a^{-2/7} \ll 1$, for large a. Equation (9) defines the inner problem in this region. It is worth noting that we would have obtained, from the very start, the same equation if we had considered the limit k going to zero, with α fixed. We apply a second time the method of Ref. 6. As before we investigate the large-r behavior of solutions of Eq. (9) with the prescription that $F \approx -i(2r)^{-1/2}$ for $\text{Re}(r) \to -\infty$. The possible exponential asymptotic correction to the algebraic series in the wedge-shaped sector $-6\pi/7 < \arg(r) < -2\pi/7$ is found to be

$$F_1 \approx \gamma r^{-3/8}\exp(2^{13/4}r^{7/4}/7). \tag{10}$$

Again γ is a constant which should be determined numerically but does not depend on a. In the outer region ($x^2 - 1 \gg a^{2/7}$), the form of transcendental corrections may be found by a WKB analysis where $1/a$ is the small parameter. We linearize Eq. (4) around its zeroth-order solution in $1/a$, $Q_0 = -1/(x^2-1)^{1/2}$, and have to solve

$$\frac{i}{a}x\frac{d}{dx}\left(x\frac{dh}{dx}\right) - 2(x^2-1)^{3/2}h = 0. \tag{11}$$

The only acceptable solution well below the segment joining $x = -1$ to $x = 1$ is

$$h = \frac{C}{(1-x^2)^{3/8}} \times \exp\int_{x_0}^{x}(-2ia)^{1/2}\frac{(x'^2-1)^{3/4}}{x'}dx', \tag{12}$$

where x_0 is an arbitrary point on the negative imaginary axis. It is straightforward to check that h matches to Q_1 of Eq. (7) for large x on their common region of validity provided that C has the same phase as $C(a)$. Now, by requiring that h matches to F_1 of Eq. (10), close to $x = -1$, we find that the phase of C is deduced from the one of γ by adding the imaginary part of

$$(-2ia)^{1/2}\int_{x_0}^{-1}(x^2-1)^{3/4}x^{-1}\,dx, \tag{13}$$

which is equal to $(a/2)^{1/2}\pi$ (this is simply obtained from the integration along an infinitesimal quarter of circle below zero with a suitable choice of integration path). Finally, we get the following asymptotic solvability condition in the large-a limit for $\alpha > 0$:

$$a = 2(n + \tfrac{1}{2} + \text{const}/\pi)^2, \tag{14}$$

where the constant is the phase of γ defined in Eq. (10). For $\alpha < 0$, the singularities are located on the imaginary axis and the phase factor linear in $a^{1/2}$, which originates from the need of a rotation in the complex plane in the preceding situation, does not exist. Therefore there is no possible solution in this case.

In conclusion, we have shown how to remove the degeneracy of the Saffman-Taylor solution at small surface tension and how to compute selected finger widths.

We are particularly indebted to Harvey Segur for having communicated and explained his results prior to publication. We also want to thank Martine Benamar, Christiane and Bernard Caroli, Chaouqui Misbah, Pierre Pelce, Bernard Roulet, and Nicolas Sourlas for interesting discussions and information. Two of us (Y.P. and A.P.) are grateful to the Center for Nonlinear Studies of the Los Alamos National Laboratory, where this work was partly initiated. The work of one of us (A.P.) has also been supported by the U. S. Department of Energy under Grant No. HDE-AC02-83-ER.

Note added.—We have now solved our nonlinear problem and the results are in agreement with known numerical results.[2,3] We intend to publish a detailed account of this work elsewhere. After this letter was sent for publication, we received a preprint from Shraiman[10] and another one from Hong and Langer[11] where approximate solvability conditions for the Saffman-Taylor problem are obtained by linear methods, the treatment essentially identical to the one given by three of us in Ref. 7.

[a]Also at Département de Physique Théorique, Centre d'Etudes Nucléaires de Saclay, F-91191 Gif-sur-Yvette, France.

[b]Also at Laboratoire de Physique Théorique et Hautes Energies, Université de Paris-Sud, Orsay, France.

[1]P. G. Saffman and G. I. Taylor, Proc. Roy. Soc. London, Ser. A **245**, 312 (1958).

[2]J. W. McLean and P. G. Saffman, J. Fluid Mech. **102**, 445 (1981).

[3]J. M. Vanden-Broeck, Phys. Fluids **26**, 2033 (1983).

[4]D. Kessler and H. Levine, unpublished; D. I. Meiron, unpublished.

[5]J. S. Langer, Phys. Rev. A **33**, 435 (1986); B. Caroli, C. Caroli, B. Roulet, and J. S. Langer, Phys. Rev. A **33**, 442 (1986).

[6]M. Kruskal and H. Segur, private communication, and to be published.

[7]T. Dombre, V. Hakim, and Y. Pomeau, C.R. Acad. Sci. **302**, 803 (1986).

[8]J. S. Langer, Rev. Mod. Phys. **52**, 1 (1980).

[9]M. Van Dyke, *Perturbation Methods in Fluid Mechanics* (Parabolic, Stanford, Calif., 1975); C. M. Bender and S. A. Orszag, *Advanced Mathematical Methods for Scientists and Engineers* (McGraw-Hill, New York, 1978).

[10]B. I. Shraiman, second preceding Letter [Phys. Rev. Lett. **56**, 2028 (1986)].

[11]D. C. Hong and J. S. Langer, preceding Letter [Phys. Rev. Lett. **56**, 2032 (1986)].

PHYSICAL REVIEW A VOLUME 30, NUMBER 5 NOVEMBER 1984

Singularities in nonlocal interface dynamics

B. Shraiman and D. Bensimon
The James Frank Institute, The University of Chicago, 5640 South Ellis Avenue, Chicago, Illinois 60637
(Received 7 May 1984)

An effective interface dynamics is derived for a Stefan-type moving-boundary-value problem corresponding to electrostatic aggregation in two dimensions. With the help of conformal-mapping techniques the problem may be reduced to a many-body-type problem described by a system of ordinary differential equations. The method allows one to show that the Mullins-Sekerka instability in this case leads to generation of cusp singularities in the interface in a finite time.

I. INTRODUCTION

Recently there has been a renewed interest in the study of dynamical structures, and in particular, processes of pattern formation in physical, chemical, and biological systems. Unfortunately the understanding of such common phenomena as solidification or dendritic growth, which has always been important to metallurgists, is hampered by the mathematical complexity of the problem. Indeed, the process of solidification is equivalent to the Stefan problem: a diffusion problem for the temperature (or impurity concentration) with boundary values specified on the moving interface, the local velocity of which is in turn determined by the heat (or impurity) flux.[1] In its general case it represents a complicated, highly nonlinear problem, which is not readily amenable even to modern numerical simulations. Thus, much of the recent work proceeds in the direction of simplifying the problem in order to gain understanding of the basic mechanisms involved and facilitate the numerical analysis. One such route is to reduce the "moving boundary value" problem to an interface dynamics problem. Such phenomenological models were proposed by Ben-Jacob, Goldenfeld, Langer, and Schon[2] and Brower, Kessler, Koplik, and Levine.[3] These models provided valuable insight into the process of dendritic growth and the importance of various physical effects, such as anisotropy and surface tension, at the expense of sacrificing some of the nonlocality of the problem. Still, the basic mechanisms at work are not fully understood.

In this paper we investigate some of the basic *nonlocal* mechanisms involved in diffusion-controlled growth. We shall derive the evolution equation for the interface for a Stefan problem in the quasistationary limit and neglecting surface tension. This will allow us to study the Mullins-Sekerka instability[4] (an instability of "bumps" on the interface) analytically beyond the linear stability analysis and show that it leads, in finite time, to the appearance of 2/3 power cusp singularities in the surface. The model that we consider is most closely related to Saffman-Taylor instability of the interface between air and water in the Hele-Shaw cell.[5] The latter in the absence of surface tension represents a realization of the model with different geometry.[6] The method used in this paper involves mapping the interface conformally onto the unit circle in the complex plane. A surprising feature of the partial differential equation (PDE) governing interface dynamics is that it may be reduced to a set of ordinary differential equations (a kind of a many-body problem) for the dynamics of the critical points of the conformal map. This facilitates analytic and numerical investigations and also points to the possibility of chaotic behavior in the evolution of the interface (the problem which requires much further study).

II. MOVING-BOUNDARY-VALUE PROBLEM AND INTERFACE DYNAMICS

Consider a cylindrical piece of "metal" with a cross-section shape given by the contour $\vec{\gamma}$. It is kept at a constant potential and is immersed in a dilute electrolyte. The cylinder grows by electrodeposition at a rate proportional to the electrostatic field. We assume that the problem is effectively two dimensional and that the growth is slow enough so that a quasistatic approximation is valid. More significant is the assumption that the electrolyte is dilute so that there is no screening of the field. The evolution of the contour is then determined by

$$\partial_t \vec{\gamma}(t) = -\vec{\nabla}\phi(\vec{\gamma}) \ , \tag{1}$$

where the electrostatic potential satisfies

$$\Delta\phi = 0, \quad \phi(\vec{\gamma}) = 0 \ . \tag{2}$$

Since we consider the model in two dimensions it is only natural to use complex variables and conformal mappings. The domain bounded by the contour $\vec{\gamma}$ is simply connected and by the Riemann theorem may be conformally mapped onto the unit disk. Let $w = x + iy$ parametrize the "physical" plane and let $z = f^{-1}(w)$ be a conformal map, mapping the exterior of $\vec{\gamma}$ onto the exterior of the unit disk in the z plane. This map provides a parametrization of the contour $\gamma(s) = f(e^{is})$ and the evolution of the contour may be specified by the evolution of the mapping.

Let us introduce a complex potential $V(w) = \phi(x,y) + i\psi(x,y) = V(f(z))$ and solve the problem outside the unit disk in the z plane

$$V(w) = \ln f^{-1}(w) - 1 \ . \tag{3}$$

The complex electrostatic field at the interface is then given by

$$E = -\left(\frac{\partial V(w)}{\partial w}\right)^{*} = -\frac{i}{(zdf/dz)^{*}} \tag{4}$$

for $z = \exp(is)$ and Eq. (1) may be rewritten as

$$\partial_t \gamma'(s) = -i\frac{\partial_s f(e^{is})}{|\partial_s f(e^{is})|^2} \ . \tag{5}$$

Reprinted with permission from *Physical Review A*, Vol. 30, No. 5, 1984.

The prime over γ in Eq. (5) is there to remind us that this equation *does not* preserve the right parametrization of the contour. The reason is that the right-hand side (RHS) of Eq. (5) represents the normal component of the velocity, which describes only the change in the geometry of the contour. The correct dynamical equation has to include a tangential component of velocity corresponding to a continuous reparametrization of the contour necessary in order to ensure that $\gamma(t,s)=f_t(e^{is})$ remains the boundary value of an analytic function (which specifies the parametrization of the contour as well as its geometry). The exact form of the reparametrization term is found by considering the infinitesimal mapping $f_{\Delta t}(z)$, induced by the evolution [Eq. (5)], and defined by $f_{t+\Delta t}(z)=f_{\Delta t}\circ f_t(z)$. We shall only give the result

$$\partial_t\gamma(t,s)=-i\partial_s\gamma(t,s)\{|\partial_s\gamma(t,s)|^{-2}+iC[|\partial_s\gamma(t,s)|^{-2}]\} . \tag{6}$$

The integral operator C is best described by its action on the Fourier transform of some real periodic function $g(s)$:

$$g(s)=a_0+\sum_{n=1}^{\infty}(a_ne^{i2\pi ns}+a_n^*e^{-i2\pi ns}) , \tag{7}$$

$$iC[g(s)]=-\sum_{n=1}^{\infty}(a_ne^{i2\pi ns}-a_n^*e^{-i2\pi ns}) . \tag{8}$$

The integral operator term in Eq. (6) reflects the *nonlocality* of the interface dynamics. This equation may easily be studied numerically; however, the analytic approach may be carried one step further.

III. DYNAMICS OF CONFORMAL SINGULARITIES

Let us interpret Eq. (6) as the evolution equation for the mapping function $f(z)$. If some contour is specified as an initial condition, one may determine the corresponding conformal map (Riemann mapping theorem). We shall consider sufficiently simple contours, such that the map is of the form

$$f(z)=a_0z\prod_{i=1}^{N}(1-a_iz^{-1}) \tag{9}$$

(While the method works for the polynomial mappings of arbitrary order, we have not succeeded in dealing with more general structures.) The requirement that $f(z)$ maps the exterior of the unit disk onto the exterior of the contour conformally demands that the critical points of the map, zeroes of its derivative, all lie within the unit disk. It is convenient therefore to deal with the factorization of $\partial_z f(z)$ directly:

$$\partial_z f(z)=\alpha_0\prod_{i=1}^{N}(1-\alpha_iz^{-1}) . \tag{10}$$

We shall now reduce the partial differential equation [Eq. (6)] to a system of ordinary differential equations governing the dynamics of the "critical points." For the purpose of the derivation that follows it is convenient to rewrite the term on the RHS of Eq. (6) in the integral form

$$g(s)+iC[g(s)]=\lim_{\epsilon\to 0}\int_0^{2\pi}\frac{d\phi}{2\pi}\,\frac{1+e^{i(\phi-s)}}{1+\epsilon-e^{i(\phi-s)}}g(\phi) . \tag{11}$$

Upon evaluating the integral one may obtain the following result:

$$|\partial_z f(z)|^{-2}+iC[|\partial_z f(z)|^{-2}]=\sum_{i=1}^{N}\frac{A_i}{|\alpha_0|^2}\,\frac{z+\alpha_i}{z-\alpha_i} \tag{12}$$

for $z=e^{is}$ with

$$A_i^{-1}=(1-|\alpha_i|^2)\prod_{j\neq i}(\alpha_i-\alpha_j)(\alpha_i^{-1}-\alpha_j^*) . \tag{13}$$

After a straightforward if somewhat lengthy computation one may rewrite Eq. (6):

$$\frac{d}{dt}\ln\partial_z f(z)=|\alpha_0|^{-2}\sum_{i=1}^{N}A_i+|\alpha_0|^{-2}\sum_{i=1}^{N}\frac{\alpha_i}{z-\alpha_i}\left[\sum_{j=1}^{N}A_j+2\sum_{j\neq i}\frac{\alpha_j(A_j+A_i)}{\alpha_i-\alpha_j}\right] . \tag{14}$$

The LHS of Eq. (14) decomposes into the sum over simple poles,

$$\frac{d}{dt}\ln\partial_z f(z)=\frac{\dot\alpha_0}{\alpha_0}-\sum_{i=1}^{N}\frac{\dot\alpha_i}{z-\alpha_i} , \tag{15}$$

and we can read off the dynamical equations for the critical points, α's, by equating the residues of the poles on both sides of Eq. (14). This yields a system of ordinary differential equations:

$$\dot\alpha_0=\alpha_0|\alpha_0|^{-2}\sum_{i=1}^{N}A_i , \tag{16}$$

$$\dot\alpha_i=-\alpha_i|\alpha_0|^{-2}\left[\sum_{j=1}^{N}A_j+2\sum_{j\neq i}\frac{\alpha_j(A_j+A_i)}{\alpha_i-\alpha_j}\right] . \tag{17}$$

This result is quite surprising: the nonlinear and nonlocal evolution equation [Eq. (6)] *conserves* the number of *critical points* and is reduced to the problem of their dynamics!

IV. MULLINS-SEKERKA INSTABILITY AND CUSPS IN THE INTERFACE

It is trivial to see that a circular contour with its radius growing as $\sqrt{t}$ is an admissible solution; however, it is linearly unstable with respect to the formation of "bumps" (the Mullins-Sekerka instability). We shall now use the dynamical equations derived above to extend the study of the interface evolution beyond the linear stability analysis.

Let us look at an n-fold perturbation of the circle:

$$\gamma(s)=re^{is}\left[1+\frac{q}{n-1}e^{-ins}\right] . \tag{18}$$

The derivative of the conformal map is then

$$\partial_z f(z)=r(1-qz^{-n}) \tag{19}$$

and the critical points are $\alpha_0=r$ and $\alpha_k=\eta\exp(i2\pi k/n)$

with $\eta^n = q$. Using Eqs. (13), (16), and (17) we get

$$\frac{d}{dt}\ln\alpha_0 = \frac{|\alpha_0|^{-2}}{1-|\eta|^{2n}} \quad (20)$$

and

$$\frac{d}{dt}\ln|\eta| = (1-2n^{-1})\frac{|\alpha_0|^{-2}}{1-|\eta|^{2n}} \ . \quad (21)$$

It follows that $q = cr^{n-2}$ and, for $q \ll 1$, $r(t) \approx \sqrt{t}$. Thus the amplitude of the n-fold mode grows as $q(t) \approx ct^{n-2/2}$. (Note that $n=2$ is a special case and shows that ellipses evolve by simple dilation, just like circles.) However, care must be taken whenever the value of $|q|$ approaches 1 and the critical points of the map hit the unit circle. This actually happens in *finite time* and corresponds to the appearance of singularities in the contour. Letting $q \to 1$ in Eq. (18) one finds that the singularities have the form of 2/3 power cusps. (It is interesting to note that the cusps appear with only few "modes" in our "analytic" parametrization, while they correspond to an infinity of "angular harmonics.") The finite time cusp formation may also be seen in a direct numerical simulation of Eq. (10): the results for a sixfold perturbation are shown in Fig. 1.

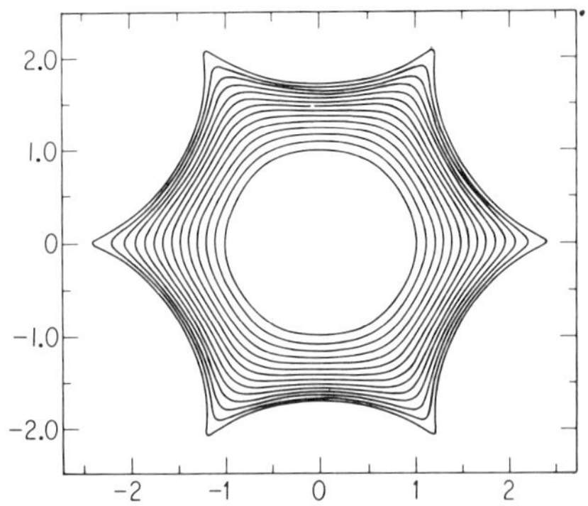

FIG. 1. The evolution of a sixfold perturbation of a circle and formation of cusps in the interface. (Result of a numerical simulation of the interface dynamics equation.)

V. CONCLUSIONS

The appearance of finite time cusp singularities in the interface is a plausible result of the nonlocal dynamics. However, while the model may provide a reasonable description of the early stages of evolution, in a real physical system singularities will not appear because of the surface tension. (Furthermore, in our "electrodeposition" model the evolution of cusps will be opposed by screening in the regions of high electrostatic field.)

We have used the mapping method to derive the interface dynamics partial differential equation for a more realistic boundary condition in Eq. (2), which includes the effects of surface tension $[\phi(\vec{\gamma}) = \sigma\kappa$, where κ is the local curvature of the interface]. The numerical simulations suggest that sharp bumps in that case undergo tip splitting rather then evolve into cusps. We shall report these results elsewhere.[6]

The conformal mapping method appears to provide a convenient tool for analytic and numerical studies. The "singularity" dynamics derived above is akin to the "pole" dynamics known in the context of the Korteweg–de Vries equation and some other conservative systems. The formulation of the problem in terms of the ordinary differential equations that it provides may allow one to study the "chaotic" aspects of the interface evolution.

Note added in proof. After this paper was accepted for publication, Nigel Goldenfeld brought to the attention of the authors an article by G. H. Meyer, in *Numerical Treatment of Free Boundary Value Problems,* edited by J. Albrecht (Birkhauser, Basel, 1982). For a model of Hele-Shaw flow with suction, G. H. Meyer using the conformal mapping method, had obtained a family of analytic solutions leading to cusp singularities.

ACKNOWLEDGMENTS

The authors would like to thank L. Kadanoff, A. Libchaber, and Steve Shenker for many helpful discussions. This work was supported in part by the U.S. Department of Energy under Grant No. DOE DEAAC02-81ER, the NSF under Grant No. NSF 80-20607, and the Materials Research Laboratory of the University of Chicago under Grant No. MRL 82-16892.

[1]J. S. Langer, Rev. Mod. Phys. **52**, 1 (1980).
[2]E. Ben-Jacob, N. Goldenfeld, J. S. Langer, and G. Schon, Phys. Rev. Lett. **51**, 1930 (1983).
[3]R. Brower, D. A. Kessler, J. Koplik, and H. Levine, Phys. Rev. Lett. **51**, 1111 (1983).
[4]W. W. Mullins and R. F. Sekerka, J. Appl. Phys. **34**, 323 (1963); **35**, 444 (1964).
[5]P. G. Saffman and G. I. Taylor, Proc. R. Soc. London, Ser. A **245**, 312 (1958).
[6]D. Bensimon and B. Shraiman (unpublished).

J. Fluid Mech. (1986), vol. 164, pp. 383–400
Printed in Great Britain

A boundary-integral method for two-phase displacement in Hele-Shaw cells

By A. J. DEGREGORIA AND L. W. SCHWARTZ
Corporate Research-Science Laboratories, Exxon Research and Engineering Co., Clinton Township, Route 22 E., Annandale, New Jersey 08801, USA

(Received 18 February 1985 and in revised form 13 August 1985)

We develop a time-dependent numerical algorithm, using a boundary-integral approach, to investigate fingering in Hele-Shaw cells. Starting from a sinusoidal variation in the initial interface, stable fingers quickly form for a wide range of the dimensionless surface-tension parameter. For very low values of the parameter, the incipient finger bifurcates. The stable fingers are clearly the same as those obtained by McLean & Saffman (1981) using a steady-state algorithm. These steady-state solutions were found to be linearly unstable. We resolve this apparent discrepancy regarding stability by tracing the fate of small disturbances placed on and about the finger tip. We show that some small disturbances do, indeed, grow initially; however, they reach a maximum amplitude and decay as they convect backward from the tip of the finger to regions where stabilizing surface tension is the major physical force. Relatively large imposed disturbances, on the other hand, cause a finger to bifurcate; the critical disturbance amplitude decreases as the surface tension is reduced,

1. Introduction

The 'fingering' phenomenon in Hele-Shaw cells has been investigated both experimentally and theoretically for the past quarter-century. Recently this interest has intensified because the phenomenon and the simplified equations that model it appear relevant to questions regarding two-phase displacement in oil reservoirs and more general investigations of pattern formation in 'noisy' systems. In both cases, the most relevant limit is when the stabilizing effect of surface tension is very small.

Saffman & Taylor (1958), in a classic paper, performed experiments displacing a viscous fluid from between two closely spaced plates, a Hele-Shaw cell, with a less viscous fluid. They point out an analogy between immiscible displacement in a Hele-Shaw cell and the equivalent problem in a two-dimensional idealization of a porous medium. In both cases, within a single-phase region of constant permeability, conservation of mass and the linear dependence of velocity on pressure gradient yield a potential-flow problem. The kinematic boundary condition, that particles on the moving interface remain there for all time, is also common to both problems. Potential differences between these problems derive from the pressure boundary condition on the moving front. In a Hele-Shaw cell, the idealized boundary condition is that the pressure jump across the front is proportional to the local curvature of the interface when the cell is viewed from above. This assumes that the interface curvature in the transverse direction, i.e. between the plates, is constant, an assumption that has been seriously questioned by Pitts (1980) and others. For porous-media flows, the situation is even less well understood. Experiments by Chouke, van Meurs & van der Poel (1959) and Peters & Flock (1981), using model porous media, suggest that acceptable results for linear stability of a slightly perturbed plane interface can be obtained by assigning an effective surface tension to this interface, proportional to the actual surface tension. Recent theoretical studies by Jeraud, Davis & Scriven (1984) suggest that the effective-surface-tension viewpoint is an oversimplification. Their more detailed theory includes pore-scale effects and leads to different values for the most unstable and neutrally stable wavelengths. Neither viewpoint has as yet been extended to the nonlinear regime when the interface is grossly distorted.

Saffman & Taylor (1958) found that a single stable finger ultimately forms in the apparatus for a wide range of experimental conditions. The experiments of Pitts (1980) confirmed their result.

Taylor & Saffman (1959) performed a linear stability analysis on steady-state finger profiles for zero surface tension, finding that the fingers are linearly unstable. McLean & Saffman (1981) extended the analysis to include the effect of finite surface tension, finding that these fingers are also linearly unstable. For non-zero surface tension, the steady-state finger profiles are determined numerically, using Newton iteration.

There is an apparent contradiction between theory and experiment – theory indicates that fingers are unstable while experiment indicates stability. To complicate the situation, Romero (1982) numerically discovered multiple steady-state solutions for a range of values of the surface tension. Indeed, Vanden-Broeck (1983) showed that there is a countable infinity of such solutions. Linear stability analyses on these additional steady-state finger profiles have not been performed to our knowledge.

Within the context of more general studies, time-dependent calculations simulating the two-phase Hele-Shaw-cell experiments have been performed by Meng & Thomson (1978) and by Tryggvason & Aref (1983, 1985) using the 'vortex-in-cell' method. Starting with a sinusoidal variation in the interface with a wavelength equal to the channel width, Tryggvason & Aref calculate fingers that evolve to a steady profile in a finger-fixed reference frame. For relatively large surface tension, these ultimate profiles agree closely with the steady-state results of McLean & Saffman. These results appear to establish that, at least, some fingers are ultimately stable in the full nonlinear sense. The vortex-in-cell method employs a fixed rectangular grid; for this reason it is not well suited to address the stability issue for low values of surface tension where high resolution is vital.

In order to help resolve the stability issue, we present a time-dependent numerical algorithm based on an accurate boundary-integral approach. We find that stable steady-state fingers can be formed for a wide range of values of surface tension. These fingers are apparently those found by McLean & Saffman (1981) using their steady-state algorithm. We provide an explanation for the apparent discrepancy between our stable fingers and the linear instability found by McLean & Saffman. Tracing small disturbances with both the full nonlinear time-dependent model and a simplified 'linear' model, we show that some disturbances initially grow, reach a maximum, and then decay. The reason for the ultimate decay is that all disturbances convect backward, away from the finger tip, to the region where stabilizing surface tension becomes the dominant physical effect acting on the disturbance.

We find that, for very low values of the dimensionless surface tension, an incipient finger, rather than progressing toward a fully developed stable finger, flattens at the tip and splits. This bifurcation is the result of a nonlinear instability. Placing a recess at the tip of a developing stable finger, we show that, above a certain amplitude, the recess causes bifurcation. As dimensionless surface tension decreases the amplitude needed for bifurcation decreases until it is within the range of 'numerical noise' that is associated with the discrete approximation of continuous functions.

Recently, branched structures have been observed in two-phase Hele-Shaw experiments by Park & Homsy (1985) and Nittmann, Daccord & Stanley (1985). The bifurcation we observe for very low surface tension may be the all-important step in this process. It is simple to show that when a finger bifurcates, the new situation of two fingers progressing up the channel side by side is also unstable. If one finger gets a little ahead of the other, it will suppress the growth of the other finger to the point where it effectively stops growing. (Saffman & Taylor 1958 observed this suppression experimentally during the initial stages of unstable displacement and it has been simulated numerically by Tryggvason & Aref 1983.) We now have a single finger progressing up the channel again and the situation can recur. From this mechanism we should expect a simple tree-like structure, where we have a main branch with single side branches at random locations on either side of the main branch. For extremely low surface tension, a finger may again bifurcate before being suppressed, so more complicated tree-like structures may also be possible. Further results, primarily focusing on the low-surface-tension limit, will be available in the near future.

Section 2 gives the basic equations and numerical method used in the fingering computations. The full equations and interfacial conditions for unsteady three-dimensional viscous flow of two immiscible fluids are given in Wehausen & Laitone (1960, p. 453 *et seq.*). Since the exact problem as posed is effectively intractable, we consider here the 'standard' simplified problem as formulated by previous workers (with the notable exception of Pitts 1980). A partial statement of the simplifying assumptions leading to these 'Hele-Shaw equations' may be found in Tryggvason & Aref (1983). See also the recent work of Park & Homsy (1984). It is important to remember that the Hele-Shaw equations do not correctly predict the experimentally observed ratio of developed finger width to channel width and the reason for the discrepancy is not well understood. When surface tension is small, however, both experiment and theory obtain values of this ratio that are close to 0.5.

In §3 we present numerical results for the growth of an imposed initial disturbance on a flat interface. For a range of values of surface tension these disturbances become steady-state fingers whose widths are in close agreement with the calculations of McLean & Saffman (1981). The effect of perturbations of these developed fingers is then explored. Two distinct effects are observed: either the disturbance ultimately decays after some period of growth, or the disturbance causes the finger to bifurcate at its tip.

A heuristic model of disturbance growth is presented in §4. For small surface tension, a perturbation is idealized as a localized pulse that passively convects backward along a developed finger. As it moves, its Fourier spectrum is altered using a local-stability argument. The resulting growth and decay are qualitatively similar to our numerical results and the apparent contradiction between predicted instability and observed stability may thus be better understood. Our simplified model is somewhat similar to the amplitude-ratio methods discussed by Reshotko (1976) in his survey of boundary-layer stability.

Results of the present study are summarized in §5.

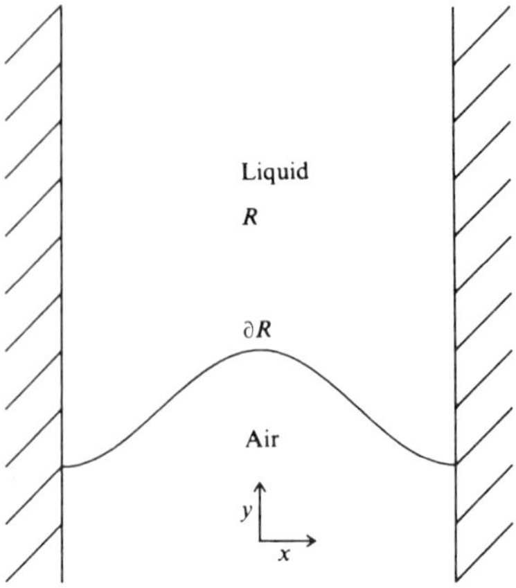

FIGURE 1. Sketch of the Hele-Shaw cell.

2. Equations and numerical method

We assume a cell of infinite length in which air is pushing a liquid, as shown in figure 1. The pressure in the air is a constant, which we may take to be zero. The pressure p in the liquid, region R, is a function of the spatial coordinates x and y. Since the plate spacing is assumed to be small compared with other lengths in the problem, we have 2-dimensional Poiseuille flow locally, in the liquid, yielding the following relationship between the depth-averaged velocity and the pressure gradient:

$$\bar{\boldsymbol{v}} = -M\boldsymbol{\nabla}p, \quad M = \frac{b^2}{12\mu}, \tag{1}$$

where b is the plate spacing and μ is the liquid viscosity. Assuming the liquid to be incompressible, the continuity equation is

$$\boldsymbol{\nabla}\cdot\bar{\boldsymbol{v}} = 0. \tag{2}$$

Combining the two equations, we obtain Laplace's equation for the pressure in the liquid:

$$\nabla^2 p = 0. \tag{3}$$

At the air–liquid boundary, denoted by ∂R in the figure, the pressure on the liquid side is augmented by surface tension from its value of zero on the air side by the expression

$$p(x, y) = \sigma\frac{\mathrm{d}\theta}{\mathrm{d}s} \quad ((x, y)\in\partial R). \tag{4a}$$

where s is the arclength along the surface, measured from some reference point, θ is the angle the tangent to the surface makes, with respect to some reference direction, and σ is the surface tension. Viewed from the liquid, p is positive at any point where the surface is concave.

On the moving interface ∂R the kinematic boundary condition, that a particle on this boundary remain on the boundary for all time, must also be satisfied. Restated, this condition is that the normal component of fluid velocity, with Cartesian components (u, v), of a particle occupying a point on the surface, is equal to the normal component of the surface velocity (x_t, y_t) at that point. Thus

$$(u, v)\cdot\bar{\boldsymbol{n}} = (x_t, y_t)\cdot\bar{\boldsymbol{n}} \quad ((x, y) \in \partial R), \tag{4b}$$

where $\bar{\boldsymbol{n}}$ is a normal vector to ∂R and subscripts signify time differentiation. Because Hele-Shaw flow is a subset of creeping motion, inertial contributions to the interface motion are neglected and surface velocity components, found by solving (3) at each instant of time, yield the instantaneous surface motion according to (4*b*).

On the sidewalls of the cell, the normal component of fluid velocity is zero, i.e.

$$\frac{\partial p}{\partial x} = 0 \quad (x = \pm\tfrac{1}{2}L), \tag{5}$$

where L is the channel width. The no-slip boundary condition for viscous flow cannot be imposed within the framework of the Hele-Shaw model.

At upstream infinity we assume constant velocity,

$$M\,\nabla p \to -\frac{Q}{bL}\bar{y} \quad (y \to +\infty), \tag{6}$$

where Q is the volumetric flow rate into the channel and $\bar{\boldsymbol{y}}$ is a unit vector in the direction of increasing $\boldsymbol{y}$.

Transforming to dimensionless length, pressure and velocity,

$$x' = \frac{x}{L}, \quad y' = \frac{y}{L}, \quad p' = \frac{Mb}{Q}p, \quad v' = \frac{bL}{Q}v, \tag{7}$$

we have the following equations and boundary conditions for the liquid:

$$\bar{v} = -\nabla p \qquad ((x, y) \in R); \tag{8}$$

$$\nabla^2 p = 0 \qquad ((x, y) \in R); \tag{9}$$

$$\frac{\partial p}{\partial x} = 0 \qquad (x = \pm\tfrac{1}{2}); \tag{10a}$$

$$\nabla p \to \bar{y} \qquad (y \to +\infty); \tag{10b}$$

$$p = \frac{\sigma Mb}{QL}\frac{\mathrm{d}\theta}{\mathrm{d}s} \quad \text{on } \partial R; \tag{10c}$$

where we have dropped the primes. The kinematic boundary condition (4*b*) remains unchanged, where, using (7), the time is measured in units of bL^2/Q. The only parameter we have is the 'dimensionless surface tension'

$$\tau = \frac{\sigma Mb}{QL}. \tag{11}$$

Because the liquid domain R in the $(z = x + \mathrm{i}y)$-plane is semi-infinite, Laplace's equation will be solved in the transformed $(\zeta = \xi + \mathrm{i}\eta)$-plane. The conformal map

$$\zeta = \mathrm{e}^{2\pi \mathrm{i}z}$$

takes the unit sink at upstream infinity into a unit sink at the origin in the ζ-plane. The moving boundary $\partial R(t)$ maps to the closed curve surrounding the origin $\partial R_\zeta(t)$. The liquid region R maps onto the annular region R_ζ bounded by ∂R_ζ and a small circle enclosing the origin and, because the flow is assumed to be bilaterally symmetric about the channel centreline in the z-plane, the complex velocity $\mathrm{d}f/\mathrm{d}\zeta$ is analytic within that annulus. Numerical values of the complex potential $f = p + \mathrm{i}\psi$ are unaffected by the mapping.

At any point within or upon ∂R_ζ, the complex potential is given by the superposition of contributions from the unit sink at the origin and distributed sources on ∂R_ζ, i.e.

$$f(\zeta) = -\frac{1}{2\pi}\ln\zeta + \frac{1}{2\pi}\int_{\partial R_\zeta} q(s_1)\ln(\zeta-\zeta_1)\,\mathrm{d}s_1. \tag{12}$$

The source line density q is a real function to be determined as part of the solution. Here s_1 may be taken as arclength on ∂R_ζ. With $p = \mathrm{Re}\{f\}$ given by (12), the integral equation for the source density is

$$\int_{\partial R_\zeta} q(s_1)\ln|\zeta-\zeta_1|\,\mathrm{d}s_1 = 2\pi p(s) - \ln|\zeta| \quad (\zeta \in \partial R_\zeta), \tag{13a}$$

where vertical bars denote the modulus of a complex number. Having determined q, the induced velocity at a point ζ, approaching ∂R_ζ from within, is given by

$$\frac{\mathrm{d}f}{\mathrm{d}\zeta} = u^{(\zeta)} - \mathrm{i}v^{(\zeta)} = -\frac{1}{2\pi\zeta} + \frac{1}{2\pi}\int \frac{q(s_1)}{\zeta-\zeta_1}\,\mathrm{d}s_1 - \mathrm{i}\,\frac{q(s)}{2}\,\mathrm{e}^{-\mathrm{i}\beta}. \tag{13b}$$

Here β is the inclination of a segment of ∂R_ζ measured counterclockwise from the direction of increasing $Re\{\zeta\}$. The integral is of Cauchy-principal-value type and the last term is the local contribution to the velocity which is most easily obtained using Gauss' theorem. The boundary-integral formulation in (13) can be shown to be equivalent to a method based on Green's formula; this relationship and the relevant existence proofs are discussed by Jaswon & Symm (1977, chap. IV).

We choose to march the solution in time using a full Lagrangian description, i.e.

$$\frac{\mathrm{d}\zeta}{\mathrm{d}t} = \frac{\mathrm{d}f}{\mathrm{d}\zeta} \quad (\zeta \in \partial R_\zeta). \tag{14a}$$

Finally, the velocities are transformed back to the z-plane by

$$\frac{\mathrm{d}z}{\mathrm{d}t} = \frac{\mathrm{d}f}{\mathrm{d}z} = \frac{\mathrm{d}f}{\mathrm{d}\zeta}\frac{\mathrm{d}\zeta}{\mathrm{d}t} \quad (z \in \partial R). \tag{14b}$$

Note that, from (4b), only the normal component of (14) need be satisfied. The full Lagrangian description allows us to monitor the surface dilation, which is sometimes also of interest.

The numerical solution is accomplished by representing the free surface in the z-plane by a set of $2N+1$ points or nodes. At any instant of time, the potential p at these nodes is calculated using a discrete version of condition (10c). After mapping the nodes to the ζ-plane, the potentials at the midpoints, between the nodes, are determined by interpolation. The source distribution is taken to be piecewise constant and the integrals in (13) are discretized, under this assumption, using formulas derived by Botha & Pinder (1983, §4.6). The resulting linear system for the source distribution from (13a) is solved and the ζ-plane velocities are calculated from (13b). After interpolation back to the nodes, the velocities are transformed to the z-plane using (14b). Note that, because of the assumption that the surface is bilaterally symmetric in the z-plane, and hence in the ζ-plane as well, the linear system to be solved is only of dimension N.

Given the node positions at any time and finding the corresponding particle velocities by the above procedure, the surface evolution reduces to the solution of a coupled system of first-order nonlinear ordinary differential equations. Since the problem becomes stiff for higher values of the dimensionless surface-tension parameter τ, we have implemented a software package that has an option for solving stiff ordinary differential equations (ODE's), LSODE (Hindmarsh 1980), finding it efficient, accurate and easy to use in this application.

Particularly at low values of τ, velocities at the surface of the moving boundary are essentially normal to the surface. Node points, at the tip of a finger propagating into the liquid, take trajectories which quickly make the spacing large near the finger tip. To maintain good coverage at the tip, we redistribute points along a cubic-spline representation of the interface at regular intervals in time. In performing the redistribution, we place more points at places of high curvature by having the grid spacing Δs inversely proportional to a linear function of the curvature

$$\Delta s \propto \left(1 + c\left|\frac{\partial \theta}{\partial s}\right|\right)^{-1}, \tag{15}$$

where c is a constant.

Since we wish to address the question of stability with the time-dependent computations, we must have sufficiently fine grid spacing to depict all potentially unstable modes. If the interface is flat and infinite, we have the following results from a linear stability analysis (Saffman & Taylor 1958, Chouke *et al.* 1959):

$$l_{cr}/L = 2\pi\tau^{\frac{1}{2}}; \tag{16}$$

$$l_{max}/L = 2\pi\,(3\tau)^{\frac{1}{2}}; \tag{17}$$

where l_{cr} refers to the critical wavelength below which modes are stable, and l_{max} refers to the wavelength of maximum growth rate. The maximum grid spacing which can adequately describe a given sinusoidal disturbance is about one quarter the wavelength. We therefore can write the following inequality for the grid spacing needed to describe the evolution of a flat interface:

$$\Delta s_{max} < 0.5\pi\tau^{\frac{1}{2}}, \tag{18}$$

where Δs is measured in units of the channel width L. As the evolving surface deviates significantly from flat, we use local conditions to determine the maximum local grid spacing

$$\Delta s_{max} < 0.5\pi(\tau/u)^{\frac{1}{2}}. \tag{19}$$

Here, u is the local interface velocity in units of Q/bL.

Unless otherwise stated, the computations used $N = 50$, which is sufficient to satisfy inequality (19). A finer spacing was sometimes used to ensure convergence. The constant c in (15) is taken to be one.

3. Results of the computations

In this paper, we wish to focus on whether a single stable finger can be formed, particularly at low values of τ. To speed up the development of a single finger, we therefore impose a sinusoidal variation in the position of the initial interface in the time-dependent computations. The wavelength of the sinusoid is one, the channel width; the phase is chosen so that the peak is at the centre of the channel; and the amplitude is 0.05 for the computations shown here.

Figure 2 shows results of the time-dependent computations for a wide range of the

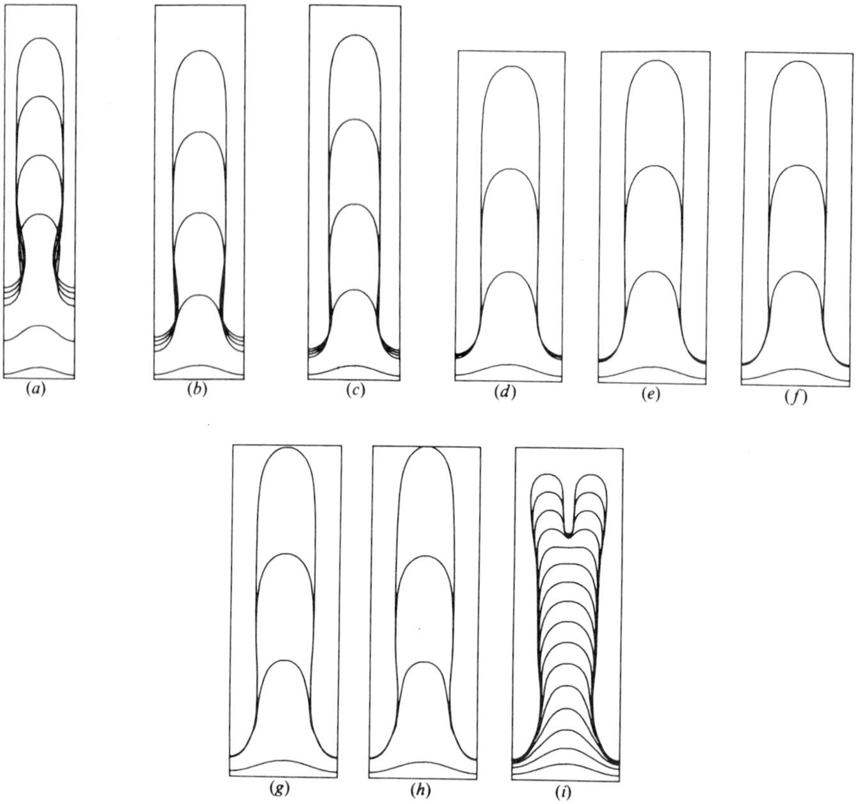

FIGURE 2. The computed time-evolution of an interface with an imposed sinusoidal variation of amplitude 0.05 (lowest curve), for different values of τ. The interface curves are shown at time intervals of 0.5. (*a*) $\tau = 0.02$, (*b*) 0.008, (*c*) 0.004, (*d*) 0.002, (*e*) 0.001, (*f*) 0.0005, (*g*) 0.0003 (*h*) 0.000 25, (*i*) 0.000 225.

surface-tension parameter τ. For all but the lowest value of τ shown, the initial small-amplitude sine wave quickly grows to form a stable finger. For high values of τ, the troughs that are formed move upward, perceptibly, and tend to round out, while, for low values of τ, the troughs, once formed, are essentially stagnant. It is easy to show, from the linear stability analysis, that, for $\tau = 0.0253$, the initial small-amplitude sine wave would be neutrally stable. We have verified this with the nonlinear time-dependent computations. For values of τ above this critical value, the small-amplitude initial sine wave decays. We cannot, therefore, generate stable fingers for $\tau > 0.0253$, by this method.

In figure 3, we show the asymptotic value of finger width λ for various values of the control parameter τ. Our results (triangles) are compared with results of Tryggvason & Aref (1985), given by circles and crosses, and the steady-state McLean & Saffman calculations (solid curve). In order to simplify direct comparison of these results, the control parameter is taken as l_{max}/L, related to τ by (17), since

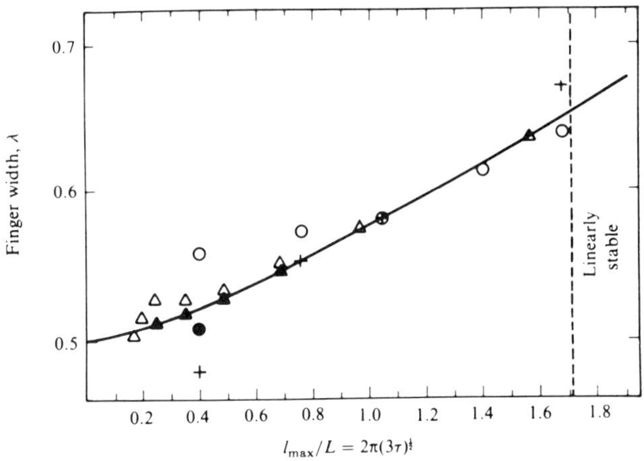

FIGURE 3. Variation of ultimate values of finger width λ with τ. The solid line is the steady-state computations of McLean & Saffman (1981); triangles denote time-dependent results from this study; circles and crosses are time-dependent results of Tryggvason & Aref (1985).

Tryggvason & Aref choose this format to present their points. Our values of τ range between 0.000 25 and 0.02 corresponding to values of l_{max}/L between 0.172 and 1.534.

For the time-dependent calculations, it is necessary to state explicitly the criteria used to establish limiting values of λ. At steady-state conditions, mass conservation yields $u_1 \lambda = 1$, where u_1 is the translational speed of the finger. In the time-dependent results, as λ approaches its limiting value, the speed at the finger tip must also reach a limit, as given by this simple relation. Our data (triangles) are calculated as the limiting value of the tip speed, i.e. the surface velocity on the channel centreline. In all cases shown, finger widths correspond to three-decimal-place convergence as the finger progresses down the channel.

Various input parameters influence the asymptotic values of λ. Specifically we monitor, by repeated runs at the same value of τ, the rate of convergence of λ with time, the accuracy of the time-stepping calculations, the number of points used to represent the profile, and the time intervals at which the points are redistributed. The first two criteria may be dismissed most simply. Three-place convergence of λ with time was achieved in all cases shown in less than 1.5 time units, although the rate of convergence decreased as τ was reduced. The LSODE package used for the time integration allows the user to specify an error tolerance; this tolerance was always set at the level required to make negligible the potential errors arising from this integration. Because the interface dilates strongly at the tip, the last two control parameters are interrelated. For the points represented by open triangles in figure 3, 50 points per half-channel width were used. This was always sufficient to satisfy the criterion of four points per unstable wavelength, using local conditions, and the time interval between point redistribution Δt_s was taken as 0.05.

Particularly for the small values of τ, the values of λ were still somewhat sensitive to the choice of Δt_s and the number of points used to describe the profile. For all λ, convergence in Δt_s is obtained using a value of 0.0125. Decreasing this parameter costs little in computation time. To test convergence with the number of grid points, we

$\tau \backslash n$	8	16	32
0.0005	splits	0.510 (79)	—
0.001	0.529 (33)	0.521 (59)	—
0.002	0.543 (26)	0.533 (46)	0.530 (81)
0.004	0.566 (21)	0.551 (35)	0.547 (65)

TABLE 1. Finger width at various values of τ for different values of n, the number of grid points per local critical unstable wavelength. Numbers in parentheses are the total number of grid points on the surface.

introduce a point-redistribution scheme that is somewhat different from the procedure described in §2. Points are redistributed along the surface with a local spacing given by

$$\Delta s = \min\left(\frac{2\pi(\tau/u)^{\frac{1}{2}}}{n}, 0.5\left|\frac{\partial\theta}{\partial s}\right|^{-1}\right),$$

where u is the local interface speed in the normal direction, and n is the number of grid points per local neutrally stable wavelength, the latter being given by (16). Using this new criterion, the number of points used to represent the profile increases as the finger grows.

Table 1 gives λ as a function of n for values of τ of 0.0005, 0.001, 0.002 and 0.004 ($l_{max}/L = 0.243$, 0.344, 0.487 and 0.688). Convergence from above is suggested with a convergence rate going approximately like $1/N^2$, where N is the total number of grid points. In the figure the solid triangles represent the values of λ obtained from table 1 using the largest value of n for which computations were performed at each τ-value. There is seen to be good correlation with the McLean & Saffman curve.

For comparison, the results of Tryggvason & Aref (1985) are shown as open circles, solid circles and crosses, corresponding to 8, 16 and 32 grid spacings per channel width respectively. Their vortex-in-cell technique works well for values of l_{max}/L in excess of 0.8; however, for $l_{max}/L = 0.4$, the smallest value for which they present results, their contention that the solution has a λ-value significantly smaller than the McLean–Saffman result is not supported by our calculations. Even for their finest rectangular grid, resolution in the critical finger-tip region is very much less than that of the boundary-integral method with periodic point redistribution. Figure 4 illustrates this by showing the point distribution of the steady-state finger for $\tau = 0.0005$ and $n = 16$. Conversely, the vortex-in-cell method may well be the algorithm of choice for larger values of τ, especially when the interface is highly convoluted.

In figure 5 we show the evolution of an interface with a sinusoidal variation, for the same parameters as those in figure 2(*b*). We show the evolution for a longer time to establish that a steady state has indeed developed. We have run this simulation, for more than twice the time shown in figure 4, with no change in the result. The value of τ of 0.008 is the value at which the channel width equals the wavelength of maximum linear growth (see (17)).

Figure 2(*i*) illustrates a very interesting phenomenon when we attempt to grow a steady-state finger for the very low value of τ of 0.000225. The incipient finger, rather than taking the path to steady state, flattens at the tip and splits in two. This finger bifurcation was observed by Paterson (1981) in circular-Hele-Shaw-cell

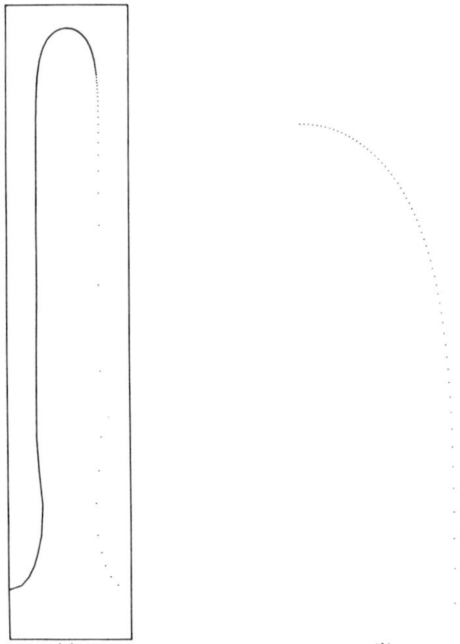

(a) (b)

FIGURE 4. (a) High-accuracy computation of a steady-state finger at $\tau = 0.0005$ ($\lambda = 0.510$) illustrating the grid-point distribution. (b) The point distribution about the tip magnified 5 times.

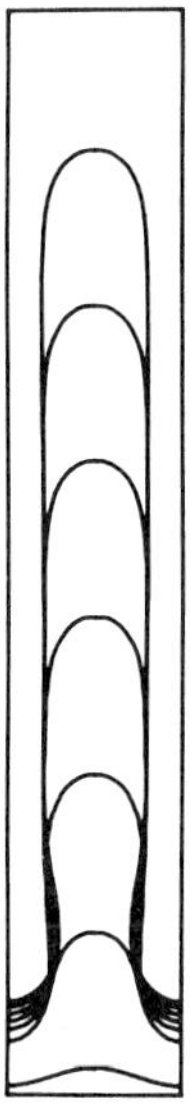

FIGURE 5. The same as figure 2(b) except the computation is shown for a longer time, indicating that a true steady-state finger has been formed.

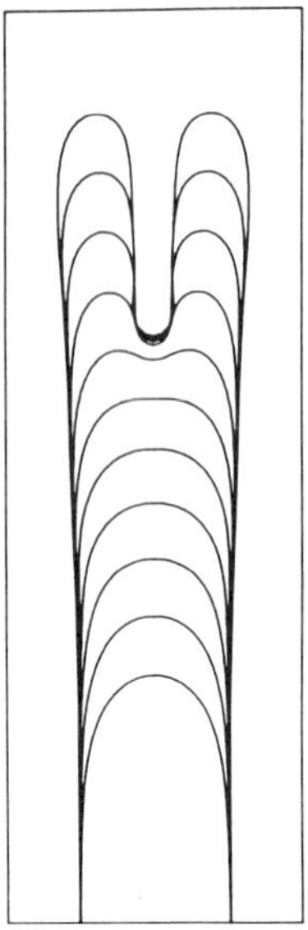

FIGURE 6. The effect of an instantaneously imposed recess of very small amplitude on the subsequent development of a finger, indicating that the fingers are linearly stable but nonlinearly unstable. The interface curves are presented at time intervals of 0.1.

experiments. (We have also simulated the circular geometry numerically and we observe finger bifurcation. These results will appear in a later paper.)

We conjecture that the splitting occurs because the fingers, for arbitrary τ, are nonlinearly unstable, i.e. developed fingers may well be stable to infinitesimal disturbances for all finite values of τ; however if sufficient noise is present the finger will split at or near the tip. This critical noise level decreases as τ is reduced. The numerical noise in our algorithm is sufficient, for the case shown in figure 2 (*i*), to cause tip splitting.

To demonstrate that fingers are nonlinearly unstable, we introduce a small-amplitude recess at the tip of the developed stable finger in figure 4 ($\tau = 0.0005$, time $t = 2.0$) and trace the subsequent evolution in figure 6. The recess can be described as follows:

$$(\Delta x, \Delta y) = A\cos(2\pi s/l_{cr})\,(n_x, n_y) \quad (|s| < \tfrac{1}{4}l_{cr}),$$
$$= 0 \quad (|s| \geqslant \tfrac{1}{4}l_{cr}),$$

where A is the amplitude of the disturbance, (n_x, n_y) is the outward normal to the finger profile, l_{cr} is the local critical unstable wavelength at the finger tip, and s is the arclength. The arclength is taken to be zero at $x = 0$, the finger tip, and is positive when x is positive. The disturbance amplitude A is -0.000225 in figure 6. As we see in the figure, this very small disturbance is sufficient to cause the finger to bifurcate. Conversely, a disturbance with an amplitude of -0.0002 gives an evolution in which the finger does not bifurcate. The finger first fattens and then necks down, ultimately returning to its steady-state shape. Thus, we demonstrate that finger bifurcation, for this value of τ, is a nonlinear instability in which a disturbance must be of sufficient

τ	Critical amplitude
0.0005	0.00021
0.001	0.00213
0.002	0.0156

TABLE 2. The critical amplitude for instability for different values of τ. The uncertainty is less than 5%.

amplitude to cause the splitting. In table 2, we list values of the critical amplitude for bifurcation for three different values of τ, determined to within 5% by a binary search. As we can see from the table, the critical amplitude drops dramatically with τ, decreasing by about an order of magnitude as τ decreases by a factor of 2. We will examine finger stability in more detail in the next section.

4. A model stability analysis

A simplified model may be constructed that exhibits the basic features leading to bounded growth of a small-amplitude disturbance. It is an heuristic model that assumes localized disturbances that are passively convected backward with the local particle velocity. For small values of the surface-tension parameter τ, the shape of a developed finger is closely approximated by the Saffman–Taylor solution for $\tau = 0$ and $\lambda = \frac{1}{2}$,

$$e^{2\pi x} = \cos 2\pi y, \tag{20}$$

in a tip-fixed coordinate system. The arclength s, measured from the tip, is given by

$$\tanh 2\pi s = \sin 2\pi y. \tag{21}$$

The surface normal at a point s is at an angle θ with respect to the oncoming stream whose velocity is U. Because the potential is approximately constant for $\tau \ll 1$, the induced velocity of the singularity distribution is essentially normal to the interface. Thus the particle velocity of the liquid, at the interface, is given by

$$u_s = U \sin\theta = U \tanh 2\pi s \tag{22}$$

and U is equal to 2 in dimensionless units for $\lambda = \frac{1}{2}$.

We now consider a localized disturbance of small amplitude that is passively convected, from its initial position s_1 at time $t = 0$, backward along the finger with the particle velocity u_s. Assuming that its breadth is small compared to the local radius of curvature, the disturbance will 'see' an essentially constant normal velocity given by

$$u_n = U\cos\theta = \frac{U}{\cosh 2\pi s}. \tag{23}$$

We identify u_n with the destabilizing velocity in the stability calculation for a plane interface translating at constant speed. A mode with wavenumber k will grow according to

$$e^{\omega t}\cos kx,$$

where

$$\omega = ku_n - \tau k^3. \tag{24}$$

The pulse initially at s_1 moves to a position s_2 on the finger in a time t given by

$$\sinh 2\pi s_2 = e^{4\pi t} \sinh 2\pi s_1 \tag{25}$$

using (20) with $U = 2$. Let l_1 be a characteristic width of the pulse when it is centred initially at s_1. As it moves backward it dilates to a dimension l_2 according to

$$\frac{l_2}{l_1} = \frac{ds_2}{ds_1} = \frac{\tanh 2\pi s_2}{\tanh 2\pi s_1} \tag{26}$$

using (25) and the assumption of small pulse width. The pulse is taken to have the bilaterally symmetric Fourier integral representation

$$f(s,t) = \frac{1}{\pi}\int_0^\infty A(k,t,l_2)\cos ks\, dk, \tag{27}$$

where s is measured from the instantaneous centre of the convected waveform. The initial spectrum is given by $A(k, 0, l_1)$ which evolves to

$$A(k,t,l_2) = A(k,0,l_2)\exp\left(\int_0^\infty \omega(t')\, dt'\right). \tag{28}$$

Using (22)–(25), the growth factor is found in closed form as

$$\exp\left(\int_0^\infty \omega(t')\, dt'\right) = \left[\frac{\tanh \pi s_2}{\tanh \pi s_1}\right]^{k/2\pi} \exp(-\tau k^3 t). \tag{29}$$

Qualitative predictions of the fate of a sufficiently small initial disturbance can now be made using this model. The centre of a disturbance will convect backwards from its initial position in a tip-fixed frame of reference. Ultimately its velocity will become constant, corresponding to the disturbance stopping at a fixed position in laboratory coordinates. At early times the disturbance amplitude will grow but must eventually reach a maximum and then decay. The magnitude of the growth depends on the initial position of the disturbance, becoming large when s_1 is small. During the growth and decay, the width of the disturbed region must increase owing both to the dilation in (26) and also to the earlier decay of the large-wavenumber components as given by the last factor in (29). As an example we calculate the time history of a trapezoidal disturbance of total width 0.06 initially centred at $s_1 = 0.13$ with τ equal to 0.0003. The evolution of the pulse shape, according to (27), is shown in figure 7. For comparison we show, in figure 8, a nonlinear time-dependent calculation with a relatively large pulse introduced at these conditions. As predicted by the linear model, we seen that the disturbance grows rapidly at first but ultimately reaches a limiting height and begins to decay. At the same time the disturbance spreads laterally, assuming a shape that is somewhat similar to those given in figure 7. The central crest of the disturbance is seen to move forward a short distance, in laboratory coordinates, before stopping. The amplitude of the initial bump, 0.003, is much larger than that required to split the finger at this low value of τ. Because it was introduced some distance from the finger tip, it produces only a transient change in finger shape.

The stability of a developed finger in a Hele-Shaw cell can now be re-examined in the light of the numerical and model results presented above. Taylor & Saffman (1959) considered the stability of fully developed fingers for $\tau = 0$ and this work has, more recently, been extended to finite surface tension by McLean & Saffman (1981). In both cases small perturbations to the steady-state solution are assumed to be of the form $\exp(\omega t)$ times a function of the space coordinates. To the extent that (eigen-)

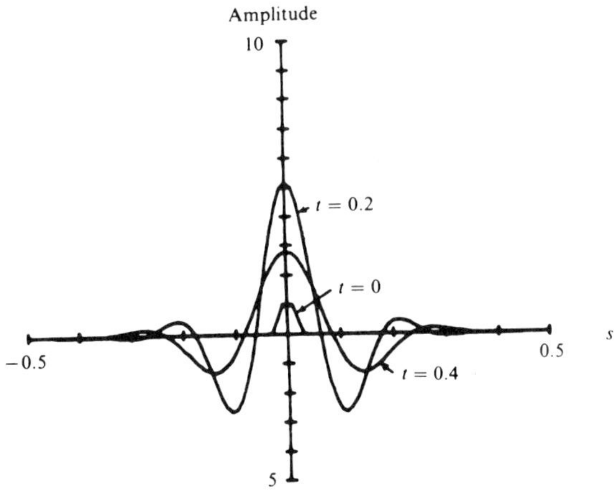

FIGURE 7. The evolution of an initial disturbance on the finger using the simplified linear model. The arclength s is measured from the centre of the disturbance, which is convecting backwards. The amplitude is the perpendicular distance from the undisturbed to the disturbed finger, and t is time.

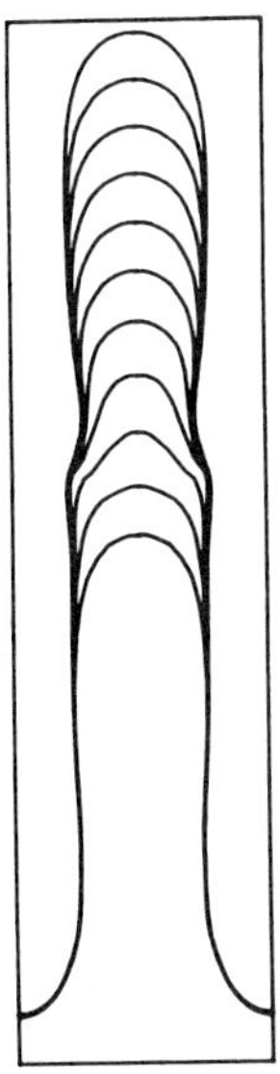

FIGURE 8. The effect of an instantaneously imposed disturbance of amplitude $+0.003$ on the subsequent development of a finger. The interface curves are presented at time intervals of 0.1. See figure 2(g) for comparison with the unperturbed evolution.

solutions of the linearized problem, corresponding to values of ω with positive real part, can be found, the underlying steady-state solution is considered to be unstable. Such eigensolutions were found to exist with or without surface tension. Taylor & Saffman (1959) conclude that the results of their stability analysis are at variance with experiment since apparently steady-state fingers can be produced in the laboratory without difficulty. In the later paper it is suggested that the resolution of this paradox is associated with the neglect of the variation of the component of interface curvature in the plane perpendicular to the plates of the cell. While this second curvature variation may explain the discrepancy between predicted values of ultimate finger width and the experimental observations (viz Pitts 1980), a simpler explanation would appear to suffice for the stability question. As the present numerical work suggests, a small-amplitude disturbance is convected backwards in a tip-fixed coordinate system. Roughly speaking, the destabilizing influence of the normal component of velocity can only produce a finite alteration of the finger profile, this alteration decreasing as the magnitude of the disturbance decreases. Provided that the disturbance is not large enough to cause the finger to split, it will ultimately decay. The presence or absence of 'instantaneous' instability would appear to be too severe a test. Only if a coordinate system translating with constant velocity could be found in which unstable eigenmodes remain stationary in space can 'operational' linear instability be inferred.

5. Conclusions

Using our nonlinear time-dependent algorithm, we have shown that a stable finger can be formed for a wide range of dimensionless surface-tension values starting from a sinusoidal variation in the interface. Comparing finger widths it is clear that the stable fingers are the same as those obtained by McLean & Saffman (1981) using a steady-state algorithm. For sufficiently large values of surface tension, our results confirm the time-dependent calculations of Tryggvason & Aref (1985) using the 'vortex-in-cell' method.

We have shown that there is no apparent contradiction between our stable fingers and the linear instability found by McLean & Saffman. There are, indeed, disturbances that grow initially, reach a maximum amplitude, and then decay. This type of linear stability analysis can only really prove the stability of a steady-state configuration. A finding of linear instability, though suggestive, can be misleading, as in this case.

We have also shown that the fingers are nonlinearly unstable. In a sense, this is an obvious result. It is clear that a deep notch placed in the tip of the finger will not 'heal' at least for smaller values of dimensionless surface tension. The principal result is that the amplitude of the notch necessary for splitting the finger is very small for small dimensionless surface tension.

We have systematically analysed the effects of recesses and protrusions at various locations on the finger. We find that a recess at the tip has the largest ultimate growth, by far. We have tried large-amplitude disturbances on the side of the finger and we have not been able to grow a vestigial finger on the side. It seems clear, therefore, that a vestigial finger on the side of a growing finger can only be generated by a bifurcation at the tip. One of the resulting fingers becomes suppressed, while the other growing finger centres and expands in the channel, leaving a vestigial finger.

While this paper was under review, our numerical algorithm was generalized by removing the requirement of bilateral symmetry. The resulting simulation of finger motion, for small τ, clearly illustrates the tip-splitting mechanism followed by a period of competition between the two resulting branches with ultimate dominance by one of them. The process repeats as the 'winner' progresses down the channel, leading eventually to a profile with multiple, essentially stagnant, side branches. Further details may be found in DeGregoria & Schwartz (1985). Recent experimental studies at small τ (Park & Homsy 1985 and Nittman *et al.* 1985), provide confirmation for this process. The fact that finger splitting had not been observed in previous experiments is probably due to the use of narrower cells yielding larger values of τ.

While agreement between computation and experiment is now good at small τ, there is still a significant discrepancy between observed and predicted finger width at large τ-values. This discrepancy is perhaps due to the variation in the radius of curvature in the third dimension with interface speed, as suggested by Pitts (1980). There has also been speculation that the neglect of this effect might account for the discrepancy between experimentally stable fingers, on the one hand, and the linear instability of theoretically computed fingers, on the other hand. The analysis of small disturbances given here seems to indicate that there really is no significant discrepancy after all.

We do not appear to have generated any of the multiple steady-state finger solutions of higher finger width that Romero (1982) or Vanden-Broeck (1983) have found. This may indicate that these other steady-state solutions are significantly less stable than the primary one, or it may simply mean that, starting from a sinusoidal variation in the interface, we achieve the steady-state finger with the smallest width for a given value of surface tension. One way to resolve this issue is to compute the profile for one of these wider steady states, start the time-dependent algorithm in this state, and see if the finger propagates unchanged. We hope to examine this in the near future.

An interesting and potentially important question concerns the stability of developed fingers at very low surface tension. Our heuristic-model calculation in §4 indicates that fingers can be linearly stable. This stability is confirmed by the numerical results for $\tau \geqslant 0.00025$. Finite-amplitude disturbances have been shown to cause tip splitting for larger τ. Thus, linear stability but nonlinear instability is demonstrated for a range of values of τ. Whether fingers remain linearly stable for arbitrarily small τ, or become linearly unstable at a small finite value of τ, is not answerable by us at present. We suspect the former, however, because it provides the simplest explanation consistent with the numerical and experimental results available to date. We have demonstrated that the noise level required to cause tip splitting decreases very rapidly as τ is reduced. Thus tip-splitting may always be expected to occur in Hele-Shaw experiments at high capillary number because some small level of noise is unavoidable. Since the Hele-Shaw model is increasingly being looked upon as a prototype case for studies of pattern formation in other nonlinear systems, the answer to this stability question may have important theoretical implications.

We are happy to acknowledge our indebtedness to Dr Ben White and Professors Greg Baker, Philip Saffman, and George Homsy for useful discussions of the fingering phenomenon.

REFERENCES

Botha, J. F. & Pinder, G. F. 1983 *Fundamental Concepts in the Numerical Solution of Differential Equations*. Wiley.

Chouke, R. L., van Meurs, P. & van der Poel, C. 1959 The instability of slow, immiscible, viscous, liquid–liquid displacements in permeable media. *Trans. AIME* **216**, 188–194.

DeGregoria, A. J. & Schwartz, L. W. 1985 Finger breakup in Hele-Shaw cells. *Phys. Fluids* **28**, 2313–2314.

Hindmarsh, A. C., *Livermore Solver for Ordinary Differential Equations* (*LSODE*) Sept. 23, 1980 version, Lawrence Livermore Laboratory, Livermore CA.

JASWON, M. A. & SYMM, G. T. 1977 *Integral Equation Methods in Potential Theory and Elastostatics.* Academic.

JERAUD, J. R., DAVIS, H. T. & SCRIVEN, L. E. 1984 *SPE/DOE Preprint No.* 12691.

McLEAN, J. W. & SAFFMAN, P. G. 1981 The effect of surface tension on the shape of fingers in a Hele-Shaw cell. *J. Fluid Mech.* **102**, 455–469.

MENG, J. C. S. & THOMSON, J. A. L. 1978 Numerical studies of some nonlinear hydrodynamic problems by discrete vortex element methods. *J. Fluid Mech.* **84**, 433–453.

NITTMANN, J., DACCORD, G. & STANLEY, H. E. 1985 Fractal growth of viscous fingers: quantitative characterization of a fluid instability phenomenon. *Nature* **314**, 141–144.

PARK, C.-W. & HOMSY, G. M. 1984 Two-phase displacement in Hele-Shaw cells: theory. *J. Fluid Mech.* **139**, 291–308.

PARK, C.-W. & HOMSY, G. M. 1985 The instability of long fingers in Hele-Shaw flows. *Phys. Fluids* **28**, 1583–1585.

PATERSON, L. 1981 Radial fingering in a Hele-Shaw cell. *J. Fluid Mech.* **113**, 513–529.

PETERS, E. J. & FLOCK, D. L. 1981 The onset of instability during two-phase immiscible displacement in porous media. *SPE J.* **21**, 249–258.

PITTS, E. 1980 Penetration of fluid into a Hele-Shaw cell: the Saffman–Taylor experiment. *J. Fluid Mech.* **97**, 53–64.

RESHOTKO, E. 1976 Boundary-layer stability and transition. *Ann. Rev. Fluid Mech.* **8**, 311–348.

ROMERO, L. 1982 Ph.D. thesis, California Institute of Technology.

SAFFMAN, P. G. & TAYLOR, G. I. 1958 The penetration of a fluid into a porous medium or Hele-Shaw cell containing a more viscous liquid. *Proc. R. Soc. Lond.* A **245**, 312–329.

TAYLOR, G. I. & SAFFMAN, P. G. 1959 A note on the motion of bubbles in a Hele-Shaw cell and porous medium. *Q. J. Mech. Appl. Maths* **12**, 265–279.

TRYGGVASON, G. & AREF, H. 1983 Numerical experiments on Hele-Shaw flow with a sharp interface. *J. Fluid Mech.* **136**, 1–30.

TRYGGVASON, G. & AREF, H., 1985 Finger interaction mechanisms in stratified Hele-Shaw flow. *J. Fluid Mech.* **154**, 287–301.

VANDEN-BROECK, J. M. 1983 Fingers in a Hele-Shaw cell with surface tension. *Phys. Fluids* **26**, 2033–34.

WEHAUSEN, J. & LAITONE, E. 1960 Surface waves. *Handbuch der Physik*, vol. 9 (ed. S. Flügge), pp. 446–778.

J. Fluid Mech. (1987), vol. 177, pp. 67–82
Printed in Great Britain

An experimental study of the Saffman–Taylor instability

By P. TABELING, G. ZOCCHI AND A. LIBCHABER
The James Franck Institute and E. Fermi Institute,
University of Chicago, 5640 S. Ellis Avenue, Chicago 60615, USA

(Received 14 August 1986)

An experiment on the Saffman–Taylor instability with wetting fluids is presented that explores a greater range of capillary numbers than did the original experiment of Saffman and Taylor. It turns out that no clear one-half plateau for the finger size is observed, and that the ensemble of experiments cannot be analysed in terms of a single control parameter. The effect of the film of oil left behind the finger is important, and we measure its thickness. A qualitative discussion of the instabilities of the fingers for large capillary numbers is presented, the first instability leading to asymmetrical fingers. Tip splitting appears for larger values of $1/B$. The $1/B$ value for the onset of instabilities is shown to be noise dependent.

1. Introduction

The penetration of a gas into a viscous fluid in a Hele-Shaw channel (Saffman–Taylor problem) is an example of the general problem of pattern formation in nonlinear systems. In the simplified approach proposed by Saffman & Taylor (1958) in their original paper, the flow of the two fluids is considered to be two dimensional and the interface is a line. In the gas the pressure is uniform, and in the viscous fluid the motion obeys a Darcy law so that all the nonlinearities of the system originate in the boundary conditions at the interface. In the absence of surface tension, a plane interface is unstable at any velocity of propagation. With a finite surface tension, the stability analysis of a flat interface gives a dispersion relation of the form (see e.g. the review paper by Bensimon *et al.* 1986):

$$\omega = \kappa U - \kappa^3 \frac{Tb^2}{12\mu}, \tag{1}$$

where the velocity of the interface U is directed from the air towards the oil, T is the interfacial tension, μ is the viscosity, κ the wavenumber and ω the exponential growth rate of an initially small sinusoidal disturbance. Since the maximum wavelength allowed by the boundary conditions corresponds to $\kappa_{\min} = 2\pi/w$, where w is the width of the cell, one obtains that the flat interface is unstable if

$$\frac{1}{B} = 12\frac{\mu U}{T}\left(\frac{w}{b}\right)^2 > (2\pi)^2. \tag{2}$$

Experimentally, Saffman & Taylor found that plane interfaces are destabilized by wavelengths scaling approximately with the maximum growth wavelength obtained from (1). The most striking observation was then that the system further evolves towards a state where a single finger occupies about one-half of the channel width at large velocities (figure 1). They could calculate analytically particular steady

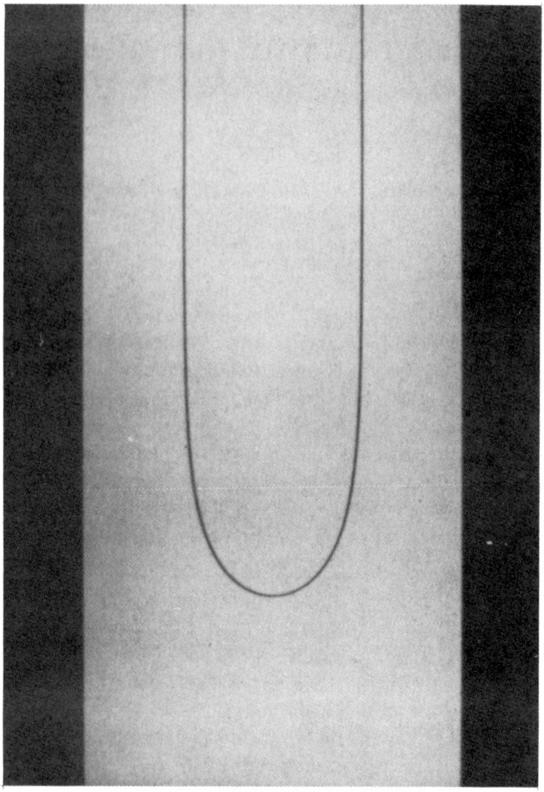

FIGURE 1. Picture of a finger close to $\lambda = \frac{1}{2}$. The dimensions of the cell are: $w = 5.4$ cm and $b = 0.8$ mm. Air penetrating into oil.

solutions of the problem in the zero-surface-tension limit. Using a conformal mapping, they found a continuous family of solutions described by a dimensionless parameter λ which represents the relative size of the finger compared to the width of the channel ($0 < \lambda < 1$). The calculated shapes agree very well with the observed ones in the regime of high capillary numbers ($Ca = \mu U/T$), provided that one chooses a value of λ equal to the observed one ($\lambda \approx \frac{1}{2}$ in this regime). On the other hand, for small capillary numbers, where surface tension is important, the computed shapes differ substantially from the actual ones.

The problem was reconsidered by McLean & Saffman (1981). Allowing a finite surface tension, they found that $1/B$ defined above is the only parameter that enters the equations of motion. Their numerical study led to values of the size of the finger close to $\frac{1}{2}$ at large velocities, which was in good agreement with the experiments. At low velocities (i.e. $1/B < 100$), the agreement with the experiments was more ambiguous since the finger sizes predicted by the theory were significantly below those actually measured (figure 7).

A more empirical approach to the problem was taken by Pitts (1980). He repeated the experiment of Saffman & Taylor and argued that the flow due to the film left behind the interface controls the boundary condition on it. He then noticed an intriguing property of self similarity for the shapes of the fingers and built from this experimental fact a theory which gives $\lambda = \frac{1}{2}$ fingers at large velocities. The agreement between theory and experiment was thus obtained at the expense of an empirical assumption.

Another question that is not yet fully understood concerns the mechanism that selects the particular solutions that are observed experimentally. It is known that the McLean–Saffman equations have more than one solution. Romero (1982) found numerically a second set of solutions to these equations, and Vanden-Broeck (1983) showed later that there exists a discrete family of solutions to this problem. On the other hand, only one solution is observed in the experiments.

As for the stability of the fingers, the recent experimental observations (Park & Homsy 1985) and computer simulations (De Gregoria & Schwartz 1985) are that the fingers become unstable at large values of $1/B$, and that the threshold for the onset of the instability depends on the noise. This can be qualitatively understood applying an argument given by Zel'dovitch *et al.* (1980) (Bensimon *et al.* 1986).

The results contained in this paper are the following: since data obtained using cells with different aspect ratios do not fall onto the same curve, a single control parameter like $1/B$ is not enough to describe the system. Also, in experiments with wetting oils no clear $\lambda = \frac{1}{2}$ plateau is observed: the λ *vs.* $1/B$ curve shows a slow and continuous decrease below this value for increasing $1/B$. For large values of $1/B$ the fingers become unstable. The first instability that we observe is an asymmetric disturbance on one side of the tip, while at higher velocities tip splitting occurs.

The paper is organized as follows: we first describe the experimental set-up in §2. In §3 we give a description of steady fingers. In §4 we study the low velocity regime. In §5, we study the film left behind the fingers. Then we analyse the large-velocity regime and the stability of the fingers.

2. The experimental set-up

Several set-ups have been built for the experiment, but all rely on the same principles. We describe here the last apparatus, from which most of the results presented in this paper have been obtained. The experimental arrangement is sketched in figure 2. The Hele-Shaw cell consists of two long rectangular glass plates separated by two aluminium spacers which define the walls of the channel. The plates are 120 cm long, 10 cm wide and 1.27 cm thick. At either end of the channel are two Plexiglas pieces in which a rectangular cavity the same size as the channel width has been machined in order to limit any possible influence of the extremities of the cell on the flow itself. Those pieces are connected to a hydraulic circuit (composed of valves and a magnetic pump), which in turn is connected to a syphon. The magnetic pump, which was driven by a d.c. motor, was used at large values of the parameter $1/B$. For small values, it turned out to be more convenient to pull the fluid by means of gravitational forces. In both cases, the experimental conditions imposed a constant fluid velocity through the channel. Experiments showed that an accurate control of the velocity was achieved; the regulation was better than 10^{-2}.

Since the homogeneity of the gap was thought to be a crucial factor for the accuracy of the experimental results (the control parameter $1/B$ depends on b^2), we used interferometry to estimate the defects of the cell and eventually reduce them. We found that the inhomogeneity of the gap did not exceed 10^{-3} cm throughout the entire channel. The largest non-uniformities were located, as expected, near the ends. In

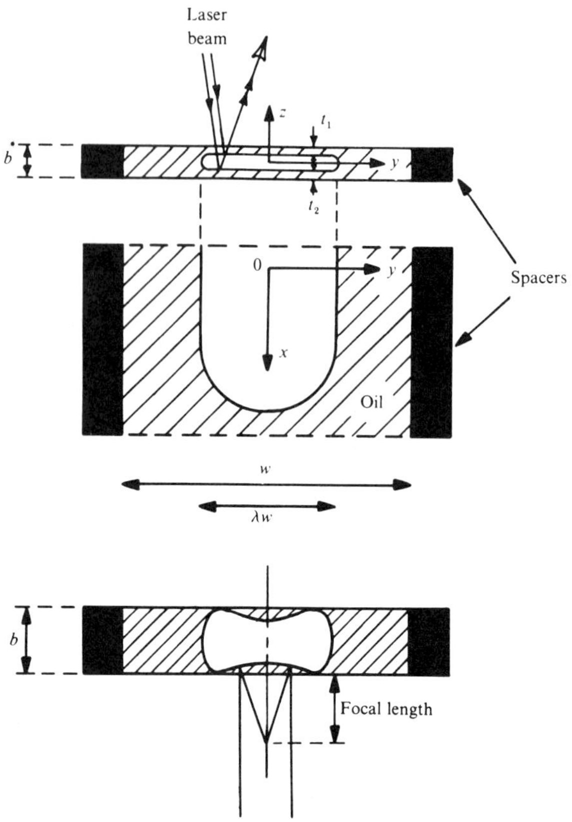

FIGURE 2. Schematic view of the finger and the oil film, which is blown up. The optical beams for the interference study are shown on the top figure. The bottom figure describes how the focal length is measured.

the central part of the cell, the size of the defects lay below 5×10^{-4} cm, and it was possible to define a large area for which this value falls to 3×10^{-4} cm. These estimates proved to be useful for studying the film left behind the fingers and the fingers' stability.

We chose oils as the viscous fluid in most of the experiments: vacuum pump oil, lubricating oils or silicone oils. Silicone oils Rhodorsil series 47V turned out to be the most convenient because their physical properties are weakly dependent on temperature and they completely wet glass. The test that we used for estimating wetting properties consisted of studying the evolution of the radius of a droplet on a clean surface and comparing it with Tanner's law (1979), which, for a wetting fluid, predicts that the radius will grow as a power of time. Some experiments have been performed with glycerol in order to visualize the flow around the fingers.

The physical parameters of the fluid were measured. The surface tension was determined by three different methods (capillary tube, pendent drop and Padday method 1969), and the viscosity was measured by the falling-ball method.

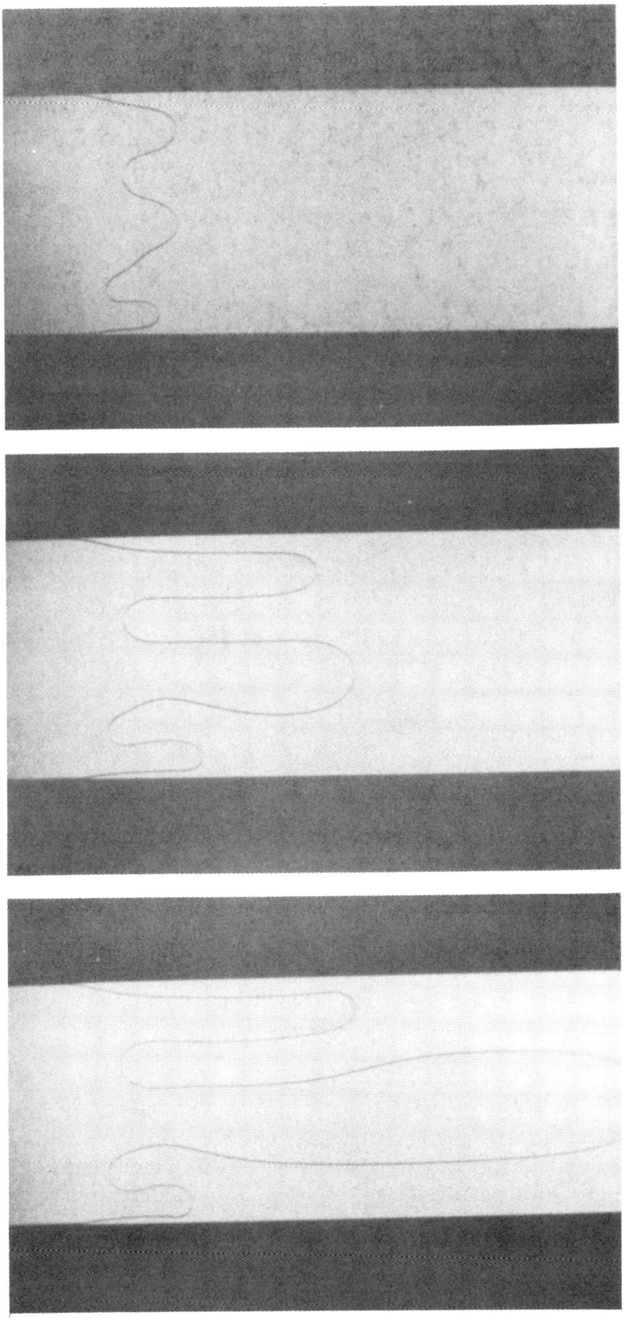

FIGURE 3. Evolution of the transient pattern as a function of time. The bottom figure shows the formation of what will become the stationary finger.

FIGURE 4. Finger of air in glycerol, seeded with small fish-skin particles. Time exposure in the frame of reference of the finger.

Several cells were studied. The gap between the plates was varied by changing the spacers, and the width of the channel could be varied by shifting them. We used three sets of spacers, corresponding to gaps of 800, 480 and 173 μm. In conclusion, we found that the quality of the experiment depends strongly on avoiding pinning effects at the walls and non-uniformities in the channel gap.

3. Description of stationary fingers

When an interface starts moving, it generates a few modes which grow and compete together, as shown in figure 3. The competition ends with a single finger occupying a part of the channel and moving steadily along it. The number of modes initially excited depends on the velocity of the interface. Experimentally, we observe that the number of modes increases more or less as the square root of the velocity, as predicted by the linear stability analysis.

An example of a steady finger is shown in figure 1. In the physical space, the dark line, which defines the interface, is a meniscus of oil that scatters the light coming

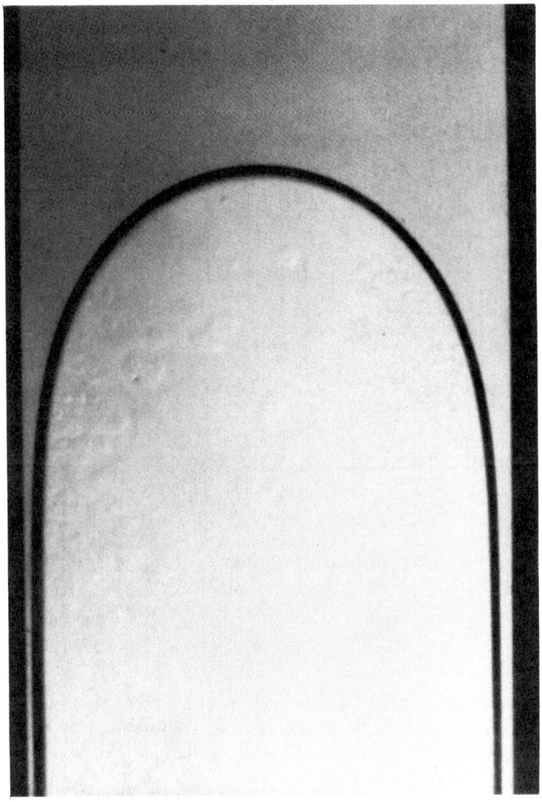

FIGURE 5. Finger shape for $1/B < 50$. At this small velocity one can observe instabilities of the film left behind.

from below. Behind the tip, the interface is parallel to the spacers. One can obtain very long fingers, almost as long as the channel itself.

How is the flow around a steady finger? We have performed experiments with glycerol in order to visualize the motion. Figure 4 shows a time exposure taken in the frame of reference of the finger. The small particles that seed the fluid show two types of trajectories: some of them, originating far ahead, deviate as they approach the finger, and are further convected along its sides. Other particles converge towards the interface, swirl on a small scale as they come close to it, and are further convected backwards. Such motion shows the existence of a pair of small eddies at the tip. Note that a few of them cross the interface and are trapped in the film of glycerine left behind the finger. This motion was previously pointed out by Pitts (1980). An interesting result is the existence of a rolling motion confined in a narrow region close to the interface. This feature has been obtained in a numerical calculation by Reinelt & Saffman (1985), and it shows that in this region the flow is three dimensional. Further away from the tip the streamlines are embedded in horizontal layers and the flow is two dimensional.

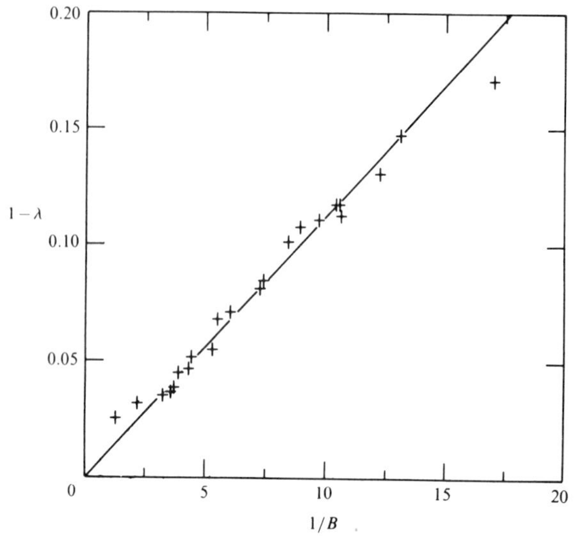

FIGURE 6. Plot of $1-\lambda$ as a function of $1/B$. $w = 3.025$ cm, $b = 0.149$ cm, $\mu = 10$ P, $T = 74$ dynes/cm. High-surface-tension oil is used, and a small-aspect-ratio cell in order to obtain a measurable velocity.

4. The low-velocity regime

Phenomenologically, the steady fingers are defined by their velocity U, their shape and their size λ. We now discuss the relation between λ and U.

The first question that we ask concerns the existence of a control parameter. The relevant pressure terms in the problem are the viscous one, which, using Darcy's law $U = -(b^2/12\mu)\,\nabla p$ and w as a typical lengthscale, gives $p \approx 12\mu U(w/b^2)$, and the one due to surface tension, which is of order T/w, so that we are indeed led to consider the dimensionless number $1/B = 12(w/b)^2\mu(U/T)$ as the control parameter of the system.

Since the size of steady fingers varies rapidly with $1/B$ in the low-velocity regime, we first restrict our study to this region. Experimentally, we observe that the finger shapes become semicircular when the velocity is decreasd to very low values, i.e. for $1/B < 50$ (figure 5). The motion of the fluid around the finger is everywhere as slow as the finger itself and no abnormal draining appears. The finger at rest is not a flat interface because of the influence of the lateral walls; it is a semicircle. Therefore, there seems to be a continuity between the finger at rest and the one moving very slowly.

Physically, it is more natural to characterize the finger sizes by $1-\lambda$, which represents the quantity of fluid left behind the finger as it moves. Our experimental results are presented in figure 6. They show a linear relation between $1-\lambda$ and $1/B$:

$$1-\lambda \approx 0.011/B. \tag{3}$$

This law holds for all values of the aspect ratio w/b measured.

Next we consider larger velocities. Figure 7 summarizes the results that we have obtained for $0 < 1/B < 250$. Experiments have been performed with a large aspect

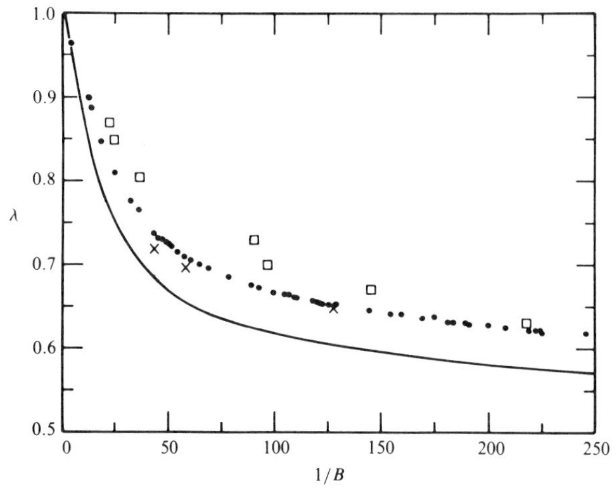

FIGURE 7. Relative size λ of the finger as a function of the parameter $1/B$. □, Saffman–Taylor experiment (aspect ratio $w/b = 31.8$); ●, experimental results obtained in cell I, with Rhodorsil oil 47V10; ×, experimental results obtained in a cell of aspect ratio $w/b = 62.3$ with Rhodorsil oil 47V50. Theoretical results of McLean & Saffman are presented as a continuous curve.

ratio $w/b = 65$, and with two oils. The purpose of this experiment was to compare our data with those of Saffman & Taylor, obtained with an aspect ratio about two times smaller.

As shown in figure 7, our data accurately follow a single line which decreases as $1/B$ is increased. The shape of the interface is circular for $1/B < 50$ (figure 5). For larger values of $1/B$ its shape is shown in figure 1. We obtain a result previously shown by Saffman & Taylor: for a given aspect ratio, the curves are universally described by a capillary number, such as $1/B$. As a matter of fact, the data obtained in two closely similar cells but with two different oils lie on a single curve. However, this curve lies significantly below that of Saffman & Taylor. Therefore, we conclude that $1/B$ is not the single control parameter of the problem. We have not found any other parameter for which all the data fit on a single curve. We thus come to the conclusion that the problem is described by two parameters.

If we now look at the theoretical results, we find that they disagree with the two experiments, as shown in figure 7. We therefore suspect that another effect, not taken into account by the theory, is involved in the experiments.

5. Influence of the film left behind the fingers

The interface between air and oil is a meniscus which lives in a three-dimensional space. When the interface is advancing, the meniscus produces a film behind it. Basically, this problem is analogous to the Landau–Levich (1942) film drained by a vertical plate. The problem is indeed well defined only when the fluid wets the surface. The theoretical study of the film drained behind a flat interface in a capillary tube has been performed by Bretherton (1961). According to this study, the film thickness follows a $\frac{2}{3}$ power law with the interface velocity, for small velocities; more

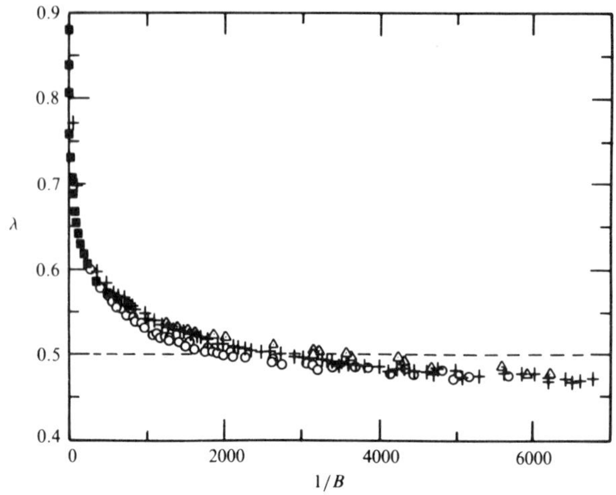

FIGURE 8. Plot of λ *vs.* $1/B$ for three different aspect ratios. △, $w/b = 112.5$; +, $w/b = 65$; ○, $w/b = 34$.

precisely, if R is the radius of the capillary tube the formula for the film thickness t reads:

$$\frac{t}{R} = 0.643\,(3Ca)^{\frac{2}{3}}. \tag{4}$$

Accordingly, the pressure drop across the interface is affected by the presence of this film. In general, if the interface is not flat, we expect the film thickness to depend on the velocity component normal to the interface.

When a finger is moving in a Hele-Shaw cell the normal velocity varies along the interface, and therefore one expects the film thickness to be non-uniform behind the interface, which means that the finger moves in a channel of variable gap. Intuitively, one feels that the physical problem may be strongly affected by this phenomenon, and this has been studied theoretically by Park & Homsy (1984).

We performed an experimental study of such a film. The results for low capillary numbers are described in a previous paper by Tabeling & Libchaber (1986). The film thickness was determined interferometrically (figure 2), and it was found that the shape of the film is of the form:

$$t = t_{\max}(U_n/U)^{\frac{2}{3}}, \tag{5}$$

where U_n is the normal velocity. This means that along the interface, the film thickness obeys a local Bretherton law. Also, the maximum thickness $t_{\max}$ as a function of Ca agrees with Bretherton's law for small Ca. In this regime the effect of the film on the finger itself can be taken into account by rescaling the parameter $1/B$ using an expression for the pressure drop across the interface derived by Park & Homsy (1984). In this way a better agreement with the McLean–Saffman theory is achieved. We thus find that the film plays a significant role in the experiments and that the previous disagreement between theory and experiments is partly due to the fact that the film had been neglected in the theoretical models.

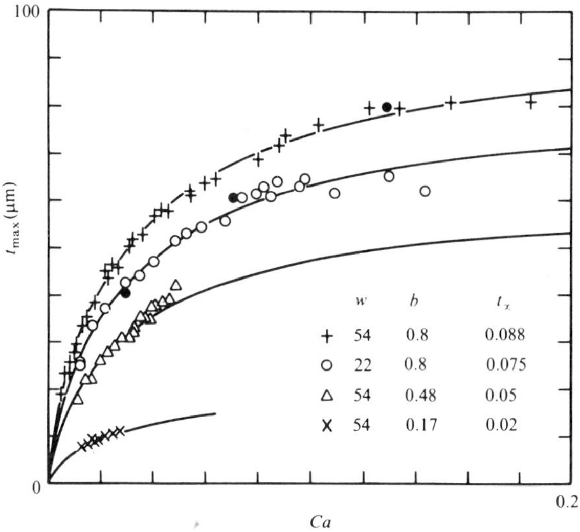

FIGURE 9. Maximum film thickness *vs.* capillary number for various aspect ratios. The solid curves represent the function $t = t_\infty(1-e^{-\beta Ca^{\frac{2}{3}}})$, with $\beta = 8.585$. ●, numerical results of Reinelt & Saffman. The unit used for w, b, t_∞ is mm.

6. The large-velocity regime

For large capillary numbers the experimental results of Saffman & Taylor seemed to indicate that the finger size saturates at the value $\lambda = \frac{1}{2}$. Although this result was also obtained in the numerical analysis of McLean & Saffman, the physical mechanism of such a behaviour remains obscure. We therefore found it useful to get accurate measurements of the finger size for different aspect ratios, in order to define the problem clearly from an experimental point of view. Our results are plotted in figure 8. Different aspect ratios led to slightly different curves. All the curves go below the value $\lambda = \frac{1}{2}$, and the differences between them are small and monotonic with the aspect ratio w/b. Hence we do not observe a $\lambda = \frac{1}{2}$ plateau and it appears that $1/B$ is not the only control parameter of the system. Therefore one suspects again that some other physical mechanism, not taken into account by the theory, plays an important role here, evidently related to the evolution of the film thickness in this regime. We thus proceeded to study it.

The interferometric method was difficult to use here, because the growing thickness of the film leads to denser fringes. Therefore we resorted to a different technique, which exploits the fact that the film can be used as a convergent mirror. By measuring its focal length, and inferring its shape from the interferometric measurements at low velocities, we can obtain the maximum thickness t_{max}. This analysis presumes that the film shape remains self-similar even at high velocities.

The results obtained for various different cells are shown in figure 9. As is predicted in his paper, Bretherton's law breaks down for values of Ca larger than about 10^{-2}, and t_{max} saturates at a value which depends on the dimensions of the cell. This technique also enabled us to carry out separate measurements for the top and bottom films, which were indeed found to be symmetric. An experimental study of the film

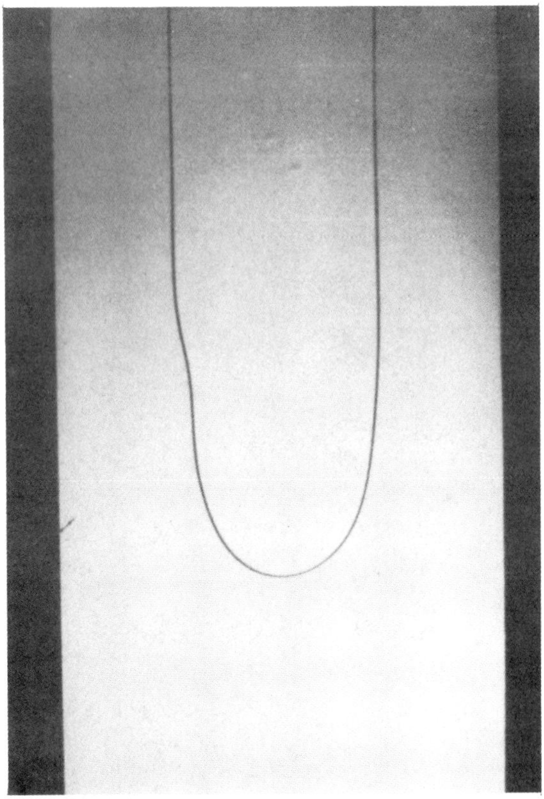

FIGURE 10. Picture of the first instability for $1/B \approx 7000$.

left behind a bubble in a capillary tube was performed by Taylor (1961), and a numerical two-dimensional calculation of the film thickness up to $Ca = 2$ has been published by Reinelt & Saffman (1985). In figure 9 we plot some points taken from figure 4 of that paper, adjusted for a cell thickness of 0.8 mm. When comparing them with the experimental data one has to take into account that in the experiments there is an effect of the aspect ratio on the film thickness, as is apparent from figure 9, while the Reinelt–Saffman calculation corresponds to the limit $w/b \rightarrow \infty$.

In passing we note that all the experimental curves can be fitted by a stretched exponential of the form:

$$t_{max} = \kappa b(1-e^{-\gamma(w/b)})(1-e^{-\beta Ca^{\frac{2}{3}}}), \tag{6}$$

with $\kappa \approx 0.119, \gamma \approx 0.038, \beta \approx 8.58$. This law reduces to Bretherton's for small values of Ca.

7. Study of the stability of the fingers

We have performed several experiments in two different cells in order to study the stability of the fingers.

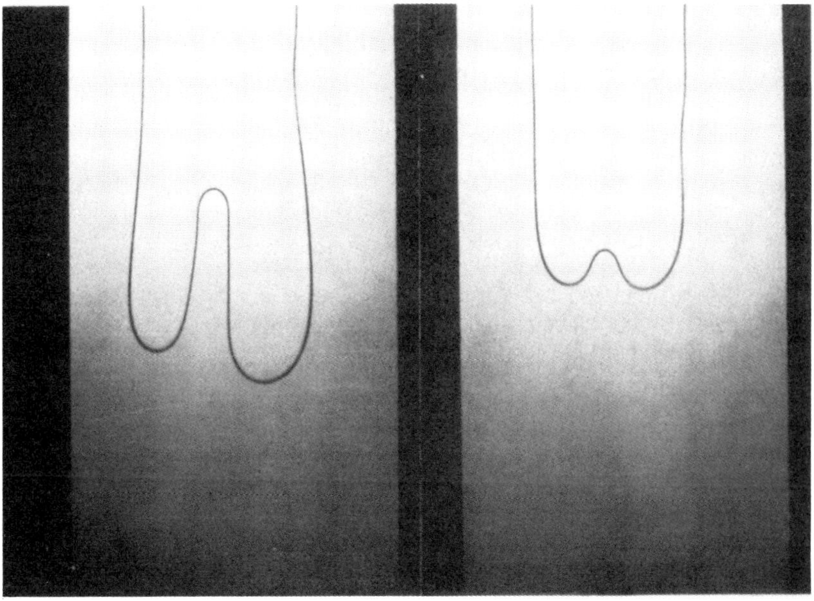

FIGURE 11. Tip splitting instability at two different times.

The first cell (cell I) had dimensions $b = 0.08$ cm, $w = 5.26$ cm and the oil used was Rhodorsil 47V500 ($T = 20.1$ dynes/cm, $\mu = 4.76$ poise). The other cell (cell II) had dimensions $w = 5.7$ cm, $b = 0.076$ cm and the oil used was a lubricating oil, whose characteristics are: $T = 30$ dynes/cm, $\mu = 3$ poises. Both cells have similar characteristics, but the gap of cell II is less uniform than that of cell I: we estimated the maximum variations $\delta b/b$ for cell II equal to 3×10^{-2}, whereas for cell I it is around 3×10^{-3}, an order of magnitude improvement in the homogeneity of the gap between the plates. The net result was that for cell II the first instabilities appeared for $1/B \approx 3000$, whereas for cell I the threshold was pushed to $1/B \approx 7000$. Apart from the onset value we have observed the same hierarchy of instabilities for both cells.

Figure 10 shows the first type of disturbance appearing beyond a critical $1/B$ value. When the finger velocity is increased, the interface becomes unstable in the way shown in the figure. A disturbance appears on one side of the tip, grows, recedes backwards and is further damped. In a frame of reference moving with the finger, the disturbance looks like a wave packet which travels backwards, while in the laboratory frame this packet looks stationary. The perturbations are stretched as they move along the interface, as a result of the velocity gradient they encounter. A particular feature of this instability is that it seems to be localized. When a disturbance grows on one side, it leaves the other side unperturbed. There is no real threshold value of $1/B$ for the instability, but in each cell, there is a narrow region in the parameter space above which the fingers are unstable. Let us call it 'threshold region'.

The physical origin of the instability is related to small non-uniformities of the cell as can be surmized from the fact that the disturbances generally appear at the same places in the channel. This feature is clearly observable when $1/B$ is close to the threshold region. In cell I, when $1/B \approx 7000$, we usually observe one or two disturbances growing along the interface and then decaying for a complete course of the finger along the channel. The number of events increases with $1/B$; for a value of $1/B$ larger than 12000, the disturbances appear almost everywhere; similar features have been observed in cell II.

Beyond the non-symmetric disturbances and for larger values of $1/B$ tip splitting occurs, as shown in figure 11. This kind of instability has also been recently observed experimentally by Park & Homsy (1985) and in computer simulations by De Gregoria & Schwartz (1985).

Let us now try to summarize the theoretical understanding; the puzzling problem was the result of McLean & Saffman (1981) showing that the fingers were linearly unstable. Then came the discovery by De Gregoria & Schwartz (1985), from computer simulations, that noise was an important factor concerning the stability. Then Bensimon (1986), Kessler & Levine (1985) showed that the fingers are linearly stable up to large values of the parameter $1/B$, and they studied the nonlinear instability of the system. There is a strong analogy between the instability of the fingers and that of the premixed flames. The latter problem was analysed by Zel'dovitch *et al.* (1980). In both cases, the non-symmetric disturbances are advected and stretched along the interface. The effect of the advection is stabilizing because the perturbation is brought to regions of lower velocities. The effect of the stretching is also stabilizing because it leads to an increase of the wavelength and hence to a decrease of the amplification rate of the perturbation. The existence of these two mechanisms may explain why the fingers can be stable. Now the instability occurs anyway if the amplitude of the disturbance is large enough. Bensimon *et al.* (1986) have obtained a relation between the amplitude of the noise ν_c necessary to obtain marginal eigenvalues and the parameter $1/B$. The striking result is that ν_c is an exponential function of $1/B$, so that the instability threshold is entirely determined by the velocity. The picture that they obtained is that of a system extremely sensitive to changes in velocity as far as the stability is concerned.

From the experimental side, many observed features are in agreement with the theory. The description of the evolution of the non-symmetric disturbances is very close to that predicted by the theory; also the shapes of the unstable fingers are in excellent agreement with the theory (see Bensimon (1986) in which a detailed comparison is made).

However, the agreement becomes more ambiguous when we look at the sensitivity of the system to noise. According to the theory, the instability of the fingers is controlled by $1/B$: significant differences between the intrinsic noise of two experiments should induce only small changes for the 'threshold' value of $1/B$. This is not what we observe. Small improvements of the quality of our experimental set-up led to dramatic changes in the threshold values of $1/B$, while the threshold value itself is not sharply defined.

We have tried to go further into this question by applying some external perturbation on the interfaces and see how they eventually become unstable. The perturbation was an electric field. Although such experiments were not entirely conclusive, it appeared that the critical value of the noise was weakly dependent on the velocity below the threshold region, which is in conflict with theory.

The origin of such a disagreement is possibly related to the existence of the film, which has been neglected by the theory. The effect of the film is stabilizing. For the case of a plane interface, the analysis shows that the presence of a film reduces the amplification rate of the disturbance and shifts the critical wavenumber towards smaller values. One feels that both effects can play a role in the nonlinear stability of the fingers.

8. Conclusion

The main conclusion of this paper is that the film plays a crucial role in the Saffman–Taylor problem. This was realized by Pitts (1980) and by McLean & Saffman, and the Reinelt–Saffman work was stimulated by this question, but no clear experimental account of its importance has been reported previously. We do not observe a $\lambda = \frac{1}{2}$ plateau in the λ *vs.* $1/B$ curve for large $1/B$, and this can be ascribed to the fact that the curvature of the meniscus keeps changing with changing Ca. Concerning the stability of the fingers, we observe that the value of $1/B$ at which the first instabilities occur depend strongly on the noise.

We thank D. Bensimon, M. Jensen, L. Kadanoff and P. Pelcé for very illuminating discussions. This work was supported by the National Science Foundation under Grant N. DMR-8316204, and also by an NSFINT-8412371 exchange award for P. Tabeling.

REFERENCES

Bensimon, D. 1986 Stability of viscous fingering. *Phys. Rev.* A (in press).

Bensimon, D., Kadanoff, L. P., Liang, S., Shraiman, B. I. & Tang, C. 1986 Viscous flows in two dimensions. *Rev. Mod. Phys.* (in press).

Bretherton, F. P. 1961 The motion of long bubbles in tubes. *J. Fluid Mech.* **10**, 166.

De Gregoria, A. J. & Schwartz, L. W. 1985 A boundary integral method for two-phase displacement in Hele-Shaw cells. *Exxon preprints.*

Hele-Shaw, H. J. S. 1898 On the motion of a viscous fluid between two parallel plates. *Nature* **58**, 34.

Kessler, D. A. & Levine, H. 1985 Theory of the Saffman–Taylor finger. Preprint.

Landau, L. & Levich, B. 1942 Dragging of a liquid by a moving plate. *Acta Phisicochimica USSR* **17**, 42.

McLean, J. W. 1980 The fingering problem in flows through porous media. PhD thesis, Caltech.

McLean, J. W. & Saffman, P. G. 1981 The effect of surface tension on the shape of fingers in a Hele-Shaw cell. *J. Fluid Mech.* **102**, 455.

Maher, J. 1985 The development of viscous fingering patterns. *Phys. Rev. Lett.* **54**, 1498.

Padday, J. F. 1969 In *Surface and Colloid Science* (ed. E. Matijevic), vol. 1, p. 39. Wiley.

Park, C. W., Gorell, S. & Homsy, G. M. 1984 Two-phase displacement in Hele-Shaw cells: experiments on viscously driven instabilities. *J. Fluid Mech.* **141**, 275.

Park, C. W. & Homsy, G. M. 1984 Two phase displacement in Hele-Shaw cells: theory. *J. Fluid Mech.* **139**, 291.

Park, C. W. & Homsy, G. M. 1985 The instability of long fingers in Hele-Shaw flows. *Phys. Fluids* **28**, 1583.

Pitts, E. 1980 Penetration of fluids into a Hele-Shaw cell. *J. Fluid Mech.* **97**, 53.

Reinelt, D. A. & Saffman, P. G. 1985 The penetration of a finger into a viscous fluid in a channel and tube. *J. Sci. Stat. Comput.* **6** (3), 542.

Romero, L. 1982 PhD thesis. California Institute of Technology.

Saffman, P. G. & Taylor, G. I. 1958 The penetration of a fluid into a porous medium or Hele-Shaw cell containing a more viscous liquid. *Proc. R. Soc. Lond.* A **245**, 312.

Shraiman, B. I. & Bensimon, D. 1984 Singularities in non local interface dynamics. *Phys. Rev.* A **30**, 2840.

TABELING, P. & LIBCHABER, A. 1986 Film draining and the Saffman–Taylor problem. *Phys. Rev.* A **33**, 794.

TANNER, L. H. 1979 The spreading of silicone oil drops on horizontal surfaces. *J. Phys.* D: *Appl. Phys.* **12**, 1473.

TAYLOR, G. I. 1961 Deposition of a viscous fluid on the wall of a tube. *J. Fluid Mech.* **10**, 161.

VANDEN-BROECK, J.-M. 1983 Fingers in Hele-Shaw cells with surface tension. *Phys. Fluids* **26**, 2033.

ZEL'DOVITCH, YA. B., ISTRATOV, A. G., KIDIN, N. I. & LIBROVITCH, V. B. 1980 Flame propagation in tubes: hydrodynamics and stability. *Comb. Sci. Tech.* **24**, 1.

Narrow fingers in the Saffman-Taylor instability

Y. Couder, N. Gérard, and M. Rabaud
Groupe de Physique des Solides de l'Ecole Normale Supérieure, 24 rue Lhomond, 75231 Paris Cedex 05, France
(Received 21 July 1986)

Saffman-Taylor fingers with a relative width much smaller than the classical limit $\lambda = 0.5$ are found when a small isolated bubble is located at their tip. These solutions are members of a family found by Saffman and Taylor neglecting superficial tension. Recent theories have shown that when capillary forces are taken into account an unphysical cusplike singularity would appear at the tip of all the fingers with $\lambda < 0.5$. Conversely, here the replacement of the tip by a small bubble makes these solutions possible. At large velocity these fingers show dendritic instability.

When between two narrowly spaced glass plates, a fluid of small viscosity drives a fluid of large viscosity, their interface is affected by the Saffman-Taylor instability. Two geometrical configurations are used for its study. In the original Saffman-Taylor experiment[1-4] the fluids are confined in a linear channel of width w. In the axisymmetric geometry introduced by Paterson[5] the viscous fluid is contained between two circular glass disks and the less viscous fluid is injected at the center. In both geometries the initial destabilization is characterized by its length scale $l_0 \sim b(T/\mu V)^{1/2}$ where T is the surface tension, μ the dynamical viscosity, b the thickness of the cell, and V the velocity of the interface. The subsequent evolution of the instability is very different in each geometry.

In the axisymmetric case[5] the fingers tend to remain a size of the order of l_0. As they move outwards they undergo a series of irregular tip splitting and side branching. In a previous experiment[6] performed in this geometry we showed that the influence of an isolated bubble at the tip created an artificial sharper point to a growing finger. Because of the higher curvature at the tip, the pressure gradient is increased and so is the finger velocity. These faster fingers take a parabolic shape which, for larger velocities, is affected by dendritic instabilities very similar to those observed in crystal growth.

In the present paper we study the influence of a perturbing bubble in the linear geometry. We will first recall the main classical results in this case. Because of the geometrical configuration, the growing fingers tend to screen each other off, so that only the fastest continue to grow. This gives rise to a stable finger shape where the length scale of the width w of the cell imposes itself. The finger is characterized by the ratio λ of its width to that of the cell. In the original paper by Saffman and Taylor λ was studied as a function of the capillary number $C = \mu V/T$. More recently a new parameter[4,7] was used, $1/B = 12[(w/b)^2 \times (\mu V/T)]$ which for equal values gave comparable λ in cells of various geometrical dimensions. In a particular cell for a given fluid the control parameter is the velocity V which can be chosen by changing the applied pressure.

The classical puzzling result is that with increasing $1/B$ the relative width λ of the finger first decreases from 1 to 0.5, at which it remains in a very large range of values of $1/B$.

Saffman and Taylor, assuming that the capillary forces should become negligible at large values of $1/B$, solved analytically the two-dimensional potential flow problem in the absence of surface tension. They found a family of solutions parametrized by λ which could take all values $0 < \lambda < 1$,

$$x = \frac{1-\lambda}{\pi} \ln \frac{1}{2}\left(1 + \cos\frac{\pi y}{\lambda}\right) . \tag{1}$$

The origin is taken at the tip of the finger; $0x$ and $0y$ are, respectively, parallel to the length and to the width of the cell and the half width $w/2$ of the cell is taken as unit length. No physical argument indicates which solution is selected. Furthermore at low velocity the observed fingers do not correspond to the predicted shape, only the particular $\lambda = 0.5$ solution corresponds to the fingers observed at large $1/B$. No reason could be found at first why the solutions $\lambda < 0.5$ were not observed at all.

Pitts,[2] using an empirical law on the local curvature at a point of the profile, found a modified relation

$$x = \frac{\lambda}{\pi} \ln \frac{1}{2}\left(1 + \cos\frac{\pi y}{\lambda}\right) . \tag{2}$$

This fitted well the observed fingers in the range $0.5 < \lambda < 0.80$, where all the experimental profiles are homothetics so that they can all be reduced to a single curve by magnification of the axes.

The first attempt to take surface tension into account in this problem was made by McLean and Saffman.[3] Following their work, recent progress in this problem has been made by Vanden-Broek[8] using numerical techniques, and by Dombre, Hakim, and Pomeau,[9] Shraiman,[10] and Hong and Langer[11] using analytical calculation. They showed that the taking into account of surface tension in the Saffman-Taylor solutions leads to finding a cusplike singularity at the tip of most of them. Only for a family of solutions the cusp's amplitude is reduced to zero. These solutions have width $\lambda > \frac{1}{2}$ and all tend toward $\lambda = \frac{1}{2}$ for vanishing surface tension.

I. EXPERIMENT SETUP

Four experimental cells were used. They were made with glass plates 1.50 m long and 1.5 cm thick separated by 0.1-cm-thick spacers. Their widths w were, respectively, 12, 6, 4, and 2 cm. The viscous fluid filling them was a silicone oil Rhodorsil 47V100 with surface tension $T=20.9$ N/m at 25° and viscosity $\mu=9.65\times10^{-2}$ kg/ms. The fluid of low viscosity was nitrogen gas. We measured finger shapes, width, and velocities either with photographs taken with a motorized camera or with video tape recordings. Small bubbles were injected into the cell before the experiment started by means of a long hypodermic needle.

II. EXPERIMENTAL RESULTS

We inject into the cell small air bubbles near the initial motionless front between nitrogen and oil. To be effective these bubbles have to be of an oblate shape between the flat plates so that their initial diameter in the cell plane must be a few times the thickness b. When pressure is applied the bubbles start moving. As their friction on the glass plate is negligible they form zones of constant pressure which distort the neighboring isobars in the oil. As the front destabilizes, one of the fingers situated behind a bubble will grow faster and catch up. The bubble then remains stable at the tip. As the velocity of this finger becomes larger (because of the increase of the curvature of the tip) it then overcomes the other fingers and forms a steady solution translating along the whole cell (Fig. 1).

Fingers with a bubble at the tip can be observed in a large range of values of velocities. However, for small velocities the bubble does not always remain stable; it drifts off on the side so that its effect vanishes. For large velocities the bubble and the fingertip are pressed against each other so that the tip of the finger and the back of the bubble are flattened and reach a stable configuration (Fig. 1).

For a given applied pressure the velocity of the new stable finger with a bubble at the tip is larger and its width λ smaller than in the usual Saffman-Taylor case. If the film that separates the bubble from the finger breaks, the finger slows down and returns to its classical shape.

The photograph of a finger with $\lambda=0.32$ obtained in the cell $w=6$ cm for $V=5$ cm/s is shown on Fig. 1. The finger profile has a well-determined shape which does not include the isolated bubble. We have superposed, excluding the bubble, points of the profiles $\lambda=0.32$ calculated, respectively, with Eqs. (1) and (2). We see immediately that this finger is not a member of the family described by Pitt's formula, but that its shape coincides with striking precision with the solution of Saffman and Taylor. The only deviation is localized behind the bubble where its influence has created a little flat part.

Figure 2 shows the evolution of the observed values of λ as a function of the capillary number in three of the four experimental cells. As in the usual fingers, λ decreases with increasing C, and a saturation occurs at a value λ_s, which depends upon the aspect ratio of the cell. We find for cells of thickness $b=0.1$ cm:

$$\lambda_s \simeq 0.46 \text{ for } w=2\text{ cm},\quad \lambda_s \simeq 0.345 \text{ for } w=4\text{ cm},$$

$$\lambda_s \simeq 0.30 \text{ for } w=6\text{ cm},\quad \lambda_s \simeq 0.225 \text{ for } w=12\text{ cm}.$$

These values are much smaller than the classical limit $\lambda=0.5$. Contrary to the classical situation, the results obtained in the various cells cannot be brought to near coincidence by using the dimensionless number $1/B$.

The mean diameter d of the efficient bubbles can be chosen in the range $b<d<8b$. For the same applied pressure the velocity of the finger will depend upon the size of the bubble. However, in a given geometry, all the ob-

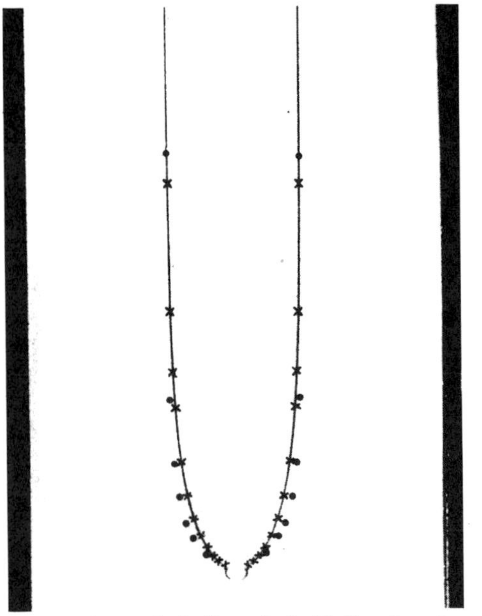

FIG. 1. Photograph of a finger obtained in the cell $w=6$ cm, $b=0.1$ cm at $V=5.3$ cm/s. The crosses are points of the theoretical profile from Saffman-Taylor Eq. (1) for $\lambda=0.32$. The dots are calculated from Pitt's Eq. (2) for $\lambda=0.32$.

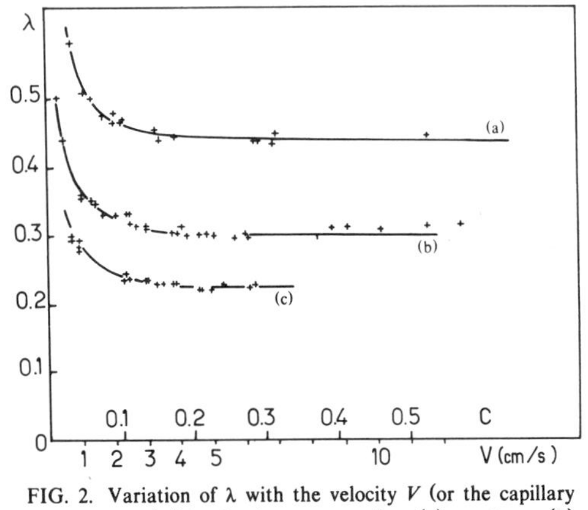

FIG. 2. Variation of λ with the velocity V (or the capillary number $C=\mu V/T$) in the three geometries: (a) $w=2$ cm, (b) $w=6$ cm, (c) $w=12$ cm.

served values of λ lie on the same curve $\lambda(V)$ (Fig. 2). This shows that the shape of a finger depends only on its velocity. This velocity is itself determined by both the applied pressure and the bubble size.

To understand these results we must remember that in the axisymmetric case the fingers, in the presence of a bubble, take a parabolic shape[6] with a curvature at the tip determined by the velocity.

In the present linear case the solutions given by Eq. (1) also have a parabolic shape near the tip. For small values of y they reduce to

$$x=\frac{\lambda-1}{4\lambda^2}\pi y^2 . \tag{3}$$

We must remember that the solutions (1) were found, $w/2$ being taken as unit lengh. In a real cell the shape of the finger near its tip can be characterized by the curvature K of the parabola at the tip

$$1/K=-\frac{\lambda^2 w}{\pi(1-\lambda)} . \tag{4}$$

From the measures of λ we can deduce the curvature K as a function of the velocity V in the different cells. The result is shown on Fig. 3. It shows that at a given velocity the curvature K is the same in all the cells (of equal thickness). Then Eq. (4) gives a relation between the observed λ and the width w of the cell:

$$\lambda=\frac{1-(1-4wK/\pi)^{1/2}}{2wK/\pi} . \tag{5}$$

Figure 4 shows the observed values of λ in the four experimental cells at a velocity $V=6$ cm/s compared to the predicted dependence $\lambda(w)$ given by the relation (5). The best fitting value of K for this velocity was $K=4.1\ \mathrm{cm}^{-1}$.

We can compare the destabilization of these new fingers to that of the classical ones. For values of $1/B>7000$ the usual Saffman-Taylor fingers destabilize when irregular side branching and tip splitting occur.[4] With a very small

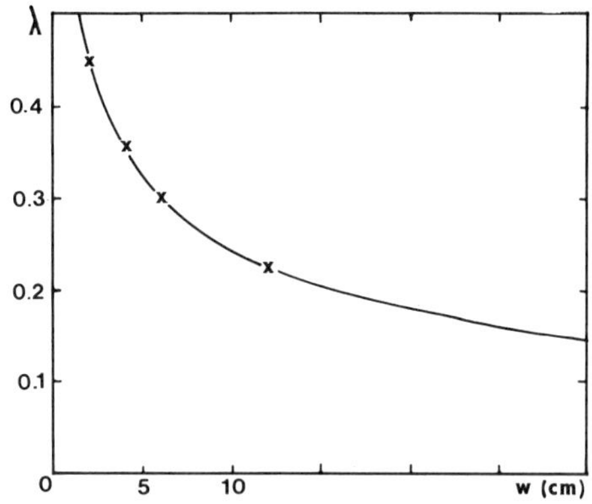

FIG. 4. Values of λ as a function of the width of the cell at velocity $V=6$ cm/s. The continuous line is calculated from Eq (5), the crosses are experimental.

bubble at their tip the fingers are more stable and can be observed to values as large as 20 000.

They also destabilize differently into one of two periodic behaviors. In the first one [Fig. 5(a)], observed for small bubbles, the curvature of the extremity of the finger varies periodically. The pulsating tip creates symmetrical lateral branches. When the bubble is large [Fig. 5(b)] the extremity of the finger oscillates transversely. We had ob-

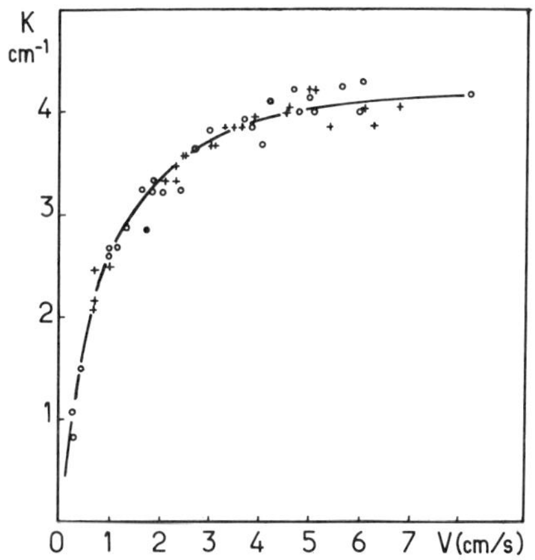

FIG. 3. The curvature of the parabolic extremity as a function of the velocity V. ○: $w=6$ cm; +: $w=12$ cm.

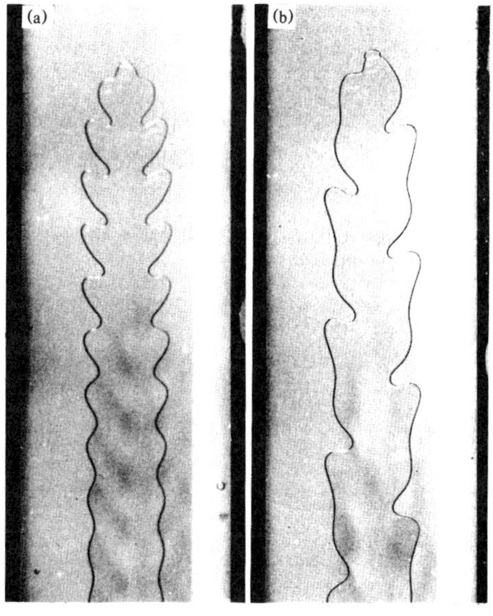

FIG. 5. (a) The pulsating tip regime of a finger with a small bubble, $w=6$ cm, $V=10$ cm/s, (b) the oscillating tip regime of a finger with a larger tip bubble, $w=6$ cm, $V=7$ cm/s.

served[6] both these instabilities in the Paterson geometry where the lateral branches could fully develop. We pointed out that the first one is identical in shape to a classical dendritic instability. In the present experiment the vicinity of the side walls limits the growth of the instability. In the widest cell, where λ is small, lateral branches can still be formed. In the narrowest the growth is practically inhibited. These instabilities will be described in more detail in a forthcoming paper.

Finally, we tried to apply to the finger another type of local perturbation of the tip. We stretched a very thin thread along the whole length of the cell and in the middle of its width. This technique was first used by Grace and Harrison[12] who showed that in a Hele Shaw cell, rising bubbles have a larger velocity when they surround a vertical wire. Except for their tip which becomes asymmetrical with respect to the wire, fingers similar to those obtained with the small bubble are observed.[13] Dendritic instability also occurs. A similar result has been obtained recently by A. Libchaber.[14]

III. CONCLUSIONS

For large velocities where viscous forces should dominate, surface tension remains important in the determination of the shape of the fingers. Recent theories have shown that this is due to its critical role in the localized region of the finger tip. Conversely, we have shown that introducing a localized perturbation at the tip suppresses the width selection described in Refs. 9–11. Other solutions (those predicted in theories neglecting surface tension) become possible. These are the parametrized Saffman-Taylor solutions here and the parabolas in the circular geometry. The curvature at the tip is determined by the velocity and selects the finger experimentally observed. Both are affected at larger velocities by dendritic instabilities.

Dendrites have usually been associated with crystalline anisotropy. In the present experiment as well as in the axisymmetric one, we show that a local perturbation of the tip creates these characteristic patterns. Anisotropy of the medium is not therefore a necessary condition to dendritic growth. It is only one of the means by which the singularity of the tip can be removed so that fingers with parabolic extremities are made possible.

Note added in proof. Recent experiments in cells of various thicknesses (b = 0.05, 0.1, and 0.2 cm) show that the relevant parameter for the cruves $\lambda(V)$ of Fig. 2 is the ratio of the width over the thickness of the cell w/b. The nondimensional version of Fig. 3 is then a plot of Kb vs C and in Fig. 4 the abscissa can be scaled adimensionally in w/b.

ACKNOWLEDGMENTS

We are very grateful to P. Tabeling for initiating us to many aspects of the Saffman-Taylor instability. We thank C. Caroli, T. Dombre, V. Hakim, A. Libchaber, and Y. Pomeau for many fruitful discussions.

[1]P. G. Saffman and G. I. Taylor, Proc. R. Soc. London, Ser. A **245**, 312 (1958).

[2]E. Pitts, J. Fluid Mech. **97**, 53 (1980).

[3]J. W. McLean and P. G. Saffman, J. Fluid Mech. **102**, 455 (1981).

[4]P. Tabeling, G. Zocchi, and A. Libchaber, J. Fluid Mech. (to be published).

[5]L. Paterson, J. Fluid Mech. **113**, 513 (1981).

[6]Y. Couder, O. Cardoso, D. Dupuy, P. Tavernier, and W. Thom, Europhys. Lett. **2**, 437 (1986).

[7]G. Tryggvason and H. Aref, J. Fluid Mech. **136**, 1 (1983).

[8]J. M. Vanden-Broeck, Phys. Fluids **26**, 2033 (1983).

[9]T. Dombre, V. Hakim, and Y. Pomeau, C. R. Acad. Sci. Ser. A **11302**, 803 (1986); R. Combescot, T. Dombre, V. Hakim, Y. Pomeau, and A. Pumir, Phys. Rev. Lett. **56**, 2036 (1986).

[10]B. Shraiman, Phys. Rev. Lett. **56**, 2028 (1986).

[11]D. C. Hong and J. Langer, Phys. Rev. Lett. **56**, 2032 (1986).

[12]J. R. Grace and D. Harrison, Chem. Eng. Sci. **22**, 1337 (1967).

[13]In the axisymmetric configuration, Ben Jacob *et al.* introduced a strong anisotropy by engraving deep grooves in the cell plates. They observed that the general anisotropy of the cell induced dendritic fingers. With a localized thread we show that a similar effect is obtained by acting on the tip of the finger only. [E. Ben Jacob, R. Godbey, N. D. Goldenfeld, J. Koplik, H. Levine, T. Muller, and L. M. Sander, Phys. Rev. Lett. **55**, 1315 (1985)].

[14]A. Libchaber (private communication).

Note added in proof: Since this publication we have shown (M.Rabaud, Y.Couder and N.Gerard, Preprint) that the relation K(V) shown on Fig. 3 is due to a crossover between two regimes. At low velocities K evolves as

$$K \approx \frac{1.0}{b}\sqrt{\frac{\mu V}{T}},$$

at high velocities the thickness of the cell becomes the dominant length scale and K saturates at

$$K \approx \frac{0.43}{b}.$$

Experimental Demonstration of the Role of Anisotropy in Interfacial Pattern Formation

E. Ben-Jacob and R. Godbey
Department of Physics, University of Michigan, Ann Arbor, Michigan 48109

and

Nigel D. Goldenfeld[a]
Institute for Theoretical Physics, University of California, Santa Barbara, California 93106

and

J. Koplik and H. Levine
Schlumberger-Doll Research, Ridgefield, Connecticut 06877

and

T. Mueller and L. M. Sander
Department of Physics, University of Michigan, Ann Arbor, Michigan 48109
(Received 4 March 1985)

We impose anisotropy in modified Hele-Shaw experiments by engraving a grid on one of the plates. Without anisotropy, the dynamics is dominated by tip bifurcations leading to a branched ramified structure. With anisotropy, a dendritic pattern forms, in qualitative agreement with the prediction of recent studies of local growth models.

PACS numbers: 61.50.Cj, 05.70.Ln, 68.70.+w, 81.30.Fb

All manner of interfacial patterns have been observed in various systems and they all seem to differ greatly from each other. For instance, dendritic structures are detected in solidification of a pure undercooled liquid,[1-3] periodic cell structures in directional solidification,[4,5] Saffman-Taylor "fingers"[6] in a Hele-Shaw cell,[7] and self-similar branching structure in electrochemical deposition.[8] In view of the variation in patterns in these systems, no universality in their interfacial dynamics appears to exist. However, we believe that underlying these seemingly unrelated phenomena is a common principle, namely that growth always occurs in two separate ways. Either parabolic dendrites propagate in a way that is governed by anisotropy, or, if the anisotropy is too small, a disorderly behavior dominated by tip splittings results.

A major step toward revealing this principle that governs the dynamics of the systems and the transition from a regime of dendritic dynamics to one of branching dynamics was the development of the geometrical model (GM)[9] and the boundary-layer model (IBM).[10] These are rather simple local growth models inspired by the models of diffusion-controlled solidification. Their simplicity has permitted not only numerical computation[11,12] of the evolution of the interface, but also the elucidation of a new selection mechanism[9,10]: For a given finite anisotropy and low enough undercooling there exists a discrete family of steady-state, shape-preserving, needle-crystal solutions. The selected dendrite is formed by side branches emerging from the needle crystal of highest velocity.[13]

Moreover, in the GM it has been shown[9] that at a critical value of anisotropy (for a given undercooling) all needle-crystal solutions are unstable except th selected one, which is marginally unstable with respec to the tip-splitting mode. At higher anisotropy (for given undercooling) this mode is stable, and it is unstable below the critical anisotropy. Dendrites are observed in the GM for this specific value of critical anisotropy (for a given undercooling). A similar analysi has not been carried out for the BLM, but there is nonzero range of anisotropy for which dendriti growth occurs in numerical simulations of the model.[1] Numerical simulations of the BLM and the GM shov that below a critical undercooling (for a given anisotropy) the tip evolves by broadening, flattening, an eventually bifurcating into two new tips which continue to evolve in this fashion, giving rise to a successio of tip bifurcations. The experiments which we repor here indicate the phase diagram presented in Fig. 1 for interfacial dynamics of diffusion-controlled processes, which is consistent with the results of the loca models.

We must bear in mind that the role of anisotrop and the phase diagram outlined above have beer demonstrated in local models which are drastic simplifications of the diffusion-controlled solidification, wit neglect of interactions between points on the interfac which are close in real space but not in terms of thei arc-length distance. The GM further neglects memor effects associated with the diffusion field. Thus th question arises: Are long-range memory effects sufficient to give rise to dendritic behavior, or is the presence of anisotropy crucial not only in the GM and th BLM, but in nature as well?[15]

The purpose of this paper is to report the first quali

Reprinted with permission from *Physical Review Letters*, Vol. 55, No. 12, 1985.

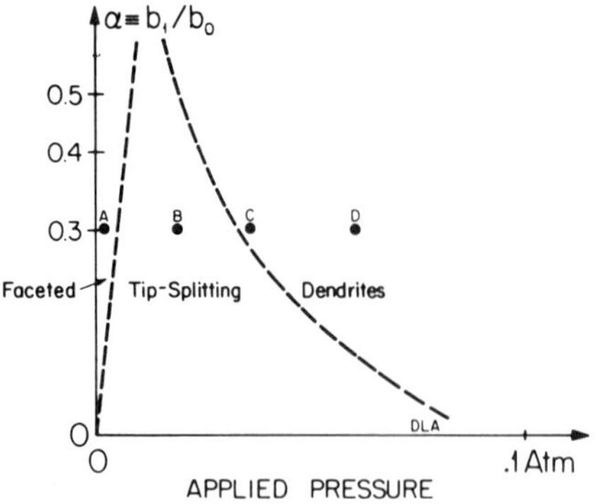

FIG. 1. Preliminary estimated phase diagram for the Hele-Shaw experiment with anisotropy. The points A,B,C,D indicate the values corresponding to the patterns shown in Fig. 2.

ative experiment to demonstrate the role of anisotropy in interfacial dynamics. Our experiment indicates that these ideas are not an artifact of the approximate nature of the GM and the BLM, but can be a universal feature of interfacial dynamics. To demonstrate the role of anisotropy in interfacial pattern formation we introduce anisotropy in a modified version of a Hele-Shaw experiment by grooving a lattice on one of the plates. Doing so, we observe the formation of dendrites with parabolic tip and side branches similar to those observed in solidification. These are not steady-state solutions, in the sense of having constant growth velocity, but their shape does seem to be time invariant. Moreover, for a given driving force, as we lower the anisotropy we observe tip bifurcations which demonstrate the transition from dendrites to branching as illustrated in Fig. 1. On the basis of our experiments, we postulate the following scenario for the transition from dendritic structure to fractal structure such as seen in diffusion-limited aggregation (DLA)[16]: For a given driving force there is a critical value of the anisotropy below which the tip of the dendrite is unstable with respect to tip bifurcation. Consequently, below the critical value the interfacial dynamics is dominated by tip splitting. As we drive the system further from equilibrium, in the absence of anisotropy, both the period between successive tip bifurcations and the minimum length scale decrease, giving rise to a DLA-like structure in the limit of a very large driving force.[17]

The Saffman-Taylor version of the Hele-Shaw experiment consisted of an interface between two immiscible fluids with different viscosity. The fluids are confined between two parallel closely separated plates. When pressure is applied to the less viscous fluid, the interface is unstable and known to develop Saffman-Taylor "fingers".[6,18]

The mathematical model describing the dynamics of the interface in a Hele-Shaw cell is the Laplace limit of the full diffusion problem: The velocity field in the viscous fluid is assumed to be two-dimensional and proportional to the pressure gradient as[6]

$$\mathbf{v} = (-b^2/12\eta)\nabla P, \tag{1}$$

where b is the plate separation and η is the viscosity. Similarly, the interface velocity along its outer normal direction (into the viscous fluid) is given by the pressure gradient at the interface. Here we consider the limit of incompressible fluids so that the pressure satisfies the Laplace equation

$$\nabla^2 P = 0. \tag{2}$$

A further assumption that the fluids are immiscible leads to the inclusion of surface tension in the pressure boundary conditions.

At infinity, in the more viscous fluid, $P_\infty = P_0$ where P_0 is the atmospheric pressure. The other boundary condition is P_s, the pressure along the moving interface:

$$P_s = P_1 - d_0\kappa, \tag{3}$$

where P_1 is the applied pressure in the less viscous fluid,[19] d_0 is the isotropic surface tension (in the appropriate units), and κ is the mean curvature of the interface given by

$$\kappa = 1/R_\perp + 1/R_\parallel. \tag{4}$$

$R_\parallel$ and $R_\perp$ are the radii of curvature of the interface parallel and perpendicular to the plane of the fluid, respectively. Equations (1) and (2), together with the boundary conditions given by Eq. (3), specify the Hele-Shaw model. Note that this system is reminiscent of DLA with finite surface tension.[20] It differs from the models of diffusion-controlled solidification[11] in three aspects: The pressure replaces the temperature field, the diffusion equation is approximated by the Laplace equation, and there is no anisotropy.

Next we describe the results of our version of the Hele-Shaw experiment. The experiment was done by application of pressure at the center of the cell in a manner similar to that of Patterson.[21] We used glycerine (94%) dyed with food color as the viscous fluid and air as the less viscous fluid. The bottom plate is circular with radius 25 cm. On this plate we engraved a regular sixfold lattice of grooves with depth $b_1 = 0.015$ in., width of 0.03 in., and edge-to-edge separation of 0.03 in. We varied the effective anisotropy, α, by changing the spacing b_0 between the two plates; α is then defined by

$$\alpha = b_1/b_0. \tag{5}$$

The typical range is from $\alpha = 0.1$ to $\alpha = 1$. The pressure was applied from a very large (5-gal.) pressure reservoir. The typical range is from $P = 10^{-3}$ atm to $P = 10^{-1}$ atm. In Fig. 1 we show a preliminary estimate of the phase diagram based on the results of the experiment. When the driving force is very small for a given anisotropy α we observe faceted growth [Fig. 2(a)]. The interface includes flat faces that advance one layer at a time via the propagation of kinks. This regime will be described in detail in a forthcoming publication.[22] In the next regime the dynamics of the interface is dominated by the tip-bifurcation instability similar to that observed in the local models [Fig. 2(b)]. We note that in the circular geometry[21] there is no stabilization effect of the walls against tip splitting, in contrast to the case of the channel geometry.[23,24] In our cell we were only able to observe a few tip-splitting bifurcations. We think that in a much larger cell the cascade of tip bifurcations will give rise to a ramified branching structure that will approach a DLA-like fractal dimension on large length scales. In order to show that, we have repeated the experiment without a grooved lattice on the bottom plate. At high pressure we have observed a ramified structure and measured a fractal dimension near to that of DLA[25] ($D \sim 1.7$). This observation gives a qualitative understanding of the fractal dimension of DLA as a limiting process of successive tip bifurcations. In Fig. 2(c) we see the symmetric "snowflake" observed above the tip splitting $\rightarrow$ dendritic transition. As we increase the pressure even further we observe the appearance of sid branches on the dendrites and the radius of curvatur of the tip of the dendrites is reduced [Fig. 2(d)], agai in agreement with the prediction of the local models We emphasize that, as seen in Fig. 2, the size of th dendrites and the spacing between the side branches i not the lattice spacing.

FIG. 2. The various patterns observed in the Hele-Shaw experiments with anisotropy. (a)–(d) correspond to points (A)–(D) in Fig. 1, respectively, and show (a) faceted growth; (b) tip splitting; (c) needle crystals; (d) dendrites. The plate is 25 cm across.

We can carry the analogy between the Hele-Shaw experiment and solidification even further. In order t find a Hele-Shaw analog of direction solidification w have introduced a "lifting version" of the Hele-Sha experiment. Here, instead of applying pressure to th less viscous fluid, we lift the upper plate at the les viscous side (air, in our experiment) at a given rate Intuitively, it is clear that by lifting we impose a pres sure *gradient* which is analogous to the imposed tem perature gradient in directional solidification. To se this, we recall that by the lifting, b in Eq. (1) become (at a given time) a function of x—the distance fror the pivot—and is of the form

$$b(x) = b_0 + b_0 x/\zeta; \quad x \ll \zeta. \tag{6}$$

Equation (2) is replaced by

$$\nabla^2 P + (2/\zeta)\partial P/\partial x = 0, \tag{7}$$

which is the equation of motion for a difussion proces in a frame moving with velocity $2D/\zeta$, where D is th diffusion constant. $R_\perp$ in Eq. (4) has the following dependence:

$$R_\perp = c/b(x) \cong (c/b_0)(1 - x/\zeta), \tag{8}$$

where c and b_0 are constants, so that the pressure the interface [Eq. (3)] becomes

$$P_s = P_1 - d_0 b_0/c - d_0/R_\parallel - (d_0 b_0/c\zeta)x. \tag{9}$$

These equations have the same form as the equatior describing directional solidification in the quasistat limit.[5] In Fig. 3 we show some results of the interfa cial dynamics in the lifting experiment.

In this experiment we have used a bottom plate wit a square lattice grooved on it, with grooves of widt 0.03 in. and depth 0.015 in., separated by 0.03 in. W control the velocity of lifting by using a variable-spee motor. In Fig. 3(a) we show the resulting "fingers" low velocity[26] (about 0.5 cm/sec; the scale is 20 cm As we increase the velocity (by increasing the rate c lifting) we observe the array of dendrites shown in Fi 3(b). The spacing of the dendrites depends on the lif ing rate—faster lifting gives smaller spacing—and o the initial plate spacing. The velocity is about 1 cm/sec and the wavelength is about 2 cm. Finally, w emphasize that, similarly to the previous experimen in the absence of anisotropy the interface dynamics dominated by tip bifurcations. The resulting pattern shown in Fig. 3(c) (the portion shown is of width 2 cm). We note that for small anisotropy (large spacing we observe patterns similar to those of the numeric simulations of Vicsek[27] and Jensen.[28] We claim th

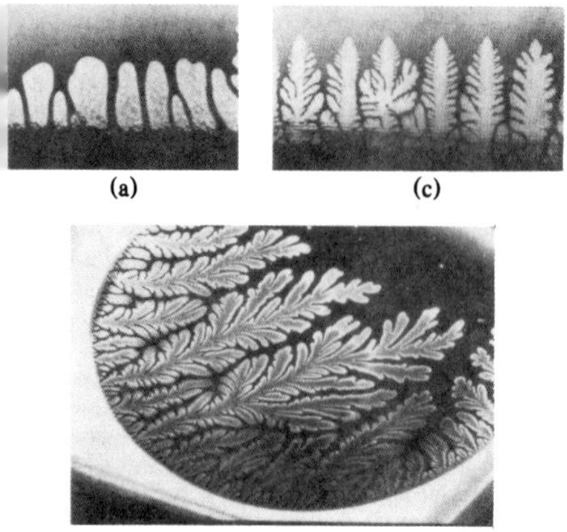

FIG. 3. The lifting Hele-Shaw experiment. (a) "Fingers" ıt low velocity. (b) Branched "fingers" at higher velocity. c) Array of dendrites with imposed anisotropy.

he use of a square lattice in the simulations in these eferences provided weak anisotropy.

In summary, we have described qualitative experinents which demonstrate the role of anisotropy in inerfacial pattern formation. This experiment proves hat anisotropy is truly essential to the formation of lendrites and not an artifact arising from the approxinate nature of the local models.[11,12] Other experinents, which study the effect of impurities on the tip ifurcation instability and impose anisotropy by use of iquid crystals as viscous fluids, are in progress. Quanitative detailed measurements of the phase diagram vill be reported in a forthcoming publication.[22]

This work was partially supported by the U.S. Department of Energy through Grant No. DE-FG0384ER45108 and the National Science Foundation hrough Grants No. DMR-82-03698 and No. PHY 82-7853, supplemented by funds from the National Aeronautics and Space Administration, and by the J.S. Army Research Office through Grant No. DAAG 9-83-K0131. One of us (E.B.-J.) acknowledges receipt of a Rackham Award, and two of us (R.G. and T.M.) acknowledge receipt of a Bendix Award. One of is (N.D.G.) thanks the University of Michigan for its iospitality during the initial stage of this project.

[a]Present address: Department of Physics, University of llinois, Urbana, Ill. 61801.

[1]M. E. Glicksman, R. J. Shaefer, and J. A. Ayers, Metall. Trans. A **7**, 1747 (1976).

[2]T. Fujioka, Ph.D. thesis, Carnegie-Mellon University, 1978 (unpublished).

[3]J. S. Langer, Rev. Mod. Phys. **52**, 1 (1980).

[4]L. A. Tarshis, J. L. Walker, and J. W. Ratter, in *Metallography, Structures and Phase Diagrams,* edited by M. B. Bauer (American Society for Metals Novelty, 1983).

[5]D. J. Wollkind and L. A. Segel, Trans. Roy. Soc. Edinburgh **51**, 268 (1970).

[6]P. G. Saffman and G. I. Taylor, Proc. Roy. Soc. London, Ser. A **245**, 312 (1958).

[7]H. S. S. Hele-Shaw, Nature **58**, 34 (1898).

[8]M. Matsushita, M. Sano. Y. Hayakawa, H. Honjo, and Y. Sawada, Phys. Rev. Lett. **53**, 286 (1984); R. Brady and R. C. Ball, Nature **309**, 225 (1984).

[9]D. Kessler, J. Koplik, and H. Levine, Phys. Rev. A **31**, 1712 (1985).

[10]E. Ben-Jacob, N. D. Goldenfeld, B. G. Kotliar, and J. S. Langer, Phys. Rev. Lett. **53**, 2110 (1984).

[11]E. Ben-Jacob, N. D. Goldenfeld, J. S. Langer, and G. Schön, Phys. Rev. Lett. **51**, 1930 (1983), and Phys. Rev. A **29**, 330 (1984).

[12]R. Brower, D. Kessler, J. Koplik, and H. Levine, Phys. Rev. Lett. **51**, 1111 (1983), and Phys. Rev. A **29**, 1335 (1984); D. Kessler, J. Koplik, and H. Levine, Phys. Rev. A **30**, 3161 (1984).

[13]For a given anisotropy, there is a critical undercooling below which there is no needle-crystal solution.

[14]The study of the stability of the needle-crystal solutions in the BLM is in progress. Preliminary results show that all solutions in the discrete family are unstable. The one with the largest velocity is the least unstable. The dendrites that have been observed are not transient since they maintain their tip velocity and tip curvature.

[15]Some additional qualitative evidence is given by a numerical simulation of the diffusion system in the zero–Peclet-number limit, where anisotropy was found to be required for tip stabilization: D. Kessler, J. Koplik, and H. Levine, Phys. Rev. A **30**, 2820 (1984).

[16]T. A. Witten, Jr., and L. M. Sander, Phys. Rev. Lett. **47**, 1400 (1981), and Phys. Rev. B **28**, 5686 (1983).

[17]The structure has a minimum length scale determined by the surface tension and the driving force. However, the skeleton has a DLA-like structure. As we drive the system further from equilibrium it shrinks to its limiting skeleton.

[18]G. Tryggvason and H. Aref, J. Fluid Mech. **136**, 1 (1983).

[19]We assume constant pressure in the less viscous fluid. This assumption is analogous to the assumption of no diffusion in the solid in diffusion-controlled solidification.

[20]L. Patterson, Phys. Rev. Lett. **52**, 1621 (1984); L. Kadanoff, to be published; C. Tang, to be published.

[21]L. Patterson, J. Fluid Mech. **113**, 513 (1981).

[22]E. Ben-Jacob, R. Godbey, N. D. Goldenfeld, and T. Mueller, unpublished.

[23]D. Kessler and H. Levine, to be published.

[24]D. Ben-Simon, private communication.

[25]Ramified structures in a Hele-Shaw experiment have been recently observed also by J. Nittmann, G. Daccord, and H. E. Stanley, Nature **314**, 141 (1985).

[26]R. J. Fields and M. F. Ashby, Philos. Mag. **33**, 33 (1976).

[27]T. Vicsek, Phys. Rev. Lett. **53**, 2281 (1984).

[28]J. Jensen, unpublished.

Dendrites

Temperature Field around a Spheroidal, Cylindrical and Acicular Crystal Growing in a Supercooled Melt

G.P. Ivantsov

We examine the growth of a single crystal in a supercooled melt under the following conditions.

The solid phase has $t = t_k$ in the entire mass. The temperature field of the liquid phase, assumed fixed, obeys the Fourier equation

$$\frac{\partial t}{\partial \tau} = a\nabla^2 t. \tag{1}$$

On the phase separation boundary

$$t = t_k, \tag{2}$$

$$-\lambda\left(\frac{\partial t}{\partial n}\right)_{t_k} = q\gamma\left(\frac{dn}{d\tau}\right)_{t_k}, \tag{3}$$

where λ is the heat conduction coefficient of the melt, q is the latent heat of crystallization, and γ is the specific weight, the same for both phases.

The melt temperature at infinity is

$$t = t_0 \qquad (t_k > t_0). \tag{4}$$

Some solutions of this problem were derived as follows.

We transform (3), replacing $dn/d\tau$ by $-\dfrac{\partial t}{\partial \tau}\Big/\dfrac{\partial t}{\partial n}$ and expanding the expression for the gradient:

$$\left[\left(\frac{\partial t}{\partial x}\right)^2 + \left(\frac{\partial t}{\partial y}\right)^2 + \left(\frac{\partial t}{\partial z}\right)^2\right]_{t_k} = \frac{q\gamma}{\lambda}\left(\frac{\partial t}{\partial \tau}\right)_{t_k}. \tag{3'}$$

We introduce in (3′) an arbitrary function of temperature $f(t)$ and view the obtained expression as a differential equation:

$$\left(\frac{\partial t}{\partial x}\right)^2 + \left(\frac{\partial t}{\partial y}\right)^2 + \left(\frac{\partial t}{\partial z}\right)^2 = f(t)\frac{\partial t}{\partial \tau}. \tag{3''}$$

Translated from the Russian by Protrans. Originally published in *Doklady Akad. Nauk SSSR*, Vol. 58, No. 4, 1947.

The problem is formulated as follows: find all surfaces of the crystallization front and the corresponding temperature fields in the liquid phase satisfying the differential equations (1) and (3″) and the boundary-value conditions (2), (4), and (3′).

The sequence for solving the problem is: find integrals of the nonlinear first-order equation (3″) in which arbitrary functions appear; determine the form of the latter with the aid of equation (1) and the arbitrary constants with the aid of (2), (4), and (3′).

In some cases, it proved to be possible to find the form of the arbitrary function and thus solve the formulated problem.

The complete integral of equation (3″) obtained by the generally known method of Lagrange and Charpit has the form

$$\frac{1 + C_1^2 + C_2^2}{C_3} F(t) = x + C_1 y + C_2 z + C_3 \tau + C_4,$$

where $F(t) = \int dt/f(t)$; hence

$$t = \Phi(x + C_1 y + C_2 z + C_3 \tau + C_4). \tag{5}$$

This form of the solution satisfies (1) but is of no interest. Eliminating from the complete integral (5) all the arbitrary constants, we obtain the singular integral

$$t = \Phi\left(\frac{x^2 + y^2 + z^2}{4a\tau}\right) = \Phi\left(\frac{r^2}{4a\tau}\right). \tag{6}$$

With the aid of equation (1), we determine the form of the arbitrary function Φ. It has the form

$$\Phi(u) = C'\left[\frac{e^{-u^2}}{u} - \sqrt{\pi}\,(1 - \operatorname{erf}(u))\right] + C'' = C'\Psi(u) + C'', \tag{7}$$

where

$$u = \frac{r}{2\sqrt{a\tau}} = \frac{R}{2\sqrt{a\tau}}\frac{r}{R} = u_0 \frac{r}{R}. \tag{8}$$

Here R is the radius of the sphere which is the phase separation surface ($r > R$).

Taking into account the boundary conditions (2), (4) and (3′) leads to an equation for determining u_0:

$$\frac{c(t_k - t_0)}{q} = 2\Psi(u_0) u_0^3 e^{u_0^2} \tag{9}$$

(here c is the heat capacity of the melt) and to the desired expression of the temperature field:

$$\frac{t - t_0}{t_k - t_0} = \frac{\Psi(u)}{\Psi(u_0)}. \tag{10}$$

The temperature field around a cylindrical crystal obtained similarly from equation (6) for the plane case is

$$\frac{t - t_0}{t_k - t_0} = \frac{\mathrm{Ei}(-u^2)}{\mathrm{Ei}(-u_0^2)}, \tag{11}$$

where u_0 is determined from the equation

$$\frac{c(t_k - t_0)}{q} = -u_0^2 e^{u_0^2} \mathrm{Ei}(-u_0^2). \tag{12}$$

We observe that in both cases the advance of the crystallization front is proportional to the root of time, which follows from expression (8) for u_0. The isothermic surfaces are spheres for the first case and cylinders for the second case.

Prescribing in the complete integral (5) arbitrary relations between the arbitrary constants and eliminating them, we find the general integrals.

Of the varied forms of investigated relations there is interest in the linear relation

$$C_3 = mC_2, \tag{13}$$

which leads to the following general integral describing the growth of an acicular crystal:

$$t = \Phi\left(-z - m\tau + \sqrt{r^2 + (z + m\tau)^2}\right). \tag{14}$$

In this case the isothermic surfaces are cofocal paraboloids of revolution, and are cofocal parabolic cylinders for the plane case.

Determining with the aid of equation (1) the form of the arbitrary function Φ in the solution (14) and the arbitrary constants with the aid of (2), (4), and (3′), we obtain an expression of the temperature field around an acicular crystal growing in a supercooled melt:

$$\frac{t - t_0}{t_k - t_0} = \frac{\mathrm{Ei}\left(-\frac{wR}{2a}v\right)}{\mathrm{Ei}\left(-\frac{wR}{2a}\right)}, \qquad \text{where } v = \sqrt{\left(\frac{r}{R}\right)^2 + \left(\frac{z - w\tau}{R}\right)^2} - \frac{z - w\tau}{R}. \tag{15}$$

Here $R = \text{const}$ is the radius of curvature of the vertex of the paraboloid that is the phase separation surface ($v = 1$), and $w = \text{const}$ is the rate of growth of the needle in the direction of its axis.

The parameter $wR/2a$ is determined by the condition

$$\frac{c(t_k - t_0)}{q} = -\frac{wR}{2a} e^{wR/2a} \mathrm{Ei}\left(-\frac{wR}{2a}\right). \tag{16}$$

We make the problem definite by prescribing the growth rate of the acicular crystal w based on some other considerations.

Acta Metallurgica Vol. 29, pp. 701 to 715, 1981
Printed in Great Britain. All rights reserved

0001-6160/81/050701-15$02.00/0

OVERVIEW 12

FUNDAMENTALS OF DENDRITIC SOLIDIFICATION—I. STEADY-STATE TIP GROWTH†

S.-C. HUANG‡[1] and M. E. GLICKSMAN[2]

[1]General Electric Company, Corporate Research & Development Center, Building 81, Room E102, Schenectady, NY 12345
[2]Materials Engineering Department, Rensselaer Polytechnic Institute, Troy, NY 12181, U.S.A.

(*Received 20 August* 1980; *in revised form* 17 *November* 1980)

Abstract—Systematic measurements of dendrite tip radius and growth velocity in succinonitrile reveal that consideration of dendrite tip stability should be incorporated into the heat transfer theory to determine the steady-state dendritic growth condition. The dendritic stability criterion measured is $2\alpha d_0/VR^2 = 0.0195$, where V is the dendritic growth velocity, R is the dendritic tip radius, α is the liquid thermal diffusivity, and d_0 is a capillary length defined in the text. Several dendritic stability models are reviewed and discussed in comparison to the present experimental results.

Résumé—Les mesures systématiques des rayons de courbure au sommet des dendrites et de leur vitesse de croissance dans le cas du succinonitrile revellent que la stabilité du sommet des dendrites doit être prise en compte dan les mecanismes de transfert de chaleur pour déterminer la condition de croissance dendritique suivant un regime stable. Le critère de stabilité des dendrites a été mesuré egal à $2\alpha d_0/VR^2 = 0.0195$, où V est la vitesse de croissance dendritique, R le rayon de courbure au sommet des dendrites, α la conductivité thermique dans le liquide, et d_0 un distance, caracteristique de l'énergie de surface, definie dans le texte. Plusieurs modèles de stabilité des dendrites sont repris et confrontés aux résultats expérimentaux presentés.

Zusammenfassung—Systematische Messungen des Radius einer Dendritenspitze und seiner Wachstumsgeschwindigkeit im System Succinonitril deuten darauf hin, dass man die Stabilität der Dendritenspitze in der Theorie des Warmeaustausches einführen soll, um die Bedingung festzustellen, wobei eine Dendrite im stationaren Zustand wachsen kann. Die gemessene Stabilitätsbedingung läutet $2\alpha d_0/VR^2 = 0.0195$, wobei V die Dendriten wachstumsgeschwindigkeit ist, R das Radius der Dendritenspitze, α die thermische Leitfähigkeit der Flüssigkeit, und d_0 die im folgenden Artikel definierte Kapillarlänge. Einige Modelle der Dendritenstabilität werden analysiert und im Vergleich mit den vorliegenden versuchs ergebnissen besprochen.

INTRODUCTION

Dendritic growth is perhaps the most common form of solidification, especially in metals and other systems that freeze with relatively low entropies of transformation. Dendritic or branched growth in alloys generates microsegregation as well as other internal defects in castings, ingots, and weldments. More subtle effects introduced by the complex dendritic microstructure in solidified materials include crystallographic texturing, hot cracking, suboptimal toughness, and reduced corrosion resistance. Moreover, the dendritic microstructure and its effects may be modified by subsequent heat treatments, but they are seldom fully 'erased.' As such, the understanding and control of dendritic growth in solidification processing is crucial in order to achieve specific material properties in final products.

Dendritic growth is a coupling of two seemingly independent growth processes: the steady-state propagation of the dendrite main stem and the non-steady-state evolution of dendrite branches. Until recently, the time-dependent features of dendrites were completely ignored and theoretical models of dendrites were limited to the mathematical description of a branchless, paraboloidal needle, which grew at a constant rate and in a shape-preserving manner. Furthermore, theoretical studies of dendritic growth have concentrated on the steady-state development of a one-component needle-dendrite growing in a pure melt, wherein the major transport process is heat conduction. Results of such theories, as reviewed in Ref. [1], express the axial dendritic growth velocity as a function of supercooling (defined as $\Delta T = T_m - T_\infty$, where T_m is the melting point and T_∞ is the temperature of the melt far from the dendritic interface), viz.,

$$V_{max} = \beta G(\Delta T)^b, \tag{1}$$

† Research supported by the Marshall Space Flight Center, through the Space Processing Applications Rocket (SPAR) Program, Office of Application, NASA.

‡ Formerly, graduate student at Rensselaer Polytechnic Institute.

Reprinted with permission from *Acta Metallurgica*, Vol. 29, 1981.

where V_{max} is the maximum growth velocity, G is a lumped material parameter, and β and b are numerical coefficients and exponents, respectively, and are determined by and specific to each theoretical model.

In a recent experiment, Glicksman *et al.* [1] measured the dendritic growth velocity, V, in succinonitrile ($CN\text{-}C_2H_4\text{-}NC$) over the range of supercooling from 1 to 10°C. The experiment was carefully designed for a critical test of the diffusion-controlled, steady-state dendritic growth theory in the form of equation (1). The results of Glicksman *et al.* revealed that, although two of the theories considered predicted the correct power-law relationship ($V \propto \Delta T^{2.65}$), *none* of the theories predicted the correct growth velocity (i.e., the coefficient β) to within two times that of the measured values.

The present study extends the investigation of dendritic growth in succinonitrile to the regime of small supercooling, from 0.1°C to 1°C. Several new phenomena arise during dendritic growth at small supercoolings. First, the dendrites grow relatively slowly ($V < 0.1$ cm/s) with concomitantly large dimensions ($R > 10^{-3}$ cm)—a favorable situation for performing precise morphology studies. As to be seen in the Background section, simultaneous measurements of tip radius and growth velocity can provide a solid basis on which improvements may be introduced into the development of dendritic growth theory. This unique way to test dendritic growth theory cannot be extended easily to high supercooling situations as in Ref. [1], where a detailed investigation of dendritic micromorphology was precluded by the then available techniques.

Another morphological aspect of dendritic growth which could be revealed at small supercoolings and which is of great interest today is that of sidebranch evolution. The nonsteady-state nature of the sidebranches controls their spacing, and the related, technically important size-scale of segregation. Also, the branching mechanism itself might influence the steady-state growth of the dendrite main stem. *In situ* observation of branch formation at small supercoolings should allow us to explore to a fuller extent the branching mechanism and, thus, understand better the overall dendritic growth process. In addition, *in situ* morphological study is extended here to include the investigation of the mechanism responsible for the branch-spacing adjustment during dendritic growth.

Second, in the range of small supercooling investigated, the natural convective fluid flow arising from the thermal gradients surrounding a dendrite will have a sizable influence on the dendritic growth velocity as well as on the dendrite morphology. Every aspect of the dendritic growth will then not only vary with supercooling but also with the dendritic growth orientation relative to the fluid flow direction.

For convenience, this paper is divided into two portions: Part I presents and discusses the results relevant to the steady-state aspect of dendritic growth, i.e., the axial propagation of the dendrite tip. Part II focuses on the nonsteady-state aspect of dendritic growth, i.e., the development of the sidebranch structure. The results on convection effects have been presented elsewhere [2, 3, 4] and will not be a subject of detailed discussion here.

BACKGROUND

A. *Experimental tests of dendritic growth theory*

Attempts to achieve experimental measurements of the growth rate coefficient β and exponent b in equation (1) have historically encountered great difficulties. One of the main experimental difficulties encountered in the past was the uncertainty in the material properties which constitute the parameter G in equation (1). For example, even in one of the most studied systems, ice/water, solid-liquid interfacial energy had not been measured to adequate accuracy until just recently [5]. Before then, arbitrary values for the ice/water interfacial energy were assigned so that a "measured" β value agreed with the prediction of a particular theory [6]. Another major experimental difficulty lies in the development of a dendritic growth technique to ensure a "free" growth condition. To be in accord with the theoretically required growth conditions, both the constraining effects of growth chamber walls and the proximity effects of neighboring dendrites must be avoided. In fact, disparate results were obtained for the ice/water system by different growth techniques [7–11]. Another serious experimental difficulty worth mentioning is the complication introduced by convection and solute redistribution—neither of which is included in any theory of the growth of thermal dendrites in pure systems.

In the experiment of Glicksman *et al.* [1], the usual pitfalls were eliminated. Specifically, a succinonitrile specimen of 99.999% purity was used to help ensure the attainment of thermal diffusion conditions assumed in the theory; reliable measurement techniques were employed to reduce the uncertainty in V and ΔT to within $\pm 5\%$; a special specimen configuration was designed to achieve free dendritic growth conditions; and, finally, the specimen properties were well-characterized, so that the lumped material parameter, G, appearing in equation (1) was known to within $\pm 10\%$. In addition, the interfacial growth kinetic coefficient of succinonitrile was estimated to be about 20 cm/s/°C [1]. Such a rapid interfacial reaction process would have had an insignificant effect on the dendritic growth in the range of growth velocity studied ($V < 10$ cm/s) and therefore could safely be ignored. Also, the anisotropy in solid-liquid interfacial energy, γ, for succinonitrile is about only 1% (deduced from the equilibrium shape of the liquid droplets, as will be discussed in Part II). Such a small anisotropy should also have but a negligible effect on the dendritic growth kinetics [1]. Thus, the theoretical assumptions of perfect isotropy and infinitely fast interfacial attachment kinetics used to obtain equation (1) are rather well-approximated in the case of pure succinonitrile freezing at $\Delta T < 10^{\circ}$C. In short, the careful techniques employed in Ref. [1] produced reliable experimental results. Consequently, the factors responsible for the failure of the dendritic growth theory to predict the correct growth kinetics must lie within the theory itself.

B. *A retest of the theory of capillary-limited dendritic growth*

As introduced in the last section, steady-state dendritic growth theory confines itself to a needle-like dendrite growing at a constant axial velocity in a supercooled, pure melt. The fully time-dependent dendritic growth problem is thus reduced to a steady-state heat diffusion problem around a smooth, shape-preserving interface. The steady-state dendritic heat-transfer problem is to be solved in the solid and liquid regions subject to two boundary conditions at the surface of the dendrite. The first condition is that the temperature at each point on the solid-liquid interface equals the *local* equilibrium freezing temperature, which is determined by the local interface curvature κ. The second boundary condition requires that the latent heat released at the interface during solidification is constantly conducted away through the adjoining solid and liquid phases. In the case of succinonitrile, the thermal conductivities of solid and liquid are within 1% of being equal to each other. This circumstance permits use of certain simplified forms of the heat transfer and stability theories.

Note that both boundary conditions are incompletely defined, insofar as in the capillary condition as well as the normal component of the temperature gradient in the energy conservation condition must be calculated from the dendrite geometry. The shape of the dendrite is not known *a priori*, however, and must be determined as part of the solution to the diffusion problem. To eliminate the nonlinear aspects of this "free boundary problem," the boundary conditions given above were usually simplified or the dendrite shape predetermined *by assumption*. Various versions of the approximate theory have been reviewed in Ref. [1]; among them are the Ivantsov model [12, 13], modified Ivantsov model [14–16], and Temkin model [17, 18], each yielding results in the form of

$$\Delta\theta = \text{Pé}\, e^{\text{Pé}} E_1(\text{Pé}) + \Delta\theta_c, \tag{2}$$

where $\Delta\theta$ is the dimensionless supercooling ($\Delta T C_p/L$); C_p is the heat capacity of the liquid phase; L is the heat of fusion; $E_1(\text{Pé})$ is the exponential integral function $\int_{\text{Pé}}^{\infty}(e^{-u}/u)du$; Pé is the Péclet number at the dendrite tip ($VR/2\alpha$); R is the dendrite tip radius; and $\Delta\theta_c$ is a term reflecting the influence of the capillary (Gibbs–Thomson) effect and is itself also a function of V and R, which in turn is a function of the velocity V.

Equation (2) reflects two major physical effects. The "point effect" (VR = constant) is the solution to the thermal diffusion model with an isothermal interface (Ivantsov model), in which $\gamma = 0$ is assumed and consequently $\Delta\theta_c = 0$. On the other hand, if $\gamma \neq 0$, then the Gibbs–Thomson effect lowers the interface equilibrium temperature by $T_m\gamma\kappa/L$, which effectively reduces by the same amount the supercooling available to "drive" the thermal diffusion. This reduction in supercooling (as reflected in the nonzero $\Delta\theta_c$ in equation (2) in both the modified Ivantsov model and in the more advanced Temkin model) prevents dendrites from growing increasingly fast as $R \rightarrow 0$, as would be required by the "point effect." Generally, the capillary-limited theory in the form of equation (2) only limits the growth velocity to being no larger than an *upper bound* value of V (V_{max}) at a given supercooling $\Delta\theta$. Unique relationships for V vs $\Delta\theta$ and R vs $\Delta\theta$ are therefore lacking in these models. To obtain a unique relationship for V vs $\Delta\theta$, it was convenient to assume that dendrites grow at the *maximum* possible velocity allowed by the capillary effect, i.e., V_{max}. The V_{max} vs ΔT relationship so-generated [equation (1)], however, disagrees with the reliable experimental results obtained by Glicksman *et al.* [1].

In summary, dendritic growth theory may be divided into two portions: first is the main body of the theory, in which the thermal diffusion problem is solved. Here, the existence of the dendrite sidebranches is neglected and the shape of the smooth needle dendrite is constrained to some *a priori* geometry. Only recently, Nash and Glicksman [19] developed a nonlinear method to solve the theory of steady-state dendritic growth as a free boundary problem. In their theory, both the capillary and the flux conditions were applied rigorously to every point on the dendrite surface, and a nonconstrained dendrite shape was determined as a part of the solution. The maximum growth velocity predicted by this exact theory, however, also disagrees with the experimental results—viz., it predicted velocities about 6.5 times larger than the velocities observed in Ref. [1]. Thus, some remaining deficiency not associated with the nature of the boundary conditions must be responsible for the 6.5-fold error remaining in the theory. Furthermore, a qualitative analysis [20] also led to the conclusion that the 6.5-fold predictive failure of the dendritic growth theory at $\Delta T > 1^{\circ}$C did not arise from the neglect of sidebranch effects in the theory. The second portion of dendritic growth theory is the "maximum growth velocity" hypothesis, which is used to select the operative growth state from among the possible (V, R, $\Delta\theta$) coordinates to yield the (V_{max}, $\Delta\theta$) relation, as expressed by equation (1). The theoretical review presented above suggests that although sidebranch effects are neglected and the capillary boundary condition is generally simplified, the main body of the steady-state theory, i.e., equation (2), should have nonetheless been substantially correct. The failure of theory in predicting the relation between V and $\Delta\theta$, i.e., equation (1), is thus likely to be related to the use of the maximum growth velocity hypothesis.

Experimentally, however, isolating the precise theoretical error remains uncertain when testing the theory only against the form of equation (1). Instead, a full test of the theory against equation (2) would be capable of locating the defect in either the main body of the theory or the V_{max} hypothesis. The latter approach requires the measurement of the tip radius, as well as the growth velocity, of a dendrite growing at a given supercooling. Based on the theoretical background reviewed above, it is expected that the (V, R) coordinate determined experimentally would fall on the solid line which describes equation (2), *but not at the point of* V_{max}.

C. *Stability criteria for a steady-state dendrite*

The needle dendrite predicted by the steady-state model in conjunction with the maximum growth velocity optimization principle was regarded to be correct to first order by some authors [21]. Second-order corrections would assumedly be needed to include the effects of the nonsteady-state sidebranch growth, which is neglected completely in the steady-state theory. For instance, Kotler and Tiller [21] applied Mullins and Sekerka's stability analysis [22] to the surface of the optimized dendrite predicted by Temkin's model. An important finding of this dendrite stability analysis was that the *whole* surface, including the tip of Temkin's optimized needle dendrite, is unstable. In particular, the greatest instability was found to lie in the region close to the dendrite tip. As a result of the nature of this instability associated with the optimized needle dendrite, branching perturbations appeared with their largest amplitudes located only $\sim 0.5R$ behind the dendrite tip [21]. The branch geometry predicted for an optimized Temkin dendrite is, however, contrary to that of a real dendrite, where large perturbations are not observed near the tip, but only away from the tip. In spite of this contradiction in the facts, their analytical result was viewed simply as evidence for a strong sidebranch influence on the dendrite tip [21].

An alternative implication of the unrealistic branching perturbation predicted by dendrite stability analyses was that the basis solution (the optimized Temkin needle dendrite) used in the analyses was incorrect. In search of a new needle dendrite, Oldfield [23] recognized the need for the dendrite tip to achieve a stable condition so that its steady-state growth could be ensured. According to this hypothesis (to be proved experimentally in this paper), the basis dendrite for a branching stability analysis should not be a needledendrite optimized to maximum growth velocity, but should be the one with a stable tip. Thus the interfacial stability, rather than providing only a secondary correction as previously thought, imposes a direct restriction on the steady-state growth. Based on this concept, several stability criteria for dendrite tips have been developed, all of which can be expressed mathematically as $VR^2 =$ constant.

(i) *Oldfield's model.* The stable size of a dendrite tip was determined, based on the idea that the destabilizing force due to thermal diffusion is equal to the stabilizing capillary force. In a simple dendrite geometry analysis, the following stability criterion was obtained (23):

$$G_{l,\mathrm{tip}} = -\frac{4T_m\gamma}{L}\frac{1}{w^2}, \tag{3}$$

where $G_{l,\mathrm{tip}}$ is the temperature gradient normal to the dendrite tip, and w is a characteristic length which describes the size of the dendrite. The relation between w and the dendrite tip radius R was evaluated by a numerical calculation; for a stable dendrite, Oldfield found that [23]

$$w = 0.2R.$$

The temperature gradient in front of a dendrite tip propagating at a steady velocity V can be expressed as

$$G_{l,\mathrm{tip}} = -\frac{LV}{\alpha C_p}. \tag{4}$$

Therefore, the stability criteria of equation (3) becomes

$$\sigma^* = 2\alpha d_0/VR^2 = 0.02, \tag{5}$$

where σ^* is the stability criterion constant, and d_0 is a capillary length defined as $T_m\gamma C_p/L^2$.

(ii) *Planar interface model.* In this model [24], the dendrite tip is first considered as a point lying on a planar interface† which propagates at the axial dendritic growth velocity, V, see Fig. [1]. The stability condition for the dendrite tip is then assumed to be the same as that of the planar interface. According to Mullins and Sekerka [25], for a crystal with a planar s/l interface to grow stably the wavelength of the perturbation must remain smaller than

$$\lambda_s = 2\pi\sqrt{2\alpha d_0/V}, \tag{6}$$

where λ_s is known as the critical perturbation wavelength. If λ_s can be related to the size of the dendrite tip such that

$$R \simeq \lambda_s, \tag{7}$$

then the criterion of equation (6) becomes

$$\sigma_s = 2\alpha d_0/VR^2 = \frac{1}{4\pi^2} = 0.0253. \tag{8}$$

(iii) *Spherical model.* The above planar interface model simulates the steady-state heat-transfer condition at the dendrite tip; however, it may not account for the proper local curvature at the dendrite tip (see Discussion). In the present spherical model, based on Ref. [26], the stability condition of the dendrite tip is assumed to be identical to that of a sphere which describes the local total curvature of the den-

† It is stressed that planar growth in an adiabatic system is unlikely to occur [25], unless perhaps when the amount of supercooling exceeds L/C_p [15]. The planar growth situation here is only invoked to simulate the heat-transfer condition at a dendrite tip.

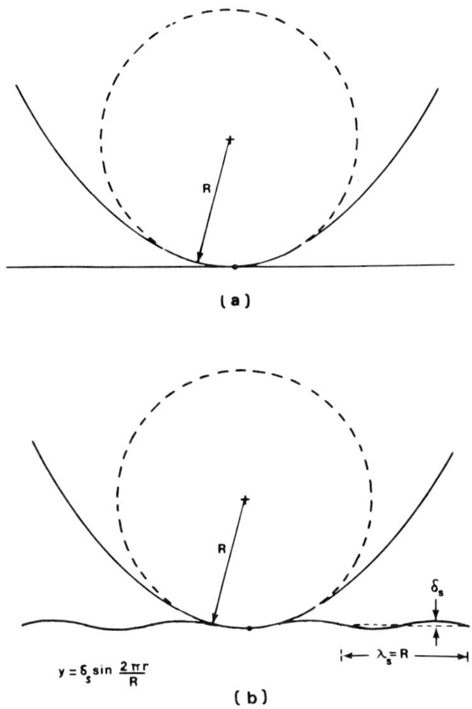

Fig. 1. Schematic showing the planar interface model for estimating dendrite tip stability: (a) before perturbation is considered; (b) the marginally stable condition, at which $\lambda_s = R$ and $\delta_s = \sigma^*\lambda_s$.

drite tip, $2/R$. The absolute stability condition for a growing sphere [22] with an instantaneous radius R has been given by

$$G_{l,\text{sphere}} = \frac{T_m \gamma}{LR^2}(n + 2)(2n + 1), \tag{9}$$

where n is the order of spherical harmonics. Note the similarity between this equation and equation (3). Note also that the growth of a sphere is indeed a nonsteady-state problem. To simulate the steady-state growth of a dendrite tip, steady-state conditions have to be imposed on the growing sphere. Under this "steady-state" condition, the temperature gradient around the sphere is again equal to $(LV/\alpha Cp)$. Now equation (9) becomes

$$\sigma^* = 2\alpha d_0/VR^2 = \frac{2}{(n + 2)(2n + 1)}. \tag{10}$$

The spherical harmonic, n, is a positive integer, and, in a model that assumes material isotropy, the choice should be independent of crystallographic symmetry. We note here that the wavelength of a perturbation wave around a sphere of radius R is a function of n, i.e., $\gamma_n = 2\pi R/n$. $n = 6$ is selected here to be consistent with the assumption used in the planar model that the dendritic tip radius approximates the perturbation wavelength, or equivalently for $n = 6 \simeq 2\pi$,

$$R = 0.95\lambda_n, \tag{11}$$

see Fig. 2. The stability criterion now reduces to

$$\sigma^* = 2\alpha d_0/VR^2 = 0.0192. \tag{12}$$

(iv) *Langer and Müller-Krumbhaar (LM-K) theory*. In this theory, rigorous mathematics and extensive numerical computation were used to study the interface behavior of the whole dendrite [24]. The procedure used by LM-K started with the derivation of a linear integro-differential equation to describe the displacement of the dendrite surface away from the unperturbed base paraboloid. The equation of motion was derived by making a quasi-stationary approximation for the time-dependent diffusion field. Also, in the derivation, thermal diffusion in the solid phase was neglected and perfect material isotropy assumed. At the limit $\text{Pé} \to 0$, an equation of motion was obtained which contains the stability parameter σ as the only system-dependent quantity. The equation of motion had no exact solution when $\sigma \neq 0$ and was studied numerically by an eigenvalue analysis. If the dendrite shape displacement had an exponential time-

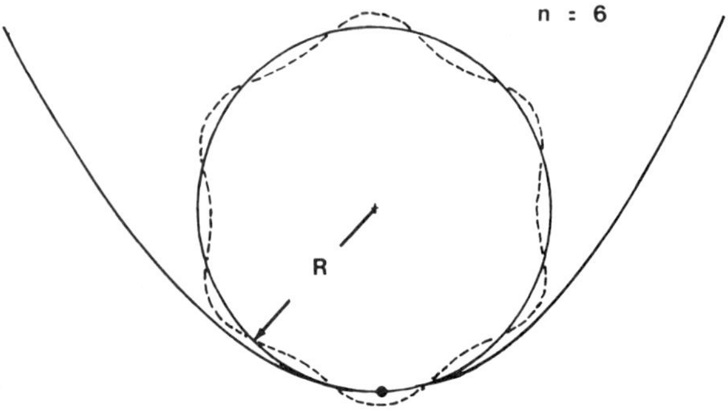

Fig. 2. Schematic showing the spherical interface model at a marginal stability condition with harmonics $n = 6$ but $\delta_n \neq 0$.

dependence, then the eigenstate found was such that when

$$\sigma^* = 2\alpha d_0/VR^2 = 0.025 \pm 0.007, \tag{13}$$

all the unstable modes at the dendrite tip vanished. Note that the $\pm 30\%$ uncertainty in σ^* shown in equation (13) reflects only the uncertainty in the numerical procedures used by LM-K to solve the equations.

All the stability criteria discussed above are in the same mathematical form, namely, VR^2 = constant. Stability criteria are thus capable of providing the conditions needed to dissociate the V–R relationship predicted by steady-state models, equation (2). Moreover, the stability constants σ^* all cluster tightly near the value 0.02, quite independently of the geometry assumed or the level of sophistication employed in the analyses. A test of the stability criteria can be performed simultaneously in a test of equation (2) in which both V and R will be measured.

EXPERIMENTAL TECHNIQUES

A. *Specimen purification and characterization*

A specimen of high purity is required in the present study of dendritic growth in a one-component system. Succinonitrile supplied by Eastman Kodak ($\sim 99\%$ pure) was therefore purified further in a series of distillation and zone refining steps [1, 27]. The purity of the specimen at each stage was characterized quantitatively by a liquidus temperature, T_l, and a solidus temperature, T_s. In order to correlate these temperatures with the final specimen purity, a preliminary series of experiments were carried out to measure T_s and T_l for succinonitrile samples extracted from various positions along a zone-refined column. These preliminary experiments allowed the calculation of some thermodynamic properties of the specimen material [27], as summarized in Table 1. When T_l of the final specimen for thermal dendritic growth studies was measured, no meaningful deviation from T_m could be resolved with our measuring equipment. As discussed in the next section, the temperature resolution was limited to 10^{-4} °C. From the material characteristics given in Table 1, this temperature resolution corresponds to a maximum detectable active impurity level of 5×10^{-7} mole fraction which corresponds to a purity of 99.99995% (6 9's plus).

B. *Temperature measurement and control*

All the temperatures were monitored with a platinum resistance thermometer, the output from which was measured with a Smith thermometer bridge. This thermometer, which has a temperature-resistance coefficient of 10 °C/ohm, was capable of a resolution of 0.0004 °C, and was calibrated by the National Bureau of Standards to an accuracy of ± 0.002 °C. Additionally, the bridge, a double Kelvin type, was calibrated by the manufacturer (Leeds & Northrup), and was capable of resolving 10^{-5} ohms. Temperature fluctuations in the observation tank were detected continuously by a null detector.

The temperature controlled observation tank was constructed from double-wall sheet glass with a well insulated exterior. A Tronac-40 temperature control system employing a nickel thermohm was used to provide nonlinear, proportioning control signals to a 220 watt, low-impedance, stainless steel, tubular heat-

Table 1. Characterization of specimen material (succinonitrile)

Segregation Coefficient k_0	Liquidus Slope m_l(K)	Solidus Slope m_s(K)	Melting Point T_m(°C)
0.124	−215.15	−1735.1	58.083 ± 0.002

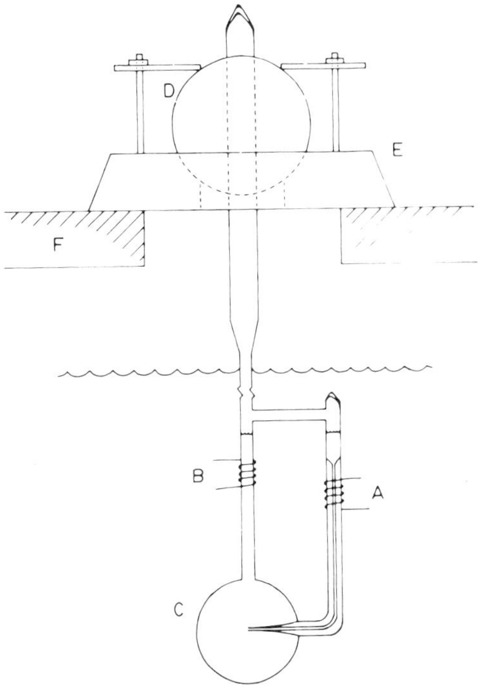

Fig. 3. Schematic of specimen configuration and support stage. A and B are control heaters; C is the crystal growth chamber; D is the tilting and secondary rotating device; E is the primary rotation and X Y translation stage; and F is the tank cover. The diameter of the crystal growth chamber C is about 4 cm.

ing coil immersed in the tank. To offset any excessive heat input from the control heater, a continuous, steady flow of water (maintained about 8 C below the control set point) was circulated through the tubular heating coil. This system provided a differential "source/sink" capable of rapidly responding to either demands for extra heating or extra cooling. Short-term temperature stability of ± 0.0004 C was achieved routinely; moreover, thermal drift of the system was found to be less than 0.001 C over a period of one week.

C. Specimen configuration and measurements of growth velocity and tip radius

Dendritic growth studies were carried out in the specimen chamber shown schematically in Fig. 3. As discussed in detail in Ref. [1], this specimen configuation was capable of achieving the "free" dendritic growth conditions over the range of supercooling we explored.

As also shown in Fig. 3, the specimen was mounted on a special supporting stage which allowed full rotation and some degree of tilt and translation. The manipulation of a growing dendrite into a desired spatial orientation with respect to the axis of observation was necessary to account for the effect of stereographic projection. More specifically, a dendrite growing in a random orientation was observed from two different directions to determine the true growth velocity and the true growth orientation with respect to the gravity vector. Since the microscope was fixed, this required rotation of the specimen chamber. Similarly, to measure the tip radius of a dendrite, a dendrite was positioned in such a way that it was observed in one of the ⟨100⟩ directions. This was based on our observation, shown in Fig. 4, that the branchless tip region as well as the branched region displayed different profiles depending on the angle of observation. As shown in the figure, the smooth region of the succinonitrile dendrite tip developed into a three-dimensional shape resembling the head of a Phillip's screwdriver. Four fin-like protrusions, which grew in the {100} planes, represent the true shape of the growing dendritic tip. To illustrate furthei the nature of this anisotropic shape development, the profiles of the dendrite tip viewed from a ⟨100⟩ direction, Fig. 4(a), and from a ⟨110⟩ direction, Fig. 4(c), were superimposed on Fig. 5. In this figure,

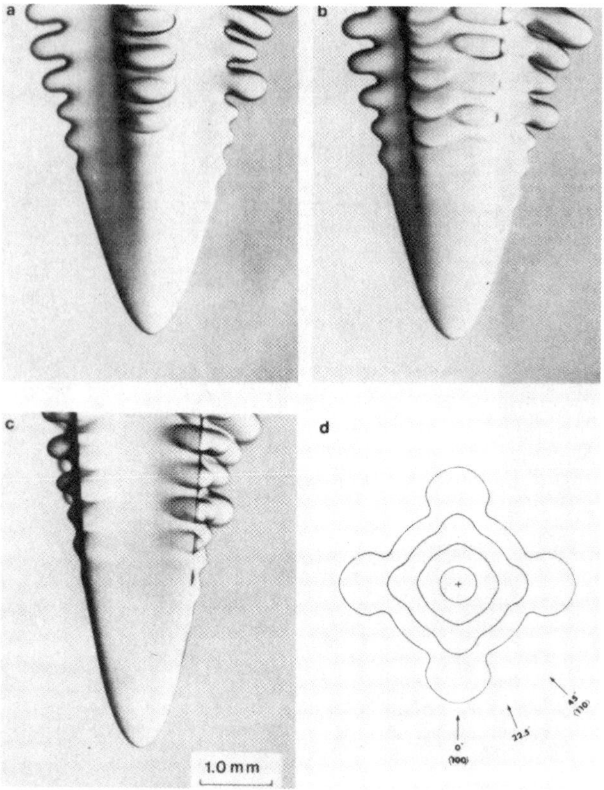

Fig. 4. Influence of crystalline anisotropy on the configuration of a succinonitrile dendrite tip, which is viewed from: (a) [100] direction, (b) a direction between [100] and [110], and (c) [110] direction; (d) shows the sequential cross sections of these dendrites, illustrating the development of sidebranches and the rotational symmetry about the growth axis close to the tip.

it is clear that the dendrite profile observed from a ⟨100⟩ direction best fits the standard parabolic curve (shown as black dots) until branch oscillations appeared. Consequently, all tip radii measurements were taken from profiles viewed from a ⟨100⟩ direction. Comparison of Figs 4 and 5 also shows that the tip regions of succinonitrile dendrites are bodies of revolution for which the total tip curvature is $2/R$.

† The speed of dendritic growth increased rapidly with supercooling and the size-scale of the morphology decreased. Consequently both the time for observation and the resolution required for photography quickly exceeded the capabilities of available microscope systems. $\Delta T = 2$ C was the practical limit.

‡ This limit occurs because the characteristic diffusion length of the dendritic heat-transfer process became comparable to the size of the growth chamber, and therefore latent heat passed through the chamber walls into the thermostat.

RESULTS AND DISCUSSION

Dendritic growth kinetics studies were carried out at 17 supercoolings. The supercooling ranged from a maximum of 2 C, where a technical limit† for morphological studies was met, to a minimum of 0.043 C, where the required adiabatic isolation of the growing crystals was no longer maintained.‡ At each supercooling studied, the dendritic growth velocity was measured as a function of the true growth angle relative to the gravity vector. The orientation dependence of the dendritic growth velocity is evident when the supercooling level is less than about 1 C. The orientation dependence reflects the influence of the convective fluid flow, and has been discussed in some detail in papers previously published [2–4]. In the present paper, discussion is limited to diffusion-controlled dendritic growth. To avoid confusion with the spatial orientation effect, only downward-growing dendrites

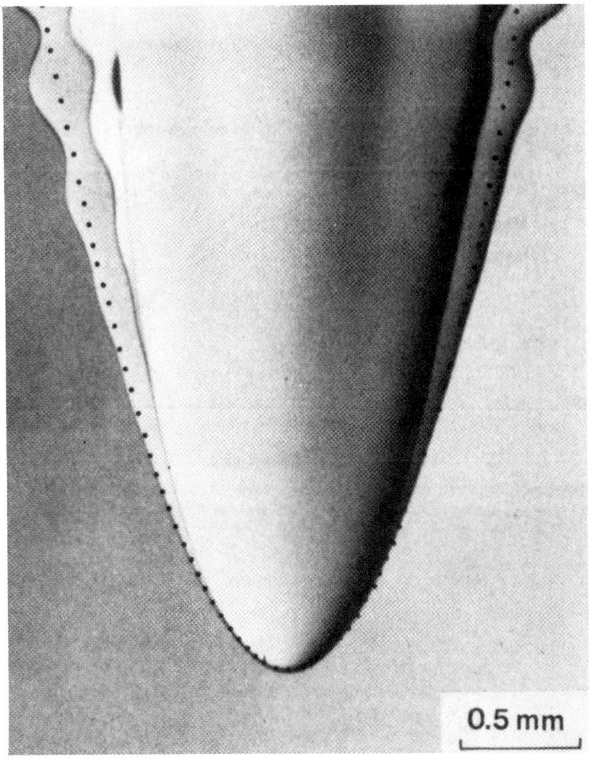

Fig. 5. Superimposition of Figs 4(a) and (c), showing fit to common parabolic curve (black dots). Dendrites of succinonitrile are bodies of revolution only in the neighborhood of the tip; progressive deviations from the paraboloidal shape occur as the fourfold symmetric branching 'sheets' and the oscillatory branching 'bumps' develop.

(i.e., dendrites growing parallel to gravity) will be considered here.

The tip radii of dendrites growing at 16 supercoolings from 0.043 C to 1.800 C were measured. Above a supercooling of approximately 2 C, the high growth velocity combined with the small size of the dendrites precludes acquisition of useful photographs of dendrites†. Thus, the tip radius analysis at and above $\Delta T = 2$ C could not be carried out with the desired accuracy. Again, *only* the tip radii for downward growing dendrites is included in the discussion which follows.

† The speed of dendritic growth increased rapidly with supercooling and the size-scale of the morphology decreased. Consequently both the time for observation and the resolution required for photography quickly exceeded the capabilities of available microscope systems. $\Delta T = 2$ C was the practical limit.

A. *A new steady-state dendritic growth theory*

Now that both the growth velocity and the tip radius of a dendrite have been measured simultaneously, a rigorous test becomes possible of steady-state dendritic growth theory in the form of equation (2). To illustrate, the coordinate (V, R) determined experimentally at $\Delta T = 1.2$ C is plotted in Fig. 6 for comparison with three of the diffusional theories. The experimental point lies at a position remote from the maxima of the curves and in a region in which the various theories tend to converge. Briefly, the following conclusions can be drawn from this result, all of which are consistent with those presented in the Theoretical Background section:

1. The simplest theory, based on the solution of the isothermal diffusion problem of Ivantsov, is substantially correct in describing the kinetics of dendritic tip growth. Thus a steady-state dendrite is essentially isothermal at T_m and effectively paraboloidal in shape.

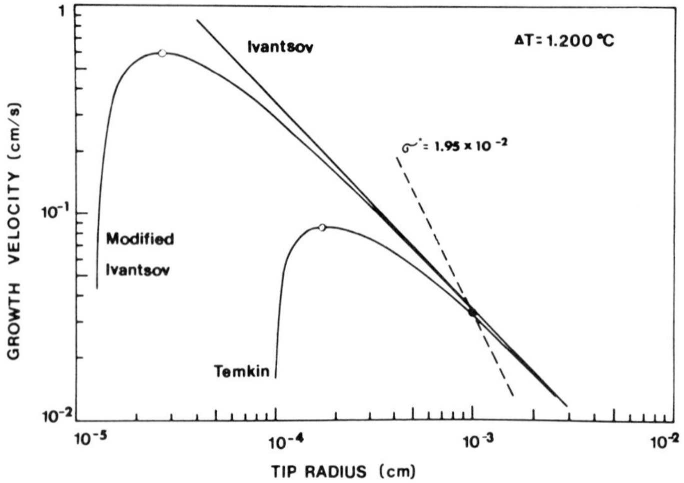

Fig. 6. Comparison of the dendrite-tip operating state measured (solid circle) to theoretical predictions. Open circles designate the maximum growth velocities predicted by two nonisothermal models. Broken line locates the operating state predicted by stability criterion, which agrees with experiment.

The inclusion of capillary effects (in all the nonisothermal models) into the isothermal theory requires only relatively small adjustments. This circumstance occurs here because the dendrites actually observed are all relatively *large* in size. For example, the interfacial temperature depression at the dendrite tip in question is only 6×10^{-4} K, which is only 0.5% of the total supercooling. This temperature depression was calculated from the well known Gibbs–Thomson relationship and from the known curvature at the dendrite tip, namely $2 \times 10^3 \text{ cm}^{-1}$. Therefore, the capillary corrections to the interfacial temperature and the shape deviations from Ivantsov's isothermal paraboloidal dendrite are essentially trivial. Various versions of dendritic growth theory developed over the last 20 years to include capiallary effects thus result in only minor differences among themselves in the *operating* region of real dendrites. The success of the Ivantsov theory also suggests that sidebranch growth has a negligible effect in as far as the growth kinetics at a dendrite tip are concerned.

2. The "maximum growth velocity" principle is incorrect. It is in fact due to adopting this *incorrect* hypothesis that the substantially *correct* (but incomplete) diffusional growth theories all fail to predict the right $V - \Delta T$ relationship. It therefore appears certain that a new physical criterion other than the "maximum velocity" principle is needed for a correct and complete theory of steady-state dendritic growth in pure materials.

3. As also shown in Fig. 6, such a criterion can be based on the stability of a dendrite tip as discussed in the Theoretical Background section (i.e., $\sigma^* = 2\alpha d_0/VR^2$ = constant). The "stability" criterion constant measured in the present study is $\sigma^* = 0.0195$ (see Fig. 7), which is within the limits of the computational uncertainty in LM-K theory, equation (13). The results of the experimental test of the "maximum growth velocity" hypothesis vs the "stability" criterion is further demonstrated in Fig. 8. In this figure, we plot as a function of supercooling the dimensionless Péclet number (Pé = $VR/2\alpha$) measured in the present experiments, along with those predicted by various dendritic growth theories which evoke either the "maximum growth velocity" principle or the "stability" criterion. When using the maximum growth velocity principle, the capillary-limited dendritic growth theories lose their predictive accuracy which is retained only when the "stability" criterion is employed.

In summary, an acceptable steady-state dendritic growth theory should, in the main, fulfill the heat diffusion condition around the whole dendrite and the marginal interfacial stability condition at the dendrite tip. Predictions of the new theory, subject to only minor adjustment from capillary effects, are shown in Fig. 9. In one case (Temkin's model with the stability criterion), agreement with experimental results is within 10%, without any adjustable parameters. The predictivity of the new theory for the dendritic tip radius is shown in Fig. 10 and essentially perfect agreement results where convection effects are small (ΔT 0.5 C).

B. *Further discussion on the dendrite tip stability models*

It has been shown above that a stability criterion in the form of VR^2 = constant is successful in selecting

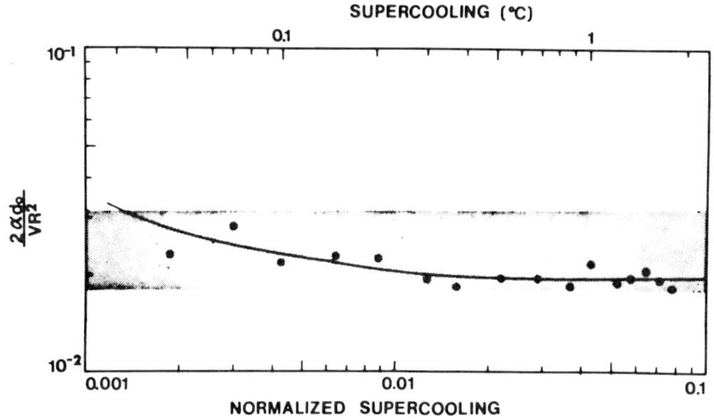

Fig. 7. Comparison of experimentally determined stability criterion constant $2\alpha d_0/VR^2$ with that predicted from the LM-K theory. Shaded band indicates the range of uncertainty within the theory. Measurements show a systematic increase in the stability constant at small supercoolings possibly due to convection effects.

the correct steady-state tip growth condition from the dendritic heat transfer solutions. The stability constant measured ($\sigma^* = 2\alpha d_0/VR^2 = 0.0195$) lies near the lower limit of the LM-K theory ($\sigma^* = 0.025 \pm 0.007$). Note again that the 35% uncertainty in the LM-K theory is only due to the uncertainty in the numerical computation used to solve the equation of interface motion. Similarly, the stability criterion constant measured in the present experiment also agreed with the stability criteria predicted by each of the three stability models presented in the Theoretical Background section. For example, the criterion given by Oldfield's model agrees to within 2.5%. However, due to the overly simple nature of that model, the stability constant predicted by the model cannot be taken too seriously. The stability constant predicted by the simple planar interface model agrees to 25%. As discussed previously, planar interface solidification is thermally unfavorable for a one-component system at supercoolings smaller than (L/C_p). Also, the reasonable assumption that $R = \lambda_s$ remains to be justified rigorously. Actually, the above assumption underestimates the R values by 15%, as shown in Fig. 11. This might have accounted for the inaccuracy of the planar interface model. However, if the planar perturbation wave is to describe the dendrite tip curvature, the arbitrary amplitude δ_s of the critically stable perturbation wave would have to be

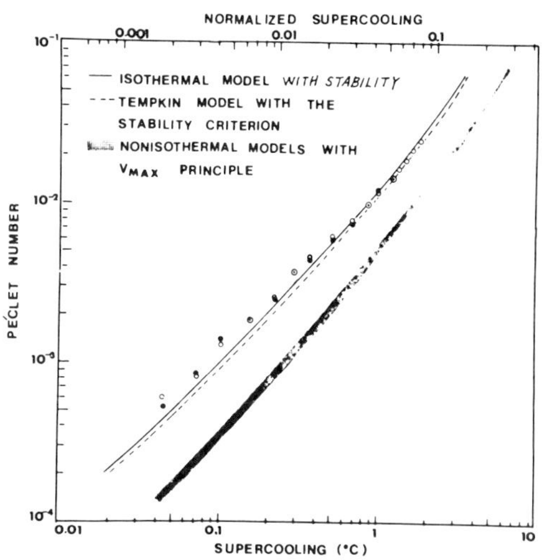

Fig. 8. Péclet number vs supercooling for pure succinonitrile.

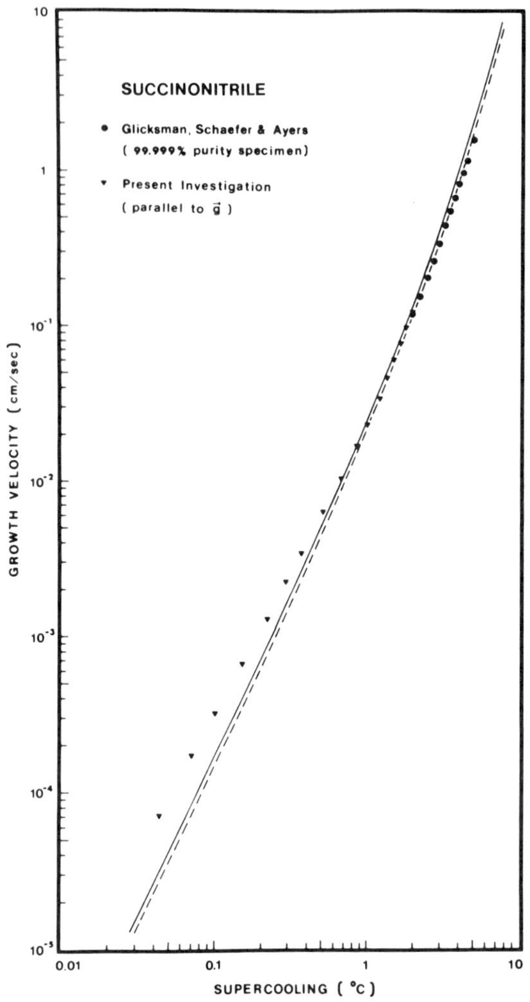

Fig. 9. Growth velocity vs supercooling. The solid curve shows prediction of Ivantsov model and the dashed curve of Temkin model, both with stability criterion ($\sigma^* = 0.0195$) applied. Systematic deviation from theoretical curves at small supercooling is due to convection effects.

confined to

$$\delta_s = \sigma^* \lambda_s. \tag{14}$$

See Fig. 2. It should be noted here that the amplitude of the perturbation wave is not actually predicted by the linear theory, but is merely a random quantity determined by the intensity of physical disturbances surrounding the tip. The limitation of equation (14) thus stands as a serious deficiency in applying the planar interface model.

In the spherical interface model, the stability constant is predicted to agree with the experimental result to within 1.5%. The steady-state thermal condition still has to be imposed at the spherical interface in order to simulate the growth condition of a dendrite tip. The dendrite tip radius is still assumed to be given approximately by the perturbation wavelength, as in the case of the planar interface model. This condition, in fact, is equivalent to selecting $n = 6$ as the dominant mode of the critical perturbation waves.

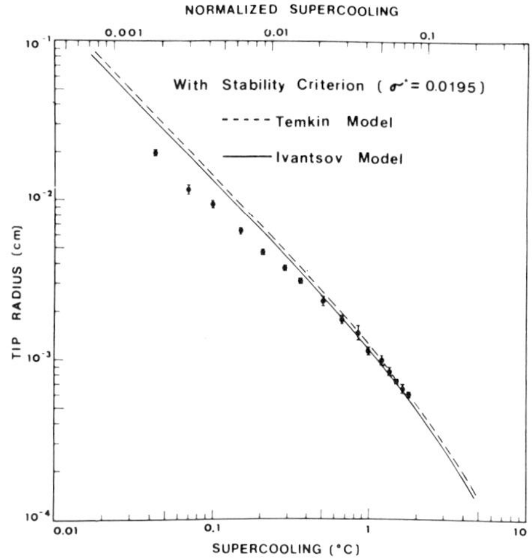

Fig. 10. Dendrite tip radius versus supercooling, comparison to diffusion-controlled steady-state theories.

However, the spherical model can account for the dendrite tip radius readily when no actual perturbation is excited, as when

$$\delta_n = 0, \tag{15}$$

where δ_n is the amplitude of the spherical perturbation wave.

Equation (15) describes the basic mode of growth, equivalent to the eigenstate solution in LM-K theory. However, even when some disturbance to growth is present and the dendrite tip is perturbed, the tip can still grow in a steady-state because of the marginal stability condition persisting there ($\dot{\delta}/\delta = 0$). In the latter growth mode, the excited dendrite tip propagates at the same velocity as an unexcited one, but at a different (constant) tip radius, see Fig. 2. The disturbance must at the same time perturb the thermal diffusion condition and thus the growth velocity. These two types of perturbations counterbalance each other and should result in continuous adjustments in the tip growth condition. The growth state variations should be microscopic in scale, assuming that the disturbance is small. On a macroscopic scale, however, the perturbed dendrite tip remains in a "steady-state," which may differ from the state of an unperturbed tip. The excited mode of dendritic growth should be the one actually observed, as disturbances of some sort are inevitable. However, it should be noted that, for analytical reasons, the solution for an unperturbed tip should be regarded as the fundamental tip growth state.

One example which might lead to the excited mode of tip growth is the original version of the spherical model [26]. In that version of the model, the temperature gradient was taken as

$$G_l = \frac{T_m\left[1 - \dfrac{2\gamma}{LR}\right] - T_0}{R}, \tag{16}$$

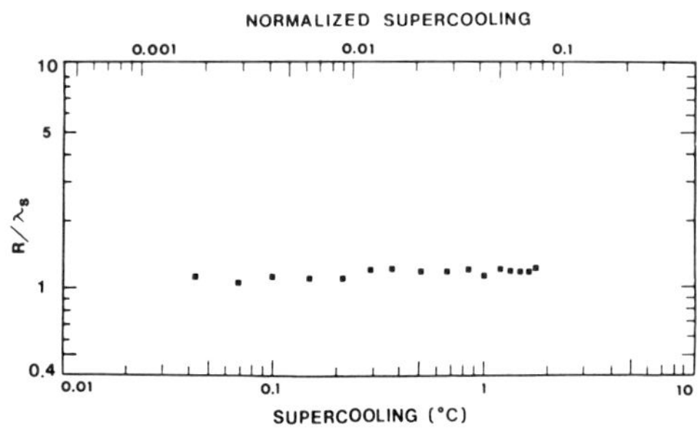

Fig. 11. Normalized dendrite tip radius vs supercooling, showing $R = 1.2\lambda_s$ over a wide range of supercooling.

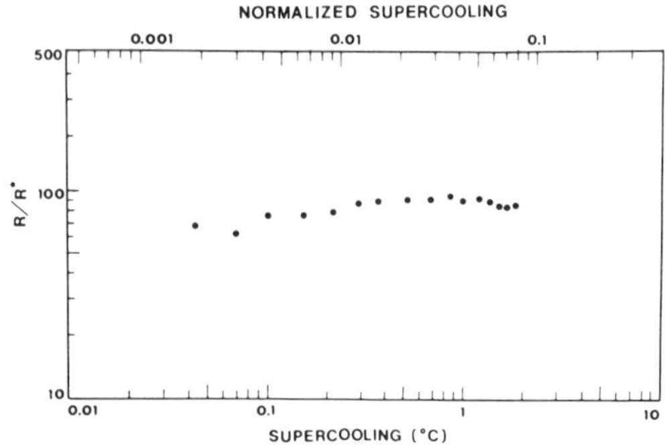

Fig. 12. Dendrite tip radius normalized to critical nucleation radius, R/R^*, as a function of supercooling.

which is the actual temperature gradient around a growing sphere. Also, the *relative* stability condition ($\dot{\delta}/\delta = \dot{R}/R$) was used [28]. Both conditions were inadequate to describe the steady-state situation at a dendrite tip. If the absolute stability condition ($\dot{\delta}/\delta = 0$) is used, equations (9) and (16) combine to give

$$R = R^* [1 + \tfrac{1}{2}(n + 2)(2n + 1)], \qquad (17)$$

where R^* is the critical nucleation radius defined as $R^* = 2T_m\gamma/\Delta S\Delta T$. The above equation compares favorably with the experimental result of $R/R^* \cong 100$ when $n = 9$ is used, Fig. 12. However, a relation between R and R^* can be obtained only when equation (16) is used, which fails to simulate the thermal condition at a steady-state dendrite tip. Therefore, the original spherical model, although resulting in a convenient relation between R and R^*, cannot be considered correct. An explanation of dendritic growth processes in terms of periodic "nucleation" of protuberances at the dendrite tip [29] is consequently of doubtful validity for freely growing dendrites.

CONCLUSIONS

1. A complete steady-state dendritic growth theory should in the main include (a) heat transfer and (b) consideration of dendrite tip stability.

2. Ivantsov's simple solution to the dendritic heat-transfer problem is substantially correct. Accordingly, the capillary effects must be trivial (since dendrites grow with tip dimensions about 100 times larger than the critical radius for nucleation). Also, the assumption of a needle dendrite shape seems consistent with the experimentally observed axial growth kinetics at the tip.

3. The hypothesis of maximum growth velocity, conventionally used in conjunction with dendritic heat-transfer theories to predict the V–ΔT relationship, is incorrect and should be replaced by a morphological stability criterion.

4. Stability criteria developed in several models all assume the mathematical form of VR^2 = constant. Among the models considered, LM-K theory is the most sophisticated one, but a modified spherical stability model developed in this study yields the criterion that provides the best agreement to our measurements in succinonitrile, i.e., $2\alpha d_0/VR^2 = 0.0192$.

5. The *fundamental* growth state of a dendrite tip, as revealed in the modified spherical interface model in accordance with the eigenstate solution of the LM-K theory, is critically stable ($\dot{\delta}_n = 0$) and unexcited ($\delta_n = 0$).

6. The physical justification of the assumption $R = \lambda_s$ used in the planar interface model and $n = 6$ in the spherical interface model remains uncertain.

Acknowledgements The authors gratefully acknowledge the assistance of Mr Keith Matuszyk in conducting the dendrite tip radius measurements and of Mr James Pan for providing technical advice on the photography. In addition, we thank Dr Robert J. Schaefer for providing guidance in designing the thermostatic observation tank and in making precision temperature measurements. Finally, the authors thank Dr John Carruthers, Dr Robert J. Naumann, and Dr Lewis Lacy for the support and encouragement provided by the Marshall Space Flight Center and the Materials Processing in Space Office of the National Aeronautics and Space Administration.

REFERENCES

1. M. E. Glicksman, R. J. Schaefer and J. D. Ayers, *Metall. Trans. A* **7**, 1747 (1976).
2. M. E. Glicksman and S. C. Huang, *Convecting Heat Transfer During Dendritic Solidification*, AIAA 16th Aerospace Science Meeting, Huntsville, Alabama, January 16–18, (1978).
3. M. E. Glicksman and S. C. Huang, to be published in the *Proceedings of the 87th National AIChE Meeting, Special Symposium on "Transport Phenomena in Phase Change Processes,"* Boston, Massachusetts, August, 1979.
4. M. E. Glicksman and S. C. Huang, *Proceedings of the 3rd European Symposium on Material Sciences in Space*, Grenoble, France, April 24–27, 1979.
5. S. C. Hardy, *Phil. Mag.* **35**, 471 (1977).

6. See, for example, J. P. Kallungal and A. J. Barduhn, *A.I.Ch.E. Jl* **23,** 294 (1977).
7. C. S. Lindenmeyer, G. T. Orrok, K. A. Jackson and B. Chalmers, *J. chem. Phys.* **27,** 822 (1957).
8. W. B. Hillig and D. Turnbull, *J. chem. Phys.* **24,** 914 (1956).
9. J. Hallet, *J. Atmospheric Sci.* **21,** 671 (1964).
10. H. R. Pruppacher, *J. Colloid Interface Sci.* **25,** 285 (1967).
11. T. Fujioka, Ph.D. Thesis, Carnegie-Mellon University, (1977).
12. G. P. Ivantsov, *Dokl. Akad. Nauk, SSSR* **58,** 567 (1947).
13. G. Horvay and J. W. Cahn, *Acta metall.* **9,** 695 (1961).
14. R. F. Sekerka, R. G. Seidensticker, D. R. Hamilton and J. D. Harrison, *Investigation of Desalination by Freezing*, Westinghouse Research Lab Report, Chap. 3, (1967).
15. M. E. Glicksman and R. J. Schaefer, *J. Cryst. Growth* **1,** 297 (1967).
16. M. E. Glicksman and R. J. Schaefer, *J. Cryst Growth* **2,** 239 (1968).
17. D. E. Temkin, *Dokl. Akad. Nauk, SSSR* **132,** 1307 1969.
18. R. Trevedi, *Acta metall.* **18,** 287 (1970).
19. G. E. Nash and M. E. Glicksman, *Acta metall.* **22,** 1283 (1974).
20. R. J. Schaefer, *J. Cryst. Growth* **43,** 17 (1978).
21. G. R. Kotler and W. A. Tiller, *J. Cryst. Growth*, **2,** 287 (1968). See also R. Trivedi and W. A. Tiller, *Acta metall.* **26,** 67 (1979).
22. W. W. Mullins and R. F. Sekerka, *J. appl. Phys.* **34,** 323 (1963).
23. W. Oldfield, *Mater. Sci. Engng* **11,** 211 (1973).
24. J. S. Langer and H. Muller-Krumbhaar, *Acta metall.* **26,** 1681 (1978).
25. W. W. Mullins and R. F. Sekerka, *J. appl. Phys.* **35,** 444 (1964).
26. R. D. Doherty, B. Cantor and S. Fairs, *Metall Trans. A* **9,** 621 (1978).
27. S. C. Huang, Ph.D. Thesis, Rensselaer Polytechnic Institute (1979).
28. S. R. Coriell and R. L. Pareker, in *Proceedings of ICCG*, Boston, June 20–24, 1966.
29. D. P. Woodruff, *The Solid–Liquid Interface*, pp. 116–117. Cambridge University Press (1973).

Acta Metallurgica Vol. 29, pp. 717 to 734, 1981
Printed in Great Britain. All rights reserved

0001-6160/81/050717-18$02.00/0

OVERVIEW 12

FUNDAMENTALS OF DENDRITIC SOLIDIFICATION—II DEVELOPMENT OF SIDEBRANCH STRUCTURE†

S.-C. HUANG‡ and M. E. GLICKSMAN
General Electric Company, Corporate Research & Development Center, Building 81, Room E102, Schenectady, NY 12345.
Materials Engineering Department, Rensselaer Polytechnic Institute, Troy, NY 12181, U.S.A.

(*Received 20 August; in revised form 17 November 1980*)

Abstract Experiments have been carried out to study the development of the dendritic sidebranch structure in succinonitrile. Results obtained provide the first insights into such fundamental problems as the origin of sidebranch perturbations, the mechanism of sidebranch evolution, the influence of anisotropy in solid liquid interfacial energy, the kinetics of isothermal coarsening, and, finally, the mechanism of dynamic sidebranch coarsening during solidification. All of the problems addressed have applications to the interpretation of microstructures arising from both conventional and rapid solidification processes.

Résumé Des experiences ont été mise en oeuvre afin d'etudier la formation de la structure dendritique à branches secondaires du succinonitrile. Les résultats obtenus donnent un premier aperçu de problèmes fondamentaux tels que l'origine des perturbations relatives aux branches secondaires, le mecanisme d'évolution de ces branches, l'influence de l'anisotropie des energies d'interface solide liquide, la cinétique de croissance isotherme et finalement la dynamique de croissance des branches secondaires au cours de la solidification. Tous les problèmes abordés s'appliquent a l'intérpretation des microstructures developpées lors de la solidification aussi bien classique que rapide.

Zusammenfassung Die Entwicklung dendritischer Sekundärarmstruktur wurde im System Succinonitril untersucht. Die erhaltenen Versuchsergebnisse liefern die erste Einsicht in viele fundamentalen Probleme, wie, z.B. Ursprung der Sekundärarmstörungen, Mechanismus der Sekundärarmentwicklung, Einfluss der Anisotropie auf die Grenzflächenenergie zwischen der festen und der flüssigen Phase, Kinetik der isothermanen Sekundärarmvergröberung, und Mechanismus der während der Erstarrung dynamischen Sekundärarmvergröberung. Alle obenerwähnten Probleme treten in der Analyse von den durch konventionelle und schnelle Erstarrungsvorgänge erzeugten Gefügen auf.

INTRODUCTION

In the preceding paper [1], we presented experimental results regarding the steady-state growth of a dendrite tip. The results were interpreted on the basis of the morphological stability at the dendrite tip. The purpose of this paper is to present a second body of the experimental results that deals with the nonsteady-state development of dendritic sidebranch structures in succinonitrile. Our main interest here is to study in detail the mechanisms of sidebranch formation and sidebranch coarsening during dendritic solidification. The results, cast as 'scaling laws,' apply to pure materials and dilute alloys.

† Research supported by the Marshall Space Flight Center, through the Spacing Processing Application Rocket (SPAR) Program, Office of Application, NASA.

‡ Formerly, graduate student at Rensselaer Polytechnic Institute.

BACKGROUND

A. *Sidebranch evolution mechanism*

The quantitative investigation of dendritic sidebranch evolution was first attempted by Tiller and co-workers [2, 3]. In an early computer model study [2], an initially smooth isothermal needle dendrite was observed to 'develop' sidebranches on a graphics terminal. It was observed that the tip radius of the initial dendrite first increased, and then decreased after a sidebranch had been formed at the 'shoulder' region just behind the dendrite tip. The formation of the second branch was also accompanied with one cycle of oscillatory change in the tip growth conditions, but the oscillation amplitude became smaller and the dendrite tip then settled down to an almost steady-state growth condition. Unfortunately, the computer simulation study was not continued with a sufficient number of time steps to investigate

Reprinted with permission from *Acta Metallurgica*, Vol. 29, 1981.

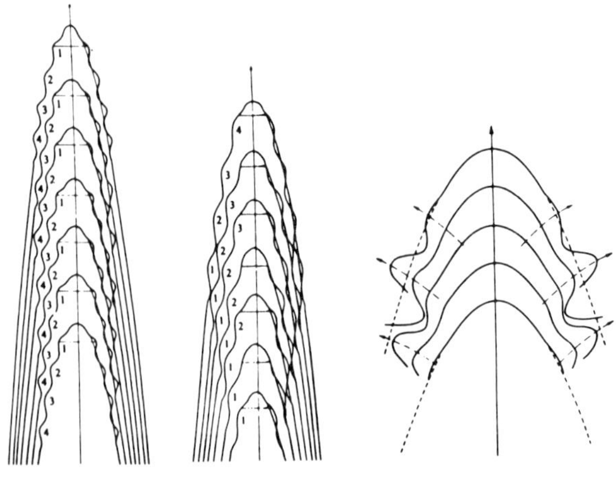

Fig. 1. Schematic of sidebranch evolution models: (a) a standing wave model [3]; (b) a traveling wave model [4]; and (c) the LM-K model [5].

the sidebranch evolution mechanism which would occur when the main dendrite tip had completed its transient growth and finally reached a steady state.

Another approach used to study the sidebranch evolution mechanism [3, 4] was to bypass the transient morphology development from an initially smooth needle dendrite to a branched dendrite. In this approach, Temkin's *optimized steady-state* solution (V_{max}, R_{opt}) was accepted (without detailed justification) as the pertinent growth conditions at the tip of a *branching* dendrite. The evolution of the branching dendrite was then investigated through a calculation of the morphological perturbation around Temkin's needle dendrite. In one of the calculations performed [3], the thermal diffusion associated with the branch perturbation was assumed to be at steady state (as in the growth of the basis needle dendrite). The result of this steady-state calculation suggested a sidebranch evolution mechanism known as the 'standing wave' model, schematically illustrated in Fig. 1(a). According to this model, the *whole* dendrite will grow at a steady state, i.e., the branched region of the dendrite as well as the dendrite tip will propagate at a constant velocity (in the direction of the dendrite main axis) preserving the undulatory shape. In another theory [4], the nonsteady-state nature of the thermal diffusion problem associated with the sidebranch perturbation was recognized and retained in the theoretical analysis. This time-dependent model yielded a 'traveling wave' model of sidebranch evolution, Fig. 1(b). It should be noted that in both calculations [3, 4], a perturbation source was believed to be necessary to initiate the branching instability.

A second time-dependent model has been developed very recently by Langer and Müller-Krumbhaar (LM-K) (5). As reviewed in Part I, the LM-K theory does not employ the 'maximum growth velocity' principle, but uses instead a marginal stability condition to select the steady state of a dendrite tip. Under this condition, the tip is the only point on a needle dendrite which is stable. Instability would still persist at *all* other portions of the dendrite surface and lead ultimately to sidebranch formation. An example of the sidebranch evolution sequence predicted by the LM-K theory is shown in Fig. 1(c). As illustrated in this figure, the LM-K theory predicts a nonsteady-state sidebranch formation mechanism which is somewhat akin to the 'traveling wave' model. Also, the LM-K theory retains the steady-state nature of the tip as does the 'standing wave' model. The LM-K theory is thus unique in its prediction that the evolution of sidebranches is a result of an intrinsic morphological instability of the needle dendrite (except at the tip) which does *not* need to be induced by growth fluctuations at the dendrite tip.

B. *Sidebranch coarsening mechanism*

Dendritic branch spacings increase with time under isothermal conditions such as exist in the mushy zone of a casting. The coarsening of the secondary sidebranches during the isothermal holding of alloy castings has been studied extensively, and at least three models of isothermal coarsening are available in the technical literature [6–8]. Two of the models [6, 8] ascribed dendrite branch coarsening to the remelting of some of the sidebranches, and the third [7] to the actual detachment of some sidebranches from the dendrite main stem. In general, it is agreed that a reduction in surface energy provides the thermodynamic driving force for the isothermal stage of the coarsening process.

On the other hand, dendrite branch spacing also continuously adjusts during the dendritic solidification process. The branch coarsening during dendritic solidification is quite complicated, and a theoretical model describing this dynamic coarsening mechanism is lacking at present. It is rather widely accepted, however, that coarsening during dendritic growth proceeds with a mechanism similar to that during isothermal holding [9]. This hypothesis was derived from the results of a series of studies [9] in which branch spacing was measured as a function of local solidification time (in interrupted solidification experiments) and coarsening time (in isothermal coarsening experiments). Data from both sets of experiments were found to lie along the same line which was predicted by two of the isothermal coarsening kinetic models [9]. It will be interesting to test whether these experimental observations for selected alloy systems remain true in a one-component system such as succinonitrile.

EXPERIMENTAL TECHNIQUES

The same high-purity succinonitrile specimen and experimental setup described in Part I [1] were used for the present study of sidebranch development. Further details of the experimental techniques have been described in Ref. [10]. It is sufficient to note the anisotropic nature of the dendritic sidebranch morphology which was delineated in Figs 4 and 5 in Part I. As shown, all the branches, resulting from the growth of perturbations along the {100} plane protrusions described in Part I, lie in {100} planes. To avoid complications arising from stereographic projection, our measurements of branch spacings, as well as qualitative studies of the branch evolution and growth mechanisms, required the same precautions as were taken in the tip radius measurements described previously. Specifically, the precaution taken was to study each dendrite by observing it from a ⟨100⟩ direction orthogonal to its main axis. A view of a dendrite in its (100) plane is shown in Fig. 2.

Generally, the branch spacing of a dendrite growing in a pure melt varies not only with the time of solidification but with the initial level of supercooling or cooling rate. The relationship between branch spacing and supercooling was studied here by measuring the branch spacing in regions near the dendrite tip. The spacing S_{tip} between sidebranches newly formed near the tip region is controlled by a perturbation wave mechanism and is only a function of supercooling. To study the subsequent coarsening effect, the initial spacings S_{tip} at several supercoolings were studied as a function of time, or more practically, as a function of the distance away from the dendrite tip. As can be seen in Fig. 2, however, each sidebranch has a different growth history, and each one at a given time might be either remelting, decelerating, or accelerating in its growth speed. The resultant overall sidebranch morphology is therefore complicated, and one needs to define carefully the 'actively growing branches' and to measure spacings only between neighboring 'active' sidebranches. In this study, we define a branch as 'active' when it is equal in length or longer than the next 'active' branch closer to the dendrite tip.

The crystal growth device shown in Fig. 3 in Part I also provides the capability of studying isothermal branch coarsening. After the dendritic structure has developed inside chamber C, control heater B in turned off, allowing the seed crystals above the heater to grow into chamber C. The crystals tend to grow along the glass walls (at a velocity approximately four times that of the free dendritic growth) thereby coating the walls with a shell of solid succinonitrile. As the solid shell continues to thicken, latent heat is released into the enclosed sphere of supercooled liquid succinonitrile. When the supercooling of the melt is eliminated by this process, a precisely isothermal condition will be established within the surrounding shell of solid. Thereafter, this solid shell serves as a thermal barrier, which actively maintains the isothermal condition of the remaining succinonitrile liquid and dendrites at the melting point. The dendritic sidebranch morphology was then observed and measured as a function of time under this isothermal situation.

RESULTS AND DISCUSSION

A. *The sidebranch perturbation sequence*

Figures 4 and 5 in Part I clearly show that the tip region of a dendrite, initially a body of revolution, bulges in four lateral {100} planes and distorts continuously into four branching sheets. Oscillating bumps then develop on the four longitudinal branching sheets. On the other hand, branching bumps cannot be seen in the troughs between branching sheets. This sequence of shape distortions is similar in certain respects to that observed in crystallizing ice cylinders, where sixfold polar instabilities (reflecting the hexagonal symmetry of ice) formed before the appearance of oscillating bumps or z-axis instabilities [11].

However, it should be noted that in Langer and Müller-Krumbhaar's theory the solidifying system is assumed to be isotropic regarding interfacial growth kinetics, and solid–liquid interfacial energy. Accordingly, the perturbation wave which arises at an isotropic dendrite tip would generate lateral, axisymmetric branching *rings* around the dendrite stem. How these rings might in turn break down into polar oscillations is not included in the LM-K theory; presumably, such oscillations could arise from yet other perturbations disturbing these rings while they are expanding in radius. This sequence of branching perturbations is obviously inconsistent with what we observe experimentally in body-centered cubic succinonitrile. To illustrate further the differences between

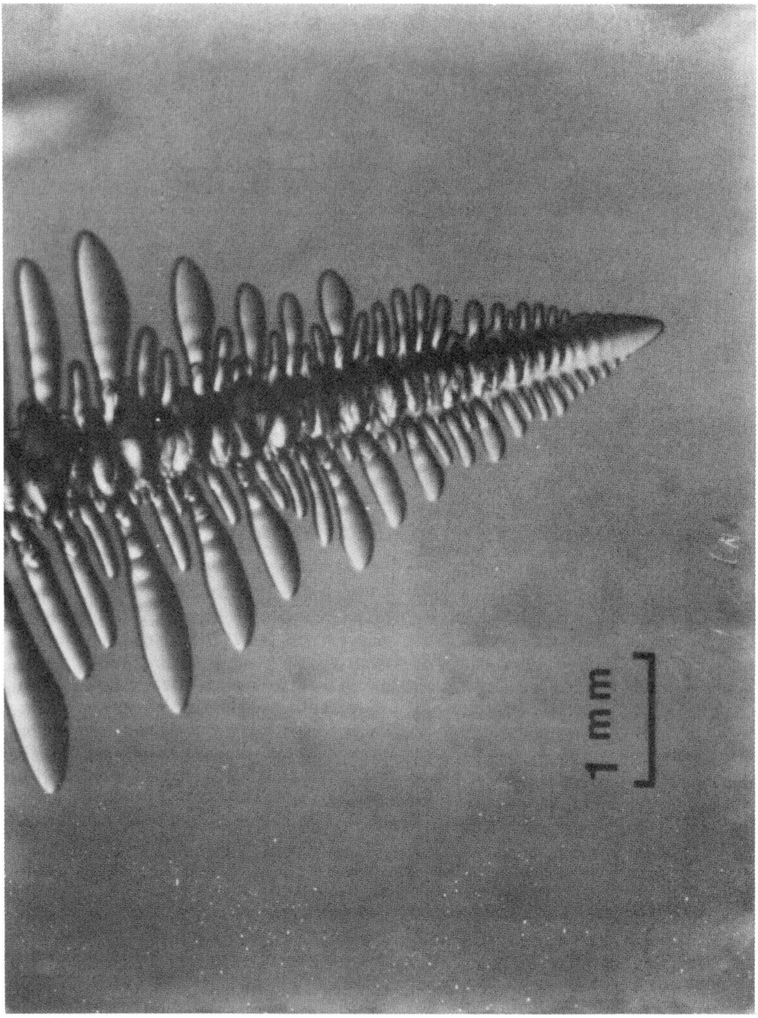

Fig. 2. An overall view of a succinonitrile dendrite, showing the variations in the sidebranch morphology with distance from the dendrite tip.

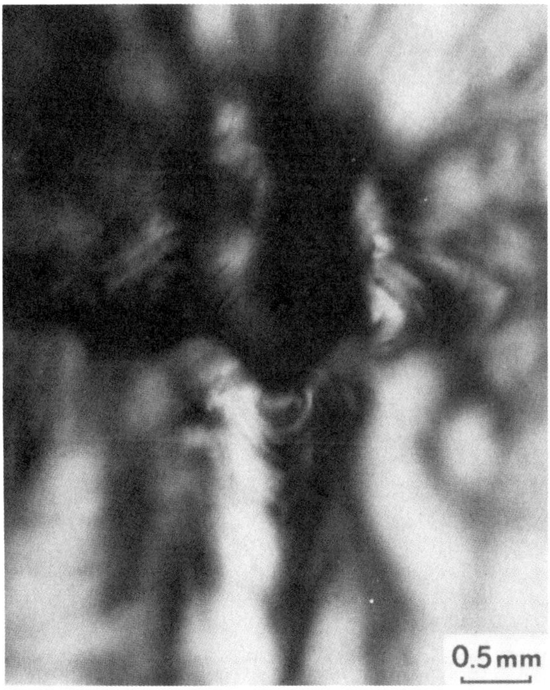

Fig. 3. Photograph of a dendrite observed in a direction nearly parallel to the dendrite axis; see Fig. 4(d) in Part I for comparison. The featureless region at the center of the photograph is the branchless, smooth tip region as can be seen in Figs 4 and 5 in Part I of the paper. $\Delta T = 0.150$ C.

the observed and the predicted sidebranch perturbation sequence, we can view a dendrite tip along its growth axis in Fig. 3. In this orientation, only perturbations oscillating along the dendrite main axis can be seen. The indentation around the dendrite tip (located near the center of the photograph) appears to be distorted from axisymmetry. This indicates that anisotropy in the dendrite cross section develops before the branching oscillation appears along the dendrite axis.

It is suggested, based on the laboratory observations described above, that the slight anisotropy in solid–liquid interfacial energy (estimated to be about 1% in succinonitrile, Fig. 4) plays an important role in the branching mechanism. Anisotropy in solid-liquid interfacial energy provides an additional source for interfacial distortion other than the interfacial instabilities associated with paraboloidal dendrites. In the early stage of branch perturbation, the slight, cubically symmetric anisotropy in solid-liquid interfacial energy seems to be the dominating force. Thereby, four branching sheets in {100} planes are always formed. It is only in latter stages that the dendritic instabilities take over, and periodic bumps evolve into branches along the branching sheets.

It is worthwhile noting that although anisotropy in solid–liquid interfacial energy has profound effects on the branch-perturbation sequence, it should have relatively little effect in changing the growth state (V,R) of the dendrite tip. The dendrite tip region approximates a body of nearly perfect axisymmetry (cf. Fig. 5 in Part I), and the usual assumptions of material isotropy made in dendritic growth theories and instability theories should apply at the tip. Another point worth noting is that anisotropy in growth kinetics may have a similar effect on the branching perturbation sequence as does the anisotropy in the solid-liquid interfacial energy. In the case of ice dendrites (12), sheet-like crystals grow from a predendritic disk in the basal plane, and branching perturbations occur only in this plane. (Interestingly, even though the anisotropy in the growth kinetics of ice crystals is very high, the dendrite-tip growth rates are still predicted closely by a three-dimensional axisymmetric model 13). In the present case of succinonitrile crystals (with b.c.c. structure), the orientation of maximum solid–

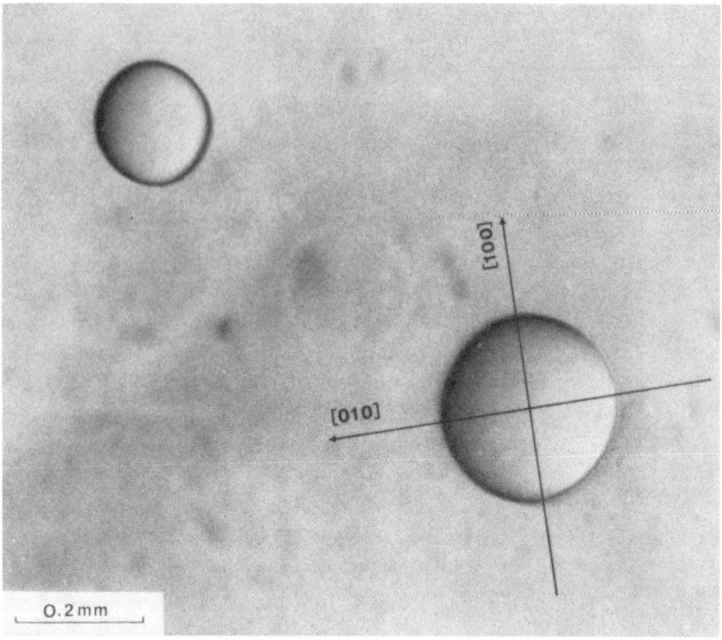

Fig. 4. Equilibrium shape of liquid droplets in a crystal of pure succinonitrile. The crystallographic directions shown were identified from the dendritic growth orientations of the crystal. The slight bulging of the droplets in the ⟨100⟩ corresponds to about a 1% anisotropy in the solid-liquid interfacial energy.

liquid interfacial energy coincides with the ⟨100⟩ 'easy growth' direction of a cubic crystal. Actually there is experimental evidence [14, 15] that an 'easy growth' direction of a cubic crystal can be a direction selected initially by the maximum interfacial energy during the early stages of growth (or nucleation) and then maintained by the 'point effect' during diffusional crystal growth. In any event, the magnitude of the anisotropy in the growth kinetics of succinonitrile is not available in the technical literature. The explanation of the formation of {100} perturbation sheets discussed above is based solely on the anisotropy of the solid-liquid interfacial energy which, for the first time, is available through our measurements, Fig. 4.

B. *Sidebranch evolution mechanism*

The sidebranch evolution resulting from the dendritic interfacial instabilities is best delineated in sequential superimposed photographs shown in Fig. 5. Note the similarity of this figure to the prediction of the LM-K theory, i.e., Fig. 1(c). The first ten or so bumps display a nearly constant spacing (a characteristic of wave-controlled mechanisms). The initial branch spacing, S_{tip}, was measured to be about three times the wavelength λ_s of the critical perturbation wave, Fig. 6. The measured ratio of S_{tip}/λ_s is higher than the prediction of the LM-K theory, i.e., $S_{tip}/\lambda_s = 2.1$. This discrepancy arises from the fact that the actual sidebranch perturbation occurs only along the edges of the branching sheets and not on a three-dimensional axisymmetric body as assumed in the LM-K theory. It can also be observed in Fig. 5 that the growth trajectory of the oscillatory instabilities points about 20° forward of the ⟨100⟩ directions orthogonal to the growth axis. This branching perturbation orientation coincides with the direction of the maximum temperature gradient around the otherwise needle-like dendrite. This observation agrees with the theoretically predicted perturbation growth directions, cf. Fig. 1(c). The non-crystallographic growth directions can also be rationalized with stability theory, by considering the growing perturbations as waves propagating away from the dendrite tip at a velocity slightly smaller than the apparent dendrite growth rate, V. A net forward shifting results such that the branches are not quite stationary in the laboratory reference frame [5]. The branches thus grow 20° out of the ⟨100⟩ directions, and the 1% anisotropy in solid-liquid interfacial energy seems to have little influence in controlling the sidebranch evolution direction. A branch, once evolved, increases its size as the dendrite tip continues

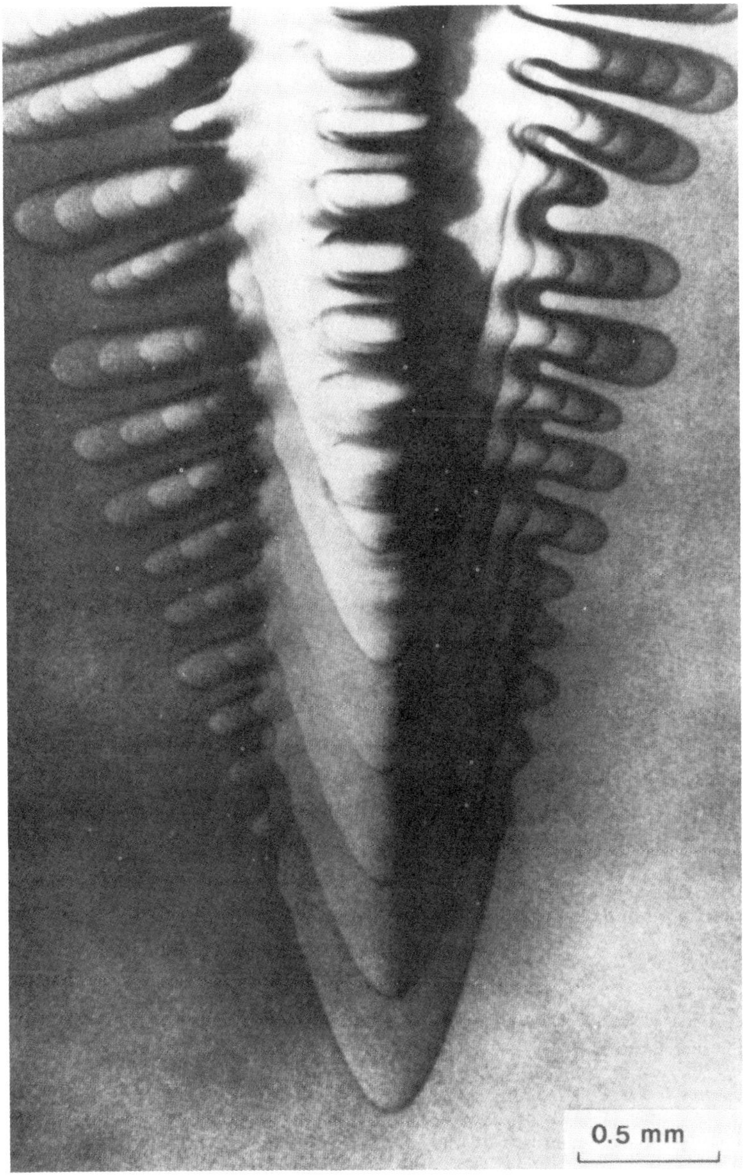

Fig. 5. Superimposition of time-lapsed photographs of a growing dendritic crystal, showing the side-branch evolution process on the {100} branching sheets.

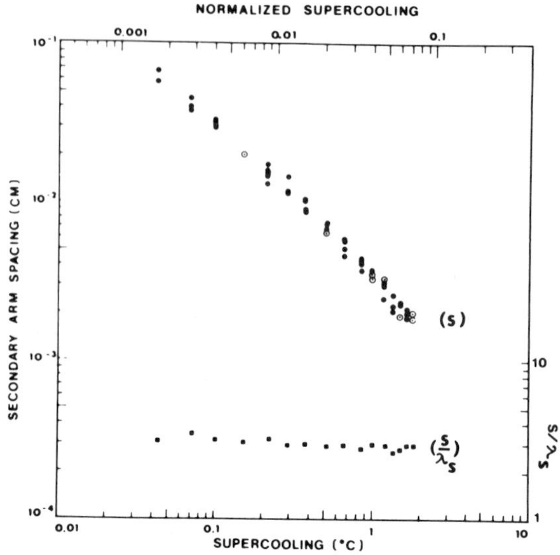

Fig. 6. Secondary branch spacing versus supercooling (● as measured; ■ as normalized to planar perturbation wavelength).

to grow away from it. Sidebranch amplification is expected from the unstable nature of the lateral surfaces of dendrites as will be discussed later. The growth rate of sidebranches is measured to be about 40% of the axial dendrite growth rate V and is predicted closely by the LM-K theory.

In short, the evolution of the first few branches must be controlled by the nonlinear development of perturbation waves. Consequently, the branch spacing is given as a unique quantity in terms of the critical instability wavelength which depends on the supercooling. The subsequent growth of the branches is due simply to the increase in the wave amplitude. The direction of the branch growth is determined by the propagation velocity of the wave which is governed by a complicated dispersion relationship, the details of which are not understood theoretically at this time.

C. *Origin of the sidebranch perturbation*

In Part I, the spherical stability model had demonstrated that an unperturbed tip ($\delta_n = 0$) is the fundamental tip growth state. The spherical model also bears upon the question concerning the origin of sidebranch perturbations. It is suggested that a sidebranch perturbation arises as a direct result of the intrinsic morphological instability existing at all regions of the needle dendrite other than at the tip. The relative magnitude and the distribution of interfacial instability are revealed in a calculation of the local stability constant ($\sigma_r = 2\alpha d_0/V_n R_r^2$) as a function of the radial distance away from the axis of the dendrite, Fig. 7. The mean radius R_r at a point (y, r) on the dendrite surface is given by

$$R_r = \frac{2R[1 + (r/R)^2]^{3/2}}{2 + (r/R)^2},$$

and the normal growth velocity V_n at the point (y, r) is given by

$$V_n = \frac{V}{[1 + (r/R)^2]^{1/2}}.$$

Therefore,

$$\sigma_r/\sigma^* = \frac{[2 + (r/R)^2]^2}{4[1 + (r/R)^2]^{5/2}},$$

where σ^* is the marginal stability constant applicable at the dendrite tip. As shown in Fig. 7, σ_r decreases rapidly to 20% of the value of the critical stability constant σ^* at a distance only one tip-radius away from the tip. According to the LM-K theory, the reduced σ_r corresponds with an increased degree of instability associated with the needle dendrite surface. Due to the generally unstable nature of the entire dendrite surface (except at the tip), any interfacial perturbation induced by an infinitesimal disturbance will continue to grow and eventually develop into sidebranches. This model of sidebranch evolution disagrees with the classical models, in which the sidebranch evolution is believed to result from oscillations in growth rate and in the size of the dendrite tip [3, 4].

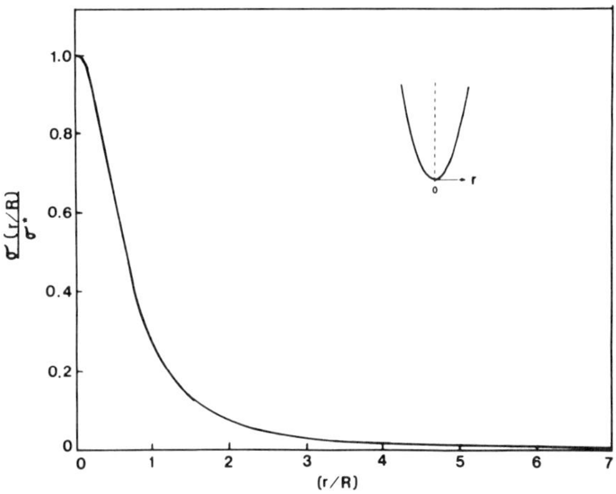

Fig. 7. Normalized local stability constant, σ_r/σ^*, as a function of the normalized radial distance from dendrite tip, r/R.

As discussed above and in Part I, the spherical stability model, despite its simplicity, retains certain important characteristics of a real dendrite, which are inherent in the more elaborate theory of LM-K. Briefly, the spherical stability model indicates that the dendrite tip grows at a marginally stable (steady-state) condition and does not itself contribute to the formation of sidebranches. Sidebranches evolve because of the morphological instability associated with the dendrite surfaces in the neighborhood of the tip. Furthermore, the experimental results presented in Part I have shown that the growth of sidebranches has a negligible effect on the axial tip growth kinetics. Therefore, it may be concluded that the dendrite's axial tip growth and the sidebranch evolution process are relatively independent, insofar as these solidification events do not interfere with each other to any significant extent.

D. *The scaling laws*

As discussed above, stability theory plays critical roles in determining both the steady-state and the nonsteady-state features of the dendrites studied here. The dendritic tip radii were measured and found to be approximately equal to the planar perturbation wavelength (Cf. Fig. 11 in Part I), whereas the branch spacing was found to be equal to three times this wavelength, Fig. 6. Since the tip radius and the branch spacing are the two characteristic dimensions of a dendrite, it follows that the general morphology of a dendrite can be scaled to λ_s. Also, since λ_s is inversely proportional to the supercooling, it follows that the general morphology of a dendrite can be scaled to ΔT^{-1}, where ΔT is the only variable in the present experiments. More specifically, the size-scaling law is such that dendrite images, when optically magnified by a factor equal to the ratio of the supercoolings, should appear to be of identical size. To demonstrate this scaling law, six dendrites solidifying at various degrees of supercooling were photographed and enlarged to produce images with the same apparent tip radius and the same branch spacing, Fig. 8. The magnifications needed so that the dendrite images appear similar to one another are as expected from the scaling law cited above.

However, we note that the position for the first distinguishable branch to appear varies slowly with the supercooling despite the scaling law of ΔT^{-1}. As shown in Fig. 8, branches appear somewhat farther back away from the dendrite tip at smaller supercoolings. Specifically, the first branch appears about ten tip radii away from the tip at $\Delta T = 0.070$°C, vs only about five tip radii at $\Delta T = 1.350$°C. Reasons remain uncertain for the non-linear supercooling dependence of the first branch location (normalized to tip radius), y_b. Thermal disturbances (normalized to supercooling) should be no larger at higher supercooling, and neither should mechanical disturbances. Therefore, the supercooling dependence of y_b cannot be attributed to the variation in these disturbance levels. The lack of scaling for the appearance of the first branch might be attributed to the non-linear amplification of the branching waves.

Not only the dendrite size but also the dendrite growth kinetics can be scaled to supercooling. Figure 9 in Part I has shown that the dendritic growth velocity is proportional to $\Delta T^{2.7}$, as predicted by the new dendritic growth theory, and as in agreement with the present experimental results. This power-law relationship between V and ΔT is particularly accurate in the dimensionless supercooling range from 5×10^{-2} to 3×10^{-1}, where unit supercooling is

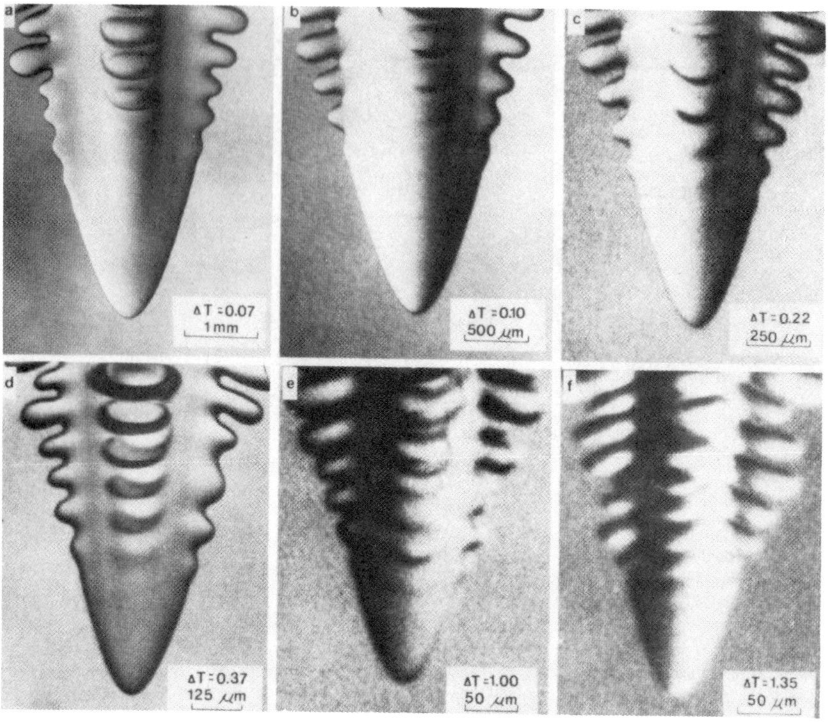

Fig. 8. Images of dendrite tips growing at different levels of supercooling. Magnifications are chosen to yield the same apparent tip radii.

defined as L/C_p. Below 0.05 (L/C_p), convective heat transfer becomes the limiting growth mechanism for dendrites of virtually all materials (16). Above 0.3 (L/C_p), molecular attachment kinetics limits the dendritic growth process in many materials (except possibly those with growth kinetic coefficients even larger than that for succinonitrile). From the practical point of view, therefore, the power law $V \propto \Delta T^{2.7}$ represents the scaling relationship between V and ΔT when *thermal diffusion* of latent heat is the growth-limiting transport process. Similarly, the sidebranch growth velocity (the perturbation amplification rate) is also scaled to $\Delta T^{2.7}$.

E. *Sidebranch competitive growth during solidification*

Figure 2 shows that dendritic sidebranches evolve in conjunction with the dendrite tip. The evolution of the sidebranches, as discussed in detail with Fig. 7, can be attributed to the needle-shaped dendrite surface which is inherently unstable (except at the tip). The perturbation mechanism alos controls the growth of the first 10 or so branches newly formed near the dendrite tip regions. As can be seen in Fig. 2, the branch spacing (or equivalently the dendritic perturbation wavelength) at this early stage of sidebranch development is constant and small (of the order of the thermal diffusion length). Thermal fields of neighboring branches thus tend to overlap and the growth of all the branches are generally retarded. On the other hand, the sidebranch growth at this stage can be explained as the increase in the amplitude of the sidebranching perturbation wave as the wave propagates away from the dendrite tip. When under the control of the perturbation wave mechanism, all the equally spaced branches tend to grow at the same velocity (about 40% of the axial dendritic growth velocity as stated earlier).

At some distance away from the dendrite tip, however, certain sidebranches experience slower growth than the others, due either to external disturbances or to statistical variations. As soon as a branch hesitates in growing, even momentarily, its further growth will be retarded rapidly because of the heat released in its proximity by neighbors. On the other hand, the leading neighbors tend to speed up their growth because they in turn achieve a larger space in which to grow.

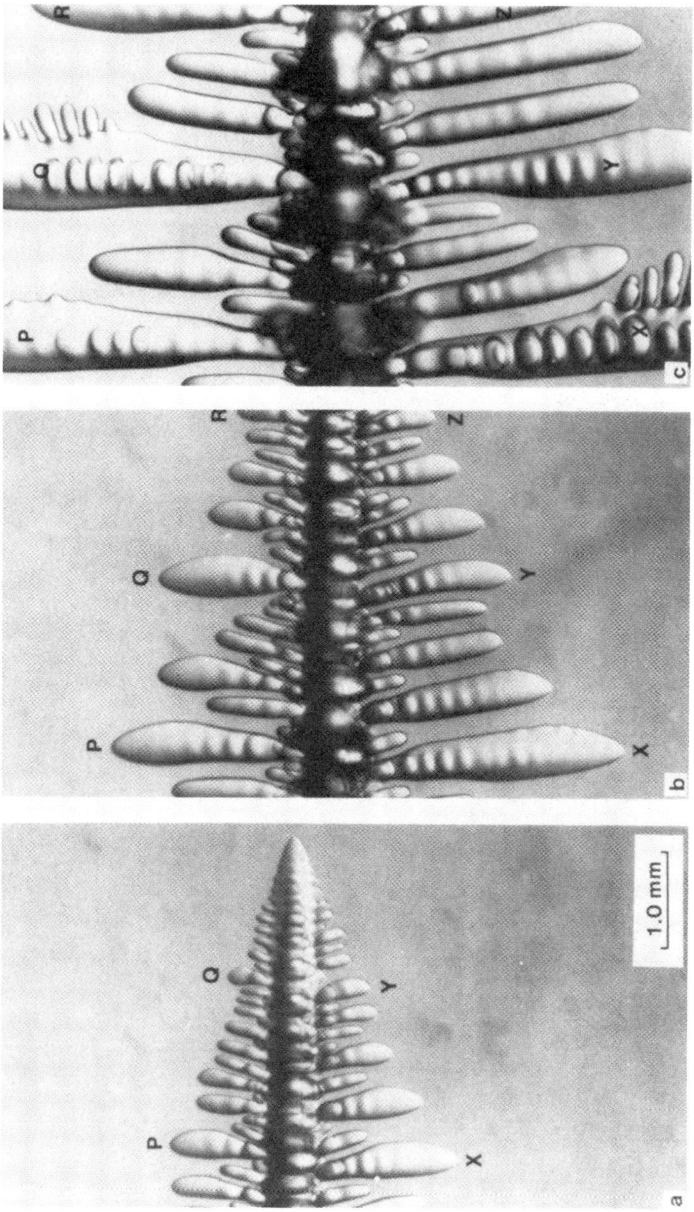

Fig. 9. Photographs illustrating the branch spacing adjustment process due to sidebranch competitive growth. 30 s between (a) and (b); 1 min between (b) and (c).

As a result, slow branches decelerate and fast branches accelerate. The slow branches soon drop out of the growth competition, and a new spacing becomes established between the leading branches. The adjustments in the sidebranch spacing due to continuous growth competition among neighboring sidebranches is quite evident in Fig. 2. As shown, fewer and fewer sidbranches survived the growth competition, and the sidebranch spacing increases rapidly with increasing distance away from the dendrite tip.

To illustrate further the dynamic branch spacing adjustment process, a dendrite was allowed to grow across the observation field of the microscope. Time-sequenced photographs of such a dendrite are shown in Fig. 9. At the time of Fig. 9a, branches between *P* and *Q* were still relatively close to the dendrite tip, and only a slight variation in sidebranch growth velocity is noticeable. A half-minute later, while the dendrite tip grew out of the observation field, branches between *P* and *Q* are clearly slowing down and dropping behind branches *P* and *Q*, Fig. 9b. One-minute later, branches *P* and *Q* continued to grow out of view while branches between them either stopped growing or actually remelted, Fig. 9c. A new branch spacing was then established between branches *P* and *Q*, which is about 15 times that of the branch spacing only one and a half minutes previously. Similarly, new spacings were established between branch *Q* and *R*, *X* and *Y*, and *Y* and *Z*. Although details of the branch spacing adjustment process were slightly different in each case, a similar spacing was established on both sides of the dendrite main stem by the time of Fig. 9c. If the dendritic growth process was further extended, sidebranches *P*, *Q*, and *R* would continue to compete against one another, leading to further increases in sidebranch spacing.

The variation of sidebranch spacing as a function of distance away from the dendrite tip was measured in Fig. 10. The distance from the dendrite tip was converted to an effective 'local solidification time' [9] by dividing the distance by the growth velocity of the dendrite stem. In measuring the branch spacing, some objective, albeit arbitrary, judgment is needed to

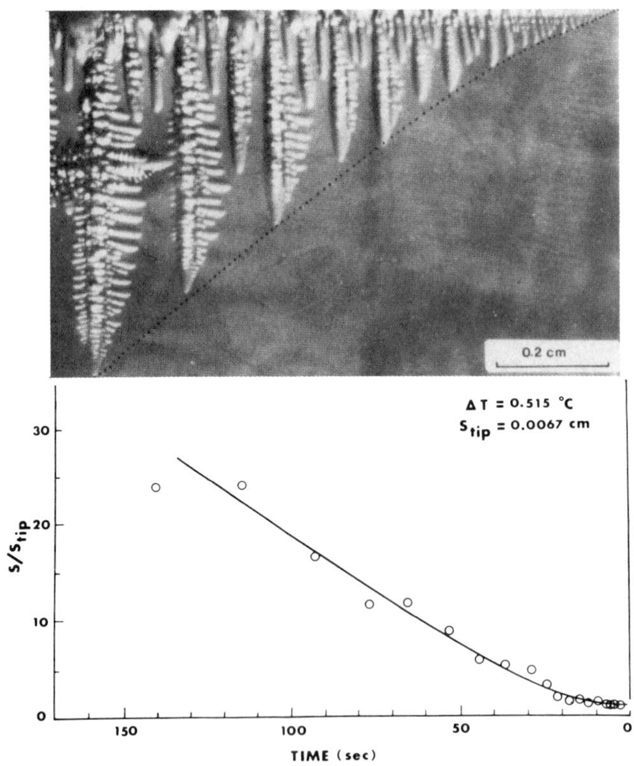

Fig. 10. Variation of sidebranch spacing as a function of distance from dendrite tip.

define the 'active' sidebranches. In the present case, a branch was judged as 'active' when it was equal to or longer than the next 'active' branch closer to the dendrite tip. Since the first 8–10 branching bumps are always 'active,' the definition is easily applied from the tip backwards. Only spacings between pairs of active branches were measured, and we ignored the presence of any intermediate 'inactive' branches. The branch spacing measured in this way was shown to increase as a power law relationship with increasing time or distance away from the dendrite tip.

While the dynamic coarsening proceeds and the growth space of surviving sidebranches increases, the surviving sidebranches do two things: first, the sidebranches originally growing 20° away from ⟨100⟩ directions gradually turn back into the proper ⟨100⟩ directions. This 'steering' effect can be seen in Fig. 2 and is attributable to the anisotropy in solid–liquid interfacial energy. Clearly the branch development at this stage is no longer controlled by the branching wave, but by a large-scale heat flow strongly influenced by local capillary processes. Second, an active sidebranch tends to grow at a higher velocity when it gets more growth space. This can be visualized in Fig. 10, in which the tangent to the curve connecting the tips of the leading sidebranches forms an increasing angle with the dendrite main stem. If the angle is increased to $\pm 45°$, then the branch must grow at a velocity equal to that of the main dendrite. This means that the particular branch must grow without the influence of its neighboring branches as an independent dendrite under an identical operating state as the tip. When this stage of branch growth is reached, competitive spacing adjustments will stop. This ultimate distribution of sidebranches leads to the structure of so-called 'superdendrites' observed in alloy systems [17]. Unfortunately, the crystal growth chamber used in the present study was not large enough for the branches to reach this ultimate growth state. As shown in Fig. 10, the maximum angle achieved between the tangent to the branch-tip envelope and the dendrite stem axis was 41°, and the process of dynamic branch coarsening was still actively underway.

The results of the dynamic branch coarsening measurements at eight levels of supercooling are compiled in Fig. 11. At each level of supercooling, branch spacings measured between neighboring active sidebranches were normalized to the wave-controlled branch spacing, S_{tip}. The normalized branch spacings were then plotted on a log–log scale against local solidification time. As shown in Fig. 11, each coarsening curve is a simple exponential function, and each has a slope of 1.3. Apparently, the same kinetic mechanism is responsible for the branch coarsening at different levels of supercooling namely, the side-branch competitive growth described above.

It should be noted that the slope found in Fig. 11 does not reflect a coarsening exponent of particularly fundamental importance. More specifically, the expression of 'coarsening rate' is arbitrary, and depends just how an 'active' sidebranch is defined. Various criteria for defining active sidebranches would yield various slopes for the logarithmic coarsening curves. One extreme criterion, for example, is to consider every existing branch, no matter whether they are growing fast, slow, or actually shrinking. This criterion would give a smaller coarsening rate which would be difficult to separate from isothermal coarsening effects driven by surface area reduction. Another extreme would be to consider as active only the leading branches (those which define the sidebranch envelope in Fig. 10). With this criterion, the responsible coarsening process is confined strictly to the branch competitive growth mechanism. This criterion would yield a relatively large coarsening rate, but it is technically difficult to apply. The criterion employed in this study as stated earlier is intermediate to these extremes and is so designed as to be able to be applied consistently and to include largely the effects of the sidebranch competitive growth mechanism.

From Fig. 11, an incubation time can be defined at $S/S_{tip} = 1$ at each level of supercooling, within which branch coarsening does not occur. The incubation time increases with decreasing supercooling. The incubation time can be changed to an incubation length, merely by multiplying it by the dendrite growth velocity, V. The supercooling can be mapped into dendrite growth velocity through the growth rate relation determined experimentally, see Fig. 9 in Part I. Therefore, a relationship between incubation length and dendrite growth velocity can be established, Fig. 12. Physically, the incubation length represents the length of dendrite near the tip region within which the branches are controlled only by the perturbation wave. Dynamic branch coarsening is not expected within the incubation length as defined here. On the other hand, the dendrite growth velocity can be related to the heat extraction rate. Thus, the experimental result of Fig. 12 could be used to estimate whether or not branches coarsen in a specimen of a given size, being quenched at a given rate. Even at high heat extraction rates, dendrite sidebranches still undergo coarsening unless the specimen size (and thus the dendrite length) is smaller than the incubation length. The coarsening mechanism in high-speed solidification processes can be due only to the sidebranch competitive growth mechanism as described above.

There have been suggestions that one can estimate cooling rates (or supercoolings) by measuring branch spacing and then referring to a 'standard' empirical relation which relates the two parameters [7]. The 'standard' reference was usually an experimentally derived relation of cooling rate versus branch spacing available in the technical literature. However, we have just shown that the branch spacing in an as-cast material depends not only on supercooling (or cooling rate in quenching experiments) but also on local

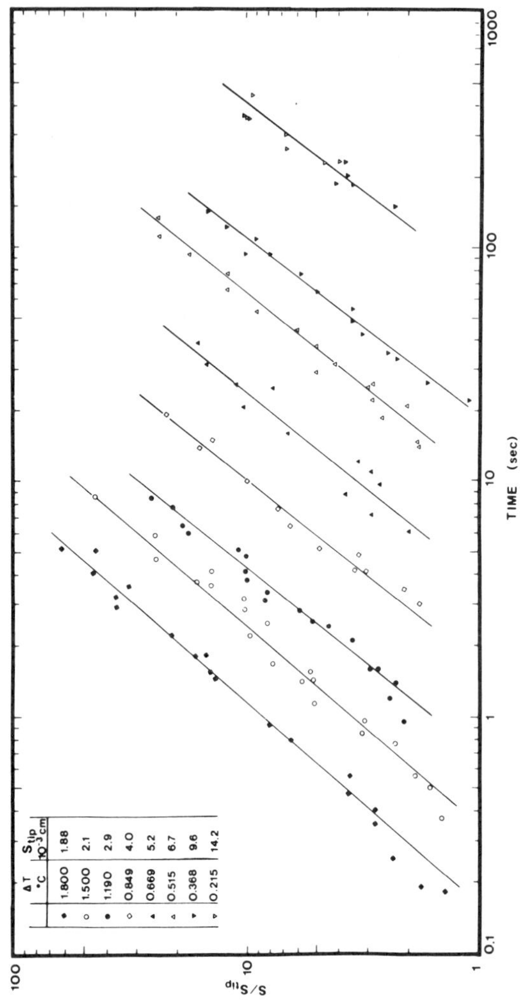

Fig. 11. Dynamic branch coarsening (normalized) measured at different levels of supercooling as a function of time.

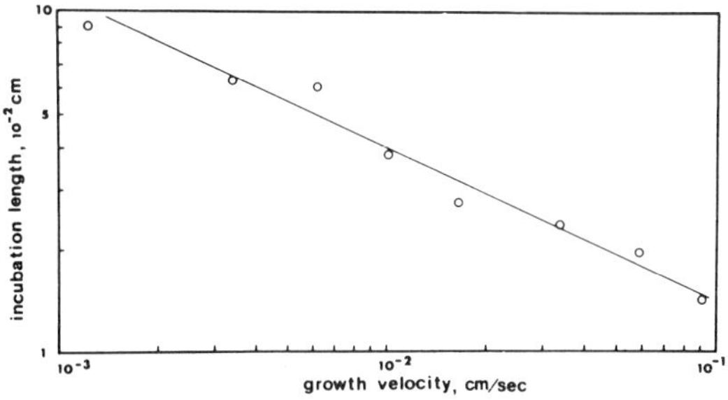

Fig. 12. Incubation length for initiation of coarsening versus growth velocity.

solidification time, Fig. 11. The method of estimating cooling rate described above appears therefore to be adequate in only two cases: 1. where the branch spacing is measured at the same local solidification time and with the same criterion of 'active' branches in all the experiments; and 2. where the branch spacing is measured before the occurrence of coarsening in all the experiments. Whether the second condition is fulfilled or not can be judged from an empirical relation involving both the heat extraction rate and the specimen size, such as the one shown in Fig. 12. If no coarsening occurs, it would be more appropriate to use the scaling law relationship for sidebranch formation, such as the one shown in Fig. 12.

F. *Isothermal coarsening*

Isothermal coarsening was studied by observing the changes in sidebranch morphology as a function of time. Three types of branch coarsening processes commonly observed in this study are shown in Figs 13–15. As shown, the detailed coarsening process is different from case to case, depending on the initial geometry of the relevant sidebranches. When the initial sidebranches are relatively long and closely spaced, they tend to touch one another, sinter, and form branch plates, Fig. 13. It is interesting to note in Fig. 13 that when branch J and I coarsened and joined together, a liquid droplet became trapped between them. Detailed equilibrium geometry of the liquid droplets allows the anisotropy in solid-liquid interfacial energy to be measured, Fig. 4. Also, the liquid droplets which develop in a coarsening process as described above are themselves the origin of some microcavities in as-cast materials. When the initial sidebranches are relatively short and widely separated, they are not able to touch one another, even though they tend to thicken in diameter during coarsening. In this case, the branches are joined together to form branch plates by filling in the liquid 'troughs' between branches, Fig. 14. For both cases of isothermal coarsening demonstrated in the last two photographs, the original sidebranches before coarsening are approximately of the same length and thickness. None of the branches is likely to remelt and allow the others to grow. As such, both of these coarsening processes result in formation of branch plates rather than remelting of some branches as usually proposed [6–8]. Only in those cases where the initial sidebranches are unequal in length and thickness might the short and thin ones be remelted, Fig. 15.

An observation common to the three different isothermal coarsening mechanisms as shown in Figs 13–15 is that of the filling in of solid in the 'trough' regions between branches. The crystal growth here indicates the existence of some slight supercooling at 'trough' regions. This local supercooling condition is not due to a lower local liquid temperature but rather to a higher interfacial melting temperature within the 'trough' region, where solid-liquid interface has a negative net curvature. In alloy systems, however, the melting temperature is a function of composition also. Usually, the microsegregation in alloys is such that the melting temperature at 'trough' regions is depressed. If the depression in melting temperature from microsegregation is larger than the elevation in melting temperature due to negative curvature, then remelting rather than freezing may occur at the root regions of branches. Remelting at the root of a branch might lead to its detachment from the main dendrite. On the other hand, branch detachment is unlikely to happen during isothermal coarsening in a one-component system. Consistently, few occasions of sidebranch detachment were ever observed in this study, except perhaps when strong melt convection occurred or a mechanical disturbance was inadvertently introduced.

Whereas the capillary-driven coarsening effect is the sole mechanism responsible for coarsening during isothermal heat treatment, a complete description of the coarsening process during solidification consists of two mechanisms. The first is a competitive sidebranch growth mechanism, described previously, by which only a selected number of branches is allowed to grow actively. The growth of the active branches then helps to set up a nearly isothermal environment for the short, inactive branches trapped between active branches. The second is the isothermal coarsening mechanism, by which the inactive sidebranches

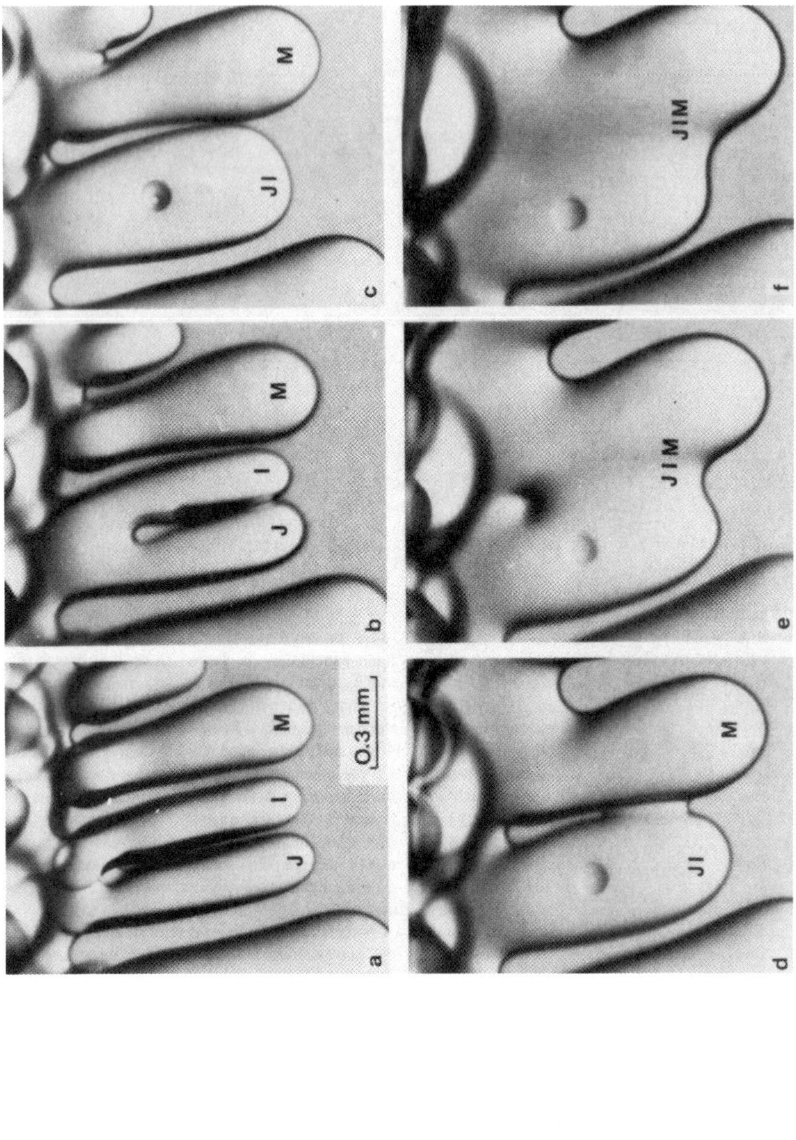

Fig. 13. Formation of a sidebranch plate during isothermal holding by a 'sintering' mechanism. Time intervals between sequential photographs are 20–30 min.

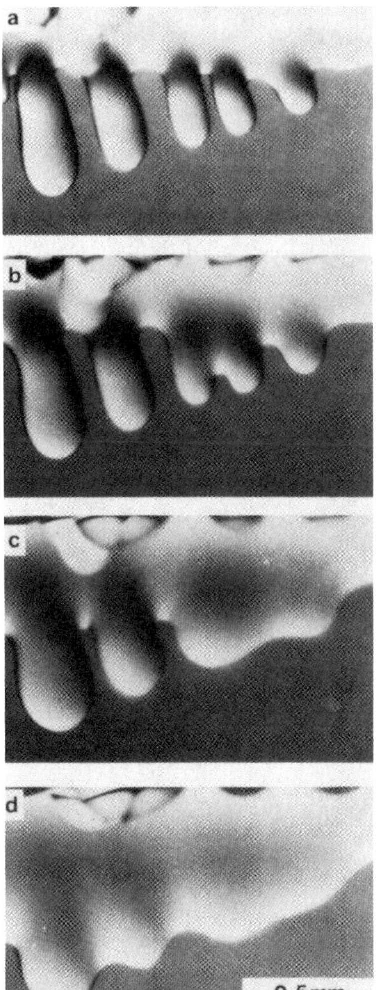

Fig. 14. Formation of branch plate during isothermal holding by a 'wave-damping' mechanism. Sequential time intervals between photographs are 20–30 min.

will eventually be eliminated. It should be emphasized, however, that it is the competitive sidebranch growth mechanism that changes the branch spacings during solidification. Isothermal coarsening appears to be an additional process which helps to establish the branch spacings set up by the branch competition mechanism. Furthermore, the time span between the sequential photographs in Figs 13–15 was typically 20–30 min. Compared with the dynamic coarsening mechanism described earlier, isothermal coarsening is a slow process. This disparity in the time scale governing the two coarsening processes adds to the doubt that a similar mechanism is responsible for both dynamic and isothermal coarsening [9].

CONCLUSIONS

On sidebranch formation

1. Branching instabilities in b.c.c. succinonitrile arise primarily from the slight anisotropy in the solid-liquid interfacial energy. The tip region, initially a body of revolution, bulges in the four longitudinal {100} planes and then distorts into branching 'sheets.'

2. The interfacial instability associated with the dendrite stem surface perturbs the branching 'sheets' into sequences of oscillating edge bumps. The branching bumps point about 20° forward of the ⟨100⟩ directions orthogonal to the dendritic growth axis. The constraining effect of anisotropy in solid–liquid interfacial energy is therefore overshadowed by the influence of the local temperature gradient which controls this early stage of sidebranch development.

3. Both the steady-state and the nonsteady-state features of dendrites are controlled largely by interfacial stability. For instance, both tip radius and branch spacing are scaled to the wavelength of the interfacial perturbation wave. Even so, the growth of a dendrite tip (steady state) and a dendrite stem (nonsteady state) do not interfere with each other and each can be considered as an essentially independent growth process.

On sidebranch coarsening

4. Dynamic coarsening during dendritic solidification occurs by competitive sidebranch growth.

5. Dynamic coarsening should occur even during high-speed solidification, unless the specimen size limits the growth to distances smaller than the incubation length.

6. Isothermal coarsening, driven by interfacial energy reduction, leads to either formation of branch plates or remelting of small branches, depending on the initial geometry of the relevant sidebranch arrays.

Acknowledgements—The authors gratefully acknowledge the assistance of Mr Keith Matuszyk in conducting the dendrite tip radius measurements and of Mr James Pan for providing technical advice on the photography. In addition, we thank Dr Robert J. Schaefer for providing guidance in designing the thermostatic observation tank and in making precision temperature measurements. Finally, the authors thank Dr John Carruthers, Dr Robert J. Naumann, and Dr Lewis Lacy for the support and encouragement provided by the Marshall Space Flight Center and the Materials Processing in Space Office of the National Aeronautics and Space Administration.

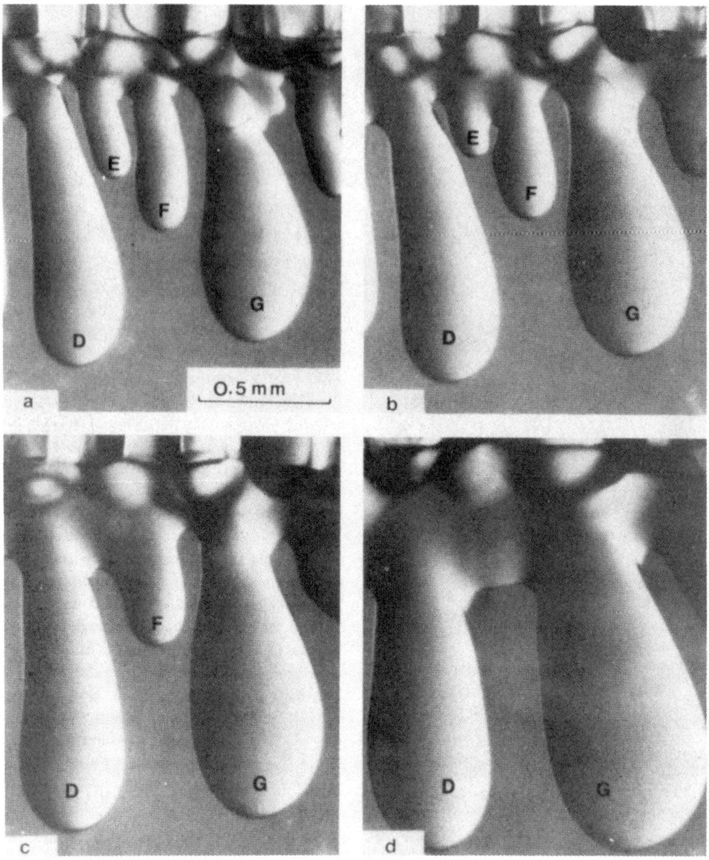

Fig. 15. Elimination of small sidebranches during isothermal holding. Sequential time intervals between photographs are 20–30 min.

REFERENCES

1. S. C. Huang and M. E. Glicksman, *Acta metall.* **29,** 701 (1981).
2. W. Oldfield and W. A. Tiller, *J. Cryst. Growth*, Library of Movie Films.
3. G. R. Kotler and W. A. Tiller, *J. Cryst. Growth*, **2,** 287 (1968).
4. Suggestion of Dr Sidney O'Hara, as cited in Ref. [3].
5. J. S. Langer and H. Müller-Krumbhaar, *Acta metall.* **26,** 1681 (1978).
6. A. A. Chernov, *Kristallografiya*, **1,** 583 (1956).
7. T. Z. Kattamis, J. C. Coughlin and M. C. Flemings, *Trans. metall Soc. A.I.M.E.* **239,** 1504 (1967).
8. M. Kahlweit, *Scripta metall.* **2,** 251 (1968).
9. T. F. Bower, H. D. Brody and M. C. Flemings, *Trans. metall. Soc. A.I.M.E.* **236,** 624 (1966).
10. S. C. Huang, Ph.D. Thesis, Rensselaer Polytechnic Institute, (1979).
11. S. C. Hardy and S. R. Coriell, *J. Cryst. Growth*, **5,** 329 (1969).
12. T. Fujioka, Ph.D. Thesis, Carnegie-Mellon University (1977).
13. J. S. Langer, R. F. Sekerka and T. Fujioka, *J. Cryst. Growth*, **44,** 414 (1978).
14. S.-K. Chan, H.-H. Keimer, and M. Kahlweit, *J. Cryst. Growth*, **32,** 303 (1976).
15. *Ibid*, **43,** 229 (1978).
16. M. E. Glicksman and S. C. Huang, *Convective Flow During Dendritic Growth*, in *NASA Workshop on Fluids Experiment System*, pp. 170–190 (1979).
17. G. S. Cole and G. F. Bolling, *Trans. metall Soc. A.I.M.E.* **242,** 153 (1968).

Acta Metallurgica, Vol. 26, pp. 1681–1687. Pergamon Press. 1978. Printed in Great Britain

THEORY OF DENDRITIC GROWTH—I. ELEMENTS OF A STABILITY ANALYSIS*

J. S. LANGER and H. MÜLLER-KRUMBHAAR†

Physics Department, Carnegie-Mellon University, Pittsburgh, PA 15213, U.S.A.

(*Received* 23 *November* 1977)

Abstract—We propose that the maximum-velocity principle previously used in theories of dendritic crystal growth be replaced by a stability criterion of the form $v\rho^2$ = constant, where v is the growth rate and ρ the tip radius of a dendrite. This form of the stability criterion is valid for small Péclet numbers (ρ much smaller than the diffusion length). A direct evaluation of the relevant constant of proportionality yields values of v and ρ, as functions of undercooling, which are in substantial agreement with the measurements of Glicksman *et al.* [3] on dendritic growth of succinonitrile crystals. In this paper, we summarize our basic ideas and describe qualitatively our theoretical results. Mathematical details are presented in two subsequent papers.

Résumé—Nous proposons de remplacer le principe de la vitesse maximale utilisé dans les théories antérieures de la croissance dendritique d'un cristal par un critère de stabilité de la forme $v\rho^2$ = constante, où v est la vitesse de croissance d'une dendrite et ρ le rayon de courbure de son extrêmité. La forme de ce critère de stabilité est valable pour les faibles nombres de Péclet (ρ beaucoup plus petit que la longueur de diffusion). L'évaluation directe de la constante de proportionalité conduit à des valeurs de v et de ρ, en fonction de la surfusion, qui sont en bon accord avec les mesures de Glicksman *et al.*, sur la croissance dendritique de cristaux de succinonitrile. Nous résumons dans cet article nos principales idées et nous décrivons qualitativement nos résultats théoriques. Les détails mathématiques sont présentés dans deux articles qui suivent.

Zusammenfassung—Wir Schlagen vor, das früher in Theorien des dendritischen Wachsens benutzte Prinzip der maximalen Geschwindigkeit durch ein Stabilitätskriterium zu ersetzen der Form $v\cdot\rho^2$ = constant, wobei v die Wachstumsgeschwindigkeit und ρ der Spitzenradius eines Dendrits ist. Diese Form des Stabilitätskriteriums gilt für kleine Péchletzahlen (ρ sehr viel kleiner als die Diffusionsweite). Eine direkte Auswertung der Proportionalitätskonstanten ergibt Werte für v und ρ als Funktion des Unterkühlungsgrades, die im wesentlichen mit den Messungen des dendritischen Wachstums von succinonitril-Kristallen von Glicksman, Schaefer und Ayers übereinstimmen. In der vorliegenden Arbeit werden die grundlegenden Vorstellungen zusammengefasst und die theoretischen Ergebnisse qualitativ dargestellt. Die mathematischen Einzelheiten werden in zwei nachfolgenden Arbeiten vorgelegt.

1. INTRODUCTION

The phenomenon of dendritic crystal growth is a particularly elegant example of a process in which an instability of a simple system generates complex but highly structured new patterns. Similar self-organizing processes occur in hydrodynamics, chemistry, biology, and perhaps even in elementary-particle physics. In few, if any, of these situations is the dynamical principle which controls restabilization understood in detail. Therefore, it seems to us that the dendrite problem is one whose importance may go well beyond its immediate relevance to metallurgical phenomena. Moreover, because there is no fundamental question about the general form of the equations of motion for solidifying systems, and the relevant measurements are conceptually—if not always practically—simple, there is a natural opportunity here for careful comparison between theory and experiment.

The dendrite problem is simply stated. Given the undercooling or super-saturation of a melt, one ought to be able to predict the growth velocity of a free dendrite and its geometry—its tip radius, the orientation and spacing of the side branches, etc. This problem is not solved by any steady-state calculation in which a nearly paraboloidal tip is found to grow at constant speed without change in shape. All versions of the steady-state theory provide only relations between velocity and tip-radius rather than definite values of these two quantities; and, therefore, these theories have to be supplemented by some subsidiary hypothesis. In the past, the relevant hypothesis always has been assumed to be a maximum-velocity condition. It is now clear, however, that velocities calculated in this way overestimate observed growth rates by almost an order of magnitude. Furthermore, the

* Research supported by AFOSR Grant F 44620-76-C-0103 and also in part by the CMU Center for the Joining of Materials.

† Permanent address: Kernforschungsanlage Jülich, Institut für Festkörperforschung, Theorie 2, D-5170 Jülich, F.R.G.

steady-state theories provide no clue about the important phenomenon of sidebranching.

Our purpose in this paper, and in the two [1, 2] which follow, is to show that the maximum-velocity condition should be replaced by a stability criterion. One of our major stimuli in this project has been the important work of Glicksman *et al.* [3], in which both the velocities and tip radii of succinonitrile dendrites have been measured as functions of undercooling and shown conclusively to be in disagreement with theoretical predictions. Using this data, Lesoult [4] has demonstrated that the discrepancy can be attributed almost entirely to the maximum-velocity hypothesis, and that the steady-state model itself seems to be qualitatively correct. The idea of using a stability criterion either to replace or to justify the maximum-velocity condition has appeared frequently in the literature. Glicksman *et al.* [3] refer to the necessity of considering time-dependent features. Holzmann [5], in the introduction to his thesis, emphasizes the potential importance of a stability analysis, and predicts that the ultimately correct mathematical model of dendritic growth will exhibit a stable limit cycle. Oldfield [6], in connection with his numerical studies of the dendrite, has proposed a simple stability criterion which turns out to be quite similar to ours in its basic idea. The calculation which is most nearly in the spirit of the one to be described in papers II and III of this series is that of Kotler and Tiller [7]. As has been pointed out by Holzmann [5], however, this calculation suffers from serious internal inconsistencies and arrives at qualitatively incorrect results.

The basic features of the results which we shall present may be understood in terms of qualitative arguments plus dimensional analysis. It is these arguments which are set forth in this first paper. Our hope is that this presentation will be complete enough to make the ideas and results accessible to readers who prefer to omit the mathematical details. A complete description of our analytic and computational procedures is contained in Papers II and III.

The scheme of this paper is as follows. In Section 2, we describe our model and define several parameters that will appear frequently throughout our analysis. Section 3 is a summary of the mathematical results, presented in detail in papers II and III, which lead to our stability criterion. In Section 4, we examine some implications of the stability theory and compare our predictions with the data of Glicksman *et al.* [3]. We conclude in Section 5 with a list of as yet unanswered but, we believe, answerable questions.

2. DEFINITION OF THE MODEL

Throughout the work to be described here, we shall confine ourselves to the simplest possible model of a dendritic process. Our system is one in which a solid phase grows in the direction of an axis of symmetry into a uniformly supercooled or supersaturated fluid. We shall assume that the rate-controlling process is either thermal or chemical diffusion, but not both, and that interface kinetics may be ignored. Further, we shall consider diffusion only in the liquid phase, and neglect any transport of heat or chemical species within the solid. One important advantage of this simplified model is that it permits us to construct a dimensionless version of the theory in terms of only a small set of physical parameters. In this way, we shall be able to see certain important scaling properties of the theory, and also might later be able to use data from both chemically and thermally controlled systems in making comparisons between experiment and theory. Another, admittedly decisive, advantage of this model is its analytic and computational simplicity. For example, we eventually shall have to account for diffusion in the solid. This is not difficult in principle; but both the mathematical complexity and the computation time will be at least twice what is required here.

In order to treat thermal and chemical systems simultaneously, we introduce a dimensionless diffusion field, u. If we are considering thermal diffusion in an undercooled pure material, then

$$u(\mathbf{r}, t) = \frac{T(\mathbf{r}, t) - T_\infty}{(\Delta T)}; \quad (\Delta T) = \frac{L}{c_p}. \tag{2.1}$$

Here T is the temperature at an arbitrary time t and position $\mathbf{r}$ within the fluid; T_∞ is the temperature infinitely far from the solidification front; and (ΔT) is the unit supercooling which is determined as shown in (2.1) by the latent heat of fusion, L, and the specific heat of the fluid, c_p. If we are considering solidification of a two-component solution in which the rate-limiting process is chemical diffusion, then

$$u = \frac{C - C_\infty}{(\Delta C)}, \tag{2.2}$$

where C denotes the concentration of one of the chemical species and (ΔC) is the miscibility gap between fluid and solid.

Let $\mathbf{v}$ be the average, constant, growth velocity of the dendrite. $\mathbf{v}$ is directed along the axis of symmetry, denoted by z in Fig. 1. Its magnitude is, of course, a quantity that we ultimately want to calculate. We write the diffusion equation in a frame of reference moving with velocity v:

$$\nabla^2 u + \frac{2}{l}\frac{\partial u}{\partial z} = \frac{1}{\alpha}\frac{\partial u}{\partial t}. \tag{2.3}$$

Here,

$$l = \frac{2\alpha}{v} \tag{2.4}$$

is the diffusion length and α is either the thermal or chemical diffusivity.

Equation (2.3) is to be solved subject to two boundary conditions at the surface of the dendrite. First, there is a thermodynamic boundary condition that we write in the form:

$$u(\text{surface}) = \Delta - d_0 \mathscr{K}(\text{surface}). \tag{2.5}$$

Here, Δ is the dimensionless undercooling or supersaturation:

$$\Delta = \begin{cases} \dfrac{T_m - T_\infty}{(\Delta T)} & \text{(thermal case)} \\[2ex] \dfrac{C_{eq} - C_\infty}{(\Delta C)} & \text{(chemical case)}. \end{cases} \tag{2.6}$$

T_m is the melting temperature of the pure substance; and C_{eq} is the equilibrium concentration in the fluid at the two-phase interface. $\mathscr{K}$ is the curvature of the surface; and d_0 is a capillarity length:

$$d_0 = \begin{cases} \dfrac{T_m \gamma}{L(\Delta T)} & \text{(thermal case)} \\[2ex] \dfrac{\gamma V C_{eq}}{(\Delta C) k_B T_m} & \text{(chemical case)}, \end{cases} \tag{2.7}$$

where γ is the surface tension, V is the atomic volume, and k_B is Boltzmann's constant. Note that, in the thermal version of (2.7), d_0 differs from the more conventionally defined capillarity length (often denoted by Γ) by the ratio of the absolute melting temperature T_m to the unit undercooling (ΔT). In this version of the formula it is also important that L be measured in units of energy per unit volume.

The second boundary condition at the surface is the continuity condition:

$$v_n = -\alpha \hat{n} \cdot \nabla u(\text{surface}). \tag{2.8}$$

Here, $\hat{n}$ is the outward-directed unit normal at a point on the surface; and v_n is the normal velocity of the surface at that point.

The equation of motion for the surface of a dendrite is to be derived from (2.3), (2.5) and (2.8) by eliminating the diffusion field u in favor of coordinates which describe the position of the moving two-phase boundary. This derivation will be described in papers II and III. For present purposes, we need only to define certain scaling parameters which will occur in this theory.

First, we define a length scale ρ, which we shall identify later as the radius of curvature of the frontmost tip of a steady-state dendrite. A second independent quantity is the diffusion length l, defined in (2.4), which appears in the diffusion equation and, effectively, in the continuity condition (2.8) if one divides through by α. The thermodynamic boundary condition introduces the dimensionless undercooling Δ and the capillarity length d_0. The explicit α remaining in (2.3) and (2.8) may be removed by an appropriate rescaling of the time t.

A properly scaled equation of motion for the surface of a dendrite will contain only three independent parameters: Δ plus two independent ratios of the lengths d_0, ρ and l. These lengths are observed to be, very roughly, in the ranges:

$$d_0(\sim 10^{-7}\ \text{cm}) \ll \rho \ll l(\sim 10^{-2}\ \text{cm}). \tag{2.9}$$

The ratio

$$p \equiv \frac{\rho}{l} \tag{2.10}$$

is known as the Péclet number. In all experimental data known to the authors, p appears to be smaller than 10^{-1}, often by one or more orders of magnitude, which means that the range of the diffusion field is much greater than the length scale of the structure at the tip of the dendrite.

It turns out that the second dimensionless ratio which appears naturally in our stability theory is

$$\sigma \equiv \frac{l d_0}{\rho^2}. \tag{2.11}$$

Note that σ can be written in the form

$$\sigma = \left(\frac{\lambda_s}{2\pi\rho}\right)^2, \quad \lambda_s \equiv 2\pi\sqrt{l d_0}, \tag{2.12}$$

where λ_s is the shortest wavelength of a disturbance which would cause a plane interface, moving at velocity v, to suffer a Mullins–Sekerka instability [8]. The parameter σ will play a crucial role in what follows.

3. SUMMARY OF THEORETICAL RESULTS LEADING TO THE STABILITY CRITERION

The first part of our theoretical development, described in paper II, is a linear stability analysis of the Ivantsov [9] paraboloidal dendrite (no capillarity) in the quasi stationary approximation and in the limit of vanishingly small Péclet number p. This analysis can be carried out without further approximation. To describe the results, we introduce the parabolic coordinate ξ, shown in Fig. 1, which measures the distance down the dendrite in units of the tip radius ρ. Specifically, for points along the unperturbed paraboloid,

$$z = \frac{\rho}{2}(1 - \xi). \tag{3.1}$$

The displacement of the surface away from the unperturbed paraboloid is given by the function $F(\xi, t)$, as shown in the figure. F is also a parabolic coordinate but, for present purposes, we may think of it as a normal displacement. The entire coordinate system is understood to be moving with constant velocity v in the z direction; thus any time-dependence of F constitutes a departure from uniform, steady-state growth.

Our procedure starts with the derivation of a linear integro-differential equation for F. (See II, equation 4.12 or 5.19.) This equation is completely scale independent—there are no system parameters whatsoever. We are then able to find exact solutions of the form

$$F(\xi,\tau) = e^{\Omega\tau} F_\Omega(\xi), \tag{3.2}$$

where $\tau = 2vt/\rho$ is a dimensionless time; Ω is any real number in the range $0 \leq \Omega < \infty$; and $F_\Omega(\xi)$ is

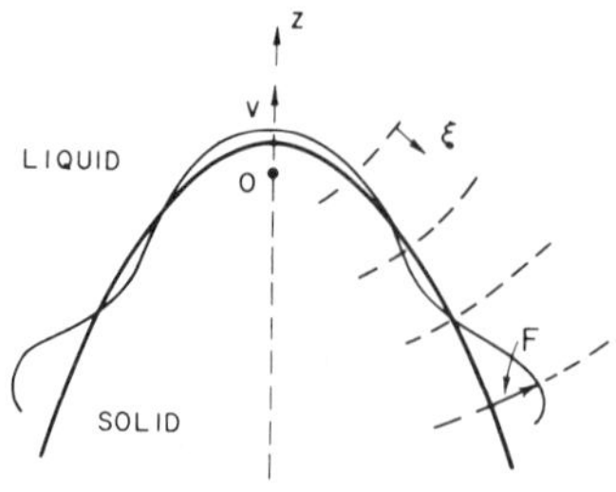

Fig. 1. Parabolic coordinates used for description of dendritic deformations.

an oscillating function which is strongly localized near the tip of the dendrite. Typical functions F_Ω are shown in II, Figs 2 and 3. Because the capillarity length d_0 is rigorously zero, there is no upper bound to Ω and no upper limit to the frequency of the spatial oscillations of F_Ω. We interpret this as a tip-splitting instability; and we emphasize the fact that the Ivantsov solution is manifestly unstable against all perturbations of this nature.

The much more interesting and difficult problem, treated in paper III, is to include the effects of capillarity in the stability analysis. The first new difficulty to arise here is that, when d_0 is finite, the steady-state shape of the dendrite is no longer exactly paraboloidal; and, in principle, we must measure perturbations F away from the properly deformed steady-state solution. However, we shall see in the next section of this paper that the capillarity corrections to the steady-state solution are very small near the real operating point of a dendrite. Therefore, when we write out the equations of motion for F, we can argue that the shape corrections are small and spatially slowly varying compared to the unstable deformations that we expect to be important.

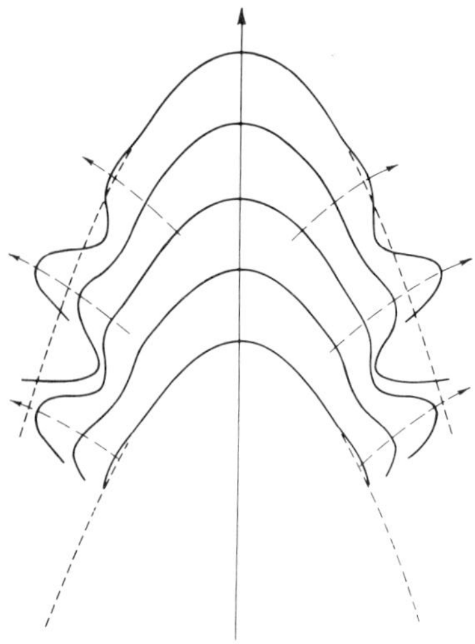

Fig. 2. Time-ordered sequence of tip shapes for a cylindrically symmetric dendrite growing at its natural velocity ($\sigma = 0.025$).

The result of this analysis, again in the limit $p \to 0$ and in the quasi-stationary approximation, is a linear equation of motion for $F(\xi, \tau)$ which contains σ as its only system-dependent parameter. (See III, equations 2.25 and 2.27.) For purposes of the stability analysis, it is useful to think of σ as identifying a point along the steady-state curve of v versus ρ. In the case $\sigma = 0$, this equation reduces to the Ivantsov limit described above; that is, the dendrite is growing very slowly with a very large tip radius and the effects of surface tension are negligable. As σ increases, v increases and ρ decreases. As long as we consider

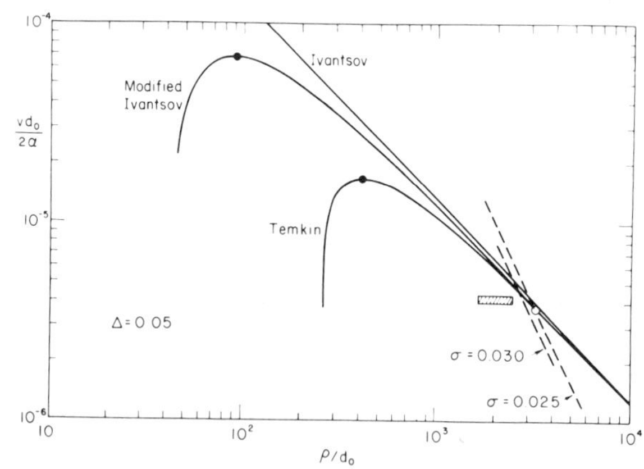

Fig. 3. Growth velocity as a function of tip radius for reduced undercooling $\Delta = 0{\cdot}05$, determined by the three steady-state approximations as indicated. The experimental point is the shaded rectangle; and the stability criterion predicts the point marked by an open circle.

only points on the large-ρ side of the maximum velocity, we may understand that larger values of σ imply sharper and faster dendrites.

Unfortunately, we have not been able to obtain much information from the general equation of motion for F by analytic methods, and have had to resort to computational techniques. We have performed two different kinds of numerical studies. First, we have integrated the equation forward in time, either starting with some initial perturbation and then letting the system move freely, or else applying a continuous random noise at the tip. A second, somewhat more sensitive, technique has been to assume a time dependence of the form $\exp(\Omega\tau)$ as in (3.2) and to look for stability eigenstates.

The most striking feature of our time-dependent results is that, for any finite σ, a perturbation at the tip always generates a train of deformations which moves down the dendrite at just such a velocity as to remain nearly stationary in the laboratory frame of reference. The size and spacing of these bumps and troughs appear to depend mainly on σ and only very little on the nature of the perturbation. In fact, the spacing between bumps is always proportional to the stability length λ_s. Invariably—even for quite large values of σ—these bumps grow in amplitude as they move. This side-branching instability is illustrated in Fig. 2, where we have shown a sequence of tip shapes computed for a cylindrically symmetric dendrite growing at what we believe to be its natural velocity. The dendrite is shown in the laboratory frame, at five equally spaced instants in time, as it grows from the bottom to the top of the page. The dashed lines indicate approximate growth paths for the secondary tips. An account of the mathematical technique used for preparing this figure will be found in III.

A more subtle effect emerges in the eigenfuction analysis. In general, the eigenvalues Ω will be complex, because the linear operator which appears in the equation of motion is not self-adjoint. Ordinarily, we think of stability as being determined by the sign of the real part of Ω, say Ω_1, with positive values being indicative of growing, unstable modes. The dominant sidebranching mode, however, has a spatial envelope which broadens with increasing ξ. To see what this means, suppose that this envelope has the form $\exp(\beta\xi)$, where β is some positive amplification factor. (We suspect that this exponential form is an oversimplification, but do not have numerical data out to large enough ξ to be able to find out what really is happening. In any case, we use the exponential here only to illustrate the effect that we observe in our calculations.) If we combine this envelope with the exponential τ dependence and look at a fixed point in the laboratory, say, $\xi = \xi_0 + \tau$, then we find an overall time dependence of the form

$$F \sim \exp[(\Omega_1 + \beta)\tau] \times \text{oscillating function of } \xi_0. \tag{3.3}$$

In the sidebranching mode, the quantity $\Omega_1 + \beta$ remains positive even when Ω_1 is negative. The crucial point is that the moving tip of the dendrite may stabilize while the sidebranching instability persists. Indeed, this is exactly what happens. For very small values of σ, the largest values of Ω_1 are positive and correspond to the tip-splitting instabilities described earlier. As σ increases, the maximum Ω_1 decreases until, at $\sigma = \sigma^* \cong 0.025$, it vanishes; but there remains an active sidebranching instability of the kind just described.

For $\sigma > \sigma^*$, there are no exponentially growing modes at the tip; however, there is an important marginal instability. An extended fluctuation in which the tip radius changes will not heal itself, but will cause the dendrite to grow at some new velocity. In this situation, the system becomes very sensitive to nonlinear effects. An obvious effect of this kind has to do with the sidebranching instability which, at finite amplitude, must develop in such a way as to maintain outward growth of the secondary dendrites while stabilizing the troughs between them. The net effect must be to thicken the dendrite, thus increasing ρ and decreasing v and σ. Moreover, the side-branching instability is stronger at smaller velocities. Thus, a downward fluctuation in velocity carries the system farther toward σ^* than an upward fluctuation carries it away. We must emphasize, however, that the effect just described is beyond the range of our linear analysis. A quantitative theory of the behavior of a dendrite for $\sigma > \sigma^*$ necessarily will involve a nonlinear analysis of the sidebranching structure, and also will require a statistical calculation of the fluctuations in the diffusion field which drive the various modes of deformation. Nevertheless, it seems plausible that the sidebranching instability must always drive the dendrite toward smaller values of σ. Because the tip becomes linearly unstable at $\sigma < \sigma^*$, we conclude that the operating point is at or near $\sigma = \sigma^*$.

In the computations reported in paper III, we find

$$\sigma^* = \begin{cases} 0.020 \pm 0.007 \text{ (2 dimensions)} \\ 0.025 \pm 0.007 \text{ (3 dimensions)}. \end{cases} \tag{3.4}$$

Both of these results are consistent with a stability criterion of the form:

$$\rho \equiv \lambda_s; \tag{3.5}$$

or

$$\sigma = \frac{2\alpha d_0}{v\rho^2} = \sigma^* \cong \frac{1}{4\pi^2} \cong 0.0253. \tag{3.6}$$

With hindsight, one realizes that this is the kind of criterion that probably must emerge from any stability theory of dendritic growth.

4. IMPLICATIONS OF THE STABILITY CONDITION

In order to explore the quantitative implications of our stability criterion, we need the steady state relation between p, σ and Δ. Unfortunately, the exact solution of Nash and Glicksman [10] is not useful for our purposes because it has built into it a maximum-velocity condition. Thus, we shall have to use one or another of the approximate versions which have been studied by Temkin [11], Trivedi [12], Sekerka [13], and others. As we shall see, these approximations turn out to be much less drastic when used in conjunction with the stability condition than when applied to a maximum-velocity theory.

Both the modified Ivantsov [13] and Temkin [11] approximations assume an exactly paraboloidal dendrite, and then adjust for this assumption by partially relaxing one or the other of the boundary conditions (2.5) or (2.8). In modified Ivantsov, one replaces the curvature $\mathscr{K}$ in (2.5) by the constant value $2/\rho$. The resulting equation, written in our notation and uncorrected for molecular attachment kinetics, is

$$\Delta - 2\sigma p = pe^{p} E_1(p), \tag{4.1}$$

where

$$E_1(p) = \int_p^\infty \frac{e^{-y}}{y} dy. \tag{4.2}$$

The Temkin approximation preserves (2.5) but imposes the continuity condition (2.8) only at the tip. This yields

$$\frac{\Delta}{1 + b\sigma} = pe^{p}E_1(p), \tag{4.3}$$

where

$$b = \tfrac{1}{2}\int_0^\infty y(1 + y)e^{-y}\frac{K_1(y)}{K_0(y)} dy \cong 1.819. \tag{4.4}$$

The quantity b is actually the value at $p = 0$ of a weakly p-dependent function [12]. It may be useful to note that, if we allow diffusion in the solid with the same diffusivity α as in the fluid, then (4.1) remains unchanged, and b is replaced in (4.3) by $b' = 2.949$. We emphasize, however, that our value of σ^* might also be changed by this modification; and we have, at present, no estimate of what this change would be.

Relations (4.1) and (4.3) are illustrated in Fig. 3 for the case $\Delta = 0.05$. Here we have plotted the dimensionless velocity $vd_0/2\alpha$ as a function of dimensionless tip radius ρ/d_0. We include in the figure the straight-line plot for the unmodified Ivantsov theory [9, 14], which can be obtained from either (4.1) or (4.3) by setting $\sigma = 0$. This particular value of Δ was chosen because it lies roughly in the middle of the rather narrow range of under-coolings within which Glicksman *et al.* [3] have measured both the radii and velocities of succinonitrile dendrites. (See Fig. 4.) The relevant experimental point is the shaded rectangle, whose dimensions indicate the apparent uncertainties in the published data. Note that the experimental point lies close to the steady-state curve at a place where the capillarity-modified theories have nearly merged with the Ivantsov line. It is this fact which has motivated our neglect of the steady-state shape correction in the stability analysis described in Section 3.

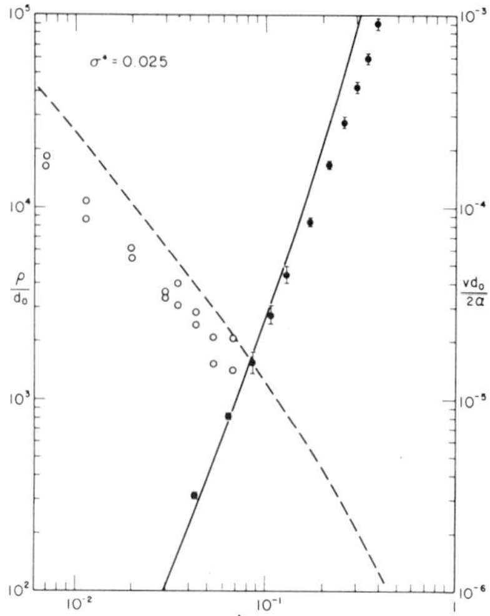

Fig. 4. Comparison between theory and experiment for growth velocity (solid curve and solid circles) and for tip radius (dashed curve and open circles) as functions of reduced undercooling Δ.

The dashed curves in Fig. 3 indicate locii along which $\sigma = 0.025$ and 0.030. The intersection between a constant-σ curve and a steady-state curve is our theoretical prediction; indicated here by an open circle for $\sigma = \sigma^* \cong 0.025$. According to the calculations discussed above, dendrites on the steady-state curve to the right of σ^* are unstable at their tips, and those to the left are driven toward σ^* by the sidebranching mechanism. The maximum-velocity predictions for the two different capillarity-corrected steady-state theories are identified by solid circles. The Temkin point, which is shown here uncorrected for diffusion in the solid, turns out to coincide almost exactly with the maximum-velocity point of Nash and Glicksman [10], whose calculations do include the latter effect. One might guess, therefore, that the large-ρ branch of the Temkin curve is a good approximation to the exact steady-state relation. The qualitative difference between maximum-velocity and stability theories is apparent in this figure. The Temkin point corresponds to a value of σ which is over an order of magnitude larger than the σ^* predicted by our stability analysis.

Figure 4 shows, on a single Δ axis, data for both $vd_0/2\alpha$ and ρ/d_0. The velocity data are indicated by solid circles with error bars. These points constitute only a selection from the published results; and the error bars are intended only to indicate the apparent scatter in the original data. The corresponding theoretical prediction for $\sigma = 0.025$ is shown by the solid curve. The open circles denote the experimental tip radii; and here we have attempted to reproduce all of the available information. The dashed curve is the theoretical ρ/d_0.

The apparent agreement between theory and experiment is encouraging, but probably should not be taken too seriously at present. There remain major uncertainties, especially in the neglect of heat flow in the solid and in the numerical evaluation of σ^*. In principle, however, this is a theory with no adjustable parameters; and we are not free to improve the fit to some of the data at the expense of others. Thus, it possibly is significant that the agreement for the velocity is considerably better at small Δ than at large Δ. This effect may be a slowing-down due to surface attachment kinetics; or it may be that we are pushing the limits of our small-p approximation. At the upper end of this curve, p is of the order 0.2. More interesting is the fact that the theory seems to predict tip radii much less accurately than velocities. We suspect that this can be explained by an argument due to Lesoult [15]: that the tip-radius ρ used in our theory ought not to be understood as the actual radius of curvature of the front-most point of the dendrite, but that ρ should be interpreted as determining some average paraboloidal envelope that includes the sidebranches which are emerging within the diffusion field of the tip. Note that the discrepancy seems greatest at the small-Δ, small-p end of the curve, where the diffusion field is very long-ranged compared to ρ. It ought to be possible to develop a quantitative version of this argument in much the same way as it ought to be possible to quantify the nonlinear sidebranching instability which drives the dendrite toward σ^*.

5. CONCLUDING REMARKS

On the basis of the analysis presented here and in the following papers, we believe that the dynamic stability condition is a significant clue to the understanding of dendritic growth. Starting with this premise, we are led to pose a number of questions about the theory and about dendrites in general. None of these questions are new; but it seems interesting to pose them now in language appropriate to the stability theory.

The most interesting question is obviously: What is the dendrite actually doing at its operating point? Our theory suggests that this point is a very active one. The sidebranching instability is as strong as it can possibly be subject to the constraint that the tip be stable against splitting or sharpening deformations. Moreover, the tip is only marginally stable against changes in radius and velocity. Does this mean that the tip undergoes some sort of erratic behaviour under the influence of thermal noise? Or is it driven into a limit cycle as suggested by Holtzman? Or does something altogether different happen?

Next, there are a number of questions which are interesting from a practical point of view. What is the effect of attachment kinetics? What happens to the analysis when one includes diffusion in the solid? What happens when one corrects for finite Péclet number in the high-velocity regime? What is the effect of crystalline anisotropy? Is the controlling instability always (or ever) cylindrically symmetric at the tip? How might one deal with strongly anisotropic situations such as growing ice crystals? These questions all appear to be answerable by means of direct, albeit elaborate, extensions of the present analysis.

Finally, there are the practical but intrinsically more difficult questions regarding the sidebranches. As we have mentioned before, we are going to need both nonlinear and statistical methods to make progress here. The question remains: What is the appropriate tip radius ρ to measure for comparison between theory and experiment? An even more interesting problem from a practical point of view is the sidebranch spacing. We predict the initial spacing to be of order ρ and thus, in accord with the stability condition, to be proportional to $v^{-1/2}$. But we know that the sidebranches coarsen as they grow. Does the restabilized, coarsened structure also scale with ρ?

We are hopeful that answers to these questions, coupled with further experiments of the quality of those of Glicksman *et al.*, will lead soon to a much improved understanding of dendritic solidification.

Acknowledgements—The authors would like to thank R. F. Sekerka and G. Lesoult for guidance and encouragement throughout the course of this work.

REFERENCES

1. J. S. Langer and H. Müller-Krumbhaar, *Acta metall.* **26**, 1689 (1978).
2. H. Müller-Krumbhaar and J. S. Langer, *Acta metall.* **26**, 1697 (1978).
3. M. E. Glicksman, R. J. Shaefer, and J. D. Ayers, *Metall. Trans.* (A) **7**, 1747 (1976).
4. G. Lesoult. To be published.
5. E. G. Holtzman. Ph.D. Thesis. Stanford University (1969).
6. W. Oldfield, *Mater. Sci. Engng* **11**, 211 (1973).
7. G. R. Kotler and W. A. Tiller, *J. Crystal Growth* **2**, 287 (1968).
8. W. W. Mullins and R. F. Sekerka, *J. appl. Phys.* **35**, 444 (1964).
9. G. P. Ivantsov, *Dokl. Akad. Nauk S.S.S.R.* **58**, 567 (1947).
10. G. E. Nash and M. E. Glicksman, *Acta metall.* **22**, 1283 (1974).
11. D. E. Temkin, *Dokl. Akad. Nauk S.S.S.R.* **132**, 1307 (1960).
12. R. Trivedi, *Acta metall.* **18**, 287 (1970).
13. R. F. Sekerka, R. G. Seidensticker, D. R. Hamilton, and J. D. Harrison: Investigation of Desalinization by Freezing, Chapter 3. Westinghouse Research Laboratory Report (1967).
14. G. Horvay and J. W. Cahn, *Acta metall.* **9**, 695 (1961).
15. G. Lesoult, To be published.

Acta Metallurgica, Vol. 26, pp. 1689 1695. Pergamon Press, 1978. Printed in Great Britain

THEORY OF DENDRITIC GROWTH—II. INSTABILITIES IN THE LIMIT OF VANISHING SURFACE TENSION*

J. S. LANGER and H. MÜLLER-KRUMBHAAR†

Physics Department, Carnegie-Mellon University, Pittsburgh, PA 15213, U.S.A.

(*Received* 23 *November* 1977)

Abstract—We present a linear stability analysis of the Ivantsov needle crystal in the limit of zero capillarity and vanishingly small Péclet number. If we make a quasi-stationary approximation for the diffusion field, the stability eigenvalue problem turns out to be exactly solvable. The result is that the paraboloidal steady-state solution is completely unstable. The instabilities are spatially oscillating modes localized at the tip.

Résumé—Nous présentons une analyse linéaire de la stabilité d'un cristal aciculaire de Ivantsov, dans la limite d'une capillarité nulle et d'un nombre de Péclet tendant vers zéro. Si nous effectuons une approximation quasi-stationnaire du champ de diffusion, le problème des valeurs propres de la stabilité peut être exactement résolu. Les résultats montrent que le solution d'un régime permanent paraboloïdal est complètement instable. Les instabilités consistent en des modes oscillant spatialement et localisés à l'extrêmité des dendrites.

Zusammenfassung—Wir legen eine lineare Stabilitätsanalyse des Ivantsov-Nadelkristalls in der Grenze der Kapillarität Null und verschwindend kleiner Pécletzahl vor. Machen wir eine quasi-stationäre Näherung für das Diffusionsfeld, dann stellt sich das Stabilitäts-Eigenwertproblem als exakt lösbar heraus. Es ergibt sich, daß die stationäre Paraboloidlösung vollständig instabil ist. Die Instabilitäten sind räumlich oszillierende Moden, die an der Spitze lokalisiert sind.

1. INTRODUCTION

In the preceding paper [1] we have described, essentially without derivation, some elements of a new theory of dendritic growth. The present paper contains the first parts of the mathematical analysis upon which that theory is based.

We shall consider here a somewhat academic problem: the linear stability of the Ivantsov [2] needle crystal in the limit of zero capillarity and vanishingly small undercooling. Our main reason for doing this is that the problem turns out to be exactly soluble, and thus provides us with a rigorous point of comparison for testing and understanding the approximations that we shall have to use for more realistic cases. Not surprisingly, the Ivantsov solution turns out to be strongly unstable. What is not quite so obvious, however, is that the instability is highly localized at the tip of the dendrite. This tip-splitting instability will serve as a useful clue in the construction of a more general theory.

The scheme of this paper is as follows. In Section 2, we define the basic model and introduce some of the notation to be used here and in a succeeding paper. Sections 3 and 4 are devoted to a detailed analysis of the two-dimensional version of this model. This version, which describes solidification of a thin layer of fluid, permits some mathematical simplification; but quantitatively reliable experiments have not yet been done in this geometry. In Section 5 we present the full three-dimensional analysis.

*Research supported by AFOSR Grant F 44620-76-C-0103 and also in part by the CMU Center for the Joining of Materials.

† Permanent address: Kernforschungsanlage Jülich, Institut für Festkörperforschung, Theorie 2, D-5170 Jülich, F.R.G.

2. DEFINITION OF THE MODEL

The model of a dendrite that we shall consider here is that which is described in Section 2 of paper I, except that the capillarity length d_0 vanishes. For completeness, we shall restate the basic equations which define this model.

Our dendrite is a solid growing in the direction of its axis of symmetry into a supercooled or supersaturated fluid. We let the average growth velocity be $\mathbf{v}$, and choose $\mathbf{v}$ to be directed along the z axis shown in Fig. 1. Diffusion, either thermal or chemical, occurs only in the fluid phase, and is described by a dimensionless field u. In the frame of reference moving with velocity $\mathbf{v}$, u satisfies:

$$\nabla^2 u + \frac{2}{l}\frac{\partial u}{\partial z} = \frac{1}{\alpha}\frac{\partial u}{\partial t} \cong 0. \tag{2.1}$$

Here, l is the diffusion length,

$$l = \frac{2\alpha}{v}, \tag{2.2}$$

and α is the diffusivity. In setting the right-hand side of (2.1) to zero, we indicate our intention of using a quasi-stationary approximation throughout this analysis. That is, we shall assume that the diffusion field responds very quickly to slow changes in the shape of the solidification front.

In the absence of capillarity, the thermodynamic boundary condition at the surface of the dendrite is simply:

$$u(\text{surface}) = \Delta = \text{constant}, \tag{2.3}$$

where Δ is the dimensionless undercooling or supersaturation. The continuity condition at the surface is

$$v_n = -\alpha\hat{n}\cdot\nabla u(\text{surface}), \tag{2.4}$$

where $\hat{n}$ is the unit, outward-pointing normal at a point on the surface and v_n is the normal velocity of the surface at the point.

As is well known, the model described here possesses a family of paraboloidal steady-state solutions. That is, for any Δ in the range $0 < \Delta < 1$, and any v, there exists a parabola in two dimensions and circular or elliptical paraboloids [3] in three dimensions which are time-independent solutions of the above equations. Our problem in what follows is to linearize these equations of motion about the steady-state solutions and to test the resulting linear system for dynamic stability.

3. TWO-DIMENSIONAL MODEL: EQUATION OF MOTION FOR THE SURFACE

To begin the above program of analysis for the two-dimensional dendrite, we transform directly into dimensionless parabolic coordinates ξ and η as shown in Fig. 1:

$$\xi = \frac{r - z}{\rho}; \quad \eta = \frac{r + z}{\rho}. \tag{3.1}$$

Here, r is the radial distance measured from 0, and ρ is a constant which we may choose to be the radius of curvature of the tip of the steady-state parabola. Only positive values of ξ and η will be considered; and the dendrite always will be assumed to be symmetric about the z axis. The surface of the dendrite will be given by

$$\eta_s(\xi, t) = 1 + F(\xi, t). \tag{3.2}$$

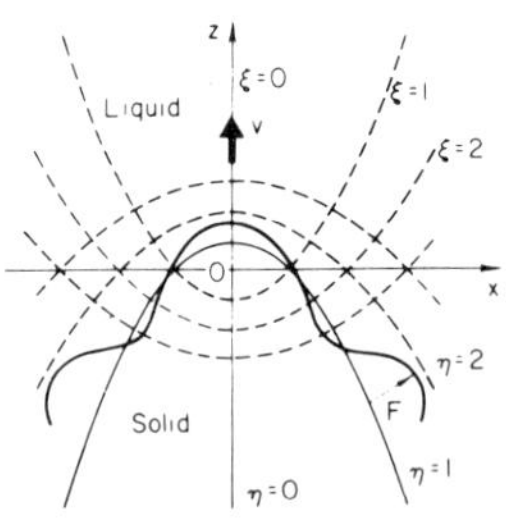

Fig. 1. Coordinate systems used for the description of dendritic deformations.

That is, $\eta_s = 1$ locates the unperturbed surface and F is the small, time-dependent deviation for which we seek an equation of motion.

In the quasi-stationary approximation for which the right-hand side of (2.1) vanishes, the diffusion equation becomes:

$$2\eta\frac{\partial^2 u}{\partial\eta^2} + (1 + 2p\eta)\frac{\partial u}{\partial\eta} + 2\xi\frac{\partial^2 u}{\partial\xi^2} + (1 - 2p\xi)\frac{\partial u}{\partial\xi} = 0, \tag{3.3}$$

where p is the Péclet number:

$$p \equiv \frac{\rho}{l} = \frac{\rho v}{2\alpha}. \tag{3.4}$$

Equation (3.3) is clearly separable. It has a general solution of the form

$$u(\eta, \xi) = \sum_n A_n V_n(p\eta)Q_n(p\xi); \tag{3.5}$$

where the A_n are arbitrary expansion coefficients, and Q_n and V_n satisfy:

$$2w\frac{d^2Q_n}{dw^2} + (1 - 2w)\frac{dQ_n}{dw} + b_nQ_n = 0, \quad w = p\xi; \tag{3.6}$$

$$2w\frac{d^2V_n}{dw^2} + (1 + 2w)\frac{dV_n}{dw} - b_nV_n = 0, \quad w = p\eta. \tag{3.7}$$

To choose the separation constants b_n, we require first that u be smooth and continuous across the z axis, that is, at $\xi = 0$. Note that the Cartesian coordinate x, shown in Fig. 1, is given in terms of ξ and η by

$$\frac{x^2}{\rho^2} = \eta\xi. \tag{3.8}$$

Along a line of constant η, ξ is proportional to x^2; therefore, any polynomial in ξ may be reflected about the z axis without producing a singularity. If we choose

$$b_n = 2n, \; n = 0, 1, 2, \ldots, \tag{3.9}$$

then equation (3.6) has solutions

$$Q_n(w) = H_{2n}(\sqrt{w}), \tag{3.10}$$

where the H_{2n} are even-order Hermite polynomials [4]. This choice of the Q_n clearly satisfies the important boundary condition at $\xi = 0$, but means that we shall have to be careful about convergence of the series in (3.5) for finite ξ. Inserting (3.9) into (3.7), we now must find solutions V_n which vanish at large w, but which are allowed to be singular at $w = 0$ because the latter point always lies inside the dendrite and therefore outside of the diffusion field. The resulting V_n are confluent hypergeometric functions whose detailed identification will not be necessary for the following analysis.

Our problem is, in effect, to determine the coefficients A_n in the expansion (3.5) by application of the boundary conditions (2.3) and (2.4). The thermodynamic boundary condition, (2.3) is

$$\sum_{n=0}^{\infty} A_nV_n\{p[1 + F(\xi)]\}Q_n(p\xi) = \Delta, \tag{3.11}$$

where we have used (3.2) to evaluate η at $\eta_s(\xi)$. The general form of the continuity condition (2.4), when rephrased in terms of our parabolic coordinates, is [5]

$$v\left(\eta_s + \xi\frac{\partial\eta_s}{\partial\xi}\right) + \frac{\rho}{2}(\eta_s + \xi)\frac{\partial\eta_s}{\partial t} = -\frac{2\alpha}{\rho}\left[\eta_s\frac{\partial u}{\partial\eta} - \xi\frac{\partial\eta_s}{\partial\xi}\frac{\partial u}{\partial\xi}\right]_{\eta=\eta_s}. \quad (3.12)$$

The steady-state solution, $\eta_s = 1$, requires only the term $n = 0$ in (3.5). We have

$$Q_0(w) = 1; \quad (3.13)$$

and

$$V_0(w) = \int_{\sqrt{w}}^{\infty} e^{-y^2}\,dy. \quad (3.14)$$

The latter result for V_0 is most easily obtained directly from (3.7). Then (3.11) is

$$u(\text{surface}) = \bar{A}_0\,V_0(p) = \Delta, \quad (3.15)$$

where $\bar{A}_0$ is the single amplitude to be determined in (3.5). By virtue of the definition of p in (3.4), the continuity equation (3.12) reduces simply to:

$$\bar{A}_0\frac{dV_0(p)}{dp} = -1. \quad (3.16)$$

Eliminating $\bar{A}_0$, and using (3.14), we find

$$2e^p\sqrt{p}\int_{\sqrt{p}}^{\infty} e^{-y^2}\,dy = \Delta. \quad (3.17)$$

Note that, in the limit of small Δ, p is very small:

$$p \approx \frac{\Delta^2}{\pi} \quad (\Delta \ll 1). \quad (3.18)$$

In order to linearize about the stationary solution, we write for the expansion coefficients in (3.5):

$$A_n = \bar{A}_0\,\delta_{n,0} + a_n(t), \quad (3.19)$$

where the a_n are slowly time-dependent quantities which are small of the order of the surface displacement function F. Expanding the argument of V_n in (3.11), and keeping only first-order terms, we obtain

$$pF(\xi) = \sum_{n=0}^{\infty} a_n V_n(p)Q_n(p\xi). \quad (3.20)$$

A similar linearization of (3.12) yields:

$$\frac{\rho}{2v}(1 + \xi)\frac{\partial F}{\partial t} = -(p + \tfrac{1}{2})F - \xi\frac{\partial F}{\partial\xi} - \sum_{n=0}^{\infty} a_n\frac{dV_n(p)}{dp}Q_n(p\xi). \quad (3.21)$$

We can formally eliminate the coefficients a_n by introducing a set of functions $\tilde{Q}_n$ which are orthonormal to the Q_n:

$$\int_0^{\infty}\tilde{Q}_n(w)Q_m(w)\,dw = \delta_{n,m}. \quad (3.22)$$

(The $\tilde{Q}_n$ are the left-eigenstates of the linear operator in (3.6).) Then from (3.20), we have

$$a_n = \frac{p^2}{V_n(p)}\int_0^{\infty}\tilde{Q}_n(p\xi)F(\xi)\,d\xi. \quad (3.23)$$

We now write (3.21) in the form

$$(1 + \xi)\frac{\partial F}{\partial\tau} = -(p + \tfrac{1}{2})F - \xi\frac{\partial F}{\partial\xi} + \int_0^{\infty} G(\xi,\xi')F(\xi')\,d\xi', \quad (3.24)$$

where

$$\tau = \frac{2vt}{\rho} \quad (3.25)$$

is a dimensionless time; and

$$G(\xi,\xi') = -\sum_{n=0}^{\infty}\frac{p^2}{V_n(p)}\frac{dV_n}{dp}Q_n(p\xi)\tilde{Q}_n(p\xi') \quad (3.26)$$

is the integral kernel through which the diffusion field enters nonlocally into the equation of motion for the interface.

4. TWO-DIMENSIONAL MODEL: CHARACTERISTIC MODES IN THE SMALL-p LIMIT

Equation (3.24) is difficult to deal with in general, but simplifies considerably in the limits $p \to \infty$ (planar interface) or $p \to 0$. It is the latter case which is of interest in the theory of dendritic growth, where Péclet numbers p are usually of order 10^{-2} or smaller. In this connection, note that relation (3.18) implies that p vanishes more rapidly than the dimensionless undercooling Δ.

Accordingly, we now focus our attention on the limit $p \to 0$ in equation (3.26) for the integral operator G. Being interested in values of ξ of order unity—because we expect important shape variations to occur on the scale of the tip radius ρ—we see that we must first find asymptotic expressions for the $Q_n(w)$ and $\tilde{Q}_n(w)$ in the limit of small values of the argument w and large values of the index n. A value of n of order $1/p$ is necessary in order that Q_n have at least one node in the interval $0 < w < p$. It is this feature which makes the polynomial expansion appear to converge poorly for small p. Most of the required asymptotic formulas can be found in the mathematical literature; but, for present purposes, it is simpler and quicker to make the analysis directly.

Returning to the variable ξ, and multiplying through by p, we rewrite (3.6) in the form

$$2\xi\frac{d^2Q}{d\xi^2} + (1 - 2p\xi)\frac{dQ}{d\xi} + 2\mu Q = 0, \quad \mu = np. \quad (4.1)$$

We then delete the explicit $p\xi$ appearing in (4.1), solve the resulting equation, and obtain:

$$\lim_{p\to 0} Q_n(p\xi) = \cos(2\sqrt{\mu\xi}), \quad \mu = np = \text{finite}. \quad (4.2)$$

Similarly, the conjugate equation for $\tilde{Q}$ is

$$2\frac{d^2}{d\xi^2}(\xi\tilde{Q}) - \frac{d}{d\xi}(1 - 2p\xi)\tilde{Q} + 2\mu\tilde{Q} = 0, \quad (4.3)$$

from which we obtain:

$$\lim_{p\to 0}\tilde{Q}_n(p\xi) = \frac{1}{\pi\sqrt{\mu\xi}}\cos(2\sqrt{\mu\xi}), \quad \mu = np = \text{finite}. \quad (4.4)$$

Repetition of this procedure for equation (3.7) yields

$$\lim_{p\to 0} V_n(p\eta) = e^{-2\sqrt{\mu\eta}}, \quad \mu = np = \text{finite}. \tag{4.5}$$

It should now be clear that, by taking the $p \to 0$ limit, we are going from a sum over n in (3.6) to an integral transform. That is, we must write

$$\lim_{p\to 0} p \sum_{n=0}^{\infty} \ldots = \int_0^{\infty} d\mu \ldots. \tag{4.6}$$

In this way we obtain

$$\lim_{p\to 0} G(\xi, \xi')\, d\xi' = \int_0^{\infty} \frac{d\mu}{\pi} \cos(2\sqrt{\mu\xi}) \cos(2\sqrt{\mu\xi'}) \frac{d\xi'}{\sqrt{\xi'}}. \tag{4.7}$$

A further simplification can be made by transforming to the variable s:

$$\xi = s^2, \tag{4.8}$$

which can be identified *via* (3.8) as the value of x (see Fig. 1), measured in units of ρ, along the parabola $\eta = 1$. We define

$$\lim_{p\to 0} G(\xi, \xi')\, d\xi' = \Gamma(s, s')\, ds' \tag{4.9}$$

so that, with $\mu = k^2/4$,

$$\Gamma(s, s') = \frac{1}{\pi} \int_0^{\infty} k\, dk \cos(ks) \cos(ks'). \tag{4.10}$$

Writing

$$F(\xi) = f(s), \tag{4.11}$$

and dropping the last explicit p in (3.24), we find the limiting equation of motion for the surface to be:

$$(1 + s^2) \frac{\partial f}{\partial \tau} = -\tfrac{1}{2} f - \tfrac{1}{2} s \frac{\partial f}{\partial s} + \int_0^{\infty} \Gamma(s, s') f(s')\, ds'. \tag{4.12}$$

Although a complete solution of (4.12) seems quite difficult, it is possible to find exact characteristic modes in which the τ-dependence is purely exponential:

$$f(s, \tau) = e^{\Omega\tau} f_\Omega(s). \tag{4.13}$$

The resulting equation for f_Ω can be solved by performing a Fourier-cosine transformation:

$$f_\Omega(s) = \sqrt{\frac{2}{\pi}} \int_0^{\infty} \phi_\Omega(k) \cos(ks)\, dk. \tag{4.14}$$

The transformed version of (4.12) is

$$\Omega\left(\phi_\Omega - \frac{d^2\phi_\Omega}{dk^2}\right) = \frac{k}{2}\left(\phi_\Omega + \frac{d\phi_\Omega}{dk}\right). \tag{4.15}$$

Before looking at the general solution of (4.15), note that, when $\Omega = 0$,

$$\phi_0(k) \propto e^{-k}; \quad f_0(s) \propto \frac{1}{1 + s^2} = \frac{1}{1 + \xi}. \tag{4.16}$$

The function f_0 describes a uniform, infinitesimal translation of the parabolic surface in the z-direction. To be precise, if the parabola is displaced a distance ϵ, then to first order in ϵ,

$$F_0(\xi) = \frac{2\epsilon}{\rho(1 + \xi)}. \tag{4.17}$$

The appearance of f_0 as a marginally stable characteristic mode is a useful check on the various approximations and transformations which have led to equation (4.15).

Next note that ϕ_0 is a solution of (4.15) for any Ω. For nonzero Ω, however, this solution is spurious; because f_0, when inserted into the left-hand side of (4.15), does not produce a well-defined Fourier-cosine transform. This observation, in effect, fixes a constant of integration for us. We write ϕ_Ω in the form

$$\phi_\Omega(k) = e^{-k} \Psi_\Omega(k), \tag{4.18}$$

and find that Ψ_Ω satisfies

$$\frac{d^2\Psi_\Omega}{dk^2} + \left(\frac{k}{2\Omega} - 2\right) \frac{d\Psi_\Omega}{dk} = 0. \tag{4.19}$$

The appropriate solution is

$$\phi_\Omega(k) = e^{-k} \int_0^k \exp\left(-\frac{k'^2}{4\Omega} + 2k'\right) dk', \tag{4.20}$$

where the requirement that ϕ_Ω contain no additive component of the displacement mode fixes the limits of the integration over k'. That is, $\phi_\Omega(k)$ vanishes at $k = 0$, and therefore vanishes sufficiently rapidly at large s to be an acceptable solution for nonzero Ω. Notice, however, that in the limit $\Omega \to 0$, ϕ_Ω approaches ϕ_0.

The most important feature of (4.20) is the fact that these solutions exist for any real, positive Ω. This means that all of the characteristic modes that we have found represent instabilities of the original parabolic dendrite. The best way to see what these instabilities look like is to consider the limit of large Ω, in which case we can make a Gaussian approximation for $\phi_\Omega(k)$:

$$\phi_\Omega(k) \approx \exp\left[-\frac{\Omega}{4}\left(\frac{k}{\Omega} - 2\right)^2\right], \quad (\Omega \gg 1). \tag{4.21}$$

Inverting the cosine transform, we find

$$f_\Omega(s) \approx e^{-\Omega s^2} \cos(2\Omega s). \tag{4.22}$$

This function is plotted in Fig. 2 as it would be seen when superimposed upon a parabolic dendrite tip.

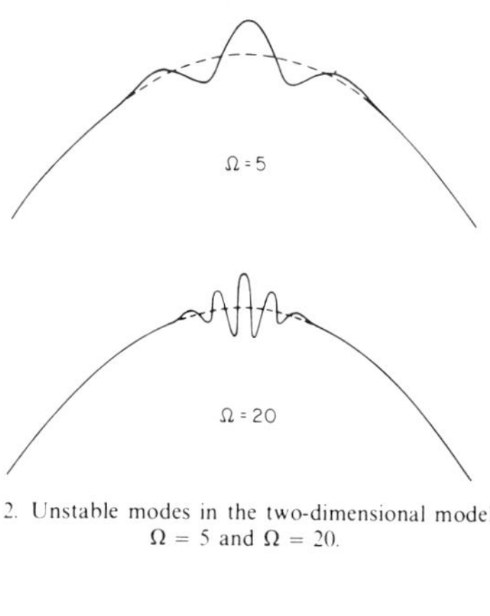

Fig. 2. Unstable modes in the two-dimensional model for $\Omega = 5$ and $\Omega = 20$.

Two cases are shown: $\Omega = 5$ and $\Omega = 20$. Note that this is a tip-splitting instability. New dendrites of relatively small tip radii appear to emerge from a region localized at the tip of the original parabola. As Ω increases, the region of instability decreases in size, but the number of secondary dendrites increases.

5. THREE-DIMENSIONAL MODEL

We turn now to the three-dimensional situation in which the Ivantsov steady-state dendrite is a paraboloid of revolution. The program of analysis is fundamentally the same as that presented in Sections 3 and 4; and the reader should consult those sections for more detailed explanations of some of the steps.

The coordinate system that we shall use is the parabolic system defined in equation (3.1) and in Fig. 1, with the addition of an azimuthal angle θ to describe orientation around the z axis. Accordingly, the surface of the dendrite is specified by

$$\eta_s(\xi, \theta, t) = 1 + F(\xi, \theta, t). \tag{5.1}$$

The diffusion equation, (3.3), generalizes to

$$\frac{4}{\eta + \xi}\left[\frac{\partial}{\partial \eta}\eta\frac{\partial u}{\partial \eta} + \frac{\partial}{\partial \xi}\xi\frac{\partial u}{\partial \xi}\right] + \frac{1}{\eta\xi}\frac{\partial^2 u}{\partial \theta^2} + \frac{4p}{\eta + \xi}\left[\eta\frac{\partial u}{\partial \eta} - \xi\frac{\partial u}{\partial \xi}\right] = 0. \tag{5.2}$$

This equation is still fully separable, and thus has solutions of the form:

$$u(\eta, \xi, \theta) = \sum_{n=0}^{\infty}\sum_{m=-\infty}^{\infty} A_{n,m}W_{n,m}(p\eta)P_{n,m}(p\xi)\,e^{im\theta}, \tag{5.3}$$

where P and W satisfy:

$$w\frac{d^2P_{n,m}}{dw^2} + (1 - w)\frac{dP_{n,m}}{dw} + \left(n + \frac{|m|}{2} - \frac{m^2}{4w}\right)P_{n,m} = 0; \tag{5.4}$$

$$w\frac{d^2W_{n,m}}{dw^2} + (1 + w)\frac{dW_{n,m}}{dw} - \left(n + \frac{|m|}{2} + \frac{m^2}{4w}\right)W_{n,m} = 0. \tag{5.5}$$

As before, the separation constants have been chosen so that the diffusion field is well behaved along $\xi = 0$. The functions $P_{n,m}$ are related to the associated Laguerre polynomials, L:

$$P_{n,m}(w) = w^{|m|/2}\,L_n^{(|m|)}(w). \tag{5.6}$$

The $W_{n,m}$ are confluent hypergeometric functions for which, as in the previous case, only asymptotic formulas will be needed.

The thermodynamic boundary condition analogous to (3.11) is essentially unchanged in three dimensions. For completeness, we state the exact form of the continuity condition (3.12):

$$v\left(\eta_s + \xi\frac{\partial \eta_s}{\partial \xi}\right) + \frac{\rho}{2}(\eta_s + \xi)\frac{\partial \eta_s}{\partial t} = -\frac{2\alpha}{\rho}\left[\eta_s\frac{\partial u}{\partial \eta} - \xi\frac{\partial \eta_s}{\partial \xi}\frac{\partial u}{\partial \xi} - \frac{\eta_s + \xi}{4\eta_s\xi}\frac{\partial \eta_s}{\partial \theta}\frac{\partial u}{\partial \theta}\right]_{\eta=\eta_s}. \tag{5.7}$$

The last two terms on the right-hand side are not needed in the linear theory.

For the steady-state Ivantsov solution, we have

$$u(\text{surface}) = \bar{A}_0W_0(p) = \Delta, \tag{5.8}$$

where W_0 is the exponential integral E_1:

$$W_0(p) = E_1(p) = \int_p^{\infty}\frac{e^{-y}}{y}\,dy. \tag{5.9}$$

The continuity condition is

$$\bar{A}_0\frac{dW_0(p)}{dp} = -1; \tag{5.10}$$

so that

$$pe^p E_1(p) = \Delta. \tag{5.11}$$

Note that p again vanishes faster than Δ at small Δ, but that this time the deviation from proportionality is only logarithmic.

The linearization about the Ivantsov solution takes place precisely as before, the only difference being that we must carry along the extra θ-dependence. However, since the angular parts of the eigenstates of the stability problem must be simply $e^{im\theta}$, it is convenient to separate these angular modes from the beginning by writing

$$F(\xi, \theta, t) = \sum_m F_m(\xi, t)\,e^{im\theta} \tag{5.12}$$

and deriving equations for the F_m directly. Then the analog of Equation (3.24) is easily found to be

$$(1 + \xi)\frac{\partial F_m}{\partial \tau} = -(1 + p)F_m - \xi\frac{\partial F_m}{\partial \xi} + \int_0^{\infty} G_m(\xi, \xi')F_m(\xi')\,d\xi', \tag{5.13}$$

where τ is given in (3.25);

$$G_m(\xi, \xi') = -\sum_{n=0}^{\infty}\frac{p^2}{W_{n,m}(p)}\frac{dW_{n,m}(p)}{dp}P_{n,m}(p\xi)\tilde{P}_{n,m}(p\xi); \tag{5.14}$$

and the $\tilde{P}_{n,m}$ are the conjugate functions which are orthonormal to the $P_{n,m}$.

The appropriate small -p limit in (5.14) is obtained by keeping np finite with fixed m. Using the same technique as in Section 4, we find:

$$\lim_{\substack{p\to 0\\ \mu = np}} P_{n,m}(p\xi) = \lim_{\substack{p\to 0\\ \mu = np}} \tilde{P}_{n,m}(p\xi) = J_m(2\sqrt{\mu\xi}); \tag{5.15}$$

$$\lim_{\substack{p\to 0\\ \mu = np}} W_{n,m}(p\eta) = K_m(2\sqrt{\mu\eta}); \tag{5.16}$$

where J_m and K_m are the ordinary and modified Bessel functions of order m respectively. Note that the small -p limits of equations (5.4) and (5.5) are self-conjugate, so that the limiting forms of P and $\tilde{P}$ are identical. This means that the limiting form of the integral kernel G has a convenient symmetry without a further change of variables as in (4.8). Specifically:

$$\lim_{p\to 0} G_m(\xi, \xi') = \int_0^{\infty} d\mu\,\Lambda_m(\mu)J_m(2\sqrt{\mu\xi})J_m(2\sqrt{\mu\xi'}) \tag{5.17}$$

where

$$\Lambda_m(\mu) = -\mu\frac{\mathrm{d}}{\mathrm{d}\mu}\ln K_m(2\sqrt{\mu}). \tag{5.18}$$

The equation of motion, (5.13), becomes:

$$(1+\xi)\frac{\partial F_m}{\partial\tau} = -\frac{\partial}{\partial\xi}(\xi F_m) + \int_0^\infty G_m(\xi,\xi')F_m(\xi')\,\mathrm{d}\xi'. \tag{5.19}$$

where G_m is here understood to have its limiting form (5.17).

Although (5.19) looks considerably more esoteric than its two-dimensional analog, (4.12), it turns out to be soluble in much the same way. We write

$$F_m(\xi,\tau) = \mathrm{e}^{\Omega\tau}F_{m,\Omega}(\xi), \tag{5.20}$$

and then perform a Fourier–Bessel transformation:

$$\left.\begin{aligned} F_m(\xi) &= \int_0^\infty \mathrm{d}\mu\,\phi_m(\mu)J_m(2\sqrt{\mu\xi}) \\ \phi_m(\mu) &= \int_0^\infty \mathrm{d}\xi\,F_m(\xi)J_m(2\sqrt{\mu\xi}) \end{aligned}\right\}. \tag{5.21}$$

(We shall drop the subscript Ω in most of what follows.) The transformed version of (5.19) is:

$$\Omega\left[1+\frac{m^2}{4\mu}-\frac{\mathrm{d}}{\mathrm{d}\mu}\mu\frac{\mathrm{d}}{\mathrm{d}\mu}\right]\phi_m(\mu) = \mu\frac{\partial\phi_m}{\partial\mu} + \Lambda_m(\mu)\phi_m(\mu). \tag{5.22}$$

As in the two-dimensional case, (5.22) has a relatively simple set of marginally stable solutions for $\Omega = 0$:

$$\phi_{m,0}(\mu) = K_m(2\sqrt{\mu}). \tag{5.23}$$

The inverse Bessel transform of $\phi_{m,0}$ is:

$$F_{m,0}(\xi) = \frac{\xi^{|m|/2}}{2(1+\xi)}, \tag{5.24}$$

which includes the translational mode for $m = 0$. We believe that the occurrence of $\Omega = 0$ solutions for $m \neq 0$ is associated with the existence of stationary states of the Horvay–Cahn [3] type which break the rotational symmetry about the z axis.

Again, the states $\phi_{m,0}$ are formal solutions of (5.22) for any Ω, but are spurious for $\Omega \neq 0$. We take the advantage of this situation to write ϕ_m in the form

$$\phi_m(\mu) = K_m(2\sqrt{\mu})\Psi_m(\mu), \tag{5.25}$$

so that (5.22) becomes:

$$\frac{\mathrm{d}^2\Psi_m}{\mathrm{d}\mu^2} + \left[\frac{1}{\mu}+\frac{1}{\Omega}+2\frac{\mathrm{d}}{\mathrm{d}\mu}\ln K_m(2\sqrt{\mu})\right]\frac{\mathrm{d}\Psi_m}{\mathrm{d}\mu} = 0. \tag{5.26}$$

Solving (5.26), we obtain for any real, positive Ω:

$$\phi_{m,\Omega}(\mu) = K_m(2\sqrt{\mu})\int_0^\mu \mathrm{d}\mu'\frac{\mathrm{e}^{-\mu'/\Omega}}{\mu' K_m^2(2\sqrt{\mu'})}. \tag{5.27}$$

These are the characteristic unstable modes of the Ivantsov dendrite.

To complete the analysis, we compute an approximation for $F_{m,\Omega}(\xi)$ in the limit of large Ω, analogous to (4.22). In this limit, the integrand in (5.27) is largest at values of μ' of order Ω^2, as can be seen by using the asymptotic expression for K_m:

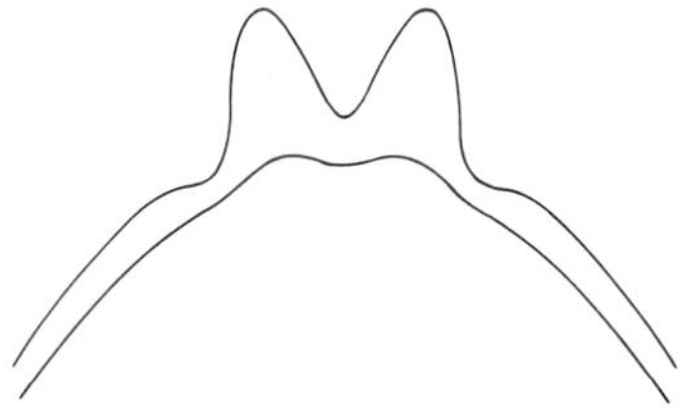

Fig. 3. An unstable mode in the three-dimensional model with $\Omega = 5$ and $m = 2$. The deformation is shown at two different times. The time interval is such that the tip has moved upward a distance of 0.2ρ between the two pictures.

$$K_m(z) \approx \sqrt{\frac{\pi}{2z}}\,\mathrm{e}^{-z}, \quad z \gg 1. \tag{5.28}$$

Inserting (5.28) into (5.27), we obtain a Gaussian approximation similar to (4.21):

$$\phi_{m,\Omega}(\mu) \approx \frac{1}{\mu^{1/4}}\exp\left[-\Omega\left(\frac{\sqrt{\mu}}{\Omega}-1\right)^2\right]. \tag{5.29}$$

Then, writing $\mu = k^2\Omega^2$ and $\xi = s^2$, we have

$$F_{m,\Omega}(s^2) \approx \int_0^\infty \sqrt{k}\,J_m(2ks\Omega)\,\mathrm{e}^{-\Omega(k-1)^2}\,\mathrm{d}k. \tag{5.30}$$

If $s\Omega \ll 1$, then the Gaussian term in (5.30) is sharply peaked in a region where the Bessel function is slowly varying, and we can write:

$$F_{m,\Omega}(s^2) \approx \sqrt{\frac{\pi}{\Omega}}J_m(2\Omega s) \quad (s\Omega \ll 1). \tag{5.31}$$

On the other hand, if $s\Omega$ is large, we may use the asymptotic formula for J_m in (5.30):

$$J_m(z) \approx \sqrt{\frac{2}{\pi z}}\cos\left(z-\frac{\pi m}{2}-\frac{\pi}{4}\right). \tag{5.32}$$

This yields:

$$F_{m,\Omega}(s^2) \approx \frac{1}{\Omega\sqrt{s}}\,\mathrm{e}^{-\Omega s^2}\cos\left(2\Omega s-\frac{\pi m}{2}-\frac{\pi}{4}\right), \quad (s\Omega \gg 1). \tag{5.33}$$

Both (5.31) and (5.33) are consistent with

$$F_{m,\Omega}(s^2) \approx \mathrm{e}^{-\Omega s^2}J_m(2\Omega s), \tag{5.34}$$

which is remarkably similar to (4.22). An illustration of (5.34) is shown in Fig. 3 for the case $m = 2$ and $\Omega = 5$. Here, the tip is shown in two successive positions as it grows in the upward direction. The time interval has been chosen such that the tip has moved a distance 0.2ρ between the two pictures. The tip-splitting deformation in the second picture is clearly well beyond the limit of validity of the linear theory. Note that, because this is the $m = 2$ mode, a section through the dendrite perpendicular to the plane of the paper would show this deformation with the opposite sign; that is, the protuberances would become indentations.

REFERENCES

1. J. S. Langer and H. Müller-Krumbhaar, *Acta metall.* **26,** 1681 (1978).
2. G. P. Ivantsov, *Dokl. Akad. Nauk. S.S.S.R.* **58,** 567 (1947).
3. G. Horvay and J. W. Cahn, *Acta metall.* **9,** 695 (1961).
4. Our basic reference for special functions, used throughout this work, is *Handbook of Mathematical Functions* (edited by M. Abramowitz and I. A. Stegun). National Bureau of Standards (1964).
5. In general, we have found it safest to derive results of this kind ourselves. A good reference for the required techniques of derivation is J. J. Stoker, *Differential Geometry*. Wiley, New York (1969).

Acta Metallurgica. Vol. 26. pp. 1697-1708. Pergamon Press. 1978. Printed in Great Britain

THEORY OF DENDRITIC GROWTH—III. EFFECTS OF SURFACE TENSION*

H. MÜLLER-KRUMBHAAR† and J. S. LANGER

Physics Department, Carnegie-Mellon University, Pittsburgh, PA 15213, U.S.A.

(*Received* 23 *November* 1978)

Abstract—The linear stability analysis of a dendrite tip, presented in the preceding paper, is generalized to include effects of surface tension. The equations of motion are studied numerically by both a forward integration method and an eigenvalue analysis. The most striking result of these computations is the appearance of sidebranching instabilities. For large enough growth velocities, the dendrite tip stabilizes, but the sidebranching instability persists. We argue that the state of dynamic stability proposed in the first paper of this series is achieved by means of a nonlinear coupling between the sidebranching mode and the extended deformations which govern the overall shape and growth velocity of the dendrite.

Résumé—Nous généralisons l'analyse linéaire de la stabilité de l'extrêmité d'une dendrite, présentée dans l'article précédent, afin d'y inclure les effets de la tension superficielle. Nous étudions numériquement les équations du mouvement par une intégration et par une analyse des valeurs propres. Le résultat le plus frappant de ces calculs est l'apparition d'instabilités latérales. Pour des vitesses de croissance suffisamment grandes, l'extrêmité de la dendrite se stabilise, mais les instabilités latérales persistent. Nous montrons que l'état de stabilité dynamique proposé dans le premier article de cette série est obtenu par un couplage non linéaire entre le mode de branchement latéral et les déformations étendues qui gouvernent la forme générale et la vitesse de croissance des dendrites.

Zusammenfassung—Die lineare Stabilitätsanalyse einer Dendritspitze, die in der vorausgehenden Arbeit behandelt wurde, wird verallgemeinert, um den Einfluß der Oberflächenspannung zu erfassen. Die Bewegungsgleichungen werden sowohl mit einer Vorwärts-Integrations-methode als auch mit einer Eigenwertanalyse numerisch untersucht. Das wichtigste Ergebnis dieser Berechnungen ist das Auftreten seitlicher Verzweigungsinstabilitäten. Bei genügend großen Wachstumsgeschwindigkeiten stabilisiert sich die Dendritspitze, jedoch bleiben die Verzweigungsinstabilitäten. Wir argumentieren, daß der in der ersten Arbeit dieser Serie vorgeschlagene dynamische Stabilitätszustand über eine nichtlineare Verkopplung zwischen der Verzweigungsmode und der ausgedehnten Verformung erreicht wird, die die Gesamtform und die Wachstumsgeschwindigkeit des Dendrits bestimmt.

1. INTRODUCTION

In this, the third in our series of papers on the theory of dendritic growth, we shall describe in some detail the calculations which have led us to the ideas presented in I [1]. Our basic problem is to include the effects of surface tension in an analysis of the stability of the tip of a growing dendrite. Surface tension is really the crucial ingredient in this theory. It is surface tension which stabilizes the tip against the instabilities described in II [2]; and it is surface tension which governs the behavior of the sidebranching instability.

The general outline of our theory has been presented in I, and a major part of the analysis needed for the discussion which follows is contained in II. In Section 2 of the present paper, we shall show how we have modified the equation of motion derived in II to account for capillarity in the case where the capillarity correction to the steady state solution is small, and where the active deformations have short characteristic lengths. We argue that this is the physically realistic regime. Our numerical analysis of this equation of motion is described in Section 3. Finally, Section 4 is devoted to a discussion of certain long-wavelength modes of deformation and the way in which these modes may couple to the sidebranching instability. Here we shall present our—still speculative—ideas about how this coupling can determine the natural dendritic growth velocity. There are two Appendices. The first contains an explicit calculation of the first-order capillarity correction to the shape of the dendrite tip; and the second reports certain technical matters concerning the computations described in Section 3.

2. EQUATION OF MOTION FOR SHORT-WAVELENGTH DEFORMATIONS

The model to be considered here has been described fully in Section 2 of I and has been studied in the case of vanishing capillarity length d_0 in II. As before, we work in a frame of reference moving

* Research supported by AFOSR Grant F44620-76-C-0103 and also in part by the CMU Center for the Joining of Materials.

† Permanent Address: Kernforschungsanlage Jülich, Institut für Festkörperforschung, Theorie 2, D-5170 Jülich, F.R.G.

at the dendritic growth velocity v. The thermal or chemical field u, outside the dendrite, obeys the diffusion equation (I-2.3) [1] and satisfies the thermodynamic boundary condition (I-2.5) and the continuity condition (I-2.8) at the solid liquid interface.

Our choice of parabolic coordinates η and ξ is defined in equation (II-3.1) [2]. In terms of these variables, the surface of the dendrite is given by

$$\eta_s(\xi, t) = 1 + E(\xi) + F(\xi, t), \tag{2.1}$$

where $1 + E(\xi)$ denotes the steady-state solution and $F(\xi, t)$ is a small time-dependent deformation. In the case of vanishing surface tension, that is, in equation (II-3.2), $E(\xi)$ is zero. When the surface tension is small but finite, $E(\xi)$ is a small ξ-dependent shape correction of order σ, where σ is the same parameter which was defined in (I-2.11) and which governs the dynamic stability of the system. We have seen in I that σ is of order 10^{-2} at the operating point of a dendrite, as opposed to order unity or greater near the maximum velocity in the steady-state solution. Thus, the actual shape correction to the steady-state paraboloid may be expected to be numerically small. Moreover, it can be seen by explicit calculation that this correction is slowly varying as a function of ξ. The details of this calculation are shown in Appendix A [for the case of a two-dimensional dendrite]. The important point is that, for purposes of investigating the stability of the dendrite tip against short-wavelength perturbations of the kind discovered in paper II, it ought to be possible to neglect the shape correction by letting $E(\xi)$ vanish.

Let us now specialize to the two-dimensional geometry considered in II, Section 3, and thus be more explicit about the nature of the approximation that we are making. We shall return later in this section to the three-dimensional situation. The diffusion equation has the same general solution as shown in (II-3.5):

$$u(\eta, \xi) = \sum_{n=0}^{\infty} A_n V_n(p\eta) Q_n(p\xi), \tag{2.2}$$

where the V_n and Q_n are the functions defined by equations (II-3.6) and (II-3.7) and in the paragraph which follows those equations. The coefficients A_n are allowed to be slowly time-dependent in the quasi-stationary approximation. The thermodynamic boundary condition, (II-3.11), becomes

$$u(\text{surface}) = \sum_{n=0}^{\infty} A_n V_n[p\eta_s(\xi, t)] Q_n(p\xi)$$
$$= \Delta - d_0 \mathscr{K}[\eta_s(\xi, t)], \tag{2.3}$$

where Δ is the dimensionless undercooling, d_0 is the capillarity length, $\mathscr{K}$ is the curvature, and $\eta_s(\xi, t)$ is given by (2.1). Similarly, the continuity condition (II-3.12), is

$$\left(\eta_s + \xi \frac{\partial \eta_s}{\partial \xi}\right) + (\eta_s + \xi)\frac{\partial \eta_s}{\partial \tau} = -\frac{1}{p}\left(\eta_s \frac{\partial}{\partial \eta} - \xi \frac{\partial \eta_s}{\partial \xi}\frac{\partial}{\partial \xi}\right)$$
$$\times \sum_{n=0}^{\infty} A_n V_n(p\eta) Q_n(p\xi)\Bigg|_{\eta = \eta_s(\xi)}, \tag{2.4}$$

where $\tau = 2vt/\rho$. In principle, we should solve (2.3) for the A_n and substitute the result into (2.4), thus obtaining an equation of motion for the interface function $\eta_s(\xi, t)$. Then $1 + E(\xi)$ is the stationary solution corresponding to vanishing τ-derivative in this equation; and the stability equation is obtained by linearizing in the deformation F. Needless to say, this would be an extremely difficult procedure if it were necessary to carry it out in detail.

A more nearly feasible approximation, which would take advantage of the expected smallness of the shape correction and would at least be systematic, would be to expand first $E(\xi)$ and then the linearized stability equation in formal powers of d_0 (or σ). Using the first-order approximation for $E(\xi)$ derived in Appendix A, we have examined a number of the associated corrections to the linear stability operator. These are integral operators which seem to be extremely cumbersome to evaluate explicitly and which do not appear to add any qualitatively new features or singular structures to the stability equation. It is at this point in our investigation that we have decided, at least for purposes of preliminary analysis, to omit the shape correction entirely—or at least almost entirely.

Note now that the only situation in which the capillarity term on the right-hand side of (2.3) becomes appreciably large is when the deformation F has a large curvature. In a linearized theory, this means that F oscillates rapidly as a function of ξ. Considering what happens in planar stability problems, we expect that the capillarity term will become large enough actually to stabilize the interface when F oscillates with wavelengths shorter than the stability length $\lambda_s \cong 2\pi\sqrt{ld_0}$. (See Eq. (I-2.12).) Because ξ is measured in units of the tip radius ρ, this means that the minimum interesting periodicity of $F(\xi)$ is small of order $2\pi\sqrt{\sigma}$. On the other hand, the shape correction $E(\xi)$ has structure only on a ξ-scale of order unity or larger. Thus, for small σ, we expect that our neglect of the shape correction should not obscure the most interesting behavior of the system. Our hypothetical operating point σ^*, where ρ is of order λ_s, appears to lie just at the limit of validity of this approximation.

Returning to (2.3), we write

$$A_n(\tau) = \bar{A}_n + a_n(\tau), \tag{2.5}$$

and

$$\bar{A}_n = \bar{A}_0 \delta_{n,0} + \tilde{a}_n \cong \bar{A}_0 \delta_{n,0}. \tag{2.6}$$

The terms $\tilde{a}_n$ denote the expansion coefficients for the shape correction, which we shall calculate explicitly in Appendix A but neglect in the present analysis. $\bar{A}_0$ is again determined by equation (II-3.16). Linearization of (2.3) yields

$$-pF(\xi) + \sum_{n=0}^{\infty} a_n V_n(p) Q_n(p\xi) = \frac{d_0}{\rho} K_2^{(1)} F(\xi), \tag{2.7}$$

where the curvature $K_2^{(1)}$ is a linear operator on F.

(The subscript 2 denotes the two-dimensional form of this operator.) To evaluate $K_2^{(1)}$, we start with the general expression for the curvature of the function $\eta_s(\xi)$ in two-dimensional parabolic coordinates:

$$\mathscr{K}_2[\eta_s(\xi)] = \frac{1}{\rho D_2^3}[\eta_s^2 - (\eta_s + \xi)\eta_s\eta_s' + (2\eta_s + \xi)\xi(\eta_s')^2 - \xi^2(\eta_s')^3 - 2(\eta_s + \xi)\xi\eta_s\eta_s''], \quad (2.8)$$

where primes denote differentiation with respect to ξ, and

$$D_2^2 = (\eta_s + \xi)[\eta_s + \xi(\eta_s')^2]. \quad (2.9)$$

Setting $E = 0$ in (2.1) and expanding in powers of F, we have

$$\mathscr{K}_2[\eta] \equiv \frac{1}{\rho}[K_2^{(0)} - K_2^{(1)}F + \ldots] \quad (2.10)$$

where $K_2^{(0)}$ is the curvature of the unperturbed parabola:

$$K_2^{(0)} = \frac{1}{(1+\xi)^{3/2}} \quad (2.11)$$

and

$$K_2^{(1)} = \frac{2\xi}{(1+\xi)^{1/2}}\frac{d^2}{d\xi^2} + \frac{1+2\xi}{(1+\xi)^{3/2}}\frac{d}{d\xi} + \frac{(2-\xi)}{2(1+\xi)^{5/2}}. \quad (2.12)$$

From (2.7) we obtain the expansion coefficients a_n:

$$a_n = \frac{p^2}{V_n(p)}\int_0^{\prime} d\xi\, \tilde{Q}_n(p\xi)[1 + \sigma K_2^{(1)}]F(\xi), \quad (2.13)$$

where the $\tilde{Q}_n$ are the orthonormal functions defined in (II-3.22). The linearized version of (2.4) is unchanged from (II-3.21). Thus the new equation of motion for F has the form

$$(1+\xi)\frac{\partial F}{\partial \tau} = -(\tfrac{1}{2} + p)F - \xi\frac{\partial F}{\partial \xi} + \int_0^{\prime} G_2(\xi, \xi'; p)[1 + \sigma K_2^{(1)}]F(\xi')\,d\xi'. \quad (2.14)$$

The integral kernel G_2 is the same as is shown in (II-3.26). In our numerical studies, we shall use G_2 only in the small-p limit given by (II-4.7). Note that (2.14) differs from (II-3.24) only by the insertion of the operator $1 + \sigma K^{(1)}$.

The analogous equation of motion for the three-dimensional situation can now be stated without detailed derivation, all steps being directly analogous to those described above. We shall restrict our attention to the case of cylindrical symmetry, and therefore drop the θ-dependence which was included in II, Section 5. The three-dimensional analog of (2.14) is then

$$(1+\xi)\frac{\partial F}{\partial \tau} = -(1 + p)F - \xi\frac{\partial F}{\partial \xi} + \int_0^{\prime} G_3(\xi, \xi'; p)[1 + \sigma K_3^{(1)}]F(\xi')\,d\xi'. \quad (2.15)$$

In this case, the kernel G_3 is given by (II-5.14) and, in the small-p limit by (II-5.17) in both cases with $m = 0$. (The present use of the subscript 3 to indicate dimensionality should not be confused with the azimuthal index m used in II.) The curvature operator is obtained from

$$\mathscr{K}_3[\eta_s] = \frac{2}{\rho D_3^3}[\tfrac{1}{2}\eta_s^2 - (\tfrac{1}{2}\eta_s + \xi)\eta_s\eta_s' + (\eta_s + \tfrac{1}{2}\xi)\xi(\eta_s')^2 - \tfrac{1}{2}\xi^2(\eta_s')^3 - (\eta_s + \xi)\xi\eta_s\eta_s''] + \frac{1}{D_3}(1 - \eta_s'), \quad (2.16)$$

with

$$D_3^2 = (\eta_s + \xi)[\eta_s + \xi(\eta_s')^2]. \quad (2.17)$$

Then, the curvature of the unperturbed paraboloid is:

$$K_3^{(0)} = \frac{2+\xi}{(1+\xi)^{3/2}} \quad (2.18)$$

and

$$K_3^{(1)} = \frac{2\xi}{(1+\xi)^{1/2}}\frac{d^2}{d\xi^2} + \frac{2+3\xi}{(1+\xi)^{3/2}}\frac{d}{d\xi} + \frac{4+2\xi+\xi^2}{2(1+\xi)^{5/2}}. \quad (2.19)$$

Equations (2.14) and (2.15) are now almost—but not quite—in the final form which we have used for numerical analyses. In II, we found it very useful to pay attention to the translational symmetry of our system. This symmetry implies that the operators on the right-hand sides of the stability equations, (2.14) and (2.15), ought each to have a zero eigenvalue. The corresponding eigenfunctions describe infinitesimal displacements of the two and three dimensional dendrites parallel to their directions of growth. In the case of vanishing surface tension, and for either dimensionality, we have seen that this translation mode is simply

$$F_T(\xi) = \frac{1}{1+\xi}. \quad (2.20)$$

This symmetry is not preserved in (2.14) or (2.15), however, because we have dropped the shape correction while retaining the capillarity term in the equation for the deformation F. Indeed, we already have pointed out that our approximate stability equations cannot be accurate for smooth, extended deformations like F_T; and we have argued that these will not be the modes of greatest interest to us. Nevertheless, we have found it convenient—if only for cosmetic reasons—to modify (2.14) and (2.15) so as to restore a translation mode. We have done this simply by adding a function, say $k(\xi)$, to the operator $K^{(1)}$ in such a way that

$$\tilde{K}^{(1)} \equiv K^{(1)} + k(\xi) \quad (2.21)$$

satisfies

$$\tilde{K}^{(1)}F_T(\xi) = 0. \quad (2.22)$$

The modified curvature operators become:

$$\tilde{K}_2^{(1)} = \frac{2\xi}{(1+\xi)^{1/2}}\frac{d^2}{d\xi^2} + \frac{1+2\xi}{(1+\xi)^{3/2}}\frac{d}{d\xi} + \frac{1-2\xi}{(1+\xi)^{5/2}}, \quad (2.23)$$

and

$$\tilde{K}_3^{(1)} = \frac{2\xi}{(1+\xi)^{1/2}}\frac{d^2}{d\xi^2} + \frac{2+3\xi}{(1+\xi)^{3/2}}\frac{d}{d\xi} + \frac{2-\xi}{(1+\xi)^{5/2}}. \quad (2.24)$$

Notice that these artificial 'shape corrections' have made only slight changes in the last, non-derivative terms of the $K^{(1)}$'s. In our various numerical studies we have not been able to detect any qualitative effects traceable to this modification of the curvature operator; and we repeat our hopeful assertion that short wavelength instabilities should not be very sensitive to this sort of long-wavelength term in the stability equation.

For ready reference, we now list the exact forms of the equations of motion which will be analyzed in the following section. These are obtained by taking the limit $p \to 0$ in (2.14) and (2.15) and inserting the modified curvature operators. In two dimensions we have:

$$(1+\xi)\frac{\partial F}{\partial \tau} = -\tfrac{1}{2}F - \xi\frac{\partial F}{\partial \xi} + \int_0^{\infty} G_2(\xi,\xi')[1+\sigma\tilde{K}_2^{(1)}]F(\xi')\,d\xi', \quad (2.25)$$

with

$$G_2(\xi,\xi') = \int_0^{\infty} \frac{d\mu}{\pi}\cos(2\sqrt{\mu\xi})\cos(2\sqrt{\mu\xi'})\frac{1}{\sqrt{\xi'}}, \quad (2.26)$$

and $\tilde{K}_2^{(1)}$ given by (2.23). For three dimensions,

$$(1+\xi)\frac{\partial F}{\partial \tau} = -F - \xi\frac{\partial F}{\partial \xi} + \int_0^{\infty} G_3(\xi,\xi')[1+\sigma\tilde{K}_3^{(1)}]F(\xi')\,d\xi', \quad (2.27)$$

with

$$G_3(\xi,\xi') = \int_0^{\infty} d\mu\left[-\mu\frac{d}{d\mu}\ln K_0(2\sqrt{\mu})\right] \times J_0(2\sqrt{\mu\xi})J_0(2\sqrt{\mu\xi'}), \quad (2.28)$$

and $\tilde{K}_3^{(1)}$ given by (2.24).

3. NUMERICAL ANALYSIS OF THE EQUATION OF MOTION

The addition of the capillarity term to the equation of motion for the deformation F has made it impossible for us to progress further by analytic methods. It makes no sense to attempt any sort of expansion in powers of σ, because σ multiplies the highest derivative in the system and thus constitutes a singular perturbation. We expect to find distinctly new behavior no matter how small the capillarity constant might be. In a qualitative sense, we can see from the sign of the curvature operator—as well as from our physical intuition—that the new term acts to stabilize short wavelength deformations. But, in order to explore the rich structure of this equation, we have had to resort to numerical techniques.

Numerical analysis of equations (2.25) and (2.27) is itself a difficult process. In the paragraphs which follow, we shall present only the general features of the methods that we have used, and shall place primary emphasis on the results. Appendix B contains a more detailed description of our computational techniques and our procedures for testing the reliability of these methods.

There are two main computational problems to be overcome. First, the function $F(\xi)$ is defined over the semi-infinite domain $0 \leqslant \xi < \infty$; and, as we shall see, interesting behavior occurs out to large values of ξ. Second, the integral kernels G_2 and G_3 are singular at $\xi = \xi'$. To deal with the first of these problems, we have had to introduce an upper cutoff, ξ_{max}, and then have checked in each calculation to see whether the results are sensitive either to the choice of ξ_{max} or to the choice of mesh-points ξ on which F has been evaluated. This remains a serious problem, however; and we shall return to it occasionally throughout the following discussion.

The singularity in the integral kernel is easier to handle. To see what is happening here, let us look at equation (2.26) for G_2, which can be evaluated analytically. As in II, we transform to the variable s:

$$\xi = s^2, \quad (3.1)$$

and write

$$G_2(\xi,\xi')d\xi' = \Gamma_2(s,s')ds' \quad (3.2)$$

so that (see II-4.10)

$$\Gamma_2(s,s') = \frac{1}{\pi}\int_0^{\infty} k\cos(ks)\cos(ks')\,dk, \quad (3.3)$$

which is clearly singular at $s = s'$. However, if Γ_2 is understood to operate only on smooth functions of s', then we are allowed to filter out the short-wavelength parts of Γ_2 by writing

$$\Gamma_2(s,s') \cong \Gamma_{2\epsilon}(s,s') \equiv \frac{1}{\pi}\int_0^{\infty} k\cos(ks)\cos(ks')e^{-\epsilon k}\,dk = -\frac{1}{2\pi}\left\{\frac{(s+s')^2-\epsilon^2}{[(s+s')^2+\epsilon^2]^2} + \frac{(s-s')^2-\epsilon^2}{[(s-s')^2+\epsilon^2]^2}\right\}. \quad (3.4)$$

$\Gamma_{2\epsilon}$ will be an accurate representation of Γ_2 as long as all other functions to be considered are smooth on a scale $\Delta s \gg \epsilon$. Then, for numerical purposes, let Δs be the interval between mesh points. Given any smooth function $f(s)$, the quantity

$$\int_{s_1-(\Delta s/2)}^{s_1+(\Delta s/2)} \Gamma_2(s,s')f(s')\,ds' \cong \left[\lim_{\epsilon\to 0}\int_{s_1-(\Delta s/2)}^{s_1+(\Delta s/2)} \Gamma_{2\epsilon}(s,s')\,ds'\right]f(s_1) \quad (3.5)$$

is perfectly well defined and trivial to evaluate even for an interval containing the point $s' = s$. Equation (3.5) provides the weight factors required for numerical evaluation of the integral on the right-hand side of (2.25). In three dimensions, we are not able to evaluate the kernel (2.28) in a closed form analogous to (3.4); but the nature of the singularity is unchanged, and our computational technique is again based on the idea of the ϵ-cutoff introduced above. Further details will be found in Appendix B.

In all of our numerical work, we have actually used the variable s instead of ξ in order to take advantage of the algebraic simplifications which were useful in II. To reduce the equation of motion to discrete form, we have chosen a range $0 \leqslant s \leqslant s_{max}$ and have divided this range into N equal intervals of width Δs. Typically, N has been chosen to be 50 or 60 and $s_{max} = N\Delta s$ to be of order 3 or 4. Thus the value of ξ_{max} is characteristically of order 10, which takes us well away from the tip. For purposes of reporting results, we shall return to the variable ξ, which is easily visualized as the distance from the tip measured back along the axis of symmetry of the dendrite. Remember that ξ is measured in units of the tip radius ρ, and that it describes a position in the frame of reference moving with the steady-state velocity of the tip. The units of time τ have been chosen such that a point which is stationary in the laboratory moves with constant velocity $d\xi/d\tau = 1$ in the ξ-frame.

Our first method of investigating equations (2.25) and (2.27) was a direct forward integration in time. We started with a perturbation of the form

$$F(\xi, 0) = \exp(-\xi/\xi_0), \tag{3.6}$$

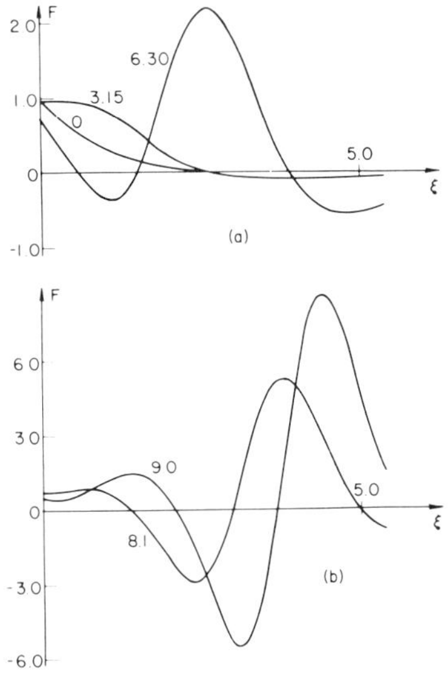

Fig. 1. Sequence of deformations $F(\xi, \tau)$ on the surface of a two dimensional dendrite. Relevant parameters are: $\sigma = 0.01$, $\Delta s = 0.0473$ and ξ_0 (equation 3.6) = 1. Numbers beside the curves indicate the times τ. Note the change of scale between graphs (a) and (b). Note also that the initial deformation ($\tau = 0$) near $\xi = 0$ is very nearly a translation mode, and that this initial displacement of the tip persists throughout the time development.

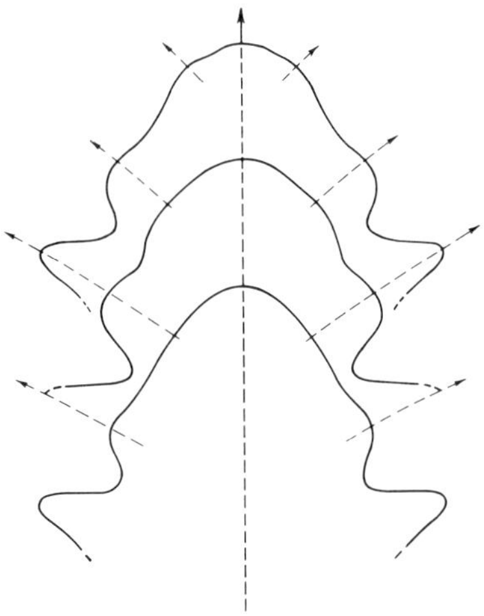

Fig. 2. Time-ordered sequence of tip shapes computed for a two-dimensional dendrite at $\sigma = 0.02$ with random noise at the tip. The dendrite grows up the page at its natural velocity as seen in the laboratory frame. The computation was begun with no deformation at time $\tau = 0$; and the shapes displayed were observed at times $\tau = 22.27$, 27.05 and 30.23. The dashed lines indicate very roughly the trajectories of secondary tips as they emerge laterally from the main body of the dendrite.

which describes a smooth bump of width ξ_0 sitting at the tip of the dendrite. A typical sequence of subsequent deformations $F(\xi, \tau)$ is shown in Fig. 1. This is a two-dimensional calculation, based on (2.25), with $\sigma = 0.01$. (Other relevant parameters are indicated in the figure caption.) As we shall see more clearly later, the tip is unstable at this value of σ. The most obvious effect is that the tip fluctuates and generates a roughly periodic train of deformations which move back along the dendrite and grow in amplitude as they move. We have repeated this experiment for a variety of different values of σ and ξ_0 and find that, after an initial transient phase, the behavior becomes independent of the initial perturbation, that is, independent of ξ_0, but is quite sensitive to σ. We shall see below that the wavelength of the deformation is proportional to $\sqrt{\sigma}$.

In a related numerical experiment, we have integrated equation (2.25) forward in time with no initial perturbation but with a random noise applied at the tip. Specifically, we have added a term to the right-hand side of (2.25) which is nonzero only at the mesh-point corresponding to $\xi = 0$, and which assumes random values at each forward step throughout the time integration. The result is shown in Fig. 2 for $\sigma = 0.02$, a value which we believe to be close to the natural operating point of the two-dimensional dendrite. Here we have shown $F(\xi, \tau)$ actually projected onto the moving parabolic dendrite as might be seen by an observer in the laboratory frame of reference.

The picture turns out to be quite similar to that shown in Fig. 2 of I, although the two figures have been constructed by very different procedures. It is clear that the train of deformations generated by the noise at the tip constitutes a sidebranching instability. The intensity of the noise source, and thus the amplitude of the sidebranches, is not determined by our theory. In preparing Fig. 2, we have chosen this intensity so that the amplitude of the outermost sidebranch (near ξ_{max}) is appreciably smaller than the thickness of the parabola at that point, and therefore is not strongly inconsistent with our linearization of the equation of motion. The amplitude is still so large that we have had to take a small liberty in drawing the outermost sidebranch in order to correct an unphysical distortion associated with the transformation of this linear approximation into parabolic coordinates. On the other hand, this amplitude is so small that the noisy transient at the tip is almost invisible. This situation in which a very regular train of sidebranches is generated by an unobservably weak noise source is in accord with the general appearance of real dendrite. Compare, for example, Fig. 9 of Glicksman *et al.* [3] with Fig. 2, or with Fig. 2 of I.

The behavior that we have observed in these time-dependent studies suggests that the motion of the dendritic surface is largely determined by one—or a few—dominant modes of instability. In order to see more precisely what is happening here, we have attempted to perform an eigenvalue analysis, that is, we have written

$$F(\xi, \tau) = e^{\Omega\tau} F_{\Omega}(\xi), \tag{3.7}$$

so that (2.25) and (2.27) have the form

$$\Omega F_{\Omega} = \mathscr{L} F_{\Omega}, \tag{3.8}$$

where $\mathscr{L}$ is the same linear integro-differential operator which appeared in the time-dependent computation. Using the same discretization in s-space as before, we have numerically diagonalized the resulting $\mathscr{L}$-matrix to obtain a spectrum of eigenvalues Ω and eigenmodes $F_{\Omega}(\xi)$. Here the problem of the upper cutoff ξ_{max} becomes particularly acute. The interesting sidebranching modes are expected to oscillate and grow without bound as ξ increases, and thus will be sensitive to the artificial boundary condition imposed at ξ_{max}. We do not have any very elegant solution to this problem. We have simply repeated the calculation for various choices of ξ_{max} and Δs and have found that, as long as we had sufficiently many meshpoints within each period of $F_{\Omega}(\xi)$ and at least three or four periods in the range $\xi < \xi_{max}$, then the physically significant modes are clearly identifiable and not strongly sensitive to the precise choice of ξ_{max}. It was this criterion which required us to use as many as $N = 60$ mesh points, and thus to diagonalize matrices $\mathscr{L}$ as large as 60×60.

In general, the result of any one of these diagonalization procedures is a spectrum of N complex eigenvalues Ω, and N corresponding complex eigenstates. Note that the operator $\mathscr{L}$ is not self-adjoint. Thus $\mathscr{L}$ permits the complex eigenvalues which are necessary in order to describe the oscillatory running waves that we have seen above. Most of the modes that we obtain by this procedure are of no special physical significance. These are highly stable modes—ones with large negative real parts of Ω—which are not smooth on the scale of the mesh spacing Δs. The interesting modes are those for which ReΩ is largest; and we find that we are always able to choose a suitable range of values of Δs within which these modes emerge unambiguously. That is, once we have identified a set of dominant modes, there never remains any unphysical mode—one which is not smooth on the scale of Δs—of a greater or even comparable degree of instability.

The results of this analysis are summarized in Figs. 3–6. Figure 3 shows spectra of ReΩ for the two-dimensional model as obtained from computations with $N = 50$ for various values of σ. The solid circles indicate sidebranching modes, that is, modes which oscillate and grow with increasing ξ and for which Ω is complex. The translation modes are indicated by open squares. In preparing this figure, we have fixed Δs for each σ so as to resolve the dominant instabilities, and then, proceding downward in ReΩ, have ceased marking modes as soon as we encountered one of doubtful validity. Note that, for $\sigma = 0.001$, a non-complex tip-splitting instability seems to have been well resolved and is indicated by the open circle. At that σ, however, the translation mode is missing because ξ_{max} is too small. Note also that, as σ increases, we are able to identify fewer modes with certainty. At the smaller σ's, where more modes can be found, there seems to be a tendency for the modes to cluster near the maximum ReΩ. (The numbers in parentheses indicate the number of modes

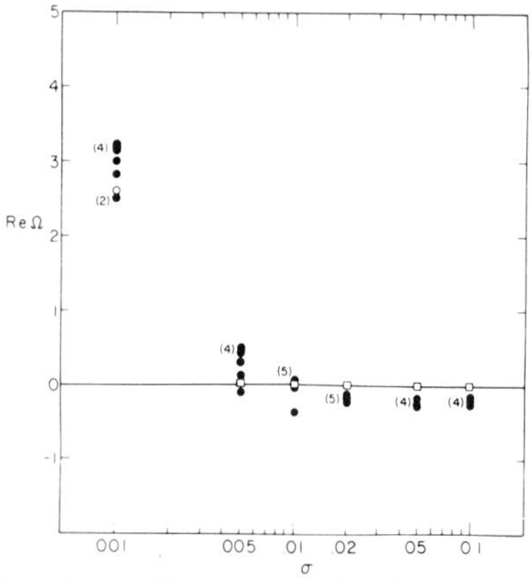

Fig. 3. Spectra of ReΩ for the two-dimensional model for various values of σ. $N = 50$ for all cases. The values of Δs used in the computations, in order of increasing σ, were: 0.0267, 0.03989, 0.04743, 0.05641, 0.07093 and 0.08435.

in a cluster.) We also have found that the density of modes increases with increasing N. Our tentative conclusion is that the dominant instability lies at the extremum of a continuous band of oscillatory sidebranching states.

In Figure 4 we have drawn the real part of $F_\Omega(\xi)$ for the most unstable mode at each of the two smallest values of σ shown in Fig. 3, that is, $\sigma = 0.001$ and 0.005. For comparison, we also have shown the translation mode $F_T(\xi)$, given by (2.20), in each of these graphs. The unstable modes are normalized so that $\mathrm{Re}F(0) = 1$. Note that, for $\sigma = 0.001$, F has grown by a factor of 1000 within its first few wavelengths, and that this activity occurs in the immediate vicinity of the tip, that is, at ξ of order unity. There appears to be a well defined wavelength, which we shall denote by the symbol Λ, and which is related to the physical wavelength λ measured parallel to the symmetry axis of the dendrite (see I-3.1):

$$\Lambda = \frac{2\lambda}{\rho}. \tag{3.9}$$

For Fig. 4(a), we estimate $\Lambda \cong 0.41$. The wave velocity v,

$$v = \frac{\Lambda}{2\pi}\,\mathrm{Im}\,\Omega, \tag{3.10}$$

would be unity for a mode which was exactly stationary in the laboratory frame. In the present case, we have $\Omega = 3.23 \pm 5.66i$, so that $v \cong 0.37$. We interpret this result to mean that the piece of the mode which

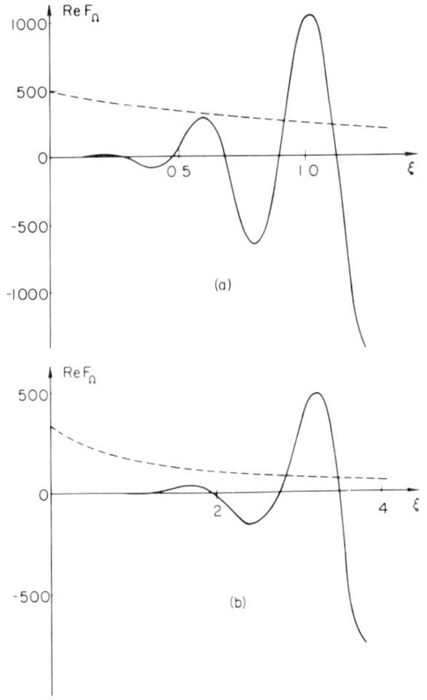

Fig. 4. Two-dimensional eigenmodes $F_\Omega(\xi)$. Dominant instabilities are shown for: (a) $\sigma = 0.001$, and (b) $\sigma = 0.005$. The dashed curves are translation modes (not normalized).

Table 1. Eigenvalues Ω, reduced wavelengths Λ, and reduced wave velocities v for two-dimensional sidebranching modes at $\sigma = 0.01$ and 0.02. Labels A–D identify modes shown in Fig. 5.

Mode	ReΩ	ImΩ	Λ	v
		$\sigma = 0.01$		
	−0.0556	1.43	4.05	0.80
	0.0168	1.65	3.37	0.89
	0.0528	1.91	2.83	0.86
A	0.0726	2.17	2.23	0.77
B	0.0425	2.43	1.75	0.68
		$\sigma = 0.02$		
	−0.233	0.940	6.81	1.02
C	−0.0924	1.17	4.77	0.89
D	−0.142	1.39	4.14	0.92
	−0.135	1.54	3.15	0.77
	−0.148	1.72	2.61	0.71

we have computed is moving out along the highly curved surface at the tip of the dendrite. At $\sigma = 0.005$, shown in Fig. 4(b), the mode extends to somewhat larger values of ξ, and the wave velocity increases. Relevant parameters are $\Lambda \cong 1.57$, $\Omega = 0.501 \pm 3.03i$ and $v \cong 0.73$.

In general, the sidebranching modes at any given σ occur in an ordered sequence with increasing values of $Im\Omega$ and decreasing Λ. As one moves through this sequence, the amplification rate, ReΩ passes through a maximum; but the size and location of this maximum varies unsystematically with the computational parameters N and Δs, and it is this variation which causes some uncertainty about intrinsic stability as a function of σ. To illustrate this pattern of behavior, we show in Table 1 the sequences of values of Ω, Λ and v for sidebranching modes at $\sigma = 0.01$ and 0.02. Selected examples of these functions $F_\Omega(\xi)$ are drawn in Fig. 5. Note that these states become large only well away from the tip, that the amplification factors are smaller than in Fig. 4, and that the velocity v is closer to unity for the dominant instability. Note also that, although the wavelength λ is quite comparable to ρ, F is relatively small and structureless throughout the region near the tip where the shape correction is appreciable. Thus the shape correction is probably not very important for these modes.

In the three dimensional model, the situation is as yet not so clear as in the two-dimensional case. A major inaccuracy occurs in the evaluation of the integral kernel G_3, as discussed in Appendix B. However, the results obtained so far are effectively indistinguishable from what we have found in two dimensions. The dominant instabilities for appreciable values of σ are sidebranching modes which are very similar to those shown in Fig. 5; and the maximum ReΩ passes through zero as σ increases, just as occurs in Fig. 3. We currently are working to improve our numerical methods for the three-dimensional case, and hope to publish a more detailed report of these

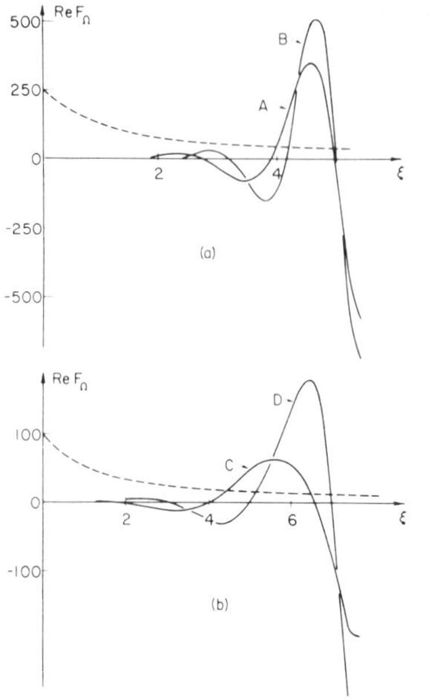

Fig. 5. Two-dimensional eigenmodes $F_\Omega(\xi)$ for (a) $\sigma = 0.01$ and (b) $\sigma = 0.02$. The labels A through D refer to the associated sets of parameters listed in Table 1. The dashed curves are translation modes (not normalized).

computations in the near future. The only preliminary data of importance for present purposes are contained in equations (3.11) and (3.13) below.

A graph of Λ vs $\sqrt{\sigma}$, for both two and three dimensional cases, is shown in Fig. 6. As mentioned above, the two cases do not appear to be distinguishable. We estimate,

$$\Lambda \cong (26 \pm 4)\sqrt{\sigma}, \quad (3.11)$$

which implies

$$\lambda \cong (2.1 \pm 0.3)\lambda_s, \quad (3.12)$$

where λ_s is the stability length defined in (I-2.12).

The main result of the eigenvalue analysis; for both two and three dimensional cases, is that the dendrite tip stabilizes at sufficiently large σ. It is apparent in Fig. 3 that the largest ReΩ vanishes at $\sigma = \sigma^*$, where, as reported in I,

$$\sigma^* = \begin{cases} 0.020 \pm 0.007 & \text{(two dimensions)} \\ 0.025 \pm 0.007 & \text{(three dimensions).} \end{cases} \quad (3.13)$$

(The stated uncertainties are themselves very uncertain, being based primarily on our estimate of the accuracy of the numerical procedure. See Appendix B.) For $\sigma > \sigma^*$, the dominant mode is the translation mode, which is only marginally unstable and is localized at the tip. The top of the sidebranch spectrum, however, does not continue to drop rapidly to negative values of ReΩ as σ increases, but appears to level off near Re$\Omega \cong -0.1$. This break in the curve of ReΩ vs σ implies the persistence of the sidebranching instability at large σ, and is also useful for purposes of identifying σ^*.

It should be emphasized that ReΩ determines apparent stability in the frame of reference moving with the dendrite tip. The sidebranching modes, even for $\sigma > \sigma^*$, remain true instabilities in the laboratory frame because of the divergence of F_Ω at large ξ. This divergence is the dominant feature in Fig. 2 which is, of course, a laboratory view of the direct time-integration of the equation of motion (2.25). Figure 2 of I is also a laboratory view, but is based directly on the eigenvalue analysis. There we have chosen just the dominant sidebranching mode for $\sigma = \sigma^* \cong 0.025$ in three dimensions, and have plotted

$$F(\xi, \tau) = Ae^{\Omega_1\tau}[(\cos\Omega_2\tau)\text{Re}F_\Omega(\xi) + (\sin\Omega_2\tau)\text{Im}F_\Omega(\xi)]. \quad (3.14)$$

where $\Omega \equiv \Omega_1 \pm i\Omega_2 \cong -0.0970 \pm 1.317i$, and the complex function F_Ω is obtained directly from the computation. In order to make this picture look as much as possible like the succinonitrile dendrite shown by Glicksman *et al.* [3], we have chosen the amplitude A to be quite small, and have used an exponential approximation to extrapolate F out to somewhat larger values of ξ than those used in the actual computation. As explained in I and in the following section, we believe that σ^* determines the natural operating point of the dendrite; thus this picture constructed from the dominant side-branching mode at $\sigma = \sigma^*$ should be an accurate representation of the tip of a physical dendrite. (Remember, however, that we have treated only rotationally symmetric perturbations here. We have not yet checked whether modes with $m \neq 0$, as in (II-5.12), might lead to larger values of σ^*.)

4. BEHAVIOR OF EXTENDED DEFORMATIONS FOR $\sigma > \sigma^*$

We turn now to the very difficult question of the behavior of a dendrite in the regime $\sigma > \sigma^*$, where

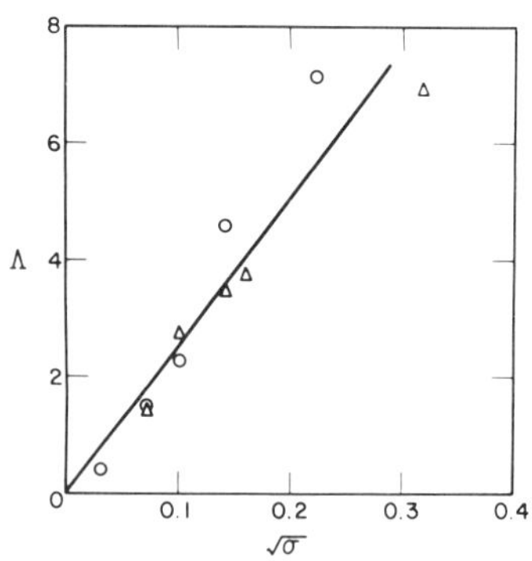

Fig. 6. Dimensionless wavelength Λ as a function of $\sqrt{\sigma}$ for the dominant side-branching instabilities in two dimensions (circles) and three dimensions (triangles).

the tip is sharp enough to be stable against localized deformations. This question is crucial to our identification of σ^* as the natural operating point of the dendrite. Our basic ideas concerning this problem are described at the end of Section 3 of I. In what follows, we shall try to make those ideas a bit more precise; but, as will be seen, we do not pretend to be able to present a systematic theory.

For $\sigma > \sigma^*$, the stability length λ_s is larger than the tip radius ρ, and thus the important deformations are those whose characteristic length scales extend well beyond the immediate vicinity of the tip, and for which our neglect of the shape correction is no longer necessarily valid. However, we do have a good idea of what these important modes of deformation must look like, and can begin to speculate about how they might couple to one another. Specifically, we expect to have to deal with a dominant sidebranching mode, the marginally unstable translation mode (2.20), and a third deformation which we shall call a 'breathing mode'. The latter is a very long-ranged deformation which has the effect of changing the tip radius ρ of the steady-state dendrite, and thus changing the steady-state velocity v.

The breathing mode, by definition, has the following property. If the deformation is a breathing mode of amplitude F_B at time t, then the change in F at time $t + dt$ is

$$dF(\xi) = \frac{1}{1+\xi}\left(\frac{dv}{d\rho}\right)F_B\,dt, \quad \left(dt = \frac{\rho}{2v}\,d\tau\right), \tag{4.1}$$

where $dv/d\rho$ is the slope of the steady-state v vs ρ curve. That is, F_B generates a translation of the form (2.20). We can demonstrate this property explicitly in a modified Ivantsov approximation, where $F_B(\xi)$ is exactly a constant because the steady-state shapes are constant-η parabolas, and where the modified thermodynamic boundary condition is

$$u(\text{surface}) = \Delta - d_0\mathscr{K}(\xi = 0). \tag{4.2}$$

For simplicity, we specialize to two dimensions, and then let $F = F_B$ in (4.2), so that

$$u(\text{surface}) \cong \Delta - \frac{d_0}{\rho}(1 - F_B). \tag{4.3}$$

Now only the $n = 0$ term in the expansion (2.2) for u is of interest. Linearization of (4.3) yields

$$a_0 \cong \frac{p^2}{V_0(p)}(1+\sigma)F_B, \tag{4.4}$$

and, substituting this into the continuity equation, we find

$$(1+\xi)\frac{\partial F}{\partial \tau} \cong \left[-p - \tfrac{1}{2} - \frac{p}{V_0(p)}\frac{dV_0}{dp}(1+\sigma)\right]F_B. \tag{4.5}$$

Self-consistency in this approximation is obtained only in the small-p limit (diffusion length l much larger than ρ). The discrepancy at large p apparently is associated with our use of the quasi-stationary approximation. From (II-3.14) we have

$$\frac{1}{V_0}\frac{dV_0}{dp} \approx -\frac{1}{\sqrt{\pi p}}. \tag{4.6}$$

Equation (4.5) becomes

$$(1+\xi)\frac{\partial F}{\partial \tau} \cong \left(-\tfrac{1}{2} + \sigma\sqrt{\frac{p}{\pi}}\right)F_B. \tag{4.7}$$

To check that the right-hand side of (4.7) agrees with (4.1), we write the modified Ivantsov approximation for the steady-state in the form (see II-3.17):

$$-\frac{V_0}{(dV_0/dp)} \cong \sqrt{\pi p} = \Delta - \frac{d_0}{\rho} \tag{4.8}$$

and differentiate with respect to ρ. The result is

$$\frac{\rho}{2v}\frac{dv}{d\rho} \cong -\tfrac{1}{2} + \sigma\sqrt{\frac{p}{\pi}}, \tag{4.9}$$

as expected.

The breathing mode constitutes another marginal instability in the sense that it drives a translation which grows linearly—not exponentially—in the time. The linear coupling between F_B and F_T vanishes only at the maximum-velocity point where $dv/d\rho = 0$; but here the sytem is strongly unstable in second order, because a finite fluctuation in ρ of either sign decreases the velocity. This simple stability consideration makes it seem extremely unlikely that a dendrite would grow at its maximum steady-state velocity.

The kinetic equations (2.25) and (2.27) are incomplete in the sense that they do not contain any information about the maximum in the v vs ρ curve. This omission can be traced back to the fact that, in going from sums to integrals in (II-3.26) and (II-5.14), we did not allow for the possibility that $F(\xi)$ might contain a long-ranged conponent of the form F_B. If we separate and retain the $n = 0$ terms explicitly (in a manner reminiscent of the way in which one treats the ground state in the Bose–Einstein condensation), we obtain an extra term on the right-hand side of (2.25):

$$\frac{p^{3/2}}{\sqrt{\pi}}\int_0^{\xi'} d\xi'\,e^{-p\xi'}(1+\sigma)F(\xi'); \tag{4.10}$$

and a similar addition to (2.27):

$$-\frac{p}{\ln p}\int_0^{\xi'} d\xi'\,e^{-p\xi'}(1+\sigma)F(\xi'). \tag{4.11}$$

For consistency with the modified Ivantsov approximation described above, we have set $\tilde{K}_2^{(1)} = \tilde{K}_3^{(1)} = 1$. All of these approximations depend on the fact that (4.10) and (4.11) are appreciably large for $p \to 0$ only when F is extremely long-ranged and non-oscillatory. An initial breathing-mode perturbation now converts (4.10) into (4.7) for σ near $\sigma(v_{\max}) \cong (\sqrt{\pi/p})/2$.

Because the behavior of the dendrite is so sensitive to F_B, it is very important to understand what drives fluctuations in this mode. Ultimately, all fluctuations must be driven by thermal noise; but thermal fluctuations are small and should couple only very weakly to an extended deformation like the breathing mode. As was stated in I, we believe that the coupling occurs nonlinearly via the sidebranching instability. We have seen quite clearly in the previous section how small

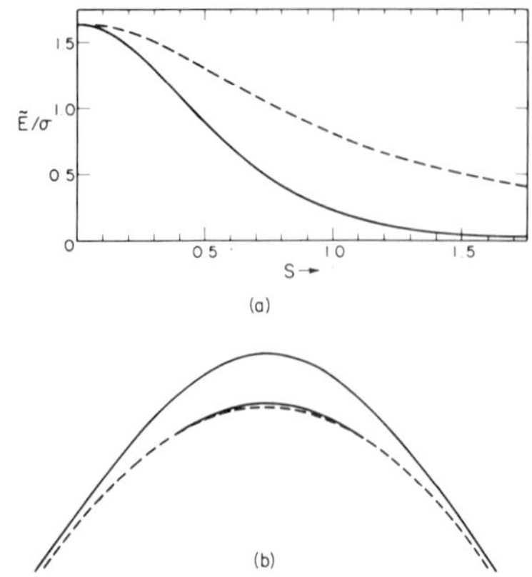

Fig. 7. Shape correction for a two-dimensional dendrite shown (a) as a function of s and (b) superimposed upon the uncorrected parabolic surface (dashed line). The dashed curve in (a) is the translation mode. The corrected shapes (solid lines) are shown in (b) for the physically realistic case $\sigma = 0.02$ (small correction) and for $\sigma = 0.4$.

fluctuations at the dendrite tip generate large sidebranches which emerge within a few tip radii—well within the diffusion field—of the front of the dendrite. In this way, the thermal noise is strongly amplified by the sidebranching instability.

The sidebranching mode should be coupled in second order to the breathing mode. Roughly speaking, this coupling occurs because the sidebranching oscillation is superimposed upon the curved surface of the dendrite, so that the curvature at the tips of the emergent secondary dendrites is somewhat less sharp than that in the troughs between them. As a result, the troughs stabilize while the tips continue to grow out, and the average thickness of the dendritic structure increases. This is certainly what one sees in real systems. To check that the observed phenomenon is consistent with our mathematical understanding of the problem, we have computed the curvature $\mathcal{K}$ out to terms of order F^2 and have used this nonlinear form in a numerical time-development calculation of the kind described in Section 3. We shall not describe these results in detail except to report that the expected outward enhancement was clearly visible. The result is effectively a breathing mode, which couples to the displacement via (4.10) or (4.11).

Our proposed picture is now the following. Suppose that we start with a paraboloidal steady-state dendrite at $\sigma > \sigma^*$. Thermal noise at the tip drives a sidebranching instability which broadens and slows the dendrite, moving it down along the v vs ρ curve. Because this new point corresponds to a smaller value of σ, the sidebranching activity must increase, and the velocity must decrease ever more rapidly. This mechanism operates along the whole right-hand (large ρ) side of the v vs ρ curve, driving the dendrite always to larger values of ρ and smaller v. When the dendrite becomes so broad that $\sigma < \sigma^*$, however, the tip splitting instability must somehow return the system to the region with $\sigma > \sigma^*$. The resulting motion remains in the neighborhood of σ^*, where the tip dynamics alternate between sidebranch-induced broadening and tip-splitting.

REFERENCES

1. J. S. Langer and H. Müller-Krumbhaar, *Acta metall.* **26,** 1681 (1978).
2. J. S. Langer and H. Müller-Krumbhaar, *Acta metall.* **26,** 1689 (1978).
3. M. E. Glicksman, R. J. Schaefer and J. D. Ayers, *Metall. Trans.* (A) **7,** 1747 (1976).

APPENDIX A

First-Order Shape Correction

Our purpose here is to compute the shape correction $E(\xi)$ to first order in the capillarity length d_0. We shall consider only the two-dimensional model.

The first step is to linearize the thermodynamic boundary condition (2.3), using d_0 as the small parameter and taking both $E(\xi)$ and the coefficients $\bar{a}_n$, defined in (2.6), to be proportional to d_0. We find (compare with 2.7):

$$-pE(\xi) + \sum_{n=0}^{\infty} \bar{a}_n V_n(p) Q_n(p\xi) = -\frac{d_0}{\rho}\frac{1}{(1+\xi)^{3/2}}, \qquad \text{(A.1)}$$

where we have used (2.10) and (2.11) for the zeroth order curvature. Next, we solve (A.1) for the $\bar{a}_n$ and insert the result into the linearized continuity condition. (The latter equation can be read directly from (II-3.21) by substituting E for F, $\bar{a}_n$ for a_n, and letting the time derivative vanish.) In this way, we obtain a linear, inhomogeneous equation for E. In the limit $p \to 0$, this is

$$\tfrac{1}{2}E(\xi) + \xi\frac{\partial E}{\partial \xi} - \int_0^{\infty} G_2(\xi,\xi')E(\xi')\,d\xi'$$

$$= -\sigma\int_0^{\infty} G_2(\xi,\xi')\frac{1}{(1+\xi')^{3/2}}\,d\xi'. \qquad \text{(A.2)}$$

where G_2 is given in (2.26).

Equation (A.2) is best rewritten in terms of the variable s ($\xi = s^2$, $E(\xi) = \tilde{E}(s)$) so that

$$\frac{1}{2}\frac{d}{ds}(s\tilde{E}) - \int_0^{\infty} \Gamma_2(s,s')\tilde{E}(s')\,ds'$$

$$= -\sigma\int_0^{\infty} \Gamma_2(s,s')\frac{1}{(1+s'^2)^{3/2}}\,ds'. \qquad \text{(A.3)}$$

where the kernel Γ_2 is given by (3.3). As in II, we solve this equation by a Fourier-cosine transformation:

$$\tilde{E}(s) = \sqrt{\frac{2}{\pi}}\int_0^{\infty} \phi(k)\cos ks\,dk. \qquad \text{(A.4)}$$

The transformed version of (A.3) is

$$\frac{d\phi}{dk} + \phi = \sigma\sqrt{\frac{2}{\pi}}\int_0^{\infty}\frac{\cos ks}{(1+s^2)^{3/2}}\,ds, \qquad \text{(A.5)}$$

for which the general solution may be written in the form:

$$\phi(k) = \sigma\sqrt{\frac{2}{\pi}}\,e^{-k}\int_0^{k} dk'\,e^{k'}\int_0^{\infty} ds\frac{\cos k's}{(1+s^2)^{3/2}} + C\,e^{-k}. \qquad \text{(A.6)}$$

The undetermined constant C appears as the amplitude of a translation mode, in accord with our expectation that we should be able to displace this solution by an arbitrary distance along the symmetry axis of the dendrite. The integration over k' in (A.6) can be performed, as can the integration over k required for transformation back to $\tilde{E}(s)$ via (A.4). The result is:

$$\tilde{E}(s) = \frac{\sigma}{(1+s^2)^{5/2}} + \frac{C'}{(1+s^2)} + \frac{2\sigma}{\pi} P \int_0^\infty \frac{s'^2\,ds'}{(s'^2 - s^2)(1+s'^2)^{5/2}}, \tag{A.7}$$

where P denotes a principal-value integration.

A natural choice for the constant C' is

$$C' = \frac{2\sigma}{\pi} \int_0^\infty \frac{s'^2\,ds'}{(1+s'^2)^{5/2}}, \tag{A.8}$$

which has the effect of cancelling the displacement mode ($\approx 1/s^2$) at large s and leaving only the localized shape correction at the tip. Our formula for $\tilde{E}$ becomes:

$$\tilde{E}(s) = \frac{\sigma}{(1+s^2)^{5/2}} + \frac{1}{(1+s^2)} \frac{2\sigma}{\pi} P \int_0^\infty \frac{s'^2\,ds'}{(s'^2 - s^2)(1+s'^2)^{3/2}}. \tag{A.9}$$

The remaining integral can be evaluated analytically by a procedure in which contour methods are used to transform to an integration over the discontinuity across a cut on the imaginary s' axis. We quote only the final result:

$$\tilde{E}(s) = \frac{2\sigma}{\pi(1+s^2)^2} + \frac{\sigma}{(1+s^2)^{5/2}} \left\{ 1 - \frac{s}{\pi} \ln\left(\frac{\sqrt{1+s^2}+s}{\sqrt{1+s^2}-s} \right) \right\}. \tag{A.10}$$

This function is shown in Fig. 7 both directly and as it appears when superimposed upon a two-dimensional parabolic dendrite with $\sigma = 0.4$ and with $\sigma \cong \sigma^* \cong 0.02$. Note that the latter, more realistic, case produces a shape correction which is essentially invisible.

APPENDIX B

Computational Methods

Our confidence in the numerical results presented in Section 3 is determined by various computational procedures and consistency checks that we have employed during the course of this investigation. We outline several of these procedures in the following paragraphs.

The integral kernel

The two-dimensional kernel, G_2, has been discussed in detail at the beginning of Section 3. For three dimensions, the equation analogous to (3.4) is

$$\int_0^\infty k^2 \frac{K_1(k)}{K_0(k)} J_0(ks) J_0(ks') \mathrm{e}^{-\epsilon k}\,dk, \tag{B.1}$$

which must be evaluated numerically. The integration routine was checked on the related integral

$$\frac{s'}{2} \int_0^\infty k^2 J_0(ks') \mathrm{e}^{-\epsilon k}\,dk = \frac{s'(2\epsilon^2 - s'^2)}{2(\epsilon^2 + s'^2)^{5/2}}, \tag{B.2}$$

which differs from (B.1) at $s = 0$ by a logarithmic correction which is important only for large values of s'. Our relative precision was better than 10^{-3} everywhere. As explained in the text, the cutoff ϵ must be chosen to be smaller than the mesh spacing Δs. The effect of this cutoff is largest at the mesh points nearest the singularity, $s' = s \pm \Delta s$. But the corresponding difference between (B.2) and (B.1) at $s = 0$ is small; and ϵ was chosen to give a relative precision better than 5×10^{-3}.

Each single integral of the form (B.1) took almost a minute of computer time; thus it was not feasible to evaluate directly all 1770 integrals for the off-diagonal triangular part of the 60×60 matrix $\Gamma(s, s')$. We evaluated only 72 integrals, equidistantly spaced in the triangular matrix, and interpolated the rest with polynomial fits in the following way. A polynomial fit is not suitable for interpolation near the singularity at $s = s'$. We therefore multiplied the sparse matrix of 72 integrals by $(s - s')^3$, then interpolated with Newton forward–backward differences up to tenth order, and finally divided the results again by $(s - s')^3$. The errors obtained were smaller than the errors already present from the numerical integration. In order to find the weight factors, analogous to the quantities in square brackets on the right-hand side of (3.5), we fit Γ at each mesh point independently to a function of the form

$$\Gamma(s, s') \cong \left[\frac{1}{(s - s')^2} + \frac{1}{(s + s')^2} \right] \times \text{constant}. \tag{B.3}$$

We then integrated (B.3) analytically in the interval $s_i - \Delta s/2$ to $s_i + \Delta s/2$ (or 0 to $\Delta s/2$ for $s_i = 0$) to obtain the weight factor associated with the ith mesh-point s_i on the s' axis. The diagonal elements of this matrix of weight factors were found by using the sum rule

$$\int_0^\infty \Gamma(s, s')\,ds' = 0. \tag{B.4}$$

Forward difference method

In performing the time integration of equations (2.25) and (2.27), a standard forward difference method was used which was stable for $\Delta\tau < 0.5(\Delta s)^2$. Two kinds of tests were made to check the numerical accuracy of this program. First, the long-wavelength behavior was tested by inserting the translation mode F_T, given in (2.20), as an initial state and checking to see that this deformation remained stationary. Because of our finite difference approximations and our necessarily finite value of s_{max}, we always found a small drift of $F(\xi, \tau)$ away from F_T. But, for parameters used in actual runs, this drift time was always longer than any other characteristic times of interest.

Second, we checked the short-wavelength behavior by testing it against the analytic results obtained in II for vanishing surface tension. We inserted as initial states the approximate eigenfunctions: (II-4.22) with $\Omega = 1, 2$ in two dimensions; and (II-5.34) with $m = 0$, $\Omega = 2, 4$ in three dimensions; and then looked to see if the computed time development was in accord with the simple exponential growth laws predicted for these four states. In order to prevent numerical instability, we used a small ($\sim 10^{-4}$ or 10^{-3}) but nonvanishing value for σ. Under these conditions, we were able to observe the initial amplitudes growing by one or two orders of magnitude before faster modes became dominant. The Ω-values were reproduced to within 2% in two dimensions and 10% in three. Considering the intrinsic instability of the system at small surface tensions, we believe this to be a very sensitive test of all aspects of the numerical method, including the representation of the integral kernel.

Eigenvalue analysis

The linear operator $\mathscr{L}$ (see equation 3.8) was taken directly from the previously tested forward difference program. It was transformed to upper Hessenberg form and diagonalized with the QR-algorithm which produces a full set of complex eigenvalues and eigenvectors.

We already have discussed, in Section 3, the sensitivity of the eigenvalue analysis to the choice of s_{max} and Δs. An additional effect which we have found to be important is that, if s_{max} is chosen too large, then the sidebranch modes are so large in amplitude and so rapidly oscillating (in s) near s_{max} that large roundoff errors occur. Our testing procedures have been most thorough and generally more successful so far for the two-dimensional model. Here we have been able to identify the translation mode (with ReΩ small of order 10^{-2}) for almost all of the cases shown in Fig. 3. We have also varied N between 35 and 60, with fixed Δs, and have found an unsystematic variation of the maximum $\mathrm{Re}\Omega$ of about 20%, with a variation of the corresponding $\mathrm{Im}\Omega$ of about 5%. These variations presumably can be attributed to edge effects at s_{max}. In all cases, however, the important features remain qualitatively insensitive to s_{max} and Δs as long as these parameters are chosen within reasonable limits as described above. In particular, the crucial changeover from instability to marginal stability is a clear property of the system, and the approximate value of σ^* at which this happens seems to be identified within roughly the errors indicated in (3.13).

PHYSICAL REVIEW LETTERS

VOLUME 51 | 26 SEPTEMBER 1983 | NUMBER 13

Geometrical Approach to Moving-Interface Dynamics

Richard C. Brower
Lyman Laboratory of Physics, Harvard University, Cambridge, Massachusetts 02138

and

David A. Kessler
Theoretical Division, Los Alamos National Laboratory, Los Alamos, New Mexico 87544

and

Joel Koplik and Herbert Levine
Schlumberger-Doll Research, Ridgefield, Connecticut 06877
(Received 7 March 1983)

A general class of models is introduced which relate the motion of a phase boundary to properties of the local interfacial geometry. These systems can undergo successive destabilizations as they grow, possibly giving rise to nonequilibrium spatial patterns. This formalism has applications to a wide variety of physical problems, especially including dendritic solidification.

PACS numbers: 05.70.Ln

Many problems of physical interest involve the question of how a nonlinear system dynamically evolves in time. This is a particularly fascinating issue whenever the initial state is prepared in a configuration which has domains that are not absolute minima of the free energy. In this case, the interface motion is necessarily a nonequilibrium problem which has the possibility of giving rise to either chaotic final states or to intricate patterns. A characteristic feature of these problems is the competition between long-wavelength instabilities and a stabilization mechanism acting at the shortest scales.

A prototypical situation is the formation of dendritic ice crystals from a supercooled melt. The equations which govern the growth of the solid phase are the heat-diffusion equation for which the interface motion, through emission of latent heat, acts as a source, and the thermodynamic boundary condition that surface tension reduces the interfacial temperature by an amount proportional to the curvature. This problem has been studied extensively by Langer and co-workers[1,2] over the past few years. There exists evidence for the operation of a pattern-selection mechanism wherein the growing dendrite undergoes a series of successive destabilizations and subsequent restabilizations by the emission of "side branches." The associated hypothesis that this sort of nonlinear limit cycle causes the system to operate at the marginal stability point seems consistent with experimental results.[2] However, the detailed workings of this mechanism as well as side-branch spacings and global symmetries are not understood. Some other systems that exhibit similar competition between stabilizing and destabilizing forces are directional solidification,[3] fluid flow in a porous material,[4] and biological growth.[5]

The purpose of this work is to introduce a simple class of models in which we believe some of the essential features of the moving-interface problem may be isolated and explored. The key idea is to reduce the complicated dynamics of the entire system which is responsible for producing the domain structure to an equation for the interface motion which is a function only of the local geometry of the interface itself. In what follows, we will show that this simplification to a geometrical surface dynamics still preserves a rich and varied phenomenology. In particular, we examine a subset of models whose behavior bears a striking resemblance to several features of the full solidification problem.

Specifically, we assume equations of the general form

$$\hat{n}\cdot\partial\vec{x}/\partial t=\mathfrak{F}(\kappa,\nabla\kappa,\dots), \tag{1}$$

where $\vec{x}$ is the interface position, κ is the local curvature, $\hat{n}$ is the outward unit normal, and ∇ is the covariant derivative on the surface. $\mathfrak{F}$ is a general function of κ and its covariant derivatives. For simplicity, we will restrict the discussion here to the two-dimensional problem where the phase boundary is just a curve. Then ∇ is just $\partial/\partial s$, where s is the arclength. An attractive choice for $\mathfrak{F}$ is

$$\mathfrak{F}=F(\kappa)+\gamma\,\partial^2\kappa/\partial s^2, \tag{2}$$

where the second term stabilizes the system at short wavelengths for $\gamma>0$.

Equation (1) has the important property of reparametrization invariance. This means that the curve $\vec{x}$ can be arbitrarily parametrized without affecting the physical content of the time evolution. In fact, the phase boundary is a purely geometrical, stringlike[6] entity, totally independent of any coordinate grid used to describe it. One consequence of this is that the tangential component of $\partial\vec{x}/\partial t$ is not determined by the physics (as contained in $\mathfrak{F}$), but instead may be chosen for convenience. An often useful choice is the orthogonal gauge,[6] $\partial\vec{x}(t,\sigma)/\partial t=\hat{n}\mathfrak{F}$, where σ parametrizes the curve.

It is useful to rewrite the time-evolution equation (1) in terms of the curvature and the arclength. The arclength is related to the parametrization σ by $ds=g^{1/2}d\sigma$, where the metric $g=\vec{x}'\cdot\vec{x}'$, and the prime means derivative with respect to σ. Defining the curvature by $\kappa=\hat{n}\cdot\partial^2\vec{x}/\partial s^2$, one can easily derive

$$\dot{\kappa}=-(\partial^2/\partial s^2+\kappa^2)\mathfrak{F}(\kappa),\quad \dot{g}=2g\kappa\mathfrak{F}. \tag{3}$$

A full discussion of these equations as well as the generalization to three dimensions will be presented elsewhere.[7]

In order to motivate a choice for F in (2), we turn to the physics of the solidification problem. It is easy to argue that the growth rate will be enhanced by increasing the curvature of the surface. Schematically, a local outward bulge will allow latent-heat diffusion to occur more rapidly than a more planar interface. Moreover, an absolutely planar interface cannot move at all, other than for certain special values of the undercooling. Thus, F must vanish at the origin. The undercooling leads to an asymmetry between positive and negative curvature, suggesting a positive term quadratic in κ. Finally, the presence of a minimum nucleation size constrains F to become negative at large κ. This leads us to suggest that solidification could be modeled by our geometrical surface dynamics with F of the form

$$F(\kappa)=\kappa+\alpha\kappa^2-\beta\kappa^3. \tag{4}$$

We view this resulting model as a suggestion in the spirit of a Landau-Ginzburg mean-field approach.

Our model allows for analytic treatment directly parallel to that applied to the solidification problem, though with significantly greater ease. As there, an exact solution of the above model is a circle whose radius obeys

$$\dot{R}_0(t)=F(1/R_0). \tag{5}$$

Because of our choice of F, large enough circles will always expand. Consider, then, the stability of such a solution. To leading order, we can set

$$R(\theta)=R_0+\textstyle\sum_m\delta_m(t)\cos(m\theta), \tag{6}$$

where the initial circle is parametrized by the polar angle θ. A standard analysis yields the instantaneous growth rate

$$\dot{\delta}_m=\frac{m^2-1}{R_0^2}\left[F'\!\left(\frac{1}{R_0}\right)-\frac{\gamma m^2}{R_0^2}\right]\delta_m. \tag{7}$$

With $\gamma=0$, stability depends on the sign of F'. This is our analog of the Mullins-Sekerka instability.[8] Nonzero γ acts as a short-distance cutoff on the fluctuation spectra, making the dominant instabilities occur at intermediate length scales. A comparison of this expression with the results of the full solidification problem[2] shows that γ is analogous to the surface tension.

The full solidification problem is known to have a family of exact solutions in the limit of zero surface tension.[8] These dendritelike structures translate at a constant velocity related to their tip radius. Similarly, for $\gamma=0$ our equations also possess such solutions. For the case $F=\kappa$, the profile of the solution is given by

$$x=\rho\theta, \quad y=\rho\ln(\cos\theta), \tag{8}$$

where the parameter θ is just the angle that the normal vector makes with the $\hat{y}$ axis. Near the tip, this curve looks parabolic with arbitrary radius ρ. The velocity of translation along the $\hat{y}$ axis is given by $1/\rho$. A stability analysis along the lines of Muller-Krumbhaar and Langer[9] leads to the eigenvalue equation

$$-\cos^2\theta(\delta''+\delta')=\omega\delta \tag{9}$$

for the distortion $\delta(\theta)$ normal to the interface. This problem can be solved exactly by transforming it to a Schrödinger equation in the presence of a kink potential.[10] The eigenfunctions are given by $\delta\sim(\cos\theta)^{1/2}P_{1/2}{}^{ik}(\sin\theta)$, $\omega=\frac{1}{4}+k^2$, where P is an associated Legendre polynomial. All modes grow in time, representing the instability of the dendrite to tip splitting in the absence of a surface-tension term.

For nonzero γ, the situation is very different. Perturbation theory seems to predict a relatively insignificant steady-state shape correction. However, a direct integration of Eqs. (4) shows that perturbation theory breaks down at large distances from the tip.[11] This manifests itself in the presence of increasingly rapid oscillations in $\kappa(\theta)$ as θ approaches $\pm\frac{1}{2}\pi$. In the strict mathematical sense, there cannot exist uniformly translating solutions. This leads us to conjecture that a similar breakdown occurs in the solidification problem, contrary to earlier expectations.[12] Nonetheless, we believe it still may be useful, even for nonzero γ, to consider the linear mode analysis around the $\gamma=0$ solution. It is straightforward to show that there is a critical value of γ/ρ^2 above which all modes become stable. This point of marginal stability may be relevant for the long-time behavior of dendrites in our model, in analogy with Langer's suggestions for the solidification problem.[9] Perhaps the lack of a steady-state solution provides a new mechanism to drive the system to the marginally stable point.

One of the most useful features of our approach is the ease of doing computer experiments. The full nonlocal diffusion-limited boundary motion is extremely difficult to treat numerically, and this has been carried out to only a limited extent.[13] Our models, on the other hand, immediately lend themselves to simulations. We have done this by integrating the curvature equations (3) and then reconstructing the curve. Our programs were checked by varying the number of points and also by comparing the results to direct integration of Eq. (1). We will discuss the detailed results of our simulation studies elsewhere,[6] but we wish to point out here that this class of models can indeed give rise to interesting structures, including dendritic growth patterns. In Fig. 1, for example, the initial pattern stabilizes to a circle as it grows, but is then destabilized by (computer) noise and emits irregular dendrites. Figure 2 exhibits tip splitting and in Fig. 3 the dendrite initiates a side-branch excitation. We emphasize that (a) these curves represent the *initial* stages of growth and that (b) we do not claim that *all* observed features of solidification are reproduced by our specific choices of $\mathfrak{F}$ and its parameters.

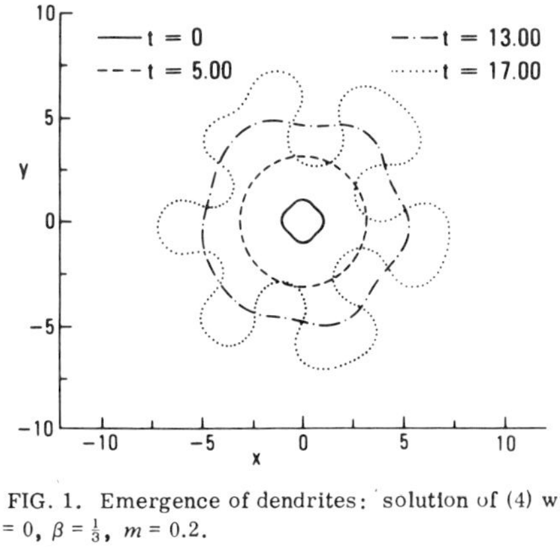

FIG. 1. Emergence of dendrites: solution of (4) with $\alpha=0$, $\beta=\frac{1}{3}$, $m=0.2$.

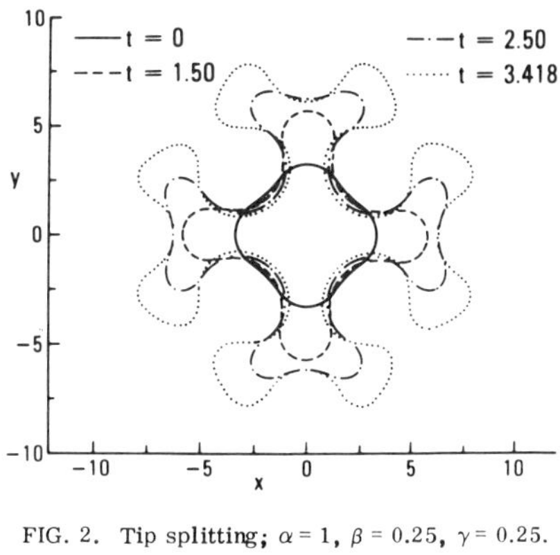

FIG. 2. Tip splitting; $\alpha=1$, $\beta=0.25$, $\gamma=0.25$.

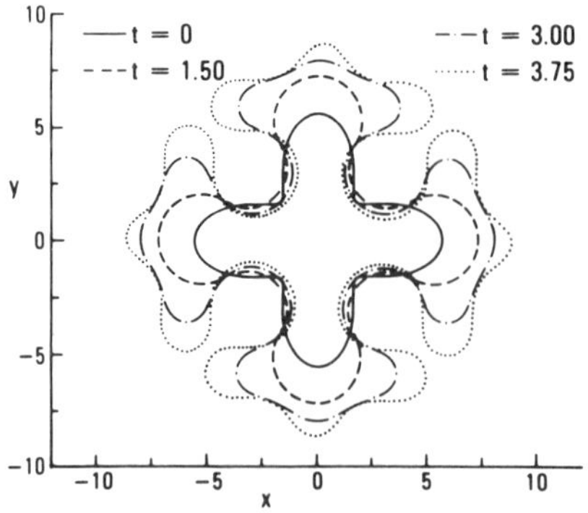

FIG. 3. Side branching; $\alpha = 1$, $\beta = 0.25$, $\gamma = 0.35$.

In fact, a crucial question in solidification is the importance of crystalline anisotropy, which may be required for repeated side branching.

The ice-crystal problem led us to consider $\mathfrak{F}$'s of the form given in (2). If we turn instead to the oscillations of a bubble of one liquid immersed in a second, immiscible liquid (e.g., oil in water), we must incorporate the fact that the area of the enclosed liquid cannot change, at least insofar as the fluid is incompressible. This will indeed be the case if we choose $\mathfrak{F}$ of the form $\partial V(\kappa)/\partial s + \gamma\, \partial^2\kappa/\partial s^2$, where V is some function of curvature. Likewise, a system whose phase boundary moves even for planar interfaces, such as a water flood displacing oil in a reservoir,[14] would necessitate a term in $\mathfrak{F}$ which would be nonzero at $\kappa = 0$. Finally, one could attempt to formulate more realistic models by allowing at least some nonlocality in $\mathfrak{F}$. Stable crystal growth has been studied in this manner,[15] and recently a model of this sort for dendritic growth has been independently proposed by Langer.[16]

In summary, our geometrical surface dynamics can be used as a starting point for investigating the question of nonequilibrium pattern formation in a variety of systems. We have demonstrated that for solidification, these models reproduce the Mullins-Sekerka instability and the tip-splitting behavior for the dendrite. We also conjecture that the lack of a steady-state solution for nonzero γ could lead to a new explanation for the marginal-stability–selection mechanism. Finally, some simple numerical experiments have convinced us that there is much to be understood here in the way of nonlinear phenomena. While this class of models cannot hope to reproduce all of the intricate details inherent in a full nonlocal approach (such as the lack of self-intersections), we are optimistic that a detailed study of these phenomena will provide much needed insight into the process of pattern formation.

We would like to thank Jim Langer for generously sharing with us his insight into the physics of these systems as well as some of his unpublished results on models which are quite similar (and in some cases identical) to the ones studied here.

[1]J. S. Langer and H. Muller-Krumbhaar, Acta Metall. 26, 1681, 1689, 1697 (1978).

[2]J. S. Langer, Rev. Mod. Phys. 52, 1 (1980), and Phys. Rev. A 27, 499 (1983); G. Dee and J. S. Langer, Phys. Rev. Lett. 50, 383 (1983).

[3]J. S. Langer, Phys. Rev. Lett. 44, 1023 (1980); G. Dee and R. Mathur, Institute for Theoretical Physics Report No. NSF-ITP 82-112 (unpublished).

[4]R. Chandler, J. Koplik, K. Lerman, and J. F. Willemsen, J. Fluid Mech. 119, 249 (1982).

[5]See, for example, *Mathematical Models of Molecular and Cellular Biology*, edited by L. A. Segel (Cambridge Univ. Press, Cambridge, England, 1980).

[6]There is an extremely large literature on reparametrization invariance in the context of the string model. For an introduction, see C. Rebbi, Phys. Rep. 12C, 1 (1974).

[7]R. C. Brower, D. A. Kessler, J. Koplik, and H. Levine, to be published.

[8]W. W. Mullins and R. F. Sekerka, J. Appl. Phys. 35, 444 (1964).

[9]H. Muller-Krumbhaar and J. S. Langer, Acta Metall. 29, 145 (1981).

[10]P. M. Morse and H. Feshbach, *Method of Theoretical Physics* (McGraw-Hill, New York, 1953), Vol. 2, Sec. 12.3.

[11]We thank J. Langer for pointing out this possibility.

[12]G. E. Nash and M. E. Glicksman, Acta Metall. 22, 1283 (1974).

[13]R. C. Brower, D. A. Kessler, and H. Levine, unpublished; J. B. Smith, J. Comput. Phys. 39, 112 (1981).

[14]For an introduction to this topic, see A. E. Scheidegger, *The Physics of Flow in Porous Media* (Univ. of Toronto Press, Toronto, 1974), 3rd ed.

[15]H. Muller-Krumbhaar, T. Burkhardt, and D. Kroll, J. Cryst. Growth 38, 13 (1977).

[16]J. S. Langer, private communication.

PHYSICAL REVIEW A VOLUME 31, NUMBER 3 MARCH 1985

Geometrical models of interface evolution. III. Theory of dendritic growth

David A. Kessler
Department of Physics, Rutgers University, Piscataway, New Jersey 08854

Joel Koplik and Herbert Levine
Schlumberger-Doll Research, Old Quarry Road, Ridgefield, Connecticut 06877-4108
(Received 26 June 1984)

We construct a theory of velocity selection and tip stability for dendritic growth in the local evolution model. We show that the growth rate of dendritic patterns is determined by a nonlinear solvability condition for a translating finger. The sidebranching instability is related to a single discrete oscillatory mode about the selected velocity solution, and the existence of a critical anisotropy is shown to be due to the zero crossing of its growth rate. The marginal-stability hypothesis cannot predict the correct dynamics of this model system. We give heuristic arguments that the same ideas will apply to dendritic growth in the full diffusion system.

I. INTRODUCTION

In a previous series of papers[1,2] (hereafter referred to as I and II) we have constructed a local geometrical evolution model for the process of dendritic crystal growth. This model assumes that the interface $\mathbf{x}(s)$ between the growing solid and the supercooled melt satisfies the equation

$$\hat{\mathbf{n}}\cdot\frac{d\mathbf{x}}{dt}=\left[\kappa+A\kappa^2-B\kappa^3+\frac{d^2\kappa}{ds^2}\right][1+\epsilon\cos(m\theta)]\ , \qquad (1)$$

where $\hat{\mathbf{n}}$ is the curve normal making an angle θ with the $\hat{y}$ axis, s is the arclength, $\kappa=d\theta/ds$ is the curvature, and A and B are parameters representing the undercooling and minimum bubble size, respectively. The term proportional to ϵ is due to crystal anisotropy, reflecting enhanced growth along the m-fold symmetry axes. In this paper we will restrict ourselves to $m=4$, but we have seen in II that the results do not change qualitatively with m. For a detailed discussion of (1), see Refs. 1 and 2.

Let us briefly review the phenomenology of Eq. (1) as it relates to dendritic growth. By a dendrite[3] we mean a moving tip which exhibits oscillatory behavior connected to the emission of sidebranches, which in turn behave as new dendritic tips. A typical example of this type of growth is depicted in Fig. 1. In II we demonstrated that this model gives rise to stable dendritic growth if and only if the anisotropy ϵ is greater than a critical anisotropy ϵ_c depending on A, B, and m, but not on the initial conditions. We discussed how the "marginal-stability hypothesis" of Langer and Müller-Krumbhaar[4] was in contradiction with our results and concluded that further theoretical understanding of dendritic growth was required.

The purpose of this paper is to present a new and comprehensive approach to the velocity selection and tip-stability problems in this type of system. We will show that the continuous family of steady-state solutions with arbitrary velocity present in the absence of surface tension breaks down to a discrete set of solutions, each with a unique velocity. (A similar behavior was discovered in a related context by McLean and Saffman, and Vanden-Broeck.[5] This will be discussed later.) The largest of these velocities is most likely to give rise to stable growth (because it can, so to speak, outrun instabilities) and it therefore is dynamically "selected." Whether or not this solution is indeed stable is then answered by a study of the eigenvalue spectrum of the linear-stability operator. As ϵ crosses from below $\epsilon_c(A,B,m)$ to above, the real part of the leading discrete complex eigenvalue becomes negative, signifying tip stability in the moving frame of reference.

The major question raised by our work is how applicable our results are to more realistic solidification models incorporating global diffusion-driven dynamics. We believe that there is an excellent chance that our selection mechanism *directly* carries over to these other systems, so

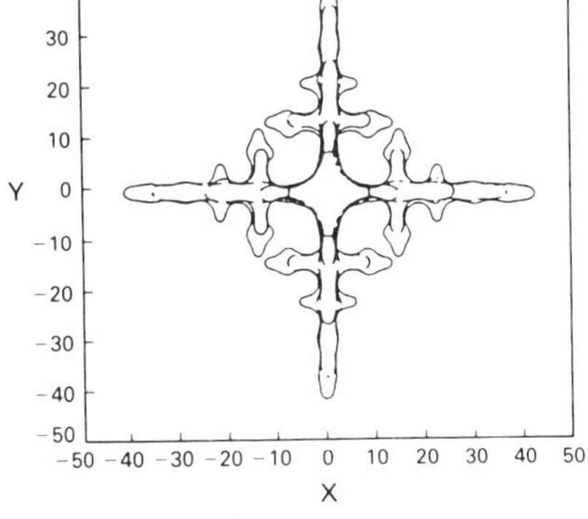

FIG. 1. Example of dendritic growth in the local evolution model ($A=4.0$, $B=1.5$, $\epsilon=0.15$).

Reprinted with permission from *Physical Review A*, Vol. 31, No. 3, 1985.

that the marginal-stability hypothesis would not apply to these cases. It would remain valid for the description of certain one space-dimensional partial differential equations,[6] and perhaps in other contexts as well. In the last section of this paper we present our reasons for believing this to be the case. We are currently attacking this problem by a combination of numerical and analytical methods and hope to report on this subject in the near future.

II. STEADY-STATE SOLUTIONS

In I we discussed how our local model, in the absence of the $d^2\kappa/ds^2$ "surface-tension" term, exhibited a continuous family of steady-state dendritic solutions corresponding to arbitrary tip velocities. To see this, note that $\hat{n}\cdot d\mathbf{x}=V\cos\theta$ for any uniformly translating solution. Equation (1) then reduces to a third-order algebraic equation for $\kappa=d\theta/ds$. We can integrate this equation from $s=0$, $\theta=0$; $\kappa(0)$ is determined algebraically and $\kappa'(0)$ equals zero automatically. For large arclength, the solution approaches

$$\frac{V\cos\theta}{1+\epsilon}=\kappa \tag{2}$$

which integrates to the asymptotic shape

$$y=\rho\ln\cos(x/\rho)\ , \tag{3}$$

where $\rho=(1+\epsilon)/V$. The dendrite approaches the fixed point $\theta=\pi/2$ exponentially, regardless of the value of the velocity. This family of solutions is the model's analog of the Ivantsov needle crystal.[7]

When we include the surface-tension term $d^2\kappa/ds^2$, we argued in I that, in general, there would be no steady-state solution at arbitrary V. This comes about because of an incompatibility of the condition $\kappa'(0)=0$ with the required approach to the $\theta=\pi/2$ fixed point. Defining $\beta=d\kappa/ds$, the evolution equation can be written as the autonomous system

$$\frac{d\kappa}{ds}=\beta\ ,\quad \frac{d\theta}{ds}=\kappa\ ,$$

$$\frac{d\beta}{ds}=\frac{V\cos\theta}{1+\epsilon\cos(m\theta)}-\kappa-A\kappa^2+B\kappa^3\ .$$

There is a fixed point at $\theta=\pi/2$, $\kappa=\beta=0$, which the solution should approach for large s. Linearizing about this fixed point and assuming that the variables behave as $e^{-\alpha s}$, we find the eigenvalue equation $V/(1+\epsilon)=\alpha+\alpha^3$. One root is real and positive, corresponding to the exponential approach to the fixed point. To find the other roots, if we let $\alpha=\alpha_r+i\alpha_i$, then

$$0=\alpha_i+\alpha_i(3\alpha_r^2-\alpha_i^2)\ ,\quad \frac{V}{1+\epsilon}=-2\alpha_r(1+4\alpha_r^2)\ .$$

These roots are complex conjugate and, since $\alpha_r<0$, they correspond to eigenvectors which exponentially diverge going down the dendrite. Thus, the requirement that κ, θ, and β approach the fixed point determines a unique trajectory, all the way back to $\theta=0$, which we hereafter label as $s=0$. One therefore expects, in general, that $\beta(0)=\kappa'(0)\neq 0$ and so a well-behaved symmetric dendrite does not exist. In I we showed that for $A=B=\epsilon=0$ there were in fact no solutions of the steady-state equation.

To further investigate the issue of steady-state solutions, we picked initial conditions such that we sat exactly on the converging eigenvector in the near vicinity of the fixed point (i.e., $\kappa=10^{-10}$). We then integrated backward in s until θ equalled zero, and determined β. If β crosses zero as the velocity is varied, we can identify a *discrete* set of possible velocities for uniformly translating shapes. A typical plot of $\beta(V)$ is shown in Fig. 2. Notice that there is, in general, a discrete set of solutions culminating in a maximal velocity solution V^*. As we will discuss later, increasing V always corresponds to increased stability; V^* is thus the most stable (but not necessarily stable) steady-state dendrite.

In Fig. 3 we plot V^* as a function of A for fixed B and ϵ, ϵ for fixed A and B, and B for fixed A and ϵ. This graph is consistent with the idea that A represents the undercooling. The larger the undercooling, the larger V^*. In fact, for $B=\epsilon=0$, $V^*\to 0$ as $A\to 0$. This is consistent with the aforementioned absence of solutions in the special case considered in I.

The key fact we wish to state in this section is that V^* is *exactly* equal to the average tip velocity as given by our numerical simulations in II. We have checked this for a wide variety of cases involving variations of all parameters A, B, and ϵ. The velocity-selection mechanism in this system is thus simply the solvability condition for the steady-state equation. The maximal velocity is selected due to its being the most stable with respect to tip-splitting perturbations. We will see this in detail in Sec. III.

Our results are in direct contrast to those predicted by the marginal-stability hypothesis as it is usually applied to dendrite-forming systems. In this approach the system "chooses" a velocity such that the steady-state solution is

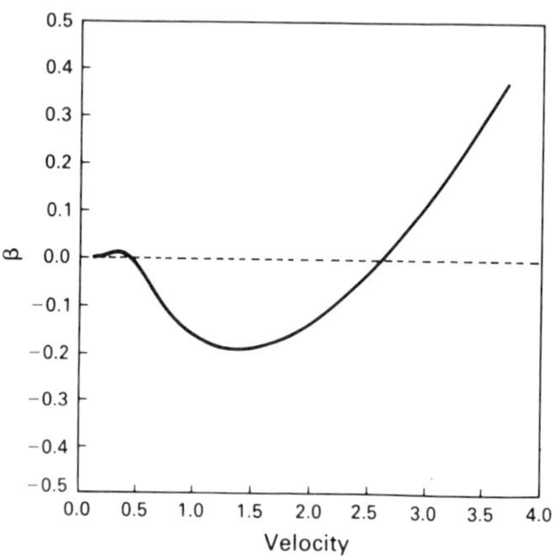

FIG. 2. $\beta(0)$ as a function of velocity for steady-state solutions.

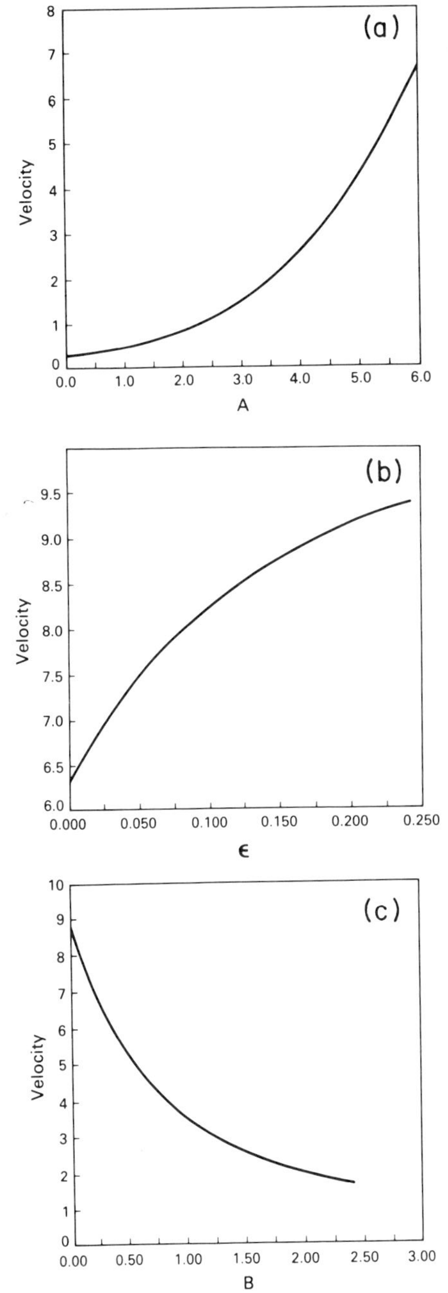

FIG. 3. Maximal velocity V^* as a function of (a) A at $B=1.5$ and $\epsilon=0.15$, (b) ϵ at fixed $A=4.0$ and $B=0.0$, and (c) B at fixed $A=4.0$ and $\epsilon=0.15$.

marginally stable to small perturbations. In our theory stability is only of concern once the velocity has been determined by a nonlinear eigenvalue constraint. Our system can and does operate away from the marginally stable point. Our analysis does, however, bear some resemblance to the ideas which underlie the work of Aronson and Weinberger.[6] In particular, we use a discrete "maximum velocity" principle, whereas previous work used a continuous version of that idea. We will later present evidence that our version stands a better chance of carrying over to more complete solutions of the dendrite problem.

III. STABILITY ANALYSIS

In II we showed that dendritic growth would be stable only for $\epsilon > \epsilon_c$. In particular, we plotted the tip velocity as a function of time and followed the change from growing to decaying oscillations as the anisotropy was increased. We now show that our solvability mechanism for velocity selection can also explain these results.

Let us assume that the curve $\mathbf{x}(s)$ is given by a small perturbation around the steady-state solution $\mathbf{x}=\mathbf{x}_0+\hat{\mathbf{n}}\delta(s)$. It is easy to see that to linear order $\theta=\theta_0-\delta'$, $\kappa=\kappa_0-\delta''-\kappa_0^2\delta$ (where the prime means d/ds_0). Substituting these expressions into (1), we derive the linear-stability equation

$$\frac{\dot{\delta}}{1+\epsilon\cos(m\theta)}=-\delta''''+C_2\delta''+C_1\delta'+C_0\delta\ , \qquad (4a)$$

where

$$C_0=-\kappa_0^2U'(\kappa_0)-4\kappa_0\kappa_0''-3(\kappa_0')^2\ ,$$
$$C_1=-5\kappa_0\kappa_0'+V[U'(\kappa_0)+\kappa_0''']/\kappa_0\ , \qquad (4b)$$
$$C_2=-U'(\kappa_0)-\kappa_0^2\ ,$$

with $U(\kappa)\equiv\kappa+A\kappa^2-B\kappa^3$. We now look for solutions of the form $\delta_k(t)\sim e^{\omega_k t}\delta_k(s)$. Stable dendrites require $\mathrm{Re}\,\omega_k<0$ for all eigenvalues k.

Fourth-order stability equations are a common feature of boundary-layer problems in fluid mechanics, and our approach to this equation will be modeled on the work of Mack[8] on the Orr-Sommerfeld equation. In such situations there can occur both discrete and continuous eigenvalue spectra, corresponding to different ways in which the eigenfunctions δ can satisfy the boundary conditions imposed on the perturbation. We will see shortly how this works in our case.

Let us consider the asymptotic (i.e., large s) form of Eq. (4), which can be written in the form

$$\frac{\omega\delta}{1+\epsilon}=-\delta''''-\delta''-\frac{V\sin\theta}{1+\epsilon}\delta'\ , \qquad (5)$$

where $\sin\theta=\pm1$ depending on which side of the tip we are on. Assuming $\delta\sim e^{qs}$, we find the asymptotic dispersion relation

$$\frac{\omega}{1+\epsilon}=-q^4-q^2+\frac{Vq}{1+\epsilon}\ .$$

We label the four roots in decreasing order of their real parts and try to form a solution which satisfies two (arbitrary) boundary conditions at $s=-L$ $(\rightarrow-\infty)$. To d this we must pick coefficients in the expansion

$$\delta = \sum_{i=1}^{4} \delta_i e^{q_i s}$$

to ensure three terms of the same approximate magnitude at $s=-L$. This requires $\delta_4 \sim e^{-(q_2-q_4)L}\delta_2$ and $\delta_3 \sim e^{-(q_2-q_3)L}\delta_2$, where $\text{Re}(q_2-q_4) \geq \text{Re}(q_2-q_3) \geq 0$.

Now, these solutions can be extended to the region around $s=0$ where $e^{q_i s}$ is replaced by an $O(L^0)$ function $\phi_i(s)$, and we must impose $\delta'(0)=\delta'''(0)=0$. In general, we only have two coefficients δ_1 and δ_2 at our disposal and this cannot be done. The only possibilities for satisfying the boundary conditions are $\text{Re}q_2=\text{Re}q_3$, or an accidental degeneracy in the $s=0$ conditions. The first gives rise to a continuous spectrum and the second to a discrete set of eigenvalues.

The first possibility was investigated in I with the following result: the most unstable mode crossed below $\text{Re}\omega=0$ at the point when $V^2(1+\epsilon)^2 \cong 2.65$. This velocity is much lower than, say, the actual velocity selected at $A=4$, $B=+1.5$, $\epsilon=0.14$ (see Fig. 3) which in II was shown to be unstable. This suggests that discrete modes exist above the continuum. For this to occur a degeneracy is needed and the above argument yields the condition

$$\Psi(\omega) \equiv \det \begin{vmatrix} \phi_1'(0) & \phi_1'''(0) \\ \phi_2'(0) & \phi_2'''(0) \end{vmatrix} = 0 . \tag{6}$$

To find Ψ we must integrate the stability equation from large negative s and find the continuation of the solutions which behave asymptotically as $e^{q_{1,2}s}$. First we found the selected velocity to 10^{-6} accuracy and identified the coefficients C_0, C_1, C_2 by spline interpolation using values every 0.01 arclength units. Then we used a predictor-corrector ordinary differential equation (ODE) solver to integrate the two solutions ϕ_1 and ϕ_2 forward from $s=10^{-10}$ towards $s=0$. In addition, after every $\Delta s=3$ step we orthogonalized ϕ_2 with respect to ϕ_1 in order to reduce the error due to an admixture of an exponentially growing ϕ_1 piece in the ϕ_2 solution. We checked the accuracy of our algorithm by using a slightly different ODE solver, and we estimate that our procedure determines Ψ to at worst a several percent error throughout the region of interest.

Typical contour plots of $\text{Re}\Psi$ and $\text{Im}\Psi$, for ϵ slightly below and slightly above ϵ_c, are presented in Figs. 4(a) and 4(b), respectively. Notice that there is only one zero in Ψ and this determines a unique eigenvalue ω^*. Of particular importance is the fact that $\text{Re}\omega^*$ moves from above zero to below zero as ϵ goes from 0.14 to 0.16. This is in agreement with an estimate of $\epsilon_c \sim 0.15$ for this set of A and B parameters made in II. In Fig. 5 we plot $\text{Re}\omega^*$ and $\text{Im}\omega^*$ versus ϵ determined by iterating towards $\Psi=0$. These numbers agree quantitatively with the growth rate and frequency of the tip velocity in our numerical simulations reported in II. Thus the stability analysis around our preselected velocity accurately describes the oscillations around constant tip translation.

Several comments are in order. Consider a typical case of $\epsilon > \epsilon_c$. The actual small oscillation eigenvector is a particular linear superposition of the ϕ_1 and ϕ_2 solutions. It turns out for reasons that we do not fully understand that the wave vector q_2 always has a small positive real part (corresponding to exponential *growth*), whereas $\text{Re}q_1$ is large and negative. Asymptotically ϕ_2 dominates, giving a sidebranch wave vector of $\text{Im}q_2$. Since $\text{Re}c_2 \sim 0$, Eq. (5) predicts

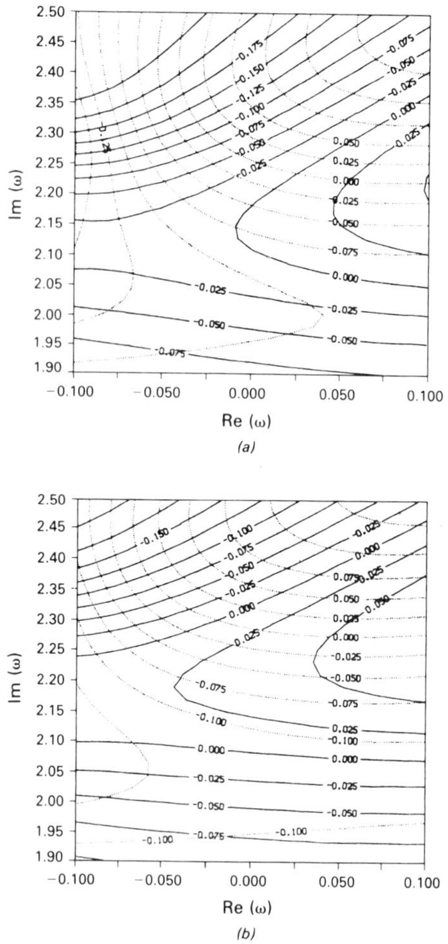

FIG. 4. Contour plot of $\Psi(\omega)$ for (a) $\epsilon=0.14$, (b) $\epsilon=0.16$. The solid lines are $\text{Re}\Psi$ and the dotted lines are $\text{Im}\Psi$.

$$V\,\text{Im}q_2 \sim -\text{Im}\omega^* . \tag{7a}$$

This is exactly the requirement that side branches are almost stationary when viewed in the laboratory frame of reference. Furthermore, it seems to be always true that

$$V\,\text{Re}q_2 + \text{Re}\omega^* > 0 \tag{7b}$$

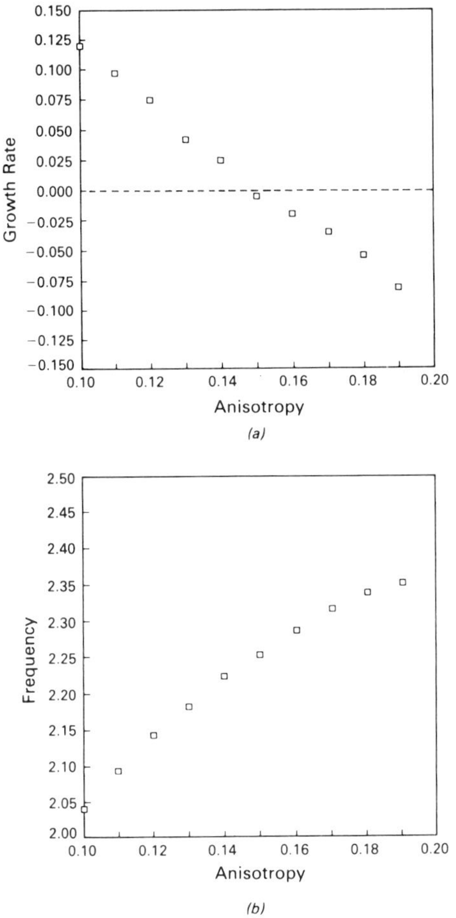

FIG. 5. Sidebranching eigenvalue as a function of ϵ; (a) real part (tip-perturbation growth rate) and (b) imaginary part (tip-oscillation frequency).

even for stable ($\mathrm{Re}\,\omega^* < 0$) tip oscillations. This predicts that in the *laboratory frame* sidebranches continue to grow even as the tip velocity spirals down to a constant value (while in the reference frame of the moving tip they are left behind). Although the velocity is selected by convergence onto the steady-state solution, the asymptotic shape always contains an infinite train of side branches. This behavior does not quite correspond to that observed in dendritic crystals,[9] however, where sidebranching always exhibits neutral stability in the tip frame remaining a constant distance behind the tip.

In the cases we have investigated so far, all of the solutions with the exception of V^* have velocities which lie well below the stability point $\mathrm{Re}\,\omega^* = 0$. It is reasonable to

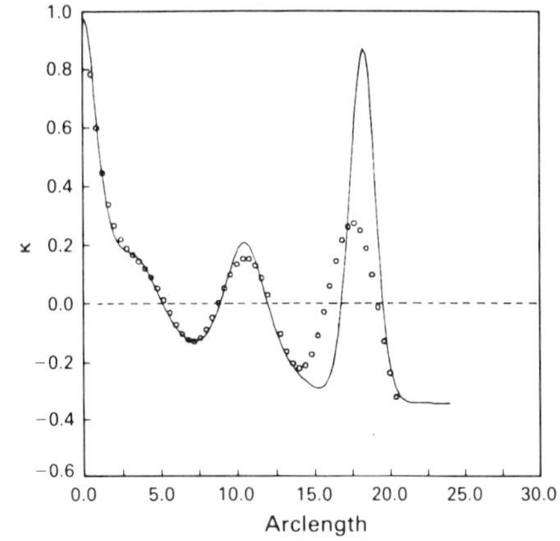

FIG. 6. Curvature plots: the solid line is the result of a direct numerical integration of (1), and the dots are the sum of the steady-state shape (2) and the leading linear perturbation mode.

speculate that this is always the case, that it is impossible for a secondary solution to be even approximately stable. If not, a system might in general be able to support several different modes of dendritic growth, the selection in a given case depending on the initial conditions.

We conclude this section with a comparison in Fig. 6 of a curvature plot generated by our numerical simulation program described in II, with a steady-state shape plus a linear perturbation approximated by its asymptotic form. We fix the complex amplitude of the perturbation by matching the first curvature minimum. Even with this rather crude approximation for the eigenvector, the system is clearly being described properly by our analysis. It is quite clear, for example, how the experimentally measurable sidebranch spacings are determined by the wave vector $\mathrm{Im}\,q_2$. At very large values of s, additional nonlinearities must be taken into account for a full description of the sidebranch dynamics.

IV. DISCUSSION

Let us summarize our findings. We have shown how the detailed phenomenology of dendritic growth in the local, geometrical evolution model can be understood as originating from a solvability condition selecting a discrete velocity and a discrete eigenmode around this velocity controlling the stability of the solution. In consequence, there is a critical anisotropy required to maintain stable dendritic growth. These findings represent a major success for the idea of using simplified models for growth processes. Because of the tractability of our equations, we have been able to propose a new approach to analyzing the dendritic growth patterns of our model.

We would now like to argue that a similar scenario should be valid for true diffusion-controlled evolution:[3] an interface coupled to a temperature field satisfying the diffusion equation and the Gibbs-Thompson boundary condition $T_I = T_M(1 - d_0\kappa)$, where T_M is the melting temperature and d_0 is the capillary length. Our prediction is that the velocity is selected by the breakdown of the Ivantsov family of steady-state solutions, under the inclusion of surface tension, to a set of discrete solutions. The repeated sidebranching about a stable tip observed experimentally would be attributed to a new discrete eigenmode of the linear-stability operator. The stability of the tip with respect to splitting may require a critical anisotropy, depending on the magnitude of the dimensionless undercooling or Peclet number. (Note, however, that the detailed sidebranch behavior in the supercritical case does not entirely correspond to that found in our local model.) We know of no experimental data that could not, in principle, be explained by this type of approach.

In Ref. 10 we showed that the usual demonstrations that the Ivantsov solution could be extended to include nonzero surface tension were in fact inadequate to answer the question. The approach used there is reminiscent of the situation here: the equation near the tip cannot in general be integrated to have the proper approach to the zero-curvature fixed point. This fact is highly suggestive of the emergence of a solvability condition. A further indication of the universality of this behavior is afforded by the boundary-layer model of Ben-Jacob *et al.*[11] An analysis similar to the one performed in Sec. II can be used to show that there exists a discrete set of steady-state solutions,[12] although numerical simulations of this model have not advanced to the stage where they could be used to test our theory. Note that there is one aspect of dendritic growth that is reproduced more accurately by this type of boundary-layer model than by the local geometrical model studied here. The dendritic solution here approaches its asymptotic shape exponentially as opposed to the power-law behavior expected for the true Ivantsov problem and the boundary-layer model. The fact that this type of solvability constraint is also valid for the boundary-layer system,[11,12] as well as our earlier remarks regarding the breakdown of the Ivantsov solution, lead us to expect that this difference is not crucial for the selection mechanism.

Another pattern-selection problem with strong similarities to dendritic growth is Saffman-Taylor fingering.[13] In that problem, when one fluid displaces a more viscous one in a narrow gap between parallel plates or a porous medium, one finds in laboratory or numerical[14] experiments that the displacing fluid eventually forms a single finger which advances with constant velocity. Theoretically, there exists an infinite family of solutions which breaks down to a discrete set[5] when surface tension is included. Recent calculations[15] have shown that the velocity is determined by a solvability condition and again, the phenomenology is quite similar to what occurs in the geometrical model.

Finally, we would like to discuss some recent numerical simulations of the diffusion equation in the zero Peclet-number limit.[16] The early time evolution of the system has a finite-sized dendrite, and aside from a slow decrease in the tip velocity with time,[17] the system should behave as a standard crystal growth system. The fact that the system did not operate at a "marginally stable point," but instead underwent a transition from unstable to stable sidebranch emission as the amount of anisotropy varied, is further evidence for the validity of the mechanism we propose here.

ACKNOWLEDGMENTS

We would like to acknowledge useful discussions with R. F. Dashen, L. Kadanoff, and Y. Pomeau and would like to thank H. Aref and Y. Pomeau for sharing some of their results regarding Saffman-Taylor fingering prior to publication.

[1]R. C. Brower, D. Kessler, J. Koplik, and H. Levine, Phys. Rev. A **29**, 1335 (1984).

[2]D. Kessler, J. Koplik, and H. Levine, Phys. Rev. A **30**, 3161 (1984).

[3]For a review, see J. S. Langer, Rev. Mod. Phys. **52**, 1 (1980).

[4]J. S. Langer and H. Müller-Krumbhaar, Acta Metall. **26**, 1681 (1978); **26**, 1689 (1978); **26**, 1697 (1978).

[5]J. W. McLean and P. G. Saffman, J. Fluid Mech. **102**, 455 (1981); J.-M. Vanden-Broeck, Phys. Fluids **26**, 2033 (1983).

[6]D. G. Aronson and H. F. Weinberger, Adv. Math. **30**, 33 (1978); G. Dee and J. S. Langer, Phys. Rev. Lett. **50**, 383 (1983); J. S. Langer and H. Müller-Krumbhaar, Phys. Rev. A **27**, 499 (1983).

[7]G. P. Ivantsov, Dokl. Akad. Nauk SSR **58**, 567 (1947).

[8]L. M. Mack, J. Fluid Mech. **73**, 497 (1976).

[9]S. C. Huang and M. E. Glicksman, Acta Metall. **29**, 701 (1981); **29**, 717 (1981).

[10]R. C. Brower, D. Kessler, J. Koplik, and H. Levine, Scr. Metall. (to be published).

[11]E. Ben-Jacob, N. Goldenfeld, J. S. Langer, and G. Schön, Phys. Rev. Lett. **51**, 1930 (1983); Phys. Rev. A **29**, 330 (1984).

[12]D. Kessler and H. Levine (unpublished).

[13]P. G. Saffman and G. I. Taylor, Proc. R. Soc. London, Ser. A **245**, 312 (1958).

[14]H. Aref (private communication).

[15]Y. Pomeau (private communication).

[16]D. Kessler, J. Koplik, and H. Levine, Phys. Rev. A **30**, 2820 (1984).

[17]J. S. Langer (unpublished).

Existence of needle crystals in local models of solidification

J. S. Langer
Institute for Theoretical Physics, University of California, Santa Barbara, California 93106
(Received 31 July 1985)

The way in which surface tension acts as a singular perturbation to destroy the continuous family of needle-crystal solutions of the steady-state growth equations is analyzed in detail for two local models of solidification. All calculations are performed in the limit of small surface tension or, equivalently, small velocity. The basic mathematical ideas are introduced in connection with a quasilinear, isotropic version of the geometrical model of Brower *et al.*, in which case the continuous family of solutions disappears completely. The formalism is then applied to a simplified boundary-layer model with an anisotropic kinetic attachment coefficient. In the latter case, the solvability condition for the existence of needle crystals can be satisfied whenever the coefficient of anisotropy is arbitrarily small but nonzero.

I. INTRODUCTION

There is growing evidence that pattern selection in local models of dendritic solidification is closely connected with the existence of so-called "needle-crystal" solutions of the steady-state equations of motion. In both the geometrical[1,2] and boundary-layer[3,4] models that have been introduced recently, as well as in more realistic nonlocal models of solidification with diffusion control,[5] there exists a continuous family of "Ivantsov"[6,7] needle crystals when surface tension is omitted. The addition of surface tension, however, is a singular perturbation of these systems which, at least in the local versions—that is, the geometrical and boundary-layer models—destroys the family of solutions and opens the possibility that sharp selection of growth rates, tip radii, etc., occurs via a solvability condition. It seems possible that a similar phenomenon occurs in the fully nonlocal problem.

The purpose of the present paper is to examine in some detail the breakdown of the Ivantsov solutions in both of the local models. The way in which this breakdown occurs provides some interesting clues about the roles played by surface tension and crystalline anisotropy in the dendrite theory. Moreover, this singular perturbation problem is of mathematical interest in its own right. As we shall see, the method of solution proposed here seems reasonable but is not rigorous or even systematic. It can, however, be tested by direct numerical computations; and the results of such tests make it plausible that the analytic approximation captures the essential features of the relevant phenomena. Indeed, as this paper is being written, it appears that there may be important new progress both in understanding the mathematical nature of the problem as posed here[8] and in showing that the basic features of this model problem also appear in the more realistic systems.[9–11] This paper has therefore been organized to serve as a starting point for study of these later developments.

We shall begin in Sec. II by looking at the simplest nontrivial version of the geometrical model in order to describe the mathematical strategy with a minimum of unnecessary complication. That strategy will be outlined in Sec. III. In Sec. IV we shall apply this method to a minimal version of the boundary-layer model, and shall see there how crystalline anisotropy may play an essential role in the selection of needle-crystal solutions. Some mathematical and computational details pertaining to both models are relegated to an appendix.

II. THE QUASILINEAR GEOMETRICAL MODEL

In all of the following, we shall restrict our attention to two-dimensional local models in which a moving one-dimensional interface is described by specifying its curvature $K=\partial\theta/\partial s$ as a function of arc length s. The relevant geometry is illustrated in Fig. 1. The basic assumption of the geometrical models[1,2] is that v_n, the normal velocity of the interface, is a function only of K and its even derivatives with respect to s. The steady-state condition is simply

$$v_n(K,d^2K/ds^2,\dots)=v\cos\theta\ , \tag{2.1}$$

so that the interface is moving at constant velocity v, without change in its shape, in a fixed direction. A needle crystal, by definition, is a solution of (2.1) that has the general form shown in Fig. 1 in which

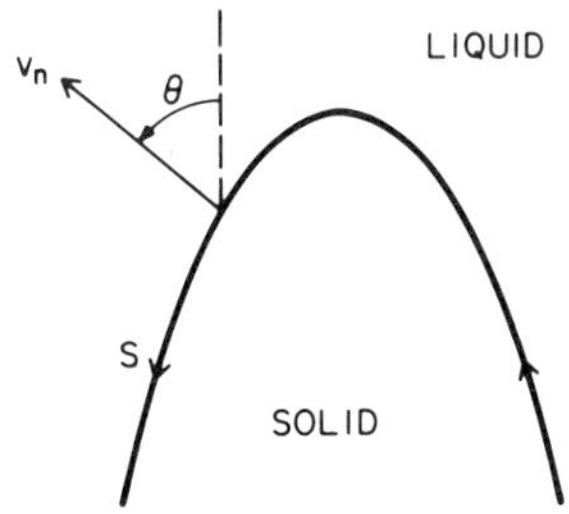

FIG. 1. Geometry of the needle crystal.

Reprinted with permission from *Physical Review A*, Vol. 33, No. 1, 1986.

$$K \to 0, \quad \theta \to \pm\pi/2 \quad \text{as } s \to \pm\infty \; . \tag{2.2}$$

The analog of the Ivantsov limit in the geometrical model occurs when we set v_n to be a function of K alone. If (2.1) is piecewise invertible for K as a function of θ, then $K(\theta)=d\theta/ds$ is a first-order ordinary differential equation which generally is solvable for θ as a function of s. We may further expect that these solutions will satisfy the needle-crystal conditions (2.2) for some continuous range of values of the parameter v. For example, the simplest possible choice is $v_n = K$, for which it turns out that

$$\theta = \cos^{-1}(\operatorname{sech} vs) \; , \tag{2.3}$$

a relation which satisfies (2.2) for all v. The family of Ivantsov solutions is conventionally characterized by the relation between v and the tip radius $R = K^{-1}(\theta=0)$, which in this case is trivially $Rv = 1$.

The above solutions are all strongly unstable. In fact, the model as it stands is not even dynamically well defined because the amplification rate of deformations is unbounded at short wavelengths. In order to produce a model whose time dependence is meaningful, it is necessary to add a term which mimics the effect of surface tension and stabilizes the system at short wavelengths. The easiest way to do this is to write

$$v_n = K + \gamma^2 \frac{d^2K}{ds^2} \; , \tag{2.4}$$

where γ plays the role of a capillary length. Equation (2.4) defines what might be called a quasilinear geometrical model; the right-hand side is linear in K. This model does not actually produce dendrites, but its dynamical behavior is well defined and quite interesting in some respects.

Our aim now is to see what effect the second derivative in (2.4) has on the needle-crystal solutions (2.3). To do this, it is convenient first to rewrite the steady-state equation (2.1) in the form

$$\kappa = \cos\theta - \nu \frac{d^2\kappa}{d\xi^2} \; , \tag{2.5}$$

where $\nu = (\gamma v)^2$, $K = v\kappa$, and $\xi = vs$. Equivalently,

$$\kappa = \cos\theta - \frac{\nu}{2}\kappa \frac{d^2\kappa^2}{d\theta^2} \; . \tag{2.6}$$

The latter form, in which ξ has been replaced by θ as the independent variable by using $\kappa = d\theta/d\xi$, is very useful but can produce spurious difficulties unless κ is everywhere non-negative.

Equations (2.5) and (2.6) can be used to illustrate some points that have been made in previous papers.[2,3] If we iterate the right-hand side of (2.6), we generate a series expansion for κ in powers of ν and $\cos\theta$. Each term in this series is consistent with the needle-crystal conditions (2.2), and the series appears to produce an accurate estimate for κ especially far down the needle where $\cos\theta$ is small. However, we know that this series can be at best asymptotic. To see this, write (2.5) as a set of three coupled equations for a ξ-dependent trajectory in the space of variables θ, $\kappa = d\theta/d\xi$, and $\lambda = d\kappa/d\xi$. As has been noted previously,[2] only one trajectory emerges from the fixed point at $\theta = -\pi/2$, $\kappa = \lambda = 0$; similarly, only one trajectory enters the reflected fixed point at $\theta = +\pi/2$. For a needle crystal to exist, these two pieces must belong to a single trajectory which joins the fixed points and, by symmetry, passes through some point on the κ axis with $\theta = \lambda = 0$. There is no special reason for this to happen. In general, the trajectories entering or leaving the fixed points, even if well approximated by some finite number of terms of the above series in powers of ν, need not reach $\theta = 0$—the tip of the needle—with $\lambda = d\kappa/d\xi = 0$. Our goal in what follows is to compute λ at $\theta = 0$ explicitly and to discover under what conditions it might vanish.

As we shall see immediately, the mathematical problem posed above is highly nontrivial. To make some progress analytically, we shall look only at small values of ν; that is, we shall look in the neighborhood of the known solution of (2.6), $\kappa_0(\theta) = \cos\theta$ at $\nu = 0$. We then shall consider solutions of (2.6) for nonzero ν that satisfy the needle-crystal conditions (2.1) for $\theta \to +\pi/2$, and shall examine their behavior near $\theta = 0$. This strategy is based on the assumption that, if ν is sufficiently small, the difference between κ and κ_0 also will be small in the interval $0 \le \theta \le \pi/2$ and may be computable by a linear approximation. For arbitrary ν, the continuation of this solution to negative θ may not remain small or even well defined, but the latter behavior need not invalidate an approximation in the region of interest. The trouble with this procedure turns out to be that $\lambda(\theta=0)$ vanishes more rapidly than any finite power of ν. Indeed, a systematic expansion in powers of ν must necessarily recover just the asymptotic series described above for which $\lambda(0)$ vanishes identically term by term. As a result, any useful scheme of approximation must go beyond a simple series in powers of the small parameter ν.

To examine this situation in greater detail, define

$$\kappa - \cos\theta \equiv \nu\kappa_1 \; , \tag{2.7}$$

and write (2.6) in the form

$$\nu \frac{d^2\kappa_1}{d\theta^2} - 2\nu\tan\theta \frac{d\kappa_1}{d\theta} + \frac{1}{\cos^2\theta}[1 + \nu(1 - 3\cos^2\theta)]\kappa_1 = -\frac{1}{\cos\theta}(1 - 2\cos^2\theta) + \mathcal{N}(\kappa_1) \; , \tag{2.8}$$

where $\mathcal{N}(\kappa_1)$ is a nonlinear term of orders $\nu^2\kappa_1^2$ and $\nu^3\kappa_1^3$ containing at most two differentiations of these quantities with respect to θ. The structure of (2.8) suggests that each differentiation with respect to θ produces a factor $1/\sqrt{\nu}$. We shall see this explicitly in what follows. Thus, a systematic approximation for (2.8) might involve dropping $\mathcal{N}$ and the term proportional to ν in square brackets on the left-hand side, or perhaps treating these terms perturbatively. The term containing $\nu\, d\kappa_1/d\theta$, which is formally of order $\sqrt{\nu}$, turns out to be essential. In Sec. III we shall examine the consequences of the above approximation. It will be useful to describe the calculation in general terms so that the technique may be applied to the boundary-layer model in Sec. IV without repetitive explanations.

III. APPROXIMATION SCHEME

The general structure of (2.8), with or without the terms to be omitted, is

$$\nu \frac{d^2\kappa_1}{d\theta^2} + \nu p(\theta)\frac{d\kappa_1}{d\theta} + q(\theta)\kappa_1 = r(\theta) . \tag{3.1}$$

The range of θ is $[0,\pi/2]$. Boundary conditions are to be imposed only at $\theta=\pi/2$. These are the needle-crystal conditions (2.2) which require that κ and $\lambda = d\kappa/d\xi = \kappa\, d\kappa/d\theta$ vanish as $\xi\to\infty$, $\theta\to\pi/2$. If we are working only to first order in κ_1, then this condition requires that κ_1 and $\kappa_0 d\kappa_1/d\theta$ vanish at $\pi/2$, where κ_0 denotes the Ivantsov solution; that is, $\kappa_0=\cos\theta$ for the geometrical model. Another way of stating this is that we want κ_1 to be consistent near $\pi/2$ with the asymptotic expansion of κ in powers of ν and $\cos\theta$. To lowest order in ν, this means

$$\kappa_1 \approx r_0(\theta)/q_0(\theta), \quad \theta\to\pi/2 \tag{3.2}$$

where r_0 and q_0 are the values of r and q at $\nu=0$. Finally, the problem that we pose for ourselves is to compute $\lambda=d\kappa/d\xi$, or simply $d\kappa/d\theta$, at $\theta=0$.

For sufficiently small ν, (3.1) can be solved by a Wentzel-Kramers-Brillouin (WKB) approximation. The homogeneous solutions are

$$\kappa_\pm^{\rm hom}(\theta) \simeq \frac{1}{q_0^{1/4}} \exp\left[\pm\frac{i}{\sqrt{\nu}}\psi(\theta) - \tfrac{1}{2}\int_0^\theta p_0(\phi)d\phi\right], \tag{3.3}$$

where

$$\psi(\theta) = \int_0^\theta q_0^{1/2}(\phi)d\phi , \tag{3.4}$$

and we have neglected terms of order $\sqrt{\nu}$ or smaller in the exponent and in the prefactor. These solutions should be accurate as long as

$$\left|\frac{1}{q_0}\frac{dq_0}{d\theta}\right| \ll \left|\frac{q_0}{\nu}\right|^{1/2}, \tag{3.5}$$

for all θ in $[0,\pi/2]$.

For the quasilinear geometrical model, $q_0=(\cos\theta)^{-2}$, and (3.5) is satisfied for $\nu \ll 1$. Moreover, $\psi(\theta)$ in (3.4) is equal to the unperturbed dimensionless arc length ξ obtained from $\kappa_0=d\theta/d\xi=\cos\theta$. Using $p_0=-2\tan\theta$, we find

$$\kappa_\pm^{\rm hom}(\theta) \simeq \frac{1}{(\cos\theta)^{1/2}} \exp\left[\pm\frac{i}{\sqrt{\nu}}\ln\left[\frac{1+\sin\theta}{\cos\theta}\right]\right] = (\cosh\xi)^{1/2}\exp\left[\pm\frac{i\xi}{\sqrt{\nu}}\right] . \tag{3.6}$$

For small ν, these are rapidly oscillating functions whose amplitudes diverge as $\theta\to\pi/2$, $\xi\to\infty$. Equation (3.6) is an analytic statement of the fact that only one trajectory in θ,κ,λ space enters the fixed point at $\theta=\pi/2$, $\kappa=\lambda=0$, and all other trajectories spiral out and away from that point. It also justifies our expectation that differentiations in (2.8) produce factors of $1/\sqrt{\nu}$.

The next step is to use the homogeneous solutions (3.3) to compute a particular solution of (3.1) by means of the formula

$$\kappa_1^{\rm part}(\theta) = \frac{1}{\nu}\int^\theta d\phi \frac{r(\phi)}{W(\phi)}\left[\kappa_+^{\rm hom}(\theta)\kappa_-^{\rm hom}(\phi) - \kappa_-^{\rm hom}(\theta)\kappa_+^{\rm hom}(\phi)\right], \tag{3.7}$$

where the Wronskian W is

$$W = \kappa_-^{\rm hom}\frac{d\kappa_+^{\rm hom}}{d\phi} - \kappa_+^{\rm hom}\frac{d\kappa_-^{\rm hom}}{d\phi} . \tag{3.8}$$

The lower limit of integration remains to be chosen. Because both of the homogeneous solutions $\kappa_\pm^{\rm hom}(\theta)$ which appear in (3.7) are inconsistent with the needle-crystal conditions, the only possible choice for this lower limit is $\pi/2$. Using (3.3), we find

$$W(\theta) \simeq \frac{2i}{\sqrt{\nu}}\exp\left[-\int_0^\theta p\, d\phi\right], \tag{3.9}$$

and

$$\kappa_1(\theta) \simeq -\frac{1}{\sqrt{\nu}}\int_\theta^{\pi/2} d\phi \frac{r_0(\phi)}{[q_0(\theta)q_0(\phi)]^{1/4}} \times \exp\left[-\tfrac{1}{2}\int_\phi^\theta p_0\, d\phi'\right] \times \sin\{(1/\sqrt{\nu})[\psi(\theta)-\psi(\phi)]\} . \tag{3.10}$$

To check that (3.10) does satisfy the asymptotic condition (3.2), integrate (3.10) twice by parts, integrating the sine as if to obtain a series in growing powers of ν. The result is

$$\kappa_1(\theta) \approx \frac{r_0(\theta)}{q_0(\theta)} + \nu M(\theta) + \cdots , \tag{3.11}$$

where M is a function of θ which is of no special interest except for the fact that it exists and is well behaved near $\theta=\pi/2$.

Our goal is to calculate $d\kappa/d\theta$ at $\theta=0$, which we denote by the symbol $\kappa'(0)$. Note that the symmetry of the system requires that q and r be even functions of θ and that p be odd. Thus

$$\kappa'(0) = -\tfrac{1}{2}[q_0(0)]^{1/4}\int_{-\pi/2}^{\pi/2} d\theta \frac{r_0(\theta)}{[q_0(\theta)]^{1/4}} \times \exp\left[\tfrac{1}{2}\int_0^\theta p\, d\phi\right] \times \cos\left[\frac{\psi(\theta)}{\sqrt{\nu}}\right] . \tag{3.12}$$

Equation (3.12) is the principal result whose validity and implications are to be explored in the remainder of this paper.

Remember that the condition for existence of a needle crystal is that $\kappa'(0)$ vanish. It is interesting, and possibly useful for future analysis, to note that (3.12) with $\kappa'(0)=0$ can be interpreted as a solvability condition for the (formally) linear inhomogeneous equation (3.1). To see this, consider (3.1) in the whole interval $[-\pi/2,+\pi/2]$ and look for solutions in the space of symmetric functions $\kappa_1(\theta)$. The function

$$\kappa_{\rm sym}^{\rm hom}(\theta)=\frac{1}{q_0^{1/4}}\exp\left[-\tfrac{1}{2}\int_0^\theta p\,d\phi\right]\cos\left[\frac{\psi(\theta)}{\sqrt{\nu}}\right] \quad (3.13)$$

is the WKB approximation for the solution of $\mathscr{L}\kappa=0$, where $\mathscr{L}$ denotes the linear operator on the left-hand side of (3.1), and

$$\begin{aligned}\tilde{\kappa}_{\rm sym}^{\rm hom}(\theta)&=\frac{1}{W(\theta)}\kappa_{\rm sym}^{\rm hom}(\theta)\\ &\simeq\frac{2i}{\sqrt{\nu}}\frac{1}{q_0^{1/4}}\exp\left[\tfrac{1}{2}\int_0^\theta p\,d\phi\right]\cos\left[\frac{\psi(\theta)}{\sqrt{\nu}}\right]\end{aligned} \quad (3.14)$$

is the solution of the adjoint equation $\tilde{\mathscr{L}}\tilde{\kappa}=0$. Thus, the solvability condition derived above has the form

$$\int_{-\pi/2}^{\pi/2}d\theta\, r(\theta)\tilde{\kappa}_{\rm sym}^{\rm hom}(\theta)=0\ , \quad (3.15)$$

which is the usual statement that the inhomogeneous term $r(\theta)$ can have no projection onto the null space of $\mathscr{L}$ if (3.1) is to have a solution. One obscure aspect of this interpretation is the definition of the function space; the divergent functions $\kappa^{\rm hom}(\theta)$ would not seem to belong to the space of acceptable needle-crystal solutions. One can make this difficulty seem less severe by working with the bounded functions $W^{-1/2}\kappa$; that is, by transforming away the first derivative in (3.1). But this mathematical point, among many such points raised in this paper, requires further scrutiny.

The general character of (3.12) can be seen by evaluating it for the quasilinear geometrical model where $q_0=(\cos\theta)^{-2}$, $p_0=-2\tan\theta$, $r_0=(2\cos^2\theta-1)/\cos\theta$, $\psi=\xi$, and $\cos\theta=\operatorname{sech}\xi$. Then, transforming to ξ as the variable of integration, we find

$$\begin{aligned}\kappa'(0)&\simeq-\tfrac{1}{2}\int_{-\infty}^{\infty}d\xi(\operatorname{sech}^{3/2}\xi)(2\operatorname{sech}^2\xi-1)e^{i\xi/\sqrt{\nu}}\\ &\approx-\frac{A}{\nu^{5/4}}\exp\left[-\frac{\pi}{2\sqrt{\nu}}\right],\end{aligned} \quad (3.16)$$

where

$$A=\frac{16\sqrt{\pi}}{15}\simeq1.8906\ . \quad (3.17)$$

The derivation of the second form of (3.16), valid in the limit $\nu \ll 1$, is summarized in the Appendix. The important point to notice is that (3.16) has an essential singularity at $\nu=0$, consistent with the failure of expansions in powers of ν.

A comparison between (3.16) and a direct numerical integration of the fully nonlinear equation (2.5) confirms the basic form of (3.16), both the exponential function and the power of ν in the prefactor. The numerically determined value of A, however, appears to be about 3.7 ± 0.2, larger than that given by (3.17) by a factor of about 2. This comparison was made for values of ν in the interval $0.002\le\nu\le0.01$, across which $\kappa'(0)$ changes by eight decades. The validity of the asymptotic evaluation of the integral in (3.16) was also checked in this interval by direct numerical integration, thus confirming that these values of ν are small enough to be in the asymptotic region.

The source of this discrepancy is related to the fact that the power of ν in the prefactor in (3.16), i.e., $\nu^{-5/4}$, is determined by the highest power of $\operatorname{sech}\xi$ in the integrand, which in turn can be traced back to the highest power of $\cos\phi$ in $r_0(\phi)$ in (3.10). (This will be seen explicitly in the Appendix, but should be apparent from (3.16) because there we are computing the high-frequency—small-ν—behavior of a Fourier transform.) Now think about a first nonlinear correction to r. That is, consider using the approximation for κ_1 given in (3.10) to evaluate $\mathscr{N}(\kappa_1)$ and including the result in r on the right-hand side of (3.1). The resulting corrections to r will be formally of relatively high order in ν but will contain various powers of $\cos\theta$, and the latter will produce inverse powers of ν in the final expression for $\kappa'(0)$. The exponential part of (3.16) will be unchanged by this procedure, but apparently the prefactor is not being computed systematically.

This line of analysis will not be pursued further in this paper. It appears that there may be a much more elegant and mathematically controlled way of arriving at formulas such as (3.16).[8] The interested reader—along with the author—should be alert for new developments in this class of problems.

IV. THE MINIMAL BOUNDARY-LAYER MODEL

The boundary-layer model[3,4] has been introduced as an attempt to include in a local description some physical features of the real solidification problem that are missing in the purely geometrical approach. In particular, the dynamics of a thermal boundary-layer field defined along the solidification front mimic some of the nonlocality and history dependence associated with a more realistic thermal diffusion field. The model has had some successes. Its steady-state solutions, including the parabolic needle crystal in the Ivantsov limit, are nearly identical to those of the full model. The boundary-layer model does produce time-dependent dendritic patterns whose growth rates and tip radii are numerically consistent with solvability conditions for needle crystals with nonvanishing surface tension. At present, however, it is still not known whether the model produces a physically realistic picture of dendritic sidebranching.

In order to make the following analysis reasonably tractable, we shall consider only a minimal version of the boundary-layer model in which the dimensionless undercooling Δ becomes vanishingly small. This is not the physically most realistic limit of the model, but it retains most of the features which seem essential for present purposes. (The opposite limit, $\Delta\to1$, will be discussed elsewhere.) The steady-state equation to be considered is the same as that of Ref. (3), Eq. (5.17), supplemented by an anisotropic kinetic attachment coefficient. Specifically,

$$\kappa=\cos^3\theta-\nu\kappa\cos^2\theta\frac{d}{d\theta}\left[\frac{\kappa}{\cos\theta}\frac{d}{d\theta}(\kappa+b_m\cos\theta)\right], \quad (4.1)$$

where the κ in Ref. 3 has been replaced by $\nu\kappa$ in order to emphasize the analogy to (2.6). Here, $\nu = d_0 v / D\Delta^5$, where d_0 is a capillary length proportional to surface tension, D is the diffusion constant in the liquid, and v is the growth velocity as before. Note that, as in the geometrical model, ν vanishes in the limit of vanishing surface tension and velocity. The quantity b_m in (4.1) is the θ-dependent kinetic coefficient. In the notation of Ref. 4, $b_m = \alpha\Delta^2(1-\cos m\theta)$. From here on, we shall specialize to the case $m=4$, and write $b_4 = 8\beta_4\cos^2\theta\sin^2\theta$.

The Ivantsov limit of (4.1), $\kappa_0(\theta) = \cos^3\theta$ is a parabola which has unit radius of curvature at its tip and which trivially satisfies the needle-crystal conditions (2.2) for all ν. In analogy to (2.7), let

$$\kappa - \cos^3\theta \equiv \nu\kappa_1 , \tag{4.2}$$

and rewrite (4.1) in terms of κ_1. The result is an equation of the form (3.1) with

$$p_0(\theta) = -5\tan\theta + \frac{8\beta_4}{\cos^3\theta}\frac{d}{d\theta}(\cos^3\theta\sin^2\theta) , \tag{4.3}$$

$$q_0(\theta) = \frac{1}{\cos^7\theta} , \tag{4.4}$$

$$r_0(\theta) = 3\cos\theta(5\cos^2-4) - \frac{8\beta_4}{\cos^2\theta}\frac{d}{d\theta}\left[\cos^2\frac{d}{d\theta}(\cos^3\theta\sin^2\theta)\right] . \tag{4.5}$$

Inserting these functions into (3.12), we obtain

$$\kappa'(0) \approx -\tfrac{1}{2}\int_{-\infty}^{\infty} d\psi\, r_0(\theta)(\cos\theta)^{\lambda} \times \exp(10\beta_4\sin^2\theta)e^{i\psi/\sqrt{\nu}} , \tag{4.6}$$

where $\lambda = \frac{31}{4} + 12\beta_4$ and

$$\psi(\theta) = \int_0^{\theta} d\phi \frac{1}{\cos^{7/2}\phi} . \tag{4.7}$$

Notice that, in contrast to the geometrical model (3.16), ψ in (4.7) is not quite the same as the unperturbed arc length ξ defined by

$$\xi(\theta) = \int_0^{\theta} d\phi \frac{1}{\kappa_0(\phi)} = \int_0^{\theta} d\phi \frac{1}{\cos^3\phi} . \tag{4.8}$$

The integral on the right-hand side of (4.6) can be evaluated asymptotically in the limit of small ν. Look, for the moment, at the case $\beta_4=0$, so that the only integrals needed are Fourier transforms of powers of the function $\cos\theta(\psi)$. An analysis which is outlined in the Appendix yields the estimate

$$\int_{-\infty}^{\infty} d\psi(\cos\theta)^{\mu} e^{i\psi/\sqrt{\nu}} \approx C(\mu)\nu^{-\alpha}\exp(-a_s/\sqrt{\nu}) , \tag{4.9}$$

where

$$a_s = \int_0^1 (1-u^2)^{3/4} du = \tfrac{3}{5}\sqrt{2/\pi}[\Gamma(\tfrac{3}{4})]^2 = 0.7189\ldots , \tag{4.10}$$

$$\alpha = \frac{\mu}{7} - \frac{1}{2} , \tag{4.11}$$

and

$$C(\mu) = (\tfrac{2}{7})^{2\mu/7}\frac{2\pi}{\Gamma(2\mu/7)} . \tag{4.12}$$

As in (3.16), the dominant exponential in (4.9) is independent of μ, but the power of ν^{-1} in the prefactor increases with increasing μ. For $\beta_4=0$, the leading μ is $\frac{43}{4}$, and

$$\kappa'(0) \approx -\frac{B}{\nu^{29/28}}\exp\left[-\frac{0.7189}{\sqrt{\nu}}\right] , \tag{4.13}$$

with

$$B = \tfrac{15}{2}C(\tfrac{43}{4}) \simeq 0.4703 . \tag{4.14}$$

Both numerical evaluation of the integral in (4.6) and numerical integration of the fully nonlinear differential equation (4.1) confirm the general form of (4.13). Specifically, a graph of

$$-2\nu^{3/2}\frac{d}{d\nu}\ln[-\kappa'(0)] \approx a - 2\alpha\sqrt{\nu} + \cdots \tag{4.15}$$

versus $\sqrt{\nu}$, evaluated for $0.0015 \le \nu \le 0.01$ and extrapolated to small $\sqrt{\nu}$, yields $a = 0.71 \pm 0.02 \simeq a_s$ and a value of α of about unity, consistent with $\frac{29}{28}$. A value of B of about 0.8 fits the integration of (4.1), whereas numerical evaluation of the approximation (4.6) confirms (4.14) in giving a value of B of about 0.5. The latter discrepancy is in the same sense and of about the same size as that which occurred for the geometrical model.

The idea that the dominant behavior of $\kappa'(0)$ at small ν is determined by the highest power of $\cos\theta$ in (4.6) provides some interesting information about the effect of crystalline anisotropy. When the anisotropy coefficient β_4 is nonzero, the dominant θ dependence in $r_0(\theta)$ comes from the second term on the right-hand side of (4.5) and enters with the opposite sign. That is, the leading term in $r_0(\theta)$ is proportional to $-\beta_4\cos^5\theta$ instead of $+\cos^3\theta$. As a result, $\kappa'(0)$ must be positive as ν approaches zero. At larger values of ν, however, the exponential factor in (4.6) is no longer oscillating so rapidly, the slower variations in $r_0(\theta)$ make the dominant contributions, and $\kappa'(0)$ becomes negative again. The situation is illustrated in Fig. 2 where $\kappa'(0)$ is drawn as a function of ν for $\beta_4=0$, 0.1, and 0.2. These graphs extend out to $\nu=0.1$, well beyond the asymptotic regime, and have been obtained by integrating the fully nonlinear equation (4.1). Equation (4.6) does seem to remain a qualitatively good approximation as shown by the dashed curves in the figure.

The important point is that, for $\beta_4 \neq 0$, $\kappa'(0)$ vanishes at a finite value of ν, indicating the existence of a steady-state needle crystal. For this particular version of the model, that is, the minimal boundary-layer model ($\Delta \to 0$) with the special form of kinetic anisotropy indicated in (4.1), such steady-state solutions occur at arbitrarily small but nonzero values of the anisotropy coefficient β_4. However, the selected value of $\nu = d_0 v / D\Delta^5$ becomes small as β_4 decreases, and it seems highly unlikely that needle crystals with very small v or, equivalently, small d_0 can be stable. Thus, it seems that the minimum value of β_4 required for dendritic behavior in this model is nonzero and is determined by a stability requirement.

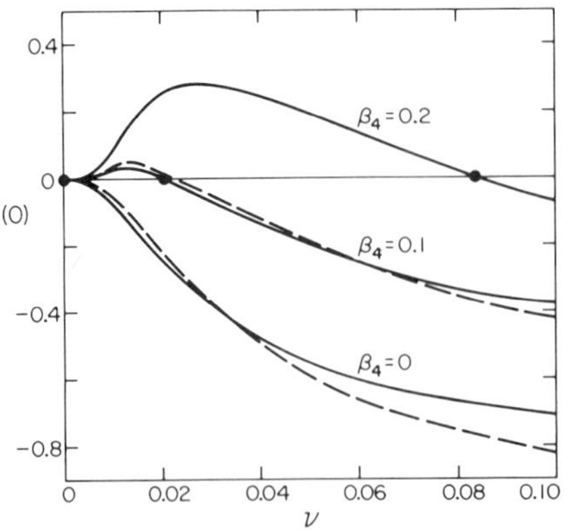

FIG. 2. $\kappa'(0)$ as a function of ν for the boundary-layer model with various values of the anisotropy parameter β_4. The solid curves have been computed by numerical integration of the nonlinear differential equation (4.1), and the dashed curves by numerical evaluation of the approximation (4.6). The heavy dots indicate values of ν at which the solvability condition is satisfied for the corresponding values of β_4.

APPENDIX: SOME MATHEMATICAL DETAILS

Asymptotic estimates for $\kappa'(0)$. To derive the asymptotic relation (3.16) for the geometrical model, we must evaluate integrals of the form

$$f_p(\nu)=\int_{-\infty}^{\infty} d\xi\,(\operatorname{sech}\xi)^p e^{i\xi/\sqrt{\nu}} \tag{A1}$$

in the limit $\nu\to 0$. This integration is most conveniently carried out by transforming to $\eta=\tan\theta$, where θ is the original angular variable that satisfies $\cos\theta=\operatorname{sech}\xi$. This technique is slightly more cumbersome than necessary for the geometrical model, but turns out to be specially suited for both the boundary-layer model to be considered next and the fully nonlocal problem.[11] In terms of η, we have

$$f_p(\nu)=\int_{-\infty}^{\infty} d\eta \frac{1}{(1+\eta^2)^{(1+p)/2}} \exp\left[\frac{i}{\sqrt{\nu}}\psi_G(\eta)\right], \tag{A2}$$

where

$$\psi_G(\eta)=\int_0^{\theta} d\theta' \frac{1}{\cos\theta'}=\int_0^{\eta} d\eta_1 \frac{1}{(1+\eta_1^2)^{1/2}} . \tag{A3}$$

Inspection of (A2) and (A3) indicates that we should deform the contour of integration so as to pass near $\eta=i$ and then carry out a steepest-descent calculation in the neighborhood of that point. (The procedure in this case requires that the path of steepest-descent pass through the branch cut and back out again.) If $\eta=i+\omega$, $|\omega|\ll 1$, then

$$i\psi_G(\eta)=i\int_0^i d\eta_1 \frac{1}{(1+\eta_1^2)^{1/2}}+i\int_0^{\omega} d\omega' \frac{1}{(2i\omega')^{1/2}} \simeq -\frac{\pi}{2}+(2i\omega)^{1/2}, \tag{A4}$$

and

$$f_p(\nu)\approx \exp\left[-\frac{\pi}{2\sqrt{\nu}}\right]\int_{-\infty}^{\infty} d\omega \frac{1}{(2i\omega)^{(1+p)/2}} \exp\left[\frac{2i\omega}{\nu}\right]^{1/2} = \frac{2\pi}{\nu^{(p-1)/2}\Gamma(p)} \exp\left[-\frac{\pi}{2\sqrt{\nu}}\right]. \tag{A5}$$

Note that, as mentioned previously, the dominant power of ν in the prefactor comes from the largest value of p; thus, in (3.16),

$$\kappa'(0)\approx -f_{7/2}(\nu) . \tag{A6}$$

The same technique works for the integral in (4.9), which we shall denote by $F_\mu(\nu)$. We have

$$F_\mu(\nu)=\int_{-\infty}^{\infty} d\eta \frac{1}{(1+\eta^2)^{(2\mu-3)/4}} \exp\left[\frac{i}{\sqrt{\nu}}\psi_B(\eta)\right], \tag{A7}$$

with

$$\psi_B(\eta)=\int_0^{\theta} d\theta' \frac{1}{(\cos\theta')^{7/2}}=\int_0^{\eta} d\eta_1 (1+\eta_1^2)^{3/4} . \tag{A8}$$

Again, writing $\eta=i+\omega$ and integrating along the path of steepest descent through $\eta=i$, we find

$$F_\mu(\nu)\approx \exp\left[-\frac{a_s}{\sqrt{\nu}}\right]\int_{-\infty}^{\infty} d\omega \frac{1}{(2i\omega)^{(2\mu-3)/4}} \times \exp\left[\frac{2}{7\sqrt{\nu}}(2i\omega)^{7/4}\right], \tag{A9}$$

which can be evaluated without further approximation to obtain the results shown in Eqs. (4.9)–(4.12).

Numerical methods. Numerical integrations of the nonlinear differential equations (2.5) and (4.1) were carried out with an implicit scheme using θ, κ, and $\lambda=d\kappa/d\xi$ as functions of arc length ξ. Initial conditions were computed using the asymptotic expansions as close to $\theta=\pi/2$ as possible, and then the equations were integrated back to $\theta=0$. Because, in practice, one can never start precisely on the trajectory which enters the fixed point, there will always be some components of the oscillating homogeneous solutions (3.3) in the function being computed. The frequency of the oscillation in (3.3) provides an estimate of the minimum step size $d\xi$ required to resolve these oscillations accurately. In addition, the rate at which the magnitudes of these homogeneous solutions decay in going toward $\theta=0$ from $\theta\simeq\pi/2$ provides an estimate of how accurately one must locate the correct trajectory near $\pi/2$ in order to obtain a desired accuracy of the solution at $\theta=0$. These estimates were confirmed numerically (roughly) and were useful for achieving convergence of the numerical procedure.

Numerical evaluations of the various forms of the solvability formula (3.12), specifically (3.16) and (4.6), were performed using the trapezoidal rule with uniformly spaced intervals in ψ (as if performing a numerical Fourier transform). Results were extrapolated to $d\psi/\sqrt{v}\rightarrow 0$ and tested for convergence in the outer cutoff.

Note added in proof. The connection between singular perturbations and solvability conditions has been discussed by Barenblatt and Zel'dovich[12,13] in the general context of similarity solutions of partial differential equations. A recent development along these lines, closely related to the work described in the present paper, is the discovery of a solvability condition for pattern selection in the theory of viscous fingering. An excellent review of the latter topic has been prepared by Bensimon, Kadanoff, Liang, Shraiman, and Tang.[14] I am grateful to L. Kandanoff for informing me about Refs. 12 and 13 and for sending me a copy of Ref. 14 prior to publication.

ACKNOWLEDGMENTS

I should like to acknowledge especially useful conversations about mathematical aspects of this problem with M. Kruskal, C. Oberman, and H. Segur. This research was supported by the U.S. Department of Energy Grant No. DE-FG03-84ER45108, and also in part by the National Science Foundation Grant No. PHY-82-17853 supplemented by funds from the U.S. National Aeronautics and Space Administration.

[1] R. Brower, D. Kessler, J. Koplik, and H. Levine, Phys. Rev. Lett. **51**, 1111 (1983); Phys. Rev. A **29**, 1335 (1984).
[2] D. Kessler, J. Koplik, and H. Levine, Phys. Rev. A **30**, 3161 (1984).
[3] E. Ben-Jacob, N. D. Goldenfeld, J. S. Langer, G. Schön, Phys. Rev. Lett. **51**, 1930 (1983); Phys. Rev. A **29**, 330 (1984).
[4] E. Ben-Jacob, N. D. Goldenfeld, B. G. Kotliar, and J. S. Langer, Phys. Rev. Lett. **53**, 2110 (1984).
[5] J. S. Langer, Rev. Mod. Phys. **52**, 1 (1980).
[6] G. P. Ivantsov, Dok. Akad. Nauk SSSR **58**, 567 (1947).
[7] G. Horvay and J. W. Cahn, Acta Metall. **9**, 695 (1961).
[8] M. Kruskal, C. Oberman, and H. Segur (private communication).
[9] D. Meiron (private communication).
[10] P. Pelcé and Y. Pomeau (unpublished).
[11] B. Caroli, C. Caroli, J. S. Langer, and B. Roulet, following paper, Phys. Rev. A **33**, 442 (1986).
[12] G. I. Barenblatt and Ya. B. Zel'dovich, Annu. Rev. Fluid Mech. **4**, 285 (1972).
[13] G. I. Barenblatt, *Similarity, Self-Similarity, and Intermediate Asymptotics* (Consultant's Bureau, New York, 1979).
[14] D. Bensimon, L. P. Kadanoff, S. Liang, B. I. Shraiman, and C. Tang (unpublished).

Dendrites in the Small Undercooling Limit

By P. Pelce and Y. Pomeau

Dendrites of a solid growing steadily in an undercooled melt have a shape given by the solution of certain nonlinear integrodifferential equations. If one does not take account of the Gibbs-Thomson condition, these equations have an exact solution due to Ivantsov and to Cahn and Horway. For elementary dimensional reasons, the tip velocity in these solutions is left undetermined. This degeneracy is removed by the effect of interface curvature on the melting temperature (Gibbs-Thomson effect). We show that, in the physically relevant limit of small undercooling, the shape of the dendrite can be deduced from parameterless similarity equations, where the Gibbs-Thomson effect is included. Every physical quantity is known in this limit, up to a nonlinear eigenvalue that has to be found numerically.

1. Introduction

When a solid grows at the expanse of an undercooled pure melt (i.e. a metastable liquid at a temperature lower than the freezing point) and when no heat can flow to the outside, this poses at the macroscopic level a complicated problem not yet completely understood. It is fair to say that the microscopic version of the problem is far from being understood too: the striking differences observed [1,2] between the interface structure at equilibrium and during crystal growth remain to be explained. A possible explanation is the formation of a glassy solid between the growing crystal and the melt. Such phenomena, among many others (convection, crystal anisotropy, etc.), are not included in the kind of theory that we shall try to develop here.

In the classical description of crystal growth, the so-called dendritic growth takes place because the latent heat generated at the moving interface has to heat a

Address for correspondence: Dr. Y. Pomeau, Orme des Merisier DPh-G/SPT, C.E.N. Saclay, 91190 Gif/Yvette Cedex, France.

STUDIES IN APPLIED MATHEMATICS 74:245–258 (1986)

Published by Elsevier Science Publishing Co., Inc.

large bulk of melt, since the temperature can nowhere exceed the freezing temperature. If one neglects surface tension effects, a famous exact solution due to Ivantsov [3] describes a dendrite growing with a constant velocity (this solution is derived in the appendices of this paper). However, the Ivantsov solution does not completely solve the problem at hand, because it is unstable against very short wavelength fluctuations (and so could lead to singularities after a finite time for smooth initial data). Another problem with the Ivantsov solution is that the velocity of the tip is left undetermined: for a given undercooling the only quantity that is fixed is the Peclet number $\mathrm{Pe} = U/aD_t$, where U is the velocity, a the curvature of the tip extremity, and D_t the thermal diffusion coefficient. It has often been suggested that the so-called Gibbs-Thomson condition, which relates the melting temperature to the curvature of the interface, should remove this degeneracy. A simple dimensional analysis indeed gives some support to this. Once the Gibbs-Thomson condition is imposed, one may construct from the physical data (undercooling, Laplace capillary constant, latent heat, thermal diffusion coefficient, etc.) quantities with the dimensions of a velocity or of a length. This is not possible for the Ivantsov solution, which basically explains its degeneracy. Thus this degeneracy should be removed at least in some sense by the Gibbs-Thomson condition. But this is not at all certain: one might think for instance of the case of the Saffman-Taylor instability. The Saffman-Taylor solution for a steady finger depends on an arbitrary real parameter and is presumably reduced to a single (or eventually a discrete set of) solution(s) if the effects of capillarity are included [5]. But that remains to be understood analytically, although some progress has been made [6] in the opposite limit (large surface tension).

The situation with the dendrite problem is quite different from that with the Saffman-Taylor instability. For the dendrite one cannot really assume that the surface tension is large or small, because in the problem without surface tension (without Gibbs-Thomson effect) there is no physical parameter for measuring this surface tension. Hence one needs another expansion parameter than the surface tension, and the Ivantsov solution cannot play the role of a "lowest order" solution. This expansion parameter is the dimensionless undercooling $\Delta = c_p(T_m - T_\infty)/L$, where T_m is the melting temperature, T_∞ the temperature far in the melt, L the latent heat per unit volume, and c_p the specific heat per unit volume. (As far as possible, our notation will be that in the review paper by Langer [10].) To get physical insight into what this number is, one might say that $\Delta = 0$ is the steady situation for a flat interface just at the melting temperature, whereas $\Delta = 1$ is the value of the undercooling such that for a plane solidification front the latent heat is just enough to heat the melt at the melting point when the front has uniform motion. In most experimental situations the undercooling is much smaller than this critical value, so that the limit of small undercooling is relevant.

We shall give first the general form of the equation to be satisfied by a curved solidification front moving steadily when the temperature at the interface is fixed by the Gibbs-Thomson condition. In the following the capillary length defined by $d_0 = \gamma T_m c_p / L^2$ and the velocity $U = D_t/d_0$ are respectively taken as the units of length and velocity, where γ is the surface tension. Let u be the dimensionless uniform velocity of the dendrite, (x, y) be the Cartesian coordinates in a plane perpendicular to the direction of propagation, and $\xi(x, y)$ be the coordinates of a point of the interface in the direction parallel to this direction.

When the thermal diffusivities of the liquid and the solid are equal, the dimensionless form of the equation to be satisfied by $\xi(\cdot, \cdot)$ is [7, 8]:

$$\Delta + \frac{(1+\xi_x^2)\xi_{yy} + (1+\xi_y^2)\xi_{xx} - 2\xi_x\xi_y\xi_{xy}}{(1+\xi_x^2+\xi_y^2)^{3/2}}$$
$$= u\int_0^{+\infty} \frac{d\tau}{(4\pi\tau)^{3/2}} \int_{-\infty}^{+\infty} dx_1 \int_{-\infty}^{+\infty} dy_1$$
$$\times \exp\left(-\frac{1}{4\tau}\left[(x-x_1)^2+(y-y_1)^2+(\xi-\xi_1+u\tau)^2\right]\right), \qquad (1)$$

where ξ stands for $\xi(x, y)$, ξ_1 for $\xi(x_1, y_1)$, ξ_x for $\partial\xi/\partial x$, etc., this sort of notation being used throughout this work.

The first term on the l.h.s. of (1) represents the undercooling; the second is the contribution from curvature to the local equilibrium condition between the solid and the liquid (Gibbs-Thomson condition). The r.h.s. represents the temperature change due to the latent heat generated at the interface.

Without the Gibbs-Thomson condition [second term on the left of (1) suppressed] one obtains the Ivantsov [3]–Cahn–Horway [4] solution. This solution is derived in Appendix A of the present paper and has the general form

$$\xi = -(ax^2 + by^2),$$

a and b being both positive and related to u and Δ by

$$\Delta = \frac{u}{\pi}\int_0^{+\infty} dX \int_0^{+\infty} dY \frac{\exp\left[-\frac{1}{4}(x^2+y^2)\right]}{aX^2+bY^2+u}. \qquad (2)$$

This reduces to the parabolic solution in $2d$ if $b = 0$ and if Δ is related to u and a by

$$\Delta = \frac{u}{\sqrt{\pi}}\int_0^{+\infty} dX \frac{\exp(-X^2/4)}{aX^2+u}; \qquad (3a)$$

in particular, the circular paraboloid corresponds to $a = b$ and implies the relation

$$\Delta = \frac{u}{2}\int_0^{+\infty} d\rho \frac{\rho\exp(-\rho^2/4)}{a\rho^2+u}. \qquad (3b)$$

In what follows we shall need the relations drawn from (3a) and (3b) in the limit

Δ close to 0. Indeed (3a) and (3b) are of the general form $\Delta = f(ua^{-1})$ and one has the estimates

$$ua^{-1} \approx 4\Delta^2/\pi$$

in 2d [Equation (3a)] and

$$ua^{-1} \approx -2\Delta/\log\Delta$$

in 3d [Equation (3b)].

Now we want to study the dendrite solution of (1) near $\Delta = 0$. More specifically we shall study the dependence of u as a function of Δ in the limit as it can be deduced from a similarity analysis, as follows. The integration over τ of the r.h.s. of (1) gives an integrand proportional to transcendental functions with an argument of order ul, l being some length. And we want to match the solution near the tip with the Ivantsov solution far from it: in this region the dendrite becomes almost flat and the curvature contribution to the temperature can be neglected, as in the Ivantsov solution. To make possible the matching with the Ivantsov solution, the length scale near the tip must be the same as the one fixed by the Ivantsov solution. Accordingly quantities of order ul, when computed near the tip, have ua^{-1} as order of magnitude, where a is the curvature parameter entering into the Ivantsov solution. But the combination ua^{-1} is precisely what is fixed by (3a) or (3b), and it has to become small when the undercooling tends to 0. Thus it is tempting to expand the transcendant functions in the integrand near their zero argument, since that is of order ul. But this cannot be done so simply, because the dominant order in this expansion leads to large distance divergences in the integral. To remove those divergences one uses the following trick: in the Ivantsov solution the integral term on the r.h.s. exactly cancels the first Δ term on the l.h.s. of (1). On the other hand we know that far from the tip the actual solution tends to the Ivantsov solution. Thus, replacing the Δ on the l.h.s. of (1) by the integral on the r.h.s. computed for the Ivantsov solution, one subtracts the result from the actual integral term. In this way one eliminates the divergences, because the large distance behavior of the integrand computed with the actual solution and with the Ivantsov solution is the same. As explained in this paper, this program works basically in the 2D case as well as for the axisymmetric dendrite in 3D. Nevertheless, this has to be done carefully when considering the order of approximation with which the Ivantsov solution is reached at large distance by the actual solution. Once the expansion of the integral term is done, one can get similarity equations by suitable scaling. This gives another relation between u and a and thus gives the low undercooling dependence of those quantities when the effect of the Gibbs-Thomson condition is included. This relation is similar in form to the one given by Langer and Muller–Krumbhaar in the framework of the marginal stability hypothesis [9]. The difference is that our similarity analysis leaves undetermined a real number (the so-called nonlinear eigenvalue of Barenblatt and Zeldovich). We give some argument supporting the view that this nonlinear "eigenvalue" has a discrete spectrum. In the LMK theory this number is such that the dendrite tip is marginally stable. Notice however that the scaling found in the present work would also follow from other stability criteria. For instance, one finds it by keeping the Ivantsov relation between radius of curvature, velocity, and undercooling, and also saying that this radius is of the order of magnitude of the most unstable wavelength in the stability analysis of the planar front in the presence of capillarity.

This paper is organized as follows: In Section 2 we consider the 2D case, the final result being the similarity equation (6). This equation is derived by scaling the velocity and the length with respect to the expansion parameter. Thus even if one cannot solve the similarity equation explicitly, one gets at least the general dependence of the dendrite velocity on the undercooling when the latter is small.

In Section 3 we consider the 3D axisymmetric case. Using the same general method as for the 2D case, we derive the similarity equation (14) as well as the scaling laws for the velocity and for the tip curvature.

2. The two dimensional case

In this section we derive the similarity equation for the 2D dendrite. As we shall see, the calculation is far from being straightforward, and much care must be exercised when performing the various approximations relevant to the small undercooling limit.

Our starting point is the integrodifferential equation (1). The 2D case is recovered by assuming that ξ is independent of y. Once the integration over the variable τ and y is done one obtains

$$\Delta + \frac{\xi_{xx}}{\left(1+\xi_x^2\right)^{3/2}} = \frac{u}{2\pi}\int_{-\infty}^{+\infty} dx_1 \exp\left(-\frac{u}{2}(\xi-\xi_1)\right) K_0\left(\frac{u}{2}|\mathbf{r}-\mathbf{r}_1|\right), \tag{4}$$

where

$$|\mathbf{r}-\mathbf{r}_1| = \left[(x-x_1)^2+(\xi-\xi_1)^2\right]^{1/2}$$

and where K_0 is the modified Bessel function of zeroth order. As said before, the Ivantsov solution (no capillarity) reads

$$\xi(x) = -ax^2,$$

a and u being both positive and related to Δ by (3a).

Now we shall be concerned with the limit as Δ goes to zero. We shall assume that in this limit the typical length scale at the tip is a^{-1} even in the presence of the Gibbs-Thomson term, and since $ua^{-1} \approx \Delta^2 \ll 1$, the arguments of the transcendental functions [i.e. $\exp(\cdot)$ and $K_0(\cdot)$ in (4)] are small, this leading to long distance divergences at the lowest order. To avoid these divergences one proceeds in two steps. First we limit the domain of integration to x_1 positive; then we notice that by inserting into the r.h.s. of (4) the function $\xi(x)$ given by the

Ivantsov solution with the same velocity u as the sought solution, we exactly cancel the Δ term in the l.h.s. Denoting by ξ^{Iv} this Ivantsov solution and by $(\xi^{\mathrm{Iv}} \leftrightarrow \xi)$ the expression obtained from the previous one by replacing ξ with ξ^{Iv}, one has

$$\frac{\xi_{xx}}{(1+\xi_x^2)^{3/2}} = \frac{u}{2\pi}\int_0^{+\infty} dx_1 \left\{\exp\left(-\frac{u}{2}(\xi-\xi_1)\right)\left[K_0\left(\frac{u}{2}\sqrt{(x-x_1)^2+(\xi-\xi_1)^2}\right) + K_0\left(\frac{u}{2}\sqrt{(x+x_1)^2+(\xi-\xi_1)^2}\right)\right] - (\xi^{\mathrm{Iv}} \leftrightarrow \xi)\right\}.$$

Now, thanks to the subtraction of the integrand computed with the Ivantsov solution, one may expand the transcendental functions for a small argument, because the solution has to tend to the Ivantsov parabola at large distances from the tip, where the radius of curvature becomes very large.

Since $K_0(\epsilon) \approx -\log\epsilon$ as ϵ goes to $0+$, one has in this limit

$$\frac{\xi_{xx}}{(1+\xi_x^2)^{3/2}} = -\frac{u}{2\pi}\int_{-\infty}^{0} d\xi_1 \left\{x_{\xi_1}\log\left(\frac{u}{2}\sqrt{(x-x_1)^2+(\xi-\xi_1)^2}\right) - (x^{\mathrm{Iv}} \leftrightarrow x)\right\}. \tag{5}$$

Note that in this equation we have used ξ as integration variable instead of x, for a reason that will appear now. We mean by $x^{\mathrm{Iv}} \leftrightarrow x$ the expression similar to the previous one where $x(\xi)$ has been replaced by the Ivantsov solution with the same velocity. Assuming that any length scale $L \approx a^{-1}$, the l.h.s. of (5) is of order a. The dimensionalization of the r.h.s. requires some more care because of the occurrence of a term with a formal logarithmic dependence on the undercooling. In fact, once the scaling of the length with a^{-1} on the r.h.s. of (5) is done, one obtains as a dominant contribution (expressed with scaled x and ξ):

$$\frac{ua^{-1}}{\pi}\int_{-\infty}^{0} d\xi_1 \left(x_{\xi_1} - x_{\xi_1}^{\mathrm{Iv}}\right)\log(ua^{-1}).$$

This term is formally of order $\Delta^2\log\Delta^2$ although the subdominant one is of order Δ^2 only (recall that $ua^{-1} \approx \Delta^2$ in 2D). But this term is constant, as it does not depend on the position on the interface. Thus if it is not zero, this implies from (5) that the curvature of the tip is constant and thus that it is a circle, which is clearly impossible because this circle cannot be matched with the Ivantsov parabola. Thus this dominant contribution must vanish. In other words, $x(\xi) - x^{\mathrm{Iv}}$ goes to zero as ξ goes to $-\infty$. This is a condition that is imposed upon the solution of the similarity equation. It is worth noticing that this (eventual) constant contribution on the r.h.s. of (5) is not obvious at all, and can be missed if one keeps the variable of integration x instead of changing to ξ as we did.

From all these considerations the similarity equation can be written as

$$\frac{\xi_{xx}}{\left(1+\xi_x^2\right)^{3/2}} = -\frac{C}{4\pi}\int_{-\infty}^{+\infty} dx_1 \log\left[\frac{(x-x_1)^2+(\xi-\xi_1)^2}{(x-x_1)^2+\left(-x^2+x_1^2\right)^2}\right], \tag{6}$$

where $C = ua^{-2}$ is a nonlinear eigenvalue (compare with Equation (4.29) of Reference [9]), the unit length a^{-1} being such that the Ivantsov parabola is $\xi^{\mathrm{Iv}} = -x^2$. As, furthermore, a^{-1} is related to u by the Ivantsov relation $ua^{-1} = 4\Delta^2/\pi$, one obtains the following formula for the dendrite velocity:

$$u = \frac{16}{\pi^2 C}\Delta^4.$$

This relation is similar to the one obtained by Langer and Muller–Krumbhaar, in whose theory the constant C satisfies a marginal stability criterion.

To simplify the writing we have used x as variable in this last expression and ξ as unknown function, although the previous considerations were made with the reverse choice. Suppose that $\xi - \xi^{\mathrm{Iv}} \approx \delta\xi(x)$ as $|x|$ goes to ∞ with $|\delta\xi(x_1)/\xi^{\mathrm{Iv}}| \ll 1$; then the integrand on the r.h.s. of (6) behaves as $\delta\xi(x_1)/\xi_1^{\mathrm{Iv}}$ for $|x_1|$ approaching $+\infty$, x fixed. Since $\xi^{\mathrm{Iv}} \approx -x_1^2$ in this limit, this excludes the possibility that $\delta\xi \approx x$ as x goes to $+\infty$, which would lead to a logarithmically diverging integral in (6). But $\delta\xi(x) \approx x$ is equivalent to $x(\xi) - x(\xi^{\mathrm{Iv}}) \approx C'$ $(C' \neq 0)$ as ξ goes to ∞. Thus the fact that the integral on the r.h.s. of (6) does not diverge at infinity is consistent with the condition derived before.

It is difficult to know from first principles if the nonlinear eigenvalue C which appears in (6) can be any positive real number or is limited to some finite interval, to discrete values, or even to some single value (to say nothing of such curious possibilities as Cantor sets). Without going into the rather complicated details, one may say that the expansion of $\xi(x)$ at $x \to \infty$ can be found term by term for an arbitrary C. This starts as $\xi(x) \approx -x^2 - (1/\pi C)\log|x|/x^2$, as shown in Appendix B. But for integral equations this kind of calculation does not help much in investigating the multiplicity of solutions. On the other hand a straight numerical approach is not without ambiguity. Let us imagine that a 1D continuum of solutions exists (i.e. solutions with different C's; the possibility of many solutions for a given C is at least unlikely, owing to the uniqueness of the asymptotic expansion for a given C). A numerical analysis of those solutions necessarily leads to a discrete and even finite number of those solutions and it does not by itself allow us to claim that a unique solution exists—or even to limit the number of solutions.

We shall sketch now a method of study for these questions. Let us assume that a solution of (6) has been found that satisfies the boundary conditions. Let us denote it as $\xi_0(x)$, and the corresponding value of C as C_0.

We shall also assume that if a continuum of solutions exists parametrized by C, this continuum is smooth with respect to C. Within this assumption, neighboring solutions can be reached by perturbation. Thus let $\xi_0(x) + \delta\xi(x)$ be the solution corresponding to the small variation of C from C_0 to $C_0 + \delta C$.

Here δC is as small as wanted. On the other hand, Equation (6) is satisfied with $\xi = \xi_0$ and $C = C_0$. The elementary linear perturbation around this solution requires $\delta\xi$ to be the solution of

$$L\left[\xi^{(0)}, x\right]\delta\xi = \frac{\delta C}{C_0}\frac{\xi^{(0)}_{xx}}{\left(1+\xi^{(0)2}_x\right)^{3/2}},$$

where $L[\ ]$ is the linear operator obtained by linearizing Equation (6) around $\xi = \xi_0$ at $C\ (=C_0)$ fixed:

$$L\left[\xi^{(0)}, x\right]\delta\xi = \frac{\delta\xi_{xx}}{\left(1+\xi^{(0)2}_x\right)^{3/2}} - 3\frac{\xi^{(0)}_{xx}\xi^{(0)}_x\,\delta\xi_x}{\left(1+\xi^{(0)2}_x\right)^{5/2}}$$
$$-\frac{C}{4\pi}\int_{-\infty}^{+\infty} dx_1 \frac{\left[\delta\xi(x)-\delta\xi(x_1)\right]\left[\xi^{(0)}(x)-\xi^{(0)}(x_1)\right]}{(x-x_1)^2+\left(\xi^{(0)}-\xi^{(0)}_1\right)^2}. \quad (7)$$

But it turns out that $L[\ ]$ has a nontrivial kernel, since it gives 0 when acting on a constant function [an obvious consequence of the fact that Equation (6) is invariant under translations with respect to ξ]. Thus (7) bears a solvability condition: its r.h.s. must be orthogonal to the adjoint kernel. In other words, one should have

$$\int_{-\infty}^{+\infty} dx \frac{\xi^{(0)}_{xx}}{\left(1+\xi^{(0)2}_x\right)^{3/2}}\delta\xi^K(x) = 0, \quad (8)$$

where $\delta\xi^K$ is the nontrivial solution of:

$$L^+\left[\xi^{(0)}, x\right]\delta\xi^K = 0,$$

L^+ being the adjoint of L. The solvability condition (8) has been written in such a way that the inner product of functions has the usual form:

$$\left(\delta\xi^a, \delta\xi^b\right) = \int_{-\infty}^{+\infty} dx_1\,\delta\xi^a(x_1)\,\delta\xi^b(x_1). \quad (9)$$

And the adjoint of L acts on a function $\delta\xi^K(x)$ according to the equation

$$L^+\left[\xi^{(0)}, x\right]\delta\xi^K = \frac{\delta\xi^K_{xx}}{\left(1+\xi^{(0)2}_x\right)^{3/2}} - \frac{3\,\delta\xi^K_x\,\xi^{(0)}_x\xi^{(0)}_{xx}}{\left(1+\xi^{(0)2}_x\right)^{5/2}}$$
$$-\frac{C}{4\pi}\int_{-\infty}^{+\infty} dx_1 \frac{\left[\delta\xi^K(x)+\delta\xi^K(x_1)\right]\left[\xi^{(0)}(x)-\xi^{(0)}(x_1)\right]}{(x-x_1)^2+\left(\xi^{(0)}-\xi^{(0)}_1\right)^2}. \quad (10)$$

As L is not symmetric for the inner product defined in (9), the adjoint kernel $\delta\xi^K$ has *a priori* no simple relation with $\delta\xi$. In particular, a constant is not an obvious solution of (10), although this is not excluded if ξ_0 has some yet unknown special property. Thus it is difficult to know (without numerical investigation) if (8) is automatically fulfilled or not. If $\delta\xi$ were a constant, for instance, it is easy to see that (8) would not be satisfied. On the other hand, both the curvature term and the adjoint kernel are even functions of x, so that a parity argument does not help. If one supposes that (8) is not satisfied, this leads one to conclude that the solutions of the similarity equation (6) are isolated (i.e. are not smoothly embedded in a continuum of solutions). In other words, the "spectrum" of this nonlinear eigenvalue is discrete.

3. The three dimensional axisymmetric case

The equation to solve is

$$\Delta + \frac{\xi_{rr} + \frac{1}{r}\xi_r^3 + \frac{1}{r}\xi_r}{(1+\xi_r^2)^{3/2}} = u\int_0^{+\infty} \frac{d\tau}{(4\pi\tau)^{3/2}} \int_{-\infty}^{+\infty} dx_1 \int_{-\infty}^{+\infty} dy_1$$

$$\times \exp\left(-\frac{1}{4\tau}\left[(x-x_1)^2+(y-y_1)^2+(\xi-\xi_1+u\tau)^2\right]\right). \tag{11}$$

As explained in the introduction, the Ivantsov solution for this equation is obtained by formally canceling the Δ term on the left hand side (this amounts to neglecting the curvature contribution to the interface temperature). This Ivantsov solution reads $\xi = -ar^2$, and a and u are related to Δ as

$$\Delta = \frac{u}{2}\int_0^{+\infty} d\rho\, \frac{\rho\exp(-\rho^2/4)}{a\rho^2+u}.$$

This gives, in the small Δ limit, $ua^{-1} \approx -2\Delta/\log\Delta$ (again this relates Δ to ua^{-1} only). As in the 2D case, one replaces the Δ term on the l.h.s. of (11) by the integral on the r.h.s. computed for the Ivantsov solution, that is, for a paraboloid in the present case. Then one subtracts the result from the r.h.s. and expands the transcendental functions for small argument, this one being of order ua^{-1}. This gives the following equation, valid in this limit of small undercooling:

$$\frac{\xi_{rr} + \frac{1}{r}(\xi_r^3+\xi_r)}{(1+\xi_r^2)^{3/2}}$$

$$= \frac{u}{4\pi}\int_0^{+\infty} dr_1 \int_0^{\pi} d\phi \left(\frac{r_1}{\sqrt{(\xi-\xi_1)^2+(r-r_1)^2+4rr_1\sin^2\phi}} - (\xi^{\mathrm{Iv}} \leftrightarrow \xi)\right). \tag{12}$$

The ϕ integration gives a complete elliptic integral since one has:

$$\int_0^{\pi} d\phi \frac{r_1}{\left[(\mathbf{R}-\mathbf{R}_1)^2+4rr_1\sin^2\phi\right]^{1/2}}$$

$$=\frac{2r_1}{\left[(\mathbf{R}-\mathbf{R}_1)^2+4rr_1\right]^{1/2}}K\left(\frac{2\sqrt{rr_1}}{\left[(\mathbf{R}-\mathbf{R}_1)^2+4rr_1\right]^{1/2}}\right). \tag{13}$$

where $K(\cdot)$ is the Jacobi complete elliptic integral (or elliptic period) and where $(\mathbf{R}-\mathbf{R}_1)^2=(\xi-\xi_1)^2+(r-r_1)^2$.

To get from (12) the similarity equation one has to scale lengths as a^{-1}, where a is the parameter that appears in the Ivantsov solution, and assume that u is related to a as specified by this Ivantsov solution. Thus (12) yields an equation where Δ has disappeared, and where the scaling leaves undetermined a "nonlinear eigenvalue" $C=ua^{-2}$. This similarity equation reads

$$\frac{\xi_{rr}+\frac{1}{r}(\xi_r^3+\xi_r)}{(1+\xi_r^2)^{3/2}}=\frac{C}{4\pi}\left\{\int_0^{+\infty}dr_1\left[\frac{2r_1}{\left[(\mathbf{R}-\mathbf{R}_1)^2+4rr_1\right]^{1/2}}\right.\right.$$

$$\left.\left.\times K\left(\frac{2\sqrt{rr_1}}{\left[(\mathbf{R}-\mathbf{R}_1)^2+4rr_1\right]^{1/2}}\right)-(\xi^{\mathrm{Iv}}\leftrightarrow\xi)\right]\right\}. \tag{14}$$

At least formally, this equation is as complicated as the similarity equation for the 2D case. An argument almost identical to the one sketched at the end of the previous section gives some support to the idea that the nonlinear eigenvalue C has a discrete spectrum.

For a given eigenvalue C, one obtains a dendrite velocity by using both the relations $C=ua^{-2}$ and the Ivantsov relation $ua^{-1}=-2\Delta/\log\Delta$. This leads to $u=(2\Delta/\log\Delta)^2/C$.

Appendix A

As shown by Ivantsov [3] and by Cahn and Horway [4], solutions of (1) can be found in the form of a parabola in 2D and of an elliptic paraboloid in 3D if one neglects the effect of surface tension. Both solutions result from the fact that the integral

$$I=\int_0^{+\infty}\frac{d\tau}{(4\pi\tau)^{3/2}}\int_{-\infty}^{+\infty}dx_1\int_{-\infty}^{+\infty}dy_1$$

$$\times\exp\left(-\frac{1}{4\tau}\left[(x-x_1)^2+(y-y_1)^2+(\xi-\xi_1+u\tau)^2\right]\right), \tag{A.1}$$

which is *a priori* a function of x and y [recall that ξ stands for $\xi(x,y)$ and ξ_1 for

$\xi(x_1, y_1)$, where $\xi(x, y)$ is the Cartesian representation of the dendrite], becomes a function of a, b, and u when computed for $\xi(x, y) = -(ax^2 + by^2)$. The 2D case is recovered with $b = 0$, and the circular paraboloid with $a = b$.

Thus we want to show that

$$I = \int_0^{+\infty} \frac{d\tau}{(4\pi\tau)^{3/2}} \int_{-\infty}^{+\infty} dx_1 \int_{-\infty}^{+\infty} dy_1$$

$$\times \exp\left[-\frac{1}{4\tau}\left((x-x_1)^2 + (y-y_1)^2 + \left[-a(x-x_1)^2 - b(y-y_1)^2 + u\tau \right]^2 \right) \right]$$

is a function of u, a, and b only. This is mainly a matter of elementary transformations of the integration variables. Let us decompose the first few ones:

Replace (x_1, y_1) by $X' = x - x_1$, $Y' = y - y_1$, then (X', Y') by $(X = X'\sqrt{\tau}$, $Y = Y'\sqrt{\tau})$, then τ by $T^* = \sqrt{\tau}$, and finally T^* by $T = T^*(aX^2 + bY^2 + u)$. This gives

$$I = \frac{2}{(4\pi)^{3/2}} \int_0^{+\infty} dT \int_{-\infty}^{+\infty} dX \int_{-\infty}^{+\infty} dY \frac{F(X, Y, T)}{aX^2 + bY^2 + u},$$

where $F(X, Y, T) = \exp\{-\frac{1}{4}[X^2 + Y^2 + (T - 2axX - 2byY)^2]\}$.

The next step consists in restricting the X and Y integrations to positive values only:

$$I = \frac{2}{(4\pi)^{3/2}} \int_0^{+\infty} dT \int_0^{+\infty} dX \int_0^{+\infty} dY \frac{[\ ]}{aX^2 + bY^2 + u}, \tag{A.2}$$

where $[\ ] = F(X, Y, T) + F(-X, Y, T) + F(-X, -Y, T) + F(X, -Y, T)$. Now the integrand is an even function of T, and one may replace $\int_0^{+\infty} dT$ by $\frac{1}{2}\int_{-\infty}^{+\infty} dT$ and perform the T integration at once. This gives

$$I = \frac{1}{\pi} \int_0^{+\infty} dX \int_0^{+\infty} dY \frac{\exp[-\frac{1}{4}(X^2 + Y^2)]}{aX^2 + bY^2 + u}.$$

This is the result sought: the x and y dependence in I has disappeared.

Appendix B

In this appendix we derive the first correction to the Ivantsov solution for the asymptotic behavior ($x \to \infty$) of solution of the similarity equation.

Two dimensional case

The similarity equation has the form

$$\frac{\xi_{xx}}{(1 + \xi_x^2)^{3/2}} = -\frac{C}{4\pi} \int_{-\infty}^{+\infty} dx_1 \log\left[\frac{(x - x_1)^2 + (\xi - \xi_1)^2}{(x - x_1)^2 + (x^2 - x_1^2)^2} \right]. \tag{B.1}$$

This is an equation for the unknown function $\xi(x)$. At large x the latter has to tend to the Ivantsov solution, so that the argument of the logarithm for large x, as a function of x_1 remains close to 1: If x_1 is close to x, then $\xi(x_1)$ is almost equal to $-x_1^2$; if x_1 is close to 1 (that is, in the region of the tip of the dendrite), then x_1 $[\xi(x_1)]$ is small with respect to x $[\xi(x)]$. So let us write $\xi(x)$ as

$$\xi(x) \approx -x^2 + \psi(x). \tag{B.2}$$

By arguments similar to the previous one, one has the following uniform estimates for large x:

$$\log\left[\frac{(x-x_1)^2+(\xi-\xi_1)^2}{(x-x_1)^2+(-x^2+x_1^2)^2}\right] \sim -2\frac{(\psi-\psi_1)(x+x_1)^2}{(x^2-x_1^2)\left[1+(x+x_1)^2\right]}$$
$$\sim \Lambda(x, x_1) \tag{B.3}$$

For large x, the integral on the r.h.s. of (B.7) is dominated by the contribution of the tip and reads

$$\int_{-\infty}^{+\infty} dx_1\, \Lambda(x, x_1) \underset{x\to+\infty}{\sim} \frac{2}{x^2}\int_{-\infty}^{+\infty} dx_1\, \psi(x_1). \tag{B.4}$$

Assume for the moment that this integral converges at large x_1, which is to be verified when we shall find the behavior of $\psi(x)$ at large x. One sees from (B.1) that the function $\psi(x)$ has to verify a kind of quantization condition, because the l.h.s. of (B.1) behaves as

$$\frac{\xi_{xx}}{\left(1+\xi_x^2\right)^{3/2}} \underset{x\to+\infty}{\sim} -\frac{1}{4|x|^3}, \tag{B.5}$$

and so cannot match the x^{-2} behavior of the r.h.s. of this equation. This quantization condition is thus

$$\int_{-\infty}^{+\infty} dx_1\, \psi(x_1) \equiv \int_{-\infty}^{+\infty} dx_1\left[x_1^2+\xi(x_1)\right] = 0, \tag{B.6}$$

which is a nontrivial output of this analysis.

The asymptotics of $\psi(x)$ is now found by matching the $|x|^{-3}$ term of the l.h.s. of the equation (B.1) with the subdominant term of the r.h.s. as $x \to \infty$. For that purpose, one subtracts from this r.h.s. the formally dominant contribution of order x^{-2}, and obtains, after a few simple manipulations,

$$\int_{-\infty}^{+\infty} dx_1\left(\Lambda(x, x_1) - \frac{2}{x^2}\psi(x_1)\right) \underset{\lambda\to\infty}{\sim} \frac{1}{x^2}\int_{-\infty}^{+\infty} dx_1\, \frac{\mu(x)-\mu(x_1)}{x^2-x_1^2}, \tag{B.7a}$$

where $\mu(x) = -2x^2\psi(x)$.

Looking now at the change in the integrand under dilatation of x, and recalling that one is looking for a $1/|x|^3$ behavior of this integral, it is convenient to take $\mu(x) = \alpha \log|x|$ (recall that μ is in general an even function), and one finds

$$\int_{-\infty}^{+\infty} dx_1 \left(\Lambda(x, x_1) - \frac{2}{x^2} \psi(x_1) \right) \sim - \frac{\alpha}{|x|^3} \int_{-\infty}^{+\infty} du \, \frac{\log|u|}{1-u^2}, \tag{B.7b}$$

from which one deduces $\alpha = +2/\pi C$. Summarizing, we have found the beginning of the large x expansion of $\xi(x)$ as

$$\xi(x) \approx -x^2 - \frac{1}{\pi C} \frac{\log|x|}{x^2} \tag{B.8}$$

It seems reasonable to expect that this expansion can be pursued at any order (even though it would be cumbersome), so that the solution is specified in this way by the single parameter C, and that this parameter can be considered as fixed by the supplementary "quantization condition" (B.6), giving some support to the idea that C belongs to a discrete set.

Three dimensional case

The similarity equation has the form

$$\frac{\xi_{rr} + \frac{1}{r}(\xi_r^3 + \xi_r)}{(1+\xi_r^2)^{3/2}} = \frac{C}{4\pi} \left\{ \int_0^{+\infty} dr_1 \left[\frac{2r_1}{\left[(\mathbf{R}-\mathbf{R}_1)^2 + 4rr_1 \right]^{1/2}} \times K\left(\frac{2\sqrt{rr_1}}{\left[(\mathbf{R}-\mathbf{R}_1)^2 + 4rr_1 \right]^{1/2}} \right) - (\xi^{\mathrm{Iv}} \leftrightarrow \xi) \right] \right\}. \tag{B.9}$$

As in the 2D case, one writes

$$\xi(r) = -r^2 + \varphi(r),$$

and one has the following estimate for large r:

$$\frac{2r_1}{\left[(\mathbf{R}-\mathbf{R}_1)^2 + 4rr_1 \right]^{1/2}} K\left(\frac{2\sqrt{rr_1}}{\left[(\mathbf{R}-\mathbf{R}_1)^2 + 4rr_1 \right]^{1/2}} \right) - (\xi^{\mathrm{Iv}} \leftrightarrow \xi)$$

$$\sim \frac{-\pi r_1 (\varphi - \varphi_1)(r_1^2 - r^2)}{\left[(r-r_1)^2 + (-r^2 + r_1^2)^2 + 4rr_1 \right]^{3/2}}. \tag{B.10}$$

As the left hand side of the equation (B.9) behaves as $-1/r$ for large r, the asymptotic behavior of φ for large r is given by the equation

$$\frac{1}{r} = \frac{C}{4\pi}\int_0^{+\infty} dr_1 \frac{r_1(\varphi-\varphi_1)(r_1^2-r^2)}{\left[(-r^2+r_1^2)^2+(r-r_1)^2+4rr_1\right]^{3/2}}. \qquad \text{(B.11)}$$

This gives the asymptotic behavior of the dendrite shape ξ for large r:

$$\xi(r) \approx -r^2 + \frac{2}{C}\frac{r}{\log r}.$$

Contrary to the 2D case, no additional "quantization condition" is required.

Acknowledgments

We are indebted to the CNLS, Los Alamos National Laboratory, University of California, Los Alamos, New Mexico, where the work was initiated, and also to Alain Pumir for fruitful discussions.

References

1. P. Boni, J. H. Bilgram, and W. Kanzig, *Phys. Rev. A* 28:2953 (1983).
2. O. N. Mesquita, D. G. Neal, M. Copic, and H. Z. Cummins, *Phys. Rev. B* (1984).
3. G. P. Ivantsov, *Dokl. Akad. Nauk SSSR* 58:567 (1947).
4. G. Horway and J. W. Cahn, *Acta Metall.* 9:695 (1961).
5. J. W. McLean and P. G. Saffman, *J. Fluid. Mech.* 102:445 (1981).
6. Y. Pomeau, *Stud. Appl. Math.*, *73*, 75–79 (1985).
7. G. E. Nash, Naval Research Lab. Reports, Nos. 7679 and 7680, 1974.
8. J. S. Langer, *Acta Metall.* 25:1121 (1977).
9. J. S. Langer and H. Muller-Krumbhaar, *Acta Metall.* 26:1681–1697 (1978).
10. J. S. Langer, *Rev. Modern Phys.* 52:1 (1980).

Université de Provence
C.E.N. Saclay

(Received May 23, 1985)

Development of Side Branching in Dendritic Crystal Growth

A. Dougherty, P. D. Kaplan, and J. P. Gollub

Physics Department, Haverford College, Haverford, Pennsylvania 19041, and
Physics Department, University of Pennsylvania, Philadelphia, Pennsylvania 19104
(Received 19 December 1986)

The growth of dendritic crystals of NH_4Br from supersaturated solution has been studied in a way that allows high-resolution measurements of the early stages of side branching. The branches are found to be nonperiodic at any distance from the tip, with apparently random variations in both phase and amplitude. Side branches on opposite sides of the dendrite are imperfectly correlated. The rms side-branch amplitude is an exponential function of distance from the tip, with no apparent onset. The implications of these results for several theories of the stability of growing dendrites are discussed.

PACS numbers: 68.70.+w, 61.50.Cj, 81.30.Fb

The problem of explaining the convoluted shapes of dendritic crystals is particularly challenging.[1-4] Dendrites growing from a supercooled melt or supersaturated solution are characterized by smooth, nearly parabolic tips growing at constant velocity with fairly regular side branches (for example, see Fig. 1). These branches compete with each other through the diffusion field in the liquid phase. Smaller ones are screened and cease growing or even begin to dissolve, while larger ones accelerate. This coarsening process is important in determining the large-scale grain structure of solidifying materials.

Much is known about the tip shape and velocity, the mean initial side-branch spacing, and the importance of crystalline anisotropy, from earlier experimental and theoretical studies.[1-5] However, the processes that determine the side-branching frequency and amplitude, and the subsequent coarsening, are still the subject of controversy. In this paper we present a study of the growth of free dendrites of ammonium bromide from supersaturated aqueous solution. Our experiment allows the side-branch amplitudes to be measured to a precision of 2% of the tip radius, and is hence particularly sensitive to the dynamics of the side-branching process. We find that the side-branch amplitude is surprisingly noisy, and comment on the implications of this fact for theoretical models.

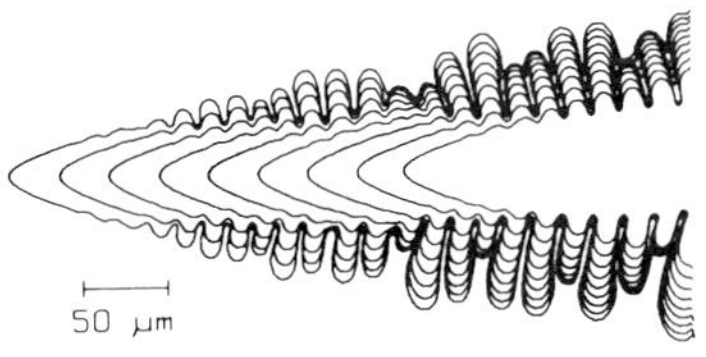

FIG. 1. Contours of an ammonium bromide dendrite growing from supersaturated solution. The contours are obtained by digital processing of microscope images taken at 20-s intervals. Supersaturation $\Delta \approx 0.007$; tip radius $\rho = 4.0$ μm; tip speed $v = 1.44$ μm/s; initial side-branch spacing $\lambda = 16$ μm.

In the standard continuum approximation, growth from solution is controlled by the diffusion of solute towards the interface. The control parameter is the dimensionless supersaturation $\Delta = (C_\infty - C_{eq})/(C_s - C_{eq})$, where C_{eq} is the equilibrium concentration of NH_4Br at the operating temperature, and C_s and C_∞ are the concentrations in the solid and in the solution far from the interface. The parameter Δ relates the concentration gradient in the solution to the concentration discontinuity across the interface. In the absence of surface tension and crystalline anisotropy, and ignoring side branching, the steady-state, or Ivantsov, solutions to the diffusion equation are paraboloids of revolution with tip radius ρ, moving at constant velocity v. These solutions fix only the Péclet number, $\mathcal{P} = \rho v/2D$, where D is the diffusion constant of the crystallizing material.[1] Experimentally, however, the dendrite grows at a unique velocity. More recent work indicates that when surface tension and anisotropy (but not side branching) are included, this continuous family breaks down into a discrete set of nearly parabolic solutions.[4] Of these solutions, only the one with the highest velocity is believed to be linearly stable.[6] However, these theoretical results have not yet been adequately tested experimentally.

To investigate the origin of side branching, which is the primary goal of this work, we look for deviations from the steady-state shape. Specifically, we measure the half-width $w_z(t)$ of the dendrite (from the center to one side) *at a fixed distance z behind the tip,* as a function of time. This amounts to viewing the dendrite in a frame of reference moving with the tip. The width oscillates as the side-branching waves pass by the observation point. Measured over long times, $w_z(t)$ gives a statistically stationary measure of the side-branching activity at that fixed location z. The measurement process is repeated for various values of z in order to characterize the initiation and subsequent development of side branches.

The experiments are conducted in a glass cell of dimensions $60 \times 6 \times 0.3$ mm^3, filled with a known concen-

Reprinted with permission from *Physical Review Letters*, Vol. 58, No. 16, 1987.

tration of NH_4Br. After the solid has been dissolved, the cell is cooled to a uniform temperature that determines Δ. A crystal nucleates at random; only experiments yielding a single free dendrite are used. The growth process is observed through a microscope with a video camera and recorded on video tape for later digital analysis.

The discussion that follows is based on a detailed analysis of a particular experimental run characterized by the following parameters: $\Delta \approx 0.007$, $\rho = 4.0 \pm 0.2$ μm, $v = 1.44 \pm 0.08$ μm/s, $D = (2.6 \pm 0.2) \times 10^{-5}$ cm^2/s, $\mathcal{P} = 0.0011$, and λ (the mean side-branch spacing near the tip) $= 16 \pm 1$ μm. (The surface tension and its anisotropy are not known for this system. However, qualitative observations of the dendritic shapes suggest that the anisotropy is somewhat larger than that of succinonitrile.) The growth was recorded for 36 min, yielding a dendrite more than 3 mm long with approximately 200 side branches. To obtain $w_z(t)$, about 14000 video frames were digitized to a resolution of 512×480 pixels, with each pixel being 0.64 ± 0.01 μm.

The character of the side branches near the tip is shown in Fig. 2, where $w_z(t)$ and its power spectrum $P_z(f)$ are shown for $z = 32$ μm. (The actual data record used to compute the spectrum is six times the length shown.) The side branching is only approximately periodic, and the amplitude has an unanticipated intermittent character. The spectrum contains a peak at the characteristic side-branching frequency of 0.09 Hz, but the peak has a standard deviation $\sigma = 0.007$ Hz when fitted by a Gaussian, indicating the presence of a range of frequencies.

This noisy behavior is also reflected in the cross-correlation function

$$C(\tau) = \langle [w_{Lz}(t+\tau) - \bar{w}_{Lz}][w_{Rz}(t) - \bar{w}_{Rz}] \rangle / \sigma_L \sigma_R,$$

where $w_{Lz}(t)$, $w_{Rz}(t)$, σ_L, and σ_Z are the width functions and their standard deviations for the two sides of the dendrite at the same z. We find that $C(0)$ is less than 0.4 for any z, and $C(\tau)$ drops to zero (within the measurement precision) for $\tau \gtrsim 66$ s, the time to nucleate six side branches. Similar behavior is seen for the autocorrelation function, so that side branches separated by more than about 6λ are essentially uncorrelated with each other. There are apparently random fluctuations in both the phase and amplitude of $w_z(t)$ that destroy any long-range correlations.

The character of the side branches for $z = 95$ μm (Fig. 3) is similar to that near the tip except for the obvious presence of competition. The side-branch amplitude is larger, but the spectral peak has essentially the same mean frequency and width as it had at $z = 32$ μm. The coarsening process has increased the overall noise level in the spectrum at all frequencies, including those below the peak. Beyond about 150 μm, the spectral peak stops growing and the background noise level continues to rise,

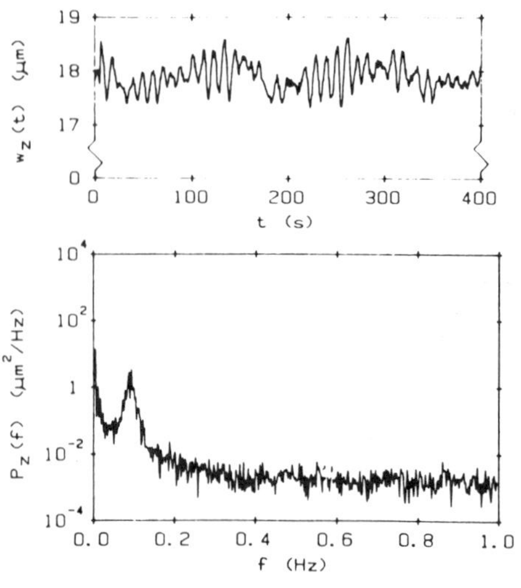

FIG. 2. Portion of the dendrite width function $w_z(t)$ and its power spectrum $P_z(f)$, for $z = 32$ μm behind the tip. The intermittency of the side-branching process is evident.

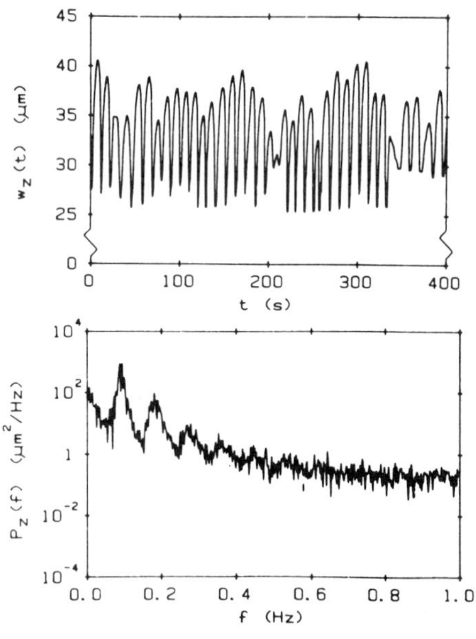

FIG. 3. Portion of $w_z(t)$ and its power spectrum $P_z(f)$ farther from the tip at $z = 95$ μm. Note the change of scales for $w_z(t)$.

yielding a relatively featureless spectrum with noise at all frequencies.

It is worth emphasizing that although these are optical measurements, they are reproducible to considerably less than the optical wavelength. The resolution limit of the microscope objective is about 1.9 μm, and the dendrite edge is actually spread over about 4 μm in the digitized image. However, interpolation over the edge region (and signal averaging) improves the effective resolution dramatically. In particular, for $z = 32$ μm, where the root mean square variation in the side-branch amplitude is only 0.32 μm, the measurement precision is about 0.05 μm.

A possible source of systematic error is the concentration gradients which locally change the index of refraction, and slightly reduce (by less than 5%) the apparent radius of curvature. This would occur even without side branches, but the distortion in that case would be static in the tip frame and thus have no effect on the spectral measurements. However, the side branches are accompanied by local enhancements of the concentration gradients, leading to a possible periodic distortion estimated to be about 0.08 μm. This is generally much smaller than the side-branch amplitudes measured.

It will be important to test whether mechanisms other than diffusion, such as convection, are significant for the dendritic growth in this system. However, preliminary estimates indicate that this is unlikely because the value for Δ obtained from the Ivantsov solution with the measured ρ and v is in approximate agreement with the measured Δ, so that diffusion is the dominant transport mechanism.

Despite the apparently restricted geometry, the dendrites in this experiment are not two-dimensional. The diffusion field around a parabolic dendrite decreases very sharply, as the exponential integral $E_1(x/l)$, where x is the distance perpendicular to the growth direction and $l = 2D/v$ is the diffusion length (about 3.6 mm in this case). Even though the cell thickness d is only about $0.08l$, the concentration deviation $[C(x) - C_{eq}]$ at $x = d/2$ is about $0.56(C_\infty - C_{eq})$. Thus the growth is not as strongly constrained in the third dimension as one might expect from the geometry. (This view is substantiated by experiments on directional solidification of a dilute binary alloy,[7] in which Somboonsuk, Mason, and Trivedi determined that the dendrite tip radius and initial side-branch spacing were independent of d for $d \gtrsim 0.06l$.)

Theoretical models at present offer little insight into the dynamics of side branching. Even for the purely local models that have been extensively studied, it is not clear whether persistent side branching occurs.[8-10] However, several possible scenarios have been suggested.

According to one hypothesis, the dendrite tip is not actually a stationary point in the tip frame of reference; rather, its position and curvature oscillate, forming a limit cycle at the side-branching frequency.[8] If this is correct, the side-branch amplitude would be periodic, and the two sides almost perfectly correlated, since noise should be ineffective in the presence of a limit cycle. Neither feature is observed in the present experiment. We have also made careful measurements of the tip velocity, and found it to be constant. In a bandwidth of 0.1 Hz around the side-branching frequency, the velocity fluctuations are less than 0.05 μm/s, or 3.5%. This is consistent with most previous experiments on free dendrites,[2] although tip oscillations have been seen in two-dimensional growth for very low supersaturation.[11] We conclude from these observations that a dynamical limit cycle involving the tip is unlikely to be an essential ingredient of the side-branching process, at least in this system.

Alternatively, the tip could be a stable structure, with side branches resulting from the amplification of finite microscopic noise.[9,12] In this view, noise (e.g., concentration fluctuations in the solution near the tip) gives rise to small interfacial perturbations. In the tip frame of reference, these perturbations propagate away from the tip and grow approximately exponentially with distance. This may be qualitatively similar to the growth of a disturbance as it is convected downstream in some hydrodynamic systems such as pipe flow.[13] In this "noise-induced side-branching" scenario, the structure is expected to be nonperiodic, and branches on opposite sides of the dendrite imperfectly correlated, as is observed. It is not implausible that concentration fluctuations could be significant, since the fractional fluctuations in Δ in a 1 $(\mu m)^3$ volume due to Poisson statistics are about 10^{-3} in this experiment.

The growth of the branches as a function of distance z from the tip is given by the square root of the area under the spectral peak (Fig. 4). This growth appears exponential at very small z, but saturates at large z. (The data are also consistent with another suggested growth law, $\exp\sqrt{s}$, where s is the arclength.[12]) This exponential growth of the side-branch amplitude in the tip frame

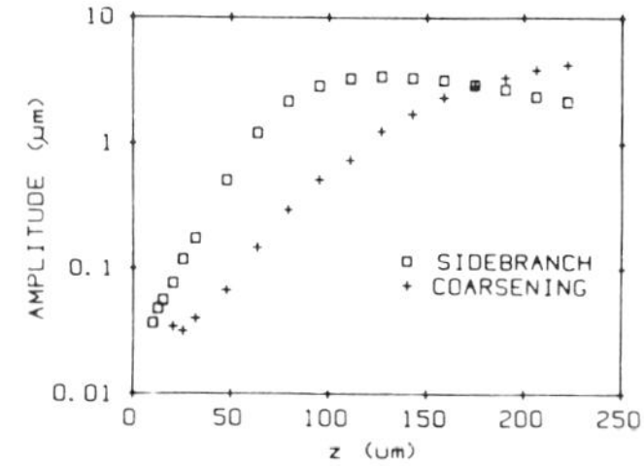

FIG. 4. Squares: rms variations in the side-branch amplitude vs distance z from the tip. Crosses: Coarsening noise (see text).

appears consistent with noise-induced side branching. There is no clear onset; the amplitude simply falls below the measurement noise level for $z < 10\ \mu m$, despite the precision of the measurements.

The development of coarsening is manifested as a rise in the low-frequency power with increasing z. Though the coarsening occurs at all wave numbers less than that of the spectral peak, we use, as a measure of the coarsening, the square root of the spectral power in a bandwidth of 0.025 Hz centered at half the side-branching frequency. Its growth as a function of z (Fig. 4) is slower than that of the side-branching instability, but is important rather close to the tip. This observation emphasizes the need to make measurements very close to the tip to explore the origin of side branching.

In summary, our measurements are consistent with the hypothesis that the amplification of microscopic noise is important to the side-branching process. To determine whether the structure is dependent on the noise intensity, the growth of larger dendrites (lower Δ) should be studied. Experiments along these lines are currently under way.

It may ultimately be necessary to consider the origin of side branches as a nucleation problem which is very sensitive to conditions on the atomic scale (for example, the presence of grain boundaries).[14] Continuum models may not be adequate to describe fully the stability of dendrites.

This work was supported by National Science Foundation Low Temperature Physics Grant No. DMR-8503543. Seed funds were previously provided by the National Science Foundation Materials Research Laboratory program under Grant No. DMR-82-16718. We appreciate helpful conversations with M. Glicksman, N. Goldenfeld, and H. Levine, and also the opportunity for extensive discussions at the Aspen Center for Physics.

[1]For a review, see J. S. Langer, Rev. Mod. Phys. **52**, 1 (1980), and references therein.

[2]S.-C. Huang and M. E. Glicksman, Acta Metall. **29**, 701, 717 (1981); H. Honjo and Y. Sawada, J. Crystal Growth **58**, 297 (1982).

[3]R. Trivedi and K. Somboonsuk, Acta Metall. **33**, 1061 (1985).

[4]D. A. Kessler, J. Koplik, and H. Levine, Phys. Rev. A **33**, 3352 (1986); B. Caroli, C. Caroli, B. Roulet, and J. S. Langer, Phys. Rev. A **33**, 442 (1986).

[5]F. Heslot and A. Libchaber, Phys. Scr. **T9**, 126 (1985).

[6]D. A. Kessler and H. Levine, Phys. Rev. Lett. **57**, 3069 (1986).

[7]K. Somboonsuk, J. T. Mason, and R. Trivedi, Metall. Trans. A **15A**, 967 (1984).

[8]See, for example, E. Ben-Jacob, N. Goldenfeld, B. G. Kotliar, and J. S. Langer, Phys. Rev. Lett. **53**, 2110 (1984); D. Kessler, J. Koplik, and H. Levine, Phys. Rev. A **30**, 3161 (1984).

[9]R. Pieters and J. S. Langer, Phys. Rev. Lett. **56**, 1948 (1986); G. Li, D. Kessler, and L. Sander, Phys. Rev. A **34**, 3535 (1986).

[10]N. Goldenfeld, private communication.

[11]H. Honjo, S. Ohata, and Y. Sawada, Phys. Rev. Lett. **55**, 841 (1985). Tip oscillations have also been reported in directional solidification: L. R. Morris and W. C. Winegard, J. Crystal Growth **1**, 245 (1967).

[12]D. A. Kessler and H. Levine, private communication.

[13]R. J. Diessler, J. Stat. Phys. **40**, 371 (1985), and Physica (Amsterdam) **18D**, 467 (1986); J. M. Chomaz, P. Huerre, and L. G. Redekopp, Bull. Am. Phys. Soc. **31**, 1696 (1986).

[14]R. J. Schaefer and M. E. Glicksman, Metall. Trans. **1**, 1973 (1970).

Directional Solidification

Reprinted from JOURNAL OF APPLIED PHYSICS, Vol. 35, No. 2, 444–451, February 1964

Stability of a Planar Interface During Solidification of a Dilute Binary Alloy

W. W. MULLINS
Department of Metallurgical Engineering, Carnegie Institute of Technology, Pittsburgh, Pennsylvania

AND

R. F. SEKERKA
Physics Department, Harvard University, Cambridge, Massachusetts
(Received 15 July 1963; in final form 9 September 1963)

The stability of the shape of a moving planar liquid–solid interface during the unidirectional freezing of a dilute binary alloy is theoretically investigated by calculating the time dependence of the amplitude of a sinusoidal perturbation of infinitesimal amplitude introduced into the planar shape. The calculation is accomplished by using gradients of the steady-state thermal and diffusion fields satisfying the perturbed boundary conditions (capillarity included) to determine the velocity of each element of interface, a procedure justified in some detail. Instability occurs if any Fourier component of an arbitrary perturbation grows; stability occurs if all components decay. A stability criterion expressed in terms of growth parameters and system characteristics is thereby deduced and is compared with the currently used stability criterion of constitutional supercooling; some very marked differences are discussed.

INTRODUCTION

IN their original experimental study of the breakdown of a planar liquid–solid interface into cellular undulations during the unidirectional freezing of a dilute binary alloy, Rutter and Chalmers[1] proposed a criterion which predicted instability of the planar interface shape whenever growth conditions produced an adjacent layer of constitutionally supercooled liquid. The conditions under which constitutional supercooling (hereafter referred to as CS) occurs were subsequently placed on a quantitative basis by Tiller, Rutter, Jackson, and Chalmers.[2] Although the resulting quantitative CS criterion has successfully predicted the dependence of the breakdown conditions observed in a number of experiments[3] on such growth parameters as the growth rate and the temperature gradient in the liquid, the theoretical status of the CS criterion has remained somewhat uncertain, since it was originally deduced by the application of equilibrium arguments concerning supercooling to the interfacial region which comprises an open system in a steady or quasisteady state.

An approach more closely related to that of the present paper has been made by Temkin[4] who attempted to analyze the stability of a planar interface by comparing the velocity of the tip of a hemispherical boss, superimposed on the plane and projecting into the liquid, with the velocity of the interface far from the boss. The velocities were calculated from the fluxes of heat and solute obtained in turn from an approximate solution to the steady-state boundary value problem corresponding to the assumed interface shape with capillarity included in the boundary conditions. His treatment, however, has the following drawbacks: (1) In place of actually obtaining the unique solution of the boundary value problem he formulates, the author, without giving any justification, imposes a guessed form for the concentration at the interface. The approximate nature of this guess introduces an over determination which, as he points out, prevents certain boundary conditions from being everywhere satisfied; for example, the velocity near the equator of the boss is not uniquely determined, having one value when calculated from solute diffusion and a second value when calculated from heat flow (the latter actually giving an infinite value). In view of the arbitrary character of the approximate solution (never actually exhibited) and of the inconsistencies to which it leads, it is difficult to assess the reliability of the author's conclusions. (2) Even if the preceding boundary value problem were rigorously solved, the results would not afford a true criterion o[f] stability since the latter must be based on whether o[r] not an *arbitrary* infinitesimal perturbation of the plana[r] shape grows with time. The hemispherical perturbatio[n] not only has a very special shape but it also gives th[e] stability problem the specious appearance of a nuclea[-]tion problem since the hemisphere, whose amplitud[e] (radius) and wavelength or lateral extent (diameter) ar[e] necessarily proportional, must always exceed som[e] critical size in order to grow owing to the dominance c[of] capillarity for sufficiently small sphere radii; by con[-]trast, we see below that instability prevails if and only [if] there are perturbations of arbitrarily small amplitud[e] which grow.

The purpose of this paper is to develop a rigorou[s] theory of the stability of the planar interface by calcu[-]lating the time dependence of the amplitude of a sinu[s-]oidal perturbation of infinitesimal initial amplitu[de] introduced into the shape of the plane; since an arb[i-]trary infinitesimal perturbation in the plane may b[e] analyzed into infinitesimal sinusoidal Fourier com[-]ponents, and since all the relevant equations are linea[r] the development of the perturbation is simply a supe[r-]

[1] J. W. Rutter and B. Chalmers, Can J. Phys. **31**, 15 (1953).
[2] W. A. Tiller, J. W. Rutter, K. A. Jackson, and B. Chalmers, Acta Met. **1**, 428 (1953).
[3] W. A. Tiller and J. W. Rutter, Can. J. Phys. **34**, 729 (1956).
[4] D. E. Temkin, Dokl. Akad. Nauk. SSSR, **133**, 174 (1960). [English transl.: Soviet Phys.—Doklady **5**, 607 (1960)].

Reprinted with permission from *Journal of Applied Physics*, Vol. 35, No. 2, pp. 444–451 (1964).

osition of the development of its components. There-
ore, the interface is unstable if any sinusoidal wave rows and is stable if none grow.

The principal physical assumptions of the present nalysis are those of isotropy of bulk and surface arameters, and of local equilibrium at the interface. 'he critical mathematical simplification is the use of teady-state values (independent of time in the frame ttached to the interface) for the thermal and diffu-ion fields. The justification for doing this is that the xperiments are usually arranged so that the interface noves with a constant velocity; if the interface shape ere also fixed, the use of steady-state values would be gorously correct. We are, however, interested in cases n which the shape of the interface is changing due to the rowth (or decay) of a perturbation in its planar nape. Clearly the steady-state field calculation gives alid results only for those perturbations whose shape nanges sufficiently slowly.[5] Nevertheless, stability riteria based on slowly changing perturbations as istinguished from the accurate calculation of the be-avior of every perturbation should be reliable for the ollowing reason: when our calculations predict in-:ability, there is always a sinusoidal perturbation of ome wavelength λ_0, above which waves grow, below hich waves decay, and at which a wave must have a onstant amplitude (Fig. 1). There is then a range of nusoidal perturbations, having wavelengths slightly reater than λ_0, which grow sufficiently slowly to be ccompanied by steady-state fields and hence to be orrectly described by the calculations[5]; therefore, nstability is rigorously predicted. When, on the other and, the calculations predict stability (for all wave-ngths) it means that no wave can grow if accompanied y a steady-state diffusion field. (The thermal field is lways in the steady state for all practical cases.) But it all the more certain not to grow if the field lags be-ind the steady-state values for it is the lateral re-istribution of solute which permits the wave to grow t all; that is, all waves decay if the solute is inhibited om moving parallel to the interface (see Sec. II). herefore, the predictions concerning stability and nstability based on the steady-state solutions should e reliable.

In the case that capillarity can be neglected (for ertain ranges of the physical parameters), the results ad to a modified CS criterion [Eq. (32)] in fairly good greement with the original one; otherwise they lead a substantially different stability criterion. In the tter case, for example, we find a critical velocity, ependent on the liquid–solid interfacial tension Eq. (27)] above which instability is impossible regard-ss of the values of the gradients or the degree of CS.

[5] The fractional change of amplitude should be small in a time ual to the relaxation time of the diffusion field (assumed larger an that of the thermal field) i.e., $|(\dot{\delta}/\delta)(\lambda^2/D)| \ll 1$, where λ is e wavelength of the sinusoidal perturbation and D is the solute fusion coefficient in the liquid.

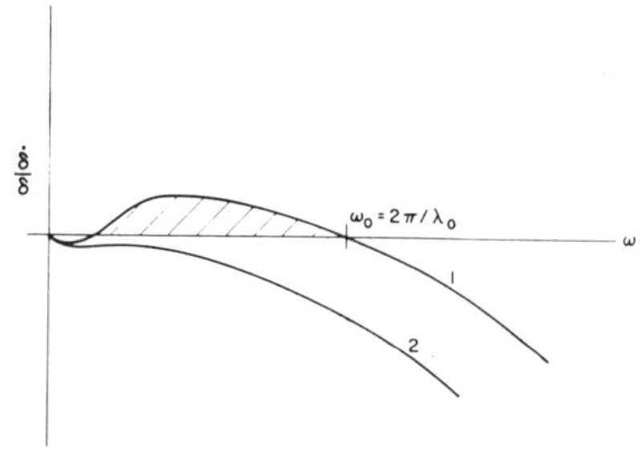

FIG. 1. $\dot{\delta}/\delta$ as a function of frequency for two cases: curve 1 unstable, curve 2 stable.

Even in the former case, the difference between the original and our modified CS should be experimentally demonstrable under the appropriate conditions [Eq. (34)]; in particular there are instances in which one can have instability without CS as well as CS without instability.

I. CALCULATION OF THE TIME DERIVATIVE $\dot{\delta}$ OF THE AMPLITUDE OF A FOURIER PERTURBATION

Let us consider a co-ordinate system attached to the planar interface moving with the constant velocity V with respect to the phases on either side[6]; let the z axis point toward the liquid and the plane $z=0$ coincide with the interface. Suppose a sinusoidal ripple, of infinitesimal amplitude, described by the equation

$$z=\phi(x,t)=\delta(t)\,\sin\omega x \tag{1}$$

is introduced into the planar interface between liquid and solid.[7] We wish to obtain an expression for $\dot{\delta} \equiv d\delta/dt$ in order to see whether the ripple grows or decays. Since this requires the calculation of the velocity $v(x)$ of each element of interface, as a function of its position x, in terms of the values of the local thermal and diffusion gradients we turn first to the determination of the thermal and diffusion fields which satisfy the required boundary conditions.

We seek expressions for the thermal and diffusion fields which:

(i) satisfy the appropriate steady-state equations in a frame moving at the constant velocity V, namely, for the liquid

$$\nabla^2 c+(V/D)(\partial c/\partial z)=0, \text{ (no convection)} \tag{2a}$$

$$\nabla^2 T+(V/D_{th})(\partial T/\partial z)=0, \tag{2b}$$

and for the solid

$$\nabla^2 T'+(V/D_{th}')(\partial T'/\partial z)=0, \tag{2c}$$

[6] Changes of density upon solidification are ignored in this treatment.

[7] For the rippled interface, the plane $z=0$ may be rigorously identified as the plane coinciding with the average position (in the z direction) of the interface and, therefore, traveling with the average velocity V (in the z direction) of the interface. The experimental conditions required to assure that V is indeed constant are discussed immediately after Eqs. (13) and (14).

where V = constant mean velocity of the interface with respect to either phase, c = concentration of solute in the liquid, T = temperature in the liquid, T' = temperature in the solid, D = diffusion coefficient of solute in the liquid, $D_{th}=\kappa_L/C_L$ = thermal diffusivity of the liquid, $D_{th}'=\kappa_S/C_S$ = thermal diffusivity of the solid, κ_L = conductivity of liquid, κ_S = conductivity of solid, C_L = specific heat of the liquid/unit volume, and C_S = specific heat of the solid/unit volume.

(ii) Satisfy the correct boundary conditions at infinity, i.e., c, T and T' should agree with the fields due to the unperturbed interface ($\delta=0$) several wavelengths from the interface.

(iii) Satisfy the following two boundary conditions at the interface $z=\phi$: first, according to the phase diagram and our assumption of local interface equilibrium the condition

$$T_\phi = mc_\phi + T_N \tag{3a}$$

must hold on the interface, where m is the slope (including sign) of the liquidus line on the phase diagram and T_N, the temperature (degrees Kelvin) in the absence of solute, is related to the melting temperature T_M (degrees Kelvin) of a flat interface by the capillary formula

$$T_N = T_M + T_M\Gamma K = T_M - T_M\Gamma\delta\omega^2 \sin\omega x. \tag{3b}$$

Here K is the average curvature at a point of the interface[8] (positive when concave toward the liquid) and $\Gamma=\gamma/L$ is a capillary constant in which γ is the specific liquid–solid interfacial free energy, assumed independent of composition, and L is the latent heat of the solvent per unit volume[8]; typically $\Gamma\simeq 10^{-8}$ cm (assuming $\gamma\simeq 30$ ergs/cm^2).

The second condition that must be satisfied at the interface is that the velocity $v(x)$ at each element of interface calculated from heat flow considerations must agree with the velocity calculated from diffusion considerations,[9] i.e.,

$$v(x) = \frac{1}{L}\left[K_S\left(\frac{\partial T'}{\partial z}\right)_\phi - K_L\left(\frac{\partial T}{\partial z}\right)_\phi\right] = \frac{D}{c_\phi(k-1)}\left(\frac{\partial c}{\partial z}\right)_\phi, \tag{4}$$

where k is the partition coefficient given by the ratio of the equilibrium concentration of solute on the solid side of the interface to that on the liquid side of the interface, and where the denominator $c_\phi(k-1)$ gives the difference in concentration between the solid and liquid sides of the interface.

The preceding conditions (i)–(iii) are most conveniently met in two stages. To this end we write for the temperature T_ϕ and concentration c_ϕ (liquid side) at the interface the expressions

$$T_\phi = T_0 + a\delta \sin\omega x = T_0 + a\phi, \tag{5a}$$

$$c_\phi = c_0 + b\delta \sin\omega x = c_0 + b\phi, \tag{5b}$$

where T_0 and c_0 are the values for a flat interface and the second terms are the first-order corrections corresponding to the infinitesimal ripple [the form of Eqs. (5) may be regarded as justified by the self-consistent values to be obtained for the coefficients a and b]. The two stages consist in first obtaining expressions for T, T' and c that satisfy conditions (i) and (ii) and that reduce to Eq. (5) on the interface and then secondly imposing condition (iii) comprising the two interface boundary conditions [Eqs. (3) and (4)] on these solutions thereby fixing a and b and yielding the complete solution.

The required expressions satisfying conditions (i) and (ii) and reducing to Eq. (5) on the interface $z=\phi$ (not $z=0$!) are the following: on the liquid side

$$c(x,z) - c_0 = (G_cD/V)(1-e^{-(V/D)z}) + \delta(b-G_c) \times \sin\omega x e^{-\omega^* z}, \tag{6}$$

$$T(x,z) - T_0 = (GD_{th}/V)(1-e^{-(V/D_{th})z}) + \delta(a-G) \times \sin\omega x e^{-\omega_{th} z}, \tag{7}$$

and on the solid side (primes denoting the solid)

$$T'(x,z) - T_0 = (G'D_{th}'/V)(1-e^{-(V/D_{th}')z}) + \delta(a-G') \times \sin\omega x e^{-\omega_{th}' z}, \tag{8}$$

where G and G' are the thermal gradients at the unperturbed flat interface ($\delta=0$) in the liquid and solid, respectively, G_c is the concentration gradient in the liquid at the unperturbed flat interface, and where

$$\omega^* = (V/2D) + [(V/2D)^2 + \omega^2]^{\frac{1}{2}}, \tag{9a}$$

$$\omega_{th} = (V/2D_{th}) + [(V/2D_{th})^2 + \omega^2]^{\frac{1}{2}}, \tag{9b}$$

$$\omega_{th}' = (V/2D_{th}') - [(V/2D_{th}')^2 + \omega^2]^{\frac{1}{2}}. \tag{9c}$$

It is easily verified by substitution that these expressions do indeed satisfy Eqs. (2a), (2b), and (2c), respectively, and that to first order[10] in δ Eq. (6) reduces to Eq. (5b), and Eqs. (7) and (8) to Eq. (5a) on the interface. Before proceeding to determine the constants a and b, we make several points concerning these solutions.

When $|z|\gg\lambda=2\pi/\omega$ the only terms in Eqs. (6)–(8) remaining are the leading ones; they describe the unperturbed fields. Whenever $\omega\gg V/2D$, $V/2D_{th}$, or $V/2D_{th}'$, respectively, Eqs. (6)–(9) show that the field in question obeys Laplace's equation for a stationary frame within a distance of several $\lambda=2\pi/\omega$ of the interface; as discussed farther on, this condition holds in al practical cases for the thermal fields.

The gradients of the thermal and diffusion fields at

---

[8] W. W. Mullins and R. F. Sekerka, J. Appl. Phys. **34**, 323 (1963).

[9] Allowing for the influence of curvature on the partition coefficient k of solute between the two phases and on the slope m leads to results which are the same as those obtained here within the approximation of dilute solution theory.

[10] Equation (4) may be viewed as the simultaneous expression o heat balance and mass balance on a differential scale at the inter face; diffusion in the solid is ignored.

the interface calculated from Eqs. (6)–(8) to the first order in δ are

$$(\partial c/\partial z)_\phi = -\omega^*(b-G_c)\delta\sin\omega x + G_c[1-(V/D)\delta\sin\omega x],$$
$$= -\omega^*\{b-G_c[1-(V/\omega^* D)]\}\delta\sin\omega x + G_c, \quad (10)$$

$$(\partial T/\partial z)_\phi = -\omega_{th}\{a-G[1-(V/\omega_{th}D_{th})]\}\delta\sin\omega x + G,$$
$$\simeq -\omega(a-G)\delta\sin\omega x + G, \quad (11)$$

$$(\partial T'/\partial z)_\phi = +\omega_{th}'\{a-G'[1-(V/\omega_{th}'D_{th}')]\}\delta\sin\omega x + G'$$
$$\simeq +\omega(a-G')\delta\sin\omega x + G'. \quad (12)$$

The second form for the temperature gradients originates from the fact that for all cases of practical interest $\lambda = 2\pi/\omega \ll D_{th}/V \simeq 10^2$ cm (typical value for a metallic system); thus $V/\omega_{th}D_{th} \simeq V/\omega D_{th} \ll 1$ and the factor multiplying G in brackets is essentially unity; the same comment applies to the solid (D_{th}'). If Eqs. (10)–(12) for the gradients are substituted into the flux boundary condition Eq. (4) [using Eq. (5b) for c_ϕ in the denominator of Eq. (4)] and then an integration with respect to x over an interval of one wavelength is performed, all the sine terms drop out and we obtain

$$\frac{1}{\lambda}\int_0^\lambda v(x)dx \equiv V = \frac{1}{L}(\kappa_s G' - \kappa_L G), \quad (13)$$

and

$$\frac{1}{\lambda}\int_0^\lambda v(x)dx \equiv V = \frac{D}{c_0(k-1)}G_c = \frac{Dk}{c_\infty(k-1)}G_c, \quad (14)$$

where the last step in Eq. (14) is required by our assumed steady-state condition which demands c_∞ (liquid) $= \langle\text{conc.}\rangle_{av}$ (solid) $= kc_0$ (liquid). Equation (13), the average heat balance condition, together with Eqs. (7) and (8) (for $z \gg \lambda$) determine the temperatures that must be maintained by a pair of furnaces traveling at the constant velocity V in order that the average plane $z=0$ of the interface may assume a fixed relative position between the furnaces and see a thermal gradient G in the liquid. Alternatively, by changing to the co-ordinate system $z' = z + Vt$, one can obtain from the same equations the temperature as a function of time that must be produced by two fixed furnaces to temporarily achieve the same result; this corresponds to the thermal value technique described by Tiller and Rutter.[3] Equation (14) originally derived by Tiller[2] *et al.* for a flat interface ($\delta = 0$), gives G_c in terms of directly measurable quantities.

Proceeding now to the second stage of the determination of the steady-state fields we substitute Eqs. (6)–(8) into Eqs. (3) and (4) [with the aid of Eqs. (10)–(12)] in order to determine a and b. Equating coefficients of like Fourier components, we obtain to first order in δ, after straightforward but tedious algebra, the expressions

$$b = \frac{2G_c T_M \Gamma\omega^3 + \omega G_c(\mathcal{G}+\mathcal{G}') + G_c[\omega^* - (V/D)](\mathcal{G}'-\mathcal{G})}{\omega^*(\mathcal{G}'-\mathcal{G}) + 2\omega m G_c - (\bar{\kappa}p/DL)(\mathcal{G}'-\mathcal{G})^2},$$
$$= \frac{2G_c T_M \Gamma\omega^3 + \omega G_c(\mathcal{G}'+\mathcal{G}) + G_c[\omega^* - (V/D)](\mathcal{G}'-\mathcal{G})}{2\omega m G_c + (\mathcal{G}'-\mathcal{G})[\omega^* - (V/D)p]}, \quad (15)$$

and

$$a = mb - T_M\Gamma\omega^2, \quad (16)$$

where we have introduced the following new symbols: $\bar{\kappa} = \frac{1}{2}(\kappa_s + \kappa_L)$, $\mathcal{G} = (\kappa_L/\bar{\kappa})G$, $\mathcal{G}' = (\kappa_s/\bar{\kappa})G'$, and $p = 1-k$. The script $\mathcal{G}$'s are thus the thermal gradients adjusted by a suitable conductivity ratio.

To determine $\dot{\delta}$, we substitute Eqs. (7) and (8) into Eq. (4) to obtain

$$v(x) = V + \dot{\delta}\sin\omega x = (\bar{\kappa}/L)(\mathcal{G}'-\mathcal{G}) + (\bar{\kappa}/L)\omega \times\{2a - (\mathcal{G}'+\mathcal{G})\}\delta\sin\omega x. \quad (17)$$

Equating like Fourier components we have

$$V = (\bar{\kappa}/L)(\mathcal{G}'-\mathcal{G}) \quad (18)$$
$$\dot{\delta} = (2\bar{\kappa}/L)\omega[a - \tfrac{1}{2}(\mathcal{G}'+\mathcal{G})]\delta. \quad (19)$$

Equation (18) is equivalent to Eq. (13). Substituting for a in Eq. (19) by using Eqs. (15) and (16), we obtain our central result

$$\frac{\dot{\delta}}{\delta} = \frac{V\omega\{-2T_M\Gamma\omega^2[\omega^* - (V/D)p] - (\mathcal{G}'+\mathcal{G})[\omega^* - (V/D)p] + 2mG_c[\omega^* - (V/D)]\}}{(\mathcal{G}'-\mathcal{G})[\omega^* - (V/D)p] + 2\omega m G_c}. \quad (20)$$

II. ANALYSIS OF THE STABILITY OF THE PLANAR INTERFACE

Stability of the interface with respect to an undulation, within the framework of our approximations, depends on the sign of $\dot{\delta}/\delta$; a positive value for any ω means growth of some undulations and hence instability of the original flat interface whereas a negative sign for all ω means decay of all undulations and hence stability.

The discussion in the introduction gave a rigorous justification for basing the prediction of instability on the steady-state solutions. It also implied that the justification for the prediction of stability on the same basis depends on the validity of the contention that a wave which is stable when accompanied by a steady-state diffusion field is even more stable (larger $|\dot{\delta}|$) if the solute has insufficient time to attain the steady-state distribution. This contention may now be supported by the preceding results. Equation (19) shows that steady-state instability can occur only if a exceeds the quantity $\frac{1}{2}(\mathcal{G}+\mathcal{G}')$ (assumed positive) and, therefore, according to Eq. (16) only if b and m have the same sign. Inspection of Eq. (5b) then shows that the concomitant alteration of concentration c_ϕ at the growing perturbation is just the opposite of that which would occur if a finite sine wave were suddenly introduced into the interface without permitting appreciable lateral diffusion of solute to occur. For example, if m were negative corresponding to $k<1$ [Eq. (28)], a growing perturbation would have a negative b and hence from Eq. (5b), a reduced concentration at the maxima and an increased concentration at the minima of the sine wave compared with values obtaining at the unperturbed interface. But these changes are the opposite of the initial changes that would occur for a finite wave suddenly introduced since in that case the rejection of solute by the freezing maxima and the liberation of solvent by the melting minima would cause the concentration to increase at the maxima and decrease at the minima; these changes taken alone would cause the temperature of the minima to exceed those of the maxima [Eq. (3a)] thereby giving rise to a flow of heat from minima to maxima favoring the disappearance of the wave. A similar argument applies to the case m>0. Therefore, we confirm the contention that if a wave cannot grow when accompanied by a steady-state diffusion field it is certainly not able to grow if the field lags behind by some transient component.

Proceeding with the analysis of Eq. (20) we note that the sign $\dot{\delta}/\delta$ depends only on the sign of the numerator since both terms in the denominator are always positive. This is true for the first term since $\mathcal{G}'>\mathcal{G}$ from Eq. (18), and since we have $\omega^*>V/D>(V/D)p$ from the definition of ω^* [Eq. (9a)] and from the fact that $p=1-k\leq 1$. The second term of the denominator is also positive since m and G_c always have the same sign [Eqs. (14) and (28)]. We, therefore, may expect instability or stability depending on whether or not the numerator ever becomes positive.

The numerator consists of three terms: the first term arises from capillarity (it vanishes if $\gamma=0$), is always negative and, therefore, favors decay of any harmonic; the second term arises from thermal gradients, is negative on our present assumption that $\mathcal{G}'$, $\mathcal{G}\geq 0$ and thus also favors decay[11]; the third term, representing the effect of solute on the equilibrium melting temperature, is always positive by the same arguments applied to the denominator and, therefore, always favors growth. Instability occurs if there is any frequency for which the magnitude of the third term exceeds that of the sum of the first two terms; otherwise stability prevails.

The frequency dependence of the terms comes about in essentially two ways: first there is an increasing dominance of the capillary effect with decreasing wavelength as evidenced by the factor ω^2 in the capillary term of the numerator; this causes stability to always prevail for sufficiently short wavelengths. The second basic frequency dependence which causes stability to prevail for sufficiently long wavelengths arises formally from the fact that for small ω the bracket (ω^*-V/D) multiplying the third term becomes approximately $(D/V)\omega^2$ [Eq. (9a)] and, therefore, small compared with the bracket $[\omega^*-(V/D)p]$ multiplying the first two terms which becomes approximately $(D/V)\omega^2+k(V/D)$. Some physical insight into this long wavelength case is provided by the observation that for sufficiently small ω, Eq. (15) combined with (18) shows that b goes to zero linearly in ω. It follows from Eq. (16) that a likewise goes to zero linearly in ω so that according to Eqs. (5) the interface tends toward a condition of constant temperature and concentration. Roughly speaking, this means that the solute redistribution becomes insignificant when the wavelength of the perturbation is sufficiently large. Equation (19) then shows, with $a\simeq 0$, that the pure thermal effect prevails and causes the perturbation to decay (we assume the wavelength is still short compared with the very large characteristic thermal distance). From the foregoing we conclude that if instability is present it must occur for a range of intermediate wavelengths. Figure 1 shows a schematic plot of $\dot{\delta}/\delta$ as a function of frequency for two cases; the curve possessing a maximum and becoming positive for a range of frequencies corresponds to an unstable interface; the other curve which is always nonpositive corresponds to a stable interface.

To study the sign of the numerator of Eq. (20) as a function of frequency, it is convenient to split out the frequency-independent terms that correspond to a modified CS criterion. For this purpose we factor out of the numerator bracket of Eq. (20) the positive quantity $2[\omega^*-p(V/D)]$ and write the remaining part upon whose sign stability depends as

$$\begin{aligned} S(\omega) &= -T_M\Gamma\omega^2-\tfrac{1}{2}(\mathcal{G}'+\mathcal{G}) \\ &\quad +mG_c\{[\omega^*-(V/D)]/[\omega^*-(V/D)p]\} \\ &= -T_M\Gamma\omega^2+mG_c g(\omega)-\tfrac{1}{2}(\mathcal{G}'+\mathcal{G})+mG_c \\ &= f(\omega)-\tfrac{1}{2}(\mathcal{G}'+\mathcal{G})+mG_c, \end{aligned} \tag{21}$$

where

$$\begin{aligned} g(\omega) &= \frac{\omega^*-V/D}{\omega^*-(V/D)p}-1=\frac{-2k}{[1+(2\omega D/V)^2]^{\frac{1}{2}}-1+2k} \\ &= -1+(D^2/V^2k)\omega^2+\cdots, \end{aligned} \tag{22}$$

and where

$$\begin{aligned} f(\omega) &= -T_M\Gamma\omega^2+mG_c g(\omega), \\ &= -mG_c+[(D^2mG_c/V^2k)-T_M\Gamma]\omega^2+\cdots. \end{aligned} \tag{23}$$

The characteristics of $f(\omega)$ [and $g(\omega)$] that are important for our purposes are the following:

(i) $f(\omega)<0[g(\omega)<0]$,
(ii) $f(0)=-mG_c[g(0)=-1]$,
(iii) $f(\omega)$ either decreases monatonically as ω increases from zero or it has one maximum for some positive value

[11] In verifying the boundary conditions expressed by Eqs. (5), the exponentials in Eqs. (6)–(8) may be expanded and all resulting terms in powers of δ higher than the first may be discarded; e.g., $\exp[-(V/D)z]_{z=\phi}=\exp[-(V/D)\delta\sin\omega x]\simeq 1-(V/D)\delta\sin\omega x$.

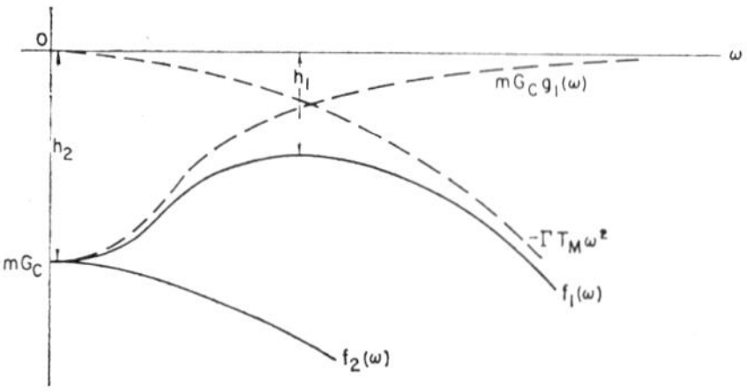

FIG. 2. Frequency-dependent part $f(\omega)$ of the stability function $S(\omega)$ for two cases; dotted curves show component terms of $f_1(\omega)$.

of ω; the first case holds when $A \equiv kT_M\Gamma V^2/mG_cD^2 \geqslant 1$ [negative or zero coefficient of ω^2 in the expansion of Eq. (23)], the second when $A<1$ (positive coefficient).

(iv) $f(\infty) = -\infty [g(\infty)=0]$.

Properties (ii) and (iv) follow immediately from the definitions given by Eqs. (22) and (23) as does property (i) when the inequality $p=1-k<1$ is taken into account; property (iii) is established in the Appendix. Figure 2 presents a schematic plot of the frequency-dependent term $f(\omega)$ with the separate contributions of the capillary term $-T_M\Gamma\omega^2$ and the term $mG_g(\omega)$ shown as dotted curves.

The negativity of $f(\omega)$ [property (i)] means that it always favors stability; evidently it poses the least barrier to instability at the frequency for which it attains its largest value (value of smallest magnitude). Let us define h to be the magnitude of this maximum value, i.e.,

$$h = |\max f(\omega)|; \tag{24}$$

It follows immediately from properties (i) and (ii) that

$$0 < h \leqslant mG_c. \tag{25}$$

The quantity h is shown on the graph of Fig. 2. With the preceding definition our fundamental stability criterion becomes

$$-\tfrac{1}{2}(\mathcal{G}'+\mathcal{G})+mG_c \begin{matrix} >h & \text{unstable} \\ <h & \text{stable.} \end{matrix} \tag{26}$$

In principle the value of h corresponding to a maximum of $f(\omega)$ can of course be determined by differentiation. The resulting equation that must be solved for ω, however, is essentially cubic.[13] A general formula for h will, therefore, not be given as it is easier in practice to simply plot $f(\omega)$ for the particular system under study, to determine h from the plot (Eq. 24) and then to apply the stability criterion given by Eq. (26). We confine the remaining discussion of this section to some qualitative remarks and to a discussion of certain special cases.

A striking new feature of the present treatment occurs for the case in which $f(\omega)$ decreases monatonically as ω increases from zero so that $|\max f| = |f(0)| = mG_c$. Since Eq. (21) then shows that $S(\omega)$ is always negative, there can be no instability under this condition. We call this case absolute stability. According to the results of the Appendix quoted above as property (iii), this absolute stability occurs when

$$A = kT_M\Gamma V^2/mG_cD^2 = k^2T_M\Gamma V/[(k-1)mc_\infty D] \geqslant 1, \tag{27}$$

where Eq. (14) is used in the last step. Absolute stability may be thought of as arising when the capillary effect completely dominates the solute effect, a situation favored by the reduced scale of transport accompanying high velocities; in line with this interpretation we note that A would vanish if Γ were zero (no capillarity). Using the Van't Hoff (4) relation for dilute solutions, namely

$$k-1 = (L/DT_M{}^2)m, \tag{28}$$

where m must be expressed in degrees per mole fraction in this formula, we obtain an alternate formula for A and another version of condition (27), namely,

$$A \simeq [k/(k-1)]^2 \cdot (L/RT_M) \cdot \Gamma \cdot (V/D) \cdot (1/c_\infty) \geqslant 1, \tag{29}$$

where the approximation sign reflects the approximate character of the Van't Hoff relation and where c_∞ must be expressed in mole fractions in this particular formula. If we take as values $RT_M/L \simeq 1$, $\Gamma \simeq 10^{-8}$ cm, and $D/V \simeq 5\times10^{-4}$ cm, then we must have $[k/(k-1)]^2 \cdot (1/c_\infty) \gtrsim 5\times10^4$ for absolute stability; this condition should be attainable in practice (values of $c_\infty \leqslant 10^{-4}$ are common and k need not be small).[14]

If capillarity is left out of the treatment, γ and Γ are zero so that $f(\omega) = mG_cg(\omega)$; since $g(\infty)=0$, [property (iv)], this means that $h = \max|f(\omega)| = mG_c \max|g(\omega)| = 0$ and that stability depends only on the sign of the frequency-independent terms $-\frac{1}{2}(\mathcal{G}'+\mathcal{G})+mG_c$ according to Eq. (25). Although the inclusion of capillarity requires in principle that some finite h always appear in the condition determining stability [Eq. (26)], it is sufficient in practice to take $h=0$ if it is small compared with the difference between the two terms $\frac{1}{2}(\mathcal{G}+\mathcal{G}')$ and mG_c; it is probably often safe to neglect h even if it is small compared with either term as the terms often differ by an appreciable percentage in a cell growht experiment. To form a rough estimate of the conditions under which the latter relation holds (e.g., $h<mG_c$), we require that the capillary term in $f(\omega)$ be small compared with mG_c at a frequency ω for which $g(\omega)$ is small in magnitude compared with 1; in other words, referring to Fig. 2, we are seeking the condition under which the dotted curve $mG_cg(\omega)$ has risen nearly to

[12] If the liquid is supercooled, $G<0$ and $\mathcal{G}+\mathcal{G}'$ may be negative leading to a thermal gradient instability. This case has been previously discussed by Mullins and Sekerka.[8]

[13] It is cubic in the variable $\{1+[2V\omega/D]^2\}^{\frac{1}{2}}$.

[14] The condition that the old CS criterion predict instability in conflict with our prediction of absolute stability is that $G<mG_c$ and $A>1$ simultaneously hold; from Eq. (27) this is equivalent to the condition $G<mG_c<k(V/D)^2T_M\Gamma \simeq 20°$K/cm assuming the same values as above and $T_M \simeq 400°$C.

zero before the capillary curve $T_M\Gamma\omega^2$ has had a chance to fall by an appreciable fraction of mG_c; this will assure that the maximum h will be small compared with mG_c.

The value ω^+ of ω required to cause $g(\omega)$ to equal a small fraction α is found by equating the denominator of $g(\omega)$ [Eq. (22)] to $\alpha^{-1}(2k)$ and solving for ω; neglecting unity in comparison with α^{-1}, the result is

$$\omega^+ = \frac{V}{D}[(k/\alpha)^2 + (k/\alpha)]^{\frac{1}{2}}. \quad (30)$$

Substituting this value into the capillary term and requiring it to be less than αmG_c, we obtain upon rearrangement and in the same approximation

$$\alpha\frac{T_M\Gamma V^2}{mG_cD^2}\left[\left(\frac{k}{\alpha}\right)^2 + \left(\frac{k}{\alpha}\right)\right] = A\left[\frac{k}{\alpha}+1\right] \leqslant \alpha^2. \quad (31)$$

If we take $\alpha = 0.05$, and if Eq. (31) holds (i.e., $A(20k+1) \leqslant 1/400$) then the magnitude of $f(\omega)$ cannot exceed 0.1 mG_c at ω^+; therefore, the value of h cannot exceed 0.1 mG_c and may be considerably less. Thus for sufficiently small values of the parameter A, as determined by Eq. (31), it is probably safe to neglect $f(\omega)$ or h in the stability criterion. [see Eq. (29) for a basic expression for A.] It would always seem best in any particular case, however, to plot $f(\omega)$, determine the size of h directly, and compare it with the difference between the frequency-independent terms to be certain whether or not it can be neglected.

When h can be neglected the stability criterion becomes

$$-\tfrac{1}{2}(\mathcal{G}'+\mathcal{G}) + mG_c < 0 \text{ stable} \quad (32)$$
$$> 0 \text{ unstable.}$$

We refer to this condition as a modified CS criterion; it may be compared with the original form of the CS criterion,[2] namely

$$-G + mG_c < 0 \text{ stable}$$
$$> 0 \text{ unstable.} \quad (33)$$

We see that the effect of the present analysis for the special case under consideration is to replace the liquid temperature gradient by a mean of conductivity weighted gradients in the liquid and solid. A better comparison between the two criteria is obtained by using Eq. (18) and the definitions of the $\mathcal{G}$'s to rewrite Eq. (32) as

$$-(\kappa_L/\bar{\kappa})G - (L/2\bar{\kappa})V + mG_c$$
$$= -G + mG_c + [(\kappa_s - \kappa_L)/(\kappa_s+\kappa_L)]G$$
$$-[L/(\kappa_s+\kappa_L)]V < 0 \text{ stable}$$
$$> 0 \text{ unstable.} \quad (34)$$

The term $[-L/(\kappa_L+\kappa_s)]V$ makes the interface more stable than the original CS criterion would predict; the term $[(\kappa_s-\kappa_L)/(\kappa_s+\kappa_L)]G$ makes it more stable if $\kappa_L > \kappa_s$ and less stable if $\kappa_s > \kappa_L$ (the usual case). Thus, assuming the present approximation applicable, it is possible to have instability without CS and CS without instability. The difference between the original CS criterion and our modified CS criterion, applicable when h can be neglected, becomes evident when either of the last two terms becomes comparable with say the first term, that is when either $|[(\kappa_s-\kappa_L)/(\kappa_s+\kappa_L)]G| \simeq |G|$ or $|[L/(\kappa_s+\kappa_L)]V| \simeq |G|$. The first case should occur often as usually $\kappa_s \sim 2\kappa_L$ (for metals). The second case should occur for fairly high velocities as equivalence of the terms occurs when $[L/(\kappa_s+\kappa_L)]V \simeq 100$ V $\simeq 10$°C/cm or $V \simeq 10^{-1}$ cm/sec.

The modified CS criterion resembles the original one[4] in giving a straight line when the c_∞ corresponding to breakdown is plotted as a function of G/V; thus substituting for G_c from Eq. (14) into Eq. (34), the equation for the breakdown line becomes

$$c_\infty = [kD\kappa_L/(k-1)m\bar{\kappa}](G/V) + (kDL/2(k-1)m\bar{\kappa}), \quad (35)$$

and differs from the corresponding original CS equation in the value of the slope $[kD\kappa_L/m(k-1)\bar{\kappa}]$ and in the possession of a finite intercept $[kDL/2(k-1)m\bar{\kappa}]$ (small on the scale of a typical plot).

We reiterate that Eqs. (32)–(35) only apply when h can be neglected and that in general Eq. (26) must be used.

Finally, any physically real interface must, in general, be described by a function of two variables rather than one variable, namely, $z = \phi(x,y)$. The Fourier components of an arbitrary infinitesimal perturbation in such a surface are of the form $\delta(t)\sin\omega_x x \sin\omega_y y$. If the preceding analysis is repeated for a perturbation of this form, it is easily seen that all of the formulae we have obtained remain valid under the substitutions

$$\partial^2/\partial x^2 \to (\partial^2/\partial x^2) + (\partial^2/\partial y^2),$$
$$\omega \to (\omega_x^2 + \omega_y^2)^{\frac{1}{2}}, \quad (36)$$
$$\lambda^{-1} \to (\lambda_x^{-2} + \lambda_y^{-2})^{\frac{1}{2}}.$$

The stability criterion and all results stemming therefrom remain the same.

III. DISCUSSION OF THE CELL SIZE

It is possible to plot $\dot{\delta}/\delta$ as a function of frequency for any particular experimental case from Eq. (20) thereby obtaining graphs similar to those appearing in Fig. 1. (Note that $\dot{\delta}/\delta$ is a considerably more complicated function of ω than is the bracket in the numerator of $\dot{\delta}/\delta$ which we have studied to determine the stability criterion.) If a plot of this type is made for typical values of the growth parameters in a cell growth experiment, it is found that the wavelength range within which $\dot{\delta}/\delta$ is positive and in particular the wavelength corresponding to the maximum of $\dot{\delta}/\delta$ is the same order of magnitude as a typical cell size, namely, $\simeq 50\ \mu$. One is tempted to press for detailed agreement between the wavelength corresponding to a maximum $\dot{\delta}/\delta$ in an unstable case and the cell size of the structure into which the unstable planar interface develops on the basis of the following rationalization: the fastest growing Fourier components of an arbitrary perturbation in the planar shape should eventually dominate causing a rippled or cellular structure with a wavelength $\simeq 2\pi/\omega_M$. This procedure carried out on the basis of Eq. (20), however, fails to give detailed agreement between the theoretically predicted and the experimentally observed relationship of the cell size and growth parameters. There are probably two main reasons for this failure: first, we have used steady-state diffusion fields in the preceding calculation. As explained in the introduction this should be a good approximation for frequencies near ω_0 where the interface shape changes very slowly. For frequencies near ω_M, however, the steady-state solutions could be seriously in error due to the possibility of a rapidly changing interface shape; that is, the true value of ω_M based on the rigorous time-dependent solutions might differ considerably from the value calculated from steady-state solutions.

The second reason is that the structure which finally develops in cell growth experiments does not have a small amplitude.[1,3] Since all of our calculations are based on the assumption of small amplitude perturbations (first order in δ), a perfectly adequate approximation for the investigation of stability, they cannot in general be expected to give reliable predictions concerning the final cell size.

The preceding two difficulties should not occur in the special case of the small amplitude perturbation developing during the initial phase of the breakdown of a planar interface into corrugations under conditions of weak instability in which the value of $\dot{\delta}/\delta$ at $\omega=\omega_M$ is sufficiently small.[1] The scale or wavelength of the small amplitude corrugations might then be expected to correspond to the predicted value of $\lambda_M=2\pi/\omega_M$ determined from a plot of Eq. (20).

ACKNOWLEDGMENT

Appreciation is expressed to Dr. W. A. Tiller for a critical reading of the manuscript.

APPENDIX. THE CRITICAL POINT OF $f(\omega)$

We first prove that the function

$$f(\omega)=-T_M\Gamma\omega^2+mG_c\frac{-2k}{[1+(2\omega D/V)^2]^{\frac{1}{2}}-1+2k}$$
$$=-mG_c+[(D^2mG_c/V^2k)-T_M\Gamma]\omega^2+\cdots \quad \text{(A1)}$$

has at the most one positive critical point (value of ω for which the derivative vanishes). Calculating $df/d\omega$ and equating it to zero one readily obtains

$$\frac{T_M\Gamma V^2}{4kmG_cD^2}\left[1+\left(\frac{2\omega D}{V}\right)^2\right]^{\frac{1}{2}}$$
$$=\left\{\left[1+\left(\frac{2\omega D}{V}\right)^2\right]^{\frac{1}{2}}+(2k-1)\right\}^{-2}. \quad \text{(A2)}$$

Introducing the substitutions

$$A=kT_M\Gamma V^2/mG_cD^2>0 \quad \text{(A3)}$$

and

$$u=[1+(2\omega D/V)^2]^{\frac{1}{2}}+2k-1 \quad \text{(A4)}$$

we may rewrite Eq. (A2) as

$$(A/4k^2)u-A[(2k-1)/4k^2]=1/u^2. \quad \text{(A5)}$$

Since ω must be real, Eq. (A4) gives the inequality

$$u-2k+1=[1+(2\omega D/V)^2]^{\frac{1}{2}}>1,$$

or

$$u>2k>0. \quad \text{(A6)}$$

Therefore, only positive roots u of Eq. (A5) can be accepted. But there is always just one positive root of Eq. (A5) as can be seen immediately by considering the equation graphically as the intersection of the straight line of positive slope $A/4k^2$ with the hyperbola u^{-2}. Therefore, there can be at most one critical point for a positive value of ω. [Not all positive roots of Eq. (A5) give real values of ω; to do so they must exceed $2k$ according to Eq. (A6).] A straightforward calculation shows $d^2f/d\omega^2<0$ at the critical point thereby proving it to be a maximum.

Since $f(\infty)=-\infty$, there must be a maximum if $f(\omega)$ increases with ω for small ω; conversely if $f(\omega)$ decreases with ω for small ω there can be no maximum as there would first have to be a minimum contradicting the proof that there is only one critical point for $\omega>0$. Therefore, from the expanded form of Eq. (A1), we deduce that there is a maximum for $\omega>0$ if the coefficient of ω^2 is positive, i.e., if

$$A=kT_M\Gamma V^2/mG_cD^2<1, \quad \text{(A7)}$$

whereas there is no maximum for $\omega>0$ and $f(\omega)$ is a monotone decreasing function of ω if the coefficient of ω^2 is nonpositive, i.e., if

$$A\geqslant 1. \quad \text{(A8)}$$

Note added in proof. In another paper to appear shortly, the quantity h [Eq. (24)] is explicitly evaluated in terms of dimensionless parameters and a stability function closely related to h is tabulated numerically. In terms of this stability function, the complete stability criterion [Eq. (26)] is studied in detail.

Interdendritic Spacing: Part II. A Comparison of Theory and Experiment

R. TRIVEDI

The primary spacing data of Part I are compared to the existing theoretical models of Hunt and of Kurz and Fisher, and a significant disagreement is found. A theoretical model based on the Hunt model is developed, and it is found that the theory adequately explains the variation in primary spacing, λ_1, with the growth rate, V. A maximum in λ_1 *vs* V is predicted and the velocity at which the maximum occurs matches with the result obtained experimentally. It is shown that the maximum in λ_1 corresponds to the dendrite-to-cell transition, and cellular structures are found to grow with much smaller spacings than dendritic structures under identical growth conditions.

I. INTRODUCTION

SINCE a theoretical analysis of the growth of a dendritic array is so complex, detailed models have been proposed only for the simple case in which the interaction between the neighboring dendrites is negligible.[1,2] Although these models have given us valuable information on the characteristics of dendrite tip, they do not enable us to understand the variation in primary dendrite spacing as a function of growth rate, temperature gradient, and alloy composition. Even though a number of empirical relationships have been proposed in the literature[3,4,5] to explain experimental observations on primary dendrite spacings, these relationships are based on experimental data obtained for a very limited range of experimental variables. Consequently, none of these models properly predicts the variation in primary dendrite spacing in different systems over a wide range of experimental conditions.

Recently, Hunt[6] and Kurz and Fisher[7] have proposed theoretical models to predict primary dendrite spacings. These models are examined in this paper to check if they explain the experimental observations reported in Part I.[8] Specifically, we determine whether these models adequately explain the following basic features of primary dendrite spacings observed experimentally:[8,9,10] (1) the primary dendrite spacing goes through a maximum as the growth rate is increased, (2) at high growth rates, the variation in primary dendrite spacing with growth rate becomes linear on a log-log plot, and (3) in the high growth rate regime, the temperature gradient affects the primary dendrite spacing significantly, whereas it has no effect on the dendrite tip radius and on the initial secondary dendrite arm spacing. We shall show that current theories do not adequately explain the growth rate dependence of primary dendrite spacing. A simplified theory will be developed which satisfactorily explains important features of the variation of primary dendrite spacing with growth rate.

R. TRIVEDI is a Senior Scientist and Group Leader at Ames Laboratory-US-DOE and Professor with the Department of Materials Science and Engineering, Iowa State University, Ames, IA 50011.

This paper is based on a presentation made at the symposium "Establishment of Microstructural Spacing during Dendritic and Cooperative Growth" held at the annual meeting of the AIME in Atlanta, Georgia on March 7, 1983 under the joint sponsorship of the ASM-MSD Phase Transformations Committee and the TMS-AIME Solidification Committee.

II. THEORETICAL MODELS

Two detailed theoretical models have been proposed in the literature[6,7] to characterize primary dendrite spacing as a function of growth rate, temperature gradient, and alloy composition. The first significant treatment was presented by Hunt.[6] His model was based on two major assumptions: (1) a dendritic interface with sidearms is approximated as a smooth steady-state interface, and (2) a constant temperature and a constant liquid composition in the direction normal to the primary dendritic growth direction are assumed. Under these assumptions, Hunt derived the shape of the cells by following the procedure developed earlier by Bower, Brody, and Flemings.[11] Since the second assumption of the Hunt model is valid only for the dendrite or cell region which is far behind the tip, the interface shape is not valid near the tip region. Hunt circumvented this problem by fitting part of a sphere to the derived shape at the growing front. Under these assumptions, he obtained the following relationships between the tip radius, ρ, and the primary dendrite spacing, λ_1:

$$G\lambda_1^2/\rho = -4\sqrt{2}[mC_t(1 - k) + DG/V] \qquad [1]$$

where G is the temperature gradient in the liquid, C_t is the solute composition at the tip, m is the slope of the liquidus, k is the equilibrium solute partition coefficient, V is the growth rate, and D is the solute diffusion coefficient in the liquid.

The relationship between the dendrite tip undercooling and the dendrite tip radius was derived for a spherical dendrite front model by using a procedure similar to that described by Burden and Hunt.[1] By substituting the value of the dendrite tip radius from Eq. [1] into this result, Hunt obtained a result which related primary dendrite spacing to tip undercooling. By assuming that the primary dendrite spacing adjusts to the minimum dendrite tip undercooling, Hunt obtained the final result as,

$$VG^2\lambda_1^4 = -(64\gamma D/\Delta S)[m(1 - k)C_\infty + kGD/V] \qquad [2]$$

where γ is the solid-liquid interfacial energy, ΔS is the entropy of fusion per unit volume, and C_∞ is the initial solute content of the alloy. When $V \gg V_{cs}$, where V_{cs} is the critical velocity at which the planar interface becomes unstable, the second term in the bracket of Eq. [2] becomes negligible. Under this condition, Hunt's result can be simplified to

$$VG^2\lambda_1^4 = -(64\gamma D/\Delta S)[mC_\infty(1 - k)] \qquad [3]$$

The right-hand side of Eq. [3] is constant for a given alloy composition. Examination of Eq. [2] shows that a maximum in λ_1 is predicted as a function of velocity when G and C_∞ are held constant. According to this theory, the maximum in λ_1 should occur at $V = 2V_{cs}$.

Another theoretical model to characterize primary dendrite spacing was developed by Kurz and Fisher.[7] They assumed that the shape of the cell or dendrite can be approximated as ellipsoids so that:

$$\rho = b^2/a = \lambda_1^2 G/3\,\Delta T' \qquad [4]$$

where a and b are minor and major axes of the ellipsoid, respectively. They can be approximated as: $a = \Delta T'/G$ and $b = \lambda_1/\sqrt{3}$, where $\Delta T'$ is the temperature difference between the tip and the base of the cell. Next, they substituted for ρ the value obtained from the marginal stability criterion for an isolated dendrite or cell. They simplified their results for the low velocity ($V < V_{cs}/k$) and for the high velocity ($V > V_{cs}/k$) regimes. For $V < V_{cs}/k$, they obtained

$$\lambda_1 = \left[\frac{6\,\Delta T'}{G(1-k)}\right]^{1/2}\left[\frac{D}{V} - \frac{k\,\Delta T_0}{G}\right]^{1/2} \qquad [5]$$

where $\Delta T_0 = -m(1-k)C_\infty/k$, in which m is the slope of the liquidus. For $V > V_{cs}/k$ they obtained

$$\lambda_1 = 4.3\,\Delta T'^{1/2}[\gamma D/\Delta Sk\,\Delta T_0]^{1/4}V^{-1/4}G^{-1/2} \qquad [6]$$

Equation [6] is very similar to Eq. [3] obtained by Hunt. Thus, in the high velocity regime, the results predicted by these two theories differ only by a constant. However, in the low velocity regime, these two theories give significantly different results in that the Kurz-Fisher model predicts a sharp increase in λ_1 with decrease in velocity at $V = V_{cs}/k$. This increase in spacing was attributed by the authors to the dendrite-to-cell transition. A comparison with experimental data in the region near $V = V_{cs}/k$ will thus enable us to assess the validity of the Kurz-Fisher model.

III. COMPARISON WITH EXPERIMENTAL DATA

A number of experimental studies on primary dendrite spacings have been carried out in metallic systems. However, in order to assess the validity of the theories, we shall use the results presented in Part I for the growth of succinonitrile dendrites in the succinonitrile-acetone system.[8] In this system, accurate measurements of primary and secondary dendrite spacing, as well as dendrite tip radius, were obtained. Furthermore, the values of all the system constants have been measured quite accurately[12] so that no uncertainty exists in the interpretation of the results. These values are given in Table I of Part I.[8] Another important point which is not always appreciated in the literature is the location at which the temperature gradient value is measured. Generally, the temperature gradient is measured (1) in the liquid ahead of the interface; (2) at the dendrite or cell tips; or (3) in the two-phase region. These temperature gradients can differ significantly in metallic systems under low growth rate conditions. For succinonitrile, the thermal conductivities of solid and liquid are nearly equal so that the temperature gradient does not vary significantly in the system under the conditions of low growth rates, *i.e.*, $V\,\Delta H \ll K_L G$, where ΔH is the enthalpy of fusion per unit volume and K_L is the thermal conductivity of liquid. Thus, the location of the temperature gradient is not as critical as in the metallic systems. The temperature gradient reported in Part I was measured at the dendrite or cell tips.

Figure 1 compares the experimental results with the theory of Hunt given by Eq. [2] and with the high growth rate regime result, Eq. [6], of Kurz and Fisher. A poor correlation is obtained between the experimental results and both theories. The maximum in primary dendrite spacing is predicted by Hunt's theory at a velocity which is more than an order of magnitude smaller than that observed experimentally. Also, in the high velocity regime, the slope of the experimental line is slightly higher than that predicted by either of these theories. The Kurz-Fisher theory also predicts cellular to dendrite transition at $V = V_{cs}/k = 0.783$ μm per second for the data in Figure 1. Our transition agrees well with this predicted growth rate. However, Kurz-Fisher predict that the primary spacing will increase sharply with decreasing velocity near $V = V_{cs}/k$. Experimental results, in fact, show just the opposite behavior in that primary spacings decrease sharply as the growth rate is decreased in the cellular region.

Both theoretical models conclude that $\lambda_1^4 G^2 V$ = constant for a given alloy composition in the high growth rate regime. Figure 2 plots $\lambda_1^4 G^2 V/k\,\Delta T_0$ *vs* velocity for the succinonitrile-acetone system. For comparison, recent experimental data by Mason *et al.*[10,13] on lead dendrite growth in Pb-Sn, Pb-Au, and Pb-Pd are also shown since much higher growth rates have been studied in these systems. All results show significant deviation from a constant value indicating that the existing theories do not adequately explain experimental results either in the low velocity regime or in the high velocity regime. Furthermore, they do not predict quantitatively the growth rate at which the maximum in primary dendrite spacing occurs. However, the growth rate at which the dendrite-to-cell transition occurs is predicted quite accurately by the Kurz and Fisher model.

IV. THEORY

We shall now present a simple model which can explain some of the characteristics of primary dendrite spacing variation with growth conditions. In order to develop a valid

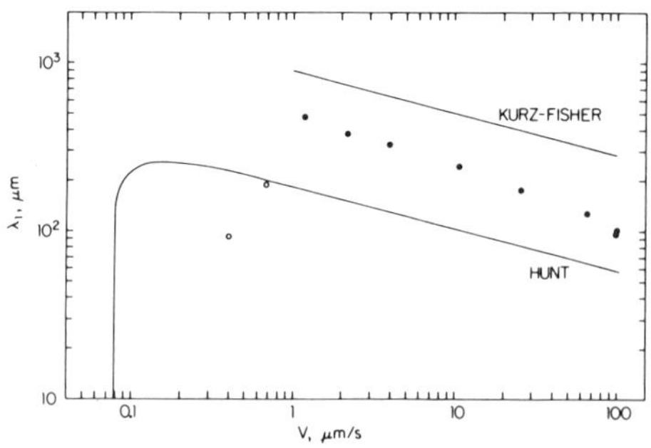

Fig. 1—Comparison of the existing theories of primary spacing with the experimental data[8] in succinonitrile-5.5 mol pct acetone system. Temperature gradient = 67 K/cm (○ cell, ● dendrite).

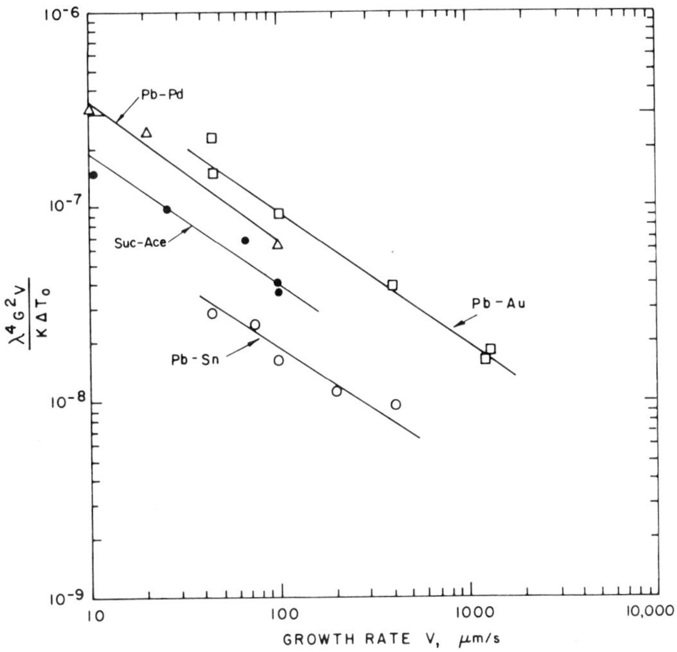

Fig. 2 — Variation of $\lambda^4 G^2 V/k\Delta T_0$ *vs* V for the high velocity data in Pb-Sn,[10] Pb-Au,[10] Pb-Pd,[13] and succinonitrile-acetone[8] system. G = 359 K/cm for the metal systems and G = 67 K/cm for the succinonitrile-acetone system.

model, we shall first examine three important conclusions reached in Part I.[8]

(1) The experimentally measured tip radii match precisely with the theoretically predicted values based on an isolated dendrite growth model. Thus, the interaction between the temperature or concentration gradient at the tip and the steady-state primary dendrite spacing is not too strong. We propose that the dendrite tip first establish its stable radius of curvature for given thermal and solute conditions, and the dendrite spacing subsequently adjusts to correspond to the radius value so that the interface can grow in a steady-state manner. Consequently, the dendrite tip radius, as well as the solute composition and the temperature at the dendrite tip can be predicted from the theory described in Part I.[8]

(2) The sharp decrease in λ_1 occurs at the dendrite-cell transition velocity. Note that the deviations from linearity in primary dendrite spacing and in tip radius (Figure 6, Part I) occur precisely at the same velocity, so that the same physical phenomenon causes a decrease in spacing and an increase in tip radius. Furthermore, the maximum in the primary dendrite spacing appears to correspond with the minimum in the solute Peclet number.

(3) In the high velocity regime, a variation in the temperature gradient significantly alters the primary dendrite spacing but has no effect on the tip radius or on the initial secondary dendrite arm spacing. Consequently, thermal diffusion plays an important part in determining the primary dendrite spacing. Thus, we define a dimensionless parameter $\Lambda = (\lambda_1/\rho)(\lambda_1/l_t)$, where l_t is the characteristic thermal diffusion length, which was defined earlier[2] as $l_t = k\,\Delta T_0/G$. Thus, the dimensionless parameter which characterizes the primary dendrite spacing can be written as

$$\Lambda = (\lambda_1^2 G/\rho k\,\Delta T_0) \qquad [7]$$

This dimensionless parameter should then be related to the dimensionless parameter which characterizes the solute diffusion length, $l_s = 2D/V$, and the capillary length, $l_c = \gamma/\Delta Sk\,\Delta T_0$. Such a dimensionless parameter, as defined earlier,[2] is:

$$\mathscr{A} = l_c/l_s = \gamma V/2\,\Delta SDk\,\Delta T_0 \qquad [8]$$

We now require a relationship between Λ and $\mathscr{A}$ to characterize the primary dendrite spacing. A number of theoretical calculations have been proposed in the literature,[6,7,14] and they all are somewhat similar to the model proposed by Hunt.[6] We shall, therefore, start out with the basic relationship between λ_1 and ρ derived by Hunt, given by Eq. [1], and then obtain a general relationship between Λ and $\mathscr{A}$. Equation [1] can be rewritten in terms of Λ as

$$\Lambda = 4\sqrt{2}[(C_t/C_\infty) - (GD/Vk\,\Delta T_0)] \qquad [9]$$

Λ must approach zero at both limits of planar interface stability. The right-hand side of Eq. [9] becomes zero at the low velocity limit of the planar interface instability, *i.e.*, at $GD/V\,\Delta T_0 = 1$. However, the value of the bracket on the right-hand side of Eq. [9] approaches $1/k$ at the velocity corresponding to the absolute stability of the planar interface.[15] We may thus modify Eq. [9] such that it reflects the appropriate limiting behavior for the absolute stability condition, *i.e.*, $\rho = \infty$ or $C_t = C_\infty/k$ when $V = \Delta SD\,\Delta T_0/\gamma k$. The right-hand side of Eq. [9] contains only one unknown, C_t. In Part I[8] we have shown that the dendrite tip radius can be predicted accurately by the isolated dendrite model in which the marginal stability criterion is employed.[2] Thus, we may substitute the marginal stability criterion in Eq. [9] to obtain

$$\Lambda = 4\sqrt{2}\mathcal{A}L/p^2 \qquad [10]$$

where p is the solute Peclet number equal to $V\rho/2D$ and $L = 1/2(l+1)(l+2)$ in which the $l = 6$ harmonic is found to be operative for the dendrite growth process.[8,16] Since the solute Peclet number approaches infinity at both the critical velocities for planar interface stability, λ_1 approaches zero at these velocities.

Equation [10] relates primary dendrite spacing to the dendrite tip radius for given conditions of solute composition, growth velocity, and temperature gradient in the liquid. In order to predict λ_1, we need to develop a relationship which determines the dendrite tip radius as a function of experimental variables. In Part I[8] we have shown that dendrite tip radius, calculated under the assumption of negligible interactions between neighboring dendrites, matches well with the experimental values obtained in the directional solidification experiments. We may thus obtain the dendrite tip radius value from the relationship developed earlier,[2] which is

$$(1/2)\mathcal{G}[1 - F_1(p)] + (\mathcal{A}L/p^2)[1 - F_2(p)] = [1 - \phi(1-k)]^{-1} \qquad [11]$$

where $\mathcal{G} = l_t/l_d = 2GD/Vk\,\Delta T_0$,

$$F_1(p) = \{\phi(1-k)/[1-\phi(1-k)]\}\{[(1-\phi)/(\phi - p + p\phi)] - (1/\phi)\},$$

$$F_2(p) = [2/L]\{\phi(1-k)/[1-\phi(1-k)]\}[1 + p - (p/\phi)],$$

and $\phi = p\exp(p)E_1(p)$.

Equation [11] can be solved to obtain the value of p for given values of V, G, and C_∞. However, it is much simpler to calculate V for a given value of p and then to use the relationship $\rho = 2pD/V$ to obtain the tip radius value. Note that Eq. [11] is quadratic in V so that for each value of p, two sets of V and ρ values are obtained. One set of values corresponds to the dendritic region whereas the other set of values corresponds to the cellular region.

Equations [10] and [11] now enable us to calculate the variation in primary dendrite spacing with the growth rate. The results of these calculations for succinonitrile-5.5 mol pct acetone system are shown in Figure 3. The correlation between the theory and the experimental data is found to be quite good. The major success of the theory appears to be in predicting the maximum in primary dendrite spacing as a function of velocity. The velocity at which the maximum in λ_1 occurs is predicted quite accurately. Also, deviations between the theoretical and experimental results in the high velocity regime are small. However, the experimental slope of log λ_1 *vs* log V plot is higher than that predicted by the theory.

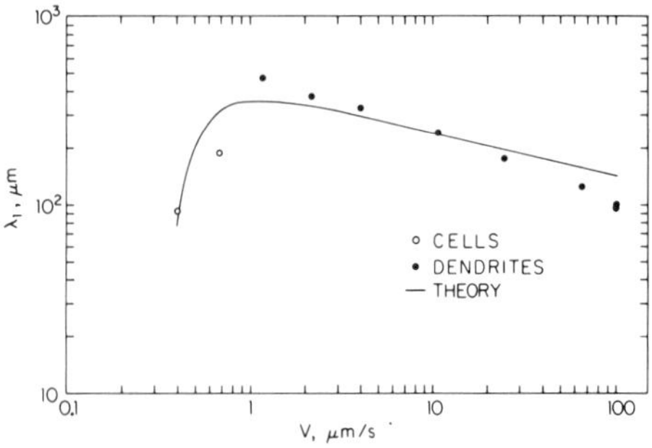

Fig. 3—Comparison of the theory and experimental data[8] on primary spacing in succinonitrile-5.5 mol pct acetone. $G = 67$ K/cm.

V. DISCUSSION

The key result of our theoretical analysis is the prediction of a maximum in primary spacing as a function of velocity. Experimental studies by Klaren *et al.*[9] had previously shown the existence of such a maximum for lead dendrite growth in Pb-Sn and Pb-Au systems. However, some doubts were raised[17] about the validity of these results since the density differences between Pb and Sn or Au are large, which could cause significant convection effects at low velocities, and thereby alter primary spacings. Recent work by Mason *et al.*[13] in the Pb-Pd system also showed a maximum in primary spacing as a function of the growth rate. Since the densities of liquid Pb and Pd are nearly identical, no appreciable convection would be present in this system. Furthermore, our results in the succinonitrile-acetone system[8] have also confirmed the existence of a maximum in λ_1 with the growth rate, and no convection effects were observed in this system.[8] A maximum in λ_1 with velocity has also been observed recently by Kurz[18] for dendrite growth in pivolic acid-naphthalene system. Our theoretical results now confirm the validity of the maximum in λ_1 with the growth rate since the theoretical model predicts quite accurately the velocity at which the maximum in primary spacing occurs.

We now examine the reasons why primary spacings increase with velocity in the low velocity regime and decrease with velocity in the high velocity regime. According to the theory, the primary dendrite spacing is inversely proportional to the square root of the solute Peclet number. Thus, a maximum in λ_1 occurs at the velocity of a minimum in the solute Peclet number, as found in Part I.[8] Note that the Peclet number decreases continuously with a decrease in velocity when the effect of the temperature gradient is negligible. However, when the effect of the temperature gradient is significant, the Peclet number increases with a decrease in the growth rate, thereby causing a minimum in the Peclet number. Consequently, it is the effect of the temperature gradient at low velocities which causes the primary dendrite spacing to decrease with the decreasing velocity. Thus, if V_m is the velocity at which the maximum in λ_1 occurs, then V_m

depends upon the temperature gradient imposed upon the system; V_m increases with an increase in the temperature gradient. This was found experimentally by Mason *et al.*[10] in the Pb-Sn system, where V_m was found to increase as the temperature gradient was increased from 66 to 107 to 350 K/cm.

The increase in the solute Peclet number at low velocities is evident in the sharp increase in the tip radius value, as seen in Figure 6 of Part I.[8] Thus, below V_m the dendrite tip radius becomes large and the transition from dendritic to cellular structure occurs. Figure 4 illustrates the interface shapes observed at the growth rates of 0.4, 0.68, and 10.0 μm per second in succinonitrile-5.5 mol pct acetone under the condition of G = 67 K/cm. These microstructures show the transition from dendritic to cellular structure as the growth rate is reduced below V_m.

One of the intriguing features that is observed experimentally is that for V just below V_m, both dendritic and cellular structures often coexist. When this happens, λ_1 between the dendrites is found to be higher than that between the cells. Figure 5 shows the interface shape for V = 0.9 μm per second and G = 77 K/cm in the succinonitrile-5.5 mol pct acetone system. Note that λ_1 (dendrites) is about 40 pct larger than λ_1 (cells). Similar observations have been made by Mason *et al.*[13] in the Pb-Pd system. These observations show that the dendrite-to-cell transition is accompanied by a decrease in the primary spacing. Furthermore, a close examination of Figure 5 shows that the dendrite front grows slightly ahead of the cellular front, so that the dendrite tip temperature is a little higher than the cell tip temperature. Also, cells have a slightly greater tip radius than do dendrites. These observations suggest that dendrites and cells should be described by different kinetic equations in this transition zone.

One of the assumptions in our model was that the dendrite tip radius was not strongly dependent on the primary spacing. This assumption was justified on the basis of the excellent agreement between the experimental measurements of tip radii with the theoretically predicted values for the isolated dendrite case. In order to check the validity of the assumption, experiments were carried out in the succinonitrile-5.5 mol pct acetone system to obtain a steady-state dendritic interface at V = 1.17 μm per second and G = 67 K/cm. The velocity was suddenly increased to 5.8 μm per second, and Figure 6 illustrates the changes that occurred initially in the shape of the interface. Note that after a very short time, the tip radius assumed the value which was characteristic of the new growth conditions, although no change in λ_1 occurred. It was also found that the primary dendrite spacing adjusted to its characteristic value only after a much longer time than that needed for the adjustment of the tip radius. Furthermore, the tip radius value did not change while the primary spacing was being adjusted. These observations also justify our assumption

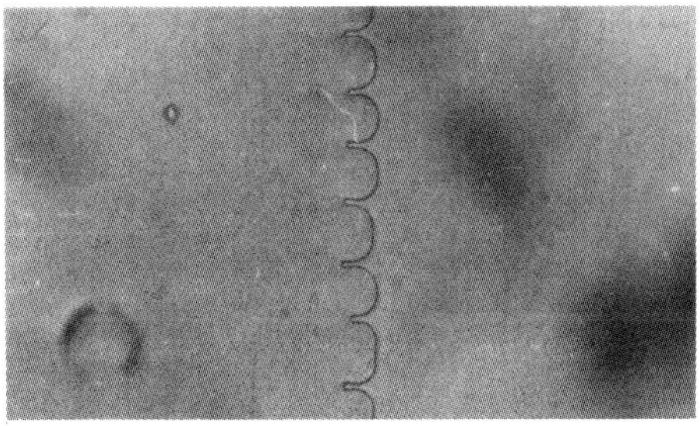

(*a*)

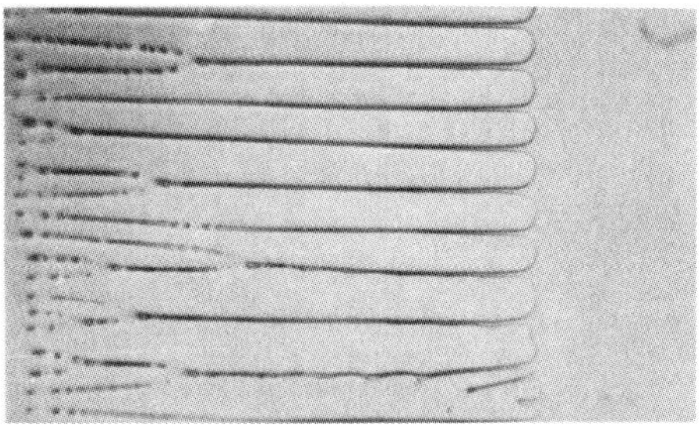

(*b*)

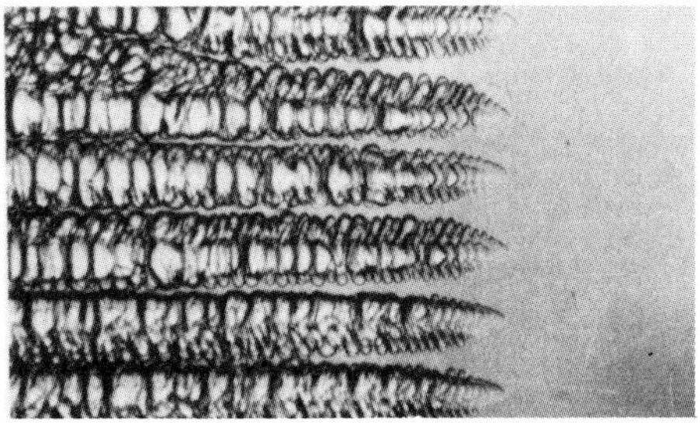

(*c*)

Fig. 4—Steady-state interface structures as a function of growth rate in succinonitrile-5.5 mol pct acetone, G = 67 K/cm. (*a*) V = 0.4 μm/s, (*b*) V = 0.58 μm/s, and (*c*) V = 10 μm/s. Magnification of (a) 56 times and of (b) and (c) 28.8 times.

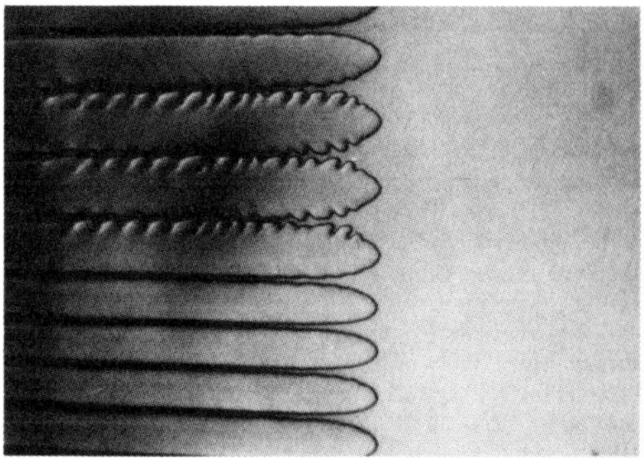

Fig. 5—Interface structure illustrating the coexistence of dendrites and cells in a directionally solidified succinonitrile-5.5 mol pct acetone system at V = 0.9 μm/s and G = 77 K/cm. Magnification 29 times.

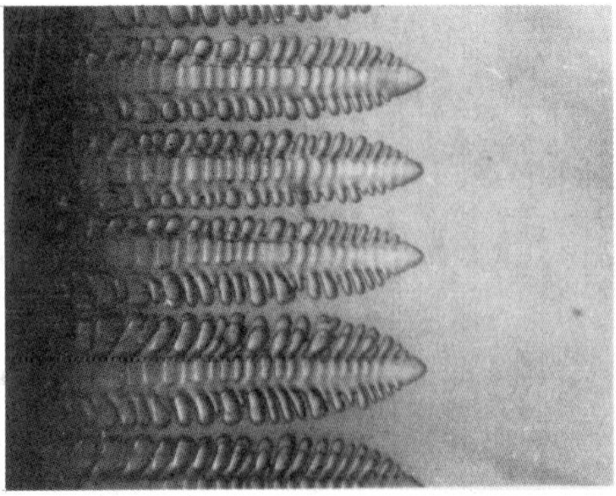

(*a*)

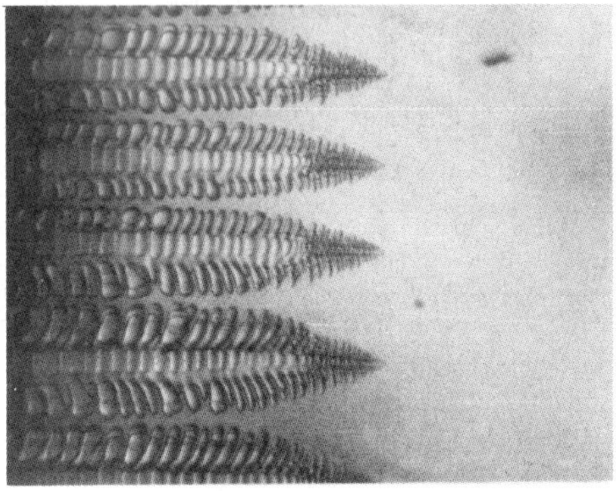

(*b*)

Fig. 6—(*a*) Steady state interface shape in the succinonitrile-5.5 mol pct acetone system for the growth conditions of $V = 1.17\ \mu m/s$ and $G = 67$ K/cm. (*b*) The change in the interface structure of (a) when the growth rate was suddenly increased to 5.8 μm/s. $t = 60$ s after the change. Magnification 24 times.

that the dendrite tip characteristics are not significantly influenced by changes in λ_1.

The theoretical model that we have developed is valid primarily for the dendritic structures since Eq. [11] is derived on the assumption of a parabolic shape front. Such an assumption has been found to be valid for all dendritic structures studied in Part I[8] and for cellular structures close to V_m. However, for $V \ll V_m$, a significant deviation from a parabolic shape occurs. Furthermore, the coupling between primary spacing and tip radius may not be negligible for cellular structures. Consequently, our model should be valid for dendritic structures in general and for cellular structures close to the V_m. Although the theoretical model proposed in this paper explains experimental results reasonably well, two important aspects still require further theoretical attention: (1) a more accurate relationship between λ_1 and ρ is needed to predict properly the slope of the log λ_1 *vs* log V plot at high velocities, and (2) a theoretical description of dendrite and cell characteristics needs to be developed in the region where both these morphologies coexist.

VI. CONCLUSIONS

The existing theories of primary spacing have been found to be inadequate in explaining experimental observations. A theoretical model based on the analysis of Hunt[6] is presented, and it is shown that the model explains reasonably well the variation of λ_1 with the growth rate. A maximum in λ_1 as a function of growth rate is predicted, and this maximum is correlated with the dendrite-to-cell transition. It is shown that the dendrite-to-cell transition occurs when the temperature gradient effects become significant and that such a transition is accompanied by a reduction in the primary spacing.

ACKNOWLEDGMENTS

Appreciation is extended to John Verhoeven and John Mason for valuable discussions and for comments on the manuscript. The author would also like to thank K. Somboonsuk for carrying out experimental work for Figures 4 to 6. This work was done at the Ames Laboratory, which is operated for USDOE by Iowa State University, under contract No. W-7405-Eng-82, supported by the Director of Energy Research, Office of Basic Energy Sciences, WPAS-KC-02-01.

REFERENCES

1. M.H. Burden and J.D. Hunt: *J. Cryst. Growth*, 1974, vol. 22, p. 109.
2. R. Trivedi: *J. Cryst. Growth*, 1980, vol. 49, p. 219.
3. J.A.E. Bell and W.C. Winegard: *J. Inst. Met.*, 1963, vol. 92, p. 357.
4. M. Ibaraki, T. Okamoto, and K. Kishitake: *J. Japan Inst. Metals*, 1969, vol. 33, p. 209.
5. G.F. Bolling and D. Fainstein-Pedraza: *Acta Metall.*, 1974, vol. 22, p. 1033.
6. J.D. Hunt: *Solidification and Casting of Metals*, The Metals Society, Book 192, London, 1979, p. 3.
7. W. Kurz and D.J. Fisher: *Acta Metall.*, 1981, vol. 29, p. 11.
8. K. Somboonsuk, J.T. Mason, and R. Trivedi: *Metall. Trans. A*, 1984, vol. 15A, p. 967.
9. C.M. Klaren, J.D. Verhoeven, and R. Trivedi: *Metall. Trans. A*, 1980, vol. 11A, p. 1853.
10. J.T. Mason, J.D. Verhoeven, and R. Trivedi: *J. Cryst. Growth*, 1982, vol. 59, p. 516.
11. T.F. Bower, H.D. Brody, and M.C. Flemings: *Trans. TMS-AIME*, 1966, vol. 236, p. 624.
12. M.E. Glicksman: private communication, R.P.I., Troy, NY, 1982.
13. J.T. Mason, J.D. Verhoeven, and R. Trivedi: *Metall. Trans.* (submitted).
14. R.A. Wasson: Ph.D. Thesis, M.I.T., Cambridge, MA, 1978.
15. W.W. Mullins and R.F. Sekerka: *J. Appl. Phys.*, 1964, vol. 35, p. 444.
16. S.-C. Huang and M.E. Glicksman: *Acta Metall.*, 1981, vol. 29, p. 701.
17. D.G. McCartney and J.D. Hunt: *Acta Metall.*, 1981, vol. 29, p. 1851.
18. W. Kurz, private communication, École Polytechnique Federale de Lausanne, Lausanne, Switzerland, 1982.

Testing shape selection in directional solidification

John Bechhoefer and Albert Libchaber
The James Franck and Enrico Fermi Institutes, The University of Chicago, Chicago, Illinois 60637
(Received 26 September 1986)

We report results from an experiment on the directional solidification of pivalic acid. In addition to the usual cellular interface patterns, we observe a new, metastable dendritic form that can be interpreted in the light of recent ideas about pattern formation.

When a thin layer of pivalic acid is drawn at a fixed velocity across a linear temperature gradient, the solid-liquid interface between crystalline pivalic acid and its melt assumes a cellular pattern. Deep liquid grooves separate periodically spaced fingers of solid. Although such patterns are familiar from previous experiments on other materials,[1] the cells in pivalic acid are unusual in that the tips of the patterns are stable and do not split, except at very low solidification velocities. We believe that this behavior is related to the large amount of surface-tension anisotropy in pivalic acid.[3] In the experiments reported below, we have taken advantage of this stability to probe patterns close to those naturally selected by the system, thereby testing recent theories concerning the mechanism of pattern formation.

In a directional solidification experiment, if we start from rest and switch on a given velocity, the system eventually selects a well-defined, stable cell wavelength that is a function of the velocity. If we start with one velocity, allow a wavelength to be selected, and then change the solidification velocity, the selected wavelength will in general also change. If the new wavelength is smaller than the old, the usual way to reduce the cell size is through tip splitting. But in pivalic acid, tip splitting is inhibited; instead, we observe hysteresis. (See Fig. 1.) When we increase the velocity, the tips sharpen and give rise to time-dependent sidebranches. Alternatively, forcibly changing the wavelength while fixing the velocity also leads to time-dependent patterns. Thus, small perturbations in the shape of a smooth, time-independent cell lead to a time-dependent metastable shape that can exist for arbitrarily long times.

Our experiment follows Jackson's design for his original work on directional solidification.[2] A thin glass cell containing pivalic acid is placed atop two temperature-controlled copper blocks that are separated by a small gap. One block is kept hotter than the melting temperature of pivalic acid, the other cooler, so that there is a solid-liquid interface that sits over the gap between the two plates. The sample cell is then drawn at a constant velocity v across the gap, from the hot to the cold side. After a sometimes slow transient, the solid-liquid interface will move with a velocity $-v$ so that its average position is maintained at a constant temperature.

In the experiments reported here, the hot block was electrically heated to 55.0 °C and regulated to ±0.1 °C. The cold block was cooled to 10.0 °C±0.1 °C by circulating water through it. The gap between the two blocks was 0.200±0.001 cm. The temperature gradient varied considerably as a function of the thickness of the sample plates and the solidification velocities and was calibrated by building a dummy cell with a thermocouple. For low velocities, the gradient was approximately 125 °C/cm, and it rose for higher velocities. The sample was pushed by a stepping-motor lead-screw combination at velocities that ranged from 0.5 μm/sec to 260 μm/sec. The sample cell was constructed from two 22-mm-square number one microscope cover glass slips glued all around, except for entrance and exit holes. Stycast, once hardened, proved to be a relatively inert glue. The cell had a thickness of approximately 20 μm and was filled with pivalic acid under vacuum. The fill holes were sealed with epoxy.

From 6 °C to 36.5 °C, pivalic acid forms a soft fcc crystal that is known as a plastic crystal phase. It is thus very similar to succinonitrile and carbon tetrabromide, which have been used in previous directional solidification experiments. Pivalic acid differs markedly from other plastic crystals in its large surface tension anisotropy. Glicksman has recently measured the relative anisotropy ϵ to be 5%, in contrast to 0.5% for succinonitrile.[3] ϵ is defined via the relation $\gamma=\gamma_0\{1+\epsilon[1-\cos(4\theta)]\}$, where γ_0 is the

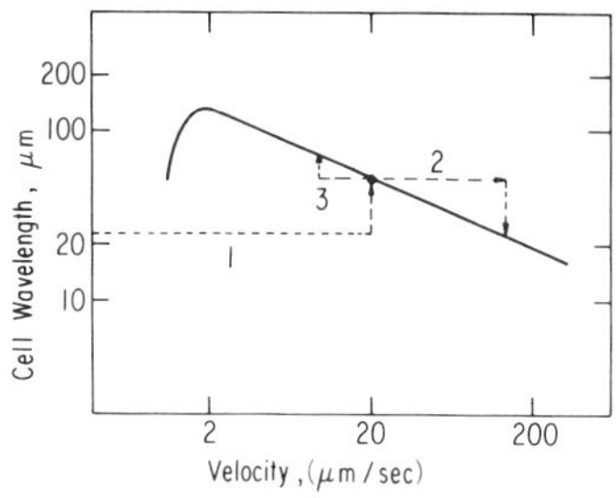

FIG. 1. Cell wavelength vs velocity for pivalic acid. The selected velocity was observed by starting from rest (line 1). Line 2 shows the hysteresis observed as the velocity is increased. The wavelength collapses when the solidification velocity is about eight times the initial velocity. Line 3 shows the smaller amount of hysteresis observed as the velocity is decreased. The system begins to choke off cells once the velocity is about half its initial value.

Reprinted with permission from *Physical Review B*, Vol. 35, No. 3, 1987.

surface tension in the ⟨100⟩ direction and θ is the angle between the orientation of the surface and the ⟨100⟩ direction.

Previous experiments in succinonitrile have shown three types of interface shapes.[4] Below a critical velocity v_c, the interface is flat. Linear stability analysis predicts that $v_c = DGk/[\,|dT/dC|\,C_0(1-k)]$, where D is the diffusion coefficient of impurities in the liquid, G is the temperature gradient, C_0 is the average concentration of impurities, dT/dC is the liquidus slope, and k is the partition coefficient. Between v_c and about $10v_c$, solid fingers freeze with impurity-rich, liquid grooves in between. The wavelength of the cells increases with velocity, while the tip curvature decreases. Above about $10v_c$, the cellular patterns develop sidebranches and become dendritic. In this regime, both the wavelength and tip curvature decrease with velocity.

Our sample of pivalic acid was distilled twice under vacuum. For the impurities, we measured $D \approx 10^{-6}$ cm^2/sec (Ref. 2). This value of D is a factor of 10 smaller than that reported for succinonitrile and acetone[6] but is consistent with other data obtained from our experiment. We also measured $(dT/dC)C_0$ to be 2.9 °C.

In pivalic acid, we observed a cellular solid-liquid interface over the entire 500-fold speed range of our driving motor. The dependence of wavelength on velocity is similar to that of succinonitrile—for $v_c < v < 10v_c$, it increases; for $v > 10v_c$, it decreases—but the shape is always cellular. Also, for $v < 10v_c$, the diffusion length D/v is larger than the cell wavelength; for $v > 10v_c$, it is smaller. We have good reason to suspect that the greatly expanded cellular regime is due to the high crystalline anisotropy of pivalic acid.[2] What is important for this study is first that there is a wide range of velocities over which one finds very stable cells. Second, the wavelength of cells of pivalic acid increases at low velocities (in our experiment, below 2 μm/sec) and decreases at higher velocities, following approximately the same laws that govern the length scales of cells and dendrites of succinonitrile. Third, and most important, where the wavelength of cells decreases with velocity, the tips of the cells were absolutely stable and were never observed to split.

The solid part of our sample was a single crystal whose ⟨100⟩ direction was aligned with the thermal gradient. When the crystallographic axes are not so aligned, the fingers grow at an angle to the thermal gradient, as described previously.[7]

In our experiments, we started from rest and switched on a given velocity. After the system had evolved to a stable system of cells, we increased the velocity. Figure 2(a) shows stationary cells 10 μm/sec. The velocity is then doubled successively in Figs. 2(b), 2(c), and 2(d). The overall size of the cell is fixed by the initial conditions, and the shape that is observed has the same wavelength as the original cells. The solution, however, is no longer stationary, for in the frame in which the tip of the dendrites is fixed, the sidebranches advect to the rear.

Figure 2 shows that the wavelength remains fixed as the velocity is increased eight fold. As the velocity is further increased, the sidebranches will themselves sprout sidebranches. Note that the system remains dendritic, even after the wavelength has decreased. The wavelength after collapse is still 50 to 100% larger than the selected wavelength. This collapse of metastable dendrites is shown in Fig. 3.

Figure 2 also reveals that the tip is sharpening up as the velocity is increased. Because the tip is the "farthest out" part of the cell, its curvature depends only on *local* conditions (velocity, temperature gradient, etc.) and not on the global structure of the fingers, as determined by initial conditions. To test this hypothesis, we made another series of runs where we started from different initial conditions and increased the velocity to the same final value.

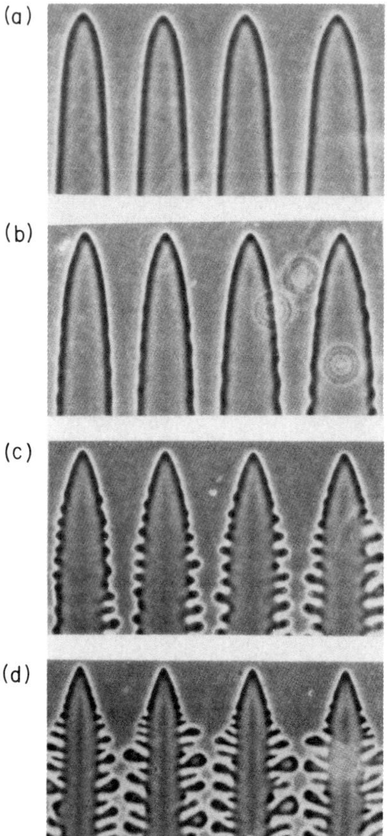

FIG. 2. Constant wavelength, variable tip curvature. (a) shows the initial condition: stable cells solidifying at 10 μm/sec. (b) shows the same sample after the velocity has been increased to 20 μm/sec. In (c), the velocity is 40 μm/sec and (d), it is 80 μm/sec. All of the pictures reflect steady state conditions. The cell wavelength is 75 μm.

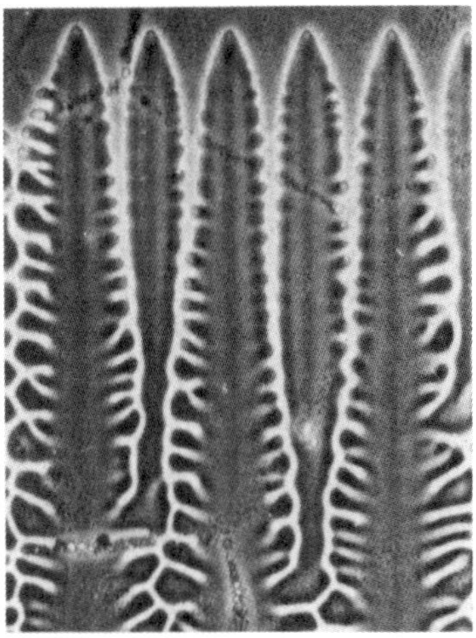

FIG. 3. Collapsing dendrites. The initial velocity was 5 μm/sec. That shown is 40 μm/sec. At 5 μm/sec, the wavelength was 107 μm; at 40 μm/sec, the wavelength is 45 μm.

Figure 4 shows one stable cell at 40 μm/sec. Figure 4(b) shows a dendrite at 40 μm/sec that came from a stable cell at 30 μm/sec, and so forth. Although the differing initial conditions lead to differing wavelengths for each of the fingers, each one has the same tip curvature, as expected.

Finally, Fig. 5 shows the result of starting from a high velocity and reducing it. The wavelength of the cells increases by a "choking off" mechanism. One cell lags slightly and then is choked off by its two neighbors. It is because the choking off mechanism operates in pivalic acid that one is able to observe the selected wavelength. When a flat interface destabilizes, the initial wavelength is always two or four times smaller than the selected cell size. The selection always goes from smaller to larger wavelengths and can thus be observed in pivalic acid.

To summarize, we have independently changed the tip curvature $1/\rho$ and the wavelength λ, only one combination of which, for a given velocity, gives time-independent patterns. Thus, the stationary cellular shape of directional

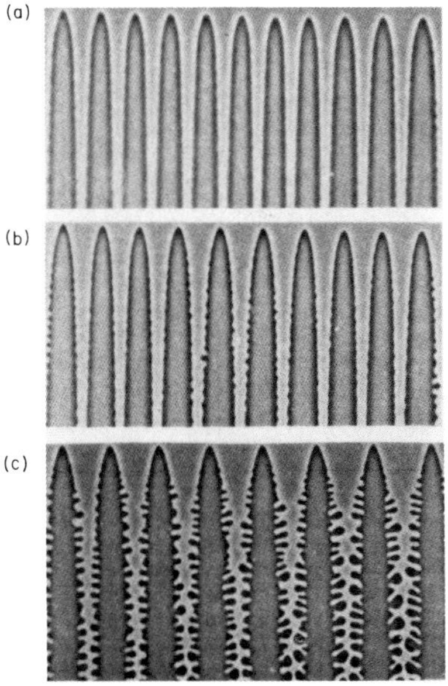

FIG. 4. Constant tip curvature, variable wavelength. The solidification rate is 40 μm/sec for all photos. In (a), the initial velocity is also 40 μm/sec (after starting from rest). In (b), the initial velocity was 30 μm/sec, and in (d) it was 15 μm/sec.

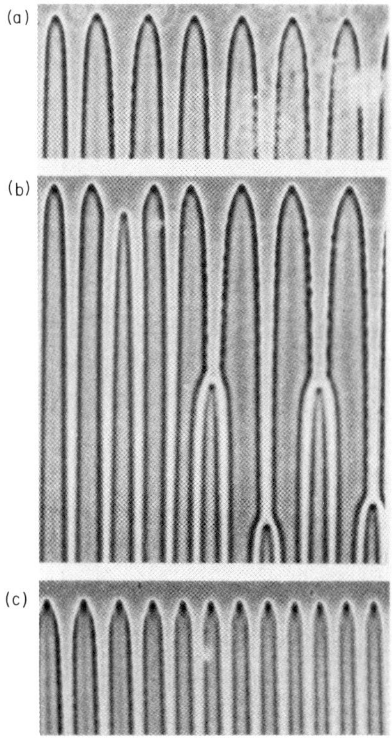

FIG. 5. Choking off mechanism. (c) shows cells at 80 μm/sec. (b) shows the choking off process that occurs when the velocity is abruptly reduced to 40 μm/sec. (a) shows the final state at 40 μm/sec. At 80 μm/sec, the wavelength was 87 μm, while at 40 μm/sec, it was 144 μm.

solidification is an isolated solution in the space of nearby shapes. By forcibly changing the shape at the tip or at the far end, we immediately observe time-dependent patterns. We are currently studying the dependence of the wavelength and growth rate of the sidebranches on the mismatch and the velocity.

These observations are relevant to recent theories of pattern selection in diffusive processes. In the absence of surface tension, such processes generally admit a continuous family of stationary solutions; however, only one shape is observed in experiments. If we look only for stationary solutions, the theory of "microscopic solvability" predicts that the introduction of surface tension along the moving interface will break up the continuous family into a discrete, countably infinite set of solutions. The theory has been applied to numerical analysis of the Saffman-Taylor problem of viscous fingering in fluids,[8,9] dendritic crystal growth into an undercooled melt,[6,10,11] and directional solidification.[12] It has also been used to solve the Saffman-Taylor fingerselection problem analytically.[13-16]

Our observation that the cellular shape is an isolated solution in the space of nearby shapes thus agrees with a prediction of microscopic solvability. We are currently looking for the other solutions.

Finally, if the slight irregularities in the sidebranches are ignored, then the metastable dendrites in Figs. 2 and 4 would be periodic solutions. A natural question to ask then is "Does a slight mismatch in the microscopic solvability condition lead to a periodic solution?" If so, then the qualitative test of microscopic solvability presented here can be made quantitative.

This work was supported by the National Science Foundation, under Grant No. NSF DMR83-16204. One of us (J.B.) received support from AT&T Bell Labs.

[1]J. S. Langer, Rev. Mod. Phys. **52**, 1 (1980).
[2]K. A. Jackson and J. D. Hunt, Acta. Metal. **13**, 1212 (1965).
[3]M. Glicksman, Mat. Sci. Eng. **65**, 45 (1984).
[4]R. Trivedi, Metal. Trans. **15A**, 977 (1984). Cf. W. Kurz and D. J. Fisher, Acta Metal. **29**, 11 (1981); and H. Esaka, Ph.D. thesis, Ecole Polytechnique Federale de Lausanne, 1986.
[5]J. Bechhoefer and A. Libchaber (unpublished).
[6]M. Glicksman, R. Schaeffer, and J. D. Ayers, Metal. Trans. **7A**, 1747 (1976).
[7]F. Heslot and A. Libchaber, Phys. Scr. **T9**, 126 (1985).
[8]P. G. Saffman and G. I. Taylor, Proc. R. Soc. London, Ser. A **245**, 312 (1958).
[9]J.-M. Vanden-Broeck, Phys. Fluids **26**, 2033 (1983).
[10]D. Meiron, Phys. Rev. A **33**, 2704 (1986).
[11]D. Kessler, J. Koplik, and H. Levine, Phys. Rev. A **33**, 3352 (1986).
[12]A. Karma, Phys. Rev. A **34**, 4353 (1986).
[13]B. Shraiman, Phys. Rev. Lett. **56**, 2028 (1986).
[14]D. C. Hong and J. S. Langer, Phys. Rev. Lett. **56**, 2032 (1986).
[15]R. Combescot, T. Dombre, V. Hakin, Y. Pomeau, and A. Pumir, Phys. Rev. Lett. **56**, 2036 (1986).
[16]M. D. Kruskal and H. Segur, Aeronautical Res. Associates of Princeton Technical Memo, 85-25 (1985).

Lamellar and Rod Eutectic Growth

K. A. Jackson and J. D. Hunt

A general theory for the growth of lamellar and rod eutectics is presented. These modes of growth depend on the interplay between the diffusion required for phase separation and the formation of interphase boundaries. The present analysis of these factors provides a justification for earlier approximate theories. The conditions for stability of rod and lamellar structures are considered in terms of the mechanisms by which the structure can change. The mechanisms considered include both small adjustments to the lamellar spacing due to the motion of lamellar faults, and catastrophic changes due to instabilities. It is concluded that stable growth occurs at or near the minimum interface undercooling for a given growth rate. The consequences of the existence of a diffusion boundary layer at the interface are discussed. The experimental results for the variation of growth rate, undercooling, and lamellar spacing are compared with the theory. We believe that the theory presented in this paper provides an adequate basis for understanding the complex phenomena of lamellar and rod eutectic growth.

THE growth of lamellar eutectics has been the subject of several theoretical and many experimental studies. The foundations for the theoretical work were laid by Zener[1] and Brandt[2] in their analyses of the growth of pearlite. Zener estimated the effect of diffusion, and took into account the surface energy of the lamellar structure. He found that the lamellar structure could grow in a range of growth rates at a given undercooling provided the lamellar spacing was appropriate for the growth rate. Since pearlite grows with only one growth rate and one lamellar spacing at a given undercooling, there is clearly an ambiguity in the theory. Zener removed this ambiguity by postulating that the growth rate was the maximum possible at the given undercooling. He predicted then that the product of the growth velocity v and the square of the lamellar spacing λ should be constant, *i.e.*, $\lambda^2 v$ = const. Brandt[2] started out by assuming that the interface between the lamellae and austenite was sinusoidal. Because of this, the ambiguity encountered by Zener did not arise. Brandt was able to obtain an approximate solution to the diffusion equation, but, since he did not take into account the surface energy, his considerations are incomplete. Tiller[3] applied some of these ideas to the growth of eutectics, and proposed a minimum undercooling condition to replace the maximum velocity condition used by Zener. These conditions are formally identical. Hillert[4] extended the work of Zener. He found a solution to the diffusion equation assuming the interface to be plane. Taking surface energy into account, and applying Zener's maximum condition, he was able to calculate an approximate shape of the interface. Jackson *et al.*[5] used an iterative method employing an electric analog to the diffusion problem to refine the calculation of interface shape. It was found that the interface shape calculated from a plane-interface solution to the diffusion equation was negligibly different from the exact solution. The method provided an analog only for eutectics for which the volumes of the two phases are equal, growing from a melt of exactly eutectic composition.

There has also been considerable experimental work on eutectics. Several experimenters[6-10] found that $\lambda^2 v$ is constant as predicted by Zener.[1] Hunt and Chilton[10] demonstrated that $\Delta T/v^{1/2}$ is also a constant for the Pb-Sn system as predicted. Lemkey *et al.*[11] have recently found $\lambda^2 v$ to be constant for a rod eutectic.

In the present paper, we present the steady-state solution for the diffusion equation for a lamellar eutectic growing with a plane interface, for the general case, that is, with no restriction on the relative volumes of the two phases, and with the melt on or off eutectic composition. A similar solution is also found for a rod-type eutectic. Expressions are obtained for the average composition at the interface and the average curvature of the interface. These equations for the average composition and curvature are similar in form to those derived by Zener[1] and Tiller,[3] and provide a justification for some of the approximations made by these authors.

The mechanisms by which the spacing in a lamellar structure can change are considered. The important mechanism for small changes in lamellar spacing involves a lamellar fault. Examination of the stability of lamellar faults leads to the conclusion that the growth occurs at or near the extremum.* The instabilities which can develop in a rodlike structure are also discussed, resulting in the conclusion that this structure also grows at or near the extremum. Comparison of the conditions for rod and lamellar growth permits a prediction of the surface-energy anisotropy required to produce rods or lamellae for various volume-fraction ratios.

The diffusion equation predicts the existence of a diffusion boundary layer at the eutectic interface unless the eutectic has 0.5 volume fraction of each phase and is growing into a liquid of eutectic composition. This boundary layer is such as to make the composition in the liquid at the interface approximately equal to the eutectic composition. This boundary layer permits changes in composition during the zone refining of eutectics.

Photographs of the eutectic interface of a growing transparent organic eutectic system have been made. Both the components of this eutectic are transparent organic compounds that freeze as metals do.[12] The in-

*Since the minimum undercooling for growth at a given velocity coincide with the maximum growth rate for a given undercooling, such growth will be termed "growth at the extremum".

K. H. JACKSON, Member AIME, is Member of the Technical Staff, Bell Telephone Laboratories, Murray Hill, N.J. J. D. HUNT, formerly with Bell Telephone Laboratories, is now with AERE, Harwell, Didcot, Berks, England.

Manuscript submitted December 31, 1965. IMD

terface shape calculated from the diffusion solution agrees well with observed shape.

Catastrophic breakdown of the lamellar structure is also considered. By catastrophic breakdown we mean either that each lamella develops a shape instability or that one of the phases is cut off by the other. These occur when the growth conditions are suddenly changed, and the gradual adjustment by lamellar faults does not have time to operate. These catastrophic mechanisms define the upper limit to the departure from the extremum conditions, and can occur for departures of as little as a factor of two from the extremum. These catastrophic effects are important in practice where growth-rate fluctuations may occur.

Finally, calculations are compared with measurements made on the lamellar growth of the Pb-Sn system for which both λ vs v and ΔT vs v data are available. The results give a diffusion coefficient in excellent agreement with the best available experimental value, and predict a value of the surface free energy for the Pb-Sn interphase boundary.

I) DIFFUSION

In this section the solution to the diffusion equation for lamellar growth with a plane interface is presented. The solution is similar to that presented by Hillert.[4] The solution to the diffusion equation for rod-type growth is also presented. The average values of the composition at the interface are calculated.

I.1) Diffusion in Lamellar Growth. Consider a plane interface, with S_α and S_β respectively half the widths of α and β phases as shown in Fig. 1. The origin of the coordinate system will be taken to be at the center of the α phase, with x and z as shown. y is parallel to the lamellae in the interface. The problem is essentially two-dimensional (independent of y) and, because of the symmetry of the problem, a solution is necessary only for the region $0 \le x \le S_\alpha + S_\beta$. The interface is assumed to be advancing in the z direction with velocity v, at steady state. For steady-state growth with the coordinate system moving with velocity v in the z direction, the diffusion equation becomes:

$$\nabla^2 C + \frac{v}{D}\frac{\partial C}{\partial z} = 0 \qquad [1]$$

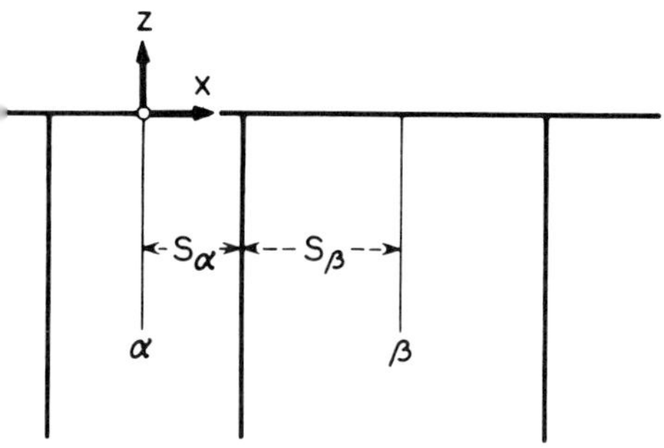

Fig. 1—Planar lamellar eutectic interface showing definition of S_α and S_β and the coordinate system.

with the boundary conditions:

$$C = C_E + C_\infty \quad \text{at} \quad z = \infty$$

$$\frac{\partial C}{\partial x} = 0 \quad \text{at} \quad x = 0 \quad \text{and} \quad x = S_\alpha + S_\beta$$

The conservation of matter at the interface requires:

$$\left(\frac{\partial C}{\partial z}\right)_{z=0} = -\frac{vC_0^\alpha}{D} \qquad 0 \le x < S_\alpha$$

$$\left(\frac{\partial C}{\partial z}\right)_{z=0} = \frac{vC_0^\beta}{D} \qquad S_\alpha < x \le S_\alpha + S_\beta \qquad [2]$$

where D is the diffusion coefficient, C_E is the eutectic composition, and C_∞ is the difference between the eutectic composition and the actual composition far from the interface. C_0^α and C_0^β are the amounts of B and A, respectively, rejected when unit volume of the α and β phase freeze, as shown in Fig. 2.*

*Note that C_0^α and C_0^β should be expressed as $C_0^\alpha = (1-k^\alpha)\,C(x, 0)$ and $C_0^\beta = (1-k^\beta)[1-C(x,0)]$, where k is a distribution coefficient defined as the ratio of the equilibrium concentration of the minor component in the solid to that in the liquid. $C(x, 0)$ is the composition at the interface as given by Eq. [3] with $z = 0$. C_0^α and C_0^β thus depend implicitly on x. The variation of composition with position across the interface is small, but must be considered where departures from the eutectic composition are important, as in the calculation of the interface shape. However, the total composition enters in Eqs. [2], not just the departure from C_E. The summation term in Eq. [3], which contains the x dependence of $C(x, 0)$, is small compared to C_E. It will be shown later that $C_\infty + B_0$ is also small compared to C_E. The value of $C(x, 0)$ can therefore be replaced by C_E with quite good accuracy for most cases. C_0^α and C_0^β can thus be determined from the phase diagram.

The solution is

$$C = C_E + C_\infty + \sum_{n=0}^{\infty} B_n \cos\left(\frac{n\pi x}{S_\alpha + S_\beta}\right) \times e^{\left(-\frac{v}{2D} - \sqrt{\left(\frac{v}{2D}\right)^2 + \left(\frac{n\pi}{S_\alpha + S_\beta}\right)^2}\right)z}$$

which reduces to

$$C = C_E + C_\infty + B_0 e^{-\frac{v}{D}z} + \sum_{n=1}^{\infty} B_n \cos\left(\frac{n\pi x}{S_\alpha + S_\beta}\right) e^{-\frac{n\pi z}{S_\alpha + S_\beta}} \qquad [3]$$

since

$$\frac{n\pi}{S_\alpha + S_\beta} \gg \frac{v}{2D} \quad \text{for} \quad n > 0$$

Using Eqs. [2] to evaluate the Fourier coefficients in Eq. [3] gives:

$$B_0 = \frac{C_0^\alpha S_\alpha - C_0^\beta S_\beta}{S_\alpha + S_\beta} \qquad [4a]$$

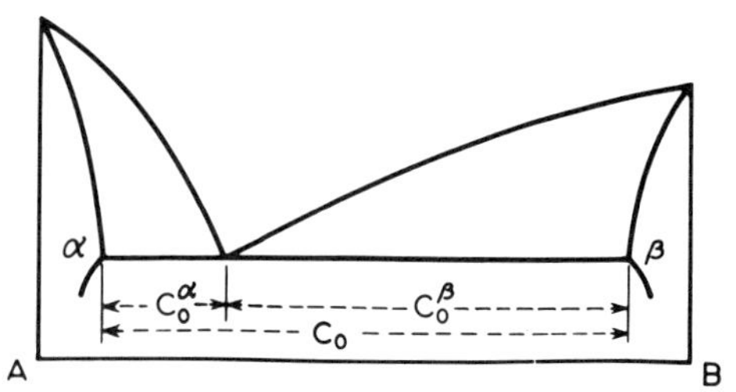

Fig. 2—Phase diagram showing definition of C_0, C_0^α, and C_0^β.

and

$$B_n = \frac{2}{(n\pi)^2}(S_\alpha + S_\beta)\frac{v}{D}C_0 \sin\left(\frac{n\pi S_\alpha}{S_\alpha + S_\beta}\right) \quad [4b]$$

where $C_0 = C_0^\alpha + C_0^\beta$, as shown in Fig. 2.*

*Note that Eq. [4a] is a statement of the lever rule.

The average composition in the liquid at the interface in front of the α phase ($z = 0$) is given by

$$\overline{C_\alpha} = C_E + C_\infty + B_0 + \frac{1}{S_\alpha}\int_0^{S_\alpha}\sum_{n=1}^{\infty} B_n \cos\frac{n\pi x}{S_\alpha + S_\beta}dx$$

$$= C_E + C_\infty + B_0 + \frac{2(S_\alpha + S_\beta)^2}{S_\alpha}\frac{v}{D}C_0 P \quad [5a]$$

and similarly, in front of the β phase, by:

$$\overline{C_\beta} = C_E + C_\infty + B_0 - \frac{2(S_\alpha + S_\beta)^2}{S_\beta}\frac{v}{D}C_0 P \quad [5b]$$

where

$$P = \sum_{n=1}^{\infty}\left(\frac{1}{n\pi}\right)^3 \sin^2\left(\frac{n\pi S_\alpha}{S_\alpha + S_\beta}\right)$$

P depends only on the ratio S_α/S_β and is tabulated in Table I.

I.2) Diffusion in Rod Growth. The diffusion solution for rod eutectics is similar to that for lamellae, except that the cosine series is replaced by a Bessel function series. The formal problem may be written as:

$$\nabla^2 C + \frac{v}{D}\frac{\partial C}{\partial z} = 0 \quad [6a]$$

with

$$C = C_E + C_\infty \quad \text{at} \quad z = \infty \quad [6b]$$

$$\frac{\partial C}{\partial r} = 0 \quad \text{at} \quad r = 0 \quad \text{and} \quad r = r_\alpha + r_\beta \quad [6c]$$

$$\left(\frac{\partial C}{\partial z}\right)_{z=0} = -\frac{v}{D}C_0^\alpha \quad 0 \le r < r_\alpha \quad [6d]$$

$$\left(\frac{\partial C}{\partial z}\right)_{z=0} = \frac{vC_0^\beta}{D} \quad r_\alpha < r \le r_\alpha + r_\beta \quad [6e]$$

In this case, we are replacing a polygonal boundary at $r = r_\alpha + r_\beta$ by a circle of equivalent area, as shown in Fig. 3.

The solution to Eqs. [6] is:

$$C = C_E + C_\infty + A_0 e^{-\frac{vz}{D}} + \sum_{n=1}^{\infty} A_n J_0\left(\frac{\gamma_n r}{r_\alpha + r_\beta}\right)e^{-\frac{\gamma_n z}{r_\alpha + r_\beta}} \quad [7]$$

where γ_n is a root of $J_1(x) = 0$. γ_n is approximately $n\pi$. With

$$A_0 = \left(\frac{r_\alpha}{r_\alpha + r_\beta}\right)^2 C_0 - C_0^\beta$$

$$A_n = \frac{2vr_\alpha}{D}C_0\frac{J_1\left(\frac{\gamma_n r_\alpha}{r_\alpha + r_\beta}\right)}{[\gamma_n J_0(\gamma_n)]^2} \quad [8]$$

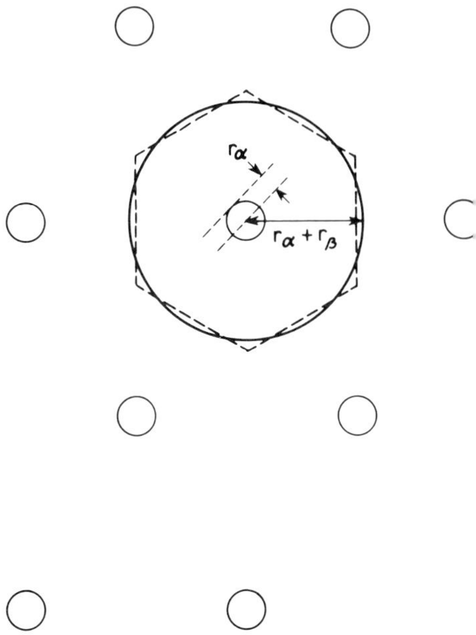

Fig. 3—Schematic drawing of a rod structure viewed normal to the interface showing definition of r_α and r_β.

Table I. Calculated Values of the Sums P and M Which Appear in the Expressions for the Average Composition in the Liquid at the Interface, Eqs. [5] and [9].

$\frac{S_a}{S_a + S_\beta} = \frac{1}{1+\zeta} = \left(\frac{r_a}{r_a + r_\beta}\right)^2$	$P = \sum_{n=1}^{10}\left(\frac{1}{n\pi}\right)^3 \sin^2\left(\frac{n\pi S_a}{S_a + S_\beta}\right)$	$M = \sum_{n=1}^{10}\frac{1}{\gamma_n^3}\frac{J_1^2\left(\frac{r_a\gamma_n}{r_a + r_\beta}\right)}{J_0^2(\gamma_n)}$
0	0.0	0.0
0.1	0.00620	0.0395
0.2	0.01627	0.0430
0.3	0.02545	0.0385
0.4	0.03167	0.0330
0.5	0.03383	0.0260
0.6	0.03167	0.0190
0.7	0.02545	0.0125
0.8	0.01627	0.0065
0.9	0.00620	0.0025
1.0	0.0	0.0

The average compositions in front of the α and β phase are given by

$$\overline{C_\alpha} = \frac{4v}{D}(r_\alpha + r_\beta) C_0 M + C_E + C_\infty + A_0 \qquad [9a]$$

and

$$\overline{C_\beta} = -\frac{4 r_\alpha^2 (r_\alpha + r_\beta)}{(r_\alpha + r_\beta)^2 - r_\alpha^2} \frac{v}{D} C_0 M + C_E + C_\infty + A_0 \qquad [9b]$$

where

$$M = \sum_{n=1}^{\infty} \frac{1}{\gamma_n^3} \frac{J_1^2\left(\frac{r_\alpha \gamma_n}{r_\alpha + r_\beta}\right)}{J_0^2(\gamma_n)}$$

M is tabulated in Table I.

II) AVERAGE CURVATURE OF THE INTERFACE

An explicit expression can be obtained for the average curvature of a lamellar or rod-type interface. The average curvature depends on the angles at the phase boundaries and on the lamellar or rod spacing. These average values can be combined with the expressions for the average composition at the interface, Eqs. [5] or [9], to obtain the total average interface undercooling.

II.1) Average Curvature of a Lamellar Interface. Taking the origin at the center of a lamella, so that the slope is zero at $x = 0$, the average curvature is given by

$$\left\langle \frac{1}{r(x)} \right\rangle = +\frac{1}{S}\int_0^S \frac{dx}{r(x)} = \frac{1}{S}\int_0^S \frac{-\frac{d^2y}{dx^2}}{\left(1 + \left(\frac{dy}{dx}\right)^2\right)^{3/2}} dx$$

$$= -\frac{1}{S} \sin \tan^{-1}\left(\frac{dy}{dx}\right)\Bigg|_0^S = \frac{1}{S}\sin\theta \qquad [10]$$

where S and θ are as shown in Fig. 4.

II.2) Average Curvature of a Rod-Type Interface. For a point P on a surface of revolution obtained by rotation about AB, see Fig. 5, one of the principal radii of curvature, ρ_1, is in the plane of the paper; the other, ρ_2, is in the plane normal to the paper which contains the normal to the surface at P. The radius of curvature in the plane of the paper is given by:

$$\frac{1}{\rho_1} = \frac{-\frac{d^2z}{dr^2}}{\left[1 + \left(\frac{dz}{dr}\right)^2\right]^{3/2}}$$

where $z(r)$ is the equation of the surface in the plane of the paper.

From Meunier's theorem,* for a surface of revolution,

$$\frac{1}{\rho_2} = -\frac{\sin\theta}{r}$$

where

$$\tan\theta = \frac{dz}{dr}$$

The total curvature at P is $1/\rho = 1/\rho_1 + 1/\rho_2$. The average curvature of the surface between r_1 and r_2 is

$$\left\langle \frac{1}{\rho} \right\rangle = \frac{1}{\pi[r_2^2 - r_1^2]}\int_{r_1}^{r_2} 2\pi r\left(\frac{1}{\rho_1} + \frac{1}{\rho_2}\right)dr$$

$$= -\frac{2}{r_2^2 - r_1^2}\int_{r_1}^{r_2} \frac{r\left(\frac{d^2z}{dr^2}\right)dr}{\left[1 + \left(\frac{dz}{dr}\right)^2\right]^{3/2}} - \frac{2}{r_2^2 - r_1^2}\int_{r_1}^{r_2} \sin\theta \, dr$$

Using $t = dz/dr$, $\sin\theta = t/\sqrt{1 + t^2}$, and integrating the first term by parts, we obtain for the average curvature of a surface of revolution between radii r_1 and r_2, see Fig. 5:

$$\left\langle \frac{1}{\rho} \right\rangle = \frac{-2}{r_2^2 - r_1^2}(r_2 \sin\theta_2 - r_1 \sin\theta_1) \qquad [11]$$

*See Salmon: *Analytic Geometry of Three Dimensions*, 7th ed, p. 306.

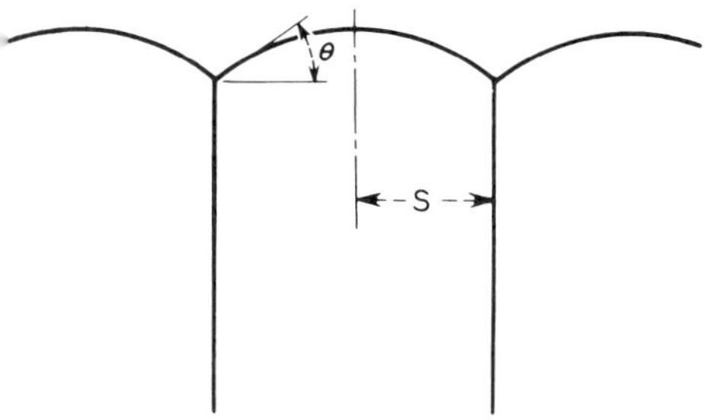

Fig. 4—Schematic drawing of lamellar interface showing definition of S and θ.

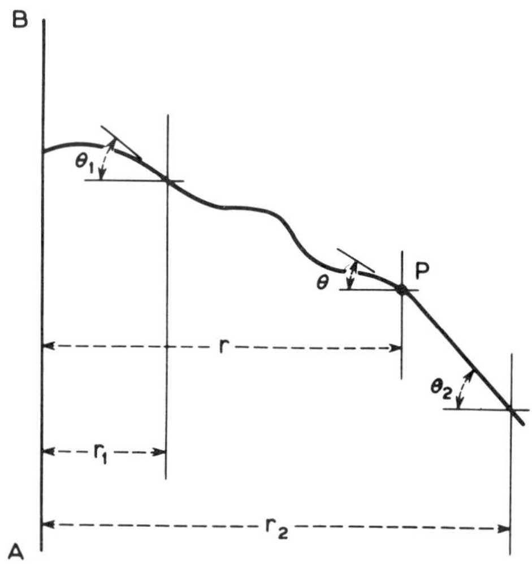

Fig. 5—The average curvature of a surface of revolution generated by rotating any curve about the line AB depends only on the limiting radii r_1 and r_2 between which the average is to be computed, and on the angles θ_1 and θ_2 at these limits.

For α rods in a β matrix, we have for the α phase:

$$r_2 = r_\alpha, \; r_1 = 0, \; \theta_2 = \tan^{-1}\left(\frac{dy}{dx}\right)_2 = -\theta_\alpha, \; \theta_1 = 0$$

so that

$$\left\langle\frac{1}{\rho_\alpha}\right\rangle = \frac{2}{r_\alpha} \sin\theta_\alpha \qquad [12a]$$

and for the β phase:

$$r_2 = r_\alpha + r_\beta, \quad r_1 = r_\alpha, \quad \theta_1 = \theta_\beta, \; \theta_2 = 0$$

so that

$$\left\langle\frac{1}{\rho_\beta}\right\rangle = \frac{2r_\alpha}{(r_\alpha + r_\beta)^2 - r_\alpha^2} \sin\theta_\beta \qquad [12b]$$

III) AVERAGE UNDERCOOLING AT THE INTERFACE AND THE EXTREMUM CONDITION

In this section expressions are developed for the average undercooling at the interface, in terms of the various factors which contribute to it. The extremum condition is then developed, followed by a discussion of how the lamellar spacing can be maintained or changed in a growing eutectic. This involves the motion of lamellar faults, and the relative stability of the lamellar and rod growth modes.

III.1) Interface Temperature. The stability of the interface depends on its temperature, and more specifically on the difference between the eutectic temperature T_E and the local actual temperature of the interface T_I. For a liquid of eutectic composition, in equilibrium with the two eutectic solids across a plane interface, $T_E = T_I$ by definition. In general for a nonplanar interface not of eutectic composition and not at equilibrium, the difference between T_E and T_I is made up of three parts. The first of these is ΔT_C due to the departure of the local composition from the eutectic composition. ΔT_C can be written as:

$$\Delta T_C = m[C_E - C(x)]$$

where m is the slope of the liquidus and $C(x)$ is the composition of the interface at x. m of course depends on which solid phase is in contact with the liquid at x. The second contribution due to a nonplanar interface is ΔT_σ, which may be written as:

$$\Delta T_\sigma = \frac{a}{r(x)}$$

where "a" is a constant given by the Gibbs-Thompson relationship, and $r(x)$ is the local curvature of the interface. The above two terms, ΔT_C and ΔT_σ, correspond to changes in the local equilibrium temperature from T_E, and may be derived from thermodynamics for an equilibrium system from the requirement that the chemical potentials must be equal. The third contribution to $\Delta T = T_E - T_I$ is a kinetic term which is specifically a nonequilibrium term. This term is due to the chemical potential difference that is driving the freezing process. Adding this term permits writing an equation which is analogous to the normal equilibrium equation and which reduces to it at zero growth rate:

$$\Delta T = T_E - T_I = \Delta T_C + \Delta T_\sigma + \Delta T_k$$

For metallic eutectic systems, however, ΔT_k can usually be ignored compared with ΔT_C and ΔT_σ (which are equal for a eutectic growing at the extremum). For example, at 10^{-2} cm per sec growth rate, the kinetic undercooling for most metals is less than 10^{-2}°C* (see, for example, the measurements of Kramer and Tiller[14] on tin). The measured value of ΔT for the Pb-Sn system at this growth rate is 1°C.[10] The ΔT_k term will therefore be ignored in the rest of this paper, although it cannot be ignored in eutectic systems which contain nonmetallic components.[15]

*The kinetic undercooling should not depend strongly on composition.[11]

It is important to note that T_I is almost constant. For typical temperature gradients (~10°C per cm) and typical lamellar spacings (~1 μ), the temperature with a few lamellar spacings (several microns) of the interface is constant to within one hundredth of a degree or less. Since this temperature variation is small compared to the observed[10] $T_E - T_I$, which is of the order of a degree, $T_E - T_I$ can be considered to be a constant over several lamellar distances. Although differences in T_I may exist between widely separated points on the interface, this need not concern us in calculating the shape of the interface or considering the local stability of lamellae.

Although both ΔT_C and ΔT_σ vary across a lamellar width in the interface, their sum, expressed as

$$T_E - T_I = \Delta T = \Delta T_C + \Delta T_\sigma$$

is constant.

III.2) The Average Undercooling of the Interface. The undercooling at any point on the interface is given by*

*The equivalence of this treatment to Zener's treatment[1] is demonstrate in Appendix B of Ref. 5.

$$\Delta T = m[C_E - C(x)] + a/r(x) \qquad [13]$$

where m is the slope of the liquidus line, C_E is the eutectic composition, $C(x)$ and $r(x)$ are the composition and curvature at a point x on the interface, and a is a constant given by the Gibbs-Thompson relationship. The average undercooling at the interface for a lamellar eutectic is obtained by inserting average values for $C(x)$ and $1/r(x)$ from Eqs. [5] and [10] into Eq. [13]:

$$\Delta T_\alpha = m_\alpha\left[C_\infty + B_0 + \frac{2v}{D} C_0 \frac{(S_\alpha + S_\beta)^2}{S_\alpha} P\right] + \frac{a_\alpha^L}{S_\alpha} \qquad [14a]$$

$$\Delta T_\beta = m_\beta\left[-C_\infty - B_0 + \frac{2v}{D} C_0 \frac{(S_\alpha + S_\beta)^2}{S_\beta} P\right] + \frac{a_\beta^L}{S_\beta} \qquad [14b]$$

where m_α and m_β are the slopes of the α and β liquidus lines, respectively (defined so that both are positive*), and from the Gibbs-Thompson equation,

*m may be positive or negative in Eq. [13]. Defining both m's to be positive introduces a minus sign into Eq. [14a].

and Eq. [10], $a_\alpha^L = (T_E/L)_\alpha \sigma_\alpha^L \sin\theta_\alpha^L$, and $a_\beta^L = (T_E/L)_\beta \sigma_\beta^L \sin\theta_\beta^L$. T_E is the melting point and L is the heat of fusion per unit volume of the appropriate phase, and σ_α and σ_β are the specific surface free energies of the α-liquid and β-liquid interfaces, respectively. The superscript L denotes the values for amellar, as opposed to R, for rod growth, the latter o be discussed in the next section.

The B_0 term appears explicitly in Eqs. [14]. B_0 ould be eliminated using Eq. [4a]. It is apparent from Eq. [4a], however, that a large change in B_0 is necessary to appreciably alter $\zeta = S_\beta/S_\alpha$, the ratio of the amellar spacings. For example a change in B_0 from to 0.01 would change ζ approximately 1 pct. This hange represents an appreciable change in the B_0 erm in Eqs. [14], but a negligible change in the other erms involving S_α and S_β. It will therefore be assumed that ζ is constant, and λ and B_0 will be reated as the independent variables in Eqs. [14] ather than treating S_α and S_β as independent variables. This approximation makes the equations ractable.* We therefore define $\lambda = 2(S_\alpha + S_\beta)$, and

*The complete form of Eqs. [14] has been obtained without this and earlier approximations. In order to proceed further, it was necessary to drop the second-order terms and retain only the lower-order terms. Dropping the second-order terms is equivalent to making the assumption that ζ is constant and also that C_0^α and C_0^β are constants (see footnote to Eq. [2]) for a given utectic.

ise $S_\alpha = \lambda/2(1+\zeta)$, $S_\beta = \zeta\lambda/2(1+\zeta)$ in Eqs. [14]. Eqs. [14] can be combined to eliminate $B_0 + C_\infty$, since $\Delta T_\alpha = \Delta T_\beta$. The resultant equation contains only the variables λ, v, and ΔT, and may be written as:

$$\frac{\Delta T}{m} = v\lambda Q^L + \frac{a^L}{\lambda} \qquad [15]$$

where m, Q^L, and a^L are constants given by:

$$\frac{1}{m} = \frac{1}{m_\alpha} + \frac{1}{m_\beta} \qquad [16a]$$

$$Q^L = \frac{P(1+\zeta)^2 C_0}{\zeta D} \qquad [16b]$$

$$a^L = 2(1+\zeta)\left(\frac{a_\alpha^L}{m_\alpha} + \frac{a_\beta^L}{\zeta m_\beta}\right) \qquad [16c]$$

Eq. [15] is shown schematically for a given growth velocity in Fig. 6. The lamellar spacing, λ, may be adjusted arbitrarily for Eq. [15] to be satisfied. An additional condition must be specified to solve for λ

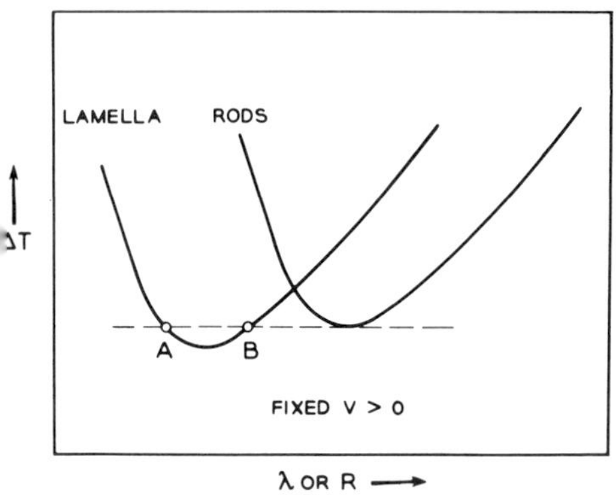

Fig. 6—Schematic drawing of the variation of interface undercooling with lamellar spacing, λ, or rod spacing, R, for a given growth rate v.

in terms of one other variable. The simplest condition is that proposed by Zener[1] and adopted by Tiller[3] and Hillert,[4] in which is was assumed that the solid grows at the extremum. Eq. [15] has a minimum at constant growth velocity given by

$$\lambda^2 v = \frac{a^L}{Q^L} \qquad [17a]$$

or

$$\frac{\Delta T^2}{v} = 4m^2 a^L Q^L \qquad [17b]$$

or

$$\Delta T\lambda = 2ma^L \qquad [17c]$$

This minimum also corresponds to the maximum in the v vs λ curve, plotted at constant ΔT.

This extremum condition, as it stands, is unsatisfactory because it is not based on a mechanism which can produce changes in lamellar spacing. As will be seen later, catastrophic changes can take place; however, these cannot account for the small adjustments in spacing which must occur in practice. These small changes are the result of the motion of faults across the eutectic grain.

Two types of fault occur during lamellar growth. Both are shown in Fig. 7. In this work, the fault which is shown schematically in Fig. 8 will be termed a lamellar fault. The other type of fault, which crosses several lamellae, will be termed a fault line.

It has been proposed[16] that lamellar spacings are determined by the stability of the lamellar fault. When the lamellar fault, Fig. 8, moves to the left the average spacing decreases and when it moves to the right it increases. The condition for stable lamellar growth therefore occurs when the fault remains stationary.

The lamellar fault or termination can be treated as part of a rod.[16] Since the termination can take up any shape, or have any effective radius R, regardless of the lamellar spacing, λ, it probably grows with the minimum rod undercooling. The minimum undercool-

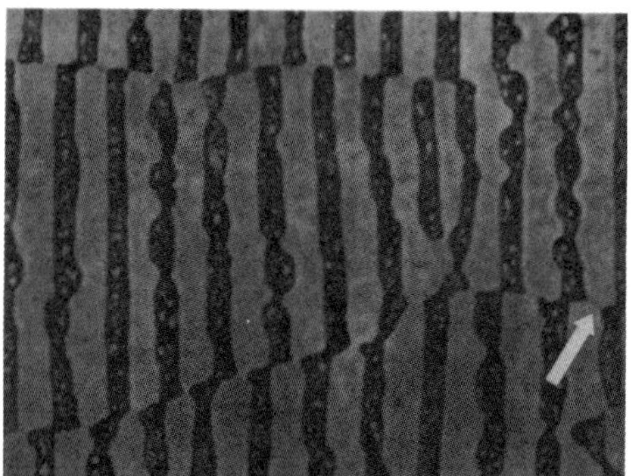

Fig. 7—Pb-Sn eutectic, electropolished in perchloric acetic acid, alcohol solution. Growth direction out of paper. Showing the two types of fault in lamellar structures. The faint line indicated by the arrow provides evidence that the fault line is a eutectic subgrain boundary. X1500. Reduced approximately 67 pct for reproduction.

ing for rod growth therefore defines the conditions for stable lamellar growth. Equations similar to Eqs. [14] but for the rod structure are obtained by combining Eqs. [9], [12], and [13]:

$$\Delta T_\alpha = m_\alpha\left[C_\infty + A_0 + \frac{4v}{D}C_0(r_\alpha + r_\beta)M\right] + \frac{2a_\alpha^R}{r_\alpha} \qquad [18a]$$

$$\Delta T_\beta = m_\beta\left[-C_\infty - A_0 + \frac{4v}{D}C_0\frac{(r_\alpha + r_\beta)r_\alpha^2}{(r_\alpha + r_\beta)^2 - r_\alpha^2}M\right] + \frac{2a_\beta^R r_\alpha}{(r_\alpha + r_\beta)^2 - r_\alpha^2} \qquad [18b]$$

where

$$a_\alpha^R = \left(\frac{T_E}{L}\right)_\alpha \sigma_\alpha^R \sin\theta_\alpha^R \quad \text{and} \quad a_\beta^R = \left(\frac{T_E}{L}\right)_\beta \sigma_\beta^R \sin\theta_\beta^R$$

The superscript R refers to the rod structure since in general σ and θ will be different for lamellar and rod interfaces. As in the lamellar analysis, it will be assumed that the ratio of the volumes of the two phases is a constant ζ and R is defined by

$$R = r_\alpha + r_\beta = r_\alpha\sqrt{1+\zeta}$$

so that

$$r_\alpha = \frac{R}{\sqrt{1+\zeta}} \quad \text{and} \quad r_\beta = R\left(1 - \frac{1}{\sqrt{1+\zeta}}\right)$$

Eqs. [18] can be combined giving

$$\frac{\Delta T}{m} = vRQ^R + \frac{a^R}{R} \qquad [19]$$

where

$$a^R = 2\sqrt{1+\zeta}\left(\frac{a_\alpha^R}{m_\alpha} + \frac{a_\beta^R}{\zeta m_\beta}\right) \qquad [20a]$$

and

$$Q^R = \frac{4(1+\zeta)}{D\zeta}C_0 M \qquad [20b]$$

Eq. [19] is formally identical to Eq. [15]. The extremum is therefore given by Eqs. [17] with R, a^R, and Q^R replacing λ, a^L, and Q^L, respectively.

Eq. [19] is shown schematically for constant v in Fig. 6. If the stability of the fault tip determines the growth conditions, the lamellar structure should grow near the undercooling shown by the dashed line. There are other considerations which influence the stability of the lamellar structure. Cahn[17] has pointed out that the growth conditions to the left of the minimum on the lamella curve, Fig. 6, are inherently unstable. This is illustrated in Fig. 9. For growth to the right of the extremum lamellar spacing, Fig. 6, if a local decrease in spacing occurs, then the local undercooling decreases, and the lamellae are stabilized by the resulting change in interface shape as shown in Fig. 9(a). The stabilization occurs because the lamellae tend to grow normal to the existing local interface. However, if the growth conditions are to the left of the extremum, a local decrease in spacing results in a depression in the interface, Fig. 9(b), and the lamellae are not stabilized by further growth, since the lamellae at the bottom of the depression get progressively narrower, because the lamellae tend to grow normal to the interface. A fluctuation which increases the local lamellar spacing is also unstable to the left of and stable to the right of the extremum in Fig. 6. Thus lamellae of smaller than the extremum spacing are inherently unstable to fluctuations in spacing or in interface shape. This instability will tend to make the stable growth conditions somewhere to the right of the minimum in Fig. 6.

These same effects are also important in motion of lamellar faults. In Fig. 8, the lamellae to the right of the fault have a slightly wider spacing than those to the left. This will tend to make a depression in the interface. If the average spacing is larger than the minimum, then the depression will be deeper to the left of the fault in Fig. 8. This will tend to make the fault

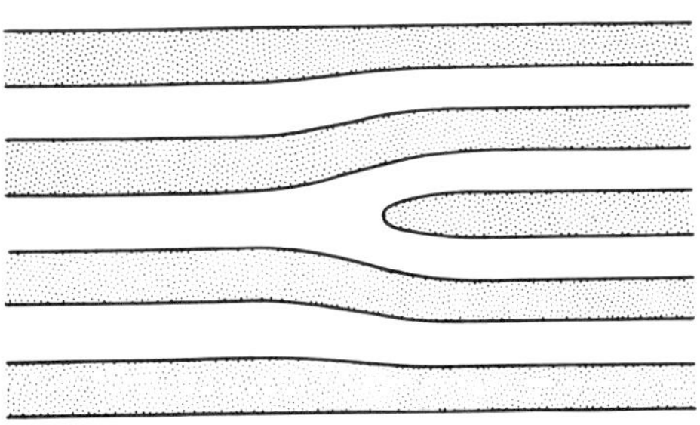

Fig. 8—Schematic drawing of a lamellar fault.

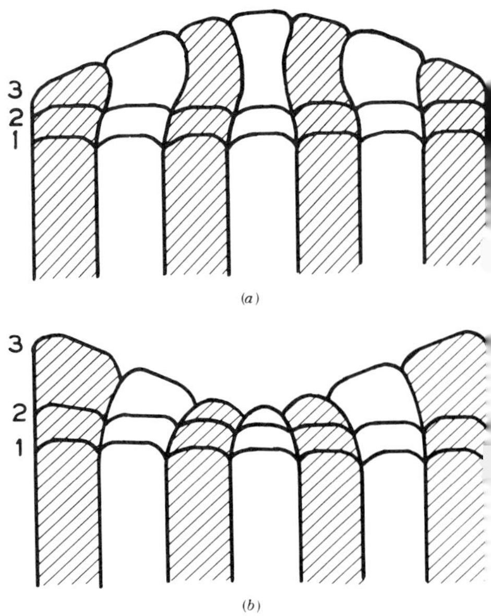

Fig. 9—The lamellae in (a) are wider than the extremum spacing, in (b) narrower than the extremum spacing. Interface position 1 shows the originally regular lamellae, position 2 the formation of narrower lamellae at center, position 3 the change in interface shape, due to the change in local undercooling. For (a) the structure is restabilized by the change in interface shape; in (b) it is not. In the latter case, a lamella will grow out.

move to the left, thereby decreasing the average spacing. This will tend to shift the stable growth conditions towards the minimum from the condition given by the intersection of point B in Fig. 6. Thus the average spacing should be within the small range of spacings between point B, Fig. 6, and the extremum spacing. It will be shown later, Table II, that the minimum undercoolings for rods and lamellae are usually close together. Consequently the lamellar structure should grow near the extremum spacing even though the spacing is actually controlled by the motion of the lamellar faults.

The formation of new faults must be considered as well as the motion of existing faults. The formation of faults may in some cases be the limiting factor. In the usual lamellar structure, the lamellae are not completely regular. The lamellar spacing is thus not a single value, but rather a range of values. The formation of new lamellar faults will be easier in the regions where the local lamellar spacing differs from the average. If the spacing is too narrow, faults will form in regions where the spacing is much smaller than the extremum spacing. If the spacing is too wide, new lamellae will form in the regions where the spacing is wider than the extremum spacing. This occurs by the formation of a hollow in the center of the wider phase, as in the catastrophic breakdown process which is discussed in detail below.

The fault lines shown in Fig. 7 may also play a part in lamellar spacing changes. The fault line is thought to delineate a low-energy eutectic grain boundary. Evidence for the boundary is seen in the area indicated by an arrow, Fig. 7. The boundary is visible as a faint line between two (light etching) tin lamellae. The fault line will grow as a groove in the solid-liquid interface. When there are different numbers of lamellae on either side of the fault, the motion of the fault line will result in a change in average spacing over the specimen. However, the amount of change will be limited since the spontaneous formation of a complete new fault line is unlikely. It is more likely that a new lamellar fault will be developed since in this case a much smaller region of the interface is involved. It should be noticed, however, that the fault lines are ideal regions for lamellar faults to develop. Once formed in a fault line, the lamellar fault can move rapidly across the grain until it runs into another fault line where it will be pinned by the eutectic subgrain boundary. Further spacing changes can then take place by the repetition of this process.

III.3) Rod Formation. When the rod minimum undercooling is lower than that for lamellae, the lamellar structure will not be formed. By comparing Eqs. [17] with the corresponding ones for rods, it is found that at a given growth velocity the minimum for rods is below that for lamellae if

$$\frac{\left(\frac{a_\alpha^L}{m_\alpha} + \frac{a_\beta^L}{\zeta m_\beta}\right)}{\left(\frac{a_\alpha^R}{m_\alpha} + \frac{a_\beta^R}{\zeta m_\beta}\right)} > \frac{4M}{P}\,\frac{1}{(1+\zeta)^{3/2}} \qquad [21]$$

For isotropic solid-liquid and α-β interfacial free energies the left-hand side of this equation is equal to one. The right-hand side is given in Table II. It is evident from the table that, if the surface energies are isotropic, rods have the lower minimum (and therefore the rodlike structure should be the stable form, since a structure which grows with smaller undercooling will grow ahead of, and ultimately eliminate, a structure which grows with larger undercooling) when $1/(1+\zeta) < 1/\pi$. This result has been obtained previously.[18,19] It is also evident from Table II that a 25 pct effective anisotropy in surface energy will be enough to stabilize a lamellar structure for any ratio of spacing $[1/(1+\zeta)]$. If there is a low-energy α-β plane, the lamellae can form in that plane, and the rods must perforce make a boundary with a variety of planes. The lamellar structure can therefore occur even in materials where $[1/(1+\zeta)$ $< 1/\pi$ provided the surface energies are anisotropic. This is probably the case in Sn-Cd, Pb-Cd, Al-Zn, and Sn-Zn where it has been shown that lamellae break down to rods, when the eutectic growth direction is suddenly altered.[19]

It is evident from Table II that the rod and lamellar structures grow with rather similar undercoolings. When the lamellar spacing on a specimen is momentarily too wide or too narrow it is possible that the lamellar undercooling is above the minimum undercooling for rods. This will cause incipient breakdown of the lamellae to rods. This breakdown can produce new lamellar faults. The faults can then move across the interface, changing the spacing. Thus the rod minimum undercooling represents a stabilizing influence on the lamellar structure. When the rod structure is the stable form, the lamellar minimum provides a similar stabilizing influence for rod formation. If there are too few rods a short lamella can form which can then break up into several rods. The two structures are thus in constant interplay with one another, producing the required changes in lamellar or rod spacing when growth conditions change.

Table II. Comparison of the Stability of Rods and Lamellae*

$\frac{S_\alpha}{S_\alpha + S_\beta} = \frac{1}{1+\zeta}$	$\frac{4M}{P}\left(\frac{1}{1+\zeta}\right)^{3/2}$
0	–
0.01	(0.8)**
0.02	0.75
0.05	0.74
0.1	0.806
0.2	0.946
0.3	0.994
0.4	1.054
0.5	1.088
0.6	1.116
0.7	1.152
0.8	1.142
0.9	1.374
1.0	–

*For isotropic interfacial free energies, rods grow with lower undercooling for values of ζ such that the number in the right-hand column is less than one. A lamellar structure grows with lower undercoolings for the cases where it is greater than one, see Eq. [21].

**The series defining M and P become difficult to sum for small values of $1/(1+\zeta)$.

IV) THE BOUNDARY LAYER

The B_0 term in Eq. [3] and the A_0 term in Eq. [7] represent boundary layers. This means that the average composition at the interface in the liquid is different than the average composition in the liquid far from the interface. The boundary layer must be present for eutectics where $C_\infty \neq 0$, since the eutectic composition is the only one which can be close to equilibrium with both the α and β phases. The solid can freeze out with any average composition $C_E + C_\infty$, produced not by changing the composition and therefore the equilibrium temperature of either phase, but by a small change in the relative widths of the two solid phases, as indicated by Eq. [4a].

The boundary layer is like the diffusion boundary layer found at a plane solid-liquid interface in an impure material in that its characteristic thickness is D/v. This is a distance which is usually large compared to the lamellar spacing. The magnitude of the boundary layer can be found by substituting Eqs. [17] into Eqs. [14], giving:

$$(B_0 + C_\infty) = \frac{\Delta T}{2m}\left\{\frac{m_\beta - m_\alpha}{m_\beta + m_\alpha} + \frac{1}{2}\frac{1-\zeta}{1+\zeta} - \frac{1}{2}\left(\frac{\dfrac{a_\alpha^L}{m_\alpha} - \dfrac{a_\beta^L}{\zeta m_\beta}}{\dfrac{a_\alpha^L}{m_\alpha} + \dfrac{a_\beta^L}{\zeta m_\beta}}\right)\right\} \qquad [22]$$

An identical equation can be obtained by combining the corresponding equations for the rod case, except that A_0 replaces B_0, and the superscripts L are changed to R. That the right-hand side of Eq. [22] is small can be seen as follows: $\Delta T/2m$ is the composition change corresponding to ΔT. Since ΔT is a few degrees at most, $\Delta T/2m$ is typically a few percent at most. The expression in parentheses is zero for a completely symmetric eutectic, *i.e.*, for $m_\alpha = m_\beta$, $\zeta = 1$, $a_\alpha^L = a_\beta^L$; it may also be zero in other fortuitous cases. It is always less than unity. $B_0 + C_\infty$ is therefore small.

For the usual case, therefore, $B_0 \approx -C_\infty$, which implies, see Eq. [3], $C(x, 0) \approx C_E$, independent of C_∞. The steady state boundary layer at the interface is such as to make the composition in the liquid at the interface almost equal to the eutectic composition, regardless of the starting composition.

The presence of the boundary layer is important in the zone refining of a eutectic. During zone refining, the composition of the sample approaches the value corresponding to no boundary layer, *i.e.*, $B_0 = 0$. It should be noted, however, that this condition does not imply that the composition approaches exactly C_E, but rather that it approaches $C_E + C_\infty^0$, where C_∞^0 is given by Eq. [22] with $B_0 = 0$. C_∞^0 depends on ΔT and hence on growth rate. For most growth rates, however, C_∞^0 is small, and so the terminal composition will differ from the eutectic composition by a small amount, a percent or less.

Eq. [22] assumes steady-state growth. A change in growth rate will change the amount of material in the boundary layer, as has been described for single-phase growth.[20] Such a change in growth rate produces an excess of one phase or the other (depending on the boundary layer) across the interface.[15] An example of this is shown in Fig. 10, where one phase has widened out at the expense of the other as a result of a growth-rate fluctuation. In an extreme case, one phase can cut off the other entirely as shown in Fig. 11. In this sample, the composition varies from left to right and the corresponding change in the boundary layer can be seen. On the right, the composition was far from the eutectic composition, so there was a large boundary layer (large B_0); on the left, the composition was close to the eutectic composition, and therefore there was practically no boundary layer. The overgrowth of one phase amounts to a catastrophic change in lamellar spacing, which will be discussed in more detail below.

The boundary layer permits a lamellar eutectic structure to grow from liquid having a range of starting compositions. In most systems the eutectic structure will grow over a range of compositions of the order of 1 or 2 pct. Without the boundary layer, the eutectic structure could only grow at one precise liquid composition for a given growth rate. The growth of a lamellar structure over a range of composition (of a few percent) is shown in Fig. 12. Note the change in $\zeta = S_\beta/S_\alpha$ from left to right. It is very apparent in this system because C_0 is very small.

V) CALCULATION OF INTERFACE SHAPE

The equations for the composition across the interface, Eq. [3], can be combined with the interface equilibrium equation, Eq. [13], using the extremum condition, Eqs. [17], to calculate the curvature at any point on the solid-liquid interface. The curvature can be integrated twice to calculate the shape of the solid-liquid interface. The formalism for this process will be presented and then used to calculate the shape of a

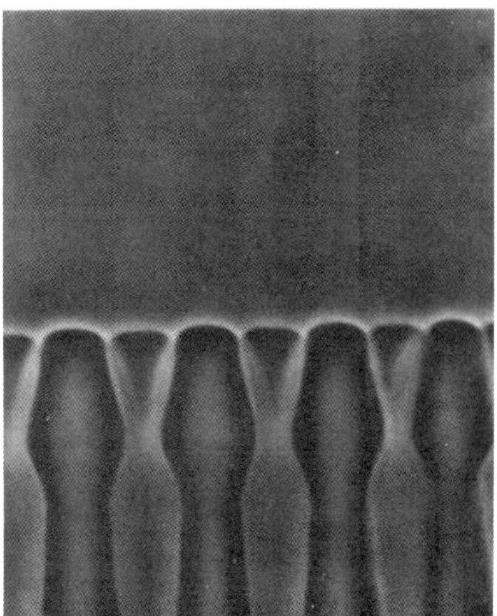

Fig. 10—Carbon tetrabomide-hexachloroethane eutectic contained in a thin cell viewed during growth in transmitted light (phase contrast). Growth direction from bottom to top of figure. Showing effect of a small growth-rate fluctuation on the eutectic structure. X2000. Reduced approximately 57 pct for reproduction.

carbon tetrabromide-hexachloroethane eutectic interface for comparison with photographs.

The undercooling at any point on the interface is obtained by combining Eqs. [3], [4], and [13]:

$$\Delta T = m_\alpha \left\{(B_0 + C_\infty) + \frac{C_0}{D}\lambda v\, f(x)\right\} + \frac{a'_\alpha}{r} \qquad 0 \le x < S_\alpha$$

$$= m_\beta \left\{-C_\infty - B_0 - \frac{C_0}{D}\lambda v\, f(x)\right\} + \frac{a'_\beta}{r} \qquad S_\alpha < x \le S_\beta + S_\alpha \qquad [23]$$

where

$$a' = \frac{T_E}{L}\sigma$$

and

$$f(x) = \sum_{n=1}^{\infty} \frac{\sin\left(\frac{n\pi}{1+\zeta}\right)\cos\left(2n\frac{\pi x}{\lambda}\right)}{(n\pi)^2} \qquad [24]$$

Inserting the definition of $1/r$, and integrating with respect to x and applying the boundary conditions:

$$\frac{dz}{dx} = 0 \quad \text{at} \quad x = 0, \quad \text{and} \quad x = S_\alpha + S_\beta$$

$$\frac{dz}{dx} = -\tan\theta_\alpha \quad \text{at} \quad x = S_\alpha^-$$

$$\frac{dz}{dx} = +\tan\theta_\beta \quad \text{at} \quad x = S_\alpha^+$$

and noting that

$$\int_0^{S_\alpha} f(x)\,dx = \frac{\lambda P}{2}$$

and

$$\int_{S_\alpha}^{S_\alpha+S_\beta} f(x)\,dx = -\frac{\lambda P}{2}$$

we obtain for the slope of the interface

$$\frac{dz}{dx} = \tan\sin^{-1}\left\{-\frac{x}{\lambda}\sin\theta_\alpha 2(1+\zeta) - m_\alpha x\frac{C_0}{D}\frac{\lambda v P(1+\zeta)}{a'_\alpha} + m_\alpha\frac{C_0}{D}\frac{\lambda v}{a'_\alpha}\int_0^x f(x)\,dx\right\} \qquad 0 \le x < S_\alpha \qquad [25a]$$

$$= \tan\sin^{-1}\left\{\frac{(\lambda - 2x)}{\lambda\zeta}\sin\theta_\beta(1+\zeta) + m_\beta\frac{(\lambda - 2x)}{2}\frac{C_0}{Da'_\beta}\frac{\lambda v P(1+\zeta)}{\zeta} - \frac{m_\beta C_0 \lambda v}{Da'_\beta}\int_0^x f(x)\,dx\right\} \qquad S_\alpha < x \le S_\alpha + S_\beta \qquad [25b]$$

Using the extremum condition, Eqs. [17], and noting

$$a_i = a'_i \sin\theta_i$$

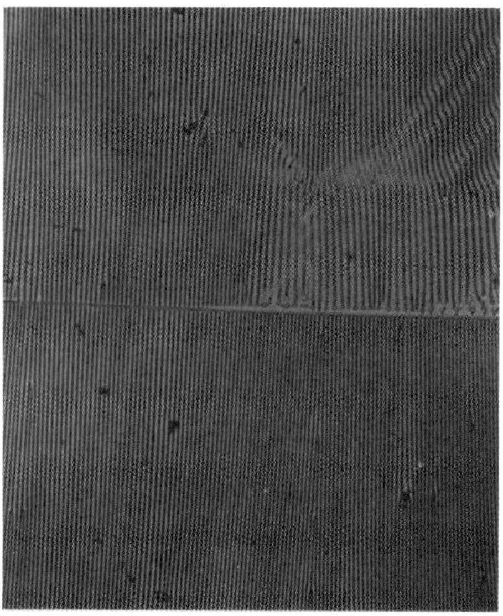

Fig. 11—Carbon tetrabomide-hexachloroethane eutectic contained in a thin cell. Growth direction from bottom to top. The thin cell was filled on the right-hand side with an alloy slightly rich in carbon tetrabromide and on the left with an alloy rich in hexachloroethane. Mixing produced a small concentration gradient in the cell from left to right. The figure shows the influence of the different liquid compositions on banding produced by a growth-rate fluctuation. X100. Reduced approximately 55 pct for reproduction.

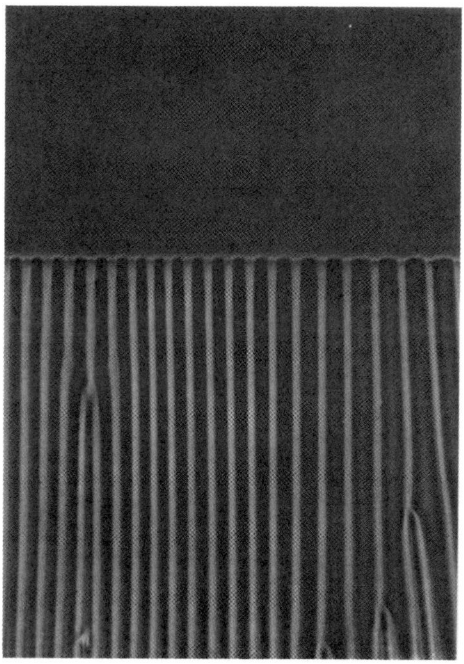

Fig. 12—Carbon tetrabromide-hexachloroethane eutectic. Growth direction from bottom to top. The sample was prepared as in Fig. 11 with the alloy rich in carbon tetrabromide on the right. The figure shows a variation in the ratio of S_α to S_β due to the different liquid compositions. X200. Reduced approximately 48 pct for reproduction.

leads to

$$\frac{dz}{dx} = \tan \sin^{-1}\left\{-\sin \theta_\alpha\left[\frac{x}{\lambda} 2(1+\zeta) + (\zeta+\eta)\left(\frac{2x}{\lambda} - \frac{2}{(1+\zeta)P}\frac{1}{\lambda}\int_0^x f(x)dx\right)\right]\right\} \quad [26a]$$

$$0 \leq x < S_\alpha$$

$$= \tan \sin^{-1}\left\{+\sin \theta_\beta\left[\left(\frac{\lambda - 2x}{\lambda}\right)\left(\frac{1+\zeta}{\zeta}\right) + \left(\frac{1}{\zeta} + \frac{1}{\eta}\right)\left(\frac{\lambda - 2x}{\lambda} - \frac{2\zeta}{(1+\zeta)P}\frac{1}{\lambda}\int_0^x f(x)dx\right)\right]\right\} \quad [26b]$$

$$S_\alpha < x \leq S_\alpha + S_\beta$$

where

$$\eta = \frac{a_\beta m_\alpha}{a_\alpha m_\beta}$$

The function $f(x)$ is known, and given by Eq. [24]. This function of x depends only on ζ, the ratio of the lamellar widths. Eqs. [26] can therefore be integrated readily to obtain the interface shape. The interface shape depends only on the groove angle at the phase boundary through the θ's, on the ratio of the lamellar spacings, ζ, and on one material parameter, η, which contains the ratios of the slopes of the liquidus lines, and of the surface energies. The diffusion coefficient, the initial composition of the melt, the undercooling at the interface, and so on, have all dropped out of the equations. It should be emphasized that this is a result of the assumption that growth is occurring at an extremum. If growth is not occurring at the extremum, then the shape of the interface can be different.

Eqs. [26] have been used to calculate the shapes of four lamellar interfaces. These are compared with photographs of the carbon tetrabromide-hexachloroethane eutectic. The shape of this eutectic interface can be photographed since this system has an unusually large lamellar spacing. This is because the shape of the eutectic diagram for this alloy is such

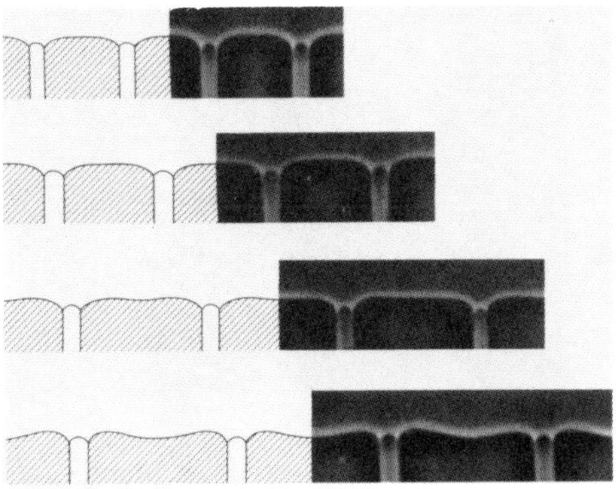

Fig. 13—Comparison of calculated and observed interface shapes. Carbon tetrabromide-hexachloroethane eutectic viewed as in Fig. 10. Growth rate ~2×10^{-4} cm per sec. X2000. Reduced approximately 61 pct for reproduction.

that $C_0^\alpha + C_0^\beta$ is very small. This means that Q, Eqs. [16], is small, and λ, Eqs. [17], is large.

The value of η can be estimated for this alloy by assuming $a_\beta = a_\alpha$. The melting point of C_2Cl_6 is 187°C, and of CBr_4 is 90.1°C. The eutectic point is at 8.6 wt pct C_2Cl_6 and 84°C. From this we deduce

$$\eta = \frac{m_{CBr_4}}{m_{C_2Cl_6}} \approx 1.58$$

θ was estimated to be 70 deg from the photographs. λ, the lamellar repeat distance, and ζ, the ratio of the widths of the two phases, were also measured from the photographs. These values give quite good agreement with the observed interface shape for Fig. 13(*a*), (*b*), and (*c*). These photographs are of lamellae which had been growing at steady state for some distance. The variation in lamellar spacing, λ, is due to different growth rates and the variation in the ratio of the spacings, ζ, is due to differences in composition.

The lamellae in Fig. 13(*d*) were photographed after a change in growth rate. The interface had time to change its shape, but the lamellar spacing λ had not had time to change. This eutectic is not growing at the extremum as assumed in deriving Eqs. [26]. The departure from the extremum condition can be taken into account by using an effective value of η, different than the one used for growth at the extremum. The calculated curve in Fig. 13(*d*), using a value of $\eta = 1.15$, gives quite good agreement with the photographed shape.

VI) CATASTROPHIC CHANGES IN LAMELLAR SPACING

There are two types of catastrophic change in lamellar spacing which we will consider. The first is illustrated in Fig. 14, and occurs when a pocket forms in the center of one phase. In such a case the pocket drops progressively further back from the interface, becomes unstable laterally, and ultimately growth of the other phase occurs in it. A photograph of this is shown in Fig. 15. On a perfect lamellar structure (*i.e.*, no faults), this type of catastrophic change of lamellae spacing can occur due to very slow changes in growth rate and in the absence of a boundary layer. This will be called a shape instability.

The other type of catastrophic change was mentioned

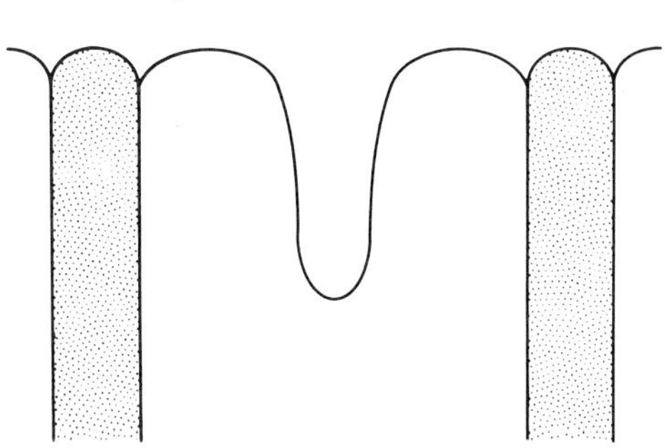

Fig. 14—Schematic drawing of a pocket developed in the center of the wider phase, leading to a shape instability.

previously. It is due to a change in composition at the interface, resulting from the effect of a growth-rate change on the boundary layer, and will therefore be termed a boundary-layer instability. This can manifest itself in the overgrowth and pinching off of one of the phases at the growth front.

In this section we will estimate the departure of growth condition from steady growth at the extremum condition which will produce these catastrophic effects.

VI.1) Shape Instability. The development of a pocket as illustrated in Fig. 14 requires that the slope of the interface, dz/dx in Eqs. [25], become infinite. This condition is simple, but tedious to evaluate numerically because of the series in Eqs. [25]. We have calculated the critical condition for two ratios of lamellar width, $\zeta = 1$ and $\zeta = 9$. The results are displayed in Table III in terms of the value of the product $\lambda^2 v$ required for instability. This condition could be converted using Eq. [17] to be expressed in terms of any other pair of variables. The corresponding values of $\lambda^2 v$ at the extremum are also given in Table III.

For $\zeta = 1$, the instability occurs simultaneously in both phases. For $\zeta = 9$, the instability occurs only in the wider phase. From Table III, for $\zeta = 1$ and $\sin\theta = 1$, $\lambda^2 v$ for instability is about ten times that of the extremum. This means that if λ remained constant while the growth rate increased, this instability would set in when the growth rate increased by ten. For the case of $\zeta = 9$, however, the instability would set in when the growth rate had increased by about a factor of two. We conclude from these calculations that this type of instability may be an important mechanism for decreasing the lamellar spacing when there is a sudden change in growth rate, particularly if there is a large difference between the widths of the two phases.

The shape instability will be even more important in a three-dimensional system than in the case we have considered, which is essentially two-dimensional. The above analysis applies to the simultaneous instability of each lamella all along its length. In a real case, the lamellar spacing is not completely regular; there are depressions in the interface, grain boundaries, eutectic colony boundaries, and fault lines in the eutectic structure. At any of these irregularities in the structure, there will be lamellae which are wider than the average, and these wide lamellae are more susceptible to shape instabilities than the average as analyzed above. The creation of lamellar faults and thereby the control of the lamellar spacing will occur at departures from the extremum condition which are much smaller than those indicated in Table III.

Table III. Comparison of $\lambda^2 v$ for Instability and at Extremum for Two Ratios of Lamellar Width, $\zeta = 1$ and 9; $\xi' = \dfrac{a'_\beta D}{PC_0 m_\beta}$ and $\xi'' = \dfrac{D}{PC_0}\left(\dfrac{a'_\alpha}{m_\alpha} + \dfrac{a'_\beta}{\zeta m_\beta}\right)$

	$\zeta = 1$	$\zeta = 9$
Value of $\lambda^2 v$ for instability	$12.5\,(1 + 0.584 \sin\theta)\,\xi'$	$2.02\,(1 + 0.851 \sin\theta)\,\xi'$
Value of $\lambda^2 v$ at extremum	$2\xi'' \sin\theta$	$1.8\xi'' \sin\theta$

The converse process, when a lamella is pinched off to increase the lamellar spacing, is shown in Fig. 16. This process will not occur simultaneously for all lamellae of one phase without the presence of a compositional boundary layer as discussed in the next section. In the absence of a boundary layer, pinch-off may occur at irregularities in the lamellar structure.

VI.2) Boundary-Layer Instabilities. The effect of changing the growth rate on the boundary layer at the interface can be estimated making use of an approximate form of the equations developed by Smith *et al.*[21] for the change in composition produced by a change in the growth rate at a plane solid-liquid interface. We find that the maximum composition difference produced by a growth-rate change Δv is approximately:

$$(\Delta C)_{\max} = \frac{C_\infty}{2}\left(1 + \frac{C_\infty}{C_E}\right)\frac{\Delta v}{v} \qquad [27]$$

where v is the final growth velocity and C_∞ is the departure of the starting composition from the eutectic composition C_E. The interface temperature during lamellar growth is a degree or so below the eutectic temperature. At this temperature, a composition fluctuation of only a few percent will move the average composition at the interface across one of the liquidus lines on the phase diagram, so that one of the phases becomes unstable and will stop growing. For a reasonably large value of C_∞, say 0.1, a change in growth rate $\Delta v/v$ of $\frac{1}{2}$ will produce, according to Eq. [27], a composition change of a few percent: enough

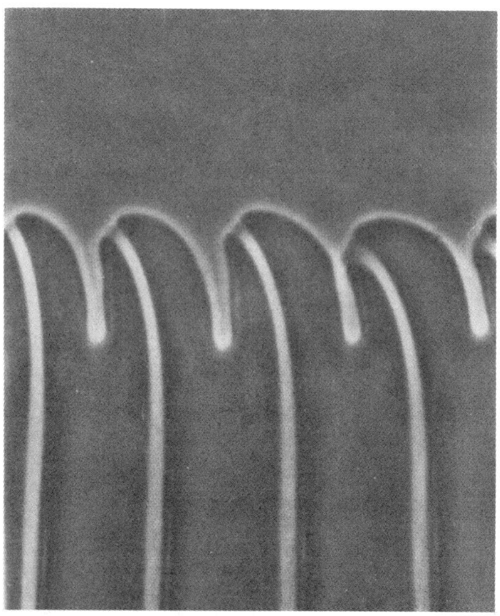

Fig. 15—Carbon tetrabromide-hexachloroethane eutectic viewed as in Fig. 10. Showing the shape instability produced by an increasing growth rate. In this case the lamellar spacing did not change by the motion of faults because faults did not exist in the very thin cells which were used to hold the eutectic. The growth of the lamellar tips to the left is an indication of the instability of the system. X2000. Reduced approximately 55 pct for reproduction.

to cut off one phase. For smaller departures from the eutectic temperatures, the growth-rate change must be larger to produce similar effects.

Fig. 10 shows a growth-rate fluctuation in an off-composition eutectic which produced changes in the relative widths of the two phases. Fig. 11 shows a growth-rate fluctuation which cut off one phase completely.

The sign in Eq. [27] should be observed. An increase in growth rate tends to cut off one phase, a decrease, the other.

VII) COMPARISON WITH EXPERIMENT

Many experimenters[6-10] have investigated the relationship between λ and v, and found, for the most part, the expected

$$\lambda^2 v = \text{const}$$

Hunt and Chilton[10] have also investigated the relationship between ΔT and v and found, for the Pb-Sn system,

$$\Delta T = 12.9\sqrt{v} \text{ deg} \qquad [28]$$

The corresponding experimental equation for Pb-Sn[10] is

$$\lambda^2 v = 3.30 \times 10^{-11} \text{ cu cm per sec} \qquad [29]$$

These data on Pb-Sn will be compared with the theory for two points on the lamella growth curve in Fig. 6: the minimum of the lamella curve and the point B. The comparison will be made by calculating values for the diffusion coefficient D, and the α-β phase boundary free energy $\sigma_{\alpha\beta}$, for these two growth conditions, and then comparing these calculated values with the best available independent experimental data.

For growth corresponding to the minimum on the lamella curve, Eqs. [17] are combined with Eqs. [28] and [29], giving

$$\frac{D\zeta}{2P(1+\zeta)^2 C_0 m} = 4.45 \times 10^{-7} \qquad [30a]$$

and

$$4(1+\zeta)m\left(\frac{a_\alpha^L}{m_\alpha} + \frac{a_\beta^L}{\zeta m_\beta}\right) = 7.41 \times 10^{-5} \qquad [30b]$$

For the Pb-Sn system, assigning α = Sn, β = Pb, from Hansen[22] we find

m_α = 120°C per atom fraction

m_β = 350°C per atom fraction

C_0 = 0.6955 atom fraction, ζ = 0.594

Interpolating in Table I, P = 0.0303 so that Eqs. [30] become

$$D = 7.17 \times 10^{-6} \text{ sq cm per sec} \qquad [31]$$

and

$$\left(\frac{a_\alpha^L}{m_\alpha} + \frac{a_\beta^L}{\zeta m_\beta}\right) = 1.3 \times 10^{-7} \qquad [32]$$

Assuming $\sigma_\alpha^L \sin\theta_\alpha = \sigma_\beta^L \sin\theta_\beta = \frac{1}{2}\sigma_{\alpha\beta}$, we find for the α-β phase boundary free energy, from Eq. [32],

$$\sigma_{\alpha\beta} = 132 \text{ crgs per sq cm} \qquad [33]$$

Equations similar to Eqs. [30] can be readily derived for point B in Fig. 6 by combining the equations for the minimum of the rod curve with Eq. [15] and Eqs. [28] and [29]. If the growth is occurring at this point on the curve, then the values of D and $\sigma_{\alpha\beta}$ are:

$$D = 5.42 \times 10^{-6} \text{ sq cm per sec} \qquad [34a]$$

and

$$\sigma_{\alpha\beta} = 89.5 \text{ ergs per sq cm} \qquad [34b]$$

These values are to be compared with $D = 6.7 \times 10^{-6}$ cm per sec obtained by extrapolation of the results of Lommell and Chalmers[23] and with surface energies for the solid-liquid interface of 59.0 ergs per sq cm for pure tin and 33.3 ergs per sq cm for pure lead obtained from nucleation experiments by Turnbull and Cech.[24] The calculated diffusion coefficient agrees well for the minimum on the lamellar curve, but not so well for point B in Fig. 6. Both are, however, well within the probable error of the experimental value. The calculated surface energy, on the other hand, is closer for the point B than for the lamellar minimum to the experimental values for the pure materials. It is difficult to estimate how different these experimental surface energies for pure materials will be from those corresponding to the two phases in contact with eutectic liquid. The independent experimental values of D and $\sigma_{\alpha\beta}$ are thus not mutually consistent when compared with the theory of eutectic growth. However, the disagreements are by a factor of two or less, well within the uncertainty of these experimental values. In spite of these uncertainties, it is evident that the growth conditions correspond to some value or range of values between

Fig. 16—Carbon tetrabromide-hexachloroethane eutectic viewed as in Fig. 10. Showing the pinch off of a lamella which thereby increases the lamellar spacing. X2000. Reduced approximately 57 pct for reproduction.

ne minimum and point B along the lamellar curve in 'ig. 6.

III) CONCLUSIONS

A solution to the diffusion equation for rod and amellar eutectics has been presented. The solution 'as obtained for the case of a flat interface, but beause of the periodic nature of the boundary condiions, gives a good approximation to the diffusion at n actual interface. Surface energy has also been aken into account for both lamellar and rod eutectics. 'hese considerations permit the calculation of the hape of a lamellar interface, which compares favorbly with the photographed shapes.

The growth conditions have been examined and it is oncluded that lamellar and rod eutectics grow near ne extremum condition. Lamellar faults play an imortant role in controlling lamellar spacing, and the nteraction between rod and lamellar types of growth nposes limiting conditions on growth. Catastrophic nstabilities of the lamellar structure have been onsidered. It was found that a deep groove in the enter of one phase could lead to splitting of the amellae at growth rates above the extremum condiion. Sudden changes in growth rate can result in vergrowth of one phase because of the boundary layer resent at the interface. This can occur for changes n growth rate of as little as a factor of two.

The boundary layer at the interface, which is preicted by the diffusion equation, has been observed nd photographed as a band of one phase produced y a change in growth rate. The boundary layer is uch that the composition in the liquid at the eutectic nterface has the same average value for any initial omposition of the liquid. This boundary layer permits a eutectic structure to grow over a range of compositions.

Good agreement has been obtained between theory and experiment for the relationship between lamellar spacing, growth rate, and interface undercooling.

REFERENCES

[1]C. Zener: *AIME Trans.*, 1946, vol. 167, pp. 550-95.
[2]W. H. Brandt: *J. Appl. phys.*, 1945, vol. 16, pp. 139-46; *AIME Trans.*, 1946, vol. 167, pp. 405-18.
[3]W. A. Tiller: *Liquid Metals and Solidification*, pp. 276-318, ASM, Cleveland. 1958.
[4]M. Hillert: *Jernkontorets Ann.*, 1960, vol. 144, p. 520.
[5]K. A. Jackson, G. A. Chadwick, and A. Klugert: *Trans. Met. Soc. AIME*, 1964, vol. 230, p. 1547.
[6]J. P. Chilton and W. C. Winegard: *J. Inst. Metals*, 1961, vol. 89, p. 162.
[7]J. D. Hunt: Ph.D. thesis, Cambridge University, 1963.
[8]G. A. Chadwick: *Progr. Mater. Sci.*, 1963, vol. 12, p. 97-182.
[9]A. S. Yue: *Trans. Met. Soc. AIME*, 1962, vol. 224, p. 1010.
[10]J. D. Hunt and J. P. Chilton: *J. Inst. Metals*, 1963, vol. 92, p. 21.
[11]F. D. Lemkey, R. W. Hertzberg, and J. A. Ford: *Trans. Met. Soc. AIME*, 1965, vol. 233, p. 334.
[12]K. A. Jackson and J. D. Hunt: *Acta Met.*, 1965, vol. 13, p. 1212.
[13]K. A. Jackson: *Can. J. Phys.*, 1958, vol. 36, p. 683.
[14]J. J. Kramer and W. A. Tiller: *J. Chem. Phys.*, 1965, vol. 42, p. 257.
[15]J. D. Hunt and K. A. Jackson: *Trans. Met. Soc. AIME*, 1966, vol. 236, p. 843-52.
[16]K. A. Jackson and B. Chalmers: unpublished; see also Ref. 8.
[17]J. W. Cahn: M.I.T., Cambridge, Mass., private communication.
[18]F. C. Frank and K. E. Puttick: *Acta Met.*, 1956, vol. 4, p. 206.
[19]J. D. Hunt and J. P. Chilton: *J. Inst. Metals*, 1963, vol. 91, p. 338.
[20]W. A. Tiller, K. A. Jackson, J. W. Rutter, and B. Chalmers: *Acta Met.*, 1953, vol. 1, p. 428.
[21]V. G. Smith, W. A. Tiller, and J. W. Rutter: *Can. J. Phys.*, 1955, vol. 33, p. 723.
[22]M. Hansen: *Constitution of Binary Alloys*, p. 1107, McGraw-Hill Book Co., New York, 1958.
[23]J. M. Lommell and B. Chalmers: *Trans. Met. Soc. AIME*, 1959, vol. 215, p. 499.
[24]D. Turnbull and R. E. Cech: *J. Appl. Phys.*, 1950, vol. 21, p. 804.

Raising Bubbles

Cohen, S. G., Ostberg, B. E., Sparrow, D. B. & Blout, E. R. 1948 *J. Polymer Sci.* **3**, 264.
Doag, K. W. 1948 *J. Amer. Chem. Soc.* **70**, 1525.
Lewis, F. M., Mayo, F. R. & Hulse, W. F. 1945 *J. Amer. Chem. Soc.* **67**, 1701.
Lewis, F. M., Walling, C., Cummings, W., Briggs, E. R. & Mayo, F. R. 1948 *J. Amer. Chem. Soc.* **70**, 1519.
Mayo, F. R. 1948 *J. Amer. Chem. Soc.* **70**, 3689.
Mayo, F. R. & Lewis, F. M. 1944 *J. Amer. Chem. Soc.* **66**, 1594.
Mayo, F. R., Lewis, F. M. & Walling, C. 1947 *Faraday Soc. Disc.* **2**, 285.
Melville, H. W. & Valentine, L. 1949 *Trans. Faraday Soc.* (in the Press).
Norrish, R. G. W. & Brookman, E. F. 1939 *Proc. Roy. Soc.* A, **171**, 147.
Price, C. C. 1946 *J. Polymer Sci.* **1**, 83.
Price, C. C. 1946 *Reactions at carbon-carbon double bonds.* New York: Interscience.
Rodebush, W. H. & Feldman, I. 1946 *J. Amer. Chem. Soc.* **68**, 897.
Staudinger, H. P. 1947 *Chem. & Ind.* **45**, 685.

The mechanics of large bubbles rising through extended liquids and through liquids in tubes

By R. M. Davies and Sir Geoffrey Taylor, F.R.S.

(*Received* 13 *September* 1949)

Part I describes measurements of the shape and rate of rise of air bubbles varying in volume from 1·5 to 200 cm.^{3} when they rise through nitrobenzene or water.

Measurements of photographs of bubbles formed in nitrobenzene show that the greater part of the upper surface is always spherical. A theoretical discussion, based on the assumption that the pressure over the front of the bubble is the same as that in ideal hydrodynamic flow round a sphere, shows that the velocity of rise, U, should be related to the radius of curvature, R, in the region of the vertex, by the equation $U = \frac{2}{3}\sqrt{(gR)}$; the agreement between this relationship and the experimental results is excellent.

For geometrically similar bubbles of such large diameter that the drag coefficient would be independent of Reynolds's number, it would be expected that U would be proportional to the sixth root of the volume, V; measurements of eighty-eight bubbles show considerable scatter in the values of $U/V^{\frac{1}{6}}$, although there is no systematic variation in the value of this ratio with the volume.

Part II. Though the characteristics of a large bubble are associated with the observed fact that the hydrodynamic pressure on the front of a spherical cap moving through a fluid is nearly the same as that on a complete sphere, the mechanics of a rising bubble cannot be completely understood till the observed pressure distribution on a spherical cap is understood. Failing this, the case of a large bubble running up a circular tube filled with water and emptying at the bottom is capable of being analyzed completely because the bubble is not then followed by a wake. An approximate calculation shows that the velocity U of rise is $U = 0{\cdot}46\sqrt{(ga)}$, where a is the radius of the tube. Experiments with a tube 7·9 cm. diameter gave values of U from 29·1 to 30·6 cm./sec., corresponding with values of $U/\sqrt{(ga)}$ from 0·466 to 0·490.

PART I

1. Introduction and experimental method

The rise of gas bubbles in liquids has been studied by several workers (Allen 1900; Hoefer 1913; Miyagi 1925, 1929), but in all the work so far published the bubbles have been so small that the results are not applicable to the study of the rise of large

Reprinted with permission from *Proceedings of the Royal Society of London, Series A, Mathematical and Physical Sciences*, Vol. 200, pp. 375–390, 1950.

volumes of gas, such as those produced in submarine explosions. In the experiments here described, bubbles ranging in volume from 1·5 to 34 cm.^{3} were formed in nitrobenzene contained in a tank, 2 ft. × 2 ft. × 2 ft. 6 in., filled to a depth of about 2 ft. with the liquid. The bubbles were photographed by spark photography at intervals of about 10 msec. (1 msec. = 10^{-3} sec.), using a revolving drum camera, and appropriate spark timing. In some further experiments, bubbles covering a range of volume from 4·5 to 200 cm.^{3} were formed in a cylindrical tank, 2 ft. 6 in. diameter, filled with water to a depth of 3 ft. 6 in., and their mean velocity of rise over a measured distance was determined. In both sets of experiments, the air volume was determined by collecting the bubble in a graduated glass cylinder.

Considerable difficulty was found in producing single, large bubbles of gas, and the method finally adopted was to pivot an inverted beaker containing air, which was then tilted so that the air was released. In general, the air is released from the beaker in a stream of bubbles of varying sizes, but by adjusting the rate of tilting, it was found possible to arrange that the air was spilled into a single bubble.

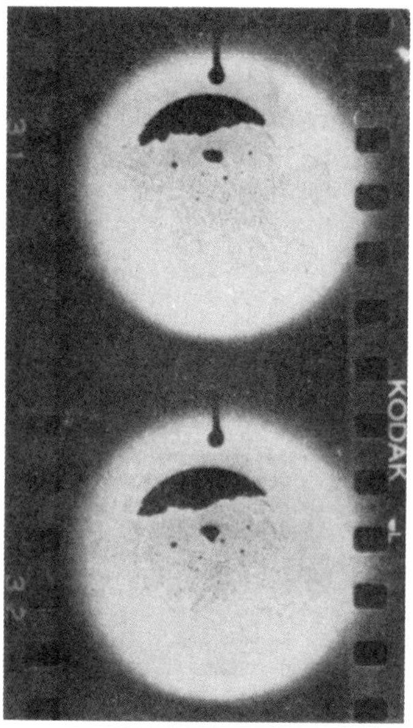

FIGURE 1. Successive spark photographs of an air-filled bubble rising in nitrobenzene. Time interval between photographs = 10·3 msec. Velocity of rise of bubble = 36·7 cm./sec. Diameter of steel ball in the upper part of the photographs = $\frac{1}{4}$ in.

Two successive photographs of a typical bubble formed in this way in nitrobenzene are shown in figure 1, the time interval between the two photographs being 10·3 msec. In addition to the bubble, the photographs show a steel ball, $\frac{1}{4}$ in. diameter, soldered at the lower end of a vertical rod immersed in the liquid; this arrangement was used to find the scale of the photographs and to give a reference mark from which the vertical displacement of the bubble could be measured.

The uniformity of the velocity of rise of the bubbles may be judged by figure 2, in which time, t, and the vertical displacement, X, of two bubbles are plotted as abscissae and ordinates, respectively. The actual measured values of X and t for the bubble of figure 1 are indicated by the circular dots in figure 2, and those for a second, larger bubble by crosses; the straight lines of closest fit drawn through the observed points are denoted by A and B respectively. It will be seen that the scatter of the experimental points is not excessive, and that the velocity of rise, U, of the two bubbles is reasonably constant over the interval measured.

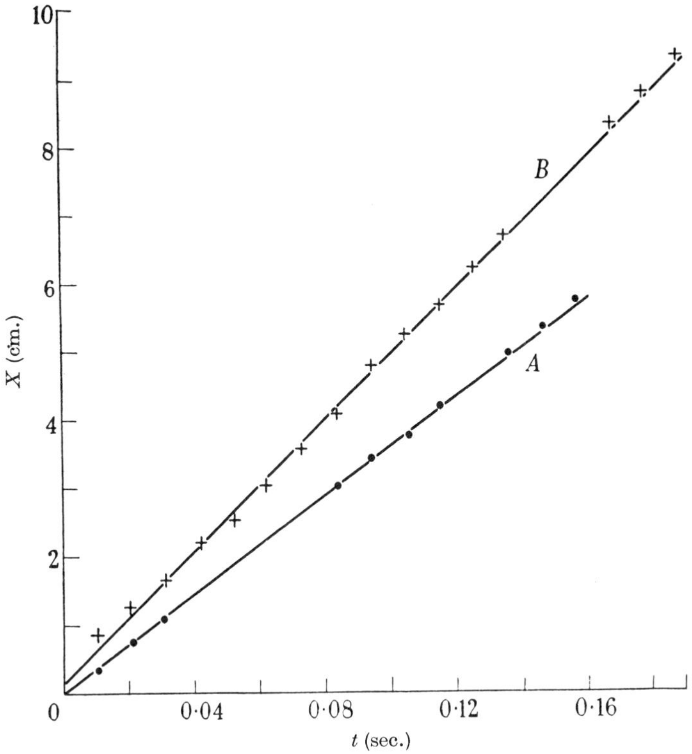

FIGURE 2. (Displacement, time) curves deduced from photographs of rising bubbles. •, + Experimental values; curve A, $U = 36{\cdot}7$ cm./sec.; curve B, $U = 48{\cdot}2$ cm./sec.

The shape of the profile of the bubbles was found by measuring the films on a travelling microscope fitted with two independent motions at right angles to one another. The results for the lower photograph of figure 1 are shown graphically in figure 3, where the circular dots represent points on the central, regular portion of the profile of the bubble, deduced from the microscope readings. In figure 3, the vertical and horizontal axes are parallel to the corresponding axes in the tank, and the origin is taken at the uppermost point on the bubble; the dimensions given refer to the actual size of the bubble. The crosses with vertical axes and with axes at 45° to the vertical in figure 3 represent points on the profile of the same bubble, obtained from measurements of photographs taken 105·7 and 132·5 msec. earlier than the lower photograph of figure 1. The agreement between the three sets of points shows that the shape of the cap of the bubble undergoes very little variation over the range of time covered by the three photographs.

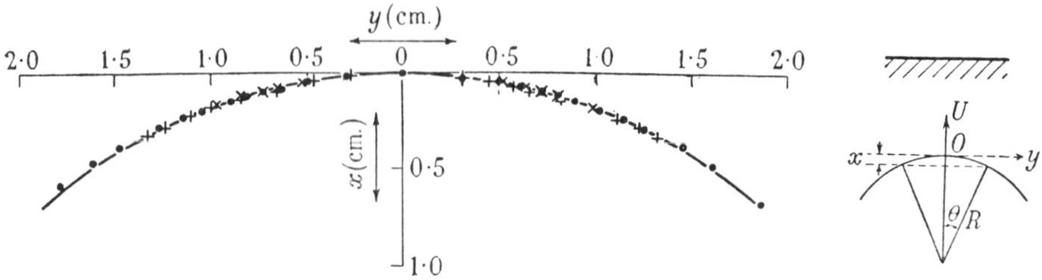

FIGURE 3. Shape of the profile of a bubble. • Bubble shown in the lower photograph of figure 1; + the same bubble 0·106 sec. earlier; × the same bubble 0·133 sec. earlier; —— an arc of circle of radius 3·01 cm.

The curve in figure 3 is an arc of circle of radius 3·01 cm., drawn to pass through the origin, and since the scatter of the observed points around this curve is within the limits of the errors made in measuring the film, the upper part of the bubble is a portion of a sphere within the experimental error. It is worth noticing that the angle subtended at the centre of the circle by the arc in figure 3 is about 75°, whilst the angular width of the whole bubble in figure 1 (referred to the centre) is about 90°.

2. Explanation of why the top of large bubbles is spherical

The perfection of the spherical shape of the top of the bubble led us to consider the condition which must be satisfied at its surface. The pressure there may be taken as constant, for the variations in pressure through the interior air must be so small as to be negligible. The pressure in the fluid outside the bubble is due to the dynamics of the flow round it, and to gravity. The condition that the pressure at the surface of the bubble is constant requires that these two causes shall neutralize one another. Applying Bernoulli's equation to steady flow relative to the bubble, which is assumed to be symmetrical so that the relative velocity at its highest point is zero, the surface condition is

$$q^2 = 2gx, \tag{2·1}$$

where x is the depth below the highest point, q the fluid velocity relative to the bubble and g is the acceleration of gravity.

The bubbles were observed to be spherical above and more or less flat beneath. It is not possible to calculate the flow round such a shape. Recourse was therefore had to experiment. Three models were made in brass, each had a spherical surface 1 in. radius and a flat under-surface. Small pressure holes were bored in both surfaces of each model. They were set up in a wind tunnel and the pressure distribution determined at a wind speed of 15 m./sec. The angles, θ_m, between the polar axes of the models and the radii to their rims were 75°, 55° and 30°, a range which more than covered the values observed in bubbles. The results are given in non-dimensional form in figure 4, when $(p_\theta - p_0)/\frac{1}{2}\rho U^2$ is plotted against θ, the angle which the radius from the centre of the sphere makes with the polar axis. Here p_θ is the pressure at angle θ, p_0 that at the vertex and $\frac{1}{2}\rho U^2$ the pitot pressure in the tunnel. It has long

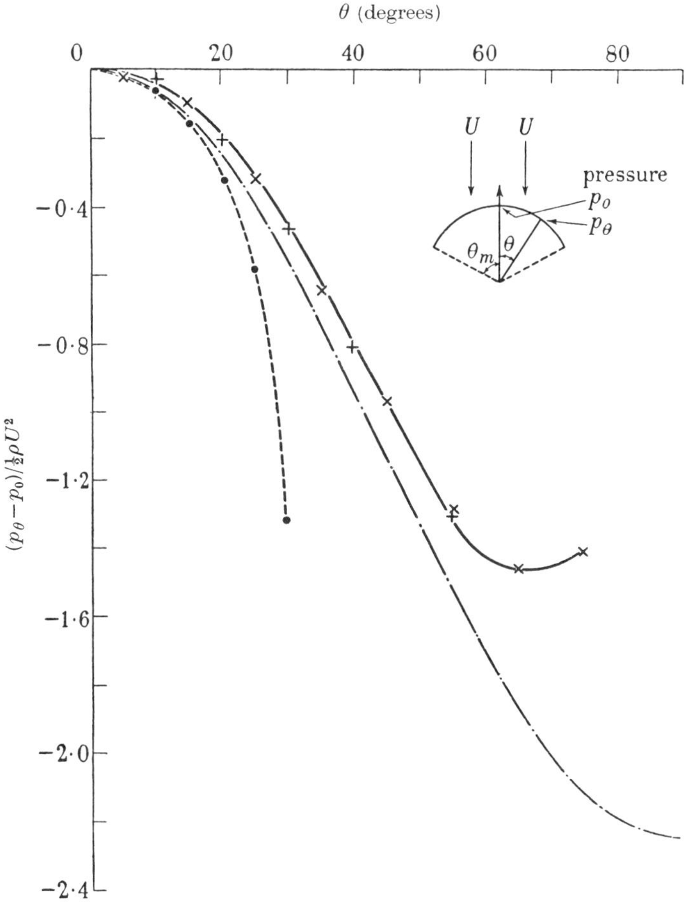

FIGURE 4. The variation of pressure over the surfaces of lenticular bodies in a wind tunnel. Experimental values: • $\theta_m = 30°$; + $\theta_m = 55°$; × $\theta_m = 75°$; ·—·—· theoretical curve for a sphere, assuming ideal fluid flow.

Mean pressures p_b on the backs of the bodies:

θ_m	30°	55°	75°
$(p_b - p_0)/\frac{1}{2}\rho U^2$	−1·32	−1·39	−1·41

been known that when a sphere is set up in a wind tunnel, the pressure distribution over a large part of the windward side is rather nearly the same as the theoretical distribution of pressure over a sphere obtained by classical methods using velocity potential. This is true even though the flow behind the sphere bears no relationship to the theoretical flow. For this reason the theoretical pressure distribution round a complete sphere was calculated from the well-known expression

$$\frac{q^2}{U^2} = \frac{p_0 - p_\theta}{\frac{1}{2}\rho U^2} = \tfrac{9}{4}\sin^2\theta, \tag{2·2}$$

and shown in figure 4. It will be seen that the measured distributions over all the lenticular bodies are rather close to the theoretical distribution for a sphere in the range $0 < \theta < 20°$. The 55° and 75° bodies retain this property nearly out to their edges, but the drop in pressure below that at the vertex at any point is rather less than the theoretical value.

The observed values of $(p_0 - p_\theta)/\frac{1}{2}\rho U^2$, which may be equated to q^2/U^2, have been taken from the faired curve for $\theta_m = 55°$ and 75° in figure 4, and their values are given in the first row of table 1. In the next row are tabulated the values of $x/R = 1 - \cos\theta$, where R is the radius of the spherical surface of the lenticular body. Below these are given the values of $q^2/U^2(1-\cos\theta) = q^2R/U^2x$. It will be seen that this ratio is nearly constant, its mean value being 3·28.

TABLE 1

θ	15°	20°	25°	30°	35°	40°	45°	50°
$\frac{q^2}{U^2} = \frac{p_0 - p_\theta}{\frac{1}{2}\rho U^2}$	0·10	0·192	0·315	0·465	0·628	0·805	0·975	1·144
$1 - \cos\theta$	0·0341	0·0603	0·0937	0·1340	0·1808	0·2340	0·2929	0·3572
$\frac{q^2}{U^2(1-\cos\theta)}$	2·94	3·18	3·25	3·44	3·47	3·44	3·33	3·20

Mean value of $\frac{q^2}{U^2(1-\cos\theta)} = 3·28$.

The condition (2·1) that bubbles of lenticular shape may have constant pressure over their spherical surfaces is satisfied if $q^2R/U^2x = 3·28$ is identical with $q^2 = 2gx$. Eliminating x/q^2

$$U^2 = 2gR/3·28 = 0·61\,gR \quad \text{or} \quad U = 0·78\sqrt{(gR)}. \tag{2·3}$$

If the pressure had been exactly the same as the calculated pressure over a complete sphere, the condition (2·1) could be satisfied over the portion near the stagnation point only, for in that case $q^2 = \frac{9}{4}U^2\sin^2\theta$ and $x = R(1-\cos\theta)$, so that $q^2/x = 2g$ if

$$\frac{U^2}{gR} = \frac{8}{9}\left(\frac{1-\cos\theta}{\sin^2\theta}\right).$$

When θ is small, $(1-\cos\theta)/\sin^2\theta \to \frac{1}{2}$, so that the pressure condition would be satisfied near the stagnation point, i.e.

$$U^2 = \tfrac{4}{9}gR \quad \text{or} \quad U = \tfrac{2}{3}\sqrt{(gR)}. \tag{2·4}$$

3. Comparison with observation

Fourteen bubbles, rising in nitrobenzene, were photographed. The results of the measurement of the films are summarized in table 2, where the first three columns give the volume V of the bubbles, the radius of curvature R and the velocity of rise U.

The fourth column gives the maximum transverse dimension $2A$, and the fifth column $\theta_m = \sin^{-1} A/R$. The sixth column gives the drag coefficient, C_D, of the bubble calculated from the equation

$$C_D \cdot \pi A^2 \cdot \tfrac{1}{2}\rho U^2 = g\rho V. \tag{3·1}$$

TABLE 2. BUBBLES IN NITROBENZENE

volume V (cm.³)	radius of curvature R (cm.)	velocity U (cm./sec.)	maximum transverse dimension, $2A$ (cm.)	$\theta_m = \sin^{-1} A/R$	drag coefficient C_D	Reynolds number Re
1·48	2·41	29·2	2·86	36·4°	0·53	2780
3·50	2·09	29·6	3·14	48·6	1·02	3090
4·06	2·04	28·9	3·48	58·3	1·00	3360
4·31	2·17	28·0	3·16	46·8	1·37	2950
6·40	2·78	35·5	4·98	63·6	0·52	5900
7·30	2·65	34·2	4·40	56·1	0·80	5010
8·02	2·67	34·0	4·23	52·5	0·97	4790
—	3·01	36·7	—	—	—	—
8·80	3·17	37·2	5·26	56·1	0·58	6520
9·18	2·77	33·0	4·53	55·8	1·00	5020
18·40	3·30	37·3	5·10	50·7	1·27	6340
21·25	3·51	38·1	5·85	56·5	1·07	7440
28·1	4·16	43·0	—	—	—	—
33·8	4·27	42·1	6·19	46·5	1·25	8700
—	4·84	48·2	—	—	—	—

The last column gives the Reynolds number Re

$$Re = UA/\nu. \tag{3·2}$$

The viscosity of nitrobenzene is 0·018 poise at 14° C and the density 1·2 g./cm.³, so that $\nu = 0{\cdot}015$ cm.²/sec.

The experimental values of U are plotted against $\surd R$ in figure 5. It will be seen that they lie closely scattered round the line $U = \frac{2}{3}\surd(gR)$ and well below the line $U = 0{\cdot}78\surd(gR)$.

It is curious that the experiments agree better with the arbitrary assumption that flow over the forward part of the bubble is the same as that calculated for a sphere moving in a frictionless liquid than with calculations based on the pressure distribution measured over the surface of a solid of nearly the same shape as the bubble.

It will be noted, however, that the flow of the liquid near the front of a bubble would be expected to be more nearly a truly irrotational one than that near a solid body, because in the latter case a boundary layer would necessarily be present, whereas in the former the air in the bubble would cause no appreciable drag so that no boundary layer would be formed. The closeness with which the observed points fit the line $U = \frac{2}{3}\surd(gR)$ is remarkable. This suggests that the flow near the front of a bubble must be very close indeed to the theoretical flow near the front of a complete sphere in an inviscid fluid.

It will be noticed in figure 5 and in table 2 that C_D is much more variable than $U/\surd R$. These values of C_D are plotted as a function of θ_m in figure 6, and those found by integrating the observed pressures on the lenticular bodies in the wind tunnel are shown on the same figure.

It will be seen that there is considerable scatter, in the case of the bubbles, from the mean line representing the variation of C_D with θ_m, and that the line representing the values of C_D for the lenticular bodies is quite different from that representing the bubbles.

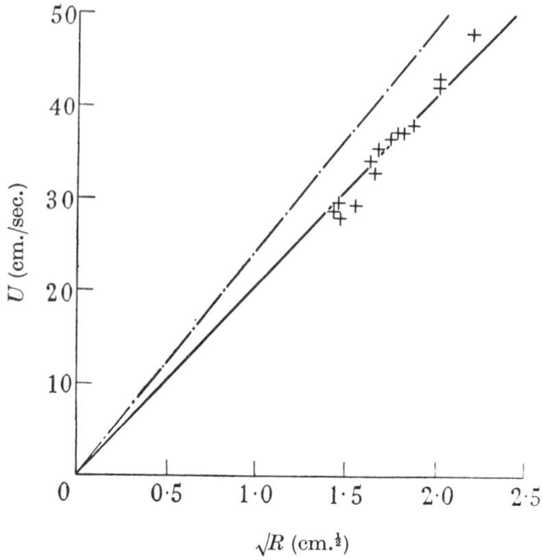

FIGURE 5. The relationship between the velocity of rise, U, of a bubble and the radius of curvature, R. + experimental values; — · — $U = 0{\cdot}78\sqrt{(gR)}$; —— $U = \frac{2}{3}\sqrt{(gR)}$.

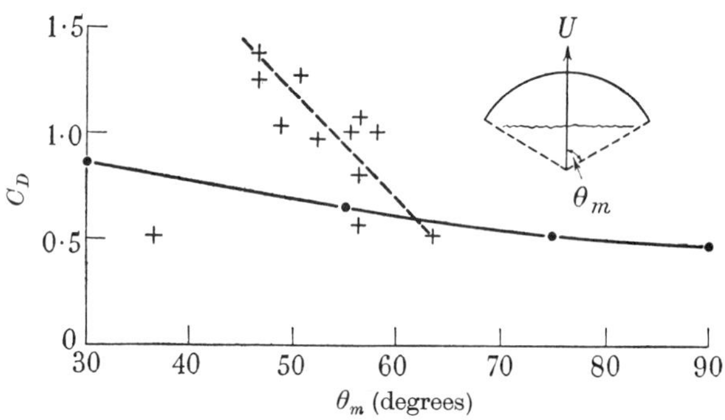

FIGURE 6. The experimental relationship between drag coefficient, C_D, and the semi-vertical angle, θ_m, subtended at the centre. + values for bubbles rising in nitrobenzene; • values for lenticular bodies in the wind tunnel.

4. The relationship between volume and rate of rise of a bubble

If all the bubbles were geometrically similar the dimension A could be expressed by

$$A = \alpha V^{\frac{1}{3}}, \tag{4·1}$$

where α would then be constant. If also the drag coefficient were constant, (3·1) shows that U would be proportional to $V^{\frac{1}{6}}$; in fact, (3·1) can be expressed in the form

$$UV^{-\frac{1}{6}} = \sqrt{(2g/\pi\alpha^2 C_D)} = 25{\cdot}0/\alpha\sqrt{C_D}\,\text{cm.}^{\frac{1}{2}}\,\text{sec.}^{-1}. \tag{4·2}$$

To test how far the assumption that α and C_D and therefore $\alpha\sqrt{C_D}$ are constant, measurements were made involving thirteen bubbles rising in nitrobenzene and seventy-five in water. The results are shown in figure 7, where $UV^{-\frac{1}{6}}$ is plotted against $V^{\frac{1}{3}}$, U being expressed in cm./sec. and V in cm.^{3}. The experimental results of Miyagi and Hoefer with smaller bubbles are also shown.

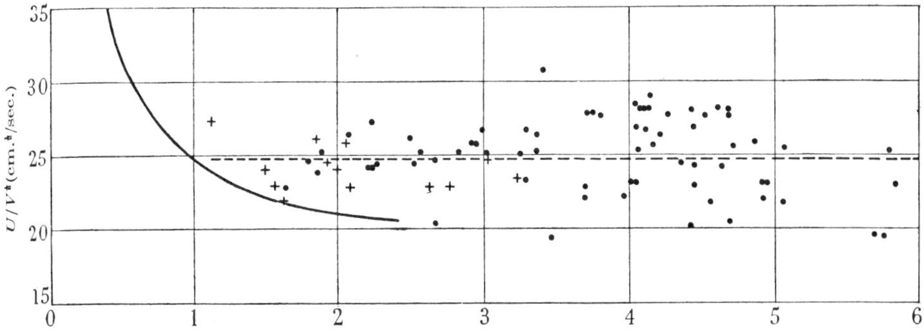

FIGURE 7. Experimental results for the velocity of rise, U, of a bubble of volume V. + bubbles in nitrobenzene; • bubbles in water; - - - $UV^{-\frac{1}{6}} = 24{\cdot}8$ (mean value); —— Miyagi and Hoefer results.

Though there is considerable scatter about the line

$$UV^{-\frac{1}{6}} = 24{\cdot}8, \tag{4·3}$$

which represents the mean value giving equal weight to all the observations, figure 7 shows that there is no systematic variation in $UV^{-\frac{1}{6}}$ with V, and that the experiments in water and nitrobenzene give the same value of $UV^{-\frac{1}{6}}$. It is of interest to compare this experimental result with (4·2). In order that bubbles may ascend with the velocity $24{\cdot}8\,V^{\frac{1}{6}}$, which represents the mean of the experimental values,

$$\alpha\sqrt{C_D} = \frac{25{\cdot}0}{24{\cdot}8} = 1{\cdot}0 \text{ approximately.} \tag{4·4}$$

It is of interest to note in table 2 that the mean value of C_D for the experiment in nitrobenzene is about 1·1. The value of α cannot be calculated unless the shape of the bottom of the bubble is known, but if the bubble were enclosed between a spherical upper surface and a flat lower one it can be shown that

$$\alpha^3 = \frac{3\sin^3\theta_m}{\pi(2-3\cos\theta_m+\cos^3\theta_m)}.$$

The values of α for a series of values of θ_m are shown in table 3.

TABLE 3

θ_m	30°	40°	50°	60°	70°	80°	90°
α	1·32	1·18	1·07	0·99	0·91	0·85	0·78

It will be seen that in range $\theta_m = 46°$ to $\theta_m = 64°$, which contains all the experimental values in table 2, except that found for the smallest bubble, α is within 10 % of 1·0.

Example. Application to the bubble of gas released in a submarine explosion

When an explosive detonates under water, a mass of gas which is known, at any rate approximately, is suddenly produced. The volume of this bubble oscillates violently at first, but after a short time this oscillation ceases and the bubble rises and finally reaches the surface. It is of interest to calculate how fast such a bubble would rise if it behaved like those described in the foregoing experiments. A charge of say 300 lb. of amatol might be expected to produce about $8{\cdot}8 \times 10^7$ cm.^{3} of gas at atmospheric pressure after the steam produced in the explosion had had time to condense.

The formula (4·3) gives then

$$U = 525\,\text{cm./sec.} = 17{\cdot}2\,\text{ft./sec.}$$

5. Turbulence in the wake behind a bubble

In the original photographs of bubbles in nitrobenzene, a region of turbulence is clearly shown behind the large bubble. This is due no doubt to some anisotropic optical property of nitrobenzene when subjected to viscous stresses. That such stresses exist could be inferred from the fact that, in the photograph shown in figure 1, the largest of the small bubbles in the wake of the large one is not spherical and is rapidly changing in shape. This bubble has a diameter of about 6 mm. Still smaller bubbles are less distorted, and one of diameter about 2 mm., seen to the left side of the 6 mm. bubble, is distorted so that its length to diameter ratio is about 1·1.

The rate of shear which might be expected to produce a distortion of this amount has been calculated by Taylor (1934). In the field of flow represented by the equations

$$u = Cx, \quad v = -Cy, \quad w = 0, \tag{5·1}$$

an air bubble of mean radius a would be pulled out so that

$$\frac{L-B}{L+B} = \frac{2C\mu a}{T}, \tag{5·2}$$

where L and B are the length and breadth of the bubble and μ and T the viscosity and the surface tension of the liquid. For nitrobenzene, $\mu = 0{\cdot}018$ poise, $T = 43{\cdot}9$ dynes/cm., so that for the 2 mm. bubble, $a = 0{\cdot}1$ cm., $L/B = 1{\cdot}1$ and $(L-B)/(L+B) = 0{\cdot}05$, giving

$$C = \frac{0{\cdot}05 \times 43{\cdot}9}{2 \times 0{\cdot}018 \times 0{\cdot}1} = 6{\cdot}1 \times 10^2\,\text{sec.}^{-1}.$$

The rate W of dissipation of energy per cm.^{3} in the flow represented by equation (5·1) is

$$W = \mu\left\{\left(\frac{\partial u}{\partial x}\right)^2 + \left(\frac{\partial v}{\partial y}\right)^2\right\} = 2\mu C^2 = 1{\cdot}34 \times 10^4\,\text{ergs/cm.}^3\text{/sec.}$$

If the rate of dissipation were constant through the wake, and if wake extends over the whole of the region which appears disturbed in figure 1, namely, through a diameter of 5·9 cm. and a length of 3·8 cm., the total rate of dissipation in the wake is

$$\begin{aligned} & 1{\cdot}34\times 10^4\times (\text{volume of wake}) \\ &= 1{\cdot}34\times 10^4\times \tfrac{1}{4}\pi\times (5{\cdot}9)^2\times 3{\cdot}8 \\ &= 1{\cdot}4\times 10^6 \text{ ergs/sec.} \end{aligned}$$

The total rate of dissipation would be known if the drag coefficient, C_D, of the large bubble were known. Since the density ρ of nitrobenzene is 1·2 g./cm.³, whilst the velocity of rise, U, of the large bubble in this experiment was 36·7 cm./sec. and its maximum transverse dimension, $2A$, was 5·1 cm., the total rate of dissipation was

$$\begin{aligned} & C_D\,.\,\tfrac{1}{2}\rho U^2\,.\,\pi A^2\,.\,U \\ &= C_D\{\tfrac{1}{2}\times 1{\cdot}20\times 36{\cdot}7^3\times \pi\times (2{\cdot}55)^2\} \\ &= 6{\cdot}1\times 10^5\, C_D \text{ ergs/sec.} \end{aligned}$$

Since C_D is found to be of the order of 1·0 (see table 2), it will be seen that the rate of dissipation which would distort the bubbles by the observed amount is of the same order as that deduced from the rate of rise.

For the largest of the small bubbles in the wake, viscous stresses would produce such a distortion that the formula (5·2) would not be expected to apply.

PART II. EMPTYING WATER FROM A VERTICAL TUBE

6. Theoretical discussion

It has been seen that large bubbles in water assume a form which is very nearly the lenticular volume contained between a sphere and a horizontal plane cutting it above its centre. The pressure distribution over the spherical surface is found experimentally to be approximately the same as that calculated for a complete sphere moving in an ideal fluid. This pressure, together with the hydrostatic pressure due to gravity, leads to a uniform surface pressure when the velocity of rise, U, is

$$U = \tfrac{2}{3}\surd(gR),$$

R being the radius of the upper surface of the bubble.

It has been found that when the rate of rise and radius of curvature of large bubbles have been measured simultaneously, these quantities do in fact very nearly satisfy this relationship.

This result depends on the observed fact that the pressure distribution over the forward portion of a lenticular shaped body is nearly the same as the theoretical distribution over a complete sphere. This empirical observation makes it possible to have a partial understanding of why it is that the upper surface of a large rising bubble is so nearly spherical, but it does not lead to a complete understanding because the relationship between the shape of the bubble and the pressure distribution over its surface is not understood. The main difficulty in this matter is to obtain a correct description of the currents in the wake which follows the nearly flat lower surface of the bubble.

For this reason we may consider the case where the bubble is confined inside a circular tube. In this case the fluid can run round the outside of the bubble and remain as a layer running down the surface of the tube and falling freely under gravity. In this case it is not necessary for the bubble to have a lower surface. The tube, in fact, is open to the atmosphere at its lower end. The problem thus posed is capable of being solved completely. The question to be answered is 'How fast will the air column rise in a vertical tube with a closed top when the bottom is opened?' Or, alternatively, 'How fast will a vertical tube with a closed top empty itself when the bottom is opened?'

If U is the velocity of rise of the air column in a tube of radius a, the flow can be brought to a steady motion by giving the whole system a downward velocity U. The top of the air column is now at rest, and if x is the depth below this, the condition which must be satisfied at points on the free surface of the air column is

$$q^2 = 2gx, \tag{6·1}$$

where q is the velocity of the liquid and g is the acceleration due to gravity.

The problem is, therefore, to find the shape of a body of revolution which if inserted in a circular tube will leave a space through which a perfect fluid could flow so that the velocity at its surface would be proportional to $\sqrt{x}$.

This problem could be solved by relaxation methods, but a rough approximation to the flow near the top of the air column may be obtained as follows:

The velocity potential

$$\phi = e^{K_n x/a} J_0(K_n r/a) \tag{6·2}$$

represents a flow contained in a tube of radius a provided K_n is a root of the equation $J_1(z) = 0$. Here J_0 and J_1 are Bessel functions. The flow has the property that it dies away to zero, when x has large negative values.

The flow represented by

$$\phi = -Ux + \sum_n A_n e^{K_n x/a} J_0(K_n r/a) \tag{6·3}$$

and by the Stokes stream function

$$\psi = -\tfrac{1}{2}Ur^2 + r\sum_n A_n e^{K_n x/a} J_1(K_n r/a) \tag{6·4}$$

has the property that it has uniform velocity U for large negative values of x, and that the radial velocity is zero at $r = a$. The pressure condition which must be satisfied on the surface $\psi = 0$ is

$$\left(\frac{\partial\phi}{\partial x}\right)^2 + \left(\frac{\partial\phi}{\partial y}\right)^2 = 2gx. \tag{6·5}$$

It seems unlikely that an analytical solution of this problem could be found but a rough approximation to the flow near the top of the air column might be obtained by using only one term of the series in (6·3). The lowest root of $J_1(z) = 0$ is 3·832, so that as a first approximation (6·3) may be taken as

$$\phi = -Ux + A_1 e^{3·832x/a} J_0(3·832r/a), \tag{6·6}$$

and (6·4) as

$$\psi = -\frac{Ur^2}{2} + A_1 r\, e^{3·832x/a} J_1(3·832r/a). \tag{6·7}$$

The surface of the bubble is $\psi = 0$. In order that $x = 0$ may be the vertex of the air column it is necessary that the coefficient of r^2 in the expansion of ψ near the origin shall be zero. This condition is satisfied if

$$A_1 = \frac{Ua}{3{\cdot}832}. \tag{6·8}$$

It is clearly impossible to satisfy the pressure condition (6·1) at more than one point when only one term in the series expansion (6·3) is used. Assume that (6·5) is satisfied at the point $\psi = 0$ when $r = \frac{1}{2}a$. The value of x at this point is found by setting

$$A_1 = Ua/3{\cdot}832, \quad r/a = \tfrac{1}{2}, \quad \psi = 0 \text{ in } (6{\cdot}5).$$

The value of $J_1[\frac{1}{2}(3{\cdot}832)]$ is 0·580, so that (6·7) gives

$$e^{3{\cdot}382x/a} = \frac{3{\cdot}832}{4(0{\cdot}580)} = 1{\cdot}65, \quad \text{and} \quad \frac{x}{a} = 0{\cdot}131.$$

Since $J_0[\frac{1}{2}(3{\cdot}832)] = 0{\cdot}273$, (6·5) gives

$$\frac{2ga}{U^2}(0{\cdot}131) = \{1-(1{\cdot}65)(0{\cdot}273)\}^2 + \{(0{\cdot}580)(1{\cdot}65)\}^2,$$

hence

$$\frac{U^2}{ga} = 0{\cdot}215$$

or

$$U = 0{\cdot}464\sqrt{(ga)}. \tag{6·9}$$

This very rough approximation to the flow might be improved by using more terms in the series in (6·3). It would then be possible to satisfy the pressure condition at as many points as the number of terms taken in the series of (6·3), the values of x/a would be calculated for each assumed value of r/a, and the final equation for determining $U/\sqrt{(ga)}$ would be left. The numerical work, however, would be very heavy, and a relaxation method would probably be more satisfactory.

7. Experiments on emptying vertical tubes

To find out whether the flow postulated in the above calculations can actually occur, a long glass tube was erected vertically over a sink in the Cavendish Laboratory. It was filled by placing the lower end in a basin of water and applying suction at the upper end through a rubber tube till the water-level reached the top. The rubber tube was then closed. To make an experiment the basin at the lower end was suddenly removed and a bubble was seen to run up the tube, the water emptying itself by running down the wall and pouring out from the bottom in the form of a tubular curtain of water. The photograph (figure 8), shows a flash-picture of the bubble running up a tube 7·94 cm. diameter. This tube was surrounded by a flat-sided glass box filled with water in order to reduce distortion by refraction.

It will be seen that the flow assumed in the analysis actually takes place. The thickness of the sheet of water that runs down the wall can be seen in the photograph, and measurements show that except close to the vertex this thickness is, as would be expected from (6·1), inversely proportional to the square root of the depth below the vertex.

Experiments were made with three tubes each about 180 cm. long and of diameters 1·23, 2·16 and 7·94 cm. The upward velocities of the tops of the bubbles were observed by timing their passage past horizontal marks with a stop-watch. The observed velocities together with the corresponding values of $U/\sqrt{(ga)}$ are given in table 4.

FIGURE 8. Emptying a glass tube 7·9 cm. diameter.

It will be seen that the values of $U/\sqrt{(ga)}$ are nearly constant but tend to rise slightly with diameter. This is probably an effect of viscosity. The Reynolds numbers associated with the bubbles are given in table 4. It will be seen that the Reynolds number associated with the largest tube is so great that it might be expected that the effect of viscosity would be negligible in that case.

TABLE 4

$2a$ (cm.)	1·23	2·16	7·94
U (cm./sec.)	9·8 to 10·15	14·5 to 15·2	29·1 to 30·6
$U/\sqrt{(ga)}$	0·40 to 0·41	0·447 to 0·468	0·466 to 0·490
$\dfrac{Ua}{\nu}$	600	1,600	12,000

calculated value, $U/\sqrt{(ga)} = 0{\cdot}464$.

It is remarkable that the very rough theory given above yields the value 0·464 for $U/\sqrt{(ga)}$, which is very close to the range 0·466 to 0·490 which was observed experimentally with the largest tube.

8. Measured profile

The profile of the bubble taken from a photograph similar to figure 8 is shown in figure 9. After passing the rounded top of the bubble, the water concentrates into a sheet near the wall. At a depth of about $1{\cdot}5a$ below the vertex the horizontal component of flow becomes so small that the vertical component, U, may without appreciable error be taken as equal to q. The equation to the part of the profile below $x = 1{\cdot}5a$ is therefore approximately

$$U\pi a^2 = \pi q(a^2 - r^2) = \pi(a^2 - r^2)\sqrt{(2gx)},$$

or if t is written for $(a-r)$, the thickness of the layer of water running down the inside of the tube, and $U = \beta\sqrt{(ga)}$,

$$2\frac{t}{a} - \frac{t^2}{a^2} = \beta\sqrt{\frac{a}{2x}},$$

and since t is small compared with a,

$$\beta = 2\sqrt{2}\left(\frac{t}{a}\right)\sqrt{\frac{x}{a}}.$$

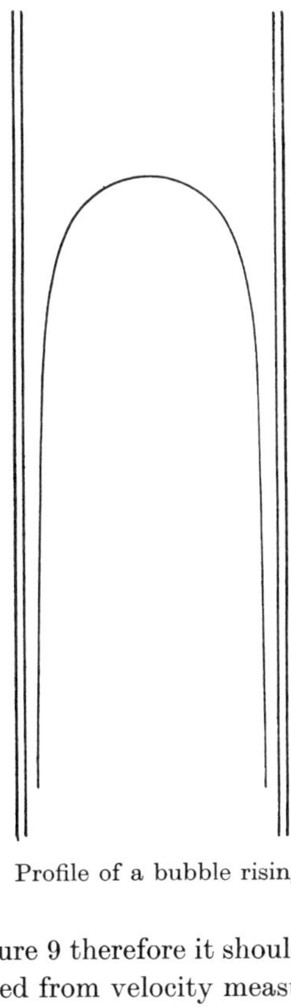

FIGURE 9. Profile of a bubble rising in a tube.

By measuring t and x in figure 9 therefore it should be possible to find β for comparison with the value deduced from velocity measurements and given in table 4. Corresponding values of t/a and x/a are given in table 5.

TABLE 5

x/a	1·71	3·43	5·14	6·17
t/a	0·171	0·123	1·103	0·093
$\beta = 2\sqrt{2(t/a)}\,\sqrt{(x/a)}$	0·63	0·64	0·66	0·65

It will be seen that the values of $t\sqrt{x}$ are nearly constant, but the value of β deduced from them is much larger than that found by measuring the velocity of the bubble. The difference may be due to the fact that the existence of the boundary layer at the inner wall of the tube was not considered, so that the downward flow estimated by measuring t and assuming that U is uniform through the layer gives rise to an overestimate. Another cause which gives rise to an overestimate of t is the refraction of the glass tube, an error which is not completely eliminated by surrounding it with a parallel-sided glass box, filled with water.

REFERENCES

Allen, H. S. 1900 *Phil. Mag.* **50**, 323, 519.
Hoefer, K. 1913 *Z. Ver. dtsch. Ing.* **57**, 1174.
Miyagi, O. 1925 *Phil. Mag.* **50**, 112.
Miyagi, O. 1925 *Tohoku Imp. Univ., Technol. Rep.*, **5**, 135.
Miyagi, O. 1929 *Tohoku Imp. Univ., Technol. Rep.*, **8**, 587.
Taylor, G. I. 1934 *Proc. Roy. Soc.* A, **146**, 501.

Polar structures in the theory of conjugated molecules. II. Symmetry properties and matrix elements for polar structures

BY D. P. CRAIG

Sir William Ramsay and Ralph Forster Laboratories, University College, London

(*Communicated by C. K. Ingold, F.R.S.—Received* 21 *December* 1948)

The inclusion of polar structures in the valence-bond theory of π-electrons entails some additions to the usual formalism, and these are given in this part. The symmetry properties of sets of structures, both non-polar and polar, and the matrix elements that come into energy calculations, are dealt with.

Using the work of part I, and the conclusions of this part, the energy parameters for work with polar structures are evaluated.

INTRODUCTION

In part I a discussion has been given of the excited states of ethylene (Craig 1949), the object being to identify in the spectrum the energy levels whose term values can be used to give the parameters required for calculations with polar structures in the valence-bond treatment.

On steady-state bubbles generated by Taylor instability

By P. R. Garabedian

(*Communicated by Sir Geoffrey Taylor, F.R.S.—Received 6 April* 1957)

A theory of the steady motion of a long bubble rising through an infinite plane vertical tube of liquid is presented. It is shown that, contrary to current belief, the flow is not uniquely determined by the width h of the tube and the acceleration of gravity g alone, but that the speed U of the bubble can also be prescribed. However, a criterion of stability singles out the unique physically significant flow of this type as the one which maximizes the velocity of the bubble. With the aid of a difference-differential equation derived from the free-boundary condition, the Froude number $U/(gh)^{\frac{1}{2}}$ for this latter case is estimated to exceed 0·2363, a value slightly higher than that indicated by earlier work on the problem.

1. Statement of the problem

This paper is concerned with the flow of an incompressible heavy liquid past a gas bubble in an infinitely long vertical tube. Attention is confined to plane flow, and it is assumed that the bubble extends downwards without limit, so that the motion is steady. A co-ordinate system attached to the bubble is chosen, with the liquid falling around the bubble, instead of the bubble rising in the liquid. It is supposed that the flow is irrotational, whence it can be described in terms of a complex potential $\zeta = \phi + \mathrm{i}\psi$ which is an analytic function of the complex variable $z = x + \mathrm{i}y$ in the physical plane. The gas in the bubble is considered to be at rest, in a state of constant pressure.

The vertical tube is represented in the z plane as an infinite strip $-\frac{1}{2}h < x < \frac{1}{2}h$ of width h. The complex potential $\zeta = \zeta(z)$ is normalized so that it transforms the region of flowing liquid in this tube conformally on to the infinite strip $-1 < \psi < 1$, with the free boundary of the bubble mapping on to a slit along the positive ϕ axis. If the origin is placed in the physical plane at the vertex of the bubble, Bernoulli's equation shows that the requirement of constant pressure along the free boundary of the bubble takes the form

$$\tfrac{1}{2}q^2 + gy = 0, \tag{1}$$

where $q = |\mathrm{d}\zeta/\mathrm{d}z|$ is the flow speed and g is the acceleration of gravity. This condition determines the shape of the free boundary, which is a curve descending from both sides of the origin to approach the asymptotes $x = \pm\frac{1}{2}h$. Our point of view will be that we are seeking a conformal mapping of the strip $-1 < \psi < 1$, slit along the positive ϕ axis, on to a region of the z plane such that the lines $\psi = \pm 1$ correspond to the lines $x = \pm\frac{1}{2}h$ and such that the slit along the ϕ axis corresponds to a curve satisfying the non-linear free boundary condition (1).

A characteristic dimensionless quantity to be associated with such a flow is the Froude number $U/(gh)^{\frac{1}{2}}$, where U denotes the speed at the upper end of the tube, which has the special value $U = 2/h$ under our present normalizations. It is pointed out that the flow can be reflected repeatedly across the vertical walls of the tube in the physical plane, and that thus extended it describes the motion of a heavy liquid through a periodic family of parallel tongues of gas. If now the liquid is imagined to

Reprinted with permission from *Proceedings of the Royal Society of London, Series A, Mathematical and Physical Sciences*, Vol. 241, pp. 423–431, 1957.

be at rest with the gas accelerated into it, $U/(gh)^{\frac{1}{2}}$ can be interpreted as a dimensionless measure of the velocity with which the tongues of gas penetrate the liquid.

Gibson (1913), Dumitrescu (1943) and Davies & Taylor (1950) have introduced the above model for axially symmetric flow and have compared experimental and theoretical values for the velocity of a large bubble rising in a cylindrical tube. However, no exact solutions have been given for non-linear free surface problems when the effect of gravity is taken into account, although an example has been presented by Vitousek (1954) and Garabedian, McLeod & Vitousek (1954) which sheds some light on the motion of an isolated bubble in an infinite liquid. Further interest in the motion of a bubble in a tube has been stimulated by the connexion which this question has with a treatment of advanced stages of Taylor instability (Taylor 1950; Lewis 1950). With this application in mind, Layzer (1955) and Birkhoff (1954, 1955) have investigated our model of plane flow in some detail.

The contribution of this paper consists in building up a more satisfactory and penetrating theoretical analysis of the problem. We shall study first the asymptotic behaviour of the flow in a freely falling vertical jet of the type separating adjacent gas bubbles in the periodic arrangement described above. The tool that enables us to derive the asymptotic development appropriate for this portion of the flow is a difference-differential equation, based on the free-boundary condition (1), which is suggested by work of Lewy (1950, 1952). Our attention is also focused on the question of existence and uniqueness of the flow under discussion. We are led to the conclusion that a one-parameter family of essentially different flows of this type exists, with each flow determined uniquely by the value of the Froude number $U/(gh)^{\frac{1}{2}}$ associated with it. This result is at odds with the hypothesis in earlier work that there would be only one single bubble shape, but the discrepancy here is explained by the fact that in nature one observes only the stable flow corresponding to the largest possible choice of the dimensionless speed $U/(gh)^{\frac{1}{2}}$.

We use an approximate free boundary flow composed of the first few terms in the asymptotic expansion for heavy jets to establish that the physically relevant maximum value of $U/(gh)^{\frac{1}{2}}$ exceeds 0·2363. This lower bound lies above the interval in which Birkhoff & Carter (see Birkhoff 1954, 1955) place the Froude number, but the deviation brings the answer nearer to the experimental data which they present. The discrepancy in Birkhoff (1954, 1955) undoubtedly stems from the incorrect assumption that the free-boundary problem has a unique solution when g and h alone are given, and that the observed value of U can be calculated without the additional characterization of the actual stable flow by an extremal property.

2. The difference-differential equation

This section investigates the asymptotic behaviour of the jets which separate adjacent bubbles in the above model of protracted Taylor instability. The discussion applies more generally to any plane jet which falls freely under the influence of gravity and possesses a vertical line of symmetry. The line $x = \frac{1}{2}h$ is such a line of symmetry for our tube flow.

In order to exploit this symmetry, we introduce the new analytic function

$$w(\zeta) = g^{\frac{1}{3}}\mathrm{i}[z(\zeta+\mathrm{i}) - \tfrac{1}{2}h], \tag{2}$$

which is real for all real values of the complex potential ζ. In terms of w, the free boundary condition (1) takes the form

$$2\mathscr{R}\{w(\phi-\mathrm{i})\}\,|\,w'(\phi-\mathrm{i})\,|^2 = 1, \tag{3}$$

valid for all $\phi > 0$. Since $w(\zeta)$ is real for real ζ, we can use the Schwarz reflexion principle to derive

$$[w(\phi+\mathrm{i})+w(\phi-\mathrm{i})]\,w'(\phi+\mathrm{i})\,w'(\phi-\mathrm{i}) = 1 \tag{4}$$

from (3). Lewy (1952) has shown that the free boundary is an analytic curve, and hence (4) is an identity which holds by analytic continuation for complex values of the variable ϕ. Thus, after the flow has been extended analytically beyond the free boundary, $w(\zeta)$ satisfies in the right half-plane $\phi = \mathscr{R}\{\zeta\} > 0$ the non-linear difference-differential equation

$$[w(\zeta+\mathrm{i})+w(\zeta-\mathrm{i})]\,w'(\zeta+\mathrm{i})\,w'(\zeta-\mathrm{i}) = 1. \tag{5}$$

The ideas of reflexion and analytic continuation which permit us to derive (5) from (3) have been developed in detail by Lewy, who proves (1952) that a condition of which (3) is a special case is sufficient to ensure analyticity of the function w across the free boundary, and who establishes (1950) a difference-differential equation analogous to (5) in order to obtain from it an asymptotic expansion of the type required in our analysis.

In the left half-plane $\phi = \mathscr{R}\{\zeta\} < 0$ it is easy to derive from the mapping properties of $w(\zeta)$ the difference equation

$$w(\zeta+\mathrm{i})-w(\zeta-\mathrm{i}) = \mathrm{i}hg^{\frac{1}{3}}. \tag{6}$$

A real solution of (5) and (6) whose only singularities are simple branch points at the odd integral multiples of i would yield the flow we are trying to find. Difficulties are encountered in solving (5) and (6) simultaneously because, due to the presence of these branch points, (5) is valid for a single branch of w only in the right half-plane and (6) is valid for a single branch of w only in the left half-plane. However, we shall confine ourselves to an analysis of the asymptotic development of solutions of (5) for large positive values of ϕ.

If we use ζ as a first-order approximation to both $\zeta+\mathrm{i}$ and $\zeta-\mathrm{i}$ for large values of ϕ, (5) reduces to the ordinary differential equation $2w(\zeta)\,w'(\zeta)^2 = 1$, which has the solution $\frac{1}{2}(3\zeta)^{\frac{2}{3}}$. Thus

$$w(\zeta) \sim \tfrac{1}{2}(3\zeta)^{\frac{2}{3}}. \tag{7}$$

To find the next term in the asymptotic expansion of w, we set $w = \frac{1}{2}(3\zeta)^{\frac{2}{3}} + W$ and substitute into (5). Neglecting higher-order terms involving two or more factors of W, we obtain

$$\frac{3^{\frac{1}{3}}}{2}\left[\frac{(\zeta+\mathrm{i})^{\frac{2}{3}}}{(\zeta-\mathrm{i})^{\frac{1}{3}}}+(\zeta-\mathrm{i})^{\frac{1}{3}}\right]W'(\zeta+\mathrm{i})+\frac{3^{\frac{1}{3}}}{2}\left[\frac{(\zeta-\mathrm{i})^{\frac{2}{3}}}{(\zeta+\mathrm{i})^{\frac{1}{3}}}+(\zeta+\mathrm{i})^{\frac{1}{3}}\right]W'(\zeta-\mathrm{i})$$
$$+\frac{W(\zeta+\mathrm{i})+W(\zeta-\mathrm{i})}{3^{\frac{2}{3}}(\zeta^2+1)^{\frac{1}{3}}} = 1-\frac{1}{2}\left(\frac{\zeta+\mathrm{i}}{\zeta-\mathrm{i}}\right)^{\frac{1}{3}}-\frac{1}{2}\left(\frac{\zeta-\mathrm{i}}{\zeta+\mathrm{i}}\right)^{\frac{1}{3}} = \frac{2}{9}\frac{1}{\zeta^2}+\dots. \tag{8}$$

If we replace $\zeta+\mathrm{i}$ and $\zeta-\mathrm{i}$ again by ζ, we are led this time to an ordinary differential equation for the first term in the asymptotic expansion of W, which turns out to have the form $C/\zeta^{\frac{1}{3}}$, where C is a constant. However, (7) shows that this term can be eliminated from the expansion of w by making a suitable translation in the ζ plane, and we lose no generality this way because one solution of (5) is transformed into another by such a translation. With $C = 0$, we deduce from (8) that $W \sim -3^{\frac{2}{3}}/27\zeta^{\frac{4}{3}}$, and hence

$$w(\zeta+K) = \frac{(3\zeta)^{\frac{2}{3}}}{2} - \frac{1}{3(3\zeta)^{\frac{4}{3}}} + \dots, \tag{9}$$

where K is an appropriate constant.

A more careful inspection of (5) shows that we can continue the above process to find in succession an infinite series of terms in the expansion (9), each composed of a real coefficient multiplied by a power of ζ whose exponent differs from $\frac{2}{3}$ by an even integer. The resulting sum, which has the form

$$w_0 = \zeta^{\frac{2}{3}} \sum_{n=0}^{\infty} \frac{a_n}{\zeta^{2n}}, \tag{10}$$

must represent one particular heavy jet corresponding to the simplest conceivable solution of (5). However, we should expect further terms to occur in the asymptotic expansion (9) to account for a quite general family of possible jets.

To find these later terms, we set $w = w_0 + W$, substitute into (5), and again neglect products involving more than one factor of W. The most important terms in the resulting linear difference-differential equation for W can be read off from (8) and have already led us to a solution asymptotic to $C/\zeta^{\frac{1}{3}}$, with C constant. This suggests that we set $W = C(\zeta)/\zeta^{\frac{1}{3}}$ with $C(\zeta)$ an undetermined function of ζ which approaches zero more rapidly than any power of ζ as $\phi \to \infty$. From (8) we deduce that $C(\zeta)$ satisfies the difference equation

$$C'(\zeta+\mathrm{i}) + C'(\zeta-\mathrm{i}) = 0, \tag{11}$$

with an error of the same order of magnitude as $C(\zeta)/\zeta^2$. The exponentials $\exp\{-\frac{1}{2}n\pi\zeta\}$, with n odd, form a complete set of solutions of (11). From this result we proceed readily to the conclusion that the full asymptotic expansion of w is a double series in powers of ζ and $\exp\{-\frac{1}{2}\pi\zeta\}$ of the general form

$$w(\zeta+K) = \frac{3^{\frac{2}{3}}}{2}\zeta^{\frac{2}{3}} + \sum_{n=0}^{\infty}\sum_{m=0}^{\infty} a_{nm}\zeta^{-\frac{1}{3}-m}\exp\{-\tfrac{1}{2}n\pi\zeta\}. \tag{12}$$

For this representation to have meaning, the inner sum over the index m must converge for each fixed value of n, whereas the outer sum can be interpreted as an asymptotic expansion.

It comes as no surprise that the general solution (12) of the difference-differential equation (5) should be a double series involving, in addition to powers of ζ, powers of an exponential of the type encountered in solving linear difference equations (cf. Lewy 1950). Once the correct form for the solution has been found, it is an easy matter to substitute the result back into (5) and to derive the recursive relations which must be fulfilled by the coefficients a_{nm}. In particular, one uses the symmetric cubic character of the left-hand side of (5) to verify that for each $n \geqslant 1$ the index m is restricted to non-negative values. Furthermore, we saw that all the a_{0m} with m even vanish for an appropriate choice of K. As was indicated by our discussion of (11), the a_{n0} with n odd can be chosen arbitrarily, and together with K they determine the remaining coefficients uniquely. For example, when K is chosen so that $a_{00} = 0$ we have

$$a_{01} = -1/3^{\frac{7}{3}}, \quad a_{11} = -4a_{10}/9\pi, \quad a_{20} = -3^{\frac{1}{3}}\tfrac{1}{8}\pi a_{10}^{2}. \tag{13}$$

For the case of the flow past a bubble in a tube, which is our principal concern, it is desirable to have a representation of $w(\zeta)$ which will describe the behaviour of this function throughout the entire ζ plane. It is easy to check here that $w(\zeta)$ has simple branch points at the odd integral multiples of i, is real for real values of ζ, and has the constant imaginary part $\pm \frac{1}{2}hg^{\frac{1}{2}}$ for $\psi = \pm 1$, $\phi < 0$. As a consequence of (6), we find that $w'(\zeta)$ is a regular function of $e^{\pi\zeta}$ for $\phi < 0$, and thus it turns out to be expedient to write down a uniformly valid asymptotic expansion for the derivative $w'(\zeta)$, rather than for $w(\zeta)$ itself. An appropriate representation is

$$w'(\zeta) = \frac{1}{(1+e^{\pi\zeta})^{\frac{1}{2}}} \sum_{n=0}^{\infty} \sum_{m=0}^{\infty} b_{nm} \left\{ \frac{(1+e^{\pi\zeta})^{\frac{1}{2}}}{\ln\left[1+(1+e^{\pi\zeta})^{\frac{1}{2}}\right]} \right\}^{m+\frac{1}{3}} \frac{1}{\left[1+(1+e^{\pi\zeta})^{\frac{1}{2}}\right]^{m+n-\frac{2}{3}}}, \tag{14}$$

where the coefficients b_{nm} are related in an obvious way to the coefficients a_{nm} and the constant K appearing in (12).

While the derivation of the basic formulas (12) and (14) leans heavily on heuristic arguments, the results have been presented in a manner quite adequate for the applications proposed for them.

3. Failure of uniqueness

If no criterion of stability is imposed, it is easy to give an example which shows that the steady flow past a bubble in an infinite tube is not uniquely determined by the dimensions of the tube and the acceleration of gravity alone. For consider a vertical rectangular tube of infinite extent in three-dimensional space whose sides are of widths h_1 and h_2, with, say, $h_1 < h_2$. A plane bubble flow in a strip of width h_1 generates a conceivable three-dimensional flow in such a rectangular tube whose velocity distribution is independent of the horizontal co-ordinate measured parallel to those sides of the tube which have the width h_2. But a quite different three-dimensional steady bubble flow, with velocities independent of the horizontal co-ordinate measured in the two remaining faces of the rectangular tube, is generated by a plane bubble flow in a strip of width h_2. Both these flows are evidently subject to Taylor instability caused by disturbances running parallel to the straight lines which lie at the top of their free surfaces. An actual bubble in such a rectangular tube has a rounded, stable free surface and moves more rapidly than either of the above plane bubbles. Its shape is more nearly approximated by the plane flow corresponding to the wider strip, since this one rises faster than the other one for equal values of the Froude number.

Among the variety of equilibrium shapes that a bubble in such a tube might assume, we can characterize the stable one by the property that it yields the most rapid motion of the bubble through the liquid. This is established by the following considerations based on the principle that the potential energy is minimized by a configuration in stable equilibrium. Suppose that the liquid flows past the bubble at a rate slower than the maximum possible speed, and suppose that the bubble is at rest, with a free surface shaped so that, in spite of all, the pressure distribution along it remains constant. Then by altering its shape slightly in such a way as to better approximate the form which penetrates the liquid at a maximum rate, the bubble can contrive to rise gradually through the falling liquid, losing potential energy in the process. Since such a change of shape results in a loss of potential energy, the original free surface must have been unstable and will not recur, whence the shape yielding the maximum speed of penetration of the bubble must be the only one that is stable and must be the one observed in nature, as was our contention.

For plane flow we can at least prove that the free boundary is uniquely determined when h, g and U are all prescribed, and therefore only a one-parameter family of essentially distinct shapes can occur, with a different choice of the Froude number $U/(gh)^{\frac{1}{2}}$ appearing in each new case. Indeed, the free boundary Γ of a plane bubble in an infinite slot is a symmetric curve with the vertical asymptotes $x = \pm\frac{1}{2}h$ and a maximum at the origin. Suppose that Γ^* is another free boundary curve of this same general character corresponding to the same values of h, g and U as Γ. We denote by $\zeta^* = \phi^* + i\psi^*$ the complex potential for the new flow, and we denote by q^* its speed at the point $z^* = x^* + iy^*$ on Γ^*, while q stands for the speed of the original flow at the point $z = x + iy$ on Γ. If Γ^* does not coincide with Γ, we can lower one of the two curves, say Γ^*, until it lies underneath the other, but still touches it at some point $z_0 = x_0 + iy_0$ with $x_0 > 0$. The asymptotic expansion (12) shows that it is possible to separate the free boundaries in this manner. We translate the flow ζ^* downwards parallel to the y axis by an equal amount and we consider the difference $\psi^* - \psi$ between the translated stream function ψ^* and the unaltered stream function ψ. Since in its new position Γ^* lies below Γ, it is clear that $\psi^* - \psi \geqslant 0$ on the arc of Γ located in the strip $0 < x < \frac{1}{2}h$, while for $x = 0$ and $x = \frac{1}{2}h$ we have $\psi^* - \psi = 0$. Hence, by the maximum principle, the harmonic function $\psi^* - \psi$ is positive at points above Γ in the strip $0 < x < \frac{1}{2}h$. At z_0 both ψ^* and ψ vanish, and it follows that the inner normal derivative of $\psi^* - \psi$ must be non-negative there. Thus $q^* \geqslant q$ at z_0, and it can even be shown that equality holds only when Γ^* is identical with Γ. Now, we have lowered Γ^* in order to make it touch Γ at z_0, and therefore if y^* denotes the ordinate of the corresponding point on Γ^* in its original position and y denotes the ordinate of the point on Γ directly underneath, then $-y^* \leqslant -y_0 = -y$. We conclude that

$$-q^2/2y \leqslant -q^{*2}/2y^* \tag{15}$$

at this pair of points on Γ and Γ^*, with equality holding only when $\Gamma^* = \Gamma$. But (15) yields the contradiction $g < g$, whence the desired uniqueness of Γ in dependence on h, g and U is established.

The familiar comparison argument just presented also shows that as the Froude number $U/(gh)^{\frac{1}{2}}$ decreases, or, for example, as g increases with h and U fixed, the corresponding free-boundary curve Γ must rise monotonically, so that two separate curves of this type can intersect only in the stagnation point at the origin. As $g \to \infty$, the free boundary approaches the segment $-\frac{1}{2}h \leqslant x \leqslant \frac{1}{2}h$ of the x axis plus the two sections of the asymptotes $x = \pm\frac{1}{2}h$ which lie in the lower half-plane. Thus the motion degenerates into a flow in a semi-infinite strip with two sinks at the corners, and the free boundary degenerates into a system of three line segments. For large g, we can convince ourselves of the existence of the bubble flow by thinking of it as a perturbation performed on this limiting configuration. Indeed, our task is to adjust the free boundary Γ so that the ratio $-q^2/2y$ becomes identically equal to g along it. Near the x axis, slight alterations in the shape of Γ modify the numerator $-q^2$ relatively little, but yield any desired value of the denominator $2y$. On the other hand, since the flux through the jets at the sides of the bubble remains fixed, it is possible to arrange the widths of the jets so that q^2 assumes a prescribed value for any fixed choice of y well below the x axis. If we grant that the free boundary can also be shaped in an appropriate way in the neighbourhood of the corners at $x = \pm\frac{1}{2}h, y = 0$, we are led heuristically to the conclusion that for fixed h and U and large enough g, the free boundary problem should have a solution.

Like the ones introduced earlier, the flat bubbles obtained in this manner exhibit Taylor instability at the uppermost portion of the interface, and it is instead the case for which g is a minimum, or, in other words, for which the Froude number $U/(gh)^{\frac{1}{2}}$ is a maximum, that is of physical significance. This latter free-boundary flow is characterized by an extremal property within the class of all flows through an infinite strip past obstacles bounded by arbitrary symmetric curves Γ descending from a maximum at the origin to approach the asymptotes $x = \pm\frac{1}{2}h$. To be precise, we contend that for given h and U the stable free boundary is the curve Γ of this type such that the maximum value of the ratio $-q^2/2y$ along Γ becomes a minimum (cf. Garabedian & Spencer (1952) where a similar maximum-minimum principle is established rigorously for wakes and cavities when the effect of gravity is neglected). If, for example, the ratio $-q^2/2y$ were less than its maximum along an arc of the extremal choice for Γ, then, according to the comparison theorem (15), an infinitesimal shift of this arc towards the flow region would generate a new curve along which the maximum of $-q^2/2y$ would be smaller than its presumed minimum value. This contradiction establishes, at least intuitively, that a solution of the above extremal problem yields a free-boundary flow which must be the desired stable motion.

Our extremal characterization of the stable bubble by the property that its free boundary minimizes the maximum value of $-q^2/2y$ serves as a tool for the estimation of the Froude number $U/(gh)^{\frac{1}{2}}$ of the observed physical flow in addition to providing the best available lead on a method of proof of the existence of a solution of the mathematical problem. For if we take any approximation to the required free boundary and, for given h and U, determine g as the value of $-q^2/2y$ at an arbitrary point z of the approximation, we obtain values for $U/(gh)^{\frac{1}{2}}$ whose minimum with variable z must be a lower bound on the true value of the maximum Froude number. Applications of this technique will be discussed in the next section. Meanwhile, we note that most of the results of this section can be generalized to include the case of the axially symmetric model.

4. Lower bound on the bubble velocity

In order to apply the theory of the preceding section to the numerical calculation of the maximum Froude number $U/(gh)^{\frac{1}{2}}$, we must construct flows which give a reasonably good approximation to the actual free-boundary flow. We can do this by using a few early terms in the uniformly valid asymptotic expansion (14). We choose for this purpose the truncated representation

$$w'(\zeta) = \frac{A_1(1+t)^{\frac{2}{3}}}{t^{\frac{2}{3}}[\ln(1+t)]^{\frac{1}{3}}} + \frac{A_2 t^{\frac{1}{3}}}{(1+t)^{\frac{1}{3}}[\ln(1+t)]^{\frac{4}{3}}} + \frac{A_3 t^{\frac{4}{3}}}{(1+t)^{\frac{4}{3}}[\ln(1+t)]^{\frac{7}{3}}} + \frac{A_4}{t^{\frac{2}{3}}(1+t)^{\frac{1}{3}}[\ln(1+t)]^{\frac{1}{3}}}, \tag{16}$$

where

$$t = (1+e^{\pi\zeta})^{\frac{1}{2}}.$$

In order to study the approximation (16), we introduce the notations

$$f(\theta; j, m, n) = \frac{2}{\pi}\frac{\sin^{m+\frac{1}{3}}\theta\cos^{n-1}\theta}{[\theta^2+(\ln\sec\theta)^2]^{\frac{1}{2}m+\frac{1}{6}}}\sin\left[(\tfrac{2}{3}-m-n)\theta+(m+\tfrac{1}{3})\tan^{-1}\frac{\ln\sec\theta}{\theta}+\tfrac{1}{2}j\pi\right], \tag{17}$$

$$I(\theta; m, n) = \int_0^\theta f(\theta; 0, m, n)\,d\theta, \tag{18}$$

and we set

$$\begin{aligned}
&F_1(\theta) = f(\theta;0,0,0), && F_2(\theta) = f(\theta;0,1,0), && F_3(\theta) = f(\theta;0,2,0), && F_4(\theta) = f(\theta;0,0,1),\\
&G_1(\theta) = f(\theta;1,0,0), && G_2(\theta) = f(\theta;1,1,0), && G_3(\theta) = f(\theta;1,2,0), && G_4(\theta) = f(\theta;1,0,1),\\
&I_1(\theta) = I(\theta;0,0), && I_2(\theta) = I(\theta;1,0), && I_3(\theta) = I(\theta;2,0), && I_4(\theta) = I(\theta;0,1),
\end{aligned}$$

with $\sec\theta = e^{\frac{1}{2}\pi\phi}$, $\phi = \mathscr{R}\{\zeta\}$. If, as described in the previous section, we substitute for g values of the ratio $-q^2/2y$ along the approximate free-boundary curve defined by (16), we find for the maximum Froude number the estimate

$$\frac{U}{(gh)^{\frac{1}{2}}} \geqslant \max_{A_k}\min_{\theta}\frac{\pi(\ln 2)^{\frac{1}{2}}\left[\sum_1^4 A_k I_k(\theta)\right]^{\frac{1}{2}}\left\{\left[\sum_1^4 A_k F_k(\theta)\right]^2+\left[\sum_1^4 A_k G_k(\theta)\right]^2\right\}^{\frac{1}{2}}}{4[A_1+A_2(\ln 4)^{-1}+A_3(\ln 4)^{-2}+\frac{1}{2}A_4]^{\frac{3}{2}}\tan\theta}, \tag{19}$$

where θ varies in the interval $0 \leqslant \theta \leqslant \frac{1}{2}\pi$ and where the range of A_1, A_2, A_3 and A_4 must be restricted so that the conformal mapping $w(\zeta)$ given by (16) remains univalent in the strip $-2 < \psi < 0$, slit along the segment $\psi = -1$, $0 \leqslant \phi \leqslant \infty$.

The technique for applying (19) to derive a numerical lower bound on $U/(gh)^{\frac{1}{2}}$ for the unique stable bubble was to tabulate the right-hand side as a function of θ for a given choice of the A_k, to locate the minima in the table, and to readjust the parameters A_k by means of a gradient method so that these minima would climb toward a common largest value. About eight repetitions of this process led to the optimal case

$$A_1 = 1, \quad A_2 = 1{\cdot}5086111, \quad A_3 = -4{\cdot}9230056, \quad A_4 = 3{\cdot}8179281$$

of the parameters, which yields by (19) the estimate

$$U/(gh)^{\frac{1}{2}} \geqslant 0{\cdot}2363 \tag{20}$$

for the physically significant maximum value of the Froude number. The table of $U/(gh)^{\frac{1}{2}}$ corresponding to this choice of the parameters exhibited appreciable deviations from the constant arrangement to be expected for the exact solution and rose, indeed, to a maximum of 0·28. Other combinations of the A_k giving a smaller interval of variation of $U/(gh)^{\frac{1}{2}}$, such as, for example,

$$A_1 = 1, \quad A_2 = 0{\cdot}877117, \quad A_3 = -1{\cdot}56682, \quad A_4 = 1{\cdot}289516,$$

which restricted the Froude number to the interval $0{\cdot}23 \leqslant U/(gh)^{\frac{1}{2}} \leqslant 0{\cdot}245$, do not challenge the validity of (20), but serve rather to verify the failure of uniqueness of the solution of the free-boundary problem when the Froude number is not prescribed in advance.

While (20) is the only rigorous result we were able to obtain using such a crude approximation as the four terms in (16), we can probably conclude from the numerical data compiled during our derivation of (20) that the maximum of the Froude number does not exceed 0·25. Our best guess of the true value based on all the evidence available at present is

$$U/(gh)^{\frac{1}{2}} = 0{\cdot}24. \tag{21}$$

The Froude number has thus been located significantly higher than previously; this is due to the awareness that the equilibrium form of the steady bubble in an infinite tube is not unique and that a criterion of stability must be introduced in order to single out the physically significant flow. It is worth mentioning, finally, that these results call for a new evaluation of the numerical solutions that have been given for the axially symmetric problem, an evaluation which will probably lead to larger estimates of the bubble speed which agree better with experimental data.

References

Birkhoff, G. 1954 *Rep. Los Alamos Sci. Lab.* no. LA-1862.

Birkhoff, G. 1955 *Rep. Los Alamos Sci. Lab.* no. LA-1927.

Davies, R. M. & Taylor, Sir Geoffrey 1950 *Proc. Roy. Soc.* A, **200**, 375–390.

Dumitrescu, D. T. 1943 *Z. angew. Math. Mech.* **23**, 139–149.

Garabedian, P. R., McLeod, E. Jr. & Vitousek, M. 1954 *Amer. Math. Mon.* **61**, part II, 8–10.

Garabedian, P. R. & Spencer, D. C. 1952 *J. Rat. Mech. Anal.* **1**, 359–409.

Gibson, A. H. 1913 *Phil. Mag.* **26**, 952–965.

Layzer, D. 1955 *Astrophys. J.* **122**, 1–12.

Lewis, D. J. 1950 *Proc. Roy. Soc.* A, **202**, 81–96.

Lewy, H. 1950 *Univ. Calif. Publ. Math.* **1**, 247–280.

Lewy, H. 1952 *Proc. Amer. Math. Soc.* **3**, 111–113.

Taylor, Sir Geoffrey 1950 *Proc. Roy. Soc.* A, **201**, 192–196.

Vitousek, J. 1954 *Tech. Rep. Stanford Univ.* no. 25.

Premixed Flames

On the Theory of Slow Combustion

By ***L. Landau***

Physical theories of slow combustion are usually based on the conception that the transfer of heat to the unburned gaseous mixture from the products of combustion, heated owing to the reaction, is a result of simple thermal conduction. The thickness of the «combustion layer» separating the region of the initial mixture from that of the products of combustion is then determined completely by thermal conduction in the gas and by purely chemical properties of the given reaction. It is essential that this thickness is independent of the characteristic dimensions of the problem (dimensions of bodies in contact with the gas as, *e. g.* diameter of the tube if combustion occurs in the latter).

We shall assume the characteristic dimensions of the problem to be large in comparison with the thickness of the combustion layer. Then in determining the hydrodynamical movement of the gas accompanying the combustion process, we may consider the whole transition layer as a surface separating the burned and unburned gas. In the considered «thermal conduction» regime of slow combustion the rate of propagation of the combustion is certainly small in comparison with the velocity of sound.

A necessary condition for the existence of the combustion regime just described is its stability. The analysis of this problem is the purpose of the computations presented below.

Let us choose a small portion of the combustion layer (considered as a discontinuity surface), which may be regarded as plane, and consider the motion of the gas near it. We shall select a coordinate system with respect to which this is at rest (the plane of the latter will be taken as plane yz). The unperturbed movement of the gas is stationary and the velocity of the gas is directed normally to the discontinuity surface (the positive direction of the x axis is chosen so as to coincide with that of the motion of the gases).

Reprinted from *Acta Physicochimica U.R.S.S.*, Vol. XIX, No. 1, 1944.

The unperturbed velocity of the gas before the discontinuity (unburned gas, $x<0$) will be denoted by v_1 that of the gas behind the discontinuity (burned gas, $x>0$) by v_2. Inasmuch as v_1 and v_2 are small compared with the velocity of sound, we may regard the gas as incompressible. Suppose now that a small perturbation v′, periodical with respect to time and to the coordinate y, is superimposed on the unperturbed motion with constant velocities v_1, v_2. This perturbation may be determined from the continuity:

$$\frac{\partial v_x'}{\partial x}+\frac{\partial v_y'}{\partial y}=0 \tag{1}$$

and Euler's equations

$$\frac{\partial v_x'}{\partial t}+v\frac{\partial v_x'}{\partial x}=-\frac{1}{\rho}\frac{\partial p'}{\partial x}; \qquad \frac{\partial v_y'}{\partial t}+v\frac{\partial v_y'}{\partial x}=-\frac{1}{\rho}\frac{\partial p'}{\partial y} \tag{2}$$

v and ρ designating v_1, ρ_1, or $v_2\ \rho_2$ according to the gas for which the equations are written. Differentiating the first equation with respect to x, the second with respect to y and adding we obtain

$$\frac{\partial^2 p'}{\partial x^2}+\frac{\partial^2 p'}{\partial y^2}=0. \tag{3}$$

The boundary conditions to be satisfied on the discontinuity surface are the following. Firstly, the tangential velocity must be continuous. Let $\zeta\ (y,\ t)$ be a small displacement of the discontinuity surface along the x axis during the perturbation; $d\zeta/dy$ will be its angle of inclination to the y axis. The component of the gas velocity tangent to the discontinuity surface consist of two parts—the projection of the velocity v′ on this surface, which, in first approximation, equals simply v_y', and the projection of the unperturbed velocity $v_x=v$, which in the same approximation equals $v\frac{d\zeta}{dx}$. Thus the following condition must be fulfilled:

$$v_{1y}'+v_2\frac{\partial\zeta}{\partial y}=v_{2y}'+v_2\frac{\partial\zeta}{\partial y} \tag{4}$$

(for $x=0$).

Further, incompressibility of the gas means, in particular, that those changes in pressure which arise during motion are so small that the variations in the rate of combustion due to them can be neglected. In other words, the velocity of propagation of the discontinuity should remain constant during the perturbation.

In our system of coordinates this means that variations in the relative velocities of the gas and the discontinuity surface must be zero. We obtain, therefore, $v'_x - \frac{d\zeta}{dt} = 0$ and arrive at the condition (for $x=0$)

$$v'_{1x} = v'_{2x} = \frac{\partial \zeta}{\partial t} . \tag{5}$$

Finally, in the same approximation the pressure on both sides of the discontinuity surface should be identical and this yields the last condition (for $x=0$):

$$p'_1 = p'_2. \tag{6}$$

We shall look for a solution of equations (1)—(3) in which p', v'_x, v'_y are proportional to the factor $e^{iky+\Omega t}$. From (3) we find that in the zone of gas 1 ($x<0$)

$$p'_1 = \text{const}\, e^{iky+kx}\, e^{\Omega t} .$$

Writing v'_x and v'_y in the same form and choosing the coefficients so that equations (1)—(3) be satisfied, we obtain

$$v'_{1x} = Ae^{iky+kx+\Omega t}, \quad v'_{1y} = iAe^{iky+kx+\Omega t}, \tag{7}$$

$$p'_1 = -A\rho_1\left(\frac{\Omega}{k} + v_1\right) e^{iky+kx+\Omega t} .$$

In gas 2 ($x>0$), besides the solution in the form const $e^{iky-kx+\Omega t}$, another particular solution of equations (1)—(3), in which the dependence on y and t is determined by the same factor $e^{iky+\Omega t}$, should also be considered. This solution can be obtained by putting $p'=0$, then the right hand sides of equations (2) disappear and the remaining homogeneous equations have solutions for which v'_x and v'_y are proportional to $e^{iky+\Omega t-\frac{\Omega}{v}x}$. The reason for this solution being considered only for gas 2 consists in the following. Our ultimate purpose is to determine the possibility of existence of such values of Ω for which the real parts are positive. The existence of such values for Ω would signify the unstability of the basic movement. For such values of Ω the factor $e^{-\frac{\Omega}{v}x}$ would indefinitely increase with increasing $|x|$ if $x<0$ and, therefore, this solution must be rejected in gas 1. Correspondingly choosing the constant coefficients we shall look for the solution for $x<0$ in the form

$$v'_{2x} = Be^{iky-kx+\Omega} + Ce^{iky+\Omega t-\frac{\Omega}{v_2}x},$$

$$v'_{2y} = -iBe^{iky-kx+\Omega t} - \frac{i\Omega}{kv_2} e^{iky+\Omega t-\frac{\Omega}{v_2}x}, \tag{8}$$

$$p'_2 = -B\rho_2\left(v_2 - \frac{\Omega}{k}\right) e^{iky-kx+\Omega t}.$$

Putting

$$\zeta = De^{iky+\Omega t} \tag{9}$$

and substituting these expressions in (4)—(6) we obtain four homogeneous equations for the coefficients A, B, C, D. A simple calculation leads to the following condition for the consistency of these equations

$$\Omega^2(v_1+v_2) + 2\Omega v_1 v_2 k + k^2 v_1 v_2 (v_1 - v_2) = 0 \tag{10}$$

(in computing it should be remebered that $(\rho_1 v_1 = \rho_2 v_2)$. If $v_1 > v$ this equation will have either two negative real roots or two complex conjugated roots with negative real parts. If $v_1 < v_2$ both roots will be real and have different signs. Thus for $v_1 > v_2$ we shall always have $\mathrm{Re}\{\Omega\} < 0$ and the motion is stable. However, if $v_1 < v_2$ there will be such values of Ω for which $\mathrm{Re}\{\Omega\} > 0$, *i.e.* the primary motion will be unstable. Since $\rho_1 v_1 = \rho_2 v_2$ it then follows from $v_1 < v_2$ that $\rho_1 < \rho_2$. We thus arrive at the conclusion that the discontinuity surface under consideration is unstable if

$$\rho_1 > \rho_2. \tag{11}$$

Owing to considerable heating during combustion the density of the combustion products (ρ_2) is practically always less than that of the initial gas mixture (ρ_1), so that condition (11) is practically always fulfilled and the combustion regime considered will be unstable.

Inasmuch as the viscosity of the gas is neglected in the computations, the result just obtained will be valid in those cases for which Reynold's numbers $\frac{av_1}{\nu_1}$. $\frac{av_2}{\nu_2}$ are large compared to unity (a is characteristic for the dimensions of the system; $\frac{1}{a}$ determines the minimum value of the wave vector k; ν_1, ν_2 are the kinematic viscosities of the initial gas mixture and combustion products, respectively. Thus a stationary regime of slow combustion can exist only under such conditions of motion which essentially depend on the viscosity and, consequently, also on the heat conduction at the gases. Generally speaking, the «combustion layer» cannot in this case be considered thin. Our results are thus in conflict with the usual point of view.

Unstability of the «discontinuity surface» should lead to turbulence and instead of a narrow transition layer there will be a diffuse combustion zone in which the movement of the gas will be turbulent. As usual turbulence will cause an intense convective mixing of the gas. This intermixing causes a transfer of heat, the velocity of which considerably exceeds that due to true thermal conduction. The propagation of combustion is, therefore, determined not by thermal conduction in the gas, as was assumed above, but by turbulent convection processes; one may speak of a «convective» regime of combustion.

It would now be incorrect to assert that the velocity of propagation of combustion and the width of the combustion zone are independent of the characteristic dimensions of the problem since turbulent motion depends on these dimensions in an essential manner.

Let us examine the burning of a gas in a long tube. The burned gas is separated from the unburned gas by a turbulent «combustion zone», which gradually moves forward in the tube. Let l be the order of magnitude of the combustion zone thickness and u — the combustion propagatlon velocity. The dependence of l and u on the diameter of the tube d can easily be determined directly from dimensional considerations. Inasmuch as thermal conduction and viscosity are not important for turbulent motion (assuming Reynold's numbers to be large) the coeffcients of viscosity and heat conduction cannot enter l or u. Further, the purely chemical kinetics of the reaction is characterized by a certain quantity of the dimension of time, which determines the velocity of the reaction. The only parameter of the dimension of length is the diameter of the tube d. From considerations of dimensions it is thus evident that

$$l \sim d, \tag{12}$$

i. e. the width of the combustion zone should be proportional to the diameter of the tube. The following relation then exists for the combustion velocity

$$u = \frac{d}{\tau} \qquad (13)$$

where τ is the time characterisic for the given reaction. The velocity of propagation of the combustion along the tube is thus proportional to the diameter of the latter. The following point should be noted. Since u is now dependent on the diameter of the tube, which is an arbitrary quantity, it cannot be asserted that the combustion velocity is always small as compared with the velocity of sound c. However it must be remembered that the relations (12) and (13) are valid only if $u \ll c$. Otherwise the gas could not be considered incompressible and there would be no reasons to assume that l and u cannot depend on the compressibility of the gas (as it was tacitly assumed above). Moreover, these results are true for tubes that are not very narrow *viz.* Reynolds numbers $\mathrm{Re} \sim \frac{ul}{\nu}$ must be sufficiently large.

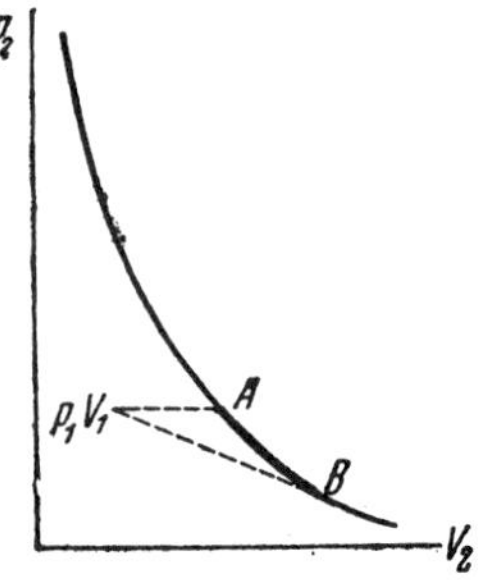

Fig. 1.

For the further investigation of the properties of slow combustion it will be more convenient to make use of a graphical representation with the aid of Hugoniot's adiabatic curve. The latter, drawn for a definite combustion process is shown in Fig 1 (p_2, V_2 are the pressure and specific volume of the combustion products; p_1, V_1—the corresponding quantities for the initial gasr mixture). Only the part AB of the curve which actually correspond to slow combustion will be of interest to us (the straight line drawn from the point p_1,V_1 to A is parallel to the axis of abscissae and that drawn to point B is tangent to the adiabatic curve). Well known properties of Hugoniot's adiabatic curve easily yield the result that on AB the propagation velocities of the combustion layer (considered as a discontinuity surface) relative to the burned and unburned gases,v_1 and v_2 respectively, are less than the corresponding velocities of sound c_1 and c_2 ($v_1 < c_1$, $v_2 < c_2$).

The combustion velocity can be described by the «material flow» $j = \rho_1 v_1 = \rho_2 v_2$ through the discontinuity surface or, in other words; by the amount of matter burned in one second per unit area of the surface. In point A the flow j and the velocities v_1, v_2 are zero; j increases along the direction from A to B. For a given reaction the material flow j depends also on the diameter of the tube (to be more definite, let us consider the combustion in a tube), increasing with increasing diameter, which corresponds to displacement along the adiabatic curve from A towards B.

It can be shown that in these conditions no combustion regime will correspond to the points on the adiabatic curve lying below point B. The discontinuity surface may possess a peculiar type of instability as a result of which sound waves will be radiated spontaneously; obviously, no such discontinuity can exist in reality. The «number» of various sound waves which could be radiated from the discontinuity is mainly limited by purely geometrical circumstances, *viz.* by the relative magnitudes of velocities v_1, v_2, and the velocities of sound c_1, c_2; this relation should be such that the discontinuity could not during its motion «overtake» the radiated wave. For $v_1 < c_1, v_2 < c_2$ (which corresponds to the section AB on the adiabatic curve) two waves could be radiated — one of them from the discontinuity towarls gas 1 and the other in the direction of gas 2. For $v_1 < c_1, v_2 > c_2$ (corresponding to the part of the adiabatic curve lying below point B) three waves are possible: one moving from the discontinuity into gas 1 and two moving in gas 2, one of them towards and the other away from the discontinuity surface. Besides its purely geometrical possibility, another necessary condition for the radiation to exist in fact is that the corresponding boundary conditions on the discontinuity surface itself may be satisfied. Computations show that two waves are not sufficient for this purpose; three waves are, however, enough to satisfy these boundary conditions. Thus the discontinuities corresponding to points lying below point B are unstable and therefore this part of the adiabatic curve does not in general correspond to any real discontinuities whatsoever.

Point B corresponds to a certain «limiting» regime of slow combustion. As to the combustion of gases in tubes with diameters larger than that corresponding to point B, the only possible stable regime will be detonation.

The usual equation of the Hugoniot adiabatic curve and the fact that the straight line drawn from p_1, v_1 to point B is tangent to the curve enable one to obtain easily an expression for the

*6

limiting velocity of propagation of «slow combustion». Calculations which will not be given here yield the following result

$$v_1 = \sqrt{\frac{\gamma_2 - 1}{2}[(\gamma_2+1)q+(\gamma_1+\gamma_2)c_{v_1}T_1]} - \sqrt{\frac{\gamma_2+1}{2}[(\gamma_2-1)q+(\gamma_2-\gamma_1)c_{v_1}T_1]} \tag{14}$$

where q is the heat of reaction per unit mass; c_v — the heat capacity at constant volume, $\gamma = \frac{c_p}{c_v}$, T — the temperature. If the heat of reaction is great ($q \gg c_{v_1}T_1$), as is usually the case, this formula can be simplified to

$$v_1 = \frac{\gamma_2(\gamma_1 - 1)c_{v_1}T_1}{\sqrt{2(\gamma_2^2 - 1)q}}. \tag{15}$$

The pressure and temperature of the gaseous product of combustion will be

$$p_2 = p_1 \frac{1}{\gamma_2+1}, \qquad T_2 = \frac{2q}{\gamma_2(\gamma_2+1)c_{v_2}}. \tag{16}$$

Let us now discuss the combustion of liquids. Only reactions of spontaneous decomposition will be considered, *i. e.* reactions which take place without the interference of foreign substance as, for instance, of oxygen from air. Combustion takes place in the layer of saturated vapour over the surface of the liquid. If the whole combustion zone be considered as a discontinuity surface, separating the initial liquid (density ρ_1) from the gaseous products of combustion (density ρ_2), a surface tension coefficient α must be ascribed to this surface, which should equal the coefficient of surface tension at the boundary between the liquid and its saturated vapour. When investigating the stability of such a combustion regime the gravitation should also be considered.

All the computations for this case are exactly similar to those carried out for combustion in gaseous mixtures, except of condition (6) (equality of pressures) which must be replaced by the condition

$$p_2' - p_1' = -g(\rho_2 - \rho_1)\zeta - \alpha \frac{\partial^2 \zeta}{\partial y^2}. \tag{17}$$

Substituting expressions (7)—(9) into conditions (4), (5) and (17) we again obtain four equations for the coefficients A, B, C, D. Instead of (10) the condition of consistency of these equations will be given by the following quadratic equation:

$$\Omega^2 (v_1 + v_2) + 2\Omega\, k v_1 v_2 + k^2 v_1 v_2 (v_1 - v_2) - kg (v_1 - v_2) - \alpha k^3 \frac{v_1 v_2}{j} = 0.$$

For stability of the motion it is necessary that $\mathrm{Re}(\Omega) < 0$, *i. e.* the equation must have either two negative real roots or two complex-conjugated roots with negative real parts. This will be the case if the free term in equation (16) is positive:

$$\alpha k^2 \frac{v_1 v_2}{j} - k v_1 v_2 (v_2 - v_1) + g (v_2 - v_1) > 0. \qquad (18)$$

On the other hand, this inequality will be true for any positive k if

$$v_1 v_2 (v_1 - v_2)^2 + 4\alpha \frac{v_1 v_2}{j} g (v_1 - v_2) < 0.$$

Inasmuch as $v_2 > v_1$, (which follows from $j \equiv \rho_1 v_1 = \rho_2 v_2$ since $\rho_1 > \rho_2$, *i. e.* the density of the liquid is greater than that of the vapour) the inequality above will be equivalent to the following one

$$\frac{4\alpha g}{j} - v_1 v_2 (v_2 - v_1) > 0.$$

Finally, expressing v_1 and v_2 in terms of the material flow j (amount of matter burned per unit time) we obtain the following inequality for the condition of stability of the combustion regime in the liquid

$$j^4 < \frac{4\alpha g \rho_1^2 \rho_2^2}{\rho_1 - \rho_2} \cong 4\alpha g \rho_1 \rho_2^2. \qquad (19)$$

In this form this inequality is referred to an infinite liquid surface for which vibrations of any wave length are possible. For liquids contained in small vessels (19) is not a necessary condition for the stability of slow combustion. A sufficient condition is that for all k greater than a certain value of the order of $1/a$, where a is the dimension of the vessel, expression (18) could not be negative. Accordingly, if condition (19) is not fulfilled a necessary condition for stability will be the inequality $a < a_0$ where

$$a_0 \sim \frac{\alpha}{j (v_2 - v_1)} \sim \frac{\alpha}{j v_2} = \frac{\alpha \rho_2}{j^2} \qquad (20)$$

Academy of Sciences of the USSR,
Institute for Physical Problems.

Received
November 26, 1943.

Experimental and Theoretical Studies of Flame-Front Stability*

GEORGE H. MARKSTEIN†
Cornell Aeronautical Laboratory, Inc.

ABSTRACT

Rich hydrocarbon flames burning in tubes were found to assume a cellular structure in the absence of turbulence in the approach stream. The effects of type of fuel, mixture composition, and pressure on the phenomenon have been studied. A first-order perturbation treatment of flame-front stability gave results in qualitative agreement with experiments when the dependence of burning velocity on flame-front curvature was taken into account.

NOTATION

a = acceleration
$A_1 \ldots A_4$ = constants
c_p = specific heat at constant pressure
d = average cell size of cellular flames
D_i = diffusion coefficient of ith species
F = $aL/(S_u^0)^2$
$f(x)$, $g(y, t)$ = functions
h = $2\pi/\lambda$ = wave number
i = $\sqrt{-1}$
k = thermal conductivity
L = characteristic length of order of flame front thickness
m = $\rho_u^0 S_u$ = mass flow
M = molecular weight of fuel
M_i = molecular weight of ith species
n_i = concentration of ith species
N_{ij} = $M_i(\nu_{ij}/\rho D_i)$
p = pressure
Q_j = heat released in jth reaction
r = radial coordinate
R = radius of curvature of flame front
S_u = burning velocity
t = time
T = absolute temperature
u, v = velocity components
w_j = rate of jth reaction
x, y = coordinates
X = x/L
Y_0 = T
Y_i = n_i, $i = 1, \ldots, n$
α = $\delta/u_b^0 h$
β_i = $k/c_p \rho D_i$
γ = $a/h(S_u^0)^2$
δ = stability parameter
ϵ = ρ_u^0/ρ_b^0
φ = angle between tangent of flame front and y-axis
χ = $Lh = 2\pi L/\lambda$
ν_{ij} = number of molecules of ith species generated in the jth reaction (ν_{ij} is negative for those consumed)
μ = parameter that determines curvature dependence of burning velocity
λ = wave length
$\lambda_{max.}$ = wave length for maximum instability
ρ = density
σ = $\mu L h$
τ = parameter that determines curvature dependence of temperature of burned gas
ξ, η = coordinates of an element of flame front
$\propto$ = proportional to

Subscripts

u = unburned
b = burned

Superscripts

0 = zero order
1 = first order

INTRODUCTION

RECENT STUDIES OF FLAME PROPAGATION in high-velocity gas streams[1] have shown that the intensity of turbulence in the combustion zone, computed from the observed flame speeds on the basis of existing theories of turbulent flame propagation,[2, 3] exceeded the level of turbulence in the approach stream by more than an order of magnitude. The high-intensity flow disturbances were assumed to originate in shear flow instabilities in the burned gas.

However, based on experimental studies of the effects of artificial disturbances on laminar Bunsen burner flames, instability of the flame front itself as a source of flow disturbances has been suggested.[4]

Subsequently, the observation of spontaneous breakdown of rich propane flames into a number of separate cells[5] in the absence of turbulence in the approach stream provided direct evidence for instability of flame fronts. This phenomenon was later found also to occur in many other hydrocarbon fuels. It appeared thus to be of sufficiently general significance to justify a detailed investigation. The experimental part of the present paper is primarily concerned with results of this study obtained with slow-burning mixtures of air, nitrogen, and hydrocarbons.

Various authors have described phenomena, such as polyhedral[6–9] and truncated pyramid[8] burner flames,

---

Presented at the 1950 Heat Transfer and Fluid Mechanics Institute, Los Angeles, June 28–30, 1950. Received August 24, 1950.

* This work was supported by the Office of Naval Research and the Office of Air Research through Project SQUID. The author is indebted to L. M. Somers for assistance with the experimental work and to H. Einbinder and W. Squire for valuable discussions in connection with the theoretical work.

† Head of Combustion Section.

FIG. 1. Setup for study of cellular flames.

rising hemispherical flamelets,[10] filamentous and corrugated flame fronts,[7] and distorted flame fronts in spherical bomb explosions,[9] which seem closely related to cellular flame structure. The majority of these workers[7, 9, 10] and others[11] favored an interpretation in terms of diffusion processes, while flow phenomena were considered to be only incidental. On the other hand, the only previous theoretical study of flame-front stability[12] was based exclusively on hydrodynamic reasoning. In the theoretical part of the present paper, it is shown that, under certain simplifying assumptions, the physicochemical properties of the combustion zone may be introduced formally into the hydrodynamic treatment in the form of appropriate boundary conditions at the flame front.

EXPERIMENTS

Apparatus and Procedure

The experiments were performed in vertical Pyrex glass tubes, open at the top and closed at the bottom except for the gas inlet. The flow velocity of the premixed gases was chosen low enough to ensure laminar flow in the tube; the range of Reynolds Numbers was about 400 to 600. The mixture compositions were selected so that the flames remained almost stationary within the tube, about 2 to 3 cm. below the top. In order to accomplish this over a wide range of fuel concentrations, the burning velocity was adjusted by adding suitable amounts of nitrogen. The proper compositions were found by starting with rich mixtures that gave diffusion flames burning on top of the tube and then gradually reducing the fuel concentration until the flames entered the tube.

Average cell sizes of the cellular flames[5] were determined by counting the number of cells over the whole tube cross section on flame photographs. Average cell size was computed as tube inner diameter divided by the square root of the number of cells. The flame photographs were taken under an oblique angle from below.

In order to obtain good cell-size averages, it appeared desirable to take a fairly large number of photographs during each run, particularly since the flames were frequently distorted by downdrafts from the top of the tube. This was done by taking motion pictures instead of stills; good exposures were obtained with a rate of 8 frames per sec. and f/2.3 on 35-mm. Eastman Linagraph Pan film.* The experimental arrangement, consisting of a flame tube of 15.25-cm. inner diameter and the movie camera, is shown in Fig. 1. The shield on top of the tube served to protect the flame against drafts. The figure shows the setup in the Altitude Chamber for work at reduced pressures. Ignition at low pressures was accomplished by a pilot flame burning from the nozzle visible on the right of the figure.

Results

With the exception of methane, all the hydrocarbons that were investigated gave cellular flames on the rich side of stoichiometric composition.† The phenomenon was, in general, already visible in rich flames that burned on top of the tube as one or more approximately hemispherical cusps, convex toward the unburned gas, near the apex of the "inner cone." As soon as the flames entered the tube, the whole primary combustion zone disintegrated into these cusps or cells. They were generally of fairly uniform size, convex toward the unburned gas and separated by dark narrow ridges, and filled the tube cross section in a single horizontal layer. The cells were always in incessant irregular motion, which was too fast to be analyzed visually or on the movies taken at low frame rate. High-speed shadowgraph movies taken axially through the tube, which was, for this purpose, closed by a glass window at the bottom, revealed that the larger cells grew continuously at the expense of their smaller neighbors and split when they reached excessive size.

Modifications of this structure were observed when the tube walls became heated in the course of consecu-

* The author is indebted to F. Stratton for suggesting this photographic technique.

† With methane, extremely shallow cells were occasionally observed in lean mixtures.

tive runs. Moderately hot tube walls gave a structure characterized by a ring of larger cells somewhere between the wall and the center of the tube. Extremely hot walls gave a structure of a few large cells arranged in a symmetrical pattern and rotating around the tube axis. These phenomena have not yet been studied in detail. They were avoided during the determinations of average cell size by inserting cooling periods of at least 30 min. between runs.

Cellular flame structures for various compositions and fuels differed not only in average cell size but also in what may be called their "strength." This property can be described at present only qualitatively. In strong structures, the cells were almost hemispherical, completely separated by dark ridges, and in lively motion. Weak structures had shallow cells, the separating ridges were often almost as luminous as the cells, and the motion was more sluggish. Cell size was more uniform in weak structures.

Extremely rich cellular flames, obtained without or with small amounts of added nitrogen, were characterized by yellow radiation due to carbon formation, which was emitted partly by the outer mantle above the tube and partly in vertical streaks or bands that originated in the ridges between the cells. This phenomenon has also been observed by others[7] and is analogous to the yellow radiation above the "open tip" of rich hydrocarbon flames. Except with benzene (the only aromatic hydrocarbon studied), addition of moderate amounts of nitrogen and corresponding reduction of fuel concentration suppressed the yellow radiation completely. In benzene, however, it persisted almost until stoichiometric composition was reached.

The strength of the structures appeared to remain constant over a wide range of rich compositions. As stoichiometric compositions were approached, the strength decreased rather abruptly within a narrow range of fuel concentrations, until the structure disappeared completely and the flame assumed the shape of a smooth disc. The limit compositions at which cell structure disappeared were determined for all the fuels investigated except for the higher hydrocarbons that are liquid at normal pressure and temperature. It was found that the limit air-fuel ratios were independent of the amount of nitrogen added within the range over which the latter could be varied without causing flashdown or blowout of the flame. Limit fuel concentrations in terms of stoichiometric concentration are given in Table 1.

The deviations from stoichiometric composition were within the errors of measurement. Thus, it seems plausible that the transitions took place exactly at stoichiometric composition.

Contrary to a statement in the preliminary note,[5] it was found that, with each fuel, average cell size varied somewhat with composition. However, for fuel concentrations within the range of about 1.20 to 1.40 times stoichiometric, average cell size remained practically constant. This range, which also gave flame photographs of best quality, was therefore chosen for comparing various fuels and for studying the effect of pressure on average cell size. Beyond this range, cell size increased gradually toward richer compositions. In the case of *n*-butane, this increase was from 1.07 to 1.28 cm. between 1.40 and 1.80 times stoichiometric fuel concentration.

The appearance of cellular flames of various fuels in a tube of 10 cm. inside diameter at atmospheric pressure is shown in Fig. 2. These photographs were taken with fuel concentrations of about 1.20 times stoichiometric and about 30 per cent added nitrogen in the atmosphere. They show that ethylene and ethane gave comparably weak structures. The flames of the

TABLE 1
Limit Compositions of Cellular Structure

Hydrocarbon	Fuel Concentration, Fraction of Stoichiometric
Ethylene	0.983
Ethane	1.030
Propylene	0.970
Propane	1.012
Butadiene 1,3	1.019
Butene 2	1.042
Isobutylene	1.020
N-butane	1.002
Isobutane	0.997

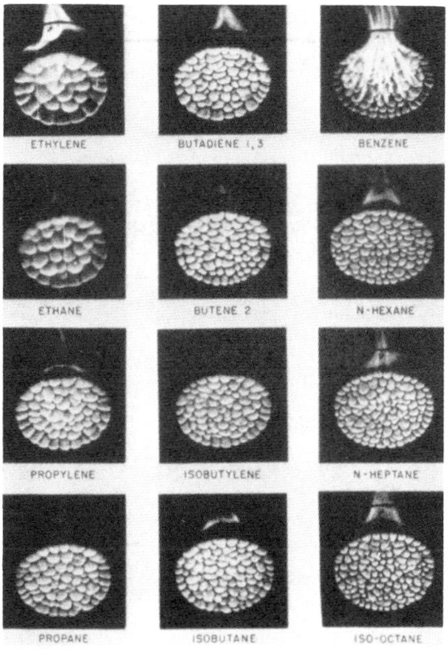

FIG. 2. Cellular flames in hydrocarbon-air-nitrogen mixtures at atmospheric pressure. Tube inner diameter, 10.0 cm.

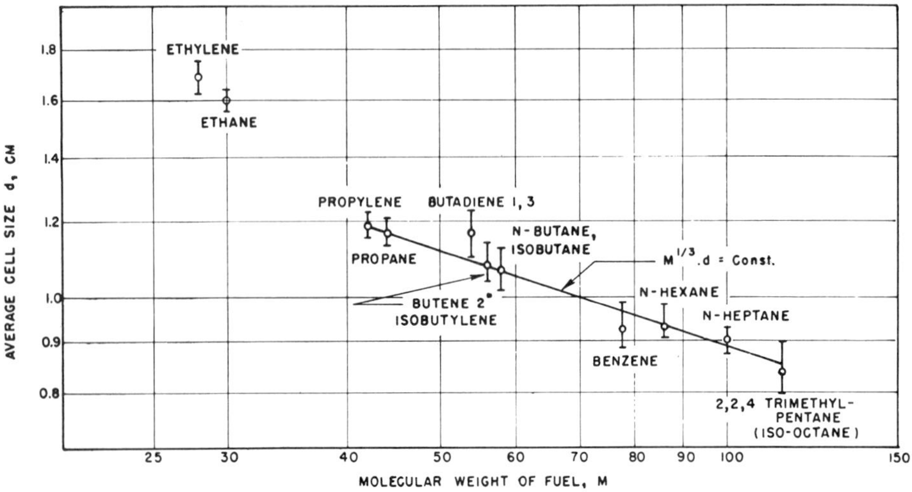

FIG. 3. Average cell size of hydrocarbon-air-nitrogen flames at atmospheric pressure vs. molecular weight of fuel.

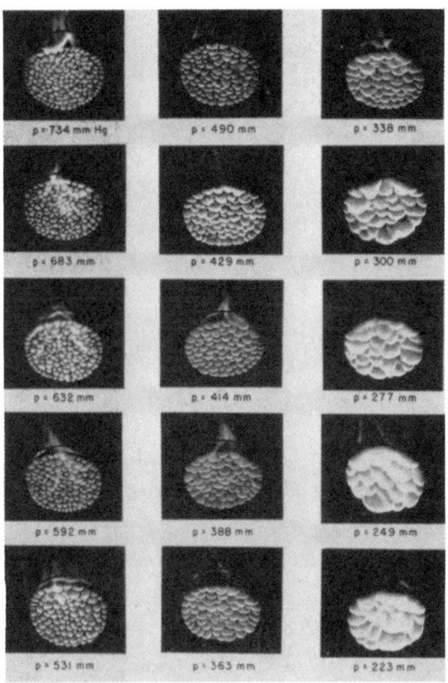

FIG. 4. Cellular flames in *n*-butane-air-nitrogen mixtures at various pressures. Tube inner diameter, 15.25 cm.

other fuels did not differ appreciably with respect to strength of structure. The photograph of the benzene flame clearly shows the streaks of yellow radiation, although it was taken with a dark-blue filter.

Average cell size decreased markedly with increasing number of carbon atoms and decreased slightly with increasing number of hydrogen atoms in the fuel molecule. No difference between straight-chain and branched molecules was found. A logarithmic plot of average cell size vs. molecular weight of fuel, Fig. 3,* showed that, with few exceptions, the data correlated remarkably well with the relation $M^{1/3}d$ = const. Excluding the values for ethylene, ethane, butadiene 1,3, and benzene, $M^{1/3}d$ values differed from their mean by only 1.7 per cent. No theoretical explanation for this empirical result can be given at this time.

The effect of reduced pressure on average cell size was studied by conducting the experiments in the Cornell Aeronautical Laboratory altitude chamber. *N*-butane was selected as fuel. Fig. 4 shows the appearance of the flames in a tube of 15.25 cm. inside diameter at various pressures. A logarithmic plot of average cell size vs. pressure, Fig. 5, showed good agreement of the data obtained with tubes of 10.0 and 15.25 cm. inside diameter and indicated good correlation with the relation $p^{3/4}d$ = const. The deviations of the values for the lowest pressures were not surprising, since considerable difficulty to stabilize the flames and obtain good photographs was experienced at these pressures. Furthermore, because of the small numbers of cells at low pressures, wall effects may have falsified the results. Significantly, the deviation started at a higher pressure with the 10.0-cm. tube than with the larger one. The significance of the result $p^{3/4}d$ = const. will be discussed in the theoretical part.

* In Figs. 3 and 5, the circles and crosses represent average values calculated from at least five photographs; the vertical lines indicate the spread of values from individual photographs.

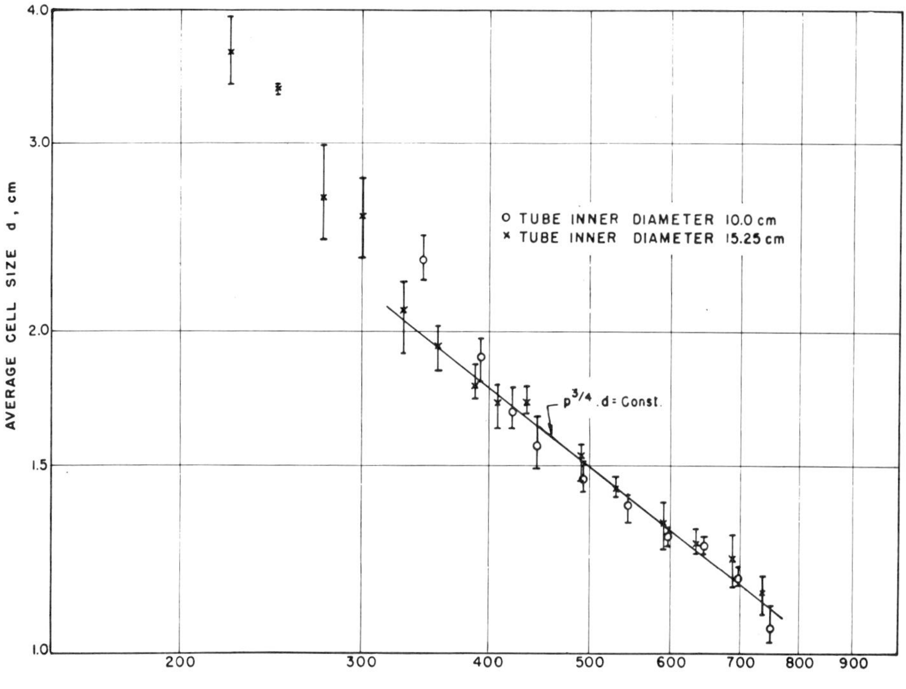

FIG. 5. Average cell size of *n*-butane-air-nitrogen flames vs. pressure.

Exploratory experiments were performed with gas mixtures containing a fourth component in addition to hydrocarbon fuel, oxygen, and nitrogen. It was found that addition of small amounts of hydrogen to *lean* mixtures of all hydrocarbons, including methane, caused strong cell formation. This is in agreement with Behrens' findings.[7] Ammonia caused formation of weak large cells in lean methane mixtures but caused an increase of cell size and decrease of strength if added to rich mixtures of higher hydrocarbons. Carbon monoxide had a similar effect on rich cellular flames. Use of carbon dioxide instead of nitrogen as diluent gave slightly smaller cells in butane-air mixtures. Helium as diluent gas gave larger cells and suppressed cell formation completely if added in larger amounts.

Studies of flame structure in fast-burning mixtures by means of high-speed shadowgraph and schlieren movie techniques are under way. A detailed discussion of their results is beyond the scope of this paper. The work done hitherto has indicated that there is a close connection between the phenomenon of vibratory flame motion[13] and cellular flame structure. Regardless of the type of combustible mixture, cells were found to appear and disappear periodically in the rhythm of the oscillation during the vibratory stage. Fig. 6 is a sequence of high-speed shadowgraph movie frames, taken axially through the flame tube, which show this phenomenon. However, a significant effect of composition was found: In those mixtures that were originally capable of forming cells, vibrations started at an earlier stage of flame travel and with a higher harmonic of the tube, while in originally noncellular mixtures, they started later and with the fundamental frequency.

It can hardly be doubted that the phenomenon of cell structure, observed to occur in the absence of stream turbulence, must be interpreted as an instability of the flame front. Flow phenomena undoubtedly play an important role; the resemblance to the phenomenon of cellular convection[14,15] is striking. In analogy to that phenomenon, one may regard each cell as a vortex ring, which, in the case of the flame front, however, is superposed over the uniform flow through the flame front. In the following theoretical treatment, it is shown that at least qualitative agreement with experimental results can be obtained by considering both the hydrodynamic and physicochemical aspects of the phenomenon.

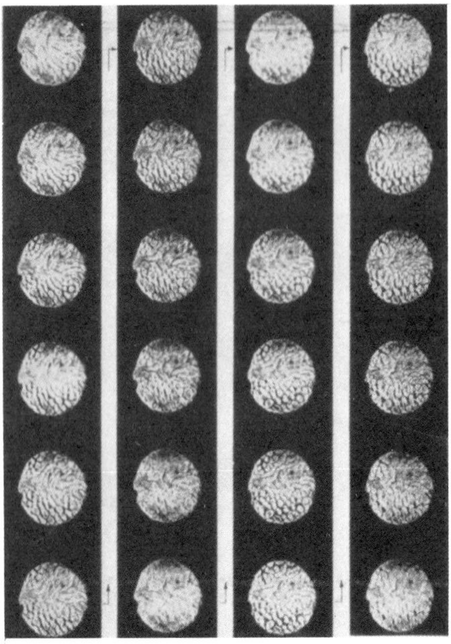

FIG. 6. High-speed shadowgraph movie of vibratory flame motion in 10.0-cm. inside diameter tube. Approximately 1,500 frames per sec.

THEORY

Various problems of stability of phenomena that are governed by nonlinear partial differential equations have been attacked successfully by the method of first-order infinitesimal perturbations. Typical examples of applications of this method are stability problems of parallel viscous flows[16] and of cellular convection.[17-19]

A rigorous formulation of the problem of stability of a plane flame front[20] must start with the Eulerian equations, the equation of continuity, the equation of energy balance, and one diffusion equation for each chemical species participating in the reaction. Perturbations of velocity components, pressure, temperature, density, and concentrations of chemical species must be considered.

A difficulty peculiar to this problem arises from the fact that even the zero-order problem—i.e., the calculation of the burning velocity of a steady plane flame front—has, in general, not yet been solved successfully. This is primarily due to lack of knowledge of the mechanisms of combustion reactions. Apart from mathematical difficulties, the presence of reaction-kinetic terms that form an integral part of the rigorous formulation thus constitute a serious obstacle for a successful solution.

Landau[12] evaded this difficulty by disregarding the physicochemical aspects of the problem completely. He regarded the flame front as a surface of discontinuity separating regions of constant steady-state velocity, density, and temperature. Perturbations of velocity and pressure only were considered in this purely hydrodynamic treatment. Thus, it is not surprising that the results obtained by Landau disagree with the experimental results given above. While he found instability for all flames (inasmuch as $\rho_u > \rho_b$ in all normal flames), the experiments gave limit compositions beyond which the flames were stable. Furthermore, he predicted a linear increase of instability with the wave number $2\pi/\lambda$, whereas the observed uniformity of cell size suggests a maximum of instability for a certain wave number.

The treatment presented below occupies an intermediate position between a rigorous treatment and Landau's. The relative simplicity of Landau's approach has been retained by restricting the validity of the treatment to wave lengths of the disturbance which are large compared to the thickness of the combustion zone. It appears permissible, then, to assume that the flow disturbance causes merely a bending of the reaction zone without affecting its internal structure directly.

Thus, the flame front may be still regarded as a surface of discontinuity, as far as the hydrodynamic part of the treatment is concerned. The physicochemical processes taking place within the reaction zone were, however, no longer disregarded; they were incorporated into the hydrodynamic treatment by the boundary conditions at the flame front. A completely determined set of boundary conditions is obtained by specifying at each element of the flame front the local instantaneous burning velocity and the temperature rise through the flame front. In view of the above restriction on the wave length of the disturbance, it was assumed that, at least in first order, local instantaneous burning velocity and temperature rise are functions of the local instantaneous curvature of the flame front only. Thus, one may write

$$S_u = S_u^0[1 + \mu(L/R) + \ldots] \qquad (1)$$

and

$$T_b = T_b^0[1 + \tau(L/R) + \ldots] \qquad (2)$$

where S_u is the burning velocity and T_b is the temperature of the burned gas, respectively, S_u^0 and T_b^0 are their values for the plane flame front, R is the radius of curvature of the flame front, and L is a characteristic length of the order of the thickness of the combustion zone. For the purpose of the first-order perturbation treatment, only the linear terms in the expansions need be retained.

The remaining difficult problem consists of calculating the parameters μ and τ. Except under special assumptions, this cannot be done at present because of lack of knowledge of the reaction mechanisms. In the Appendix, a general formulation of this problem is given, and its solution for a special case is indicated.

The hydrodynamic treatment, on the other hand, can now be carried out by introducing expressions (1) and (2) formally without further reference to chemical kinetics, diffusion, and heat conduction phenomena in the reaction zone. In order not to obscure the physical significance of the treatment by too lengthy derivations, the further assumption $\tau = 0$ has been made in the following: This corresponds to the assumption of complete combustion, since, in this case, the temperature of the burned gas must be independent of flame-front curvature. The more general case $\tau \neq 0$ offers no particular additional difficulties; its treatment will be published elsewhere.

The treatment follows Landau except for the boundary conditions. Thus, the steady-state flame front is taken as the y-z plane, separating the region of unburned gas $x < 0$ from the region of burned gas $x > 0$. The undisturbed flow velocities have the constant values u_u^0 and u_b^0 in the unburned and burned gases, respectively. Compressibility and viscosity are neglected. Thus, taking into account the continuity relation $\rho_u^0 u_u^0 = \rho_b^0 u_b^0$, the temperature ratio ϵ across the flame front is equal to the velocity and inverse density ratios

$$\epsilon = T_b^0/T_u^0 = u_b^0/u_u^0 = \rho_u^0/\rho_b^0 \tag{3}$$

Furthermore, in order to maintain a steady flame front, the conditions

$$u_u^0 = S_u^0, \quad u_b^0 = S_b^0 = \epsilon S_u^0 \tag{4}$$

must be fulfilled. The steady-state pressure drop across the flame front is given by

$$p_u^0 - p_b^0 = \rho_u^0 (S_u^0)^2 (\epsilon - 1) \tag{5}$$

Over this steady state, small time-dependent two-dimensional velocity perturbations u^1, v^1 and pressure perturbations p^1 are superposed. Neglecting second-order terms, the Eulerian and continuity equations for the perturbations become

$$\left.\begin{aligned} \frac{\partial u_u^1}{\partial t} + u_u^0 \frac{\partial u_u^1}{\partial x} &= -\frac{1}{\rho_u^0}\frac{\partial p_u^1}{\partial x} \\ \frac{\partial v_u^1}{\partial t} + u_u^0 \frac{\partial v_u^1}{\partial x} &= -\frac{1}{\rho_u^0}\frac{\partial p_u^1}{\partial y} \\ (\partial u_u^1/\partial x) + (\partial v_u^1/\partial y) &= 0 \end{aligned}\right\} \tag{6u}$$

for the unburned gas and an analogous system (6b) for the burned gas.

Following the usual procedure,[12, 16, 17] a particular solution periodic in y and exponential in t is sought; thus, all perturbations depend on y and t by

$$g(y, t) = \exp(ihy + \delta t) \tag{7}$$

where h is the wave number, connected with the wave length λ by

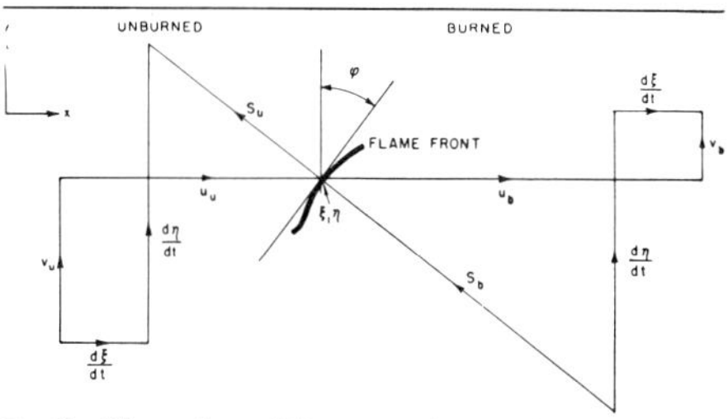

FIG. 7. Kinematic conditions at an element of the flame front.

$$h = 2\pi/\lambda \tag{7a}$$

and where i is the imaginary unit. The parameter δ plays the role of an eigenvalue of the problem. Stability of the system is determined by the real part of δ. If any nontrivial solutions of Eqs. (6), subject to certain boundary conditions, correspond to positive values of the real part of δ, the system is unstable.

Because Eqs. (6) have constant coefficients, the solutions are also exponential in x. By inserting solutions of the form of Eq. (7), it is easily seen that they must be of the form $\exp(hx)$, $\exp(-hx)$, or $\exp[-(\delta/u^0)x]$.

Of these solutions, only the first can be retained in the unburned gas and only the other two in the burned gas, since the perturbations must vanish at $x = \pm\infty$. Thus, with

$$\alpha = \delta/u_b^0 h = \delta/\epsilon S_u^0 h \tag{8}$$

The solutions of Eq. (6) can be written

$$\left.\begin{aligned} u_u^1/u_u^0 &= A_1 f_1(x) g(y, t) \\ v_u^1/iu_u^0 &= A_1 f_1(x) g(y, t) \\ p_u^1/\rho_u^0(u_u^0)^2 &= -(\alpha\epsilon + 1) A_1 f_1(x) g(y, t) \\ u_b^1/u_u^0 &= [A_2 f_2(x) + A_3 f_3(x)] g(y, t) \\ v_b^1/iu_u^0 &= [-A_2 f_2(x) - \alpha A_3 f_3(x)] g(y, t) \\ p_b^1/\rho_u^0(u_u^0)^2 &= (\alpha - 1) A_2 f_2(x) g(y, t) \end{aligned}\right\} \tag{9}$$

where

$$\left.\begin{aligned} f_1(x) &= \exp(hx) \\ f_2(x) &= \exp(-hx) \\ f_3(x) &= \exp(-\alpha hx) \end{aligned}\right\} \tag{9a}$$

The A_i are constants that have to be determined from the boundary conditions at the flame front. In order to establish the boundary conditions for the general case $S_u \neq S_u^0$, one starts best with the rigorous kinematic conditions at an element of the flame front, Fig. 7. By definition of the burning velocity, the element at point ξ, η propagates relative to the unburned gas with a velocity S_u normal to its surface and relative to the burned gas with velocity $S_b = \epsilon S_u$.

Thus, the components of velocity of motion of the element ξ, η relative to the coordinate system are

$$\left.\begin{aligned} d\xi/dt &= u_u - S_u \cos\varphi = u_b - S_b \cos\varphi \\ d\eta/dt &= v_u + S_u \sin\varphi = v_b + S_b \sin\varphi \end{aligned}\right\} \tag{10}$$

where φ is the angle between the tangent to the flame front and the y-axis.

In the first-order perturbation treatment, the displacements of the flame front are caused by the flow perturbations and are, thus, themselves small quantities. Therefore, one may write, with sufficient accuracy,

$$\cos \varphi \approx 1, \quad \sin \varphi \approx \varphi \approx \partial\xi/\partial y \tag{11}$$

The displacement ξ must depend on y and t in the same way as the other perturbations.

$$\xi = (1/h)A_4 g(y, t) \tag{12}$$

In Eq. (1) for the burning velocity, the radius of curvature may be approximated by*

$$1/R \approx -\partial^2\xi/\partial y^2 \tag{13}$$

The burning velocity is thus given by

$$S_u = S_u^0[1 + \sigma A_4 g(y, t)] \tag{14}$$

with

$$\sigma = \mu L h \tag{15}$$

Finally, the pressure drop across the flame front is given by

$$p_u - p_b = \rho_u^0 S_u^2(\epsilon - 1) \tag{16}$$

Inserting Eq. (9) for $x = 0$ and Eqs. (11), (12), and (14) into the boundary conditions (10) and (16), one obtains, after canceling zero-order terms on the basis of Eqs. (4) and (5) and neglecting second-order terms, the following system of linear homogeneous equations for the constants A_1, A_2, A_3, and A_4:

$$\left.\begin{aligned} A_1 - (\alpha\epsilon + \sigma)A_4 &= 0 \\ A_2 + A_3 - \epsilon(\alpha + \sigma)A_4 &= 0 \\ A_1 + A_2 + \alpha A_3 - (\epsilon - 1)A_4 &= 0 \\ (\alpha\epsilon + 1)A_1 + (\alpha - 1)A_2 + 2\sigma(\epsilon - 1)A_4 &= 0 \end{aligned}\right\} \tag{17}$$

The system has a nontrivial solution only if the condition

$$\begin{vmatrix} 1 & 0 & 0 & -(\alpha\epsilon + \sigma) \\ 0 & 1 & 1 & -\epsilon(\alpha + \sigma) \\ 1 & 1 & \alpha & -(\epsilon - 1) \\ \alpha\epsilon + 1 & \alpha - 1 & 0 & 2\sigma(\epsilon - 1) \end{vmatrix} = 0 \tag{18}$$

is fulfilled. This is a cubic equation for α with the roots

$$\left.\begin{aligned} \alpha_{1,2} &= \frac{1}{1+\epsilon} \times \\ &\left[-(1 + \sigma) \pm \left(1 + \epsilon - \frac{1}{\epsilon} + \sigma^2 - 2\sigma\epsilon\right)^{1/2}\right] \\ \alpha_3 &= 1 \end{aligned}\right\} \tag{19}$$

The root α_3 must be discarded, since it leads to vanishing solutions (9). The negative sign of the square root in $\alpha_{1,2}$ corresponds to a stable solution.

Thus, stability will be determined by the solution

$$\delta = \frac{S_u^0}{L}\frac{\chi\epsilon}{1+\epsilon} \times \left\{\left[1 + \epsilon - \frac{1}{\epsilon} + (\mu\chi)^2 - 2\epsilon\mu\chi\right]^{1/2} - (1 + \mu\chi)\right\} \tag{20}$$

where

$$\chi = Lh = 2\pi L/\lambda \tag{21}$$

and Eqs. (8) and (15) have been taken into account.

This solution is plotted in Fig. 8 for $\epsilon = 5$. Landau's solution corresponds to $\mu = 0$. It is seen that the effect of curvature on burning velocity is stabilizing if $\mu > 0$—i.e., if the burning velocity is larger when the flame front is concave toward the unburned gas. This is the type of dependence one would normally expect on the basis of simple thermal theories of flame propagation. The stabilizing effect increases with increasing wave number, and, thus, a maximum of instability for a certain wave number is obtained in qualitative agreement with the experimental results.† It is shown in the Appendix that for the special case

$$D_i = k/c_p\rho \quad (i = 1, \ldots, n) \tag{22}$$

where D_i is the diffusion coefficient of the ith chemical species, k is the thermal conductivity, and c_p is the specific heat at constant pressure; one obtains $\mu = 1$ with

$$L = k/c_p\rho_u^0 S_u^0 \tag{23}$$

Fig. 8 shows that, for this case, maximum instability occurs for $2\pi L/\lambda_{max.} = 0.2$. With the reasonable value $L = 0.03$ cm. at atmospheric pressure, a wave length of about 1 cm. is obtained in good order-of-magnitude agreement with the experimental average cell sizes.

As to the dependence of average cell size on pressure, under the assumption that ϵ, μ, and k/c_p are independent of pressure, the theory predicts

$$\lambda_{max.} \propto L \propto (\rho_u^0 S_u^0)^{-1} \propto (p S_u^0)^{-1}$$

The experimental result $d \propto p^{-3/4}$ would thus yield the relation $S_u^0 \propto p^{-1/4}$. The same relation has been theoretically predicted[21] and experimentally verified for various hydrocarbons.[22] However, work on the pressure dependence of the burning velocity of butane-air mixtures[23] did not substantiate this relation and gave results that depended critically on fuel concentration. Further experimental work on burning velocities at reduced pressure is thus needed before the result $d \propto p^{-3/4}$ can be interpreted without ambiguity.

Eq. (20) still does not explain the existence of completely stable flames, since a range of instability exists for any value of μ. However, for large positive values

* Thus, curvature is defined as positive when the flame front is convex toward the burned gas.

† This maximum is actually a saddle point for the real part of δ in the complex χ plane. A discussion of the physical significance of this result will be published elsewhere.

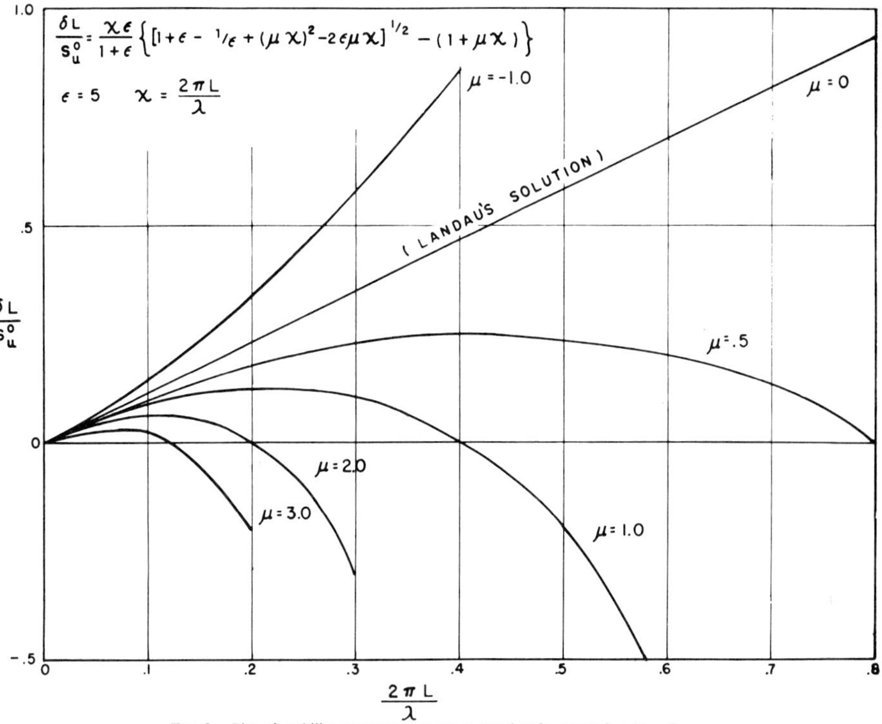

FIG. 8. Plot of stability parameter vs. wave number for several values of μ.

of μ, this range could easily correspond to cell sizes larger than the tube diameters used in the experiments, and instability could thus have escaped observation. Furthermore, the effect of gravity has been neglected so far in the treatment. The influence of gravity acting in the direction from the burned to the unburned gas is equivalent to an acceleration a in the opposite direction. It causes an additional pressure drop across the flame front proportional to the displacement ξ and the difference of densities. Thus,

$$(p_u^1 - p_b^1)_{\text{acceleration}} = a(\rho_u^0 - \rho_b^0)\xi \qquad (24)$$

This gives an additional term in the last of Eqs. (17), which becomes

$$(\alpha\epsilon + 1)A_1 + (\alpha - 1)A_2 + (\epsilon - 1) \times [2\sigma + (\gamma/\epsilon)]A_4 = 0 \qquad (17a)$$

where

$$\gamma = a/h(S_u^0)^2 \qquad (25)$$

The final result is

$$\delta = \frac{S_u^0}{L}\frac{\chi\epsilon}{1+\epsilon}\left\{\left[1 + \epsilon - \frac{1}{\epsilon} + (\mu\chi)^2 - 2\epsilon\mu\chi - \left(1 - \frac{1}{\epsilon^2}\right)\frac{F}{\chi}\right]^{1/2} - (1 + \mu\chi)\right\} \qquad (20a)$$

where

$$F = aL/(S_u^0)^2 \qquad (25a)$$

Fig. 9 shows a plot of this expression for $\epsilon = 5$ and $\mu = 1$. It shows that the effect of gravity is stabilizing and increases with the wave length. Thus, a finite range of wave lengths for instability is now obtained, and, for sufficiently large values of μ, stability over the whole range is possible. It is interesting to note that the acceleration term in Eq. (20a) is identical with the expression for the effect of acceleration on the stability of an interface between two liquids.[25]

Finally, under conditions of vibratory flame motion, alternating accelerations are superposed over the effect of gravity. Thus, instability would increase and decrease periodically corresponding to the family of curves, Fig. 9. The results of the high-speed movie studies, Fig. 6, are in agreement with this conclusion.

APPENDIX

It is desired to calculate the dependence of burning velocity and temperature of the burned gas on curvature of the flame front. By restricting the treatment to small values of L/R, it can be formulated as an eigenvalue perturbation problem.

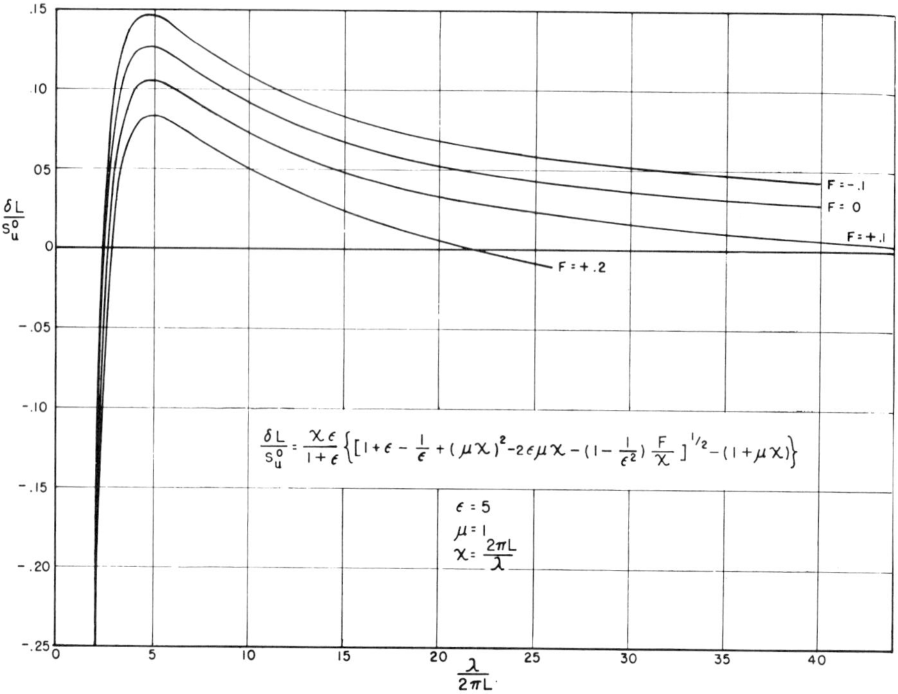

FIG. 9. Plot of stability parameter vs. wave length for several values of F.

The basic equations for a plane steady flame front are

$$\left.\begin{aligned} &k\frac{d^2T^0}{dx^2} - c_p m^0 \frac{dT^0}{dx} + \sum_j Q_j w_j(T^0, n_1^0, \ldots, n_n^0) = 0 \\ &\rho D_i \frac{d^2 n_i^0}{dx^2} - m^0 \frac{dn_i^0}{dx} + M_i \sum_j \nu_{ij} w_j(T^0, n_1^0, \ldots, n_n^0) = 0 \\ &m^0 = \rho_u{}^0 S_u{}^0 \end{aligned}\right\} \quad (26)$$

where $n_i{}^0$ is the concentration of the ith chemical species, Q_j is the heat released in jth reaction, w_j is the rate of the jth reaction, ν_{ij} is the number of molecules of ith species generated in the jth reaction, and M_i is the molecular weight of the ith species. The heat conductivity k and the product ρD_i have been assumed to be constant.* The mass flow m^0, which determines the burning velocity, is an eigenvalue of the nonlinear system (26).

* Eqs. (26) lack generality in other respects.[24] They are used in this form here rather to show the method of attack than to lead to a rigorous result.

Introducing the parameters

$$\left.\begin{aligned} &\mu^0 = (c_p m^0 L)/k, \quad N_{j0} = Q_j/k \\ &N_{ij} = M_i \frac{\nu_{ij}}{\rho D_i}, \quad \beta_i = \frac{k}{c_p \rho D_i} \quad (i = 1, \ldots, n) \end{aligned}\right\} \quad (27)$$

and replacing the independent variable by

$$X = x/L \quad (28)$$

and the dependent ones by

$$Y_0{}^0 = T^0, \quad Y_i{}^0 = n_i{}^0 \quad (i = 1, \ldots, n) \quad (29)$$

Eqs. (26) can be written

$$\frac{d^2 Y_i{}^0}{dX^2} - \beta_i \mu^0 \frac{dY_i{}^0}{dX} + \sum_j N_{ij} w_j(Y_0{}^0, \ldots, Y_n{}^0) = 0 \quad (i = 0, \ldots, n) \quad (26a)$$

The general equations for a cylindrically curved flame front are

$$k\left(\frac{d^2T}{dr^2} + \frac{1}{r}\frac{dT}{dr}\right) - c_p m \frac{dT}{dr} + \sum_j Q_j w_j(T, n_1, \ldots, n_n) = 0 \quad (30)$$

(and similar equations for the n_i).

If the range of the radial coordinate r over which Eqs. (30) have to be integrated is small compared to its average value R, it is permissible to replace $1/r$ by $1/R$ in the second term. Furthermore, the dependent variables and the eigenvalue may be developed in terms of L/R and the variable r replaced by $X = (r - R)/L$.

If this is carried through, canceling of zero-order terms and neglecting higher-order terms in L/R leads to the system

$$\frac{d^2 Y_i^1}{dX^2} - \beta_i \mu^0 \frac{dY_i^1}{dX} + \sum_j N_{ij} \times \sum_l \left(\frac{\partial w_j}{\partial Y_l}\right)_{Y_i = Y_i^0} Y_l^1 = (\beta_i \mu^1 - 1) \frac{dY_i^0}{dX} \quad (31)$$

where the reaction rates have been developed into Taylor series with respect to the zero-order values and where substitutions analogous to Eqs. (27) and (29) have been performed.

Assuming that the solution of Eqs. (26) is known, the problem consists of finding a solution to Eqs. (31) which fulfills the proper boundary conditions. μ^1 is the first-order perturbation of the eigenvalue and is thus the parameter that determines the dependence of burning velocity on curvature. The value that Y_0^1 assumes at the upper boundary determines the dependence of temperature of the burned gas on curvature.

Solution of this system is, in general, a difficult problem. For the special case of complete combustion, a solution can be found if, in addition, the conditions

$$\beta_i = 1 \quad (i = 1, \ldots, n) \quad (32)$$

are fulfilled. In this case, the right sides of Eqs. (31) vanish for $\mu^1 = 1$. By differentiating Eqs. (26a) with respect to X, it is seen that the derivatives

$$Z_i^0 = dY_i^0/dX \quad (33)$$

fulfill the equations

$$\frac{d^2 Z_i^0}{dX^2} - \beta_i \mu^0 \frac{dZ_i^0}{dX} + \sum_j N_{ij} \sum_l \left(\frac{\partial w_j}{\partial Y_l}\right)_{Y_i = Y_i^0} Z_l^0 = 0 \quad (34).$$

which are formally identical with the left sides of Eqs. (31). The Z_i^0 vanish at the boundaries, since the Y_i^0 are constant in the unburned and burned gases. In the case of complete combustion, temperature and concentrations in the burned gas must be independent of curvature; thus, the Y_i^1 also vanish at the boundaries.

Thus $Y_i^1 \propto Z_i^0$ is a valid solution of Eqs. (31) in the special case $\beta_i = 1$ with the eigenvalue $\mu^1 = 1$. By selecting the value of L so that $\mu^0 - 1$, this solution becomes

$$S_u = S_u^0 \left[1 + \frac{L}{R} + \ldots\right] \quad \left(L = \frac{k}{c_p \rho_u^0 S_u^0}\right) \quad (35)$$

for

$$D_i = k/c_p \rho \quad (i = 1, \ldots, n) \quad (32a)$$

This result is physically reasonable, since under conditions (32a) the concentration and temperature distributions will not be displaced relative to each other by the effects of curvature on heat conduction and diffusion. Thus, under these conditions the effect of curvature is independent of the reaction kinetics.

References

[1] Williams, G. C., Hottel, H. C., and Scurlock, A. C., *Third Symposium on Combustion, Flame and Explosion Phenomena*, p. 21; Williams and Wilkins, Baltimore, 1949.

[2] Damköhler, G., Z. Elektrochemie, Vol. 46, p. 601, 1940; N.A.C.A. T.M. No. 1112, 1947.

[3] Shchelkin, K. I., J. Tech. Phys. (U.S.S.R.), Vol. 13, p. 520, 1943; N.A.C.A. T.M. No. 1110, 1947.

[4] Markstein, G. H., *Third Symposium on Combustion, Flame and Explosion Phenomena*, p. 162; Williams and Wilkins, Baltimore, 1949.

[5] Markstein, G. H., J. Chem. Phys., Vol. 17, p. 428, 1949.

[6] Smith, F. A., and Pickering, S. F., J. Research Nat. Bur. Stand., Vol. 3, p. 65, 1929.

[7] Behrens, H., Naturwiss, Vol. 32, pp. 297, 299, 1944.

[8] Wohl, K., Kapp, N. M., and Gazley, C., *Third Symposium on Combustion, Flame and Explosion Phenomena*, p. 3; Williams and Wilkins, Baltimore, 1949.

[9] von Elbe, G., *Research on Flame and Ignition Phenomena*, Progress Report No. 9, U.S. Bureau of Mines, October–December, 1949.

[10] Böhm, G., and Clusius, K., Z. Naturforsch, Vol. 3a, p. 386, 1948.

[11] Jost, W., Z. Phys. Chemie, Vol. 193, p. 332, 1944.

[12] Landau, L., Acta Physicochim., U.R.S.S., Vol. 19, p. 77, 1944.

[13] Coward, H. F., Hartwell, F. J., and Georgeson, E. H. M., J. Chem. Soc., 1937, Pt. 2, p. 1482, and many other papers.

[14] Bénard, H., Revue générale des Sciences, Vol. 12, pp. 1261, 1309, 1900.

[15] Bénard H., and Avsec, D., J. de Physique, Vol. 9, p. 486, 1938.

[16] Lin, C. C., Quart. Appl. Math., Vol. 3, pp. 117, 218, 277, 1945–1946.

[17] Lord Rayleigh, Phil. Mag., Vol. 32, p. 529, 1916.

[18] Jeffreys, H., Phil. Mag., Vol. 2, p. 833, 1926; Proc. Roy. Soc. (London), Vol. A118, p. 195, 1928.

[19] Low, A. R., Proc. Roy. Soc. (London), Vol. A125, p. 180, 1929.

[20] Einbinder, H., unpublished work.

[21] Tanford, C., and Pease, R. N., J. Chem. Phys., Vol. 15, p. 861, 1947.

[22] Badin, E. J., Stuart, J. G., and Pease, R. N., J. Chem. Phys., Vol. 17, p. 314, 1949. Wheatley, P. J., and Linnett, J. W., Trans. Faraday Soc., Vol. 45, p. 1152, 1949. Garner, F. H., Ashforth, G. K., and Long, R., Nature, Vol. 164, p. 884, 1949.

[23] Wohl, K., and Kapp, N. M., Meteor Report UAC-42, University of Delaware, 1949.

[24] Boys, S. F., and Corner, J., Proc. Roy. Soc. (London), Vol. A197, p. 90, 1949.

[25] Taylor, G. I., Proc. Roy. Soc. (London), Vol. A201, p. 192, 1950.

J. Fluid Mech. (1982), *vol.* 124, *pp.* 219–237
Printed in Great Britain

Influence of hydrodynamics and diffusion upon the stability limits of laminar premixed flames

By PIERRE PELCE AND PAUL CLAVIN
Laboratoire de Dynamique et Thermophysique des Fluides (LA 72), Université de Provence, Centre St Jérôme, rue H. Poincaré, 13397 Marseille cedex 13, France

(Received 26 October 1981)

An analytical theory is developed for the stability properties of planar fronts of premixed laminar flames freely propagating downwards in a uniform reacting mixture. The coupling between the hydrodynamics and the diffusion process is described for an arbitrary expansion of the gas across the flame. Viscous effects are included with an arbitrary Prandtl number. The flame structure is described for a large value of the reduced activation energy and for a Lewis number close to unity. The flame thickness is assumed to be small compared with the wavelength of the wrinkles of the front, this wavelength being also the characteristic lengthscale of the perturbations of the flow field outside the flame. A two-scale method is then used to solve the problem. The results show that the acceleration of gravity associated with the diffusion mechanisms inside the front can counterbalance the hydrodynamical instability when the laminar-flame velocity is low enough. The theory provides predictions concerning the instability threshold. In particular, the dimensions of the cells are predicted to be large compared with the flame thickness, and thus the basic assumption of the theory is verified. Furthermore, the quantitative predictions are in good agreement with the existing experimental data.

The bifurcation is shown to be of a different nature than predicted by the purely diffusive–thermal model.

The viscous diffusivities are supposed to be independent of the temperature, and then the viscosity is proved to have no effect at all on the dynamical properties of the flame front.

1. Introduction

Owing to heat release Q, the temperature T increases inside a premixed flame of thickness d from $T_{-\infty}$ to T_b. According to a practically isobaric condition, the corresponding density decrease from $\rho_{-\infty}$ to ρ_b can be expressed in terms of a parameter γ, defined by $\gamma = (\rho_{-\infty} - \rho_f)/\rho_{-\infty} = 1-(T_{-\infty}/T_f)$, $0 < \gamma < 1$, describing the gas expansion through the flame. The subscripts $-\infty$ and b refer to conditions in the fresh mixture and in the local burned gases respectively. Subscript f refers to the burned gases in the case of planar and adiabatic flames, i.e. $T_f = T_{-\infty} + Q/C$, where C is the specific heat of the mixture. When the activation energy E of the chemical reaction is large, $\beta = (E/RT_f)(T_f - T_{-\infty}/T_f) > 1$, the reaction takes place only in a thin zone of vanishing thickness d/β located close to the maximum of temperature T_b, while the combustion rate is a strongly increasing function of the temperature of combustion T_b. In the planar case, the gradients of the temperature and of species concentration are oriented only in the direction perpendicular to the front, and the temperature of combustion is equal to T_f. The thickness d and the velocity u_L of the front are determined by a simple balance of the corresponding fluxes of heat and mass with the chemical production characterized by the reaction time τ_r:

$$\tau_r(T) \sim e^{E/kT}, \quad u_L^2 \sim \frac{\rho D_{th}}{\rho_f \tau_r(T_f)}, \quad d = \frac{\rho D_{th}}{\rho_{-\infty} u_L}, \tag{1}$$

where D_{th} is the thermal diffusivity of the gas mixture. For a wrinkled front, the transverse gradients modify the local balance of energy and affect the temperature of combustion so that $T_b \neq T_f$, and the local velocity u of the front is also modified, so that $u \neq u_L$. Thus the dynamic behaviour of a wrinkled front must depend on the diffusive and convective transport of energy and mass, which control T_b.

However, apart from these transport mechanisms occurring inside the flame, the motion of a wrinkled front depends also on hydrodynamical effects that are developed on a scale larger than d (see figure 1). In fact, the combustion rate controls only the front velocity u relative to the local upstream flow velocity $u_{-\infty}$, which produces a convective displacement of the front. However, owing to hydrodynamics, the upstream flow field $u_{-\infty}$ is itself modified by the wrinkling of the front on a distance of the order of magnitude of the wavelength Λ of the wrinkles. Thus a feedback mechanism is developed in the motion of the wrinkled fronts where hydrodynamic effects are coupled to the transport processes taking place inside the flame. The origin of this feedback lies in the fact that the gas expands through the flame thickness $\gamma \neq 0$, and so the conservation of mass and momentum causes a deflection of the streamlines of the gas flow at the tilted front. This local deflection induces a modification of the flow, which, as a consequence of negligibly small Mach number, can be considered as instantaneous and essentially incompressible. To point out the typical difficulty involved in the study of the dynamic properties of a wrinkled front, let us recall that this coupling between hydrodynamics and transport processes is not involved in the calculation of the steady and unidimensional basic solution where the planar front, perpendicular to a uniform flow (with a velocity equal to u_L), is not affected by any deflection of the streamlines. The solution for this freely propagating planar flame, which is characterized by (1), has been obtained a long time ago by Zel'dovich & Frank-Kamenetzki (1938). However, for non-steady wrinkled fronts the above-mentioned coupling causes great difficulties in analytical studies, and, to the best of our knowledge, a complete and satisfactory analytical solution of the stability of the planar flame has not yet been obtained.

It is the purpose of the present work to provide such an analysis on the basis of recent results of Clavin & Williams (1982) (referred to hereinafter as I). Attention is focused on a realistic situation where γ is not small. In this case, according to the pioneering studies of Darrieus (1938) and Landau (1944), the hydrodynamic effects are known to be of particular importance because they produce a strong mechanism of instability of the front, which leads one to question the physical relevance of the basic planar solution. The method employed in the present study is based on the difference of scale between the flame thickness d and the relevant wavelengths Λ; $\epsilon = d/\Lambda < 1$. A two-scale method is used that identifies the scale d for the longitudinal variation in the temperature and the scale Λ for the transverse variation. The flow is then separated into two parts: a far flow field, which varies on scale Λ, and the flow inside the flame thickness, which varies on scale d. This method allows us to obtain local relations between the flame velocity and the characteristic properties of the far flow field at the front. Thus it is shown, as was anticipated in the pioneering studies quoted above, that the problem is reduced to the solution of an incompressible-flow problem outside the flame thickness with boundary conditions specified on both sides of the front. The final result is sensitive to these boundary conditions, which can be specified only by a detailed analysis of the structure of the wrinkled flame. Such an analysis is completely absent not only in the first studies mentioned above but also in the more recent phenomenological description reported in Markstein (1964) and in Zel'dovich *et al.* (1980). The phenomenological boundary conditions used there limit the validity of the final result. Even in the simplifying approximation $\gamma = 0$ it is only recently that the structure of the wrinkled front has been completely described by Sivashinsky (1977*a*) and by Joulin & Clavin (1979), who have used an asymptotic expansion in $\beta \to \infty$ for solving the purely thermal diffusive model introduced and first studied by Barenblatt, Zel'dovich & Istratov (1962). When the gas expansion is retained, $\gamma \neq 0$, this local structure is not only of a diffusive nature but also involves transverse convective transport, which evolves within the flame as a consequence of the expansion. In addition to the hydrodynamic instability described by Darrieus and Landau, which is associated with a longitudinal convection, the modification of the flame structure by the transverse convective transports is a supplementary effect of the gas expansion. This particular aspect has been partially solved in I, where it is shown how the flame structure can be determined for small values of $\epsilon = d/\Lambda$ for any arbitrary expansion, $0 < \gamma < 1$. In this case the front is only weakly wrinkled and the deflection of the streamlines produces only weak effects on the flame structure even for strong expansion ($\gamma \sim 1$). But the corresponding modifications are found to be of the same order of magnitude as the diffusive effects. The study of Clavin & Williams (I) was concerned primarily with the establishment of the local equation for the evolution of the front expressed in terms of the value of the upstream far-flow velocity at the front, which is considered as a given quantity. In the present paper we are concerned with the calculation, in the laminar case, of this flow field produced by the wrinkling of the front. In contrast to I, where the determination of the local temperature of combustion T_b is sufficient, the flow calculation requires further computations of the jump conditions across the flame for flow velocity and pressure. These relations are derived in §3, and associated with the evolution equation of I they constitute the boundary conditions necessary to solve the fluid-mechanical problem involved in the study of the front stability developed in §4. As reported in the review monograph of Markstein (1964), the size of the cellular structures appearing at the instability threshold of premixed flames are observed to be large (~ 1 cm) compared with the flame thickness d ($\sim 10^{-2}$ cm) for most of the reactive mixtures he studied. Thus the perturbation analysis in small values of $\epsilon = d/\Lambda$ used herein is expected to be good for the study of the limit of stability of planar flames.

As suggested by Landau (1944) for the combustion of liquids and by Einbinder (1953) in gases, the hydrodynamical instability can be damped out for all wavelengths when the acceleration due to gravity g is added to the transport mechanisms. The corresponding criterion is that the Froude number $F = u_L^2/(1-\gamma)gd$, must be smaller than some critical value F_c. This corresponds to the experimental observations (see Lewis & von Elbe 1961, p. 390; Markstein & Somers 1953), which show that it is only for slow-burning flames that planar freely propagating fronts can be stabilized, in a uniform laminar flow, provided that they propagate in the downward direction. The expression for F_c is derived in this paper in terms of the thermal expansion γ and of the diffusive properties of the reactive mixture. The corresponding instability threshold is predicted by this expression for F_c to be experimentally observable in usual hydrocarbon–oxygen mixtures diluted by nitrogen when the flame velocity u_L of the planar front is between 5 cm/s and 17 cm/s. This threshold is also predicted to concern only the wavelengths close to $\Lambda_c = \pi u_L^2/(1-\gamma)g$, and to correspond to a fuel-rich composition for slow-burning flames and to a lean composition for fast flames. This is in qualitative and quantitative agreement with the experimental results reported by Markstein (1951, 1964) concerning slow-burning freely propagating flames. The predicted bifurcation is of a different nature than those described previously, either by the phenomenological theory of Markstein (1970) or by the purely thermal-diffusive model studied by Sivashinsky (1977*a*) and by Joulin & Clavin (1979), even when the modifications proposed by Matkowsky & Sivashinsky (1979) are introduced to take into account the acceleration due to gravity in the presence of a negligible expansion.

In addition to the considerations concerning the acceleration due to gravity, the results obtained here provide the two first orders in the development in powers of ϵ of the exact dispersion relation associated with the instability of the basic planar solution,† the first term corresponding to the analyses of Darrieus and Landau.

The viscous effects cannot be neglected inside the flame, since the Prandtl number is of order unity. Nevertheless, it turns out that for temperature-independent diffusivities the viscous effects balance themselves, and do not play any role in the dynamical properties of the flame front. A similar result has been independently obtained by Frankel & Sivashinsky (1982). This clarifies the controversy concerning the stabilizing or destabilizing influence of viscosity reported, for example by Markstein (1964) and by Zel'dovich *et al.* (1980).

The mathematical formulation is given in §2. The analysis is presented in §3, which can be omitted in a first reading. In §4 the results are discussed and compared with the previous studies. A quantitative discussion and the concluding remarks are presented in §5.

2. Formulation

The flame model is identical with that of I, and the formulation is similar. The reader is referred to I for more details. Slight modifications are introduced here to take into account the acceleration due to gravity. Furthermore, the laminar character of the upstream flow and the linearization appropriate to the stability analysis introduce further simplifications.

Attention is restricted to an overall exothermic reaction controlled by one limiting reactant. An Arrhenius rate is adopted with a large activation energy $\beta \gg 1$; d and d/u_L are the units of length and time respectively, $x = \alpha(y,t)$ defines the location of the reactive zone and the moving coordinates‡ $\xi = x-\alpha$, $\eta = y$, $\tau = t$ are introduced, $r = \rho/\rho_{-\infty}$ denotes the density ratio, U and v are the non-dimensional longitudinal (x-direction) and transverse (y-direction) components of the flow velocity. The subscript L refers to the one-dimensional steady-state solution, and the non-subscripted variables refer to the perturbation. The linearized versions of the conservation equations are written in the moving system of coordinates; we introduce

$$s_L = r_L U_L, \quad s = r_L\left(U - \frac{\partial \alpha}{\partial t}\right) + rU_L, \tag{2}$$

† According to Zel'dovich (1981) there may exist steady but non-planar flame fronts propagating in tubes, which could be intrinsically stable with respect to hydrodynamical effects. In such cases, which are beyond the scope of the present paper, the stability analysis is expected to be of a different nature. Such a situation cannot be excluded also for planar fronts in 'curved flows' such as a stagnation-point flow.

‡ In order to simplify the notation, only one transverse coordinate is specified.

where $r_L = 1$, $U_L = 1$, $r = 0$, $U = 0$ when $\xi \to -\infty$. Here s identifies the non-dimensional longitudinal mass flux in the moving coordinate system. The mass conservation gives $s_L = 1$ for the steady-state solution and

$$\frac{\partial r}{\partial \tau} + \frac{\partial s}{\partial \xi} + \frac{\partial (r_L v)}{\partial \eta} = 0 \tag{3}$$

for the perturbation.

Energy conservation is written with the non-dimensional temperature

$$\theta = (T - T_{-\infty})/(T_f - T_{-\infty}).$$

When the Mach number $M = [\rho_{-\infty} u_L^2 / p_{-\infty}]^{\frac{1}{2}}$ is negligibly small, the conservation of energy outside the reaction zone gives

$$\theta_L = \begin{cases} e^{\xi} & (\xi < 0, \\ 1 & (\xi > 0) \end{cases}$$

for the steady-state solution and

$$r_L \frac{\partial \theta}{\partial \tau} + \frac{\partial \theta}{\partial \xi} + \left(s + \frac{\partial^2 \alpha}{\partial \eta^2}\right) \frac{\partial \theta_L}{\partial \xi} = \left(\frac{\partial^2 \theta}{\partial \xi^2} + \frac{\partial^2 \theta}{\partial \eta^2}\right) \tag{4}$$

for the perturbation, with the boundary conditions

$$\theta = 0 \quad (\xi \to -\infty),$$

$$\theta = 0 \quad (\xi \to +\infty).$$

r is obtained from the solution of (4) by using the nearly isobaric conditions ($M \ll 1$)

$$r_L + r = \left[1 + \frac{(\theta_L + \theta)\gamma}{1 - \gamma}\right]^{-1}, \tag{5}$$

to give for the steady-state solution

$$r_L = \begin{cases} \left[1 + \dfrac{e^{\xi}\gamma}{1-\gamma}\right]^{-1} & (\xi < 0), \\ 1 - \gamma & (\xi > 0), \end{cases}$$

and $r = -(\gamma/(1-\gamma)) r_L^2 \theta$ for the perturbation.

Outside the reaction zone, species conservation gives

$$\psi_L = \begin{cases} 1 - e^{L\xi} & (\xi < 0), \\ 0 & (\xi > 0), \end{cases}$$

and

$$r_L \frac{\partial \psi}{\partial \tau} + \frac{\partial \psi}{\partial \xi} + \left(s + \frac{1}{L} \frac{\partial^2 \alpha}{\partial \eta^2}\right) \frac{\partial \psi_L}{\partial \xi} = \frac{1}{L} \left(\frac{\partial^2 \psi}{\partial \xi^2} + \frac{\partial^2 \psi}{\partial \eta^2}\right), \tag{6}$$

with

$$\psi = 0 \quad (\xi \to -\infty, \ \xi > 0).$$

In (4) and (6) the thermal and molecular diffusivities have been considered as temperature-independent. The Lewis number L is defined by $L = D_{th}/D_{mol}$, where D_{mol} is the molecular diffusivity of the reactive species that is in deficit, so that it controls the reaction rate.

Following Joulin & Clavin (1979), whenever the reduced activation energy β of the exothermal reaction is large, the chemical production is located at $\xi = 0$ in a thin reaction zone of thickness d/β and this production term induces jump conditions for the space derivative of ψ and θ:

$$\left[\frac{\partial \theta}{\partial \xi} + \frac{1}{L}\frac{\partial \psi}{\partial \xi}\right]_{\xi = 0^-}^{\xi = 0^+} = 0, \quad \left.\frac{\partial \theta}{\partial \xi}\right|_{\xi = 0^-} = \tfrac{1}{2}\beta\theta(\xi = 0). \tag{7}$$

$\psi(\xi=0)=0$ defines the origin ($\xi=0$) and $\theta(\xi=0)=O(1/\beta)$. These relations are valid asymptotically in the limit $\beta\rightarrow\infty$ up to $O(1/\beta)$ with the restriction $1-1/L=O(1/\beta)$. This is not too restrictive for gaseous mixtures where the Lewis number is close to one. In this case by inspection of (4) and (6), the difference $\psi-\theta$ appears to be of order $1/\beta$ and $\theta(\epsilon=0)$ is seen to be proportional to $1-1/L$ in such a way that the relevant parameter of order unity in the limit $\beta\rightarrow\infty$ is $l=\beta(1-1/L)$. In the following, unless explicitly specified, all the quantities should be understood as being $O(1)$ in the limit as $\beta\rightarrow\infty$. Furthermore, at order ϵ^2, as will be explained later, (4) and (6) reduce to quasi-planar and quasi-steady equations, $\partial\theta/\partial\tau$, $\partial^2\theta/\partial\eta^2$, $\partial\psi/\partial\tau$, $\partial^2\psi/\partial\eta^2=o(\epsilon^2)$, and thus can be solved very easily in terms of s and $\partial^2\alpha/\partial\eta^2$. Then, by prescribing the boundary condition (7), one obtains an integral relation for $s(\xi)$ which, at this order ϵ^2 reduces to

$$\int_{-\infty}^{0}\left(s+\frac{\partial^2\alpha}{\partial\eta^2}\right)e^{\xi'}d\xi'=\frac{1}{2}l\left[\int_{-\infty}^{0}s(\xi')(1+\xi')e^{\xi'}d\xi'-\frac{\partial^2\alpha}{\partial\eta^2}\right]. \tag{8}$$

At this stage let us outline the method. As soon as v is known inside the flame thickness d, a direct integration of (3) gives the variation of s through d, and (8) provides its upstream value ($\xi\rightarrow-\infty$). Then, s being completely known inside the flame thickness, (2) gives the flame velocity relative to the upstream flow as well as the modification to the longitudinal gas velocity U inside d. For the planar solution where $s_L=1$ and $U_L(\xi)=1/r_L$ one has

$$U_L(\xi)=\begin{cases}1+e^{\xi}\dfrac{\gamma}{1-\gamma} & (\xi<0), \quad (9a)\\[2ex] \dfrac{1}{1-\gamma} & (\xi>0). \quad (9b)\end{cases}$$

The non-dimensional pressure, i.e. the ratio of the pressure to its value in the fresh mixture, is denoted by $1+M^2p$. p is given by solving the longitudinal momentum equation, which, when the viscous diffusivities are supposed temperature-independent, can be written as

$$r_L\frac{\partial U}{\partial\tau}+\left(s+P\frac{\partial^2\alpha}{\partial\eta^2}\right)\frac{\partial U_L}{\partial\xi}+\frac{\partial U}{\partial\xi}=-r\frac{gd}{u_L^2}-\frac{\partial p}{\partial\xi}+P\left(\frac{\partial^2}{\partial\xi^2}+\frac{\partial^2}{\partial\eta^2}\right)U+P'\frac{\partial}{\partial\xi}\left(\frac{\partial U}{\partial\xi}+\frac{\partial v}{\partial\eta}\right), \tag{10}$$

with $p=0$, $U=0$ as $\xi\rightarrow-\infty$.

The pressure in the steady-state solution is given by the corresponding equation for $p_L(\xi)$:

$$\frac{\partial U_L}{\partial\xi}=-r_L\frac{gd}{u_L^2}-\frac{\partial p_L}{\partial\xi}+(P+P')\frac{\partial^2U_L}{\partial\xi^2}, \tag{11}$$

with $p_L=0$, $U_L=1$ as $\xi\rightarrow-\infty$: P and P' are the Prandtl numbers based on the first and second coefficients of viscosity respectively.

v is given by the transverse component of momentum conservation:

$$r_L\frac{\partial v}{\partial\tau}+\frac{\partial v}{\partial\xi}=-\frac{\partial p}{\partial\eta}+\frac{\partial\alpha}{\partial\eta}\frac{dp_L}{d\xi}+P\left(\frac{\partial^2}{\partial\xi^2}+\frac{\partial^2}{\partial\eta^2}\right)v+P'\frac{\partial}{\partial\eta}\left(\frac{\partial U}{\partial\xi}+\frac{\partial v}{\partial\eta}\right)-P'\frac{\partial\alpha}{\partial\eta}\frac{\partial^2U_L}{\partial\xi^2} \tag{12}$$

with $v=0$ as $\xi\rightarrow-\infty$.

The momentum equations (10) and (12) are valid everywhere in the gas flow, and, since they do not contain the chemical-production term, the jump conditions through the reactive zone concerning p, U and v are obtained directly on integration over this zone. Because U and v are continuous on the reactive zone, it is found that

$$[p]_{\xi=0_-}^{\xi=0_+} = (P+P')\left[\frac{\partial U}{\partial \xi}\right]_{\xi=0_-}^{\xi=0_+}, \tag{13a}$$

$$[p_L]_{\xi=0_-}^{\xi=0_+} = (P+P')\left[\frac{\partial U_L}{\partial \xi}\right]_{\xi=0_-}^{\xi=0_+}, \tag{13b}$$

$$-\left(\frac{\partial \alpha}{\partial \eta}\right)\left[\frac{\partial U_L}{\partial \xi}\right]_{\xi=0_-}^{\xi=0_+} = \left[\frac{\partial v}{\partial \xi}\right]_{\xi=0_-}^{\xi=0_+}. \tag{13c}$$

The system of linear equations (3), (4), (6), (10), (12) with the jump conditions (7) and (13) is solved by perturbation in $\epsilon = d/\Lambda$, up to the second order ϵ^2, by using, as explained in §2, a two-scale method. All the transverse derivatives $\partial/\partial\eta$ are by definition of order ϵ. Inside the flame the longitudinal derivative $\partial/\partial\xi$ is of order unity. Outside the flame, θ and ψ are constant (at the dominant order in the asymptotic limit $\beta \rightarrow \infty$), and, the spatial scale in the variation of the flow field being Λ, the spatial derivatives, including $\partial/\partial\xi$, are of order ϵ. Furthermore, as is shown by the analyses of Darrieus and Landau, the reduced flow perturbation has an intensity of order ϵ and the evolution is on a reduced timescale of order $1/\epsilon$, $\partial/\partial t = O(\epsilon)$. Thus, according to (3), $\partial s/\partial\xi = O(\epsilon^2)$ inside the flame, and, according to (4) and (6), $\theta = O(\epsilon^2)$, $\psi = O(\epsilon^2)$. So, as was anticipated in writing (8), the equations (4) and (6) appear to be quasi-steady and quasi-planar at order ϵ^2. Additionally, one introduces $G = O(1)$, defined by

$$\frac{gd}{u_L^2} = \epsilon G. \tag{14}$$

The reason is that the critical wavelength Λ_c is predicted by a phenomenological analysis to be given by $\Lambda_c = \pi u_L^2/(1-\gamma)g$, which greatly exceeds the flame thickness d as soon as $u_L > 5$ cm/s. Furthermore, the final result will show that for very slow-burning flames $u_L < 5$ cm/s, and for the diffusive properties of the usual reactive components the front is always stable, so that (14) can be used in the stability analysis without loss of generality.

3. Analysis

The flame position is developed in powers of ϵ:

$$\alpha = \alpha_0(Y, T) + \epsilon\alpha_1(Y, T) + o(\epsilon) \tag{15}$$

where $Y = \epsilon\eta$, $T = \epsilon\tau$. In addition to the reactive zone of thickness d/β, which is completely described by the jump conditions (7) and (13), three more zones are considered: the thermal diffusive zone of thickness d, and, outside, the two hydrodynamical zones, denoted by the subscripts $-\infty$ and $+\infty$ for the fresh mixture and the burnt gas respectively (see figure 1). In these two hydrodynamical zones of thickness Λ the temperature and the concentration are constant, and the solutions are written in the form

$$\left.\begin{aligned} u_{\pm\infty} &= \epsilon U_{\pm\infty}(X, Y, T), \quad v_{\pm\infty} = \epsilon V_{\pm\infty}(X, Y, T), \\ p_{\pm\infty} &= \epsilon P_{\pm\infty}(X, Y, T), \quad s_{\pm\infty} = \epsilon S_{\pm\infty}(X, Y, T), \\ \psi_{\pm\infty} &= \theta_{\pm\infty} = 0, \end{aligned}\right\} \tag{16}$$

where $X = \epsilon x$, $Y = \epsilon\eta$, $T = \epsilon\tau$. Further, $r_L = \psi_L = 1$, $\theta_L = 0$ and $\partial p_L/\partial X = -G$ in the upstream zone $-\infty$, and $r_L = 1-\gamma$, $\psi_L = 0$, $\theta_L = 1$ and $\partial p_L/\partial X = -(1-\gamma)G$ in the downstream zone $+\infty$.

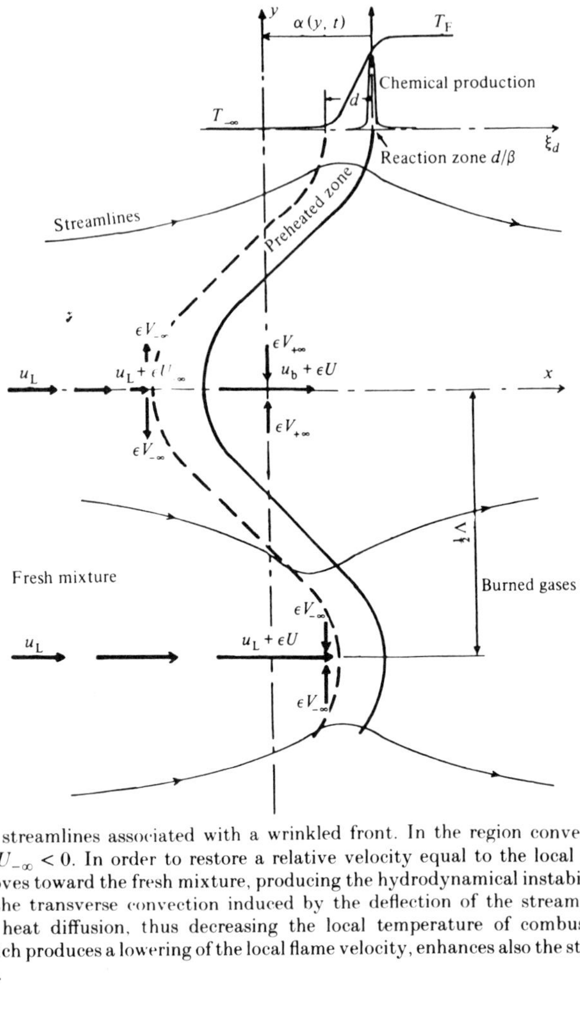

FIGURE 1. The streamlines associated with a wrinkled front. In the region convex towards the fresh mixture, $U_{-\infty} < 0$. In order to restore a relative velocity equal to the local flame velocity u_L, the front moves toward the fresh mixture, producing the hydrodynamical instability. However, in this region the transverse convection induced by the deflection of the streamlines enhances the transverse heat diffusion, thus decreasing the local temperature of combustion T_b. This mechanism, which produces a lowering of the local flame velocity, enhances also the stabilizing effect of the diffusion.

The approximation (5) associated with (16) implies that $r_{\pm\infty} = 0$, and thus these two zones are simply controlled by incompressible fluid mechanics:

$$(2) \Rightarrow S_{+\infty} = (1-\gamma)\left(U_{+\infty} - \frac{\partial \alpha}{\partial T}\right), \quad S_{-\infty} = \left(U_{-\infty} - \frac{\partial \alpha}{\partial T}\right), \tag{17a, b}$$

$$(3) \Rightarrow \frac{\partial U_{+\infty}}{\partial X} + \frac{\partial V_{+\infty}}{\partial Y} = 0, \quad \frac{\partial U_{-\infty}}{\partial X} + \frac{\partial V_{-\infty}}{\partial Y} = 0, \tag{17c, d}$$

$$(10) \Rightarrow \begin{cases} (1-\gamma)\dfrac{\partial U_{+\infty}}{\partial T} + \dfrac{\partial U_{+\infty}}{\partial X} = -\dfrac{\partial P_{+\infty}}{\partial X} + \epsilon P\left(\dfrac{\partial^2 U_{+\infty}}{\partial X^2} + \dfrac{\partial^2 U_{+\infty}}{\partial Y^2}\right), & (17e) \\[2ex] \dfrac{\partial U_{-\infty}}{\partial T} + \dfrac{\partial U_{-\infty}}{\partial X} = -\dfrac{\partial P_{-\infty}}{\partial X} + \epsilon P\left(\dfrac{\partial^2 U_{-\infty}}{\partial X^2} + \dfrac{\partial^2 U_{-\infty}}{\partial Y^2}\right), & (17f) \end{cases}$$

$$
(12) \Rightarrow \begin{cases} (1-\gamma)\dfrac{\partial V_{+\infty}}{\partial T} + \dfrac{\partial V_{+\infty}}{\partial X} = -\dfrac{\partial P_{+\infty}}{\partial Y} - \dfrac{\partial \alpha}{\partial Y}(1-\gamma)\,G + \epsilon P\left(\dfrac{\partial^2 V_{+\infty}}{\partial X^2} + \dfrac{\partial^2 V_{+\infty}}{\partial Y^2}\right), & (17g) \\[2ex] \dfrac{\partial V_{-\infty}}{\partial T} + \dfrac{\partial V_{-\infty}}{\partial X} = -\dfrac{\partial P_{-\infty}}{\partial Y} - \dfrac{\partial \alpha}{\partial Y}\,G + \epsilon P\left(\dfrac{\partial^2 V_{-\infty}}{\partial X^2} + \dfrac{\partial^2 V_{-\infty}}{\partial Y^2}\right). & (17h) \end{cases}
$$

Using the normal-modes decomposition

$$
\left.\begin{aligned} \alpha &= \mathscr{C} e^{iKY+\Sigma T}, \quad U_{\pm\infty} = \mathscr{U}_{\pm\infty}(X)\,e^{iKY+\Sigma T}, \quad V_{\pm\infty} = \mathscr{V}_{\pm\infty}(X)\,e^{iKY+\Sigma T}, \\ P_{+\infty} &= (1-\gamma)\,\alpha G + \mathscr{P}_{+\infty}(X)\,e^{iKY+\Sigma T}, \quad P_{-\infty} = -\alpha G + \mathscr{P}_{-\infty}(X)\,e^{iKY+\Sigma T}, \end{aligned}\right\} \quad (18)
$$

and assuming bounded values everywhere for $\Sigma > 0$, the solution of the system (17) can be written in the following form:

$$
\begin{aligned}
&\mathscr{P}_{+\infty}(X) = \mathscr{P}_{+\infty}(0)\,e^{-KX}, \quad \mathscr{P}_{-\infty}(X) = \mathscr{P}_{-\infty}(0)\,e^{+KX}, \\
&\mathscr{U}_{+\infty}(X) = \mathscr{A} \exp\left\{\frac{1-[1+4\epsilon P((1-\gamma)\,\Sigma+\epsilon PK^2)]^{\frac{1}{2}}}{2\epsilon P}x\right\} + \frac{K}{(1-\gamma)\,\Sigma - K}\,\mathscr{P}_{+\infty}(0)\,e^{-KX}, \\
&\mathscr{U}_{-\infty}(X) = -\frac{K}{K+\Sigma}\,\mathscr{P}_{-\infty}(0)\,e^{KX}, \quad \mathscr{V}_{+\infty}(X) = \frac{i}{K}\frac{\partial}{\partial X}\,\mathscr{U}_{+\infty}(X), \quad \mathscr{V}_{-\infty}(X) = i\mathscr{U}_{-\infty}(X).
\end{aligned} \quad (19)
$$

These linear solutions are valid at all orders of the development in powers of ϵ. Notice that there is another root on the $-\infty$ side that does not appear in (19) because it corresponds to an exponential decrease on the x-scale of the preheated zone of the flame.

The four constants of integration, $\mathscr{C}$, $\mathscr{A}$ and $\mathscr{P}_{\pm\infty}(0)$, have to be determined afterwards. In order to carry out this determination, the flame structure has to be investigated to provide the jumps through the flame thickness, $\mathscr{P}_{+\infty}(0) - \mathscr{P}_{-\infty}(0)$, $\mathscr{U}_{+\infty}(0) - \mathscr{U}_{-\infty}(0)$, $\mathscr{V}_{+\infty}(0) - \mathscr{V}_{-\infty}(0)$. This will be done up to the second order in the development in powers of ϵ.

In the third zone, corresponding to the preheated zone of the flame, the solutions are written in the following form:

$$U = \epsilon U_{-\infty}(\epsilon x, Y, T) + \epsilon \hat{u}_1(\xi, Y, T) + \epsilon^2 \hat{u}_2(\xi, Y, T) + o(\epsilon^2), \quad (20a)$$

$$v = \epsilon V_{-\infty}(\epsilon x, Y, T) + \epsilon \hat{v}_1(\xi, Y, T) + \epsilon^2 \hat{v}_2(\xi, Y, T) + o(\epsilon^2), \quad (20b)$$

$$p = \epsilon P_{-\infty}(\epsilon x, Y, T) + \epsilon \hat{p}_1(\xi, Y, T) + \epsilon^2 \hat{p}_2(\xi, Y, T) + o(\epsilon^2), \quad (20c)$$

$$p_{\mathrm{L}} = p_{\mathrm{L}0}(\xi) + \epsilon p_{\mathrm{L}1}(\xi, Y, T), \quad (20d)$$

$$s = \epsilon S_{-\infty}(\epsilon x, Y, T) + \epsilon \hat{s}_1(\xi, Y, T) + \epsilon^2 \hat{s}_2(\xi, Y, T) + o(\epsilon^2), \quad (20e)$$

$$\theta = \epsilon^2 \hat{\theta}_2(\xi, Y, T) + o(\epsilon^2), \quad \psi = \epsilon^2 \hat{y}_2(\xi, Y, T) + o(\epsilon^2), \quad (20f, g)$$

where $\hat{\theta}_2(\xi = 0) = O(\beta^{-1})$, and thus should be considered as zero. Furthermore, all the quantities $\epsilon U_{-\infty}(\epsilon x)$, $\epsilon V_{-\infty}(\epsilon x)$, $\epsilon S_{-\infty}(\epsilon x)$, $\epsilon P_{-\infty}(\epsilon x)$ have to be developed in powers of ϵ.

It will be shown in the following that, when $\xi \to -\infty$, all the quantities tagged with the circumflex ^ go to zero like $\xi^n e^{\xi}$. Thus the expressions (20) valid in the preheated zone are automatically matched with expressions (16) corresponding to the upstream hydrodynamical region labelled by $-\infty$.

The solution of (3) gives directly

$$\hat{s}_1 = 0, \quad (21a)$$

$$\hat{s}_2 = \frac{\partial V_{-\infty}}{\partial Y} \int_{\xi}^{-\infty} (r_{\mathrm{L}}(\xi') - 1)\, d\xi' + \int_{\xi}^{-\infty} r_{\mathrm{L}}(\xi') \frac{\partial \hat{v}_1}{\partial Y}(\xi')\, d\xi'. \quad (21b)$$

When the first orders of s given by (21) are introduced in (8) one obtains, at the order ϵ, $S_{-\infty,0}(X=0)=0$ for $\xi \to -\infty$, which gives, according to (15) and (17b).

$$\frac{\partial \alpha_0}{\partial T} = U_{-\infty,0}(X=0). \tag{22}$$

Then (2) and (10) give

$$\hat{u}_1 = 0, \quad \hat{p}_1 = 0.$$

According to (11),

$$\frac{\partial p_{L0}}{\partial \xi} = -\frac{\partial U_L}{\partial \xi} + (P+P')\frac{\partial^2}{\partial \xi^2} U_L.$$

Then, using (9), a direct integration of (12) provides the expression for $\hat{v}_1$:

$$\hat{v}_1 = -[\partial \alpha_0/\partial Y]\frac{\gamma}{1-\gamma} e^{\xi}, \tag{23}$$

where the last jump condition of (13) has been used to determine the constant of integration. When the result (23) is introduced in (21), an expression for $\hat{s}_2$ is obtained, and (8) with (20) leads to the expression for $S_{-\infty}(X=0)$. One obtains

$$S_{-\infty 1}(X=0) = \frac{\mathscr{L}}{d}\left[-\frac{\partial^2 \alpha_0}{\partial Y^2} + \left.\frac{\partial U_{-\infty,0}}{\partial X}\right|_{X=0}\right], \tag{24a}$$

$$\hat{s}_2(\xi=0) = \left[-\left.\frac{\partial U_{-\infty,0}}{\partial X}\right|_{X=0} + \frac{\partial^2 \alpha_0}{\partial Y^2}\right] \ln \frac{1}{1-\gamma}, \tag{24b}$$

where

$$\frac{\mathscr{L}}{d} = \frac{1}{\gamma}\ln\frac{1}{1-\gamma} + \tfrac{1}{2}\beta\left(1-\frac{1}{L}\right)\frac{1-\gamma}{\gamma}D(\gamma), \tag{25a}$$

with

$$D = \int_0^{\gamma/(1-\gamma)} dx\, x^{-1} \ln(1+x). \tag{25b}$$

Recalling that $\hat{\theta}_2(\xi=0) = O(\beta^{-1})$, (2) specifies the value of $\hat{u}_2(\xi=0)$:

$$\hat{u}_2(\xi=0) = \frac{\gamma}{1-\gamma} S_{-\infty 1}(X=0) + \frac{1}{1-\gamma}\hat{s}_2(\xi=0). \tag{26}$$

The integration of the order ϵ^2 of (10) yields the value of $\hat{p}_2$:

$$\begin{aligned}\hat{p}_2(\xi=0) = {}& -2\frac{\gamma}{1-\gamma} S_{-\infty 1}(X=0) - \frac{2}{1-\gamma}\left(\frac{\partial^2 \alpha_0}{\partial Y^2} - \left.\frac{\partial U_{-\infty,0}}{\partial X}\right|_{X=0}\right)\ln\frac{1}{1-\gamma} \\ & + \frac{\gamma}{1-\gamma}(1-P-P')\frac{\partial^2 \alpha_0}{\partial Y^2} + \left(\ln\frac{1}{1-\gamma}\right)\frac{\partial U_{-\infty,0}(X=0)}{\partial T} + (P+P')\left.\frac{\partial \hat{u}_2}{\partial \xi}\right|_{\xi=0}\end{aligned} \tag{27}$$

and the integration of (12) provides the value of $\hat{v}_2(\xi=0)$:

$$\begin{aligned}\hat{v}_2(\xi=0) = {}& -\frac{\gamma}{1-\gamma}\frac{\partial \alpha_1}{\partial Y} + P\left(\frac{\partial V_{+\infty,0}}{\partial X} - \frac{\partial V_{-\infty,0}}{\partial X}\right)_{X=0} \\ & + \left[\frac{\partial V_{-\infty,0}(X=0)}{\partial T} + \frac{\partial^2 \alpha_0}{\partial T \partial Y} + G\frac{\partial \alpha_0}{\partial Y}\right]\ln\frac{1}{1-\gamma},\end{aligned} \tag{28}$$

where the jump conditions (13) through the reactive zone have been used to determine the constant of integration.

4. Results and discussion

The definitions (20) and the results (22)–(28) specify the boundary and jump conditions across the flame up to the order ϵ^2, providing the necessary information for the solution of the hydrodynamical problem (17), (19). To begin with, let us consider the modification of the flame velocity obtained from (22), (24*a*) by using the definitions (16) and (17*b*):

$$\frac{\partial \alpha}{\partial t} = u_{-\infty}(X=0) - s_{-\infty}(X=0),$$

$$s_{-\infty}(X=0) = \frac{\mathscr{L}}{d}\left(\left.\frac{\partial u_{-\infty}}{\partial x}\right|_{x=0} - \frac{\partial^2 \alpha}{\partial y^2}\right) + o(\epsilon^2), \tag{29}$$

associated with the following modification $T_b - T_f$ of the temperature of combustion:

$$(T_b - T_f)/(T_f - T_{-\infty}) = +\left(1-\frac{1}{L}\right)\left(\frac{1-\gamma}{\gamma}\right) D(\gamma)\left(\left.\frac{\partial u_{-\infty}}{\partial x}\right|_{x=0} - \frac{\partial^2 \alpha}{\partial y^2}\right) + o(\epsilon^2), \tag{29'}$$

where $\mathscr{L}/d$ and $D(\gamma)$ are given by (25*a*, *b*).

Equations (29) and (29′) are valid in the limit $\beta \to \infty$ for $1 - L = O(b^{-1})$, and correspond to the linearized version of the result obtained in I. $u_{-\infty}(X=0)$ represents the convective effect upon the flame movement produced by the modification of the upstream gas flow induced by the front wrinkling. This was the only effect retained in the early analyses of Darrieus (1938) and Landau (1944), where $s_{-\infty}$ was set equal to zero. $s_{-\infty}(X=0)$ represents the change of the front velocity relative to the upstream gas flow produced by the modification of the flame structure involved in the front wrinkling. Such a modification was first introduced by Markstein (1951) through a phenomenological relation

$$\frac{\partial \alpha}{\partial t} = u_{-\infty} - \frac{\mathscr{L}}{R}, \quad \frac{\mathscr{L}}{d} > 0, \tag{30}$$

where $1/R$ is the curvature of the front, which, in the weakly curved case, is approximated by $-(\partial^2\alpha/\partial y^2)\,d^{-1}$. Except for a supplementary term $\partial u_{-\infty}/\partial x|_{x=0}$, which will be commented on later in the paper, (29) is similar to (30), and provides the expression of the 'Markstein phenomenological length' $\mathscr{L}$. This modification of the flame structure is produced by the transverse diffusive flux of mass and energy appearing when the front is wrinkled, and is also affected by the corresponding transverse convective flux produced by the deflection of the streamlines. This last effect disappears when the gas expansion is neglected, $\gamma = 0$, as in the purely 'diffusive-thermal' model first studied by Barenblatt *et al.* (1962). In this case, the hydrodynamical effects also disappear completely, $u_{-\infty}(X) = 0$, and the equation (29) for the evolution of the front reduces to a simple diffusion equation

$$\frac{\partial \alpha}{\partial t} = \frac{\mathscr{L}}{d}\frac{\partial^2 \alpha}{\partial y^2}, \tag{31}$$

with a 'Markstein length'

$$\frac{\mathscr{L}}{d} = 1 + \tfrac{1}{2}\beta\left(1 - \frac{1}{L}\right) \tag{32}$$

given by (25*a*) for $\gamma = 0$. These results, valid in the limit $\beta \to \infty$ for $l \equiv \beta(1 - 1/L) = O(1)$, are in agreement with that of Barenblatt *et al.* (1962) and can be easily interpreted from a physical point of view (Clavin 1982). The term 1 in (32) represents the contribution of the thermal relaxation of a wrinkled front associated with a constant value of the maximum of the temperature of combustion. The second term on the right-hand side of (32) describes the effect of the modification of the flame velocity induced by the change of the temperature of combustion resulting in the competitive diffusion processes of heat and mass developed in the transverse direction by the front wrinkling. Further analytical studies of the 'diffusive-thermal' model have been recently carried by Sivashinsky (1977*a*) and by Joulin & Clavin (1979) to get a better understanding of the dynamical properties of the front. One of the motivations for these works was that, according to (32), the diffusion coefficient $\mathscr{L}/d$ in (31) may easily change sign for the diffusive properties of the usual rich reactive mixture, providing an interpretation of the instability threshold for the appearance of the cellular structures. But, when the transverse convection is taken into account, the result (25*a*) shows that $\mathscr{L}/d$ can no longer easily reach zero, and indeed is strictly positive even for the lightest limiting component involved in the usual reactive mixtures (an exception is possible for hydrogen). Thus, in agreement with the assumption (30) used by Markstein but contrary to the prediction of the 'diffusive-thermal model', the modification of the flame structure by the front wrinkling produces, for the usual hydrocarbon reactive mixtures, a stabilizing effect ($\mathscr{L}/d > 0$). Nevertheless, this 'diffusive-thermal model' is still interesting, because it allows us to investigate the diffusive phenomena in the complete range of wavenumbers (the assumption $\epsilon \ll 1$ is not necessary). Thus, for example, an interesting diffusive mechanism of instability leading to travelling waves has been pointed out by Sivashinsky (1977*a*) and Joulin & Clavin (1979) for lean mixtures of heavy hydrocarbons.

A comparison between (29) and (30) reveals that, as shown previously in the study of Eckhaus (1961), it is not only the curvature of the front that is involved in the modification of the flame structure, but also the quantity $\partial u_{-\infty}/\partial x$, which is related to the inhomogeneities of the upstream flow. This effect has been known for a long time to exist, and it was referred to as the 'Karlowitz effect' in the traditional literature of combustion (see e.g. Lewis & von Elbe 1961). Eckhaus (1961) interprets (29) as the curvature of the front relative to the deformation of the streamlines. The recent result by Clavin & Joulin (1982), concerning the nonlinear case of finite amplitudes of deformation, provides the precise formulation of this effect, which is shown to be more complex.

It is clear from (29) that the complete description of the dynamical properties of the front requires the solution of the purely hydrodynamical problem for determining the modification of the upstream flow field $u_{-\infty}(X)$, $v_{-\infty}(X)$. This was tentatively undertaken by different early analytical studies, for example those of Eckhaus (1961), Chu & Parlange (1962), Maxworthy (1962) and Istratov & Librovich (1966). However, in the absence of a systematic technique to solve the structure of the wrinkled flame, these analytical studies lead to results that are contradictory and not completely satisfactory. For example, they do not reduce, in the limit $\gamma = 0$, to the rigorous result expressed by (31) and (32). More interesting is the work of Sivashinsky (1977*b*), but this is exact only for small values of γ. There is also an intermediate study by Lazarev & Pleshanov (1980), in which an incomplete analysis of the flame structure is carried out in such a way that the solution requires the introduction of a phenomenological relation such as (30), but in which the 'Karlowitz effect' is overlooked. More relevant are the results obtained by Markstein (1964) in his early phenomenological analysis, in which the study of the flame structure is not addressed at all, but is replaced by relevant phenomenological relations on the flame front considered as a surface of discontinuity. On both sides of the front, the flow is treated as in the pioneering works of Darrieus (1938) and Landau (1944) in the inviscid and incompressible approximation.

Here, the problem is completely solved analytically for small values of $\epsilon = d/\Lambda$ up to order ϵ^2, but for an arbitrary gas expansion γ and for arbitrary Prandtl numbers. It is proved that the problem is reduced, effectively, to solving for an incompressible flow in the two hydrodynamical zones, with the boundary and jump conditions determined by the analysis of the flame structure of §3. In fact, Markstein (1964) used a phenomenological relation similar to (29), which is proved here and in I to be exact to order ϵ^2, but the jump conditions used for the pressure and the flow field were incomplete.

Thus (26) gives the jump condition, through the flame thickness, for the longitudinal flow velocity:

$$u_{+\infty}(0) - u_{-\infty}(0) = \left[\frac{\gamma}{1-\gamma}\frac{\mathscr{L}}{d} - \frac{1}{1-\gamma}\ln\frac{1}{1-\gamma}\right]\left[\frac{\partial u_{-\infty}}{\partial x}\bigg|_{x=0} - \frac{\partial^2\alpha}{\partial y^2}\right] + o(\epsilon^2). \quad (33)$$

The first term $[\gamma/(1-\gamma)]\mathscr{L}/d$ corresponds to the modification of the longitudinal velocity associated with a constant value $\epsilon S_{-\infty}(X=0)$ of the longitudinal mass flux across the flame. This term is the only one appearing in the semi-phenomenological analysis of Markstein (1964) (see also Zel'dovich *et al.* 1980). But an additional term appears in (33), corresponding to the modification $\hat{s}$ of s produced by the transverse convection due to the deflection of the streamlines through the flame thickness.

Equation (27) lead to the modification of the pressure through the flame thickness:

$$\begin{aligned} p_{+\infty}(0) - p_{-\infty}(0) = &-2\frac{\gamma}{1-\gamma}\left[\frac{\mathscr{L}}{d} - \frac{1}{\gamma}\ln\frac{1}{1-\gamma}\right]\left[\frac{\partial u_{-\infty}}{\partial x}\bigg|_{x=0} - \frac{\partial^2\alpha}{\partial y^2}\right] \\ &+ \frac{\gamma}{1-\gamma}(1-P-P')\frac{\partial^2\alpha}{\partial y^2} + (P+P')\left[\frac{\partial u_{+\infty}}{\partial x}\bigg|_{x=0} - \frac{\partial u_{-\infty}}{\partial x}\bigg|_{x=0}\right] \\ &+ \left(\ln\frac{1}{1-\gamma}\right)\frac{\partial u_{-\infty}(0)}{\partial t} + o(\epsilon^2). \end{aligned} \quad (34)$$

The first term corresponds to the linearized Bernoulli law associated with (33). Only a part of this term was retained in the phenomenological analysis mentioned above. Among the additional terms, $[\gamma/(1-\gamma)]\partial^2\alpha/\partial y^2$, which produces an effect similar to surface tension, will be seen to be one of the most important. According to (18), the jump condition associated with $\epsilon\mathscr{P}_{\pm\infty}$ is obtained by adding $-\epsilon\gamma G\alpha$ to (34). Finally, (23) and (28) specify the jump condition for the transverse velocity:

$$\begin{aligned} v_{+\infty}(0) - v_{-\infty}(0) = &-\frac{\gamma}{1-\gamma}\frac{\partial\alpha}{\partial y} + P\left(\frac{\partial v_{+\infty}}{\partial x} - \frac{\partial v_{-\infty}}{\partial x}\right)_{x=0} \\ &+ \left[\frac{\partial v_{-\infty}(X=0)}{\partial t} + \frac{\partial^2\alpha}{\partial t\,\partial y} + G\frac{\partial\alpha}{\partial y}\right]\ln\frac{1}{1-\gamma} + o(\epsilon^2). \end{aligned} \quad (35)$$

When the solutions (19) are introduced into (29), (33)–(35), the four constants of integration $\mathscr{C}$, $\mathscr{A}$, $\mathscr{P}_{\pm}(0)$ are given by the solution of four homogeneous linear equations. In order to obtain a non-trivial solution the corresponding 4×4 determinant must be zero. This leads to the following dispersion relation:

$$A(k)\sigma^2 + B(k)\sigma + C(k) = 0. \quad (36)$$

where $\sigma = \epsilon\Sigma$, $k = \epsilon K$, and where the coefficients A, B, C are given by

$$A = (2-\gamma)+\gamma\left[\frac{\mathscr{L}}{d}-\frac{1}{\gamma}\ln\frac{1}{1-\gamma}\right]k, \tag{37a}$$

$$B = 2k+\left[\frac{2}{1-\gamma}\frac{\mathscr{L}}{d}-\frac{2}{1-\gamma}\ln\frac{1}{1-\gamma}\right]k^2, \tag{37b}$$

$$C = \frac{\gamma}{1-\gamma}k\left\{\frac{gd}{u_L^2}(1-\gamma)-k\left[1+\epsilon G(1-\gamma)\left(\frac{\mathscr{L}}{d}-\frac{1}{\gamma}\ln\frac{1}{1-\gamma}\right)\right]\right.$$
$$\left.+k^2\left[1+\frac{2+\gamma}{\gamma}\frac{\mathscr{L}}{d}-\frac{2}{\gamma}\ln\frac{1}{1-\gamma}\right]\right\}. \tag{37c}$$

Equations (36) and (37) correspond to the three first orders of the expansion in ϵ of the exact dispersion relation.

The first point of interest is that the Prandtl numbers P and P' do not appear in the final result (36), (37). This is surprising, because they are present at all the stages of the calculation, and the cancellation mechanism is by no means trivial and has not yet received a simple physical interpretation. For example, the viscous effects inside the thermal diffusive zone, described by the term

$$-\frac{\gamma}{1-\gamma}(P+P')\frac{\partial^2\alpha}{\partial y^2}$$

of (34), are exactly counterbalanced in the last term C of the dispersion relation (36) by the viscous effects inside the reactive zone described by (13a, b), which induces the term

$$(P+P')\left[\left.\frac{\partial u_{+\infty}}{\partial x}\right|_{x=0}-\left.\frac{\partial u_{-\infty}}{\partial x}\right|_{x=0}\right]$$

of (34). *Thus, when the diffusivities are supposed to be temperature independent, the dynamical properties of the flame front are found to be not affected at all by the viscous effects.* This result answers the open question concerning the influence of the viscosity, which was found to be stabilizing by some authors and destabilizing by others (see Markstein 1964, 1970).

It is interesting to rewrite the coefficient C of (36) in the following form:

$$C = \frac{\gamma}{1-\gamma}k\left[\underset{\text{gravity}}{\tfrac{1}{2}k_c}-k\left(\underset{\text{hydrodynamics}}{\delta}-\underset{\text{diffusion}}{\frac{k}{k_n}}\right)\right], \tag{38}$$

where k_c varies with the flame speed according to

$$k_c = 2(1-\gamma)\frac{gd}{u_L^2}, \tag{39}$$

and where k_n depends on the equivalence ratio of the reactive mixture through the Lewis number L appearing in $\mathscr{L}/d$ (25),†

† Far from the stoichiometric composition, L is defined with the binary molecular diffusivity of the limiting component (oxygen in the rich mixtures, fuel in the lean ones) in the neutral component (nitrogen in the air). As the analysis of Joulin & Mitani (1981) has shown, for mixtures close to stoichiometric composition, L appears to be a function of the equivalence ratio (ratio of the amount of fuel to the amount of oxygen) such that L varies continuously from its value based on the oxygen for rich mixtures to its value based on the fuel for lean mixtures.

$$\delta = 1+\frac{gd}{u_L^2}(1-\gamma)\left(\frac{\mathscr{L}}{d}-\frac{1}{\gamma}\ln\frac{1}{1-\gamma}\right).$$
$$k_n = \left[1+\frac{2+\gamma}{\gamma}\left(\mathscr{L}/d\right)-\frac{2}{\gamma}\ln\left(\frac{1}{1-\gamma}\right)\right]^{-1} \tag{40}$$

where the second term of δ can be easily neglected compared with unity because its numerical value is of order 10^{-2} in the range of interest for the flame velocity ($u_L > 5$ cm/s). For ordinary reactive mixtures k_n is positive, making the diffusion a stabilizing mechanism.

When the acceleration of gravity is not considered, the dispersion relation (36), (37) reduces to an expression similar to the one obtained in the inviscid approximation by the phenomenological work of Markstein (1964), but where supplementary terms, involving $\ln(1-\gamma)^{-1}$, appear in A and in B. Furthermore, the expression (40) for k_n is slightly different from $k_n = \{[(2+\gamma)/\gamma]\mathscr{L}/d\}^{-1}$ obtained by Markstein. The additional terms, involving $\ln(1-\gamma)^{-1}$, come from the modification of the longitudinal mass flux s associated with the transverse convection produced, inside the flame thickness, by the deflection mechanism of the streamlines. The supplementary term 1 in the denominator of (40) comes from the additional 'surface-tension effect' already mentioned in (34). Without the acceleration of gravity, $k_c = 0$, and there is always a range of unstable wavenumbers located around zero. This can be clearly understood because, as the pioneering works of Darrieus (1938) and Landau (1944) have shown, the hydrodynamical instability produces in (38) a k^2 term. The stabilizing diffusive effects being associated with a k^3 term, they cannot overcome the hydrodynamical instability near $k = 0$. As Markstein (1970) pointed out himself, this cannot explain the existing experimental results, which show that the cellular structures fade away at an intrinsic size independent of the dimension of the burner.

Because of a mechanism similar to the gravity waves, the acceleration due to gravity in flames propagating downwards induces a positive k term in C. Thus the criterion of stability given by $C > 0$ can possibly be satisfied for all wavelengths provided the acceleration due to gravity and the diffusive effects are strong enough. *Thus, infinite planar fronts of slow-burning flames associated with a limiting component sufficiently heavy are predicted to be stable when they propagate downwards.*

The critical stability condition is given by

$$k_n = 2k_c \tag{41}$$

where k_c and k_n are given by (39) and (40) respectively.

For $k_n > 2k_c$ the instability occurs only for a finite range of wave vectors centred on k_c and vanishing at the critical condition (41). This unstable domain is limited by the two marginal wavenumbers $k_{\pm}$ given by

$$k_{\pm} = \tfrac{1}{2}k_n \pm k_c\left(1-\frac{2k_c}{k_n}\right)^{\frac{1}{2}}. \tag{42}$$

Using (39), (40) and (25) the critical stability condition is presented in figure 2 on the plane (u_L, L) for different values of γ. *These curves are universal and can be used for any flame.* The parameter L is directly related to the equivalence ratio, and the flame velocity u_L is controlled in the experimental situation by the composition of the reactive mixture, including the dilution but also the equivalence ratio. For each specific reactive mixture, the transcription of the limits of stability on the usual plane (equivalence ratio, dilution) is straightforward as soon as the dependence of the burning velocity of the mixture studied is known.

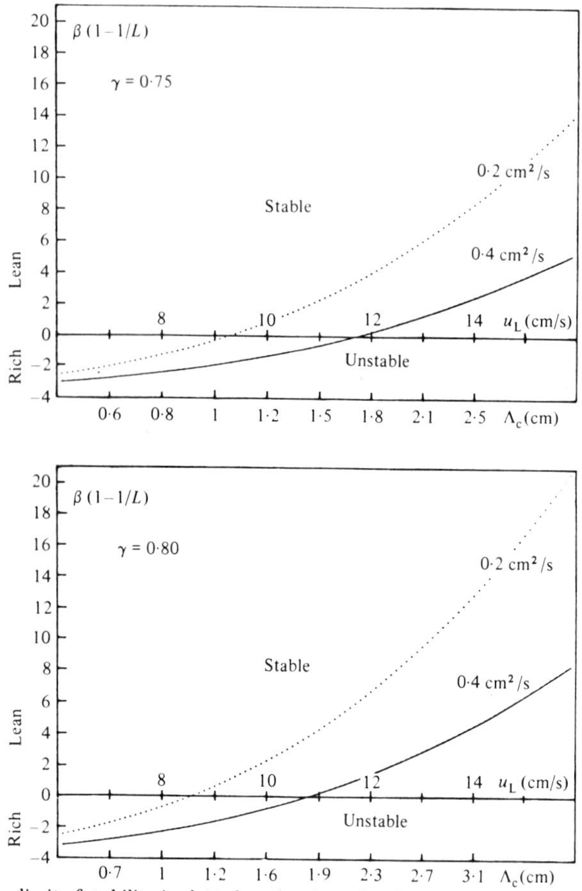

FIGURE 2. The limit of stability is plotted on the plane ($l = \beta(1-1/L), u_L$) for two values of $\rho D/\rho_{-\infty}$ corresponding to the fresh mixture, 0·2 cm²/s, and the burned gases, 0·4 cm²/s. This last curve is expected to be the more relevant because the reaction zone is at the temperature of the burned gases. The effect of the temperature dependence of ρD is described by Clavin & Garcia (1982). The two figures correspond to different values of the gas expansion, $\gamma = 0{\cdot}75$ and 0·8, which are realistic for conventional hydrocarbon flames in the range of flame velocities considered. For hydrocarbons heavier than oxygen, the lean mixtures correspond roughly to higher values of l, $l > 1$, and the rich to $l < 1$.

The present analysis is limited to the range of small values of k. According to (39), the critical value k_c is effectively found to be small as soon as u_L is large enough ($u_L > 5$ cm). For lower flame velocities the experimental values of k_n corresponding to the usual reactive mixtures are expected to be smaller than the corresponding value of $2k_c$. Such flames are thus predicted to be stable at any composition. Thus the limits of stability predicted by the present study is found to cover all the range of the usual reactive mixtures.

5. Concluding remarks

The analytical study developed here provides a rigorous description of the coupling between the hydrodynamical and diffusive effects occurring in the dynamical properties of premixed flame fronts. The corresponding description of the limits of stability is in good qualitative agreement with the existing experimental data (Markstein 1951, 1964; Markstein & Somers 1953). As for the quantitative comparison with these data, the main defect of the model is the use of an over-simplified chemical kinetic description, which impairs the prediction with an insufficient precision of the effective Lewis number L corresponding to each composition of the actual reactive mixture used in the experiments. Nevertheless, one can obtain a good idea of the order of magnitude of this number by looking at the values of the different diffusivities of the reactive mixture. Let us for example consider the case of a propane–oxygen mixture highly diluted with nitrogen. At the normal conditions, the thermal diffusivity of the nitrogen is $D_{th} \approx 0{\cdot}19\ cm^2/s$ and the binary molecular diffusion coefficients of propane and oxygen in nitrogen are $D(C_3H_8,N_2) \approx 0{\cdot}11\ cm^2/s$ and $D(O_2,N_2) \approx 0{\cdot}22\ cm^2/s$ respectively (Fristrom & Westenberg (1965)). As the activation energy E of propane is about 55 kcal/mol (Kaskan 1957). β is close to 15 for the flame temperature of 1500 K, and the corresponding values of the quantity $l = \beta(1-1/L)$ appearing in (14) are $l(C_3H_8) = 6{\cdot}3$ and $l(O_2) = -2{\cdot}4$.

Thus, when the composition varies from lean to rich, one can expect that the effective value of $l \equiv \beta(1-1/L)$ appearing in the expression (25) for $\mathscr{L}/d$ varies from 6·3 to −2·4. In fact one can expect that, for most of the time, the influence of the chemical kinetics enhances the stability of the fronts and causes a slight shift of these limits towards higher values. This is due to the fact that the extremely mobile intermediate species involved in chemical reaction attain, as a rule, maximum concentration at the reaction zone. Since the flame velocity increases with their concentration, the molecular diffusivity of these species is expected to *reinforce* the stabilizing influence of the heat diffusion. Nevertheless, for the usual hydrocarbons, to attain experimental values of l larger than 10 seems unlikely. Thus the results shown in figure 2 demonstrate that freely propagating planar flames with a *laminar velocity* u_L *larger than* 17 cm/s *cannot be stabilized in a uniform flow*. On the other hand, it seems very difficult to encounter values of l smaller than −3 (except for hydrogen). Hence our theory predicts that slow flames with *laminar velocity* u_L *smaller than* 5 cm/s *should be stable for all compositions. Thus, the instability threshold should be experimentally observable between* 5 cm/s $< u_L <$ 17 cm/s, *at a fuel-rich composition for the lower values of* u_L *and at a lean composition for higher values of* u_L. Furthermore, independently of chemical-kinetics effects, the size of cells at the bifurcation is predicted by (39) and (42) to vary from 0·5 cm to 3·5 cm, as the *square of the laminar velocity of propagation* u_L^2. It must be emphasized that these considerations apply only to the size of the cells at the critical threshold of the instability, and not to the evolution of this size by the nonlinear effects occurring beyond the threshold.

It should be also noted that for the rich mixtures close to the instability threshold, the temperature is predicted by (29′) and (19) to be higher in the parts of the front concave towards the burnt gases. According to the result of our analysis, this is particularly the case at the threshold for cell formation on flame fronts with slow enough laminar velocity ($u_L < 12$ cm/s, cf. figure 2).

In the usual hydrocarbon mixtures, highly diluted with nitrogen, these low velocities correspond to situations that are not very far from the thermal-extinction limit (~ 8 cm/s). According to Joulin & Clavin (1979), one may expect that close to the extinction limit, the above predictions could be affected by the heat losses. The model used in the present paper can be made more realistic by introducing such effects as the influence of heat losses, the temperature variation of the diffusivities, or a more detailed kinetic scheme. The corresponding development is tractable and will be published in the near future (see Clavin & Garcia 1982; Clavin & Nicoli 1983). We have deliberately omitted such complications here, in order to emphasize the important basic phenomena.

In conclusion, when the expansion of the gas is taken into account in adiabatic flames, the diffusive effects appear to be stabilizing ($\mathscr{L}/d$, $k_n > 0$) at all compositions for the usual hydrocarbons, and thus cannot be responsible for the cellular structures of the planar fronts. Nevertheless, very light limiting components such as hydrogen require the more-detailed study quoted above. An exception is also still possible for very heavy limiting components, which can possibly produce, as in solid combustion, spinning waves (cf. the discussion in §4). Apart from these extreme cases, which are beyond the scope of the present study, the cellular structures appearing on planar fronts of usual hydrocarbon adiabatic flames are shown to be described by the competition between the hydrodynamical instability and two stabilizing phenomena: acceleration due to gravity and diffusive effects. The corresponding threshold is predicted to be experimentally observable for ordinary hydrocarbon flames propagating downwards with a flame speed between 5 cm/s and 17 cm/s.

The authors are greatly indebted to G. Searby, A. Liñan and A. K. Oppenheim for fruitful discussions. This work was supported in part by the C.N.R.S. under A.T.P. 3966 and by the D.R.E.T. under grant no. 80/434.

REFERENCES

Barenblatt, G. I., Zel'dovich, Ya. B. & Istratov, A. G. 1962 On diffusional thermal stability of laminar flame. *Prikl. Mekh. Tekh. Fiz.* **2**, 21–26.

Chu, B. T. & Parlange, J. Y. 1962 On the stability of laminar flame. *J. Méc.* **1**, 293–312.

Clavin, P. 1982 Dynamical behavior of premixed flame fronts in laminar and turbulent flows. *Prog. Energy Combust. Sci.* (to appear).

Clavin, P. & Garcia, P. 1982 The influence of the temperature dependence of diffusivities on the dynamics of flame fronts. *J. Méc.* (to appear).

Clavin, P. & Joulin, G. 1982 Premixed flame propagation in high intensity turbulent flow. *J. Physique* (to appear).

Clavin, P. & Nicoli, C. 1983 Effects of the heat losses on the limits of stability of premixed flames propagating downwards. *Combust. Flame* (submitted).

Clavin, P. & Williams, F. A. 1982 Effects of molecular diffusion and of thermal expansion on the structure and dynamics of premixed flames in turbulent flows of large scale and low intensity. *J. Fluid Mech.* **116**, 251–282.

Darrieus, G. 1938 Propagation d'un front de flamme. Essai de théorie des vitesses anormales de déflagration par développement spontané de la turbulence. Unpublished works presented at La Technique Moderne 1938 and at Le Congrès de Mécanique Appliquée 1945.

Eckhaus, W. 1961 Theory of flame-front stability. *J. Fluid Mech.* **10**, 80–100.

Einbinder, H. 1953 The hydrodynamic stability of flame fronts. *J. Chem. Phys.* **21**, 480–489.

Frankel, M. L. & Sivashinsky, G. I. 1982 The effect of viscosity on hydrodynamic stability of a plane flame front. *Combust. Sci. Tech.* (to appear).

Fristrom, R. M. & Westenberg, A. A. 1965 *Flame Structure*. McGraw-Hill.

Istratov, A. G. & Librovich, V. B. 1966 The effect of transport processes on the stability property of a plane flame front. *Prikl. Mat. Mekh.* **30**, 451–466 (in Russian).

JOULIN, G. & CLAVIN, P. 1979 Linear stability analysis of nonadiabatic flames: diffusional–thermal model. *Combust. Flame* **35**, 139–153.

JOULIN, G. & MITANI, T. 1981 Linear stability of two reactants flames. *Combust. Flame* **40**, 235–246.

KASKAN 1957 The dependence of flame temperature on mass burning velocity. In *Proc. 6th Symp. of Combustion Institute*, pp. 134–143.

LANDAU, L. 1944 On the theory of slow combustion. *Acta Physicochimica URSS* **19**, 77–85.

LAZAREV, P. P. & PLESHANOV, A. S. 1980 Linear stability of a laminar flame front. *Fiz. Gor. Vzryva* **16**, 45–52 (English transl. 1981, Plenum).

LEWIS, B. & VON ELBE, G. 1961 *Combustion, Flames and Explosions of Gases*. Academic. (2nd edn 1967.)

MARKSTEIN, G. H. 1951 Experimental and theoretical studies of flame front stability. *J. Aero. Sci.* **18**, 199.

MARKSTEIN, G. H. 1964 *Nonsteady Flame Propagation*. Pergamon.

MARKSTEIN, G. H. 1970 Flames as amplifiers of fluid mechanical disturbances. In *Proc. 6th U.S. Nat. Congr. of Appl. Mech., Cambridge, Mass.*, pp. 11–33.

MARKSTEIN, G. H. & SOMERS, L. M. 1953 Cellular flame structure and vibratory flame movement in *n*-butane methane mixtures. In *Proc. 4th Symp. on Combustion*, pp. 527–535. Williams & Wilkins.

MATKOWSKY, B. J. & SIVASHINSKY, G. I. 1979 Acceleration effects on the stability of flame propagation. *SIAM J. Appl. Math.* **37**, 669–685.

SIVASHINSKY, G. I. 1977*a* Diffusional–thermal theory of cellular flames. *Combust. Sci. Tech.* **15**, 137–145.

SIVASHINSKY, G. I. 1977*b* Nonlinear analysis of hydrodynamic instability in laminar flames. I. Derivation of basic equations. *Acta Astronautica* **4**, 1177–1206.

ZEL'DOVICH, YA. B. & FRANK-KAMENETZKI, D. A. 1938 A theory of thermal propagation of flame. *Acta Physicochimica URSS*, **9**, 341–350.

ZEL'DOVICH, YA. B., BARENBLATT, G. I., LIBROVICH, V. B. & MAKHVILADZE, G. M. 1980 *Mathematical Theory of Combustion and Explosion*. Nauka. (In Russian.)

ZEL'DOVICH, YA. B. 1981 Structure and stability of steady laminar flame at moderately large Reynolds numbers. *Combust. Flame* **40**, 225–234.

Stability Limits and Critical Size of Structures in Premixed Flames

J. Quinard,* G. Searby,* and L. Boyer†
Université de Provence—Centre St. Jérôme
Marseille, France

Abstract

An apparatus for the study of the instabilities of a premixed flame in a laminar flow is described. In this burner, the gas flow is laminar, uniform, and homogeneous. The flame front is 9 cm in diameter and, below the instability threshold, is flat to within 1%. This work concerns a propane/air mixture diluted with nitrogen. The flame speeds are measured by laser anemometry close to the front and range from the extinction limit (6 cm/s) to the autoturbulent regime (∿11 cm/s). The dilution ranges from 12% to 16% and the equivalence ratio from 0.8 to 1.2. The isodilution curves and the stability limits are presented and compared with theoretical predictions. The critical size of the cellular structures is obtained from an appropriate analysis of photographic images of the flame front. The experimental results agree fairly well with the relation:

$$\Lambda_c = \pi U_{fc}^2/(1-\gamma)g$$

where Λ_c is the critical wavelength, U_{fc} is the flame velocity, γ is the thermal expansion ratio, and g is the acceleration of gravity.

Presented at the 9th ICODERS, Poitiers, France, July 3-8, 1983.

*Chargé de Recherche au C.N.R.S. Laboratoire de Dynamique et Thermophysique des Fluides.

†Professor, Laboratoire de Dynamique et Thermophysique des Fluides.

Introduction

The study of the stability threshold of a premixed flame front should provide fundamental information about the combustion process, including the coupling of reactive, diffusive, and convective mechanisms. The results of such a study may be used to check theoretical models before they are applied to more realistic situations in combustion. As in other fields, the onset of instability is revealed by the appearance of structures that are stationary at the threshold and that become turbulent as the system becomes more unstable.

Since the first observations of Smithells and Ingle (1892) and the quantitative study of Markstein (1964), which led to a phenomenological insight into the instability process, the understanding of this phenomenon has remained incomplete. The three fundamental mechanisms controlling stability --diffusion, hydrodynamics, and buoyancy-- had been recognized and described independently (Darrieus 1938, 1945; Landau 1944; Markstein 1964; Barrenblatt et al. 1962). Recently Pelcé and Clavin (1982) have developed a model which accounts for these three effects simultaneously and have shown that they are intimately and nontrivially coupled.

This analytical theory is concerned with the stability of an initially flat flame front propagating downwards and in which the reactive properties are approximated by an overall onestep reaction. In this case and for wavelengths large compared to the flame thickness d_f (this assumption is realistic as the cell size is about 1 cm, while the flame thickness is 0.01 cm), instability results from the competition between three mechanisms, two of which are stabilizing:

1) The flame front separates the "heavy" unburnt mixture (at the bottom of the burner) from the "light" hot gas; thus, the force of gravity tends to flatten any deviation from horizontality.

2) Ahead of a wrinkled flame front, the transverse diffusion of heat and reactants modify the local flame speed (Barenblatt et al. 1962) in such a way that an initial perturbation of the flame front is flattened by heat diffusion, whereas it is amplified by the mass diffusion of the reactants. However, advective fluxes created ahead of a wrinkled front interfere with the diffusive fluxes, so that the net effect is globally stabilizing for usual hydrocarbon flames (Clavin 1983).

The third mechanism is destabilizing: The normal acceleration of the gas across the flame front induces a deflection of the streamlines and gives rise, in turn, to a modification of the gas velocity ahead of the flame front. As a consequence, the incoming velocity of the unburnt gas is increased where the flame lags locally behind the average front position. This is the well-known hydrodynamic instability (Darrieus 1938, 1945; Landau 1944).

The strength of these effects is essentially controlled by three parameters:

1) The expansion coefficient,

$$\gamma = 1 - \rho_b/\rho_o \simeq 1 - T_o/T_b \simeq 1 - U_o/U_b$$

where ρ, T, U are, respectively, the density, temperature, and gas velocity in the fresh mixture (indexed with o) and in the burnt gas (indexed with b). This parameter controls simultaneously the strength of the buoyancy forces and of the hydrodynamic instability.

2) The normal flame speed, taken equal to the normal gas velocity U_o just ahead of the flame front. Increasing the normal flame velocity increases the strength of the hydrodynamic instability.

3) The Lewis number, $L_e = D_{th}/D_m$, which measures the relative efficiency of the diffusion of heat (D_{th} is the thermal diffusivity) to the diffusion of mass (D_m is the molecular diffusivity). In the present work, the mass fractions of the reactants are small, and the thermal diffusivity is essentially that of the dilutant (N_2). The effective Lewis number, which controls the actual way in which the flame speed depends on the different species present is not necessarily given simply by the diffusivity of the limiting component but may also depend on the concentration of intermediate species produced in the reaction layer. However, we may still expect the flame to be more stable when the concentration of the heavier reactant (propane in this work) is decreased.

The main conclusions of this linear analysis are shown on Fig. 1, where the amplification rate σ of an initial perturbation is plotted as a function of its wavenumber k (the deviation of the flame front position from the planar case is taken as $a(y,t) = a\, e^{iky+\sigma t}$). The stabilization by buoyancy effects is dominant at low wavenumbers, while transverse diffusion gives a contribution proportional to k^2 that stabilizes the perturbations at large wavenumbers. Independently of the equivalence ratio, an increase in the flame speed increases the positive contribution of hydrodynamical effects, and the amplification rate becomes positive for a critical flame speed U_{fc} related to the diffusive properties of the mixture and the expansion coefficient, γ. This instability threshold may be represented in a universal form in the plane of reduced Lewis number‡ ℓ vs flame speed (for a given value of γ). The original "universal" result of Pelcé and Clavin (1982) for adiabatic flames with constant transport properties has been extended by Clavin and Nicoli (1984) to include the effects of heat losses; and by Clavin and Garcia (1983) to include the effects of temperature-dependent transport coefficients. These latter considerations modify the form of the theoretical instability limit; however, one major conclusion of these authors is that to first-order analysis, the size of the structures appearing at the instability threshold can be expressed in a form that does not depend on the transport coefficients or the chemical kinetics, but is simply given by

$$\Lambda_c = \pi U_{fc}^2/(1-\gamma)g \qquad (1)$$

where g is the acceleration of gravity. This provides an opportunity to check this model without further assumptions concerning the actual chemical kinetics or the transport properties of the mixture.

Apparatus

The stability properties of a premixed flame must be studied in a flowfield that is uniform, homogeneous, and laminar on both sides of the flame front. This requirement eliminates devices commonly used for combustion studies, such as flame holders or porous plugs and also flames stabilized by stagnation flows, which strongly modify the flow conditions and the expected stability limits (Clavin 1983). For this study a flat flame burner (11 cm in diameter) has been constructed in which the flame is far from any obstacle but is dynamically stabilized by adjustment of the mean flow velocity (Fig. 2). The mass flow of each gas (fuel, air, nitrogen) is regulated by a sonic throat following two pressure regulators that reduce the fluctuations to about 0.1%.

‡ $\ell = \beta(1-1/L_e)$, where β is the reduced activation energy related to the actual activation energy E_a through $\beta=(E_a/kT_b)(T_b T_o)/T_b$. The Boltzmann constant is k.

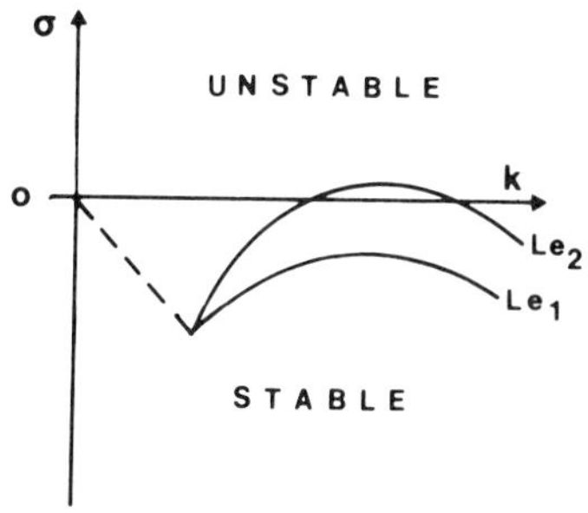

Fig. 1 Amplification rate σ vs wavenumber k for identical hydrodynamic conditions but with different Lewis numbers ($L_{e1} > L_{e2}$).

The secondary flow is regulated with an electrovalve controlled by the tomographic signal (Boyer et al. 1983). The flow velocity can be measured at any point in the burner using a laser Doppler anemometer. Oil droplets are used for tomography measurements and for velocity measurements in the unburnt gas, whereas alumina particles are used for velocity measurements in the burnt gas.

The main experimental difficulty is to maintain a uniform velocity field ahead of the flame front while the upstream part of the burner is nonuniformly heated by heat radiation from the downstream walls. This heating can induce convective flows with velocities of the same order of magnitude as the mean flow velocity. In these experiments, a nearly uniform temperature field is ensured up to 5 cm ahead of the flame front via differential wall heat exchangers (see Fig. 2). The temperatures of three independent cooling circuits are adjusted so that the wall temperature, seen by the reactive flow, is maintained as close as possible to the temperature of the flow itself. Because of thermal conduction along the burner walls, a thin convective boundary layer is established 4 to 5 cm ahead of the flame front, but the velocity field in the center of the burner does not vary more than 1% over a diameter of 9 cm and from 3 cm upstream of the flame front to the preheat zone (Fig. 3).

When the flame front is flat and stable, the normal flame speed is taken as the gas velocity just before the increase of velocity in the preheat zone. When the flame is unstable, the normal flame speed is taken as the mean velocity measured 0.3 to 0.5 cm upstream of the luminous zone of the flame. This measurement procedure is the only one that yields reproducible results for

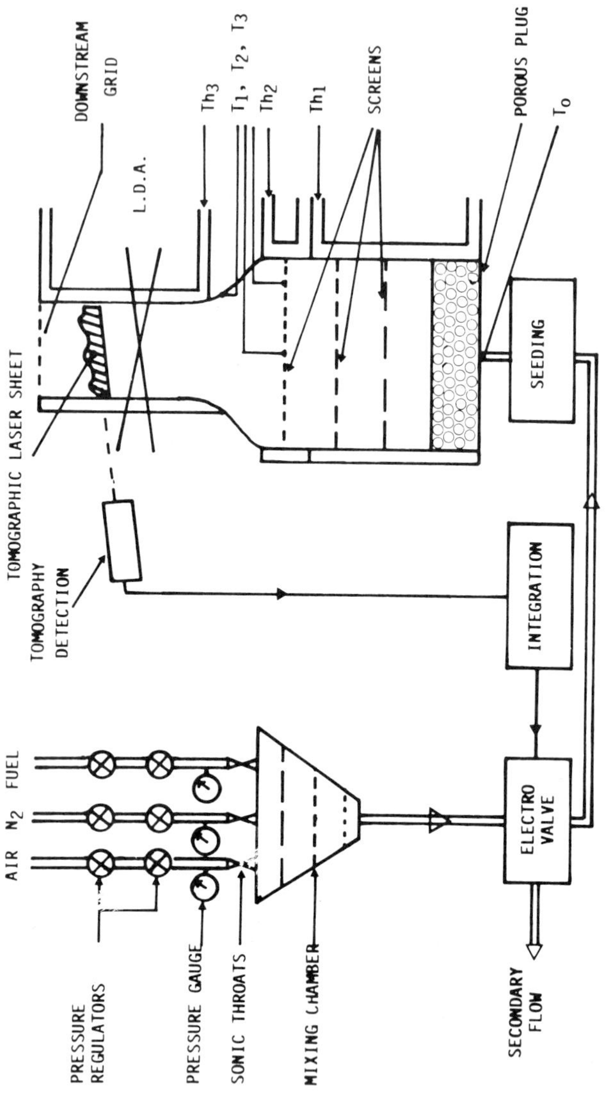

Fig. 2 Apparatus. Th1, Th2, Th3: Water flows. T_o, T_1, T_2, T_3: temperature probes. LDA: laser Doppler anemometry setup.

a given composition in nonuniform flowfields. In this latter case, the velocity field induced by the presence of the quasiplanar front is nonzero (see Searby et al. 1983). In this work, situations have been observed for which the induced field causes the total velocity field to vary by more than 50% in the distance 1 cm upstream of the preheat zone. These experiments suggest the basis of a standard protocol for the measurement of the "fundamental normal flame velocity", which can be defined uniquely as the gas velocity observed at the intersection between the fast varying region (preheat zone) and slow varying region (approach zone) for a free flame stationary in the laboratory frame. The results presented here were obtained at the center of the burner, but it has been checked that the normal flame speed is constant to within 1% over a diameter of 9 cm.

Close to the instability threshold, the wrinkles of the flame front are of small amplitude (less than 0.1 cm in height). The most sensitive technique for measuring the cell size is to produce photographs with a high contrast obtained by full development of a Kodak Tri-X Pan film (400 ASA) up to 1600 ASA. The views are taken from below, at an angle of about 45 deg. The photographs are then reduced to digital form for analysis with an HP85 computer to compensate the perspective effects and the optical deformations due to the cylindrical shape of the burner (Fig. 4). With this technique, it was possible to obtain reproducible results (flame speed, stability limit, critical cell size) to within 1%.

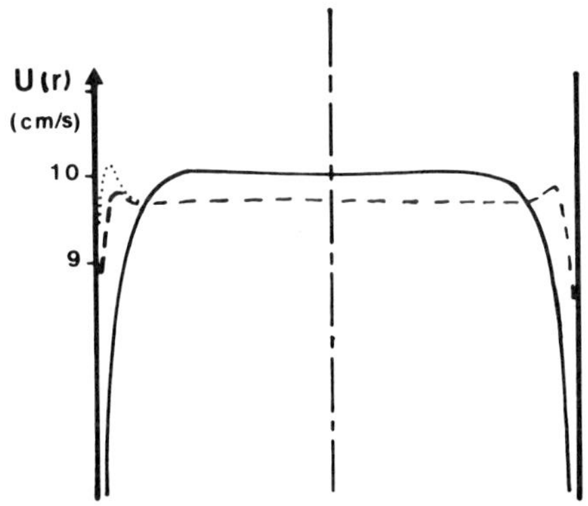

Fig. 3 Radial velocity profile in the presence of a flame (burner diameter = 11 cm). (——): 6 cm upstream of the flame front; (----): 2 cm upstream of the flame front; (····): 0.5 cm upstream of the flame front.

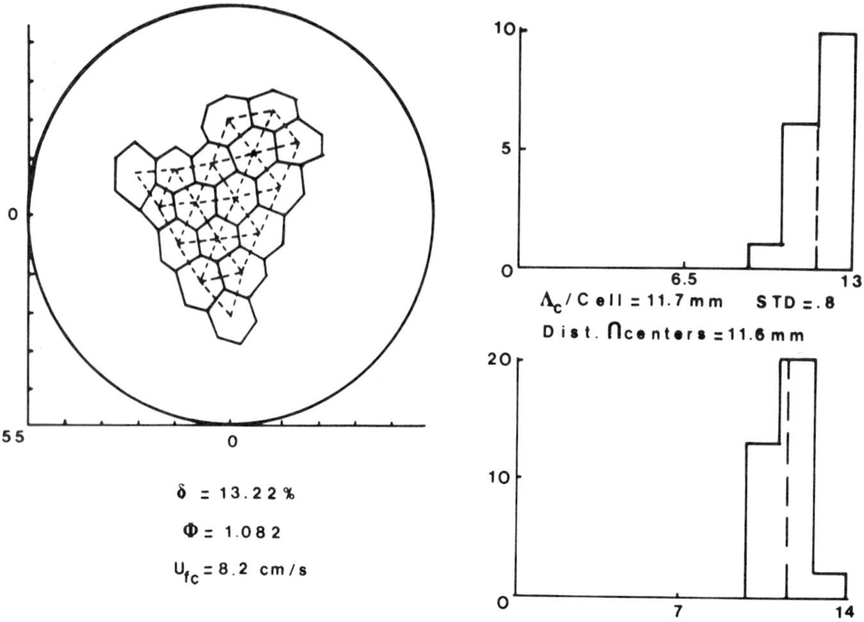

Fig. 4 Left side: reconstitution of the flame front structure. Upper right: histogram of the mean wavelength per cell (Λ_c in mm). Lower right: histogram of the mean distance between centers.

Results

The present work concerns propane/air mixtures diluted with nitrogen. The normal flame speed has been measured as a function of the equivalence ratio (between 0.8 and 1.2) and for various dilutions (from 12% to 16% of oxygen in the mixture) (see Fig. 5). The flame speed at the extinction limit was about 6 cm/s in this burner, and the temperature of the fresh mixture was maintained at 300 ±1 °K.

The stability limit was determined when the flame front showed the first perceptible signs of structure, but the cell size could be measured with sufficient precision only when the amplitude of the corrugation reached about 5% of its characteristic length, that is, when the flame speed was 1% or 2% larger than the critical flame speed at the threshold. However, experimental observations suggest that the averaged cell size remains relatively constant near the threshold. Figure 6 shows the measured cell size for cell amplitudes ranging from

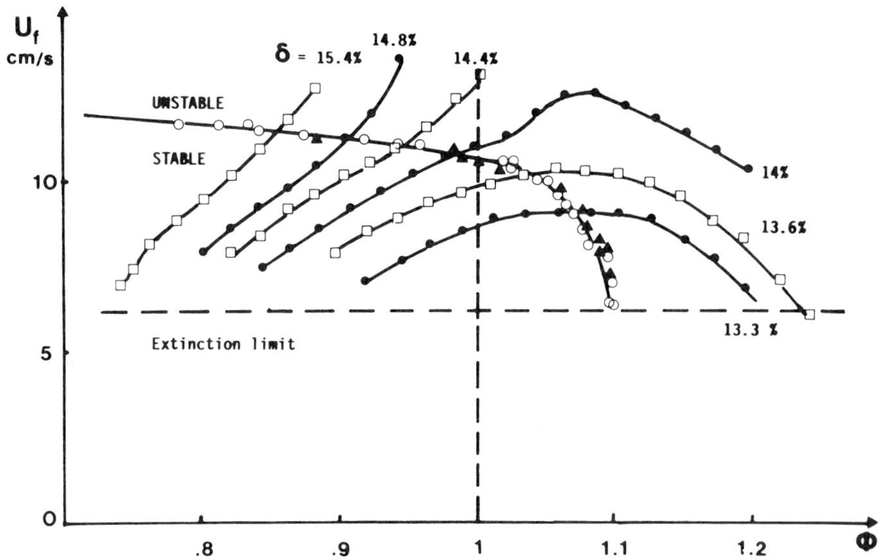

Fig. 5 □, ●: Flame velocity of $C_3H_8/O_2/N_2$ gas mixtures. Dilution $\delta = O_2/(O_2+N_2)$ by volume; ϕ = equivalence ratio; ○, ▲ = stability limit.

5% to 50% of the cell size. In the light of these remarks, these results should be comparable to the results obtained from the linear stability analysis established for the size of the structures at the threshold.

The representation of the stability limit (Fig. 5) is not universal, as it depends upon the diffusive properties of the mixture and the kinetic characteristics of the reactions. However, the comparison with the theoretical predictions (Fig. 7) evaluated for a simplified chemical scheme and with constant diffusion coefficients shows good qualitative agreement, particularly when heat losses are taken into account: The negative slope of the stability limit on the lean side of the mixture can be explained by considering the evolution of the Lewis number from a high value ($L_e(C_3H_8 \rightarrow N_2) \simeq 1.7$) towards smaller values ($L_e(O_2 \rightarrow N_2) \simeq 0.9$). Close to the stochiometric composition, the critical velocity decreases faster due to the joint effect of the decrease in Lewis number and the relative increase of heat losses.

The experimental values of the critical cell size are plotted as a function of the critical velocity in Fig. 8 along with the theoretical values calculated from

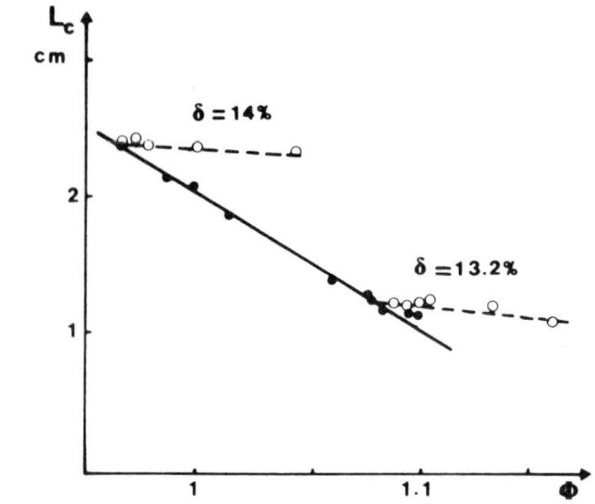

Fig. 6 Cell size as a function of the equivalence ratio. ●: averaged values at the threshold; ○: measured values when moving away from the threshold.

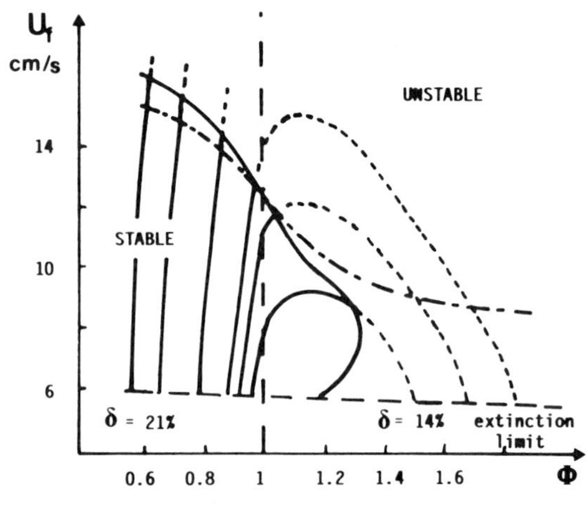

Fig. 7 Calculated values of the flame velocity and the stability limits (Clavin and Nicoli 1983). (— - —): stability limit in the adiabatic case; (———): stability limit with heat losses leading to an extinction velocity of 6 cm/s.

Eq. (1) and the experimental value of the expansion coefficient (Fig. 9). The mean cell size is in quantitative agreement with Eq. (1), and, moreover, it increases effectively with the critical flame speed as predicted. However, the expected parabolic behavior is not well verified. Below 10 cm/s, the critical wavelength tends to a limiting value of about 1 cm. This occurs on the

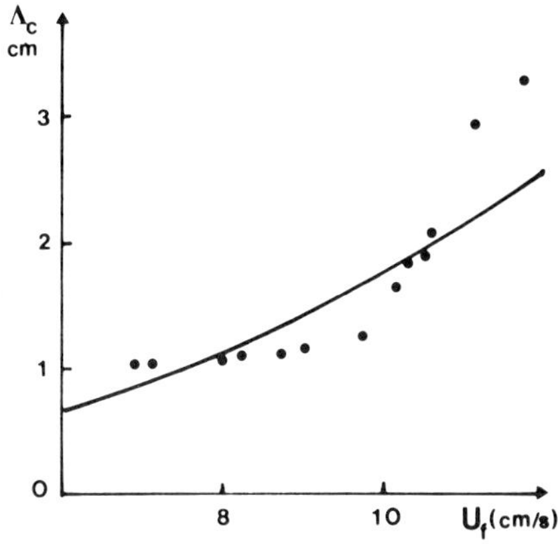

Fig. 8 Experimental values of the critical wavelength vs critical flame velocity. Full line: theoretical prediction with the experimental value of γ.

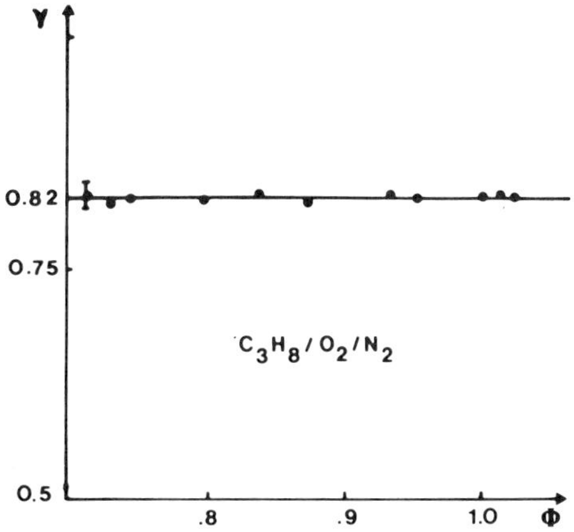

Fig. 9 Measured expansion coefficient as a function of the equivalence ratio along the stability limit.

rich side of the marginal stability curve close to the extinction limit, where heat losses become important and may affect the validity of the linear analysis. The stabilizing higher-order terms in the development of the linear analysis would also lead to a relative increase of the critical wavelength particularly at low flame speeds.

When the critical flame speed is greater than 10 cm/s, the slope is steeper than predicted and varies roughly as U_{fc}^3. Small terms neglected in the derivation of relation (1), would produce a similar deviation but their magnitude is too small to explain the observed discrepancy. However, it is not certain that the finite size of the burner (and hence of the flowfield) has no effect on the size of larger structures. Further experiments are planned in different sized burners to elucidate this point.

Finally, it is possible that these discrepancies arise from nonlinear effects, absent from the theoretical models, but which are necessarily present and limit the amplitude of the cell height in the unstable domain.

Conclusion

These experiments, performed in conditions as close as possible to those considered in the theoretical models, confirm the existence of a stability threshold with the appearance of structures of finite size at the threshold. This finite size results from the competition between three fundamental mechanisms controlling the stability of a planar premixed flame propagating downwards: buoyancy, hydrodynamics, and diffusion. For a given equivalence ratio, that is, for given diffusive properties of the reactants, the instability threshold is crossed when the strength of the hydrodynamic instability is increased following an increase in the normal flame speed. The critical velocity decreases with the effective Lewis number, owing to the strengthening of destabilizing mass fluxes. As a first approximation, the cell size is given by $\Lambda_c = \pi U_{fc}^2/(1-\gamma)g$, in qualitative agreement with the theory; but further experiments are needed with other fuels and in different burners to establish whether the observed deviations of critical cell sizes are caused by finite burner size and heat losses or by transport effects arising from chemical kinetic schemes that cannot be considered to be confined to a thin reaction zone.

Acknowledgments

Thanks are due to J. P. Pahin and F. Abetino for their technical assistance. This work was supported in part by the CNRS and by the Direction des Recherches et Etudes Techniques under Contract No. 80/434.

References

Barenblatt, G. I., Zel'dovich, Y. B., and Istratov, A. G. (1962) On diffusional thermal stability of laminar flame. Prikl. Mekh. Fiz. 2, (n°4) 21-26.

Boyer, L., Searby, G., and Quinard, J. (1983) Some recent applications of laser tomography in combustion. Proceedings of XV ICHMT Symposium on Heat and Mass Transfer Measurement Techniques, Dubrovnik, Hemisphere publishing corporation (to appear).

Clavin, P. (1984) Dynamical behavior of premixed flame fronts in laminar and turbulent flows. Progress in Energy and Combustion Science, Pergamon Press, New York (to appear).

Clavin, P. and Garcia, P. (1983) The influence of the temperature dependence of diffusivities on the dynamics of flame fronts. J. Mec. Théo. Appl. 2, (n°2) 245-263.

Clavin, P. and Nicoli, C. (1984) Effect of the heat-losses on the limits of stability of premixed flames propagating downwards. Combust. Flame (to appear).

Darrieus, G. (1938, 1945) Propagation d'un front de flamme. Essai de théorie des vitesses anormales de déflagration par dévelopement spontané de la turbulence. Presented at La Technique Moderne (1938) and at Le Congrès de Mécanique Appliquée (unpublished).

Landau, L. (1944) On the theory of slow combustion. Acta Physicochim. URSS 19 (n°1), 77-85.

Markstein, G. H. (1964) Nonsteady Flame Propagation (p.15-103). Pergamon Press, New York.

Pelcé, P., and Clavin, P. (1982) Influence of hydrodynamics and diffusion upon the stability limits of laminar premixed flames. J. Fluid Mech. 124, 218-237.

Searby, G., Sabathier, F., Monreal, J., Clavin, P. and Boyer,L. (1983) The feed-back of a flame front on turbulent flows. IXth International Colloquium on Dynamics of Explosions and Reactive Systems, Poitiers, Ed.AIAA Progress in Astronautics and Aeronautics.

Smithell, A. and Ingle, H. (1982) J. Chem. Soc. 61, 204.

Acta Astronautica. Vol. 4, pp. 1177–1206. Pergamon Press 1977. Printed in Great Britain

Nonlinear analysis of hydrodynamic instability in laminar flames—I. Derivation of basic equations

G. I. SIVASHINSKY

Department of Mathematical Sciences, Division of Applied Mathematics, Tel-Aviv University, Ramat-Aviv, Israel

(*Received* 24 *February* 1977; *revised* 16 *June* 1977)

Abstract—An asymptotic nonlinear integrodifferential equation is derived for spontaneous instability of the plane front of a laminar flame.

If the combustible mixture is deficient in the light component, spontaneous instability will lead to self-turbulization of the flame, and the flame front assumes a strongly nonstationary cellular structure.

If there is an excess of the light component, spontaneous instability produces stationary, irregular wrinkles on the flame front, and the flame continues to propagate in a laminar regime.

It is shown that in all cases spontaneous instability of the flame implies an increase in its propagation velocity.

Introduction

It is well known that the difficulties involved in setting up a theory of flame propagation in turbulent flow of a gaseous combustible mixture are due largely to the problem of spontaneous instability of laminar flames. This question was discussed in detail by Markstein (1964), and also in a recent survey of turbulent combustion.†

Owing to the intrinsic instability of a laminar flame front, the turbulent field of the incoming flow causes (in the linear approximation) unlimited amplification of long-wave disturbances of the front. This effect is a serious obstacle to developing a general theory of flame propagation by direct asymptotic expansion in powers of the amplitude of the external turbulent pulsations.

In the context of a flame progagating freely in *open space*, nonlinearity is evidently the only factor inhibiting unlimited amplification of disturbances. Therefore, any consistent theory of turbulent flame propagation must take it into consideration.

At the present time, effective nonlinear analysis of hydrodynamic instability is feasible only in cases that can be formulated in terms of bifurcation theory. In such cases one can often accomplish asymptotic analysis in the neighborhood of a bifurcation point (a certain value of the parameter, representing the transition from absolute stability to instability), and this makes it possible to describe certain essentially nonlinear effects.

†See Libby and Williams (1976); also Andrews, Bradley and Lwakabamba (1975); Williams (1970).

As we shall see later, spontaneous instability of a laminar flame does indeed fall into this category of problems.

It is now known that spontaneous instability of a flame is due primarily to two destabilizing effects: (i) thermal expansion of the gas passing through the flame front, and (ii) interaction of the diffusion and heat-conduction processes within the flame front structure.

The destabilizing effect of thermal expansion was noticed by Landau (1944), who treated the flame as a density discontinuity in an ideal fluid, traveling at constant velocity relative to the gas. It is clear that this effect is always present in flame of exothermal reactions.

The second destabilizing effect was described in a qualitative framework by Zel'dovich (1944), and later subjected to a quantitative analysis by Barenblatt, Zel'dovich, Istratov (1962) and Sivashinsky (1977). It is found that the destabilizing effect occurs if the diffusivity of the component limiting the reaction exceeds the thermal conductivity of the mixture. In other words, the second effect appears only in mixtures in which the more mobile (i.e. more easily diffusing) component is present in small concentrations.

These two destabilizing effects have been corroborated by a great number of experiments studying laminar flame stability.†

If the mixture is deficient in the heavy component (e.g. lean mixtures of heavy hydrocarbons with air), the flame will experience only hydrodynamic instability. In that case the flame front is criss-crossed by wrinkles. The distribution of these wrinkles over the surface of the flame front is extremely nonuniform, depending on the initial disturbance imparted to the smooth flame front. If there are no persistent external disturbances of sufficiently high amplitude (e.g. turbulence in the incoming flow), the resulting corrugated flame front is highly stable to small disturbances and the initial pattern of the wrinkles is well preserved. In a freely propagating spherical flame in a stationary gas, the wrinkles grow and deepen as the flame front expands, though their number remains the same. Such corrugated flames have been produced artificially by passing flames through wire gauze (Markstein, 1964). When the flame passes through the gauze, the latter leaves a characteristic periodic pattern of wrinkles on the front, and these are clearly preserved in further propagation of the flame.

The appearance of inhomogeneity on the flame front is accompanied by a marked increase in the mean propagation velocity of the flame, compared with the case of a smooth plane, flame front. Note that if there is no external turbulence a corrugated flame propagates in a clear-cut laminar regime.

If there is an excess of the heavy component (e.g. rich mixtures of heavy hydrocarbons and air) the effect of hydrodynamic instability is superimposed on that of diffusional-thermal instability. Considering a freely propagating, expanding spherical flame in an initially stationary gaseous mixture, one discerns three main stages of spontaneous instability (Istratov and Librovich, 1969). At first, the spherical flame front is covered with widely spaced, irregular wrinkles. Only

†See, e.g. Markstein (1964), Schelkin and Troshin (1964), Istratov and Librovich (1969), Palm-Leis and Strehlow (1969).

hydrodynamic instability of the type described above can arise at this stage. Then, when the front exceeds a certain critical radius, its entire surface becomes covered with uniformly distributed weak protuberances, somewhat reminiscent in appearance of regular hexagons of equal size. With further increase in flame radius, the amplitude of the protuberances increases, their regular hexagonal shape disappears, but their typical characteristic dimension remains unchanged. This stage is that of cellular flame structure. The size of the cells depends only on the chemical composition of the combustible mixture and may vary over a wide interval: from fractions of a millimeter to a centimeter.

As in the case of noncellular corrugated flames, the wrinkles in a cellular flame front are sharper toward the combustion products. Since the characteristic dimension of the cells remains the same as the flame propagates, the number of cells increases continually as the radius of the sphere increases. For some time following the appearance of cells, the flame continues to propagate in a quasi-stationary laminar regime. With further increase in radius, the flame becomes unstable and gradual self-turbulence sets in. The turbulent flame front shows the same dimension of inhomogeneity as the quasistationary cellular front; in fact, it may be regarded as an essentially nonstationary cellular flame in a state of constant, irregular pulsation, with the cells breaking up and merging at random. The turbulent cellular structure gradually destroys the initial wrinkles, which were due to hydrodynamic instability.

We emphasize that under conditions of diffusional-thermal instability the flame spontaneously becomes turbulent even when the incoming flow shows no signs of turbulence far from the flame.

The transition of the flame from the corrugated noncellular to the turbulent regime of propagation is characterized by a constant increase in its mean propagation velocity. With increase in the radius of the sphere, this velocity gradually overshoots a certain maximum corresponding to saturation in the development of turbulence. This maximum may apparently be treated as the normal velocity of a "plane" turbulent flame front propagating in a self-turbulent regime.

The purpose of study in this paper is nonlinear analysis of hydrodynamic flame instability due to thermal expansion of the gas and to transport effects.

Adopting the simplest possible assumptions concerning diffusion and chemical kinetics, the following asymptotic equation describing the evolution of a disturbed plane flame front was derived:

$$F_t + 4(1+\epsilon)^2\nabla^4 F + \epsilon\nabla^2 F + \frac{1}{2}(\nabla F)^2 = \frac{(1-\sigma)}{8\pi^2}\int_{-\infty}^{\infty} |\mathbf{k}|\, e^{i\mathbf{k}(\mathbf{x}-\mathbf{z})}F(\mathbf{z},t)\,d\mathbf{k}\,d\mathbf{z} \qquad (1)$$

Here $F = F(\mathbf{x}, t)$ is the dimensionless perturbation of the flame front surface in units of the width l_T of thermal structure of flame, $\mathbf{x}$ = dimensionless space coordinates in units of l_T, t = dimensionless time in units of l_T/U_b, U_b = normal velocity of the flame (velocity of a plane flame front relative to the combustion products), $\epsilon = (L_0 - L)/(1 - L_0)$, $1/2N(1-\sigma)(1-L_0) = 1$, N = dimensionless activation energy in units of R^0T_b, R^0 = universal gas constant, T_b = adiabatic temperature of combustion products, σ = coefficient of thermal expansion of the gas ($\sigma < 1$), L = Lewis number for the component of the mixture limiting the reaction.

Equation (1) covers the entire range of diffusional, thermal and hydrodynamic effects in the flame. At $L > L_0$ one can also consider an asymptotic situation that leads to the equation

$$F_t + \epsilon\nabla^2 F + \frac{1}{2}(\nabla F)^2 = \frac{(1-\sigma)}{8\pi^2}\int_{-\infty}^{\infty} |\mathbf{k}|\, e^{i\mathbf{k}(\mathbf{x}-\mathbf{z})} F(\mathbf{z}, t)\, d\mathbf{k}\, d\mathbf{z}, \qquad \epsilon < 0. \tag{2}$$

Asymptotic analysis is based on the conditions that (i) the Lewis number is near a certain critical value $L_0(L_0 < 1)$, (ii) the thermal expansion of the gas is weak and (iii) the nondimensional activation energy is a large number.

Flame model and fundamental equations

Despite the fact that real combustion mixtures contain more than two components and the reactions involve many steps—these effects play a secondary role in the formation of cellular (or noncellular) flame structure and the onset of self-turbulence; they may therefore be disregarded in a qualitative explanation of the relevant physical mechanism. Below we shall adopt the simple hydrodynamic flame model proposed by Istratov and Librovich (1966) in their work on the effect of transport processes on the hydrodynamic stability of a plane flame front.

In this model the flame is described exclusively by the diffusion equation for the concentration of the component M limiting the combustion reaction, by the heat conduction equation and the Navier–Stokes equations. The transport coefficients are assumed to be constant.

The reaction zone is assumed to be narrow, concentrated on a certain surface (the flame front) where the heat and mass flows are discontinuous and the concentration of the limiting component vanishes.

A model of this type may be derived rigorously from the general equations of combustion theory, for the case, say, of a bimolecular reaction with high activation energy and high heat release, when one of the reactants is present in minute concentrations. One must also assume in that case, that the initial fresh mixture consists entirely of reacting components, or that the reactants are weak traces in some homogeneous, "inert" gas (Fristrom and Westenberg, 1965). The flame is assumed to propagate slowly in open space (quasi-isobaricity). The condition that the transport coefficients are constant is not essential for the subsequent analysis, and is easily eliminated if necessary.

Such an elementary model is of course incapable of describing the possible effects of multicomponent diffusion or the transition from combustion to detonation. Nevertheless, for all its simplicity, the model of Istratov and Librovich (1966) proves adequate to describe cellular structure, wrinkles, self-turbulence and the resultant increase in the mean propagation velocity of the flame. In a full gas-dynamical model the latter effect should ultimately produce a detonation wave.

Below is the framework of the above model, we shall present a non-linear analysis of the stability of a propagating plane flame front. Subject to suitable choice of dimensionless variables (see Introduction), the system of equations for the flame may be written as follows:

Diffusion:

$$\rho \frac{\partial C}{\partial t} + \rho U_i \frac{\partial C}{\partial x_i} = \frac{1}{L} \frac{\partial}{\partial x_i} \frac{\partial C}{\partial x_i} - Q\delta_F. \tag{3}$$

Heat conduction:

$$\rho \frac{\partial T}{\partial t} + \rho U_i \frac{\partial T}{\partial x_i} = \frac{\partial}{\partial x_i} \frac{\partial T}{\partial x_i} + (1-\sigma)Q\delta_F. \tag{4}$$

Continuity:

$$\frac{\partial \rho}{\partial t} + \frac{\partial \rho U_i}{\partial x_i} = 0. \tag{5}$$

Momentum:

$$\frac{\partial \rho U_i}{\partial t} + \frac{\partial \rho U_i U_k}{\partial x_k} = \frac{\partial P_{ik}}{\partial x_k}, \tag{6}$$

where

$$P_{ik} = \nu\left(\frac{\partial U_i}{\partial x_k} + \frac{\partial U_k}{\partial x_i} - \frac{2}{3}\delta_{ik}\frac{\partial U_m}{\partial x_m}\right) - P\delta_{ik}.$$

Dynamic incompressibility condition:

$$\rho = \frac{1}{T} \tag{7}$$

$$Q = \exp \frac{1}{2} N(T-1), \tag{8}$$

where δ_F is the surface delta-function

$$\delta_F = \sqrt{[1+(\nabla F)^2]}\,\delta(x_1 - F), \tag{9}$$

$x_1 = F(x_2, x_3, t)$ is the perturbation of the plane flame front ($x_1 = 0$).

T is the dimensionless temperature, referred to adiabatic temperature T_b of combustion products; C = dimensionless concentration of the limiting component, referred to its initial concentration C_0; ρ = dimensionless density, referred to density ρ_b of combustion products; P-dimensionless pressure, referred to $\rho_b U_b^2$; U_i = gas velocity referred to U_b; ν = Prandtl number. For a justification of the eqn (8) defining the reaction rate intensity, we refer the reader to Sivashinsky (1975).

Far ahead of the flame front ($x_1 \to -\infty$) the temperature of the gas is equal to the temperature T_0 of the fresh mixture, while the concentration of the combustible component is constant and equal to C_0. Beyond the flame front ($x_1 \to +\infty$) the concentration is zero, and the temperature of the gas is equal to the adiabatic temperature T_b of the combustion products. In the dimensionless variables introduced above, these conditions may be written thus

$$\begin{gathered} T(-\infty, x_2, x_3, t) = \sigma, \qquad C(-\infty, x_2, x_3, t) = 1, \\ T(+\infty, x_2, x_3, t) = 1, \qquad C(x_1, x_2, x_3, t) \equiv 0 \quad \text{for} \quad x_1 \geq F(x_2, x_3, t). \end{gathered} \tag{10}$$

The following relations are satisfied on the flame front ($x_1 = F$) (see (3), (4), (8)–(10))

$$\begin{gathered} [\nabla T \cdot \mathbf{n}] + (1 - \sigma)Q = 0, \qquad [\nabla C \cdot \mathbf{n}] - LQ = 0, \\ [T] = 0, \qquad C = 0 \end{gathered} \tag{11}$$

where $\mathbf{n}$ is the normal to the surface $x_1 = F(x_2, x_3, t)$. For the undisturbed plane flame front ($F = 0$) we have

$$\begin{gathered} Q^0 = 1; \qquad T^0 = \sigma + (1 - \sigma) \exp x_1, 1; \qquad C^0 = 1 - \exp Lx_1, 0; \\ P^0 = 2 - \sigma + (1 - \sigma)\left(\frac{4}{3}\nu - 1\right) \exp x_1, 1, \end{gathered} \tag{12}$$

for $x < 0$ and $x > 0$, respectively

$$U_1^0 = T^0, \qquad \rho^0 = (T^0)^{-1}, \qquad U_2^0 = U_3^0 = 0,$$

(The superscript 0 indicates the solution of the one-dimensional stationary problem.)

For the further analysis, it is convenient to transform to a coordinate system attached to the surface of the disturbed flame front

$$x_1 = x + F(y, z, t), \qquad x_2 = y, \qquad x_3 = z. \tag{13}$$

Auxiliary model problem. Linear stability analysis

Before discussing nonlinear instability of a plane flame front, we shall find it extremely useful to carry out the classical, linear analysis of the stability of one-dimensional propagation of a plane flame front to infinitesimal periodic disturbances. Solution of this relatively simpler problem will provide us with valuable information for the subsequent asymptotic nonlinear analysis.

However, it is well known that linear stability analysis in the general case involves a mathematical problem of great difficulty: to determine conditions for the existence of a nontrivial solution to a system comprising a large number of ordinary linear differential equations with variable coefficients. The variability of the coefficients is due to the temperature-dependence of the density, and this constitutes the main difficulty, precluding reduction of the problem to the investigation of a certain algebraic equation.†

However, since the aim of our linear analysis is only to obtain some idea of the orders of magnitude involved, it will be quite sufficient to consider a certain approximate flame model, which is amenable to analytical solution. This model will be a "brute-force" combination of the hydrodynamic model of Landau (1944) and the diffusional-thermal model of Zel'dovich *et al.* (1944) and Barenblatt *et al.* (1962). In other words, we shall treat the flame as a discontinuity jump in a non-viscous, incompressible fluid, but without neglecting the effects of diffusion, heat conduction and temperature dependence of the reaction rate. The model will clearly show both hydrodynamic and diffusional-thermal destabilizing effects.

As the density is now constant both ahead of and behind the front, it is easy to derive an algebraic relation between the stability index ω, the disturbance wavenumber k, and the parameters σ, L and N. Indeed, set

$$\rho_- \equiv 1/\sigma \quad \text{for} \quad x<0, \qquad \rho_+ \equiv 1 \quad \text{for} \quad x>0, \qquad \text{and also} \quad \nu = 0. \tag{14}$$

In the linear approximation, the solution to the perturbed system of equations (transformed) (3)–(8) is sought in the form‡

$$\begin{aligned} F &= D \exp(iky + \omega t), & C &= C^0 + F\tilde{s}(x), \\ T &= T^0 + F\tilde{\theta}(x), & P &= P^0 + F\tilde{p}(x), \\ U_1 &= T^0 + F\tilde{u}(x), & U_2 &= F\tilde{v}(x). \end{aligned} \tag{15}$$

where D is the amplitude of small periodic perturbations. From (20), (22)–(25), (30), (31) we obtain the system of ordinary differential equations for the functions $\tilde{s}(x)$, $\tilde{\theta}(x)$, $\tilde{p}(x)$, $\tilde{u}(x)$, $\tilde{v}(x)$:

$$A_i(\tilde{s}(x), \tilde{\theta}(x), \tilde{p}(x), \tilde{u}(x), \tilde{v}(x)) = 0 \qquad i = 1, 2, 3, 4, 5 \tag{16}$$

For brevity's sake, we shall not give the explicit form of (16).

The mathematical problem of stability may be stated thus: determine the spectrum $\omega = \omega(k)$ in such a way that the problem (16) has a nontrivial solution which goes to zero as $x \to \pm\infty$.

We shall seek conditions under which $Re\,\omega$ is nonnegative, i.e. the flame is unstable. The solution of system (16) presents no mathematical difficulties, and we shall therefore skip the intermediate steps and present the final result:

The rate of stability parameter $\omega = \omega(k)$ is determined from the algebraic

†A survey of the research on stability of flames under small perturbations, may be found in Istratov and Librovich (1965) (see also Andrews *et al.* (1975)).

‡To simplify matters, we shall confine ourselves to the case of two-dimensional disturbances, i.e. the derivatives with respect to z are assumed to vanish everywhere.

equation

$$\frac{1}{2}N(1-\sigma)=\frac{kB[R(1,\sigma)+R(1,1)]}{(B-\omega+k\sigma)[L-1+R(1,\sigma)-R(L,\sigma)]}+\frac{[L-R(L,\sigma)][R(1,\sigma)+R(1,1)]}{2[L-1+R(1,\sigma)-R(L,\sigma)]}, \tag{17}$$

where

$$R(L,\sigma)=\sqrt{(L^2+4k^2+4k\omega/\sigma)}.$$

For $x<0$

$$\tilde{u}(x)=B\exp kx, \tag{18}$$

where

$$B=\frac{(1-\sigma)(\omega^2+2k\omega+\sigma k^2)}{2(k+\omega)}.$$

We shall not present the cumbersome expressions for the remaining quantities $\tilde{p}$, $\tilde{v}$, $\tilde{u}(x>0)$. In the next section we shall give asymptotic expressions for them—the only information necessary for the subsequent analysis.

The first term on the right of (17) clearly represents the perturbation of the flame due to thermal expansion, while the second is due to the diffusional-thermal effect.

We first consider the case in which the concentration and temperature fields are similar, $L=1$. This classical case is of particular interest, since when $L=1$ the temperature at the exit from the reaction zone is not affected by the disturbance of the front but remains equal to T_b. The coefficient of the activation energy vanishes (see (17)) and no longer figures in the solution. On the other hand (Zel'dovich and Barenblatt, 1959), when $L=1$ the diffusion and conduction processes exercise only a stabilizing effect on the flame front. Thus, in the case of similarity the hydrodynamic instability effect appears in its simplest and purest form (i.e. uncontaminated by diffusional-thermal instability or by temperature-dependence of the reaction rate).

When $L=1$ and $\omega=0$, we obtain from (17)

$$k_0=\frac{(1-\sigma)(3-\sigma)}{4(2-\sigma)}. \tag{19}$$

Even without the effect of high activation energy and viscosity, this value of the critical wavenumber testifies to the powerful stabilizing effect of heat conduction. Indeed, for a typical thermal expansion coefficient, say $\sigma=0.2$, we have

$$k_0\simeq 0.31, \qquad \Lambda_0=2\pi l_T/k_0\simeq 20.2 l_T.$$

Thus, near the stability limit the characteristic disturbance wavelength considerably exceeds the width l_T of the thermal structure of the flame. Here thermal expansion of the gas is the sole destabilizing factor. As $\sigma\to 1$, therefore, the instability index $Re\,\omega$ will tend to zero. Thus $\sigma = 1$ is a bifurcation point of the problem.

On the other hand, we see from (19) that as $\sigma \to 1$

$$k_0 \simeq \frac{1}{2}(1-\sigma). \tag{20}$$

Thus, in the neighborhood of the bifurcation point the structure of the disturbed flame is not only quasistationary ($\omega \ll 1$) but also quasi-one-dimensional ($k \ll 1$).

By comparing (19) and (20), one can derive a rough estimate of the error in the principal term of the asymptotic expansion with respect to $(1-\sigma)$. For example, for $\sigma = 0.2$ the relative error is $\sim 30\%$. To improve the approximation, we must of course proceed to higher approximations relative to $(1-\sigma)$.

Consider the second extreme case

$$L = 1, \qquad k \sim \omega \ll 1.$$

From (17) we obtain the asymptotic case of Landau (1944)

$$\omega \simeq \frac{k\{\sqrt{[\sigma + \sigma^2(1-\sigma)]} - \sigma\}}{1+\sigma}, \tag{21}$$

or as $\sigma \to 1$

$$\omega \simeq \frac{1}{2}(1-\sigma)k. \tag{22}$$

Equations (20) and (22) give an indication of the correct scaling factor for ω and k in the case $\sigma \to 1$:

$$k = (1-\sigma)K, \qquad \omega = (1-\sigma)^2\Omega. \tag{23}$$

From this and (17) we obtain the following asymptotic equation:

$$\Omega = \frac{1}{2}K - K^2 \qquad \text{or} \qquad \omega = \frac{1}{2}(1-\sigma)k - k^2. \tag{24}$$

Equations (24) provides a corroboration of the well-known Markstein (1964) condition, which relates the propagation rate of the flame to the curvature of the front.

In the general case, $L \neq 1$, we can also scale the variables in accordance with (23) and isolate the principal term of the asymptotic expansion as $\sigma \to 1$, $N \to \infty$, $(1-\sigma)N < \infty$. By (17),

$$\frac{K^2 - 1/2KL + \Omega L}{K^2} = \frac{1}{2}N(1-\sigma)(1-L),$$

or

$$\omega = \frac{1}{2}(1-\sigma)k - \frac{1}{L}\left[1 - \frac{1}{2}N(1-\sigma)(1-L)\right]k^2. \tag{25}$$

Like (24), eqn (25) may be written as a linearization of Markstein's condition (1964):

$$\frac{\partial F}{\partial t}=\frac{1}{L}\left[1-\frac{1}{2}N(1-\sigma)(1-L)\right]\frac{\partial^2 F}{\partial y^2}+u'(0, y, t), \tag{26}$$

where $u'(x, y, t)$ is the perturbation of the x-component of the velocity of the gas.

For condition (26) to be well-posed, the coefficient of the second derivative must be positive. In other words, the Lewis number L should satisfy the condition

$$L>L_0\equiv 1-\frac{2}{N(1-\sigma)}. \tag{27}$$

Thus, for finite K, Ω, $N(1-\sigma)$, the asymptotic treatment at $\sigma\to 1$, $N\to\infty$ is admissible only for sufficiently large Lewis numbers ($L>L_0$).

To describe the case $L<L_0$, we begin with the asymptotic case $\sigma\to 1$ when ω, k and $N(1-\sigma)$ are finite.

In the first approximation, the term representing hydrodynamic instability drops out of the expression on the right of eqn (17). We obtain an equation describing only the effect of the diffusion and conduction effects on flame stability:

$$\frac{1}{2}N(1-\sigma)=\frac{R(1, 1)[L-R(L, 1)]}{L-1+R(1, 1)-R(L, 1)}. \tag{28}$$

At $\omega=0$, we obtain from (28) an expression for the critical wavenumber:

$$k_0=\frac{1}{2}\sqrt{\left[\frac{1}{2}N(1-\sigma)(1-L)-1\right]}. \tag{29}$$

Thus, in direct opposition to the previous asymptotic case, neutral stability is possible here only at sufficiently small Lewis numbers ($L<L_0$).

At $\omega\sim k^2\ll 1$, we obtain from (28)

$$\omega\simeq\left[\frac{1}{2}N(1-\sigma)(1-L)-1\right]k^2. \tag{30}$$

Thus, at $L<L_0$ a plane flame front is unstable to long-wave disturbances. $L=L_0$ is clearly a bifurcation point for the diffusional-thermal problem.

We consider the situation when L is close to L_0, or

$$\frac{1}{2}N(1-\sigma)(1-L)-1=\epsilon, \quad (\epsilon\ll 1). \tag{31}$$

Let us see what happens to eqn (28) in this asymptotic case. Equations (29) and

(30) indicate the correct scaling factor for the parameters ω and k:

$$k = \sqrt{\epsilon}\tilde{K}, \qquad \omega = \epsilon^2\tilde{\Omega}. \tag{32}$$

From this and (28) we obtain the following asymptotic relation between $\tilde{\Omega}$ and $\tilde{k}$:

$$\tilde{\Omega} = \tilde{K}^2 - 4\tilde{K}^4,$$

or

$$\omega \simeq \left[\frac{1}{2}N(1-\sigma)(1-L) - 1\right]k^2 - 4k^4. \tag{33}$$

At $L < L_0$ the flame is stable only to short-wave disturbances ($k > k_0$), and unstable to long-wave disturbances. Here there is a number k_c corresponding to the most rapid amplification of disturbances (maximum ω):

$$k_c = \frac{\sqrt{2}}{4}k_0. \tag{34}$$

In terms of the disturbed flame front, eqn (33) may be written as

$$\frac{\partial F}{\partial t} = -4\frac{\partial^4 F}{\partial y^4} + \left[1 - \frac{1}{2}N(1-\sigma)(1-L)\right]\frac{\partial^2 F}{\partial y^2}. \tag{35}$$

Condition (35) is well-posed for any values of L, since the coefficient of the fourth derivative is always negative.

The essential limitation of the result (35) is that the destabilizing effect of thermal expansion does not appear in the asymptotic relations (33), (35). In order to bring this effect into play, while retaining the advantages of eqn (35), we must consider the asymptotic situation as $\sigma \to 1$, $\epsilon \to 0$ simultaneously. To find a suitable uniform asymptotic relationship, we use the scaling factors of (32) and eqn (17). Letting $\epsilon \to 0$, $\sigma \to 1$ in (17), we obtain

$$-\frac{(1-\sigma)}{2\epsilon\sqrt{\epsilon}}\frac{1}{\tilde{K}} + \frac{\tilde{\Omega} + 4\tilde{K}^4}{\tilde{K}^2} = 0. \tag{36}$$

Hence it is evident that if the hydrodynamic term is to be regained as $\sigma \to 1$, the relevant orders of magnitude must satisfy the condition

$$(1-\sigma) \sim \epsilon\sqrt{\epsilon} \qquad \text{or} \qquad \frac{1}{2}N(1-\sigma)(1-L) - 1 = \alpha(1-\sigma)^{2/3}. \tag{37}$$

α is a certain constant of the order of unity. In terms of the original variables ω and k, eqn (36) may be written as

$$\omega + 4k^4 + \left[\frac{1}{2}N(1-\sigma)(1-L) - 1\right]k^2 - \frac{1}{2}(1-\sigma)k = 0, \tag{38}$$

or, in terms of the disturbed flame front,

$$\frac{\partial F}{\partial t} + 4\frac{\partial^4 F}{\partial y^4} + \left[\frac{1}{2}N(1-\sigma)(1-L) - 1\right]\frac{\partial^2 F}{\partial y^2} = u'(0, y, t). \quad (39)$$

Equation (39) is a generalization of Markstein's condition (26); this uniform asymptotic treatment includes both hydrodynamic and dissipative effects.

Choice of characteristic dimensions for nonlinear asymptotic analysis of flame stability

The results of the preceding section amount to two different asymptotic situations, which reduce to eqns (26) and (39) on the flame front. The first case (26) holds when $N \to \infty$, $\sigma \to 1$ in such a way that the product $N(1-\sigma)$ remains finite. Here the characteristic scaling factors of the time t and space variable y (along the flame front) are seen from (23) to be $(1-\sigma)^{-2}$ and $(1-\sigma)^{-1}$, respectively.

Using (16), we have the following asymptotic expression for the perturbations of the hydrodynamic quantities, the temperature and the concentration as $\sigma \to 1$:

$$\tilde{u}_- = -\tilde{p}_- = -i\tilde{v}_- = \frac{1}{2}(1-\sigma)k \exp kx. \quad (40)$$

$$\tilde{u}_+ = -\tilde{p}_+ = i\tilde{v}_+ = \frac{1}{2}(1-\sigma)k \exp -kx. \quad (41)$$

$$\begin{aligned} \tilde{s}_- &= k^2 x \exp x, \qquad \tilde{\theta}_- = -(1-\sigma)k^2 x \exp x - (1-\sigma)(1-L)k^2 \exp x \\ \tilde{s}_+ &= 0, \qquad \tilde{\theta}_+ = -(1-\sigma)(1-L)k^2 \exp[-(\omega + k^2)x]. \end{aligned} \quad (42)$$

Note that if the incoming hydrodynamic flow to the flame front is irrotational (see (40)), the flow will remain irrotational behind the flame front too, in the first approximation relative to $(1-\sigma)$ (see (41)). Generally speaking, this property does not hold for arbitrary $(1-\sigma)$ (Landau, 1944).

By (40)–(42), along the normal to the flame front there are three characteristic zones (layers), whose widths are of orders 1, $(1-\sigma)^{-1}$ and $(1-\sigma)^{-2}$. The first zone corresponds to the thermal structure of the flame, of width l_T. The second corresponds to the depth of hydrodynamic disturbances due to distortion of the flame front. In the third zone, the disturbances are dissipated by the effect of transport processes. We use the results obtained for the model problem and introduce suitable characteristic time and space variables:

$$\begin{aligned} &\tau = (1-\sigma)^2 t,\ \eta = (1-\sigma)y,\ \zeta = (1-\sigma)z, \\ &x = x \text{ (thermal structure of flame)}; \\ &\xi = (1-\sigma)x \text{ (hydrodynamic structure of flame)}; \\ &\chi = (1-\sigma)^2 x \text{ (dissipation zone of thermal and hydrodynamic disturbances)}. \end{aligned} \quad (43)$$

We shall use the variables (43) for an asymptotic nonlinear analysis of the stability problem, based on the exact eqns (3)–(8).

By contrast, the information gained from the linear analysis is no longer sufficient to enable us to choose the characteristic scaling factor for the amplitude of the disturbed flame front. In this case Landau's model (1944) turns out to be useful, provided thermal expansion of the gas is excluded. In a model of this kind, the flame front is simply a surface moving at a constant unit velocity relative to the stationary gas. Thus the disturbed surface $F(y, z, t)$ of the plane flame front satisfies the equation

$$1 - F_t = \surd(1 + F_y^2 + F_z^2). \tag{44}$$

Using (43), we can transform eqn (44) to the form

$$1 - (1-\sigma)^2 F_\tau = \surd[1 + (1-\sigma)^2 F_\eta^2 + (1-\sigma)^2 F_\zeta^2]$$

or

$$(1-\sigma)^2 F_\tau + \frac{1}{2}(1-\sigma)^2 F_\eta^2 + \frac{1}{2}(1-\sigma)^2 F_\zeta^2 = o((1-\sigma)^2). \tag{45}$$

Hence it is clear that in the nonlinear asymptotic analysis we must treat F as a quantity of the order of unity as $\sigma \to 1$. We shall assume henceforth that this result, based on a Landau-type model, remains valid for our fundamental problem, which is based on eqns (3)–(8):

$$F(\tau, \eta, \zeta) \sim 1. \tag{46}$$

Now, on the basis of (46) and the results (40)–(42) of the linear analysis, we can estimate the orders of magnitude of the perturbations of velocity, pressure temperature and concentration:

$$u' \sim v' \sim w' \sim p' \sim (1-\sigma)^2, \qquad s' \sim (1-\sigma)^2; \qquad \theta' \sim (1-\sigma)^3. \tag{47}$$

We now turn to the second asymptotic situation (39), again considering the limit as $N \to \infty$, $\sigma \to 1$ with $N(1-\sigma)$ remaining finite. At the same time, we assume that the Lewis number L is close to a certain L_0 such that the order relation (37) is satisfied. The time t and space variable y (along the front) have characteristic scales

$$t \sim \epsilon^{-2} \sim (1-\sigma)^{-4/3} \qquad \text{and} \qquad y \sim z \sim \epsilon^{-1/2} \sim (1-\sigma)^{-1/3}$$

respectively (see 32).

As in the first asymptotic situation, there are again three characteristic zones along the normal to the flame front; their widths are of orders 1, $(1-\sigma)^{-1/3}$, $(1-\sigma)^{-4/3}$, respectively. Thus the characteristic time and space variables are:

$$\begin{aligned}&\tau=(1-\sigma)^{4/3}t,\ \eta=(1-\sigma)^{1/3}y,\ \zeta=(1-\sigma)^{1/3}z;\\&x=x\ \text{(thermal structure)};\\&\xi=(1-\sigma)^{1/3}x\ \text{(hydrodynamic structure)};\\&\chi=(1-\sigma)^{2/3}x\ \text{(dissipation zone)}.\end{aligned} \tag{48}$$

To determine the characteristic scaling factors for the amplitude of the disturbed front, we again use eqn (44). Using (48), we can write eqn (44) as

$$(1-\sigma)^{4/3}F_\tau+\frac{1}{2}(1-\sigma)^{2/3}F_\eta^{\,2}+\frac{1}{2}(1-\sigma)^{2/3}F_\zeta^2=o((1-\sigma)^{2/3}). \tag{49}$$

Thus, in order to retain the nonlinear terms as $\sigma\to 1$ we must stipulate that $F(\tau,\eta,\zeta)$ is of the order of $(1-\sigma)^{2/3}$:

$$F\sim(1-\sigma)^{2/3}. \tag{50}$$

Using this and eqns (40)–(42), we have the following relations for the perturbations of hydrodynamic variables, temperature and concentration:

$$u'\sim v'\sim w'\sim p'\sim(1-\sigma)^2;\qquad s'\sim(1-\sigma)^{4/3};\qquad \theta'\sim(1-\sigma)^{7/3}. \tag{51}$$

Derivation of nonlinear equation for disturbed flame front. First asymptotic case

In the first asymptotic case, in view of (47) we set

$$\begin{aligned}&T=T^0+(1-\sigma)^3\theta;\qquad C=C^0+(1-\sigma)^2S;\qquad U_1=T^0+(1-\sigma)^2u;\\&U_2=(1-\sigma)^2v;\qquad U_3=(1-\sigma)^2w;\qquad P=P^0+(1-\sigma)^2p\end{aligned} \tag{52}$$

$$\frac{1}{2}N(1-\sigma)=\beta \tag{53}$$

where

$$\begin{aligned}&T^0=1+(1-\sigma)I(x),\qquad P^0=1+(1-\sigma)\Pi(x),\\&I(x)=\exp x-1,\qquad \Pi(x)=1+\left(\frac{4}{3}\nu-1\right)\exp x\qquad \text{for } x<0,\\&I(x)=\Pi(x)=0\qquad \text{for } x>0.\end{aligned} \tag{54}$$

We shall look for asymptotic expansions of θ, S, u, v, w, p, F in the form

$$\begin{aligned}&\theta=\theta_0+(1-\sigma)\theta_1+\cdots,\qquad S=S_0+(1-\sigma)S_1+\cdots,\qquad u=u_0+(1-\sigma)u_1+\cdots\\&v=v_0+(1-\sigma)v_1+\cdots,\qquad w=w_0+(1-\sigma)w_1+\cdots,\qquad p=p_0+(1-\sigma)p_1+\cdots,\\&F=F_0+(1-\sigma)F_1+\cdots.\end{aligned} \tag{55}$$

In the thermal structure zone of the flame (x), the first approximation that follows from eqns (3)–(8) and from (43), (52)–(55) is

$$[u_0 - (F_0)_\tau]I_x + (\theta_0)_x = (\theta_0)_{xx} - [(F_0)_{\eta\eta} + (F_0)_{\zeta\zeta}]I_x + [(F_0)_\eta{}^2 + (F_0)_\zeta^2]I_{xx} + \left[\beta\theta_0 + \frac{1}{2}(F_0)_\eta{}^2 + \frac{1}{2}(F_0)_\zeta^2\right]\delta(x). \tag{56}$$

$$L[u_0 - (F_0)_\tau]C_x^0 + L(S_0)_x = (S_0)_{xx} - [(F_0)_{\eta\eta} + (F_0)_{\zeta\zeta}]C_x^0 + [(F_0)_\eta{}^2 + (F_0)_\zeta^2]C_{xx}^0 - L\left[\beta\theta_0 + \frac{1}{2}(F_0)_\eta{}^2 + \frac{1}{2}(F_0)_\zeta^2\right]\delta(x) \tag{57}$$

$$[u_0 - (F_0)_\tau]_x = 0 \tag{58}$$

$$[2u_0 - (F_0)_\tau]_x = \left[-p_0 + \frac{4}{3}\nu(u_0)_x\right]_x \tag{59}$$

$$(v_0)_{xx} = \nu(v_0)_x - (F_0)_\eta\left(\frac{1}{3}\nu I_x - \Pi\right)_x \tag{60}$$

$$(w_0)_{xx} = \nu(w_0)_x - (F_0)_\zeta\left(\frac{1}{3}\nu I_x - \Pi\right)_x \tag{61}$$

By (58)–(61),

$$u_0(x, \eta, \zeta, \tau) = f(\eta, \zeta, \tau) \qquad \text{for } -\infty < x < \infty \tag{62}$$

$$v_0(x, \eta, \zeta, \tau) = g(\eta, \zeta, \tau) \qquad \text{for } x > 0$$

and (63)

$$v_0(x, \eta, \zeta, \tau) = g(\eta, \zeta, \tau) + (F_0)_\eta - (F_0)_\eta \exp x \qquad \text{for } x < 0$$

$$w_0(x, \eta, \zeta, \tau) = h(\eta, \zeta, \tau) \qquad \text{for } x > 0$$

and (64)

$$w_0(x, \eta, \zeta, \tau) = h(\eta, \zeta, \tau) + (F_0)_\zeta - (F_0)_\zeta \exp x \qquad \text{for } x < 0$$

$$p_0(x, \eta, \zeta, \tau) = e(\eta, \zeta, \tau) \qquad \text{for } -\infty < x < \infty. \tag{65}$$

The actual form of the functions f, g, h, e will depend on the matching conditions which link (62)–(65) to the external solution describing the hydrodynamic structure of the disturbed flame (variable ξ):

$$u_0(\xi, \eta, \zeta, \tau) \to f(\eta, \zeta, \tau) \qquad \text{for } \xi \to \pm 0 \tag{66}$$

$$\begin{aligned} v_0(\xi, \eta, \zeta, \tau) &\to g(\eta, \zeta, \tau) && \text{for } \xi \to +0 \\ v_0(\xi, \eta, \zeta, \tau) &\to g(\eta, \zeta, \tau) + (F_0)_\eta && \text{for } \xi \to -0 \end{aligned} \tag{67}$$

$$\begin{aligned} w_0(\xi, \eta, \zeta, \tau) &\to h(\eta, \zeta, \tau) && \text{for } \xi \to +0 \\ w_0(\xi, \eta, \zeta, \tau) &\to h(\eta, \zeta, \tau) + (F_0)_\zeta && \text{for } \xi \to -0 \end{aligned} \tag{68}$$

$$p_0(\xi, \eta, \zeta, \tau) \to e(\eta, \zeta, \tau) \qquad \text{for } \xi \to \pm 0 \tag{69}$$

we now obtain from (56), (57), (62)

$$S_0(x, \eta, \zeta, \tau) = [L(F_0)_\eta^2 + L(F_0)_\zeta^2 - (F_0)_{\eta\eta} - (F_0)_{\zeta\zeta} + L(F_0)_\tau - Lf]x \exp Lx$$
$$\text{for } x < 0 \quad \text{and} \quad S_0(x, \eta, \zeta, \tau) \equiv 0 \quad \text{for } x > 0 \tag{70}$$

$$\theta_0(x, \eta, \zeta, \tau) = \theta_0(0, \eta, \zeta, \tau) \exp x - [(F_0)_\eta^2 + (F_0)_\zeta^2 - (F_0)_{\eta\eta} - (F_0)_{\zeta\zeta} + (F_0)_\tau - f]x \exp x$$
$$\text{for } x < 0 \quad \text{and} \quad \theta_0(x, \eta, \zeta, \tau) \equiv \theta_0(0, \eta, \zeta, \tau) \quad \text{for } x > 0. \tag{71}$$

Integrating (56), (57) in the neighborhood of $x = 0$, we obtain the following expressions for the discontinuities of the derivatives:

$$\begin{aligned} [(S_0)_x] &= \beta L\theta_0 - \frac{1}{2} L(F_0)_\eta^2 - \frac{1}{2} L(F_0)_\zeta^2, \\ [(\theta_0)_x] &= -\beta\theta_0 + \frac{1}{2}(F_0)_\eta^2 + \frac{1}{2}(F_0)_\zeta^2. \end{aligned} \tag{72}$$

In view of (66), (70), (71), eqns (72) form a system of two equations for $\theta_0(0, \eta, \zeta, \tau)$ and $F_0(\eta, \zeta, \tau)$. Eliminating $\theta_0(0, \eta, \zeta, \tau)$, we obtain

$$(F_0)_\tau = \frac{1}{L}[1 + \beta(L-1)][(F_0)_{\eta\eta} + (F_0)_{\zeta\zeta}] - \frac{1}{2}(F_0)_\eta^2 - \frac{1}{2}(F_0)_\zeta^2 + f \tag{73}$$

where $f(\eta, \zeta, \tau) = u_0(0, \eta, \zeta, \tau)$ (see 66).

To determine $u_0(0, \eta, \tau)$, we consider the zone of hydrodynamic structure (variable ξ). By (4)–(7), the principal term of the asymptotic expansion (55) gives

$$\begin{aligned} (u_0)_\xi + (p_0)_\xi &= 0, \qquad (v_0)_\xi + (p_0)_\eta = 0 \\ (w_0)_\xi + (p_0)_\zeta &= 0, \qquad (u_0)_\xi + (v_0)_\eta + (w_0)_\zeta = 0. \end{aligned} \tag{74}$$

From (66)–(69) we obtain the conditions that must hold at $\xi = 0$.

$$\begin{aligned} u_0(+0, \eta, \zeta, \tau) &= u_0(-0, \eta, \zeta, \tau), \\ p_0(+0, \eta, \zeta, \tau) &= p_0(-0, \eta, \zeta, \tau), \\ v_0(+0, \eta, \zeta, \tau) &= v_0(-0, \eta, \zeta, \tau) - (F_0)_\eta, \\ w_0(+0, \eta, \zeta, \tau) &= w_0(-0, \eta, \zeta, \tau) - (F_0)_\zeta. \end{aligned} \tag{75}$$

We see that the hydrodynamic variables may be expressed in terms of the flame front disturbance $F_0(\eta, \zeta, \tau)$ by solving the simple linear problem (74), (75). We require a solution of (74), (75) that is bounded for $|\xi| < \infty$.

As we are interested in this paper in *intrinsic* instability of the flame, caused by initial disturbances of the front, we must exclude all *external* disturbances; to this end, we assume that the incoming flow to the flame is irrotational, i.e. for $\xi < 0$:

$$(u_0)_\eta - (v_0)_\xi = 0, \qquad (u_0)_\zeta - (w_0)_\xi = 0, \qquad (v_0)_\zeta - (w_0)_\eta = 0. \tag{76}$$

Solving problem (74)–(75) by Fourier transforms, we obtain

$$(u_0)_\mp = \frac{1}{8\pi^2}\iiiint_{-\infty}^{\infty} \sqrt{(k^2+l^2)}\, e^{ik(\eta-\eta')+il(\zeta-\zeta')\pm\sqrt{(k^2+l^2)}\xi} F_0(\eta', \zeta', \tau)\, d\eta'\, d\zeta'\, dk\, dl. \tag{77}$$

The signs (+) and (−) refer to the zones $\xi > 0$ and $\xi < 0$, respectively. Note that the flow at $\xi > 0$ is irrotational (for the principal term of the asymptotic expansion).

Using (77), we can write (73) as an integrodifferential equation for F_0:†

$$F_\tau = \frac{1}{L}[1+\beta(L-1)]\nabla^2 F - \frac{1}{2}(\nabla F)^2 + \frac{1}{8\pi^2}\int_{-\infty}^{\infty} |\mathbf{k}|\, e^{i\mathbf{k}(\boldsymbol{\eta}-\boldsymbol{\eta}')} F(\boldsymbol{\eta}', t)\, d\mathbf{k}\, d\boldsymbol{\eta}' \quad \boldsymbol{\eta} = (\eta, \zeta). \tag{78}$$

This equation is clearly meaningful only when $L > L_0$. Otherwise, the sign of the coefficient of the second derivative is negative and the corresponding initial-value problem is not well-posed.

Nonlinear equation for flame front. Second asymptotic case

In the first asymptotic case, the first-approximation equations were sufficient to yield an equation of type (2) for the flame front. In the second asymptotic case the situation is less trivial. Here the first approximation enables us to find only a branch point of the general bifurcation problem. The equation of the flame front is then derived as the condition for solvability of the second-approximation system of equations.

In view of (51), in the second approximation we must set

$$\begin{gathered} T = T^0 + (1-\sigma)^{7/3}\theta, \qquad C = C^0 + (1-\sigma)^{4/3}S, \\ U_1 = T^0 + (1-\sigma)^2 u, \qquad U_2 = (1-\sigma)^2 v, \qquad U_3 = (1-\sigma)^2 w, \\ P = P^0 + (1-\sigma)^2 p, \qquad F = (1-\sigma)^{2/3}\Phi \end{gathered} \tag{79}$$

where C^0, T^0, and P^0 are defined in (54). In view of (37), we express the Lewis number L as

$$L = L_0 + \alpha(L_0 - 1)(1-\sigma)^{2/3}. \tag{80}$$

In the thermal structure zone (variable x), the system of equations for heat conduction and diffusion (3), (4) may be written, up to terms of order $o((1-$

†We have omitted the index "0", which is not necessary here.

$\sigma)^{2/3}$), as follows:

$$\theta_x - \theta_{xx} + I_x(\Phi_{\eta\eta} + \Phi_{\zeta\zeta}) = (1-\sigma)^{2/3}(\theta_{\eta\eta} + \theta_{\zeta\zeta}) + (1-\sigma)^{2/3} I_{xx}(\Phi_\eta{}^2 + \Phi_\zeta{}^2) + (1-\sigma)^{2/3}(\Phi_\tau - u)I_x + \left[\beta\theta + \frac{1}{2}(1-\sigma)^{2/3}(\Phi_\eta{}^2 + \Phi_\zeta{}^2)\right]\delta(x) \quad (81)$$

$$LS_x - S_{xx} + C_x^0(\Phi_{\eta\eta} + \Phi_{\zeta\zeta}) = (1-\sigma)^{2/3}(S_{\eta\eta} + S_{\zeta\zeta}) + (1-\sigma)^{2/3} C_{xx}^0(\Phi_\eta{}^2 + \Phi_\zeta{}^2) + L(1-\sigma)^{2/3}(\Phi_\tau - u)C_x{}^0 - L\left[\beta\theta + \frac{1}{2}(1-\sigma)^{2/3}(\Phi_\eta{}^2 + \Phi_\zeta{}^2)\right]\delta(x). \quad (82)$$

A solution of system (81), (82) will be sought as an asymptotic expansion in powers of $(1-\sigma)^{2/3}$, i.e.

$$\Phi = \Phi_0 + (1-\sigma)^{2/3}\Phi_1 + \cdots, \qquad \theta = \theta_0 + (1-\sigma)^{2/3}\theta_1 + \cdots,$$
$$S = S_0 + (1-\sigma)^{2/3}S_1 + \cdots, \qquad u = u_0 + (1-\sigma)^{2/3}u_i + \cdots. \quad (83)$$

By virtue of (80), the undisturbed concentration distribution C^0 may also be expanded in powers of $(1-\sigma)^{2/3}$

$$C^0 = C_0{}^0 + (1-\sigma)^{2/3}C_1{}^0 + \cdots. \quad (84)$$

It is readily seen that solution of the hydrodynamic problem for the second asymptotic case reduced in the first approximation to solution of eqns (58)–(61), (74), as done in the first asymptotic case.

Clearly, the orders of magnitude of Φ_0, ξ, η, τ are determined by eqns (48), (50) and (79). Thus, as in the preceding section,

$$u_0(x, y, z, \tau) = f(\eta, \zeta, \tau) \quad (85)$$

where $f(\eta, \zeta, \tau)$ is given in terms of $\Phi_0(\eta, \zeta, \tau)$ by

$$f(\eta, \zeta, \tau) = \frac{1}{8\pi^2}\int_{-\infty}^{\infty} |\mathbf{k}|\, e^{i\mathbf{k}(\boldsymbol{\eta}-\boldsymbol{\eta}')}\Phi_0(\boldsymbol{\eta}', \tau)\, d\boldsymbol{\eta}'\, d\mathbf{k}. \quad (86)$$

In the first approximation, system (81), (82) may be written as

$$(\theta_0)_x - (\theta_0)_{xx} + I_x[(\Phi_0)_{\eta\eta} + (\Phi_0)_{\zeta\zeta}] = \beta\theta_0\delta(x)$$
$$L_0(S_0)_x - (S_0)_{xx} + (C_0{}^0)_x[(\Phi_0)_{\eta\eta} + (\Phi_0)_{\zeta\zeta}] = -\beta L_0\theta_0\delta(x). \quad (87)$$

The bounded solution of system (87) satisfying the continuity condition at $x = 0$ has the form

$$S_0 = -[(\Phi_0)_{\eta\eta} + (\Phi_0)_{\zeta\zeta}]x \exp L_0 x \qquad \text{for } x < 0$$

$$\theta_0 = [(\Phi_0)_{\eta\eta} + (\Phi_0)_{\zeta\zeta}]\left(x + \frac{1-L_0}{L_0}\right)\exp x \qquad \text{for } x < 0$$

$$\theta_0 = \frac{1-L_0}{L_0}[(\Phi_0)_{\eta\eta} + (\Phi_0)_{\zeta\zeta}], \qquad \text{for } x > 0$$

$$S_0 \equiv 0 \qquad \text{for } x > 0. \tag{88}$$

The conditions for discontinuity of the derivatives at $x = 0$ imply a necessary condition for the existence of a nontrivial solution:

$$\beta(1 - L_0) = 1. \tag{89}$$

Thus, we have obtained the critical Lewis number L_0, corresponding to a bifurcation point of the problem. Note that the fact that the disturbance θ_0 is not damped out as $x \to +\infty$ is due to the existence of an external region $x \sim (1-\sigma)^{-2/3}$ (with characteristic space variable $\chi = (1-\sigma)^{2/3}x$) in which θ_0 is exponentially damped (see (42)).

In the second approximation, system (81), (82) becomes

$$(\theta_1)_x - (\theta_1)_{xx} + I_x[(\Phi_1)_{\eta\eta} + (\Phi_1)_{\zeta\zeta}] = (\theta_0)_{\eta\eta} + (\theta_0)_{\zeta\zeta} + I_{xx}[(\Phi_0)_\eta^{\,2} + (\Phi_0)_\zeta^{\,2}]$$
$$+ I_x[(\Phi_0)_\tau - u_0] + \left[\beta u_1 + \frac{1}{2}(\Phi_0)_\eta^{\,2} + \frac{1}{2}(\Phi_0)_\zeta^{\,2}\right]\delta(x) \tag{90}$$

$$L_0(S_1)_x - (S_1)_{xx} + (C_0{}^0)_x[(\Phi_1)_{\eta\eta} + (\Phi_1)_{\zeta\zeta}] = \alpha(1-L_0)(S_0)_x - (C_1{}^0)_x[(\Phi_0)_{\eta\eta} + (\Phi_0)_{\zeta\zeta}]$$
$$+ (S_0)_{\eta\eta} + (S_0)_{\zeta\zeta} + (C_0{}^0)_{xx}[(\Phi_0)_\eta^{\,2} + (\Phi_0)_\zeta^{\,2}] + L_0(C_0{}^0)_x[(\Phi_0)_\tau - u_0]$$
$$- \left[L_0\beta\theta_1 + \frac{1}{2}L_0(\Phi_0)_\eta^{\,2} + \frac{1}{2}L_0(\Phi_0)_\zeta^{\,2} + \alpha\beta(L_0 - 1)\theta_0\right]\delta(x). \tag{91}$$

The second-approximation solution of (90), (91) satisfying the continuity condition at $x = 0$ is

$$\theta_1 = \frac{1}{2}[(\Phi_0)_{\eta\eta\eta\eta} + 2(\Phi_0)_{\eta\eta\zeta\zeta} + (\Phi_0)_{\zeta\zeta\zeta\zeta}]x^2 \exp x + \left\{\left(2 - \frac{1}{L_0}\right)[(\Phi_0)_{\eta\eta\eta\eta}\right.$$
$$\left. + 2(\Phi_0)_{\eta\eta\zeta\zeta} + (\Phi_0)_{\zeta\zeta\zeta\zeta}] - (\Phi_0)_\eta^{\,2} - (\Phi_0)_\zeta^{\,2} - (\Phi_0)_\tau + u_0 + (\Phi_1)_{\eta\eta} + (\Phi_1)_{\zeta\zeta}\right\}x \exp x$$
$$+ \theta_1(0)\exp x, \qquad \text{for } x < 0 \tag{92}$$

$$S_1 = \frac{1}{2L_0}[(\Phi_0)_{\eta\eta\eta\eta} + 2(\Phi_0)_{\eta\eta\zeta\zeta} + (\Phi_0)_{\zeta\zeta\zeta\zeta}]x^2 \exp L_0 X$$
$$+ \frac{\alpha(1-L_0)}{2L_0}[(\Phi_0)_{\eta\eta} + (\Phi_0)_{\zeta\zeta}]x^2 \exp L_0 x$$
$$+ [L_0(\Phi_0)_\eta^{\,2} + L_0(\Phi_0)_\zeta^{\,2} + L_0(\Phi_0)_\tau - L_0 u_0 - (\Phi_1)_{\eta\eta} - (\Phi_1)_{\zeta\zeta}]x \exp L_0 x$$
$$- \frac{1}{L_0{}^2}[(\Phi_0)_{\eta\eta\eta\eta} + 2(\Phi_0)_{\eta\eta\zeta\zeta} + (\Phi_0)_{\zeta\zeta\zeta\zeta}]x \exp L_0 x, \qquad \text{for } x < 0 \tag{93}$$

$$\theta_1 = \theta_1(0) + \left(\frac{1}{L_0} - 1\right)x[(\Phi_0)_{\eta\eta\eta\eta} + 2(\Phi_0)_{\eta\eta\zeta\zeta} + (\Phi_0)_{\zeta\zeta\zeta\zeta}]. \tag{94}$$

$$S \equiv 0, \qquad x > 0.$$

To define $\theta_1(0)$, Φ_0 and Φ_1, we appeal to the conditions for discontinuity of the derivatives at $x = 0$. These conditions may be expressed as a system of two linear equations in $(\Phi_1)_{\eta\eta} + (\Phi_1)_{\zeta\zeta}$ and $\theta_1(0)$

$$\begin{aligned}(\beta - 1)\theta_1(0) - (\Phi_1)_{\eta\eta} - (\Phi_1)_{\zeta\zeta} &= \frac{3L_0 - 2}{L_0}[(\Phi_0)_{\eta\eta\eta\eta} + 2(\Phi_0)_{\eta\eta\zeta\zeta} + (\Phi_0)_{\zeta\zeta\zeta\zeta}] \\ &\quad - \frac{1}{2}(\Phi_0)_\eta^{\,2} - \frac{1}{2}(\Phi_0)_\zeta^2 - (\Phi_0)_\tau + u_0, \\ \beta L_0\theta_1(0) - (\Phi_1)_{\eta\eta} - (\Phi_1)_{\zeta\zeta} &= \frac{1}{L_0}[(\Phi_0)_{\eta\eta\eta\eta} + 2(\Phi_0)_{\eta\eta\zeta\zeta} + (\Phi_0)_{\zeta\zeta\zeta\zeta}] \\ &\quad - L_0(\Phi_0)_\tau + L_0 u_0 + (\Phi_0)_{\eta\eta} + (\Phi_0)_{\zeta\zeta}.\end{aligned} \tag{95}$$

As the determinant of this system vanishes (see, 89), it will be solvable at $\Phi_0 \not\equiv$ *const.* only if the following condition holds:

$$(\Phi_0)_\tau + \frac{1 + 3L_0}{L_0^{\,2}}[(\Phi_0)_{\eta\eta\eta\eta} + 2(\Phi_0)_{\eta\eta\zeta\zeta} + (\Phi_0)_{\zeta\zeta\zeta\zeta}] + \frac{\alpha}{L_0}[(\Phi_0)_{\eta\eta} + (\Phi_0)_{\zeta\zeta}] + \frac{1}{2}[(\Phi_0)_\eta^{\,2} + (\Phi_0)_\zeta^2] = u_0. \tag{96}$$

Together with (85), (86), this yields a nonlinear integrodifferential equation for $\Phi_0(\eta, \zeta, \tau)$. If $\beta \gg 1$ (or $L_0 \simeq 1$), this equation has the form:†

$$F_t + 4\nabla^4 F + \epsilon\nabla^2 F + \frac{1}{2}(\nabla F)^2 = \frac{(1-\sigma)}{8\pi^2}\int_{-\infty}^{\infty} |\mathbf{k}|\, \mathrm{e}^{i\mathbf{k}(\boldsymbol{\eta} - \boldsymbol{\eta}')}\Phi(\boldsymbol{\eta}', \tau)\, \mathrm{d}\mathbf{k}\, \mathrm{d}\boldsymbol{\eta}'. \tag{97}$$

Now in terms of the natural dimensionless variables (in units of l_T and the time interval l_T/U_b) eqn (97) may be transformed as follows:

$$F_t + 4\nabla^4 F + \epsilon\nabla^2 F + \frac{1}{2}(\nabla F)^2 = \frac{(1-\sigma)}{8\pi^2}\int_{-\infty}^{\infty} |\mathbf{k}|\, \mathrm{e}^{i\mathbf{k}(\mathbf{x} - \mathbf{z})}F(\mathbf{z}, t)\, \mathrm{d}\mathbf{k}\, \mathrm{d}\mathbf{z}. \tag{98}$$

Improvement of asymptotic expansions

As we have already mentioned, when $L = 1$ the thermal expansion coefficient σ of the gas is the only parameter in the stability problem for a plane flame front. In that case, in the vicinity of $\sigma = 1$ there is a single correct nonlinear asymptotic form

†We have omitted the index "0".

for the disturbance of the front, leading to the equation

$$F_t - \nabla^2 F + \frac{1}{2}(\nabla F)^2 = \frac{(1-\sigma)}{8\pi^2}\int_{-\infty}^{\infty} |\mathbf{k}|\, e^{i\mathbf{k}(\mathbf{x}-\mathbf{z})} F(\mathbf{z}, t)\, d\mathbf{k}\, d\mathbf{z}. \tag{99}$$

We now examine the appearance of eqn (98) when $L = 1$. As in eqn (99), the coefficient of the second derivative is -1. However, as the term involving the fourth derivative does not vanish, eqn (98) does not degenerate into eqn (99). This discrepancy is not surprising. Equation (98) was derived on the assumption that L lies near L_0. Thus, extrapolation to the value $L = 1$ ($L_0 < 1$) is generally speaking not legitimate, and may lead to significant errors.

Nevertheless, it turns out that in regard to the second asymptotic case one can modify the scaling factors (37), (48), (50) in such a way that the resulting asymptotic equation for F is a good approximation in a larger neighborhood of L_0 and, in particular, at $L = 1$. To this end, the parameter α, the normalized variables τ, η, ζ and the disturbance Φ of the flame front must be defined as follows:

$$\begin{gathered} \alpha = \epsilon(1+\epsilon)^{-2/3}(1-\sigma)^{-2/3}, \qquad \tau = (1-\sigma)^{4/3}(1+\epsilon)^{-2/3} t \\ \eta = (1-\sigma)^{1/3}(1+\epsilon)^{-2/3} y, \qquad \zeta = (1-\sigma)^{1/3}(1+\epsilon)^{-2/3} z, \\ \Phi = (1+\epsilon)^{-2/3}(1-\sigma)^{-2/3} F. \end{gathered} \tag{100}$$

These transformations again yield an equation of type (97) for the principal asymptotic term of the function Φ. Now, however, in terms of the natural dimensionless variables (in units of l_T and the time interval l_T/U_b), eqn (97) becomes an equation of type (1).

Equation (1) is correct for any Lewis number, and in particular for $L = L_0$ and $L = 1$. At $L = 1$, the coefficient of the fourth derivative vanishes and eqn (99) results.

Note that the Lewis number of a gaseous mixture never exceeds $L = \sqrt{2}$.

The relation linking the stability index ω to the wave-number k for eqn (1), in the case of small harmonic disturbances, is

$$\omega = \frac{(1-\sigma)}{2} k - \left(\frac{L - L_0}{1 - L_0}\right) k^2 - 4\left(\frac{1 - L}{1 - L_0}\right)^2 k^4. \tag{101}$$

Figure 1 illustrates the shape of the curve $\omega = \omega(k)$ for $L \gtrless L_0$ and fixed parameters N and σ.

On solution of the nonlinear equations. Some physical conclusions

To investigate the evolution in time of the disturbed flame front, we must consider an initial-value problem for eqns (1), (2), (78), (98). Although there are no effective analytical tools for solution of this nonlinear problem, it seems possible to conduct a qualitative investigation, obtaining certain estimates of physical interest. For example, it would be interesting to estimate the pro-

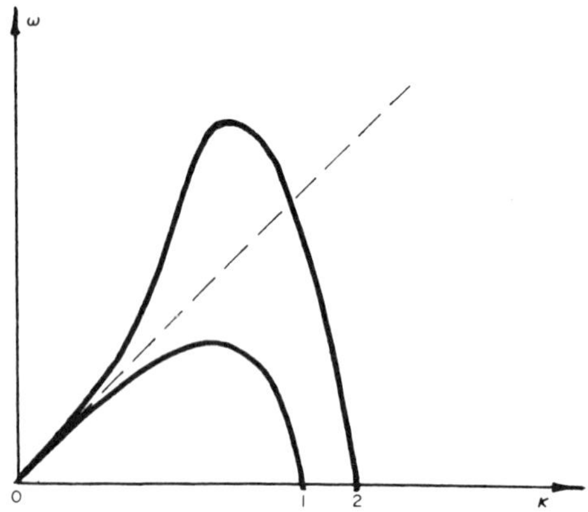

Fig. 1. Stability index ω plotted against wavenumber k. Curve 1—$L \geq L_0$, curve 2—$L < L_0$, dashed line—approximation of Landau (1944).

pagation velocity and amplitude of the disturbed flame front at times far removed from the origin, when the solution has "forgotten" the details of the initial disturbances and its behavior has become regular. This is clearly a profound mathematical problem, far beyond the scope of the present article.

As a preliminary step, numerical solution of eqns (1), (2), (78), (98) is quite admissible and at the same time provides a great deal of information. Analysis of the corresponding finite-difference scheme and computer solution of the problem will be the subject of the second part of this paper, written in collaboration with Michelson (1978). Here we shall give a brief description of the main conclusions following from the numerical work. We shall limit ourselves to the case of a two-dimensional flame, as it is the simplest case and the result is obtained most rapidly.

For the sake of brevity we shall discuss only eqn (97), since eqn (78) may be obtained from it as a long-wave asymptotic form as $\alpha \to -\infty$. Equation (97) contains a single dimensionless parameter α (see (100)) whose value governs the entire (qualitative) variety of possible situations. Linear stability analysis of a plane flame front (Fig. 1) has shown that, whatever the value of α, there is always a critical wavelength l_c, corresponding to the maximum amplification rate of the initial harmonic disturbance (i.e. maximum ω). Extrapolating this result to the general nonlinear case, one would expect that for any α the flame front will have the shape of some cellular (wrinkled) surface with cells of characteristic dimension l_c.

Solution of eqn (97) shows, however, that this is not always so. The type of instability displayed by the flame front changes entirely as the parameter α goes through a certain critical value α_c,

$$\alpha_c \simeq -0.3. \tag{102}$$

At $\alpha > \alpha_c$, following an arbitrary initial disturbance, the solution of eqn (97) ultimately settles down to a steady, stationary regime:

$$\Phi(\tau, \eta) \rightarrow \tilde{\Phi}(\eta) - V\tau, \qquad \tau > \infty. \tag{103}$$

The limiting shape $\tilde{\Phi}(\eta)$ of the flame front is a warped surface, composed as it were of pieces of parabola-like arcs convex toward the fresh mixture (Fig. 2). The velocity V is practically independent of $\alpha (\alpha < \alpha_c)$, being approximately equal to 0.18.

The positive sign of V indicates that the velocity of the corrugated front is greater than the normal velocity of the plane flame front. The shape of the steady-state front, i.e. the dimensions of the arc-shaped wrinkles and their distribution over the front, depends essentially on the initial random disturbance. The very existence of different limiting configurations of the front points to their stability under small disturbances. Numerical solution shows (Michelson and Sivashinsky, (1978) that the dimensions of the individual wrinkles may consider-

Fig. 2. General view of a wrinkled laminar flame front ($\alpha < \alpha_c$).

ably exceed the value l_c predicted by linear analysis. Here the only role of l_c is formation of the structure (radius of curvature) of the angles that arise in the front at points where contiguous wrinkles join up.

Away from these angular points, the shape of the steady stationary front may be fully described within the unstructured model of Landau (1944). As $(1-\sigma)\to 0$, the Landau model clearly leads to the equation

$$\Phi_\tau + \frac{1}{2}(\Phi_\eta)^2 = \frac{1}{4\pi}\iint_{-\infty}^{\infty} |k|\, e^{ik(\eta-\zeta)}\Phi(\zeta,\tau)\, d\zeta\, dk. \tag{104}$$

In the nonstationary case ($\Phi_\tau \neq -V$), the initial-value problem for eqn (104) is obviously not well-formed, because of the rapid amplification of high-frequency disturbances. Nevertheless, it is easy to regularize the equation, for example by expressing the flame velocity as a phenomenological function of the curvature of the front (Markstein, 1964).

This result implies that the well-known Landau paradox (1944) concerning absolute instability of a flame front may be partly explained in the framework of the original Landau model. As we now see, Landau's model allows the flame to assume a stationary form with distorted wrinkles. In the framework of the unstructured Landau model, such configurations are obviously unstable only to very short-wave disturbances; such instability is eliminated by merely including dissipative effects. Overly large wrinkles will of course be unstable even when transport effects are taken into consideration. However, thanks to the strong nonlinear effect of the disturbed-flow geometry (as shown by numerical experiments), the dimensions of the individual wrinkles may considerably exceed the stability limit ($2\pi/k_0$) predicted by linear analysis for a plane flame front. Thus, the source of the Landau paradox is the complete linearization of the equations for the disturbed flame, which eliminates a major stabilizing factor.

The inclusion of transport effects regularizes the nonstationary Landau model. However, the characteristic dimension l_c introduced by transport effects has almost no significance for the shape of the steady flame front or for the dimensions of the individual wrinkles.

In this sense, the Markstein model (1964) provides a complete solution to the problem of nonstationary formation of a corrugated noncellular flame front. However, in contrast to the predictions of the linear theory, which is based on the existence of l_c, the Markstein model does not describe the formation of cellular flame structure. This is also in full accordance with well-known experimental data. As we saw in the section on Auxiliary Model Problem, Markstein's phenomenological model is valid only for sufficiently large Lewis numbers, i.e. when flame structure is not cellular. To summarize: when $\alpha < \alpha_c$, the plane flame front is transformed as a result of *spontaneous* instability into a stationary distorted, corrugated front, but the flame continues to propagate in a laminar regime. Figure 3 is a plot of the function $\alpha = \alpha(\epsilon)$, which shows that flames with Lewis number greater than unity (i.e. $\epsilon > -1$) are always of the noncellular wrinkled type.

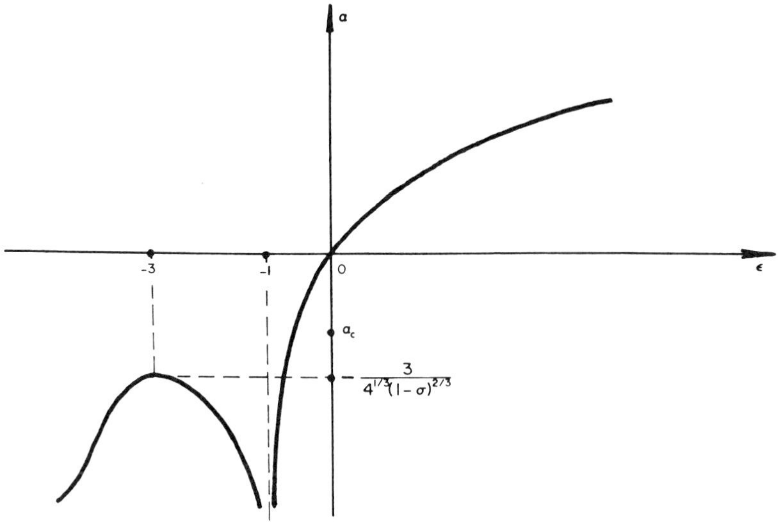

Fig. 3. Parameter α plotted against ϵ.

The situation is quite different when $\alpha > \alpha_c$. Here only the second-asymptotic eqns (1), (98) may be used. Following an initial disturbance, the solution of eqns (1), (98) ultimately settles down to a steady but essentially *nonstationary* regime. The front assumes the form of a random corrugated surface, comprising cells of more or less equal dimensions, constantly breaking up and recombining in a chaotic fashion (Fig. 4). As in the case $\alpha < \alpha_c$, there are clearly defined protuberances pointing toward the burnt-out gas. The whole of this turbulent front propagates at a certain constant mean velocity. As when $L > L_0$, the propagation velocity is greater than normal. We emphasize here that the turbulent front we have described owes its formation exclusively to spontaneous instability of the flame; it has nothing to do with external factors. According to the formulation of the problem, the incoming hydrodynamic flow far ahead of the flame front is uniform and homogeneous, with a constant flow rate.

In contradistinction to the situation for laminar noncellular flames ($\alpha < \alpha_c$), when $\alpha > \alpha_c$ the mean propagation velocity $V = V(\alpha)$ depends essentially on α:

$$V(-0.15) \simeq 0.18, \qquad V(0) \simeq 0.6, \qquad V(0.75) \simeq 1.6 \tag{105}$$

for

$$\alpha \gg 1, \qquad V(\alpha) \simeq 0.3\alpha^3. \tag{106}$$

The asymptotic relation (106) corresponds to a purely diffusional-thermal flame model, in which the effect of thermal expansion of the gas is ignored.

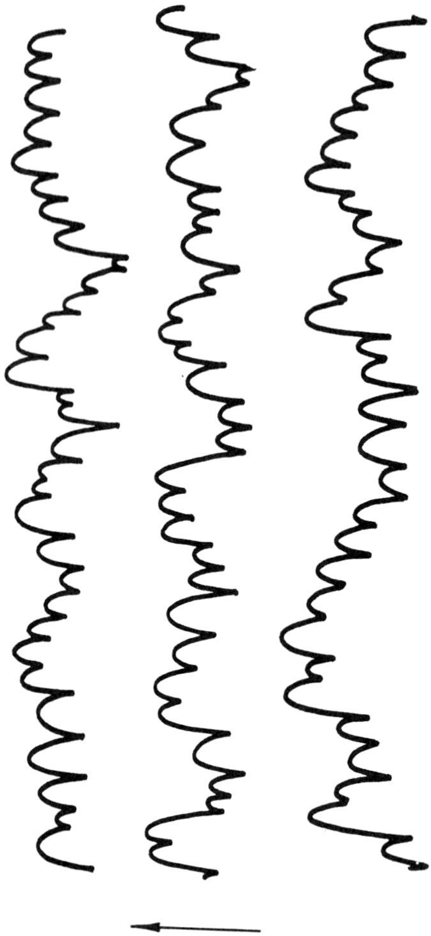

Fig. 4. General view of a turbulent cellular flame front ($\alpha > \alpha_c$).

Recall that in terms of the natural dimensionless coordinates

$$\langle F_t \rangle = -(1-\sigma)^2 V(\alpha). \tag{107}$$

An interesting outcome of the hydrodynamic model is that, if thermal expansion is taken into account ($\sigma \neq 1$), the flame may assume turbulent cellular structure even for nonpositive ϵ (i.e. $L \geq L_0$). Since the diffusional-thermal instability is possible only when $\epsilon > 0$, it follows that if $\alpha_c < \alpha \leq 0$ the onset of cellular structure depends exclusively on hydrodynamic instability. This effect is observed, for example, when $\epsilon = 0$. Equation (1) is then of the form

$$F_t + 4\nabla^4 F + \frac{1}{2}(\nabla F)^2 = \frac{(1-\sigma)}{8\pi^2}\int_{-\infty}^{\infty} |\mathbf{k}|\, e^{i\mathbf{k}(\mathbf{x}-\mathbf{z})} F(\mathbf{z}, t)\, d\mathbf{k}\, d\mathbf{z}. \tag{108}$$

In mathematical terms, the new formation of cellular structure is apparently due to the fact that the stabilizing effect of the dissipative operator $(-4\nabla^4)$ is *weaker* than the analogous effect of the operator $(+\nabla^2)$.

For sufficiently large negative α, the stabilizing effect of the operator $(+\nabla^2)$ completely suppresses the tendency for cells of dimension l_c to form, as predicted by the linear stability analysis. In this situation, only wrinkled noncellular flames are formed. At small negative α, the major factor in stabilization of the flame is the weaker operator $(-4\nabla^4)$, which preserves the tendency toward formation of cellular structure.

Conclusion

The sole aim of our analysis of *linear* stability was to determine the order of magnitude of the disturbances. We therefore confined attention to a simplified mathematical model, neglecting, among other things, the viscosity (14). This was no longer the case in the asymptotic nonlinear analysis, where viscosity was included along with thermal conduction and diffusion. For it is well known that in gases these three effects are of the same order of magnitude, and it would be physically unjustifiable to neglect viscosity compared with the other transport effects.

However, despite the appearance of viscosity in the initial equations (6) and in the disturbances of the hydrodynamic variables (59), (60), the *principal* term of the asymptotic form of the flame-front equation (eqns (1), (2)) turns out to be independent of the Prandtl number ν. Thus, in phenomena of hydrodynamic instability the viscosity effect is as it were secondary. Of course, the Prandtl number ν must appear in approximations of higher order in $(1-\sigma)$.

It is interesting to compare this conclusion with the results of the Istratov–Librovich asymptotic linear analysis (Istratov and Librovich, 1966). In that work, where the same initial model (3)–(8) was used, the stabilizing role of viscosity was evident even in the first approximation. This does not contradict our result, since the asymptotic approach of Istratov and Librovich is based on a different choice of expansion parameter. Their point of departure is the assumption that the width l_T is small compared with the wavelength of the front disturbance, the parameter $(1-\sigma)$ being assumed finite. In our analysis, however, it is assumed not only that the wavenumber k is small but also that

$$k \sim (1-\sigma) \sim N^{-1} \ll 1.$$

We also note that, although our linear analysis was based on a flame model that makes no allowance for viscosity and for the continuous temperature-dependence of density, the asymptotic relations (25), (26), (38), (39) are nevertheless correct from the standpoint of the exact flame model (3)–(8).

The integrodifferential equations derived above for the disturbed flame front comprise only the principal terms of the expansions in question. In this sense, the equations provide a good description of weak turbulence only, or of a weakly distorted flame front. To extend the range within which the results are applicable and to confirm the various extrapolations, it would be highly interesting to compute also higher approximations relative to the parameter $(1-\sigma)$.

In addition to eqns (1), (2) for F_0, such an approach would require successive introduction of linear equations for $F_1, F_2, \ldots$, with coefficients which may be treated as given functions defined by previous approximations. It is clear that the flame front may be described at each step of the approximation procedure by a certain explicit function of the space coordinates, which certainly excludes disruption of the turbulent front. Disruption of the flame front and formation of islets of combustible mixture all over the flame are familiar phenomena in turbulent propagation of flames. The regime of spontaneous self-turbulence in a quiescent gas (or laminar flow) is characterized, on the other hand, by a continuous turbulent front (Libby and Williams, 1976; Palm-Leis and Strehlow, 1969).

The nonlinear equations of spontaneous instability (1), (2) may now serve as a basis for derivation of equations for a flame front propagating in an *external* turbulent flow. Of the many situations possible here, the simplest is probably interaction of the flame front with an incoming vortex wave of low intensity (Markstein, 1964).

Note that eqns (1), (2) were not derived by considering an initial-value problem for the equation system (3)–(9), but by isolating the "low-frequency" time-asymptotic case $(t \sim (1-\sigma)^{-4/3})$. In principle, this is the only asymptotic situation of physical interest, representing as it does a limit regime. A more thorough mathematical treatment must be based on the general initial-value problem. This means that, apart from the interval $(t \sim (1-\sigma)^{-4/3})$ one should also consider the initial time interval $t \sim 1$, in which the disturbances of temperature, concentration and flame front adjust to one another. At $t = 0$ these disturbances must satisfy certain arbitrary, incompatible initial values.

In regard to the ideas discussed here, the reader is referred to the papers of Matkowsky (1970a and 1970b).

To end this article we note that spontaneous flame instability is also of interest from the general physicomathematical point of view. It turns out that a flame, though a fully deterministic physical system, is under certain conditions liable spontaneously to become turbulent, as happens in flow of viscous fluids at large Reynolds numbers. In contrast to the situation obtaining in the flow of a classical viscous fluid, in the case of eqn (1) turbulence may arise even in the one-dimensional case.

At $\sigma = 1$ the corresponding random solutions will obviously have the properties of statistical homogeneity and isotropy. Equation (1) can thus describe simpler types of turbulence than the Navier–Stokes equation, and in this sense it may provide a useful mathematical model for studies of the onset and essence of turbulence-type phenomena.

Acknowledgements—The author is deeply indebted to Prof. G. I. Barenblatt and Prof. Ya. B. Zel'dovich for initial encouragement and to Prof. B. Matkowsky for a helpful discussion of the problem. This research was supported in part by the National Science Foundation Grant MPS 75-08328.

References

Andrews G. E., Bradley, D. and Lwakabamba, S. B. (1975) Turbulence and turbulent flame propagation—a critical appraisal. *Combustion and Flame* **24**, 285–403.

Barenblatt, G. I., Zel'dovich, Ya. B. and Istratov, A. G. (1962) On heat and diffusion effects in stability of laminar flames. *J. Appl. Mech. Tech. Phys.* (*PMTF*) **4**, 21–26 (in Russian).

Fristrom, R. M. and Westenberg, A. A. (1965) *Flame Structure.* McGraw-Hill, New York.

Istratov, A. G. and Librovich V. B. (1966a) The effect of transport processes on the stability of a plane flame front. *J. Appl. Math. Mech.* (*PMM*) **30**(3), 541–547.

Istratov, A. G. and Librovich, V. B. (1966b) Stability of Flames. *Hydromechanics*—1965. Acad. Sc. USSR. Institute of Scientific Information. (English translations, FSTC-HT-23-952-68).

Istratov, A. G. and Librovich, V. B. (1969) On the stability of gasodynamic discontinuities associated with chemical reactions. The case of spherical flame. *Astronautica Acta* **14**, 453–467.

Landau, L. D. (1944) On the theory of slow combustion. *J. Exp. Theor. Phys.* (*ZhETF*) **14**(6), 240–245 (in Russian).

Libby, P. A. and Williams, F. A. (1976) Turbulent flows involving chemical reactions, *Annual Review of fluid Mechanics* **8**, 351–376.

Markstein, G. H. (Editor) (1964) *Non-steady Flame Propagation.* Pergamon Press, Oxford.

Matkowsky, B. J. (1970a) A simple nonlinear dynamic stability problem. *Bull. Amer. Math. Soc.* **76**(3), 620–625.

Matkowsky, B. J. (1970b) Nonlinear dynamic stability: a formal theory. *SIAM J. Appl. Math.* **18**(4), 872–883.

Michelson, D. M. and Sivashinsky, G. I. (1978) Nonlinear analysis of hydrodynamic instability in laminar flames. Part 2. Numerical Experiments. *Acta Astronautica* (in press).

Palm-Leis, A. and Strehlow, R. A. (1969) On the propagation of turbulent flames. *Combustion and Flame* **13**, 111–129.

Schelkin, R. I. and Troshin, Y. K. (1964) Gas Dynamics of Combustion, Nauka, Moscow (English translations, NASA TT F231).

Sivashinsky, G. I. (1975) The structure of Bunsen flames. *J. Chem. Phys.* **62**(2), 638–643.

Sivashinsky, G. I. (1977) Diffusional-thermal theory of cellular flames. *Combustion Sc. Tech.* **15**(3/4), 137–145.

Williams, F. A. (1970) An approach to turbulent flame theory. *J. Fluid Mech.* **40**(2), 401–421.

Zel'dovich, Ya. B. (1944) Theory of Combustion and Detonation of Gases. *Acad. Sc. USSR*, Moscow.

Zel'dovich, Ya. B. and Barenblatt, G. I. (1959) Theory of flame propagation. *Combustion and Flame* **3**(1), 61–74.

Appendix A

Nomenclature

- B see eqn (18)
- C_0 concentration of the limiting component
- C dimensionless concentration (referred to C_0)
- E activation energy
- e see eqn (65)
- F flame front surface (referred to l_T)
- f see eqn (62)
- g see eqn (63)
- h see eqn (64)
- I see eqn (54)
- K see eqn (23)
- $\tilde{K}$ see eqn (32)
- k dimensionless wavenumber (referred to $1/l_T$)
- L Lewis number

l_T thermal thickness of flame
N dimensionless activation energy ($=E/R^0T_b$)
P dimensionless pressure (referred to $\rho_b U_b^2$)
p see eqns (52), (79)
$\tilde{p}$ see eqn (15)
p' perturbation of pressure
Q strength of surface source
R^0 universal gas constant
R see eqn (17)
S see eqn (52), (79)
$\tilde{S}$ see eqn (15)
S' perturbation of concentration
T dimensionless temperature (referred to T_b)
T_b adiabatic temperature of combustion products
T_0 temperature of fresh mixture
t dimensionless time (referred to l_T/U_b)
U_i dimensionless gas velocity (referred to U_b)
U_b velocity of a plane flame front relative to the burned gas
u, v, w see eqns (52), (79)
$\tilde{u}, \tilde{v}$ see eqn (15)
u', v', w' perturbation of velocity
V velocity of corrugated flame front
x, y, z dimensionless space coordinates (referred to l_T)
$\mathbf{x}$ (y, z)
α see eqns (37), (100)
β see eqn (53)
ϵ see eqn (31)
ζ, η see eqns (43), (48), (100)
θ see eqns (52), (79)
ν Prandtl number
ξ see eqns (43), (48)
ρ dimensionless density (referred to ρ_b)
ρ_b density of combustion products
σ temperature (density) ratio ($= T_0/T_b$)
τ see eqns (43), (48), (100)
Φ see eqns (79), (100)
Ω see eqns (23)
$\tilde{\Omega}$ see eqn (32)
ω rate of stability parameter
Π see eqn (54)
χ see eqn (43), (48)

J. Physique **46** (1985) 1485-1494 SEPTEMBRE **1985**

Classification
Physics Abstracts
05.45 — 47.20 — 47.70F

Application of pole decomposition to an equation governing the dynamics of wrinkled flame fronts

O. Thual

CNRM, 42 avenue Coriolis, 31057 Toulouse Cedex, France

U. Frisch and M. Hénon

CNRS, Observatoire de Nice, B.P. 139, 06003 Nice Cedex, France

(*Reçu le 27 février 1985, accepté le 13 mai 1985*)

Résumé. — L'équation intégrale de Sivashinsky gouvernant certaines instabilités hydrodynamiques de fronts de flamme unidimensionnels est un cas particulier des modèles de plasma non linéaires de Lee et Chen ; en tant que telle elle possède une décomposition en pôles. Ceci explique les structures très organisées observées dans les simulations numériques. L'équation de Sivashinsky a des solutions stationnaires stables avec les pôles alignés parallèlement à l'axe imaginaire. Avec des conditions aux limites périodiques, quand le nombre de modes linéairement instables est élevé, les pôles se condensent en une distribution ln coth. Ceci est illustré par calcul numérique des positions d'équilibre des pôles. La condensation des pôles explique les plis de certains fronts de flamme. Le spectre d'énergie du déplacement du front suit une loi en $\ln^2 k$.

Abstract. — The Sivashinsky integral equation governing certain hydrodynamical instabilities of one-dimensional flame fronts is a special case of Lee and Chen's (*Phys. Scr.* **2** (1982) 41) non linear plasma models; as such it has a pole decomposition. This explains the highly organized structures observed in numerical simulations. The Sivashinsky equation has stable steady solutions with the poles aligned parallel to the imaginary axis. With periodic boundary conditions, when the number of linearly unstable modes is large, the poles condense into a ln coth distribution. This is illustrated by numerical calculations of equilibrium positions of poles. The pole condensation explains cusp-like wrinkles in certain flame fronts. The energy spectrum for the front displacement follows a $\ln^2 k$ law.

1. Introduction.

Sivashinsky [1] has shown that in a suitable asymptotic regime the dynamics of wrinkled flame fronts is governed by a non linear partial (pseudo)differential equation. In the one-dimensional case it reads

$$\partial_t u + u\,\partial_x u = \Lambda u + \nu\,\partial_x^2 u\,. \qquad (1.1)$$

Λ is a linear singular operator defined conveniently in terms of the spatial Fourier transform :

$$u(t, x) = \int_{-\infty}^{+\infty} \mathrm{e}^{ikx}\,\hat{u}(t, k)\,\mathrm{d}k\,, \qquad (1.2)$$

by

$$\Lambda : \hat{u}(t, k) \mapsto |\,k\,|\,\hat{u}(t, k)\,. \qquad (1.3)$$

We here use notation different from Sivashinsky's in order to bring out similarities with Burgers' equation

$$\partial_t u + u\,\partial_x u = \nu\,\partial_x^2 u\,. \qquad (1.4)$$

What is here denoted u is actually not a velocity but $\partial_x \phi$ where ϕ is the flame front displacement; so it is the slope of the flame front. Also we find it more convenient to vary the « viscosity » ν rather than the size of the domain.

Various studies of the Sivashinsky equation have been reported [1-4]. The results indicate that the solutions are highly organized in the form of one or several wrinkles (see Figs. 2 and 3 of Ref. [3] and also Figs. 2 and 3 of the present paper).

At the root of this simple behaviour of the Sivashinsky equation is the fact that it possesses a *pole decomposition* : equation (1.1) admits solutions of the form

$$u(t, x) = -2\nu \sum_{\alpha=1}^{2N} \frac{1}{x - z_\alpha(t)}. \qquad (1.5)$$

The z_α's are poles in the complex plane (coming in c.c. pairs) moving according to the laws of motion of poles $(\alpha = 1, 2, \ldots, 2N)$

$$\dot{z}_\alpha = -2\nu \sum_{\beta \neq \alpha} \frac{1}{z_\alpha - z_\beta} - i \operatorname{sign}(\Im(z_\alpha)), \qquad (1.6)$$

where $\Im$ denotes the imaginary part. When 2π spatial periodicity is assumed it is enough to restrict attention to poles with real parts between 0 and 2π. Instead of equations (1.5) and (1.6) one then uses

$$u(t, x) = -\nu \sum_{\alpha=1}^{2\pi} \cot\left(\frac{x - z_\alpha}{2}\right) \qquad (1.7)$$

and

$$\dot{z}_\alpha = -\nu \sum_{\beta \neq \alpha} \cot\left(\frac{z_\alpha - z_\beta}{2}\right) - i \operatorname{sign}(\Im(z_\alpha)). \qquad (1.8)$$

The existence of this pole decomposition follows from a more general result of Lee and Chen [5] for a class of non linear dynamical models arising in plasma turbulence. In our notation the governing equations for the Lee and Chen models are

$$\partial_t u + u\, \partial_x u = \nu\, \partial_x^2 u + \rho \Lambda u + \sigma \Lambda\, \partial_x u, \qquad (1.9)$$

which is a linear combination of the Sivashinsky equation and the Benjamin-Ono [6-7] equation. For the convenience of the reader a derivation of the pole decomposition for the Sivashinsky equation is given in the appendix A.

We shall not here review work on pole decompositions. A list of references up to 1981 may be found in section 4 of reference [8]. We also mention a recent application of pole decomposition to the formation of cusp singularities in interface dynamics [9].

This paper is organized as follows : section 2 is a systematic study of the two pole problem giving some qualitative insight into N-pole dynamics; in particular there is a tendency for the poles to align themselves parallel to the imaginary axis. Subsequently the paper is devoted essentially to the detailed study of such « vertical alignments ». Section 3 is about the discrete dynamics of N pairs of poles : for N large the poles condense into a continuous distribution which is determined analytically (Sect. 4). This is the analog for the Sivashinsky equation of the pole condensations for Burgers' equation [10]. From the distribution of poles we deduce the shape of flame wrinkles and the high wavenumber behaviour of the energy spectrum. In section 5 we mention some implications and open problems.

2. The two-pole problem and qualitative dynamics.

We begin by studying the simplest dynamical situation with two poles at complex conjugate locations $z(t) = a(t) + ib(t)$ and $z_* = a - ib$ with $b > 0$. From equation (1.6), we have

$$\dot{z} = -\frac{2\nu}{z - z_*} - i. \qquad (2.1)$$

It follows that a is a constant and that

$$\dot{b} = \frac{\nu}{b} - 1. \qquad (2.2)$$

Note that the $-2\nu/(z - z_*)$ term in equation (2.1), already present in Burgers' dynamics, tends to push the poles away from the real axis while the $-i$ term produces a uniform drift towards the real axis. Their competition leads for $t \to \infty$ to a stable equilibrium configuration with

$$b_{eq} = \nu. \qquad (2.3)$$

The corresponding steady solution of the Sivashinsky equation is

$$u(x) = -\frac{4\nu x}{x^2 + \nu^2}. \qquad (2.4)$$

In the periodic case equations (2.3) and (2.4) become respectively

$$b_{eq} = \operatorname{arg\,tanh} \nu, \qquad (2.5)$$

and

$$u(x) = -\frac{2\nu \sin x}{(1 - \nu^2)^{-1/2} - \cos x}. \qquad (2.6)$$

A graph of $u(x)$ for $\nu = 5 \times 10^{-2}$ is shown in figure 1. The aspect is quite similar to the localized structures appearing in figure 3 of reference [4], concerning an equation differing very slightly from the Sivashinsky equation.

Next we consider two poles z_1 and z_2 in the upper half plane $\Im(z) > 0$ which we assume to be sufficiently close to each other so that we can ignore their interactions with other poles. We then have

$$\dot{z}_1 = -\frac{2\nu}{z_1 - z_2} - i, \quad \dot{z}_2 = -\frac{2\nu}{z_2 - z_1} - i. \qquad (2.7)$$

Thus $\zeta = z_1 - z_2 = a + ib$ satisfies

$$\dot{\zeta} = -\frac{4\nu}{\zeta}. \qquad (2.8)$$

Hence

$$\zeta^2(t) - \zeta^2(0) = -8\nu t. \qquad (2.9)$$

When t varies from 0 to $+\infty$, then $\Im(\zeta^2) = 2ab$ remains constant so that the point (a, b) moves on a

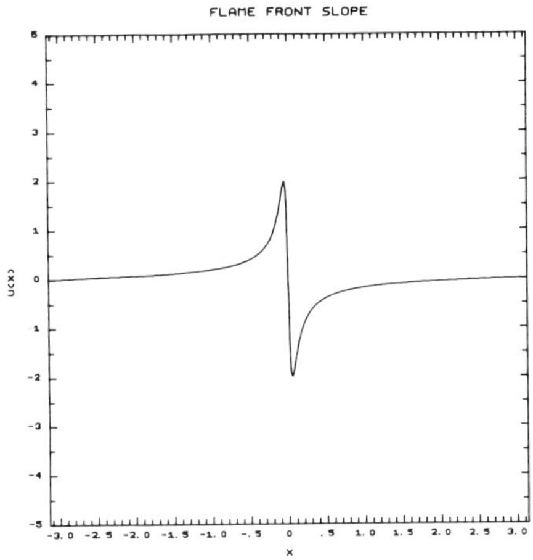

Fig. 1. — Structure of the slope of the flame front $u(x)$ for the steady space-periodic solution with two complex conjugate poles.

hyperbola. For large times $a \to 0$ and $b \to +\infty$: poles tend to align on a parallel to the imaginary axis.

The above analysis indicates that, roughly, poles tend to attract each other horizontally (parallel to the real axis) and to repel each other vertically. In addition, they are subject to a drift towards the real axis. Extrapolation of these qualitative features to a situation with many poles suggests that the poles tend to form vertical alignments, eventually coalescing into a single one. This is indeed happening in Burgers' dynamics and may be viewed as the mechanism for the formation and coalescence of shocks. According to Lee and Chen [5] this merging process is a universal property of their non dispersive models.

In the periodic case the poles are constrained to be on 2π periodic arrays or, equivalently, to be all contained in a domain of extension 2π in the real direction and to have cot rather than z^{-1} interactions. The short distance behaviour is thus unaffected, but the global dynamics are quite different. At the moment there is no result concerning the eventual fate of a system of poles (periodic or non periodic). In the periodic case numerical integrations of the Sivashinsky equation [3-4] indicate that for small ν the solutions often tend to a steady state with one or several wrinkles in the flame front displacement (in fact vertical pole condensations as we shall see). These features are reproduced in simulations of our own using a pseudo-spectral method with 512 Fourier modes. Figure 2 shows a solution with one wrinkle for $\nu = 0.1$. Figure 3 shows a solution with two wrinkles, also for $\nu = 0.1$, but restricted to the class of functions $u(x)$ which are odd (otherwise this solution would be unstable). The labels a and b correspond respectively to the front displacement $\phi(x)$ and the front slope $u(x)$.

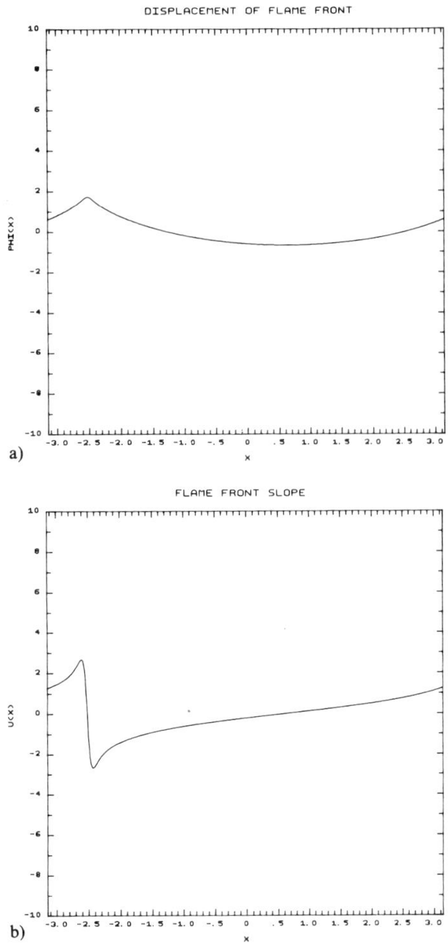

Fig. 2. — Structure of the flame front (a) and its slope (b) for a steady one-wrinkle space-periodic solution obtained by pseudo-spectral simulation of the Sivashinsky equation with 512 Fourier modes and « viscosity » coefficient $\nu = 0.1$.

Henceforth we shall consider a single vertical alignment, possibly containing a large number of poles, with or without periodicity.

3. Vertical alignments of poles : discrete dynamics.

We consider N pairs of complex conjugate poles constrained to be on a parallel to the imaginary axis, located at

$$x = x_0 , \quad y = \pm y_1, \ldots, \pm y_N . \tag{3.1}$$

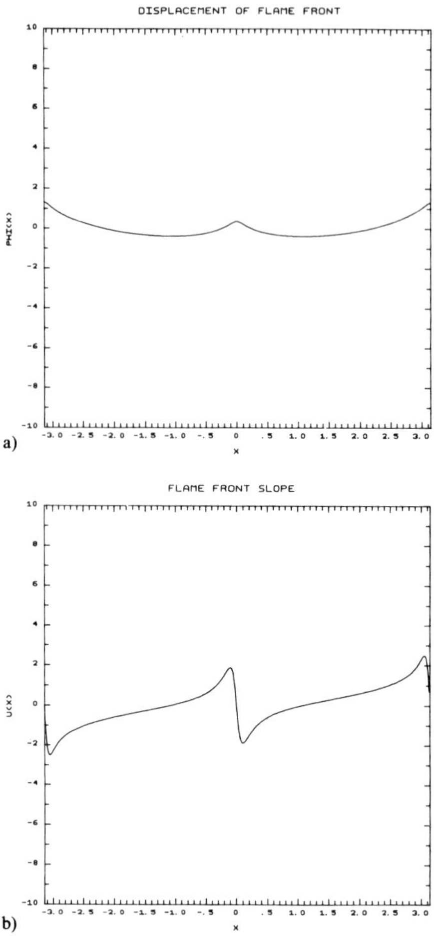

Fig. 3. — Same as figure 2, except that the solution $u(x)$ is restricted to odd functions, allowing stable coexistence of two wrinkles.

Without loss of generality, we assume that $x_0 = 0$; we also assume that the y_j's have been ordered :

$$0 < y_1 < \cdots < y_N . \qquad (3.2)$$

This order will never change since the repulsion between two poles becomes infinite when they approach each other. For the same reason, we always have strict inequalities in (3.2).

The motion of poles is given by a set of N simultaneous real differential equations :

$$\dot{y}_j = f_j(y_i, \ldots, y_N) , \qquad (3.3)$$

with, in the non-periodic case

$$f_j = \nu\left[\frac{1}{y_j} + \sum_{k \neq j}\left(\frac{2}{y_j - y_k} + \frac{2}{y_j + y_k}\right)\right] - 1 , \qquad (3.4)$$

and in the periodic case

$$f_j = \nu\left[\coth y_j + \sum_{k \neq j}\left(\coth\frac{y_j - y_k}{2} + \coth\frac{y_j + y_k}{2}\right)\right] - 1 . \qquad (3.5)$$

For $N = 1$, (3.4) reduces to (2.2). A steady state will be a solution of the N simultaneous ordinary equations

$$f_j = 0 \quad (i = 1, \ldots, N) . \qquad (3.6)$$

We note first some properties of these equations.

(i) In the non-periodic case, ν can be eliminated by the change of variables

$$y_j = \nu Y_j , \quad t = \nu T . \qquad (3.7)$$

ν is therefore an irrelevant parameter. On the contrary, in the periodic case a scale is prescribed so that the value of ν is relevant.

(ii) In the non-periodic case, multiplying (3.4) by y_j and summing over j, we obtain the virial-like relation

$$\frac{1}{2}\frac{\mathrm{d}}{\mathrm{d}t}\sum_{j=1}^{N} y_j^2 = \nu N(2N - 1) - \sum_{j=1}^{N} y_j . \qquad (3.8)$$

The right-hand side must vanish at equilibrium; this provides a test of computational accuracy.

(iii) In the periodic case, we derive from (3.5) for the highest pole $j = N$, using the fact that

$$\coth x > 1 \quad \text{for} \quad x > 0 :$$

$$f_N > \nu(2N - 1) - 1 . \qquad (3.9)$$

If the right-hand side is positive or zero, f_N is always positive, and no steady state can exist; the highest pole moves towards $y = +\infty$ in the imaginary direction with an (asymptotically) constant speed. Therefore a necessary condition for the existence of a steady state in the periodic case is

$$\nu(2N - 1) < 1 . \qquad (3.10)$$

(iv) The following properties hold in the non-periodic case, and also in the periodic case when (3.10) is satisfied :

1. *There exists one and only one steady state.*

2. *Any solution of* (3.3) *tends towards the steady state for* $t \to +\infty$. The proof of these properties is based on the existence of a Lyapunov function (see Appendix B).

(v) The equations (3.6) are easily solved for $N = 1, 2, 3$ in the non-periodic case. We take $\nu = 1$ for simplicity. For $N = 1$, the steady state solution is

$$y_1 = 1 . \qquad (3.11)$$

For $N = 2$:

$$y_1 = 3 - \frac{3\sqrt{2}}{2} = 0.878680\ldots,$$

$$y_2 = 3 + \frac{3\sqrt{2}}{2} = 5.121320\ldots. \qquad (3.12)$$

For $N = 3$, the y_j are the three roots of the equation

$$y^3 - 15\,y^2 + \frac{45 + 15\sqrt{129}}{4}\,y - \frac{135\sqrt{129} - 1\,395}{4} = 0\,, \qquad (3.13)$$

i.e.

$$y_1 = 0.818742\ldots,\quad y_2 = 4.254082\ldots,\quad y_3 = 9.927176\ldots \qquad (3.14)$$

These valus can also be used as a check on the program for the numerical solution of (3.6).

We turn now to numerical determination of the equilibrium positions. The equations (3.6) were solved numerically by a relaxation method : the position y_i of each pole in turn is adjusted to a new value

$$y_j' = y_j + \Delta y_j\,. \qquad (3.15)$$

Δy_j is computed by a one-step Newton formula :

$$\Delta y_j = -\,f_j\left(\frac{\partial f_j}{\partial y_j}\right)^{-1}, \qquad (3.16)$$

with the constraint that it should not exceed one-half of the distance to the next pole. A convenient initial state is : $y_j = 2\,\nu j$. The computation is halted when the relative change $|\,\Delta y_j\,|/y_j$ becomes less than some prescribed small number for all poles. This method was found to always converge to the equilibrium, although the convergence slows down for large N.

The value

$$\nu = 0.005 \qquad (3.17)$$

was used in all numerical computations, in order to facilitate comparisons. As a consequence of (3.10), a small value of ν is necessary in order to allow a large number of poles in the periodic case. With the value (3.17), the maximum of N in the periodic case is

$$N_{\max} = 100\,. \qquad (3.18)$$

Figures 4a and 4b show (full curves) the cumulative distribution, i.e., the number of poles between 0 and y as a function of y, for the non-periodic case and for $N = 10$ and $N = 100$ respectively. The dotted and dashed curves are analytic approximations which will be described in section 4. Table I gives in columns 2 and 3 the positions of the lowest and highest poles, y_1 and y_N, for various N. Column 4 is an analytic estimate which will also be described in section 4.

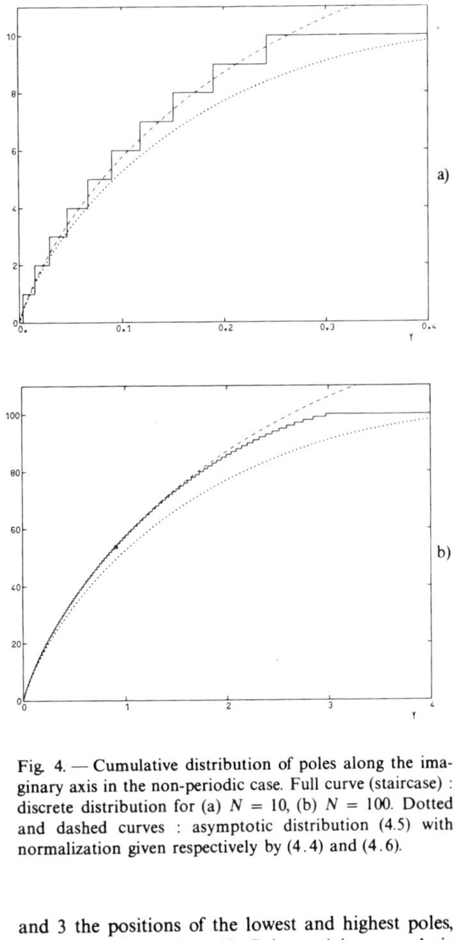

Fig. 4. — Cumulative distribution of poles along the imaginary axis in the non-periodic case. Full curve (staircase) : discrete distribution for (a) $N = 10$, (b) $N = 100$. Dotted and dashed curves : asymptotic distribution (4.5) with normalization given respectively by (4.4) and (4.6).

Table I

N	y_1	y_N	y_1^*
1	0.005000	0.005000	0.008070
2	0.004393	0.025607	0.006129
3	0.004094	0.049636	0.005417
4	0.003902	0.075278	0.005016
5	0.003764	0.10188	0.004747
10	0.003387	0.24178	0.004084
20	0.003076	0.53470	0.003596
50	0.002740	1.4409	0.003115
100	0.002531	2.9761	0.002833
200	0.002351	6.0730	0.002600

In the periodic case, when N is well below N_{max}, the distribution differs only slightly from the non-periodic case. For $N = 10$, for instance, one obtains a curve which is barely distinguishable from that of figure 4*a*. The reason is simply that for small y, (3.5) reduces to (3.4). As N increases and approaches N_{max}, however, the periodic case deviates from the non-periodic case. Figure 5 shows the cumulative distribution of the periodic case for $N = N_{max} = 100$. This distribution differs noticeably from that of figure 4*b* (the horizontal scale is not the same). The dashed curve is again an analytic approximation which will be described in section 4. Table II gives the positions of the lowest and highest poles for various N. Comparing with table I, we see again that the difference with the non-periodic case is negligible for small N but increases quickly as N approaches N_{max}.

Table II.

N	y_1	y_N
1	0.005000	0.005000
2	0.004394	0.025609
3	0.004094	0.049647
4	0.003902	0.075312
5	0.003765	0.10195
10	0.003389	0.24261
20	0.003083	0.54272
50	0.002775	1.6051
100	0.002663	9.3282

4. Continuous approximation and pole condensations. We begin with the non-periodic case. The very regular distribution of poles shown in figure 4 suggests going over to a continuous approximation, with the individual poles at locations $\pm iy_j$ replaced by a density of poles $\rho(y) \geqslant 0$ normalized to the total number of poles

$$\int_{-\infty}^{+\infty} \rho(y)\,\mathrm{d}y = 2N. \qquad (4.1)$$

Equation (3.6) is then replaced by an integral equation (the P in front of the integral means Cauchy Principal value)

$$2\nu P \int_{-\infty}^{+\infty} \frac{\rho(y')}{y - y'}\,\mathrm{d}y' - \operatorname{sign} y = 0. \qquad (4.2)$$

The general solution of this linear integral equation, which may be obtained by Fourier transformation, is

$$\rho(y) = \frac{1}{\pi^2 \nu} \ln \frac{c}{|y|}, \qquad (4.3)$$

where c is an arbitrary positive constant. This solution is neither normalized nor positive (for $|y| > c$).

This is not surprising : we have assumed that (4.2) is true for all y, but if the density drops to zero at a finite distance y_{max} then there is no need for (4.2) to be true for $y > y_{max}$. In fact (4.2) cannot be true for all y : for any finite distribution, the first term in (4.2) tends to zero for $y \to \infty$. Equation (4.3) only represents an inner expansion near the origin and c must be determined by matching.

One method for determining the constant c is to truncate the density (4.3) to zero beyond $|y| = c$, and to use the normalization condition (4.1). This gives

$$c = \pi^2 \nu N. \qquad (4.4)$$

From (4.3) we compute the cumulative distribution :

$$R(y) = \int_0^y \rho(u)\,\mathrm{d}u = \frac{y}{\pi^2 \nu}\left(1 + \ln \frac{c}{y}\right). \qquad (4.5)$$

The curve $R(y)$ is shown on figures 4a and 4b as a dotted line, for c given by (4.4). It reproduces correctly the shape of the observed distribution, but stays somewhat below. The optimal matching is obtained with

$$c = 1.28\,\pi^2 \nu N, \qquad (4.6)$$

which produces the dashed curve on figures 4a and 4b. There is now very good agreement, except in the outermost parts of the distribution for $N = 100$.

The discrete positions of the poles can be estimated by solving the implicit equation

$$R(y_j) = j - \tfrac{1}{2}. \qquad (4.7)$$

(This formula would give the exact pole positions if the curve $R(y)$ passed through the middle point of each vertical step in Fig. 4.) The solution for the innermost pole ($j = 1$), using (4.6) and (4.7) is given in table I, column 4. Agreement with the true value of y_1 is seen to improve slowly as N increases.

For $N \to \infty$, (4.7) gives $y_1 \to 0$, and more generally $y_j \to 0$: the concentration of poles near the origin increases without limit (*pole condensation*). It increases extremely slowly, however : y_j varies to leading order in N as $1/\ln N$. The innermost poles get squeezed in logarithmically with N due to the repulsion from all other poles.

Now we turn to the periodic case and reinsert ν. By (3.10), the maximum number of pole pairs that can be put in an equilibrium configuration is inversely proportional to ν. We observe that when entire initial data are prescribed for the Sivashinsky equation, poles are created at $t = 0_+$ at infinity in arbitrary large numbers by essentially the same mechanism that operates for the Burgers equation [10-11]. We shall therefore assume that the number of poles present in the steady state is equal to its maximum acceptable value, N_{max}. When ν is small we then get order $1/\nu$ very closely packed poles (*pole condensation*). The distribution within the condensation can again be obtained by going to the continuum limit; the analog of equation (4.2) is

$$\nu P \int_{-\infty}^{+\infty} \rho(y') \coth \frac{y-y'}{2} \, dy' - \text{sign}\, y = 0 \quad (4.8)$$

The general solution is

$$\rho(y) = \frac{1}{\pi^2 \nu} \ln \left(c_1 \coth \frac{|y|}{4} \right) \quad (4.9)$$

where c_1 is an arbitrary positive constant. This is again an inner expansion. Since we know that there is only a finite number of poles $O(\nu^{-1})$, we impose that the density $\rho \to 0$ as $y \to \infty$. This gives

$$c_1 = 1 . \quad (4.10)$$

Moreover ρ is then positive for all y and

$$\int_{-\infty}^{+\infty} \rho(y) \, dy = \frac{1}{\nu} . \quad (4.11)$$

Since we assume N to be large and equal to the maximum value N_{max} allowed by (3.10), we have

$$\frac{1}{\nu} \sim 2N - 1 \sim 2N , \quad (4.12)$$

so that the normalization condition (4.1) is fulfilled to leading order.

We define again the cumulative distribution as

$$R(y) = \int_0^y \rho(u) \, du . \quad (4.13)$$

For ρ given by (4.9), this integral does not appear to have a closed form, so we compute it numerically. The integrand being singular at $y = 0$, it is convenient to do first an integration by parts :

$$R(y) = \frac{y}{\pi^2 \nu} \ln \coth \frac{|y|}{4} + \frac{1}{2\pi^2 \nu} \int_0^y \frac{u \, du}{\sinh u/2} . \quad (4.14)$$

The singularity is then isolated in the first term, and the second term is easily evaluated numerically. The curve $R(y)$ is shown on figure 5 as a dashed line. The agreement with the observed distribution is excellent over the whole y range.

Note that for small y, the density is to leading order given by

$$\rho(y) \sim \frac{1}{\pi^2 \nu} \ln \frac{4}{|y|} . \quad (4.15)$$

We recover the form (4.3) of the non-periodic case, with $c = 4$. The best-fit solution (4.6) for the non-periodic case, together with (4.12) would have given

$$c \simeq 0.64 \, \pi^2 \simeq 6.32 . \quad (4.16)$$

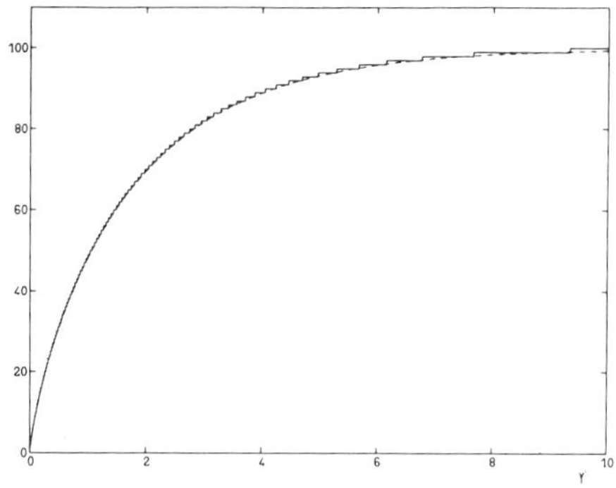

Fig. 5. — Cumulative distribution of poles along the imaginary axis in the periodic case. Full curve : discrete distribution for $\nu = 0.005$, $N = 100$, Dashed curve : asymptotic distribution (4.14).

Thus, for given large N, the periodic and non-periodic cases have the same functional form for $R(y)$ near $y = 0$ but different normalizations. We can again estimate the position of the innermost pole y_1 from (4.7) and (4.14); since y is small, the latter can be replaced by

$$R(y) = \frac{y}{\pi^2 \nu} \left(1 + \ln \frac{4}{y} \right) . \quad (4.17)$$

for $\nu = 0.005$, we obtain from this continuous approximation

$$y_1^* = 0.003012 , \quad (4.18)$$

to be compared with the value computed from the discrete case

$$y_1 = 0.002663 . \quad (4.19)$$

If ν is varied, with N simultaneously changed so that is always has its maximum value, then y_1, the distance of the innermost pole to the real axis, is given to leading order in ν by

$$y_1 \sim \frac{\pi^2 \nu}{2 \ln (1/\nu)} . \quad (4.20)$$

Using the above asymptotic expansions we can determine some key features of the small ν steady solutions in the periodic case. The solution of the Sivashinsky equation with a pole condensation on the imaginary axis is given in real physical space by

$$u(x) = -\nu \sum_{\substack{n=-N \\ n \neq 0}}^{N} \cot \frac{x - i y_n}{2} . \quad (4.21)$$

For $|x| \gg \nu/\ln(1/\nu)$ we can use the continuous density and obtain

$$u(x) \sim -\frac{1}{\pi^2}\int_{-\infty}^{+\infty} \cot\frac{x-iy}{2} \ln\left[\coth\frac{|y|}{4}\right] dy. \tag{4.22}$$

The asymptotic expansion of this integral for (not too) small x gives

$$u(x) \sim \frac{1}{2\pi} \operatorname{sign}(x) \ln|x|. \tag{4.23}$$

The corresponding expansion for the flame front displacement $\phi(x)$ is

$$\phi(x) - \phi(0) \sim \frac{1}{2\pi}|x| \ln|x|. \tag{4.24}$$

Sivashinsky (1983) refers to a similar logarithmic singularity at the cusps of the flame folds, obtained by McConnaughey (1982). Actually there is no real singularity : from equation (4.20) the closest complex singularity is within a distance $y_1 = O(\nu/\ln \nu^{-1})$; thus the cusp is slightly rounded over a distance y_1.

We now calculate the spectrum $E(k)$ of the solution, defined as the squared modulus of the Fourier transform of the steady flame front displacement. Since $u(x) = \partial_x \phi(x)$ we have

$$E(k) = k^{-2} |\hat{u}_k|^2, \tag{4.25}$$

where

$$\hat{u}_k = \frac{1}{2\pi}\int_0^{2\pi} e^{-ikx} u(x)\, dx. \tag{4.26}$$

A simple residue calculation based on the pole decomposition gives, for $k > 0$

$$\hat{u}_k = 2i\nu \sum_{n=1}^{N} \exp(-ky_n). \tag{4.27}$$

For large k we must distinguish two ranges :

(i) *Dissipation range.* When $|k| y_1 \gg 1$ only the nearest pole singularity matters and we get

$$E(k) \sim 4\nu^2 k^2 \exp(-2y_1 k), \tag{4.28}$$

with y_1 given by equation (4.20).

(ii) *Inertial range.* When $1 \ll |k| \ll y_1^{-1}$ we may replace the sum in equation (4.27) by an integral over the continuous distribution of poles; this gives to leading order (for $\nu \to 0$ first and then $k \to \infty$)

$$E(k) \sim \frac{4}{\pi^4} \ln^2 |k|. \tag{4.29}$$

Note that the energy spectrum for $u(x)$ follows a $k^{-2} \ln^2 |k|$ law; i.e. it is somewhat shallower than the k^{-2} spectrum obtained for Burgers' model. The numerical simulation of Pumir [4] (his Fig. 5b) gives for $u(x)$ an energy spectrum that approximately follows a $k^{-\alpha}$ law with $\alpha \simeq 1.23$ which is also shallower than k^{-2}. Given that Pumir's equation is not exactly the Sivashinsky equation and given also the limited range of wavenumbers involved there is probably no genuine discrepancy.

5. Discussion.

We have found that the pole decomposition provides a reduction of the Sivashinsky equation to a discrete Dynamical System. The latter admits steady state solutions with all the poles aligned parallel to the imaginary axis. In the spatially periodic case the maximum number N of complex conjugate pairs of poles in an alignment is such that $\nu(2N-1) < 1$. We conjecture more general results for analytic periodic initial data : (i) as $t \to \infty$ all the singularities (poles and others) are pushed off to infinity, except a finite number N of pairs of poles satisfying the above inequality; (ii) for real times poles stay uniformly bounded away from the real axis. The former result has been recently established for the case of initial conditions with a finite but arbitrary number of poles [13, 14]. In this context we also mention a result of Foias, Nicolaenko, Sell and Temam [17] concerning the family of equations

$$\partial_t u + u\,\partial_x u = M_\varepsilon u, \tag{5.1}$$

where the operator M_ε has the following representation in Fourier space

$$M_\varepsilon : \hat{u}_k \mapsto [\varepsilon(|k| - k^2) + (1-\varepsilon)(k^2 - k^4)]\,\hat{u}_k. \tag{5.2}$$

It is assumed that $0 \leqslant \varepsilon \leqslant 1$; $\varepsilon = 1$ is the Sivashinsky equation. For the case of space-periodic odd solutions of equation (5.2) Foias *et al.* [15] have shown that when $t \to \infty$ the solutions have a finite dimensional attractor imbedded in a finite dimensional manifold.

In the periodic case with small ν we found that the poles on a vertical alignment condense into a ln coth distribution the signature of which in real physical space is a cusp in the flame front. This suggests a real singularity. Actually, as we have seen in section 4 there are only complex poles, but the ones nearest to the real axis are at a distance $O(\nu/\ln \nu^{-1})$ so that the cusp is slightly rounded. It is noteworthy that this distance is smaller by a factor $1/\ln \nu^{-1}$ to what would have been inferred from a naive analysis based on the observation that wavenumbers in excess of ν^{-1} are linearly stable. This discrepancy is important when attempting to numerically solve the Sivashinsky equation : the number of Fourier modes that are necessary scales like $\nu^{-1} \ln \nu^{-1}$ and not like ν^{-1}. Otherwise the numerics will work well as long as the poles stay within $O(\nu)$ of the real axis (as in the two-pole solution); after some time however the poles will start piling up vertically and spurious singularities may be observed [4].

We finally mention an open problem. We have shown that the Sivashinsky equation has stable steady solutions, which can be made trivially time-dependent by a Galilean transformation. There may also be non-trivial time-dependent solutions. In one spectral simulation with 25 linearly unstable modes ($\nu = 1/25$) we have observed a complicated time-dependent régime going through a succession of single and multi-wrinkle configurations. At the moment we cannot rule out ever-lasting, possibly chaotic, time-dependent solutions [16].

Acknowledgments.

We have benefitted from discussions with Y. C. Lee, A. Pumir, B. Nicolaenko, and S. Zaleski. Some of the numerical results reported in section 3 were obtained independently by J. P. Rivet.

Appendix A.

THE POLE DECOMPOSITION. — We wish to show that equations (1.1)-(1.3) have pole solutions of the form

$$u(t, x) = -2\nu \sum_{\alpha=1}^{2N} \frac{1}{x - z_\alpha(t)}. \qquad (A.1)$$

with

$$\dot{z}_\alpha = -2\nu \sum_{\beta \neq \alpha} \frac{1}{z_\alpha - z_\beta} - i\,\mathrm{sign}\,(\Im(z_\alpha)). \qquad (A.2)$$

We note that $u(x)$ is a linear combination of terms of the form

$$p_z(x) = \frac{1}{x - z}. \qquad (A.3)$$

We claim that

$$\Lambda p_z(x) = i\,\mathrm{sign}\,(\Im(z))\,\partial_x p_z(x), \qquad (A.4)$$

where Λ is the pseudo-differential operator defined by (1.3), i.e. multiplication by $|k|$ in Fourier space. A simple residue calculation gives the following expression for the Fourier transform $\hat{p}_z(k)$ of $p_z(x)$:

$$\text{for } \Im(z) > 0, \quad \hat{p}_z(k) = iH(-k)\,e^{-ikz}; \qquad (A.5)$$

$$\text{for } \Im(z) < 0, \quad \hat{p}_z(k) = iH(+k)\,e^{-ikz}, \qquad (A.6)$$

where $H(.)$ is the Heavyside function. Thus $\hat{p}_z(k)$ has its support in the negative or positive k-axis, depending on the sign of the imaginary part of z. Since the Fourier transform of ∂_x is the multiplication by ik (see (1.2)), the operator Λ acting on the function p_z is equivalent to $\pm i\partial_x$. This proves (A.4). We may thus interpret the operator Λ in the Sivashinsky equation (1.1) as producing an advection in the complex plane, in the imaginary direction towards the real axis.

Finally, the proof that (A.1) and (A.2) satisfies the Sivashinsky equation (1.1) is obtained by substitution and straightforward algebra, using the identity

$$\sum_{\alpha,\beta} \frac{1}{x - z_\alpha}\,\frac{1}{x - z_\beta} = \sum_\alpha \frac{1}{(x - z_\alpha)^2} + \sum_{\alpha \neq \beta} \frac{2}{z_\alpha - z_\beta}\,\frac{1}{x - z_\alpha}. \qquad (A.7)$$

Appendix B.

We give here the proofs of the assertions made in section 3. We remark first that the differential equations (3.3) may be written in terms of a Lyapunov function U :

$$\dot{y}_j = f_j = \frac{\partial U}{\partial y_j}, \qquad (B.1)$$

with, in the non-periodic case :

$$U = \nu\left[\sum_j \ln y_j + 2\sum_{\substack{j,k\\ j<k}} (\ln(y_k - y_j) + \ln(y_k + y_j))\right] - \sum_j y_j, \qquad (B.2)$$

and in the periodic case :

$$U = \nu\left[\sum_j \ln\sinh y_j + 2\sum_{\substack{j,k\\ j<k}} \left(\ln\sinh\frac{y_k - y_j}{2} + \ln\sinh\frac{y_k + y_j}{2}\right)\right] - \sum_j y_j. \qquad (B.3)$$

A simple computation shows that the function $U(y_1, \ldots, y_N)$ has a negative curvature in every direction. Specifically, if we consider a linear displacement : $y_j = a_j + b_j v$, with a_j, b_j constants and v variable, then $d^2U/dv^2 < 0$. An immediate consequence is that there cannot exist more than one steady state : if two existed, on the straight-line segment joining them we would have $dU/dv = 0$ at each end, which is in contradiction with the above property. Another consequence is that if a steady state does exist, then it is a global maximum for U.

Next we show that a steady state actually exists (provided, in the periodic case, that the inequality (3.10) holds). We consider a sub-problem in which $y_{p+1}, \ldots, y_N$ have given values while $y_1, \ldots, y_p$ are free, and we look for a steady state for the free variables. Specifically, we look for values of $y_1, \ldots, y_p$ such that (3.6) is satisfied for $j = 1, \ldots, p$. The accessible phase space is :

$$0 < y_1 < \cdots < y_p < y_{p+1} . \tag{B.4}$$

The same reasoning as above shows that there cannot exist more than one equilibrium for $y_1, ..., y_p$. We show now by recursion that one exists. Consider first the case $p = 1$. The domain of variation of y_1 is $0 < y_1 < y_2$. For $y_1 \to 0$, $f_1 \to +\infty$; and for $y_1 \to y_2$, $f_1 \to -\infty$. Therefore there exists a y_1 for which f_1 vanishes, i.e., a steady state for y_1. Suppose now that we have proved the existence of a steady state for $p-1$ poles and we want to prove it for p poles, with $p < N$. We let y_p vary in the interval $0 < y_p < y_{p+1}$, with the other poles $y_1, ..., y_{p-1}$ continually adjusted to their equilibrium position. For $y_p \to 0$, $f_p \to +\infty$; and for $y_p \to y_{p+1}$, $f_p \to -\infty$ (it is easily seen that the distance $y_p - y_{p-1}$ remains finite in the limit). Therefore there exists a y_p for which $f_p = 0$.

The last case $p = N$, is treated in the same fashion but with a slight difference. The interval of variation is now $0 < y_N < \infty$. For $y_N \to \infty$, $f_N \to -1$ in the non-periodic case, and $f_N \to \nu(2N-1) - 1$ in the periodic case. Therefore there exists a y_N for which $y_N = 0$. This completes the proof of existence of a steady state.

Finally, we have from (B.1) :

$$\dot{U} = \sum_j \frac{\partial U}{\partial y_j} \dot{y}_j = \sum_j f_j^2 \geqslant 0 , \tag{B.5}$$

the equality being obtained only at a steady state. Thus U always increases with time, except at equilibrium. On the other hand, U is bounded from above by its value at the steady state, which is a global maximum. Therefore U is a genuine Lyapunov function, and every solution of (3.3) tends toward the steady state for $t \to \infty$. This completes the proof of stability for pole-displacements parallel to the imaginary axis. This result has been extended to arbitrary displacements other than the trivial ones where all the poles undergo a same real translation [17].

References

[1] SIVASHINSKY, G. I., *Acta Astronaut.* **4** (1977) 1177.
[2] SIVASHINSKY, G. I., *Ann. Rev. Fluid Mech.* **15** (1983) 179.
[3] MICHELSON, D. M. and SIVASHINSKY, G. I., *Acta Astronaut.* **4** (1977) 1207.
[4] PUMIR, A., *Phys. Rev.* **A 31** (1985) 543.
[5] LEE, Y. C. and CHEN, H. H., *Phys. Scr.* **T 2** (1982) 41.
[6] BENJAMIN, T. B., *J. Fluid Mech.* **25** (1966) 241.
[7] ONO, H., *J. Phys. Soc. Japan* **39** (1975) 1082.
[8] FRISCH, U. and MORF, R., *Phys. Rev.* **A 23** (1981) 2673.
[9] SHRAIMAN, B. and BENSIMON, D., *Phys. Rev.* **A 30** (1984) 2840.
[10] BESSIS, D. and FOURNIER, J. D., *J. Physique Lett.* **45** (1984) L-833.
[11] FRISCH, U. in Proceed. *Sixth Kyoto Summer Institute on Chaos and Statistical Methods*, ed. Y. Kuramoto, p. 211, Springer series in Synergetics, Springer (1984).
[12] MCCONNAUGHEY, H. V., *Combust. Sci. Technol.* **33** (1983) 113.
[13] LEE, Y. C., CHENG, H. H., LEE, T. T. and QIAN, S. N., *Solitons and Turbulence ; Chaos in Infinite Dimensional Systems* ; Proc. Second Intern. Workshop on Non Linear and Turbulent Processes in Physics ; Kiev, October 1983, p. 1425, Gordon and Breach.
[14] LEE, Y. C., private communication (1985).
[15] FOIAS, C., NICOLAENKO, B., SELL, G. and TEMAM, R., *Inertial Manifold for the Kuramoto-Sivashinsky equation*, preprint Indiana University, Depart. Math. (1984).
[16] MICHELSON, D. M. and SIVASHINSKY, G. I., *Combust. Flame* **48** (1982) 211.
[17] RIVET, J. P., private communication (1985).

Combustion Science and Technology, 1980, Vol. 24, pp. 1–13
0010-2202/80/2401-0001$06.50/0

Flame Propagation in Tubes: Hydrodynamics and Stability

YA. B. ZEL'DOVICH, A. G. ISTRATOV, N. I. KIDIN., and V. B. LIBROVICH *Academy of Sciences of the USSR, Moscow*

(*Received April 17, 1980*)

Abstract—The steady propagation of a convex laminar flame front in long channels is under consideration. In comparison to the previous theoretical efforts by Zel'dovich (1944), Tsien (1951), Chernyi (1954) and Borisov (1978) the existence of a stagnation zone fixed with respect to the flame front is taken into account. The flame front is supposed to be a hydrodynamic discontinuity with the known normal rate of propagation through the cold gas; the boundary surface of the stagnation zone is considered as a discontinuity of the tangential component of velocity, *i.e.* as a vortex sheet.

The upstream flow field of cold gases can be treated as a potential one; however, the downstream flow of combustion products must be rotational. Vortices generated at the curved flame front drift along the streamlines and fill the whole flow area behind the flame. The gas viscosity, thermal conductivity, and diffusivity of the reacting components and heat losses at walls are neglected in this hydrodynamic model. The integral equation deduced under the simplified flow field description in front of the flame yields the dependence of the flame propagation velocity along the channel on its normal speed and on the ratio of initial and final temperatures. The results of an analytical approximation and numerical solution of this equation are compared.

Special attention is given to the hydrodynamic instability of the curved flame front with respect to disturbances of small amplitude. It appears to be much more stable than the plane flame front because of the supplementary stabilizing effects: stretching of the wavelength of the disturbance drifting by the tangential velocity component along the flame front and quenching by the channel walls. An analysis performed shows that a critical Reynolds number corresponding to the appearance of the flame instability can attain values over a few hundred and is in accordance with the experimental observations.

INTRODUCTION

General Outline

The propagation of flames in channels has long attracted the attention of many investigators (Zel'dovich, 1944; Tsien, 1951; Chernyi, 1954; Uberoi, 1959; Maxworthy, 1962; Borisov, 1978). The photographs show a curved shape of the flame front in channels up to considerable values of Reynolds numbers, Re (Re is calculated from the flame propagation rate along the tube, the tube diameter, and the cold gas viscosity). The flame front has a stationary curved surface at Reynolds numbers equal to a few hundred, although the hydrodynamic flame instability theory developed by Landau (1944), Darrieus (1944) and Markstein (1951) predicts stable propagation at Reynolds numbers just exceeding unity. The first consideration of this anomaly was that of Istratov and Librovich (1966)†.

† It also has been under consideration in the recent paper by Zel'dovich (1980).

The photograph taken from the paper of Uberoi (1959) and presented in Figure 1 shows the shape of the flame front propagating in a plane channel. It shows stagnation zones near the channel walls where the gas rests with respect to the flame front. The stagnation zones are caused by the refraction of stream lines at the extreme points of the flame front.

We will consider the cold-gas-convex surface of the flame front as a curved cell of the flame which had been formed after the plane flame lost its stability. The steady state of the convex flame is a result of the nonlinear hydrodynamic interaction with the gas flow field (see Zel'dovich, 1966, 1979). In the linear approximation the flame perturbation amplitude grows in time in accordance with Landau theory, but this growth is restricted by nonlinear effects.

The flame propagating in channels has common features with freely propagating spherical flames which are known to be more stable to hydrodynamic disturbances (Shchelkin, Troshin, and Rakipova, 1947; Rozlovsky and Zel'dovich, 1954; Istratov and Librovich, 1969). Indeed, if, starting from one position of the front, we construct, by the

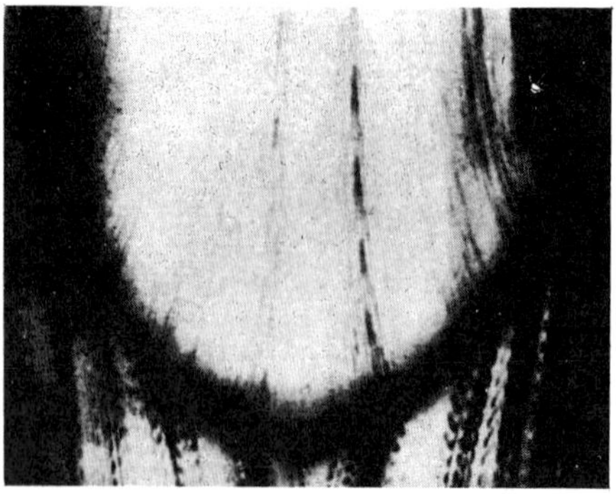

FIGURE 1 Photograph of flow field and flame front shape from the work Uberoi (1959). Flame is propagating downward in vertical tube, small particles added to gaseous mixture indicate streamlines before and behind flame front. One can see stagnation zone close to tube wall.

Huygens' principle, its next position, the flame dimensions would increase, if the tube walls were not restricting. A peculiarity is that the expanding flame front runs into the cold wall and disappears and this condition provides the steady-state regime of flame propagation.

Theoretically, there always exists a stationary plane solution for the flame front propagating in a channel in the absence of the gravitational force, but this solution is unstable with respect to small hydrodynamic disturbances. In accordance with Landau theory the rate of growth of hydrodynamic disturbances is inversely proportional to the wavelength, so that the most rapidly growing are the short-wave disturbances (Landau, 1944; Zel'dovich, 1966, 1979). But in this case the distortions in the flame front cause the changes in its structure and velocity, and the front cannot be treated as an infinitely thin hydrodynamic discontinuity. These changes in flame structure related to the transfer processes (diffusion and heat conductivity) may have a pronounced effect on the development of hydrodynamic instability. The disturbances with wave-lengths of the order of a characteristic front thickness appear to be the most dangerous and grow at the highest rate.

In a channel the maximum size of the disturbance is restricted by the channel diameter. Small-scale disturbances with a wave-length considerably smaller than the diameter may also arise at any point of the curved flame front surface. Particularly, if the disturbance wave-length is comparable with the flame front thickness, the disturbance seems to be most dangerous. Nevertheless, because of the peculiar flow field in the vicinity of the curved flame front, the steady flame propagation appears to be stable up to Reynolds numbers of the order of hundreds. There exists a velocity gradient positioned along the flame (see Figures 1, 2); *i.e.*, conditions arise for the appearance of a hydrodynamic stretching of the flame front and changing of itr propagation rate—the so called "stretch-effect" (Karlovitz, 1953; Istratov and Gremyachkin, 1972). If the effective coefficient of thermal conductivity in the flame exceeds the diffusion coefficient of the main reactant or is equal to it (which is often the case), the effect of stretching reduces the flame speed, or improves the stability to hydrodynamic disturbances. However, of greater importance is the fact that flame stretching causes stretching of the wave-length for a given disturbance.

There is another effect which we wish to take into account. The time for the development of disturbances in a channel is limited. Along the front there exists a tangential gas flow, which transfers the developed disturbance to the channel walls, where it is quenched. In contrast to the freely expanding spherical flame with disturbances developing as standing waves, disturbances in channels propagate along the flame front as running waves. But, both for the flame in the channel and the spherical flame, one observes an increase in the wave-length of the initial disturbance with time (Istratov and Librovich, 1966; Zel'dovich, 1979).

The present assumptions will be used below to interpret the steady propagation of a flame in a tube at rather large Reynolds numbers and to explain the stability of the curved flame front with respect to hydrodynamic disturbances of small amplitudes.

BASIC CONSIDERATION

Gasdynamics of Combustion in Tubes. Gas Flow before the Flame Front. Shape of the Flame Surface

Consider the gasdynamic fields ahead of and behind the flame front separately. The gas flow ahead of the flame front can be assumed to be potential. The potential function ϕ satisfies the Laplace equation:

$$\Delta\phi = 0, \mathbf{w} = \operatorname{grad} \phi. \tag{1}$$

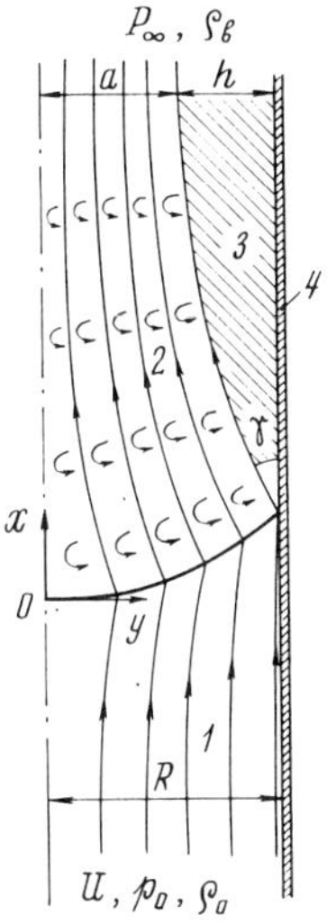

FIGURE 2 Hydrodynamic model of flow at flame propagation in a channel (Region 1—potential flow of cold gas, region 2—vortex flow of hot products, region 3—stagnation zone, region 4—wall).

Consider the flame in a plane channel, and choose the Decarte coordinate system (x, y) moving with flame at a constant velocity equal to the flame propagation along the channel (Figure 2), where y changes from 0 to R, x is read from the middle of the flame front into the combustion products side, so that $x = 0$, $y = 0$ is the leading point of the flame.

The solution of the equation in this coordinate system when the flow field is stationary may be presented in the form:

$$\phi = Ux + \sum_{n=1}^{\infty} A_n \exp(k_n x) \cos(k_n y) \qquad (2)$$

Where U is the velocity of flame propagation along the channel axis (in the moving coordinate system U is the gas velocity at $x = -\infty$), and A_n and k_n are constants. A bound of this solution far from the front (at $x = -\infty$) is taken into account in the Relation (2).

At first we shall account for the first term in the series of Eq. (2) only, namely, the function $A_1 \exp(k_1 x) \cos(k_1 y)$. This function describes the general features of the cold gas flow qualitatively correctly. The subsequent functions with more frequent zeroes specify the details of the flow, and do not change qualitatively its main properties.

In order to find the constants A_1 and k_1, we equate to zero the velocity component which is normal to the tube wall (the wall is gas-impenetrable). Due to the symmetry we also use the condition that the central streamline intersects the flame front at a right angle, and the velocity of the cold gas at that point is equal to the normal flame velocity u_n. Therefore

$$\begin{aligned} &y = R, \qquad v = \partial_y \phi = 0, \\ &y = 0, \qquad x = 0, u = \partial_x \phi = u_n \end{aligned} \qquad (3)$$

where R is the channel semiwidth, and u, v are components of velocity vector along the x, y axes, correspondingly.

From the first Condition (3) we obtain $k_1 = \pi/R$ and from the second, $A_1 = -\epsilon u_n/k_1$, where $\epsilon = U/u_n - 1$. The velocity vector components of the cold gas flow are written as

$$\begin{aligned} u/u_n &= 1 + \epsilon[1 - \exp(k_1 x)\cos(k_1 y)], \\ v/u_n &= \epsilon \exp(k_1 x)\sin(k_1 y) \end{aligned} \qquad (4)$$

As is seen, the maximum velocity of the gas flow is reached near the channel walls, where it is directed along the walls. At the channel axis the velocity is a minimum and has only the longitudinal component. At intermediate points of the channel cross-section the gas overflows from the axis to its periphery.

Knowing the flow field before the flame front, one can find the shape of its surface. For this, one should use the fact that the normal velocity component at the front must be equal to the normal flame velocity. In the simplest case, when the normal velocity of the flame is constant at all points of the surface, we obtain:

$$u \sin\theta - v \cos\theta = u_n \qquad (5)$$

where θ is the angle between the tangent to the flame surface and the channel axis and $ctg\ \theta = dx'/dy$, where $x' = x'(y)$ is the equation of the surface.

After simple transformations of (5) we have:

$$dx/dy = \pm[vu_n - u(w^2 - u_n{}^2)^{1/2}]/[uu_n + v(w^2 - u_n{}^2)^{1/2}],\ w^2 = u^2 + v^2 \qquad (6)$$

with the boundary condition $y = 0$, $x = 0$, where the velocity vector components u and v have the form as shown in Eq. (4). For a cold gas–convex front one should choose the minus sign in Eq. (6).

In extreme cases Eq. (6) can be simplified and integrated. At a U/u_n value that slightly differs from unity, $\epsilon \ll 1$ (the flame shape is close to plane), Eq. (6) yields:

$$x_f \simeq 4\cdot\epsilon^{1/2}\{1 - (0.5)^{1/2}\sin(k_1 y)/[1 - \cos(k_1 y)]^{1/2}\}/k_1 \qquad (7)$$

At $\epsilon \gg 1$ (a highly convex flame front), the integral of Eq. (6) is

$$x_f = \{\ln[k_1 y/\sin(k_1 y)]\}/k_1 \qquad (8)$$

For practically important intermediate values of ϵ, one can perform a numerical integration of Eq. (6). Results of numerical calculations of the flame shape in a channel at different values of the ratio of the flame propagation velocity along the channel to the normal flame velocity are given in Figure 3.

The pressure distribution ahead of the flame front can be found from Bernoulli's integral. Since at $x = -\infty$ the flow is homogeneous ($u = U$, $v = 0$), and the pressure is known $p(-\infty, y) = p_0$, then

$$p = p_0 + \rho_0(U^2 - u^2 - v^2)/2 \qquad (9)$$

ρ_0 is cold gas density.

In conclusion of this section it should be noted that the main dimensionless parameters of the problem considered are: the ratio of the propagation flame velocity along the channel axis to the normal velocity of flame (parameter $\epsilon = U/u_n - 1$), and the degree of thermal gas expansion ($\alpha = T_0/T_b$; where T_0 and T_b are the initial and burned-gas temperatures).

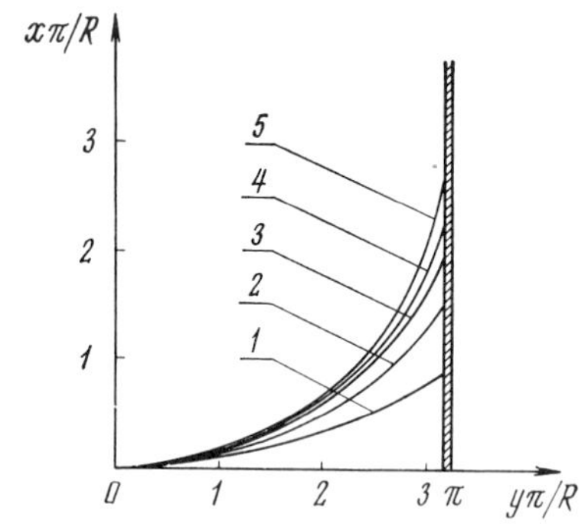

FIGURE 3 Numerical calculations of the flame front shape at different values of parameter $\epsilon = U/u_n - 1$ (1: $\epsilon = 0.05$, 2: $\epsilon = 0.15$, 3: $\epsilon = 0.25$, 4: $\epsilon = 0.35$, 5: $\epsilon = 0.45$).

The Flow of Combustion Products. Conditions at the Flame Front. Determination of the Flame Propagation Velocity

The flow of combustion products behind the flame front has a non-zero vorticity. When the gas crosses the flame front, which represents the gasdynamic discontinuity surface where the velocity, pressure, density and gas temperature are steplike changing, the vorticity of combustion products is generated (Zel'dovich, 1944, 1966, 1979, 1980; Borisov, 1978; Zel'dovich *et al.* 1979; Tsien, 1951, and Chernyi, 1954).

Starting from the flame front the intensity of the vortices remains constant along each streamline, so that the region filled by combustion products is a rotational one. In some of the previous works mentioned, however, the existence of the stagnation zone behind the flame front has not been accounted for, so that the quantitative conclusions differ essentially from those of the hydrodynamic model presently under consideration. It should be noted that the boundary streamline of the stagnation zone is a tangential velocity component discontinuity surface or a vortex sheet. As a consequence of the rotational nature of the flow of combustion products, instead of Eq. (1) one should solve the following equations for the stream function

$$\rho_b \Delta\psi = -\omega(\psi) \qquad (10)$$

and

$$\rho_b u = \partial_y \psi,\ \rho_b v = -\partial_x \psi \qquad (11)$$

where ω is the flow vorticity, which is constant along each streamline and can be calculated from the boundary conditions at the flame front and the cold gas flow field before it (discussed later), ρ_b is the hot gas density.

At the flame front the front-normal velocity component grows step-like due to variation of the gas density in accordance with the mass flow conservation law

$$\rho_0 u_n = \rho_b u_b \qquad (12)$$

(u_b is the normal flame velocity with respect to the combustion products).

But the tangential velocity component u_τ is unchanged

$$u_\tau|_{-f} = u_\tau|_{+f} \qquad (13)$$

where the signs "$-$" and "$+$" pertain to the cold gas and the hot combustion products respectively and the index $_f$ means that the values are taken at the flame front.

From the momentum conservation equation for the normal flow velocity component

$$p_{-f} + \rho_0 u_n^2 = p_{+f} + \rho_b u_b^2 \qquad (14)$$

one can find the pressure difference at the flame front, which proves to be constant along the entire surface if the normal flame velocity is constant:

$$\Delta p = p_{-f} - p_{+f} = \rho_0 u_n^2 (1-\alpha)/\alpha,$$

$$\alpha = \rho_b/\rho_0 = T_0/T_b \qquad (15)$$

The value of the stream function is the same on both sides of the hydrodynamic discontinuity:

$$\psi_{-f} = \psi_{+f} \qquad (16)$$

From Eqs. (12)–(16) one can calculate on the hot side of the flame front

$$P(\psi) = p + \rho_b(u^2+v^2)/2 \qquad (17)$$

and determine

$$\omega(\psi) = -dP(\psi)/d\psi \qquad (18)$$

The last formulae can be obtained from the motion equation written in the Gromeko–Lamb form

$$\mathbf{w} \times \boldsymbol{\omega} = \nabla(p/\rho + w^2/2), \; \mathbf{w} = \mathbf{i}u + \mathbf{j}v \qquad (19)$$

if a plane fow geometry and Relationships (11) are taken in account.

By making substitutions, the following expression for the vorticity distribution along the flame surface can be deduced:

$$\omega_{+f} = 0.5(\alpha-1)d(u^2_{-f} + v^2_{-f})/d\psi_{-f} \qquad (20$$

Consider the condition, which determines th velocity of the curved flame front propagation i the channel. Inside the stagnation zone filled b combustion products the pressure is constant an is equal to the value at infinity (when $x = \infty$ Because of Bernoulli's integral along the streamlin restricting the stagnation zone, the gas motio velocity remains unchanged. Since at $x = \infty$ th flow is plane-parallel (p_∞ = const, $v = 0$), dis tributions of velocity u and of the stream functio ψ_∞ are associated with the vorticity distribution

$$\rho_b u_\infty(y) = d\psi/dy = -\int_0^y \omega(y)dy \qquad (21$$

Consider now the channel cross sections a $x = \infty$ and $x = -\infty$ (see Figure 2), and writ the mass flow conservation integrals as

$$\rho_b \int_0^a u_\infty(y)dy = \int_0^{\psi_a} d\psi = \psi_a = \rho_0 UR \qquad (22$$

and the momentum integrals as:

$$p_\infty R + \rho_b \int_0^a u_\infty^2(y)dy = (p_0 + \rho_0 U^2)R \qquad (23$$

where a is a half-width of the cross section filled b the moving combustion products at infinity (i our coordinate system), and ψ_a is an extreme valu of the stream function. The latter integral, wit allowance for Eq. (22), can be expressed as

$$(p_0 - p_\infty + \rho_0 U^2)R = \int_0^{\psi_a} u_\infty(\psi)d\psi \qquad (24$$

By writing Bernoulli's integrals along eac streamline from the flame front to infinity, one ca obtain:

$$u_\infty^2(\psi) = 2[P_{+f}(\psi) - p_\infty]/\alpha\rho_0, \qquad (25$$

so (24) can be rewritten as

$$(p_0 - p_\infty + \rho_0 U^2)R = \int_0^{\psi_a} \{2[P_{+f}(\psi) - p_\infty]/\alpha\rho_0\}^{1/2} d\psi \qquad (26$$

'his expression, with the boundary conditions at flame front [Eqs. (12)–(16)] and Eq. (6) for the ne surface shape, determines the combustion nt propagation velocity U as a function of the mal flame velocity u_n and the ratio $\alpha = \rho_b/\rho_0$:

$$+u_*^2-2u_n^2(1-\alpha)/\alpha=(2/\alpha\rho_0)\int_0^{\rho_0 UR}[u_n^2(1-\alpha^2)+u_*^2-\alpha(1-\alpha)w_{-f}^2]^{1/2}d\psi \qquad (27)$$

ere u_* is the cold gas velocity at the contact nt of the flame front with the channel wall. It is venient to determine the solution of Eq. (27) the following manner. Starting from some value the parameter $\epsilon = U/u_n-1$, define the surface pe and u_* by integrating Eq. (6). Then from . (27) determine the corresponding value α.

An approximate analytical solution of the ations at $\epsilon \ll 1$, $1-\alpha \ll 1$ results in the depend- e:

$$U/u_n \simeq 1+(1-\alpha)^2/2 \qquad (28)$$

te, also, that for any α there exists the trivial ution $U/u_n = 1$ (plane front), which is, however, stable.] Numerical solutions of the equations illustrated in Figure 4.

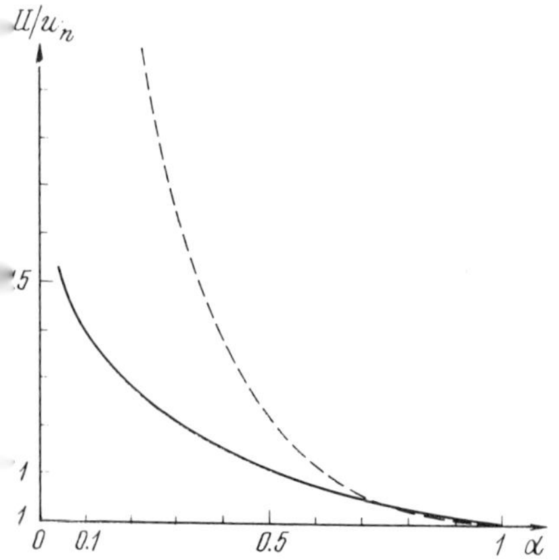

URE 4 Flame propagation velocity related to the nal flame velocity U/u_n along the channel axis as the tion of the gas expansion in the flame front.
——— numerical results of presented model.
– – – results of cellular model of Zel'dovich (1966).

The thickness of the jet of combustion products at $x = \infty$ is determined from the following integral:

$$a = \int_0^{\psi_0}[1/\rho_b u_\infty(\psi)]d\psi = \frac{1}{\rho_0}\int_0^{\rho_0 UR}1/[u_n^2(1-\alpha^2)+\alpha u_*^2-\alpha(1-\alpha)w^2]^{1/2}d\psi \qquad (29)$$

An approximate analytical solution at $\epsilon \ll 1$ yields α as

$$a \simeq R(1-\epsilon) \simeq R[1-(1-\alpha)^2/2] \qquad (30)$$

and thickness of the stagnation zone h as

$$h \simeq R\epsilon \simeq R(1-\alpha)^2/2 \qquad (31)$$

Results of numerical integration of Eq. (29) are represented in Figure 5. For ordinary flames $\alpha \sim 0.1$–0.2 the flame propagation velocity along the channel is greater than the normal flame velocity by a factor of 1.25–1.40. The stagnation zone in the combustion products occupies about a quarter of the channel cross section.

Comparison of Flame Propagation in a Channel and Cellular Flames

If one considers the disturbances of finite amplitude at the infinite thin flame surface from the point of view of Huygens' principle, then with time there appear angular points, which diminish the flame front surface, *i.e.* with time the disturbances become smoother. By allowing for this and the hydrodynamic instability effect, one can come to the conclusion that the growth of disturbances is restricted by a certain amplitude which is proportional to the wave-length and depends on the thermal expansion of the gas in the flame (Zel'dovich, 1966).

By using these results one can derive the dependence of the flame propagation velocity on the degree of gas expansion. The disturbance amplitude (see Figure 6) satisfies the following equation with negative nonlinearity:

$$dA/dt = \Omega A+\kappa A^2 \qquad (32)$$

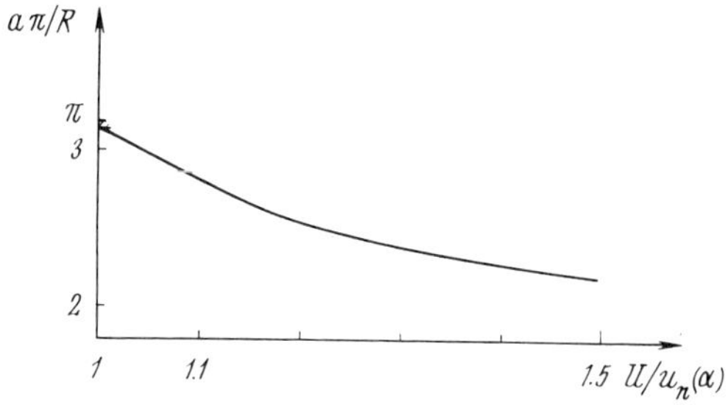

FIGURE 5 Jet thickness a at $x = \infty$ versus dimensionless flame propagation velocity U/u_n.

where $\Omega = ku_n f(\alpha)$ is linear characteristic frequency taken from the Landau dispersion equation, k is the disturbance wave number, $k = 2\pi/\lambda$, and λ is the wave-length, $\kappa = -2k^2 u_n{}^2/\pi^2$ (see Zel'dovich,

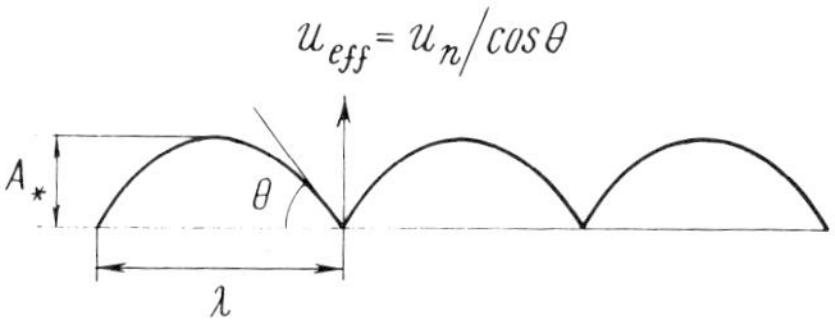

FIGURE 6 Model of the cellular flame from the work of Zel'dovich (1966).

1966). Equation (32) at $t \to \infty$ yields the stationary amplitude of the disturbance:

$$A(t \to \infty) = A_* = -\Omega/\kappa = f(\alpha)\pi^2/2k \qquad (33)$$

The relation $A_* \sim k^{-1}$ means that the cell shape depends on α only, and not on the wave-length. In order to compare this result with the channel combustion problem, we assume that $k = 1$, and then obtain the amplitude of the curved flame corresponding to the channel width 2π:

$$A_* = f(\alpha)\pi^2/2 \qquad (34)$$

where $f(\alpha) = \{-1 + [(1-\alpha)^2/\alpha + 1]^{1/2}\}/(1+\alpha)$ is the positive root in the Landau dispersive equation.

In the Zel'dovich work (1966) the cell shape was approximated by a parabola. So, to find the propagation velocity dependence on α, it is sufficient to calculate the length L of the parabola arc with the amplitude A_* and wave-length λ = (see Figure 6):

$$L = 0.5\lambda\{[1 + (4A_*)/\lambda]^{1/2} + (\lambda/4A_*)\text{arcsh}(4A_*/\lambda)\}$$

and to use the evident relation $L/\lambda = U/u_n$:

$$U/u_n = 1 + \epsilon = 0.5\{[1 + (\pi f(\alpha))^2]^{1/2} + (1/\pi f(\alpha))\text{acrsh}(\pi f(\alpha))\}$$

At $\epsilon \ll 1$ and $1-\alpha \ll 1$, Eqs. (34) and (36) yield

$$A_* = \pi(1.5\epsilon)^{1/2} \simeq 3.85\epsilon^{1/2}$$
$$\epsilon = \pi^2(1-\alpha)^2/24 \simeq 0.41(1-\alpha)^2$$

(Compare these results with $A_* \simeq 4\epsilon^{1/2}$, $\epsilon \simeq (1-\alpha$ obtained in the preceding section).

The dependence ϵ vs α derived from Eq. (36 plotted on Figure 4 as a dotted line. Solutions Eq. (36) and of Eq. (27) show good agreement o at small values ($\epsilon < 0.05$) as would be expec because the assumptions used in Eq. (32) are va only at sufficiently small amplitudes A_*/λ Note that changes of the gas motion before a after the curved front were considered indire through value $f(\alpha)$ taken from the Landau the for a small disturbance.

Correction to the Flow Field Ahead of Flame Fro Propagating in a Channel

The results of analytical and numerical calculati obtained in the previous section, and graphic 'epicted in Figures 4 and 5 were derived, as noted bove, when only one term was taken from the /hole series of Eq. (2).

Now consider a contribution of the following erms to the series of Eq. (2). From the conditions t the channel wall and axis it follows that $k_n = nk_1$. 'onsequently, neither of the harmonics, excluding he first one, can be realized singly. For instance, or $n = 2$ the value of velocities $u_* = (u_*^2 + v_*^2)^{1/2}$ t the extreme point of the flame front at the wall, urns out to be less than u_n, that is inadmissible. Iowever, to the first harmonic one can add the thers separately with a small "weight" β, in order o fulfill the conditions at all frontal points.

$$\begin{aligned} /u_n &= 1+\epsilon[1-(1-\beta)\exp(k_1x)\cos(k_1y) \\ &\qquad -\beta \exp(nk_1x)\cos(nk_1y)], \\ /u_n &= \epsilon[(1-\beta)\exp(k_1x)\sin(k_1y) \\ &\qquad +\beta \exp(nk_1x)\sin(nk_1y)] \end{aligned} \tag{38}$$

uch a "mixing in" of separate terms of the series, s shown by the calculations, insignificantly affects he shape of the flame surface, with the exception f the region adjacent to the channel wall (see 'igure 7), but changes greatly the stagnation zone hickness and the flame propagation velocity. By alculating again the flame surface shape from q. (6) for each β and chosen harmonic number, nd by solving the integral Eq. (27), one obtains or each β the dependence $\epsilon(\alpha)$. These dependencies are represented in Figure 8. For each value β, for the harmonic number n and ratio α, there exists a value at which the solution vanishes. This value corresponds to the case when the flame front touches the channel wall at right angles, and the stagnation zone disappears. The numerical calculation shows that at the given value α, the critical value ϵ_{max} is independent of the number of the added harmonic. The dependence $\epsilon_{max}(\alpha)$ is represented in Figure 8 by a dotted line. The dependence of the stagnation zone thickness on α at different values of β is shown in Figure 9.

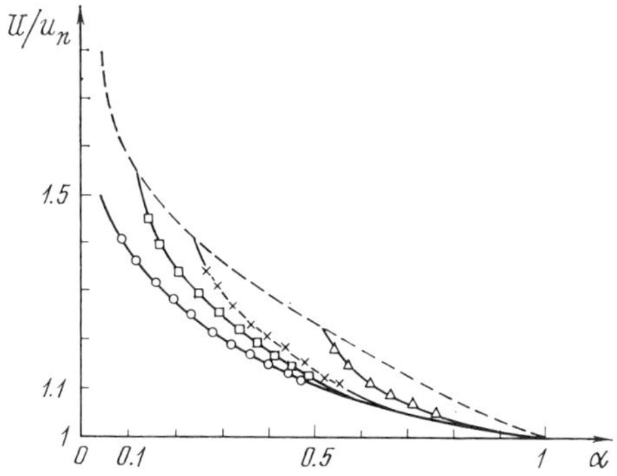

FIGURE 8 Flame propagation velocity as the function of α at different values of β (which is the "weight" of second harmonic ($n = 2$)).(○ $\beta = 0$; □ $\beta = 0.01$; × $\beta = 0.05$; △ $\beta = 0.1$).

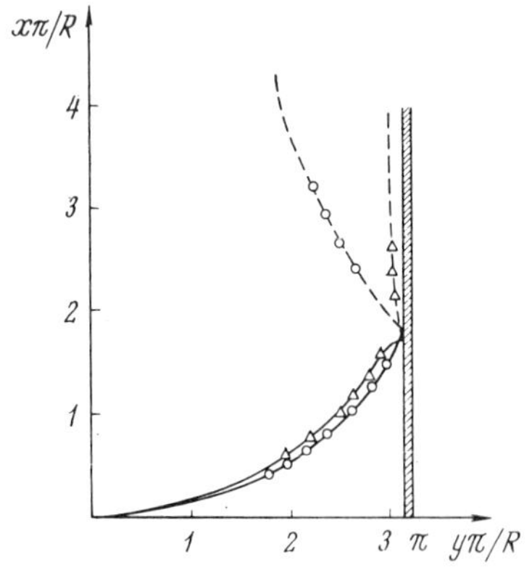

IGURE 7 Differences in the flame shape due to the dded first and second harmonics with weight β. $\beta = 0$; △ $\beta = 0.1$, $n = 2$; $\epsilon = U/u_n - 1 = 0.2$.

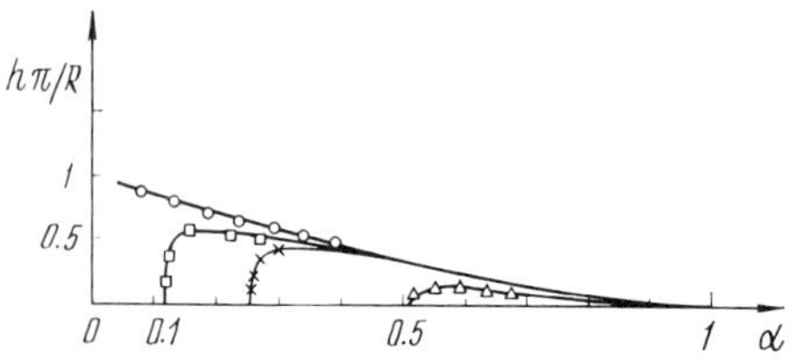

FIGURE 9 Dependence of the stagnation zone thickness on α at different values β (○ $\beta = 0$; □ $\beta = 0.03$; × $\beta = 0.05$; △ $\beta = .01$).

A similar procedure can be applied with the negative weight of harmonic added. In this case a critical value ϵ_{min} corresponds to the minimum of the propagation velocity when the jet thickness at $x = \infty$ tends to zero, and the stagnation zone occupies the whole channel cross section. One may suggest that the real values of the propagation velocity at the given u_n and α lie between the extreme curves corresponding to the maximum and minimum of the propagation velocity.

The Stability of the Convex Flame to Small Disturbances

As noted above, the convex flame front in the channel can be considered as one cell arising due to the plane flame hydrodynamic instability up to the limit determined by the nonlinear stabilization. But for such a flame one may investigate the stability of the curved shape to a disturbance with a smaller wave-length, that is, with higher increment of growth. Suppose that at the flame surface one may locate many of such waves. We will follow their evolution by using an approximate approach, which is an analog of the geometrical optics method or of the method by Ventzel–Kramers–Brillouin used in quantum mechanics (for details see the book by Landau and Lifshitz, 1954). In the inhomogeneous gas flow with the oblique curved flame surface one can consider the trajectories (beams) of the propagation of disturbances (see Figure 10). They are moving along the channel together with the flame propagation and approach the channel walls due to the tangential component of velocity at the flame front. Because of the variation of the tangential velocity the disturbance wave-length does not remain constant; that is there exists an effect of the stretching of the initial disturbance.

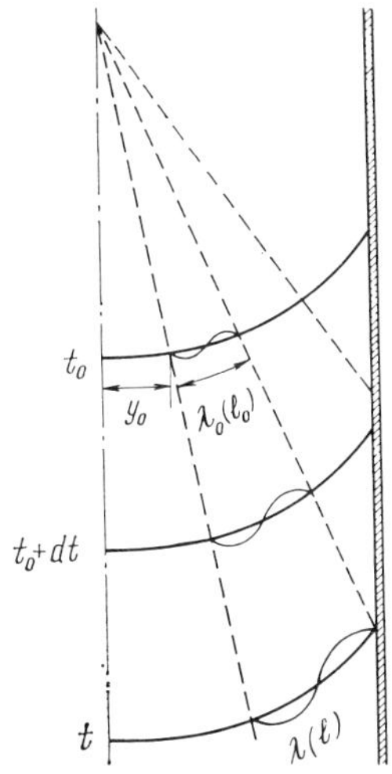

FIGURE 10 Evolution of disturbances along the flame surface with time (t)—Ventzel–Kramers–Brillouin approximation.

Thus, consider the disturbance of the wave-length λ_0 located not far from the channel axis at $y = y_0$ and determine how it would grow up until the moment of contact with the wall. Let A_0 be an initial amplitude of the disturbance, and A_w the finite amplitude near the wall. The shortwave perturbations are assumed to be small in order to use the linear approach. Since in accordance with Landau theory

$$dA/dt = \Omega A, \; dl = u_\tau dt \tag{39}$$

(l is the coordinate along the flame surface), then the maximum A_w reached at the wall is expressed through the initial A_0 as

$$A_w/A_0 = \exp(\Gamma), \Gamma = \int_{l_0(y_0)}^{L(R)} (\omega/u_\tau)dl \tag{40}$$

In the approximation of the Ventzel–Kramers–Brillouin method, we take $\Omega = k(l)u_n f(\alpha)$ from the Landau theory on the hydrodynamic instability of the plane flame. The relative change in the disturbance because of the stretch-effect is equal to that in the tangential velocity component

$$d\lambda/\lambda = du_\tau/u_\tau \tag{41}$$

Taking into account Eq. (41) and $k(l) = 2\pi/\lambda(l)$ we obtain for Γ:

$$\Gamma = 2\pi f(\alpha)(u_n u_{\tau 0}/\lambda_0) \int_{l_0(y_0)}^{L(R)} [u_\tau(l)]^{-2} dl \tag{42}$$

It should be noted here that $\Gamma \to \infty$ and, consequently, $A_w \to \infty$ if we take $l_0 = 0$, because the tangential velocity component $u_\tau = 0$ at the channel axis. The real disturbances, however, hav finite dimensions and cannot be placed exactly at the leading point of the flame front. The disturbances with wave-length smaller than the thermal thickness of the flame decay in time and do not provoke the flame instability. The most rapidly growing disturbances, following Markstein (1951) theory, have a wave-length $\lambda_* = 2\kappa/[u_n(1-\alpha)]$ (κ is thermal diffusivity); and for moderately large Reynolds numbers of the flame front propagation along the channel (Re of the order of hundreds), they are much smaller than the channel width.

For small l_0 one may write

$$\psi(y_0) \simeq [(1+\epsilon)y_0 - \epsilon \sin(k_1 y_0)]\rho_0 u_n \simeq \rho_0 u_n y_0 \quad (43)$$

In order to look for the most dangerous disturbance let us also take $y_0 = 4\pi\kappa/u_n(1-\alpha)$. The integration of Eq. (42) at $\epsilon \ll 1$ yields

$$\Gamma = \epsilon^{1/2}\mathrm{Re}(y_0/4\pi)[\epsilon\pi(1-y_0/R) + \mathrm{ctg}(y_0\pi/2R)] \quad (44)$$

which, under the condition $y_0/R \ll 1$ approximately equals

$$\Gamma = (\epsilon)^{1/2}\mathrm{Re}/2\pi \quad (45)$$

where the Reynolds number $\mathrm{Re} = u_n R/\kappa$ is introduced. The results of the numerical integration of Eq. (42) at different ϵ, and at the corresponding values α (see previous sections) are represented in Figure 11. The main part of the integral Γ is gained near the lower integration boundary. On other words, the disturbances grow significantly near the channel axis, but not at the channel wall.

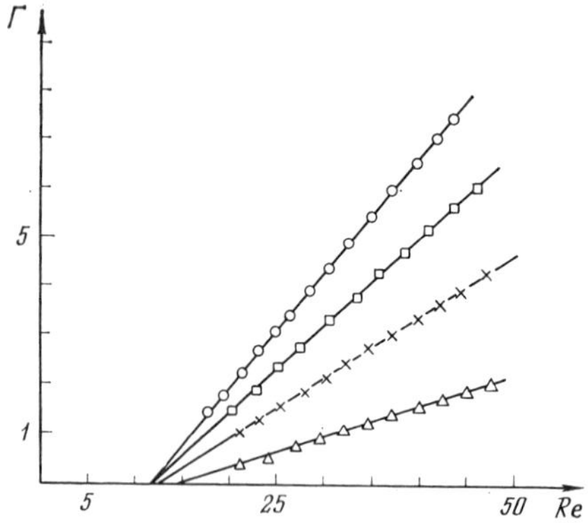

FIGURE 11 Numerical calculations of the integral Γ as the function of Reynolds number Re at different values of the flame propagation velocity along the channel axis U/u_n (○ $U/u_n = 1.05, \alpha = 0.675$; □ $U/u_n = 1.15, \alpha = 0.4$; × $U/u_n = 1.25, \alpha = 0.23$; △ $U/u_n = 1.35, \alpha = 0.12$).

Let us estimate the increase of the disturbance amplitude at the channel wall with respect to its initial value taken from the thermodynamic fluctuation theory. This theory predicts the initial disturbance of the order of $\lambda_0/(N)^{1/2}$, where N is the number of gas molecules in the volume λ_0^3. Therefore, the condition for reaching a noticeable value of the disturbance amplitude leads to the relation

$$\ln(N) = 2\Gamma \quad (46)$$

which yields, by using Eq. (45), the following critical Reynolds number Re_*

$$\mathrm{Re}_* = \pi(\epsilon)^{-1/2}\ln N \quad (47)$$

If we take $\lambda_0 = \kappa/u_n \sim 10^{-2}$ cm and $N \sim 5.3\times10^{22}(\kappa/u_n(1-\alpha))^3 \sim 5.3\times10^{16}(1-\alpha)^{-3}$ of molecules, an estimate of the critical Reynolds number at $\epsilon \sim 0.1$ is equal to $\mathrm{Re}_* \sim 380$. The critical Reynolds number will be greater if we take into account the dependence of the characteristic frequency ω on the flame front curvature following the Markstein theory (1951).† The solution of the plane flame instability problem considering the influence of the transfer processes (heat conduction, diffusion and viscosity) by means of the matched expansion series approach leads to the following expression for Ω (Istratov and Librovich (1966)).

$$\Omega = k(l)u_n f(\alpha)[1 - k(l)g(\alpha)\kappa/u_n] \quad (48)$$

The second correction term in brackets is important for disturbance wave-lengths comparable with the thermal width of the flame, which now are under consideration. For Lewis and Peclet numbers equal unity the function $g(\alpha)$ has such a form

$$g(\alpha) = \{1-\alpha-[1+\alpha+2\alpha f(\alpha)][1+\ln\alpha/(1-\alpha)]\}/[1-\alpha-\alpha f(\alpha)] \quad (49)$$

† The Markstein's approach is the next approximation to the Landau theory accounting for the disturbance wavelengths comparable with the flame front thickness. It should also be noted that stretching flow field produces an additional effect of decreasing of the disturbance amplitude in the time that is related to the increase of the wavelength.

Let us take again the most dangerous disturbances growing most rapidly, *i.e.* $-k_0 = 0.5u_n/[\kappa g(\alpha)]$, and by substituting Eqs. (48) and (49) into the expression for Γ (42) we obtain:

$$\Gamma = 2f(\alpha)(u_n u_{\tau 0}/\lambda_0) \times \int_{\lambda_0}^{L} [1-2\pi u_{\tau 0} g(\alpha)\kappa/(u_n \lambda_0 u_\tau)]/u_\tau^2 \, dl \quad (50)$$

At $\epsilon \ll 1$, Eq. (50) yields

$$\Gamma \simeq 3/[32\pi(2)^{1/2}]\,[1-(2\epsilon)^{1/2}/6]\mathrm{Re}. \quad (51)$$

The results of the numerical integration of Eq. (50) are represented in Figure 12.

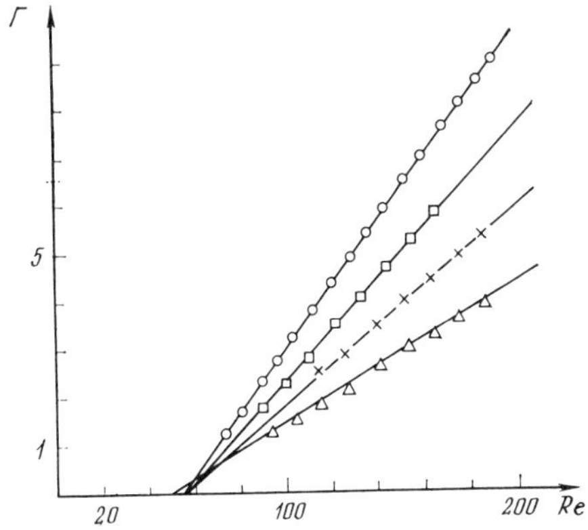

FIGURE 12 Second approximation of calculations of the integral Γ (○ $U/u_n = 1.05$, $\alpha = 0.675$; □ $U/u_n = 1.15$, $\alpha = 0.4$; × $U/u_n = 1.25$, $\alpha = 0.23$; △ $U/u_n = 1.35$, $\alpha = 0.12$).

Condition $A_w \sim \lambda_0$ yields the critical Reynolds number

$$\mathrm{Re}_* \simeq 16(2)^{1/2}(\pi/3)\ln N, \quad (52)$$

which at the same typical parameters, as those used above, equals ~ 900.

DISCUSSION

An approximate analysis of the flame propagation in the channel performed in this paper includes the consideration of the non-viscid gas flow fields before and after the curved flame front considered as a hydrodynamic discontinuity. The refraction of streamlines at the flame front demands the introduction of the stagnation zones in the combustion products where the gas is at rest with respect to the flame. The boundary between the combustion products flow and the stagnation zone, being the surface of the tangential velocity discontinuity, is unstable. The appearing small disturbances are increasing, and finally form a turbulent tail. Therefore, in reality, an ideal stagnation zone, taken into the analysis and shown in Figure 2, does not exist. However, if the disturbances are not growing very rapidly and they have time to be carried away by the flow, then the considered scheme would be justified and would yield the correct results. Besides, the turbulent mixing is accompanied by momentum conservation, and therefore, it does not upset the equation used in the determination of the flame propagation velocity.

Processes that have not been taken into account in the scheme considered were the heat removal from the flame to the channel walls, and the influence of the gas viscosity. Note that one can provide the combustion conditions when these effects may be neglected (high Reynolds numbers). And, *vice versa*, there exist such regimes when these effects determine the whole process of the flame propagation.

In the scheme considered, as shown above, the convex flame front affects the hydrodynamic of the gas flow, and forms some velocity distribution ahead of it. This is associated with the pressure difference at the flame front. In other words, it is always necessary to solve a conjugate problem on the front propagation and the gas motion. Restricting the analysis by the first term in the series describing the flow field before the flame and taking into account the corresponding shape of the flame front, as was shown, one can look for the flame propagation velocity from the integral mass and momentum conservation relations taken far from the flame in the cold and hot gases. More detailed description of the cold gas field demands, in its turn, the detailed consideration of the rotational flow of combustion products behind the flame, in order to find a relation for the flame propagation velocity.†)

† It should be kept in mind that the direct application of numerical methods is complicated by the Landau instability effect.

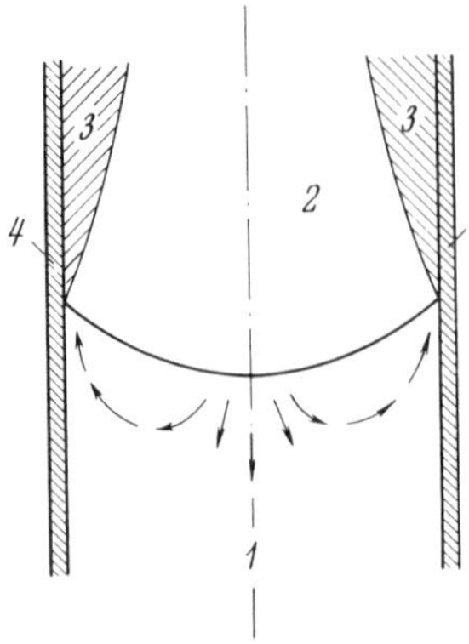

IGURE 13 Flow field before the flame front in the aboratory coordinate system. (Region 1—potential flow, egion 2—vortex flow, region 3—stagnation zone, region —channel wall).

The velocity distribution ahead of the flame ront in the room coordinate system is shown in igure 13. As seen from the figure, the gas motion s accompanied by the rotation and displacement of luid elements, since the velocities near the axis and valls are directed in opposite directions. The otation centres of each of the fluid elements are ehind the flame front. But the flow is potential nd $\omega = \text{rot}\,\boldsymbol{u} = 0$. In the coordinate system of the flame front, the velocities near the axis and walls have the same direction (see Figure 2).

By describing the flame as a gasdynamic discontinuity, one neglects the thermodiffusional flame structure which determines the normal velocity of its propagation. Such an approach supposes that the normal flame velocity is an *a priori* given parameter or taken from experiment.† The stability to the shortwave disturbances also depends on the flame structure. Actually, in the analysis of the flame propagation stability up to the moderately high Reynolds numbers performed in the present paper, the structure of the flame has been taken into consideration by selecting the most dangerous wave-length of the small disturbances.

In conclusion we would like to mention that in order to verify the assumptions and conclusions made in this paper, it would be useful to perform a set of experiments on flame propagation at zero gravity with the visualization of flow in channels and tubes and with artificially created local disturbance. Similar experiments seem to be able to verify the basic theoretical conclusions of the present paper under appropriate conditions because the buoyancy usually affects the hydrodynamics of flow during the combustion and the flame front stability in the essential way.

† To take into account the stretch effect on the flame propagation rate, the consecutive theory needs the treatment the flame structure. The phenomenological Markstein (1951) theory is the first in this direction.

Appendix

he Case of Cylindrical Symmetry

'he conclusions drawn in previous sections and the ethod for obtaining the integral relations deterining the flame propagation velocity along the lane channel are applicable to the case of the ylindric tubes also. The difference lies in the etails of the formulas and in the quantitative esults.

Equation (10) in cylindrical coordinates, allowing 1at the solution is bounded at the tube axis, can e solved as a series of Bessel's functions of the hole order

$$\phi = Uz + \sum_{i=0}^{\infty} A_i \exp(k_i z) J_i(k_i r) \qquad (53)$$

(r and z are the cylindrical coordinates).

By restricting ourselves to considering the first term in the series, and by satisfying the conditions at the tube walls and at the flame leading point

$$r = R,\ v = 0,$$
$$r = 0,\ z = 0,\ u = u_n \qquad (54)$$

we obtain $k_0 = \lambda_1/R$, $A_0 = (u_n - U)/k_0$, λ_1 is the first zero of the Bessel's function $J_1(x)$. Introducing for simplicity the dimensionless values $u' = u/u_n$, $r' = k_0 r$, $z' = k_0 z$ and omitting the primes, we have

$$\begin{aligned} u &= 1 + \epsilon[1 - \exp(z)J_0(r)], \\ v &= \epsilon \exp(z) J_1(r), \\ \psi &= (1+\epsilon)r^2/2 - \epsilon \exp(z) r J_1(r) \end{aligned} \qquad (55)$$

The equation for the flame surface takes the form

$$dz/dr = [-u+v(u^2+v^2-1)^{1/2}]/[v+u(u^2+v^2-1),]$$
$$r = 0, z = 0 \quad (56)$$

The integrals of the mass flow and of the axial impulse component conservation yield:

$$2\alpha \int_0^a ru(\infty, r)\, dr = \lambda_1^2(1+\epsilon) = 2\psi_a \quad (57)$$

$$\lambda_1^2[p_0 - p_\infty + (1+\epsilon)^2] = 2\alpha \int_0^a ru^2[\infty, r)dr$$

$$= 2\int_0^{\psi_a} \{2[P_{+f}(\psi) - p_\infty]/\alpha\}^{1/2} d\psi \quad (58)$$

The latter, after considering Bernoulli's integral, is transformed into

$$(1+\epsilon)^2/2 + u_*^2/2 - (1-\alpha)/\alpha$$

$$= (2/\alpha\lambda_1^2) \int_0^{\psi a} \{1 - \alpha^2 u_*^2 - \alpha(1-\alpha)w_{=f}^2\}^{1/2} d\psi \quad (59)$$

The integral Eq. (59) together with the surface shape Eq. (56) would determine the flame propagation velocity along the tube axis as a function of the normal flame velocity and the density ratio α. The width of the stagnation zone could be derived by integration

$$h = \lambda_1 - \{2 \int_0^{\psi a} 1/[\alpha u_\infty(\psi)]d\psi\}^{1/2} \quad (60)$$

The performed numerical calculations for the case of cylindrical symmetric yielded the same qualitative dependence of the flame surface shape, the propagation velocity, and thickness of the stagnation zone, as in the case of the plane channel. The quantitative results for the cylindrical tube are somewhat different, *e.g.* the dimensionless propagation velocity along the tube axis at a real α proves to be 50 percent higher than in the case of plane symmetry.

REFERENCES

Aldushin, A. P. *et al.* (1980). Flame propagation in the reacting gaseous mixture. *Archivum combustionis*, **1.**

Borisov, V. I. (1978). On the velocity of uniform flame propagation in a plane channel. *FGV* **14,** 26 (In Russian).

Chernyi, G. G. (1954). Gas flow in a tube in the presence of the flame front. *Theoretical Hydromechanics Moscow*, **4,** 31 (In Russian).

Darrieus, G. (1944). Propagation d'un front de flamme. Essai de theorie de vitesses anomales de deflagration par development spontane de la turbulence. *Z. Phys. Chem.*, **88,** 641.

Istratov, A. G., and Librovich, V. B. (1966). On the influence of transfer processes on the plane flame front stability, *PMM*, **30,** 451 (In Russian).

Istratov, A. G., and Librovich, V. B. (1969). On the stability of gasdynamic discontinuities associated with chemical reactions. The case of a spherical flame. *Astr. Acta*, **14,** 453.

Istratov, A. G., and Gremyachkin, N. M. (1972). On the stability of plane flame in the flow with the velocity gradient. *Combustion and Explosion, Nauka, Moscow*, pp. 305–308 (In Russian).

Karlovitz, B. *et al.* (1953). A. Flame propagation across velocity gradients. B. Turbulence measurement in flames. *Fourth Symposium (International) on Combustion*, Williams and Wilkins, Baltimore, p. 613.

Landau, L. D. (1944). On the theory of slow combustion, *Acta Phys.-Chem., URSS*, **19,** 77.

Landau, L. D., and Lifshits, E. M. (1954). *Flow Mechanics.* Gostechizdat, Moscow (In Russian).

Markstein, G. (1951). Experimental and theoretical studies of flame front stability. *J. Aero. Sci.*, **18,** 199.

Maxworthy, T. (1962). Flame propagation in tubes, *Phys. Fl.*, **5,** 407.

Rozlovsky, A. I., and Zel'dovich, Ya. B. (1954). About the conditions for the spontaneous instability of normal burning, *DAN USSR*, **57,** 365 (In Russian).

Shchelkin, K. I. *et al.* (1947). Measurements of normal velocities of acetylene–oxygen mixture flame. *JETF*, **17,** 1397 (In Russian).

Tsien, H. S. (1951). Influence of flame front on the flow field. *J. Appl. Mech.*, **18,** 188.

Uberoi, M. S. (1959). Flow field of a flame in a channel, *Phys. Fl.*, **2,** 72.

Zel'dovich, Ya. B. (1944). Comments on combustion of a rapid flow in a tube. *Journ. Techn. Phys.*, **14,** 162 (In Russian).

Zel'dovich, Ya. B. (1966). About some effects stabilizing the curved front of laminar flame. *PMTF*, **1,** 102 (In Russian).

Zel'dovich, Ya. B. (1979). Structure and stability of the stationary laminar flame at moderately large Reynolds numbers. Preprint of the Inst. of Chemical Physics, the USSR Ac. of Sci., Chernogolovka.

Zel'dovich, Ya. B. *et al.* (1979). Hydrodynamics of vortical flow behind the flame front propagating in tubes. *VIth International Symposium on Combustion Processes*. Abstracts. Karpacz, 83.

Zel'dovich, Ya. B. (1980). Structure and stability of steady laminar flame at moderately large Reynolds numbers. *Combustion and Flame* (in print).

Index

Perspectives in Physics

Huzihiro Araki, J. A. Libchaber, and Giorgio Parisi, editors

Frances Bauer, Octavio Betancourt, Paul Garabedian, Masahiro Wakatani, *The Beta Equilibrium, Stability, and Transport Codes: Applications to the Design of Stellarators*

Pierre Pelcé, editor, *Dynamics of Curved Fronts*